BASIC GEOMETRY

Ray C. Jurgensen
Richard G. Brown

Editorial Adviser: Albert E. Meder, Jr.
Teacher Consultant: Robert J. McMurray

Houghton Mifflin Company • Boston
Atlanta Dallas Geneva, Illinois
Lawrenceville, New Jersey Palo Alto Toronto

The Authors

Ray C. Jurgensen, formerly Chairman of the Mathematics Department and holder of the Eppley Chair of Mathematics at the Culver Academies, Culver, Indiana. Mr. Jurgensen has been a lecturer at the National Science Foundation institutes for mathematics teachers and meetings of the National Council of Teachers of Mathematics.

Richard G. Brown, Mathematics teacher at the Phillips Exeter Academy, Exeter, New Hampshire. Mr. Brown has taught a wide range of mathematics courses for both students and teachers at several schools and universities, including Newton (Massachusetts) High School, the University of New Hampshire, Arizona State University, and the North Carolina School for Science and Mathematics.

Editorial Adviser

Albert E. Meder, Jr., Professor of Mathematics, Emeritus, Rutgers, The State University of New Jersey.

Teacher Consultant

Robert J. McMurray, Mathematics teacher at the Albuquerque Technical Vocational Institute, Albuquerque, New Mexico. Mr. McMurray has many years of experience teaching geometry in secondary schools.

Printed in U.S.A.

ISBN: 0-395-41196-3

CDEFGHIJ-RM-9543210/898

CONTENTS

UNIT A

CHAPTER 1 Points, Lines, and Angles

1 Beginning Your Study of Geometry 4
2 Points and Lines 6
3 Angles 10
4 Classifying Angles 15
5 Vertical Angles 18
6 Angle Bisectors 20
7 Three Constructions 25
8 Postulates of Equality 29
9 Postulates of Geometry 33

FEATURES
Consumer Applications 14 Puzzle Problems 17 Experiments 22
Applications 23 Career Notes 32

REVIEWS AND TESTS
Self-Tests 14, 22, 37 Skills Review 38 Chapter Review 39
Chapter Test 41 Mixed Review 42

CHAPTER 2 Introducing Proof

1 Three Theorems 46
2 "If . . . then" Statements 50
3 Reaching Conclusions (Optional) 56
4 Writing Proofs 60
5 Parallel Lines 66
6 Proving Lines Parallel 72
7 Constructing Parallel Lines 77

FEATURES
Applications 54 Career Notes 59 Calculator Activities 71
Consumer Applications 76 Reading Geometry 81 Problem Solving Strategies 82

REVIEWS AND TESTS
Self-Tests 53, 80 Skills Review 84 Chapter Review 85 Chapter Test 87
Cumulative Review 88

UNIT B

CHAPTER 3 Triangles
1 The Angle Sum of a Triangle 92
2 Classifying Triangles 97
3 Defining Congruent Triangles 102
4 The SSS Postulate 107
5 The SAS Postulate 115
6 The ASA Postulate 121
7 The AAS and HL Theorems 126

FEATURES
Puzzle Problems 101, 112, 132 Applications 113 Experiments 120, 132
Consumer Applications 120 Career Notes 125 Reading Geometry 133
Computer Activities 135

REVIEWS AND TESTS
Self-Tests 101, 112, 125, 132 Skills Review 136 Chapter Review 137
Chapter Test 139 Mixed Review 140

CHAPTER 4 Using Congruent Triangles
1 Proving Corresponding Parts Equal 144
2 Congruent Triangles and Constructions 149
3 Segment Bisectors 154
4 Altitudes and Medians of Triangles 158
5 Inscribed and Circumscribed Circles 163
6 Triangles with Two Equal Sides 167
7 Triangles with Two Equal Angles 173

FEATURES
Consumer Applications 148 Reading Geometry 153 Puzzle Problems 157, 162
Experiments 162, 178 Career Notes 172 Applications 179
Problem Solving Strategies 180

REVIEWS AND TESTS
Self-Tests 152, 166, 178 Skills Review 182 Chapter Review 183
Chapter Test 185 Cumulative Review 186

UNIT C

CHAPTER 5 Polygons

1 Introducing Polygons 190
2 Angle Sums of Polygons 194
3 Special Quadrilaterals 198
4 Properties of Parallelograms 203
5 Properties of Special Parallelograms 209
6 Proving Figures Are Parallelograms 214
7 Properties of Trapezoids 220
8 The Midpoints Theorem 225

FEATURES
Experiments 193 Career Notes 202 Applications 208 Puzzle Problems 224
Reading Geometry 229

REVIEWS AND TESTS
Self-Tests 197, 207, 219, 228 Skills Review 230 Chapter Review 231
Chapter Test 233 Mixed Review 234

CHAPTER 6 Areas

1 Areas of Rectangles 238
2 Areas of Parallelograms 243
3 Areas of Triangles 247
4 Areas of Trapezoids 250
5 The Pythagorean Theorem 255
6 Converse of Pythagorean Theorem 259
7 Circumferences of Circles 263
8 Areas of Circles 268

FEATURES
Consumer Applications 246 Puzzle Problems 254, 261 Career Notes 262
Applications 272 Problem Solving Strategies 274 Computer Activities 276

REVIEWS AND TESTS
Self-Tests 246, 254, 262, 271 Skills Review 278 Chapter Review 279
Chapter Test 281 Cumulative Review 282

UNIT D

CHAPTER 7 Ratios and Proportions

1 Ratios 286
2 Proportions 289
3 Using Proportions 293
4 Maps and Scale Drawings 296
5 Making Maps and Scale Drawings 299

FEATURES
Calculator Activities 292 Puzzle Problems 295 Applications 302

REVIEWS AND TESTS
Self-Tests 295, 301 Skills Review 304 Chapter Review 305
Chapter Test 307 Mixed Review 308

CHAPTER 8 Similar Polygons

1 Defining Similar Polygons 312
2 The AA Postulate 317
3 A Special Case of Similar Triangles 323
4 The Triangle Proportionality Theorem 325
5 Perimeters and Areas 329

FEATURES
Career Notes 316 Calculator Activities 322, 332 Puzzle Problems 324
Experiments 333 Applications 334 Problem Solving Strategies 336

REVIEWS AND TESTS
Self-Tests 322, 333 Skills Review 338 Chapter Review 339
Chapter Test 341 Cumulative Review 342

UNIT E

CHAPTER 9 Circles

1 Basic Terms 346
2 Tangents 350
3 Arcs and Central Angles 355
4 Chords 359
5 Inscribed Angles 364
6 Other Angles 369
7 Segments of Chords 373

FEATURES

Experiments 349 Consumer Applications 349 Puzzle Problems 354, 368, 376
Applications 377 Computer Activities 378

REVIEWS AND TESTS

Self-Tests 358, 368, 376 Skills Review 380 Chapter Review 381
Chapter Test 383 Mixed Review 384

CHAPTER 10 Areas and Volumes of Solids

1 Lines and Planes in Space 388
2 Right Prisms 393
3 Right Circular Cylinders 398
4 Regular Pyramids 402
5 Right Circular Cones 406
6 Spheres 410
7 Similar Solids (Optional) 414

FEATURES

Experiments 392, 405, 409 Consumer Applications 392 Puzzle Problems 401, 405
Calculator Activities 409 Applications 413 Reading Geometry 417
Problem Solving Strategies 418

REVIEWS AND TESTS

Self-Tests 401, 412 Skills Review 420 Chapter Review 421 Chapter Test 423
Cumulative Review 424

UNIT F

CHAPTER 11 Right Triangles
1 Reviewing Right Triangles 428
2 Special Right Triangles 432
3 Using Special Right Triangles 437
4 Diagonals of Rectangular Solids 440
5 Right Triangles in Pyramids and Cones 443
6 The Tangent Ratio 447
7 The Sine and Cosine Ratios 452

FEATURES
Applications 431, 457 Calculator Activities 436 Puzzle Problems 442, 446, 456, 459
Career Notes 451

REVIEWS AND TESTS
Self-Tests 439, 446, 456 Skills Review 460 Chapter Review 461
Chapter Test 463 Mixed Review 464

CHAPTER 12 Coordinate Geometry
1 Points and Coordinates 468
2 Distance Between Two Points 472
3 Midpoint of a Segment 476
4 Slope of a Line 486
5 Parallel and Perpendicular Lines 491
6 Equations and Lines 496
7 Coordinate Geometry Proofs 501

FEATURES
Calculator Activities 475, 481 Applications 482, 505 Career Notes 500
Problem Solving Strategies 506 Computer Activities 508

REVIEWS AND TESTS
Self-Tests 480, 495, 504 Skills Review 510 Chapter Review 511
Chapter Test 513 Cumulative Review 514

TABLE OF SQUARES 515
TABLE OF SQUARE ROOTS 516
TABLE OF TRIGONOMETRIC RATIOS 517

POSTULATES 518
THEOREMS 519
CONSTRUCTIONS 522

REVIEW EXERCISES 523
APPENDIX: TRANSFORMATIONAL GEOMETRY 531

GLOSSARY 543
INDEX 551

ANSWERS TO SELF-TESTS 559
ANSWERS TO SELECTED EXERCISES 562

SYMBOLS

		Page
$\angle$, $\angle$s	angle, angles	10
$\angle ABC$	angle ABC, measure of $\angle ABC$	10
$\overset{\frown}{CD}$	arc with endpoints C and D, measure of $\overset{\frown}{CD}$	355
A	area	238
B	area of base	394
$\odot P$	circle with center P	263
C	circumference	263
$\cong$	congruent, is congruent to	102
cos	cosine	452
$^\circ$	degrees	10
d	diameter,	263
	length of diagonal,	440
	distance	472
$=$	equals, is equal to	7
$\approx$	is approximately equal to	264
$\neq$	is not equal to	59
$>$	is greater than	34
$<$	is less than	53
L.A.	lateral area	394
h	length of altitude, height	238
b	length of base	238
s	length of a side of a square	239
l	length of slant height	402
XY	length of $\overline{XY}$	7
$\overleftrightarrow{CD}$	line through points C and D	6
(x, y)	point with coordinates x and y	84
$\parallel$	parallel, is parallel to	34
$\square$	parallelogram	198
p	perimeter	240
$\perp$	perpendicular, is perpendicular to	15
π	pi	264
r	radius	264
$\frac{a}{b}$, $a:b$	ratio of a to b	286
$\overrightarrow{AB}$	ray with endpoint A	6
$\overline{AB}$	segment with endpoints A and B	6
$\sim$	similar, is similar to	312
sin	sine	452
$\sqrt{x}$	positive square root of x	256
tan	tangent	447
T.A.	total area	394
$\triangle$, $\triangle$s	triangle, triangles	15
V	volume	394

METRIC SYMBOLS

mm	millimeter
cm	centimeter
m	meter
km	kilometer
cm^2	square centimeter
m^2	square meter
km^2	square kilometer
cm^3	cubic centimeter
m^3	cubic meter
L	liter
g	gram
g/cm^3	grams per cubic centimeter
min	minute
h	hour

READING GEOMETRY

HOW TO READ YOUR GEOMETRY TEXTBOOK

When you read a story or a newspaper article, you probably read rather quickly, without giving full attention to every sentence. To get the most out of a textbook, you need to read differently. A geometry textbook calls for slow, thoughtful reading. Here are some suggestions to help you comprehend more when you read about geometry.

VOCABULARY

You will find many new words in this book. Some of them are not much used except in geometry—for example, *collinear*, *vertices*, and *transversal*. Others are everyday words that have special meanings in geometry—for example, *ray*, *acute*, and *supplement*. As you turn the pages, you will see words printed in **heavy type.** These are important words and phrases, whose meanings you will learn as you study the lessons. They are defined or explained when they first appear, and are often illustrated by diagrams. For example, see the definition and illustration of *vertical angles* on page 18. Later, if you need to refresh your memory, you can look up "vertical angles" in the Glossary at the back of the book, or in the Index. The Glossary repeats the definition and tells you the page where it is first stated in the text. The Index may give other page references if you are looking for additional information.

SYMBOLS

Almost as soon as you began to study arithmetic, you learned to use symbols for the operations. Later you became acquainted with other mathematical symbols. Symbols are especially useful in geometry because they provide a lot of information in a small amount of space. For example, look at Statement 1 in the proof of Theorem 1 on page 154, "$\overleftrightarrow{RM} \perp \overline{CD}$." In words this says, "The line through points R and M is perpendicular to the segment with endpoints C and D." Be sure to look for *all* the information given by statements in condensed form. If you have trouble reading a statement expressed by symbols, refer to the list of symbols on the facing page. Then reread the statement slowly. Saying it aloud or writing it out in words may help you. Exercises such as Classroom Practice Exercises 6–10 on page 8 will give you practice in interpreting symbols.

DIAGRAMS

Open this book almost anywhere and you will see diagrams. They are there to help you understand the explanations and the exercises, and to present additional information. You need to read the diagrams along with the words of the text. Be careful not to miss any of the information that is given. Sometimes, as on page 6, you will find red boxes and arrows below or beside the diagrams. These often provide a word statement of information given in a diagram. They may also call your attention to a particular part of the figure.

In reading a diagram, be careful not to take anything for granted that isn't actually shown. Look again at the second diagram on page 7. It wouldn't be safe to say that AM and MB are equal if the little red marks weren't there to tell you that this is true.

Sometimes you may find it helpful to make a quick sketch of your own to illustrate an explanation that you are studying. Be sure to label your drawing to show the relationships correctly.

Exercises such as Written Exercises 3–20 on page 9 and Written Exercises 8–14 on page 49 will help you gain skill in reading diagrams.

ORGANIZATION OF THE TEXTBOOK

Your textbook has been planned to make reading and learning geometry as easy as possible. New words and phrases are printed in **heavy type** and carefully explained when they first occur. Brief explanations in red boxes often appear along with diagrams and other important items. Definitions are repeated in the Glossary, with page references, and the Index lists all pages where you can find information about a particular item. Large red boxes are used to indicate theorems, postulates, and constructions. Other important information is also shown in red. All the postulates, theorems, and constructions are listed at the back of the book, with page references to where they first occur.

Each chapter contains one or more Self-Tests that you can use to check your progress. Answers are given at the back of the book. At the end of every chapter you will find a summary, review exercises, and a review test for the chapter. There is either a cumulative test or a mixed review at the end of every chapter. Reviews of arithmetic and algebraic skills are also included. You will have many opportunities to check your progress and strengthen your skills.

UNIT A

Here's what you'll learn in this chapter:

To name points, lines, rays, segments, angles, and planes.

To measure and classify angles.

To do basic constructions with a compass and a straightedge.

To use postulates of equality and postulates of geometry to justify statements.

Many examples of the geometric figures discussed in this chapter and later in the book can be found in the photograph above. Look for parts of the building that suggest points, lines and line segments, planes, and angles.

Chapter 1

Points, Lines, and Angles

1 Beginning Your Study of Geometry

For each diagram below, tell which line looks longer—the solid red one or the solid black one.

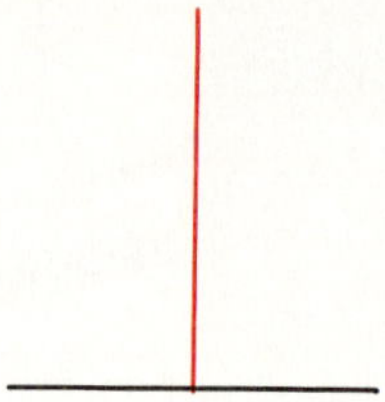

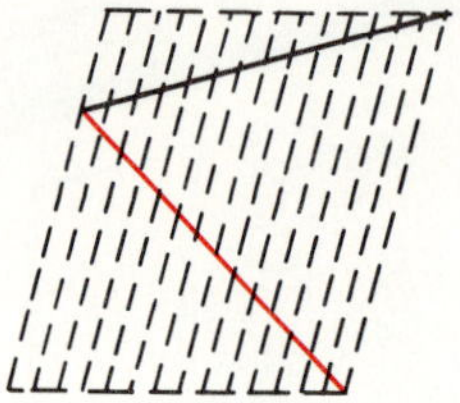

Check your answers by measuring.

Were you fooled by these optical illusions? They show that you cannot always depend on appearances. Sometimes you need to measure. In this course you will need three tools for measuring: a ruler, a protractor, and a compass.

Not all geometry problems can be solved by measuring. Many are solved by reasoning logically, starting with known geometric facts. This is the kind of reasoning you will study in this course. The following activities preview some things that you will study. Try them.

1. Draw a rectangle.
   Join the midpoints of the sides.
   What kind of figure is formed?

   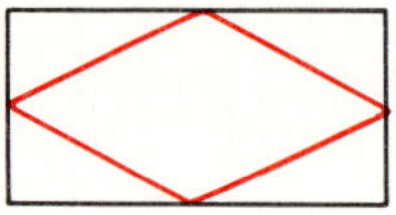

   Now join the midpoints of the "inside" figure to form a third four-sided figure. What kind of figure is formed? How do its dimensions compare with the dimensions of the original rectangle?

   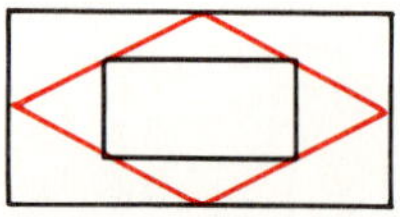

   Try this experiment again, beginning with a different-looking rectangle.

2. Imagine that the earth is a perfectly round ball.
   A giant hoop fits exactly around the earth's equator.
   The hoop is then stretched so that it is 2 m longer.
   (Two meters is about the height of a tall person.)

   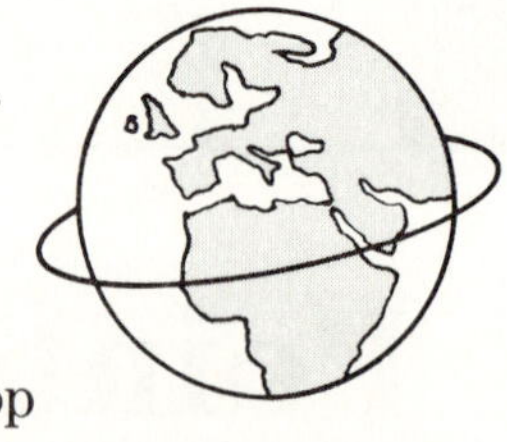

   The hoop now floats evenly around the earth.
   Is there enough room between the earth and the hoop for a flea to crawl under? a fly? a mouse? a cat?

The answers are at the top of the next page.

Activity 1: When you join the midpoints of a rectangle, you form a parallelogram with all sides equal. This figure is called a *rhombus.*

When you join the midpoints of this rhombus, you form a rectangle. The length and width of this rectangle are half those of the original rectangle.

Activity 2: All of them could crawl under the hoop. Later in the course you will be able to prove that there are about 32 cm between the earth and the hoop.

The activities should suggest that some geometric problems have surprising results. To get these results, we begin our study with some very *un*surprising ones. We will piece these together to prove more substantial results, called *theorems.* In a way, this process is like detective work. We will fit together geometric clues to arrive at conclusions. The exercises will give you a chance to do some detective work.

Exercises

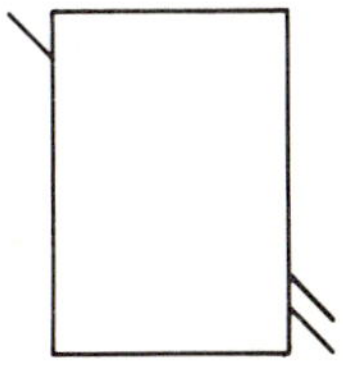

1. Which line at the right of the rectangle is the continuation of the line at the left?

2. Draw several four-sided figures. For each, join the midpoints of the sides. In each case, what kind of figure is formed?

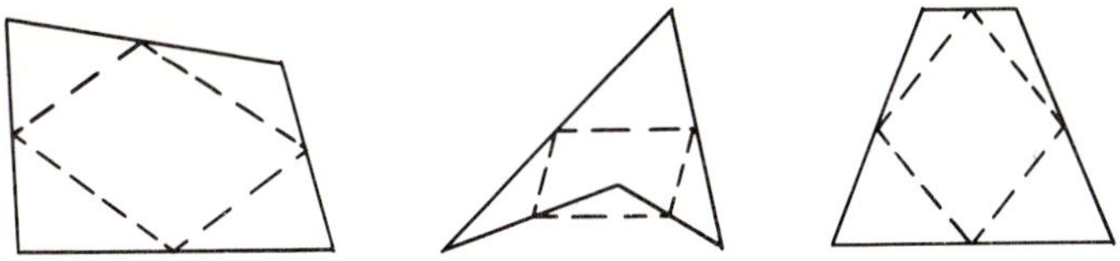

3. Wrap a piece of string tightly around a large wastebasket. Wrap another piece around a tin can. Then add two meters of string to each piece. Finally, spread the lengthened strings evenly around the wastebasket and the can as shown.

Compare the space between the string and the wastebasket with the space between the string and the can. What do you notice?

2 Points and Lines

Do you know that a color television picture is composed of thousands of dots? You don't notice the spaces between the dots because the dots are so close to each other.

The idea of tightly-packed dots is sometimes used in mathematics. For example, the picture below suggests that a **line** contains very many **points.**

Each dot represents a point of the line.

Actually, a line contains an unlimited number of points. We usually picture a line by drawing along the edge of a ruler, like this.

The arrowheads in the drawing suggest that a line extends indefinitely far in both directions.

The drawings below explain how we name points and lines.

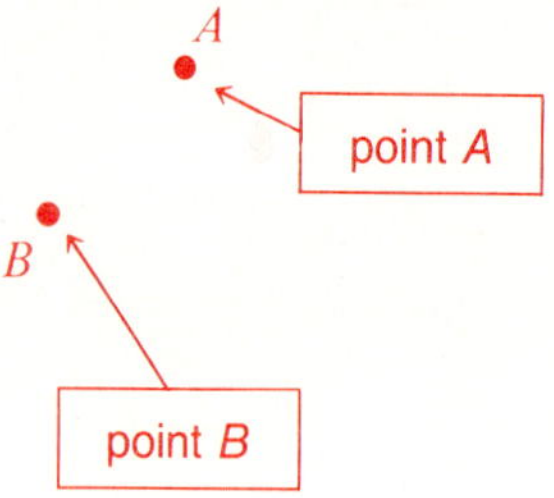

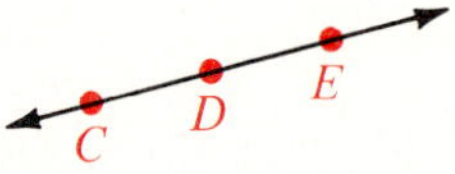

$\overleftrightarrow{CD}$ (read *line CD*) means the line through points *C* and *D*. $\overleftrightarrow{CD}$, $\overleftrightarrow{DE}$, and $\overleftrightarrow{CE}$ all refer to the same line.

We can also refer to a line using a lower-case letter. The line shown can be called line *l*.

The points *C*, *D*, and *E* shown above are called **collinear points.** This means that they are all on one line. On the other hand, the points *X*, *Y*, and *Z* pictured at the right are not collinear points. They do not lie on one line.

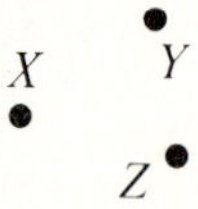

A *segment* and a *ray* are parts of a line as shown below.

$\overline{AB}$ is read **segment** *AB*.
A and *B* are called *endpoints* of the segment.

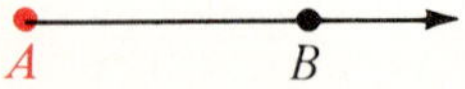

$\overrightarrow{AB}$ is read **ray** *AB*.
A is called the *endpoint* of the ray.
The endpoint is always named first.

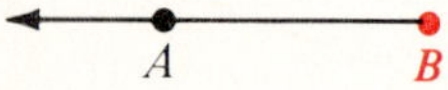

$\overrightarrow{BA}$ is read *ray BA*. A ray continues indefinitely in one direction—in this case, from *B* through *A* and beyond.

Notice that $\overrightarrow{BA}$ and $\overrightarrow{AB}$ refer to different rays. On the other hand, $\overline{BA}$ and $\overline{AB}$ refer to the same segment.

Each point of a line can be paired with a real number. Similarly, each real number can be paired with a point.

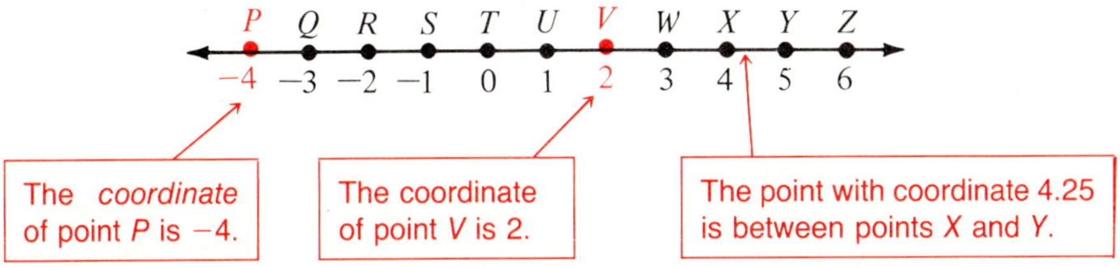

The symbol XY stands for the length of $\overline{XY}$. This length is just the distance between X and Y. It can be found by counting the number of units from X to Y. It can also be found by subtracting the smaller coordinate (the one on the left) from the larger coordinate (the one on the right). Using the figure above, we find:

$$XY = 5 - 4 = 1$$

$$SU = 1 - (-1) = 1 + 1 = 2$$

$$PS = -1 - (-4) = -1 + 4 = 3$$

In the diagram above, the *midpoint* of $\overline{SU}$ is the point T. It is exactly halfway between S and U. In general, the **midpoint** of a segment is defined in this way:

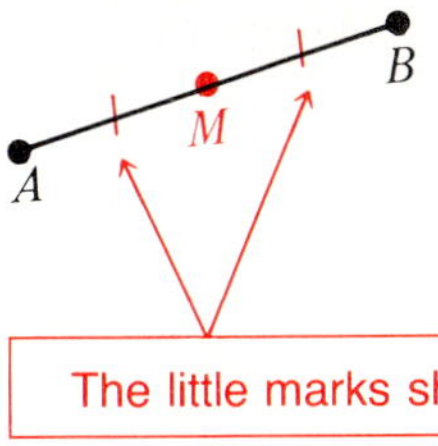

M is the midpoint of $\overline{AB}$ means that

(1) M is on $\overline{AB}$ and

(2) $AM = MB$.

From the definition of midpoint it is clear that

$$AM = \tfrac{1}{2}AB \text{ and } MB = \tfrac{1}{2}AB.$$

EXAMPLE

In the diagram, Q is the midpoint of $\overline{PS}$, $RS = 2$, and $QR = 3$. Find PS.

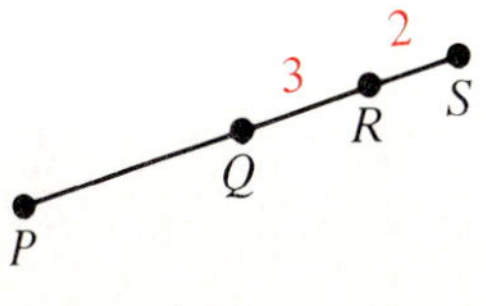

$QS = QR + RS = 3 + 2 = 5$
Since $QS = 5$, PQ also equals 5.
Then $PS = PQ + QS = 10$.

Classroom Practice

Two lines meet at O as shown. Tell whether each statement is true or false.

1. $\overleftrightarrow{OS}$ is the same as $\overleftrightarrow{OR}$.

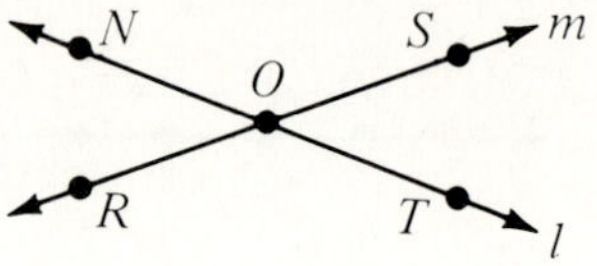

2. $\overline{OS}$ is the same as $\overline{OR}$.
3. Line m contains just three points.
4. Points R, O, and S are collinear.
5. Points N, O, and S are collinear.

Exercises 6–10 refer to the diagram at right.

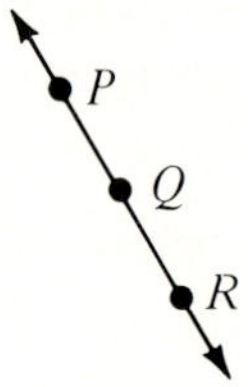

6. Read the symbol $\overline{QR}$ and explain what it means.
7. Name the endpoints of: **a.** $\overline{PQ}$; **b.** $\overline{QP}$.
8. Read the symbol $\overrightarrow{QP}$ and explain what it means.
9. Name the endpoint of: **a.** $\overrightarrow{PQ}$; **b.** $\overrightarrow{QP}$.
10. Which points are in both $\overrightarrow{QR}$ and $\overrightarrow{PQ}$?

Exercises 11–15 refer to the diagram.

11. Name the coordinate of X.

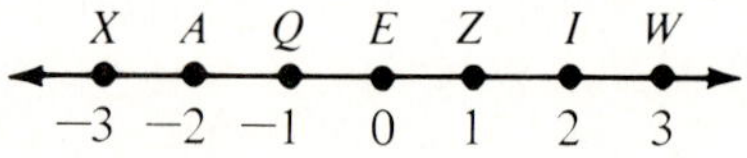

12. Name the point with coordinate 3.
13. Name the midpoint of $\overline{ZW}$.
14. Find each distance: **a.** ZW; **b.** EI; **c.** XZ.
15. How many real numbers are there between 1 and 2—a limited number or an unlimited number? How many points are there between Z and I?

Written Exercises

A

1. Suppose $AB = 4$, $BC = 3$, and $AD = 14$.
   a. Then $CD =$ __?__.
   b. Is C the midpoint of $\overline{AD}$?

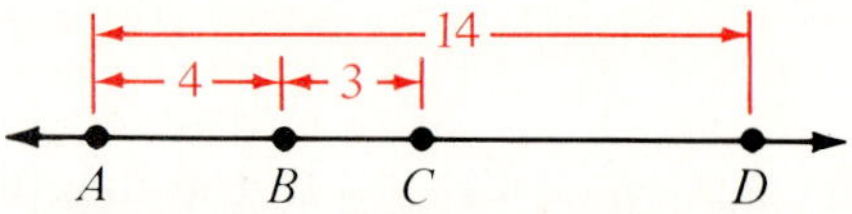

2. Suppose O is the midpoint of $\overline{HL}$.
   a. If $HL = 16$, then $OL =$ __?__.
   b. Suppose you also know that W is the midpoint of $\overline{OL}$. Then $OW =$ __?__.

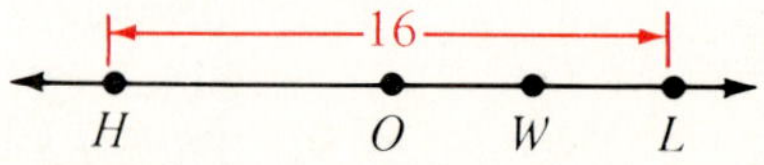

Exercises 3–10 refer to the figure.

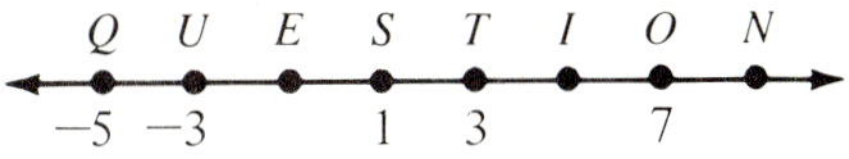

Exercises 3–10

3. The coordinate of point U is _?_.

4. If I is the midpoint of $\overline{TO}$, then the coordinate of I is _?_.

5. If U is the midpoint of $\overline{QE}$, then the coordinate of E is _?_.

6. If $ON = 2$, then the coordinate of N is _?_.

Find each distance.

7. TO 8. TU 9. QU 10. OQ

The coordinates of I and A are -10 and 20, as shown. O is the midpoint of $\overline{IA}$, $IE = 6$, and $IU = 23$. Find each of the following.

11. The coordinate of O

12. The coordinate of E

13. The coordinate of U

Exercises 11–20

14. IO 15. EU 16. EA

Using the diagram above, tell whether each statement is true or false.

17. There are an unlimited number of points between E and O.

18. $\overline{EU}$ and $\overline{UE}$ refer to the same segment.

19. $\overrightarrow{EU}$ and $\overrightarrow{UE}$ refer to the same ray.

20. $\overleftrightarrow{EU}$ and $\overleftrightarrow{AO}$ refer to the same line.

Suppose S is the midpoint of $\overline{RT}$. Complete the table.

B

	21.	22.	23.	24.	25.	26.	27.	28.
coordinate of R	8	−4	−18	5	17	−18	a	$b - 10$
coordinate of T	12	12	−2	?	?	?	$a + 6$	6
length of $\overline{RT}$	?	16	?	?	?	?	?	?
coordinate of S	10	?	?	9	29	−10	?	?

29. Look up the meaning of the prefix "geo-." Relate this meaning to the words "geometry," "geography," and "geology."

C 30. Points X, Y, and Z have coordinates 2.001, $\frac{21}{8}$, and $\sqrt{5}$. Which of these points are between points C and D?

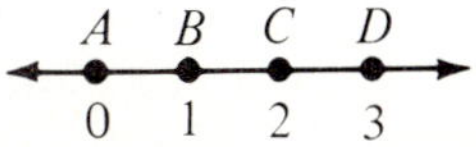

3 Angles

As you may know, an **angle** is a figure formed by two rays or segments with a common endpoint. The rays or segments are called the **sides** of the angle and the endpoint is called the **vertex** of the angle. The diagrams below show how angles are named.

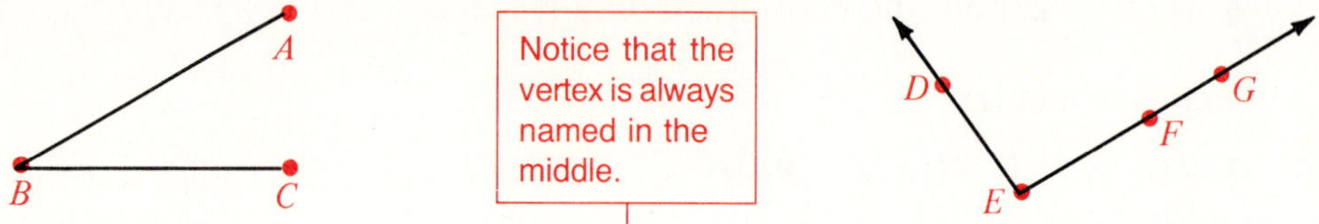

$\angle ABC$ means angle ABC.
$\angle CBA$ is another name for $\angle ABC$

$\overline{BA}$ and $\overline{BC}$ are the sides.
B is the vertex.

$\angle DEF$ means angle DEF.
$\angle DEF$ is the same as $\angle DEG$.

$\overrightarrow{ED}$ and $\overrightarrow{EF}$ are the sides.
E is the vertex.

Sometimes we use just one letter or one number to name an angle.

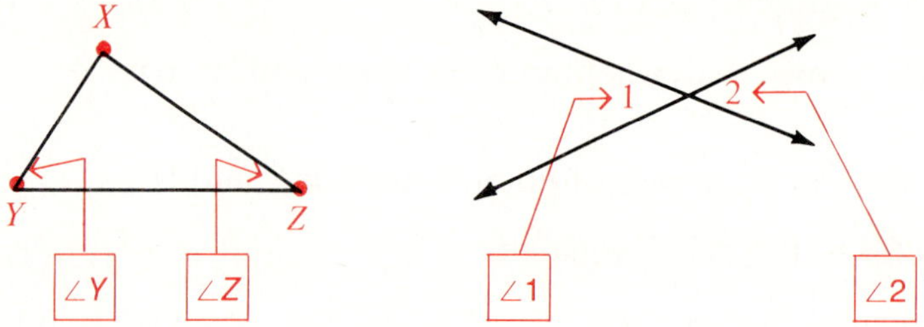

To find the *measure in degrees* of an angle, we use an instrument called a *protractor*. The diagram below shows how to do this. Can you see that the measure of $\angle AOB$ is 40°? This fact is written: $\angle AOB = 40°$.*

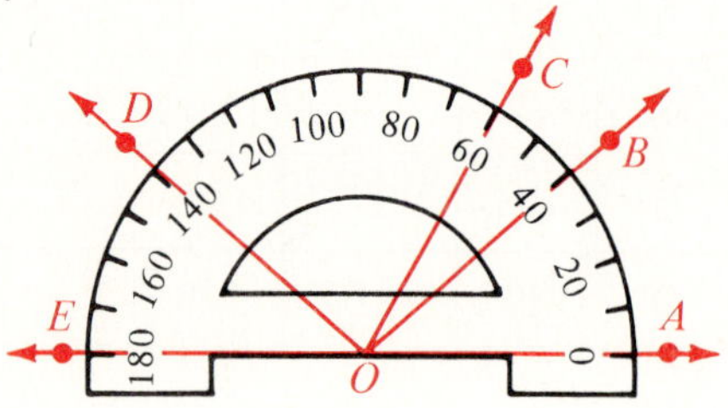

Notice that:
$\angle AOB = 40°$
$\angle AOE = 180°$
$\angle BOC = 60° - 40° = 20°$
$\angle DOB = 140° - 40° = 100°$
$\angle EOC = 180° - 60° = 120°$

*Note: Some books use the notation "$m\angle AOB = 40°$" and say, "the measure of $\angle AOB$ is 40°." In this book, we shall use the simpler notation "$\angle AOB = 40°$" and say, "$\angle AOB$ equals 40°."

Many protractors have two scales, one reading from left to right, and the other reading from right to left. To use this kind of protractor, follow these steps:

1. Estimate: Is the angle measure less than or greater than 90°?
2. Carefully line up your protractor as shown below. If the sides of the angle do not reach the scale, extend them.

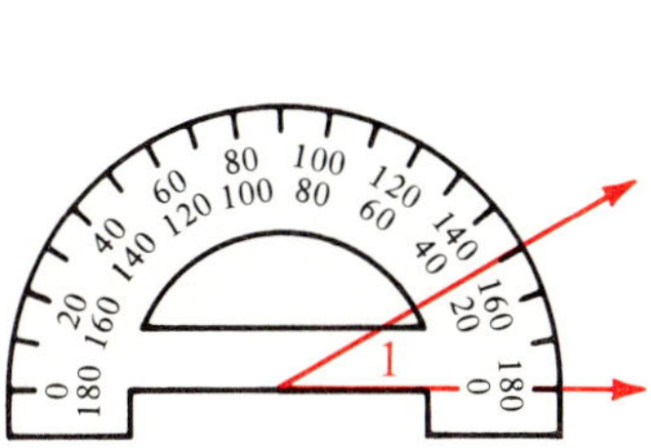

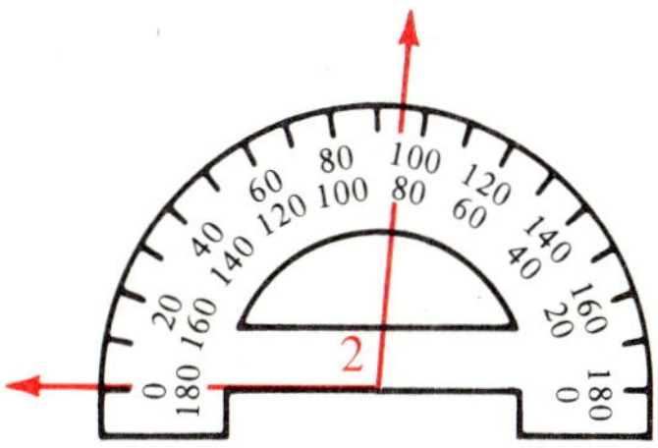

3. Choose the scale that has 0° at one side of the angle. Read the measure of the angle. Does the measure agree with your estimate?

EXAMPLE

What is the measure of $\angle 3$?

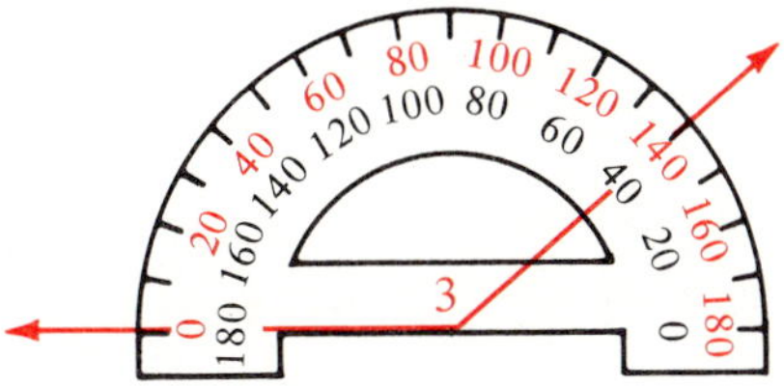

$\angle 3$ is greater than 90°.

Choose the scale on top.

$\angle 3 = 140°$

The diagram at the bottom of page 10 suggests that we can add and subtract angle measures. Here are some other examples.

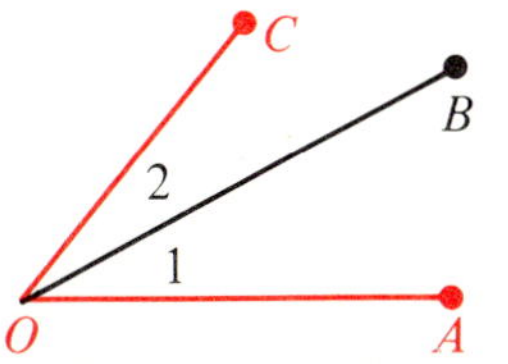

$\angle 1 + \angle 2 = \angle AOC$

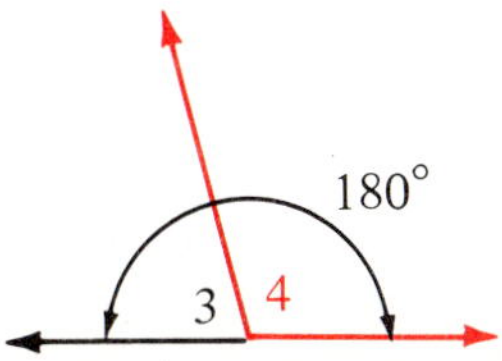

$180° - \angle 3 = \angle 4$

Classroom Practice

Exercises 1–7 refer to the diagram.

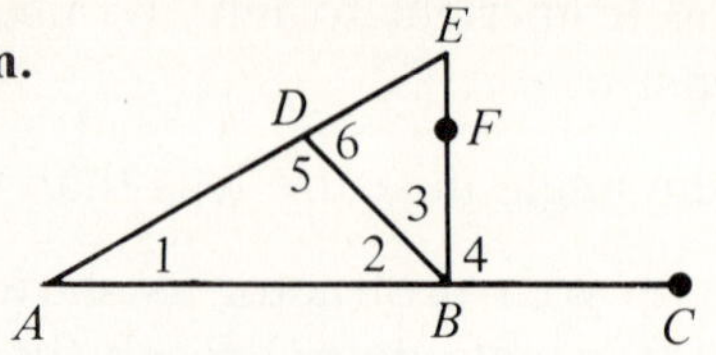

1. Name the vertex of $\angle 4$.
2. Name the sides of $\angle 2$.
3. $\angle BAE$ and $\angle 1$ are names for the same angle. Give another name for each of the following angles: $\angle ABD$, $\angle AEB$, $\angle BDE$, $\angle 3$, $\angle 5$, $\angle 4$.
4. $\angle 5 + \angle 6 = \underline{\quad ? \quad}^\circ$
5. $\angle ABE - \angle 3 = \angle \underline{\quad ? \quad}$
6. Explain why you should not refer to $\angle B$.
7. **a.** Is it correct to refer to $\angle A$?
   b. Is it correct to refer to $\angle D$?

In Exercises 8–10, use the diagram shown.

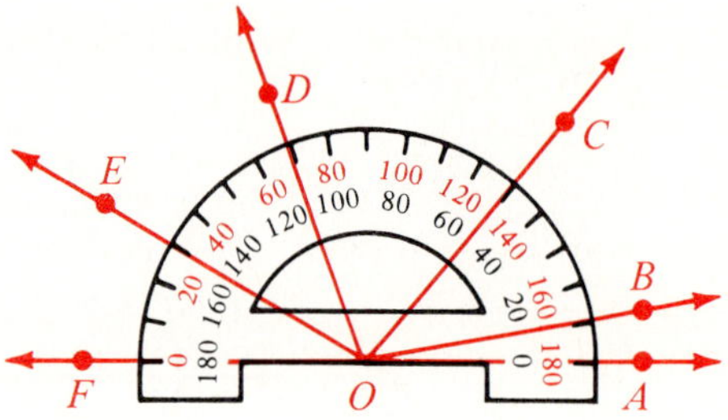

8. Using three letters for each angle, name five angles that have $\overrightarrow{OA}$ as one side.
9. Using three letters for each angle, name five angles that have $\overrightarrow{OC}$ as a side.
10. State the measure of each of the following: $\angle AOB$, $\angle AOC$, $\angle AOD$, $\angle AOE$, $\angle AOF$.
11. Estimate the measure of each angle shown.

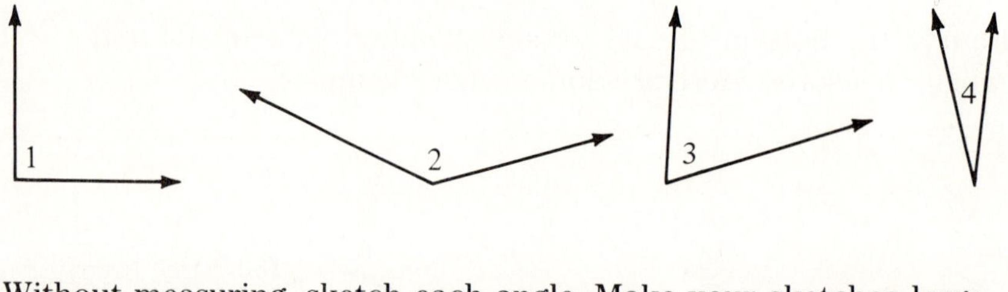

12. Without measuring, sketch each angle. Make your sketches large enough so that you can use a protractor to check the accuracy of your work.

    a. 90° angle **b.** 60° angle **c.** 10° angle **d.** 170° angle

Written Exercises

Complete each statement.

A 1. The sides of $\angle AIH$ are _?_ and _?_.

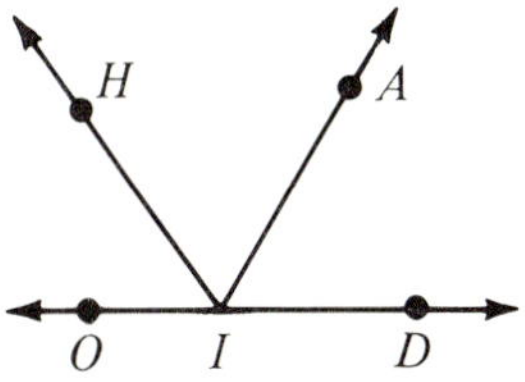

2. The vertex of $\angle AIH$ is _?_.

3. Another name for $\angle OIH$ is $\angle$ _?_.

4. $\angle DIA + \angle AIH = \angle$ _?_

5. $\angle DIH + \angle HIO = \angle$ _?_ = _?_ °

6. $180° - \angle DIA = \angle$ _?_

Find the measure of each angle named.

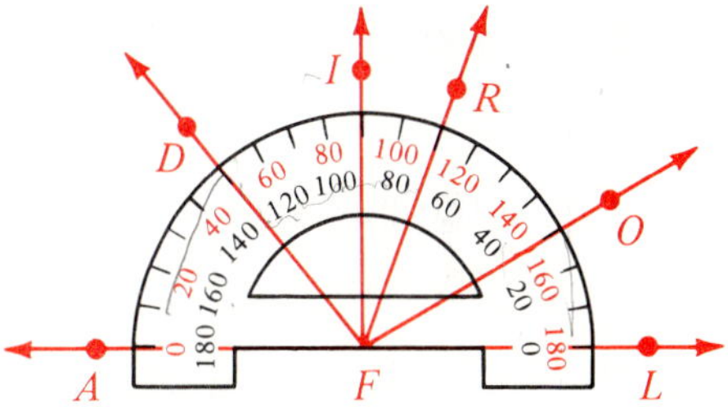

7. $\angle LFO$
8. $\angle RFA$
9. $\angle DFO$
10. $\angle AFL$
11. $\angle DFR$
12. $\angle IFL$

Use your protractor to draw an angle with the given measure.

13. 20°
14. 65°
15. 90°
16. 140°

17. Estimate the measure of each angle shown.

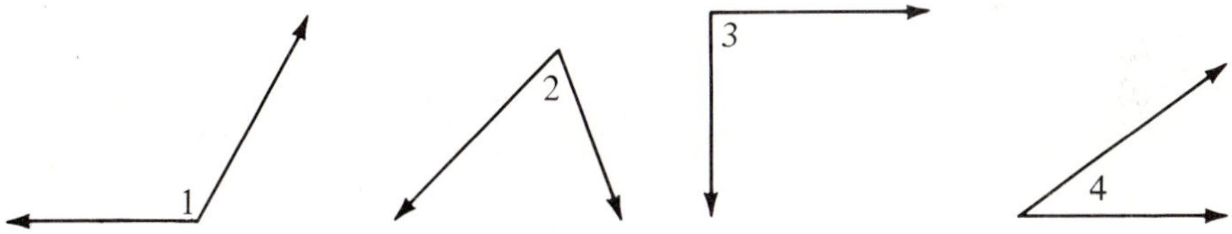

18. Without measuring, sketch each angle. Then use a protractor to check the accuracy of your work.
   a. 45° angle
   b. 15° angle
   c. 85° angle
   d. 160° angle

Complete each statement.

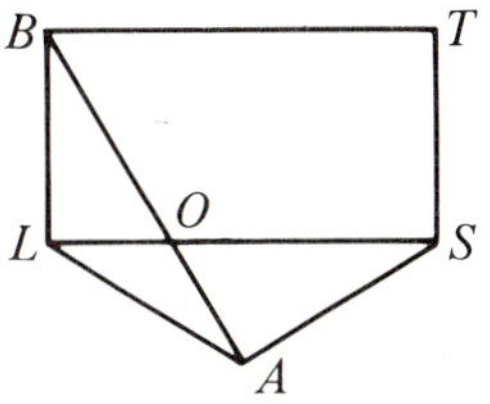

SAMPLE $\angle BLO + \angle OLA$ = _?_
$\angle BLO + \angle OLA = \angle BLA$

B 19. $\angle BAS + \angle BAL = \angle$ _?_

20. $\angle TSO + \angle OSA = \angle$ _?_

21. $\angle LAS - \angle LAB = \angle$ _?_

22. $180° - \angle BOS = \angle$ _?_ $= \angle$ _?_

SELF-TEST

Vocabulary

line (p. 6)
point (p. 6)
collinear points (p. 6)
segment (p. 6)
ray (p. 6)
midpoint (p. 7)
angle (p. 10)
sides of an angle (p. 10)
vertex of an angle (p. 10)

Chapter 1, Section 2

Tell whether each statement is true or false.

(1-2)

1. $\overleftrightarrow{AB}$ is the same as $\overleftrightarrow{BC}$.
2. $\overrightarrow{AB}$ is the same as $\overrightarrow{BA}$.
3. C is the midpoint of $\overline{BD}$.
4. $AB = 6$

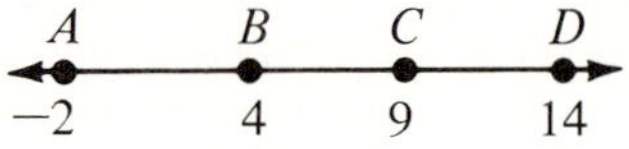

(1-3)

5. What is the vertex of $\angle PRQ$?
6. Name $\angle 2$ using three letters.
7. What angle has a measure of about 90°? Use three letters to name the angle.
8. $\angle 5 + \angle 6 =$ __?__ °

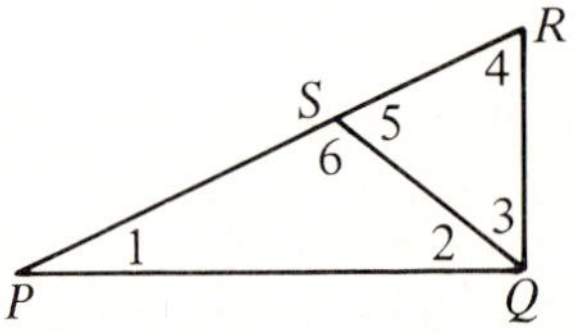

CONSUMER APPLICATIONS

TIME IS MONEY

People use the money they earn in their jobs to buy things they want and need: food, clothing, housing, books, and recreation. How much time must a person work to make these common purchases?

Suppose a worker takes home \$5.40 per hour. That is equivalent to 9¢ per minute. ($5.40 \div 60 = 0.09$) How much time will it take the worker to earn the money to pay for the following items?

a \$16.83 toaster	a monthly rent of \$360
a \$216 television set	a \$48.60 grocery bill
a \$3.78 paperback book	a 5¢ piece of gum

4 Classifying Angles

Angles are often classified by their measures.

An **acute angle** has a measure between 0° and 90°.

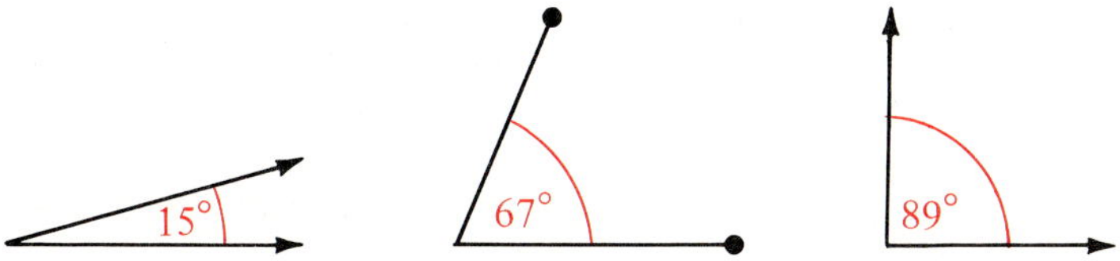

A **right angle** has measure 90°. We indicate a right angle by drawing a small square inside the angle as shown at the right.

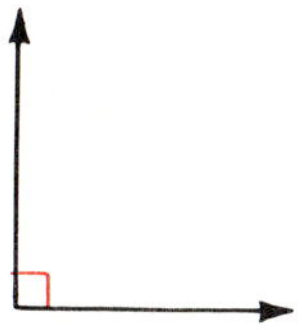

If an angle formed by two lines is a right angle, then the lines are **perpendicular.** Also, if two lines are perpendicular, they form right angles. In the diagram, line *AB* is perpendicular to line *CD*. This is written:

$$\overleftrightarrow{AB} \perp \overleftrightarrow{CD}.$$

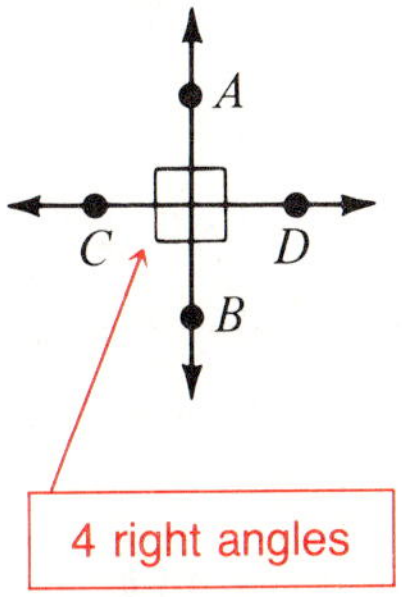

An **obtuse angle** has a measure between 90° and 180°.

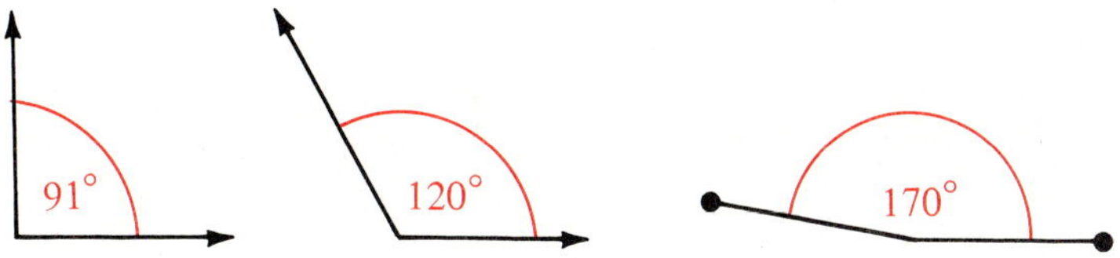

A **straight angle** has measure 180°.

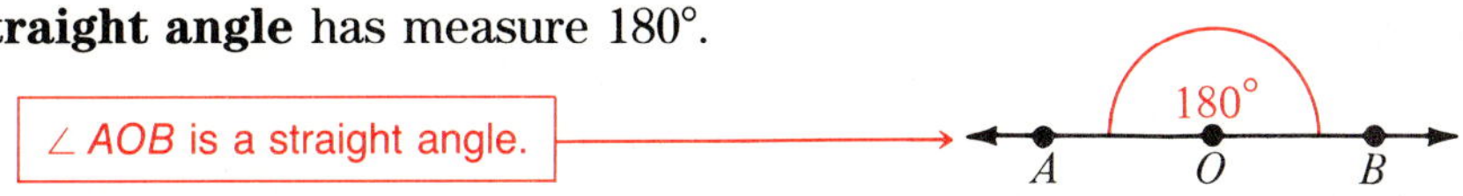

The diagram at the right shows a **triangle** with **vertices** *A*, *B*, and *C*. *Vertices* is the plural of *vertex*. The triangle pictured is called *triangle ABC* and is represented by the symbol $\triangle ABC$. Since $\angle B$ and $\angle C$ have equal measures, we say that the angles are equal and write

$$\angle B = \angle C.$$

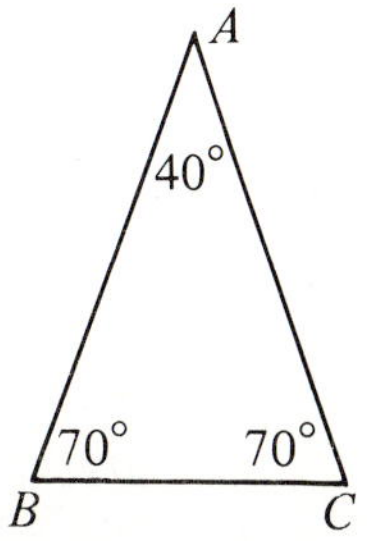

Also, since the measures of $\angle A$, $\angle B$, and $\angle C$ total 180°, we write

$$\angle A + \angle B + \angle C = 180°.$$

EXAMPLE In the diagram, $\overline{AC} \perp \overline{EB}$ and $\angle EAD = 40°$.
Find the measure of **a.** $\angle DAC$ and **b.** $\angle DAB$.

a. $\angle DAC = \angle EAC - \angle EAD$
$= 90° - 40°$
$= 50°$

b. $\angle DAB = \angle EAB - \angle EAD$
$= 180° - 40°$
$= 140°$

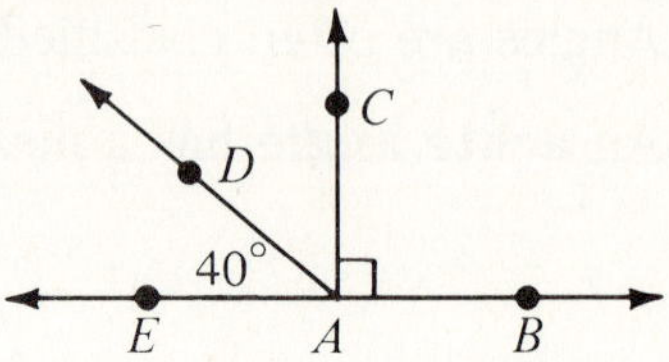

Classroom Practice

Classify each angle as acute, right, obtuse, or straight.

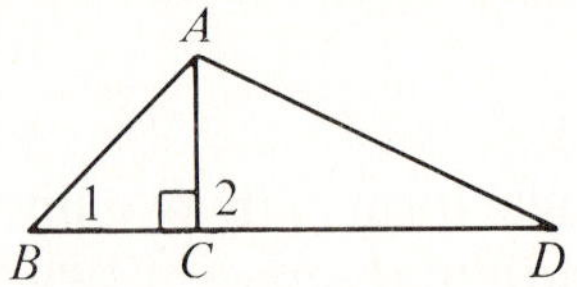

1. $\angle 1$ **2.** $\angle 2$ **3.** $\angle DAB$

4. $\angle ACB$ **5.** $\angle CAB$ **6.** $\angle DCB$

7. An angle that is equal to $\angle 2$ is $\angle$ __?__.

8. You know that $\overline{AC} \perp \overline{BD}$ because $\angle 2$ is __?__.

9. The three triangles pictured are $\triangle$ __?__, $\triangle$ __?__, and $\triangle$ __?__.

10. Name the three vertices of $\triangle ZOS$.

11. Name the three angles of $\triangle XOZ$.

12. How many angles are shown in the diagram? (Do not count straight angles.)

13. How many triangles are shown in the diagram?

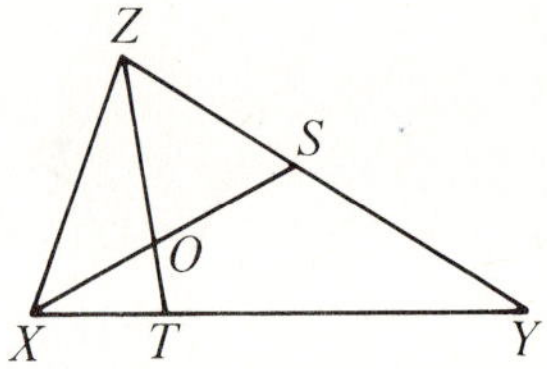

Written Exercises

The measures of several angles are given below.
Classify each angle as acute, right, obtuse, or straight.

A **1.** 18° **2.** 157° **3.** 128° **4.** 90° **5.** 180° **6.** 89°

7. Name four obtuse angles.

8. Name six acute angles.

9. Name four pairs of angles whose sum is a straight angle.

10. Name four triangles.

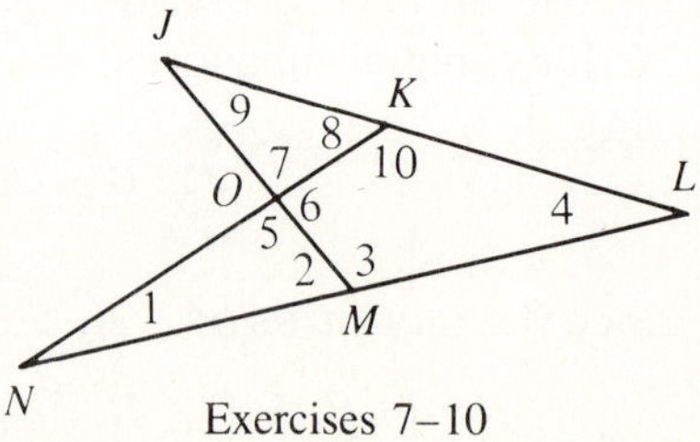

Exercises 7–10

In the figure, $\overline{AB} \perp \overline{CD}$ and $\angle SRD = 30°$.

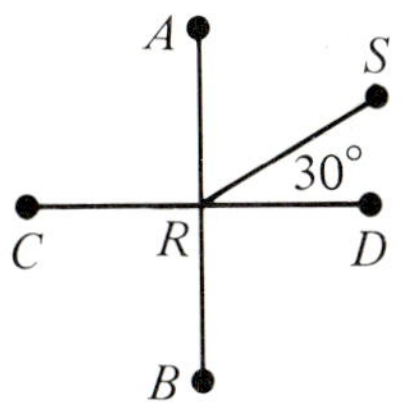

11. Name four right angles.

12. Find the measure of each angle named.
 a. $\angle ARS$ b. $\angle CRS$ c. $\angle BRS$

In the figure, $\overline{TQ} \perp \overline{VR}$ and $\overline{UQ} \perp \overline{QS}$.

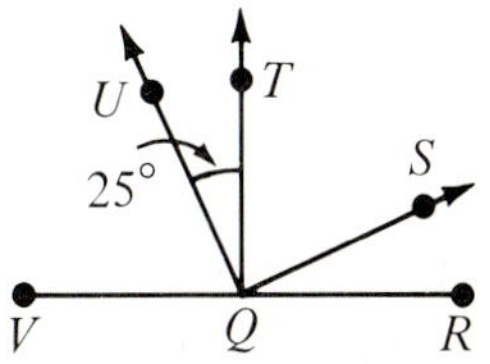

13. Name three right angles.

14. Name two obtuse angles.

15. Find the measure of each angle named.
 a. $\angle VQU$ b. $\angle TQS$ c. $\angle SQR$

B 16. a. Draw a triangle with three acute angles.
 b. Draw a triangle with just two acute angles.
 c. Can you draw a triangle with just one acute angle?

17. If the angles between the panes of the stained glass window shown are all equal, find the measure of each angle.

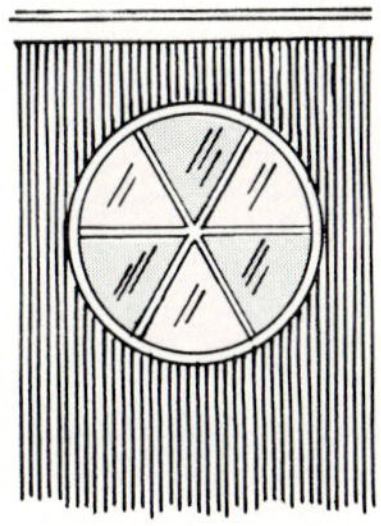

Exercise 17

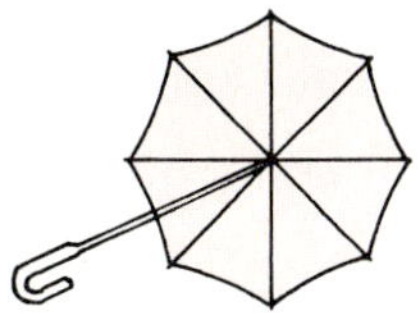

Exercise 18

18. If the angles between the spokes of an umbrella are all equal, find the measure of each angle.

19. Use a dictionary to find the meanings of an "acute" person and an "obtuse" person.

20. Study a corner of a room where two walls and the floor meet. How many right angles are formed at the corner?

PUZZLE ◆ PROBLEMS

Here are six lines.
Add five more lines to make NINE!

5 Vertical Angles

When two lines meet, they form four angles. Two angles that are opposite each other, such as $\angle 1$ and $\angle 3$ in the diagram, are called **vertical angles.** $\angle 2$ and $\angle 4$ are also vertical angles.

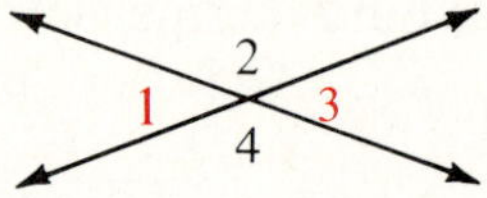

Suppose $\angle 1 = 30°$. Do you see that $\angle 2 = 150°$? (Remember that $\angle 1 + \angle 2 = 180°$.) Now do you see that $\angle 3 = 30°$? How large is $\angle 4$? This information is given in the first row of the table at the right. Copy and complete rows 2 and 3.

$\angle 1$	$\angle 2$	$\angle 3$	$\angle 4$
30°	150°	30°	150°
20°	?	?	?
25°	?	?	?

You can probably guess that vertical angles are always equal. Before reading further, see if you can explain why this is so.

The following steps show that *vertical angles are always equal.* Refer to the diagram above.

Step 1 $\angle 1 = 180° - \angle 2$ because $\angle 1 + \angle 2 = 180°$.
Step 2 $\angle 3 = 180° - \angle 2$ because $\angle 2 + \angle 3 = 180°$.
Step 3 $\angle 1 = \angle 3$ because both angles equal $180° - \angle 2$.

In the same way, you can show that $\angle 2 = \angle 4$.

EXAMPLE

Study the diagram and find the measure of
a. $\angle DOE$ **b.** $\angle BOC$ **c.** $\angle COD$.

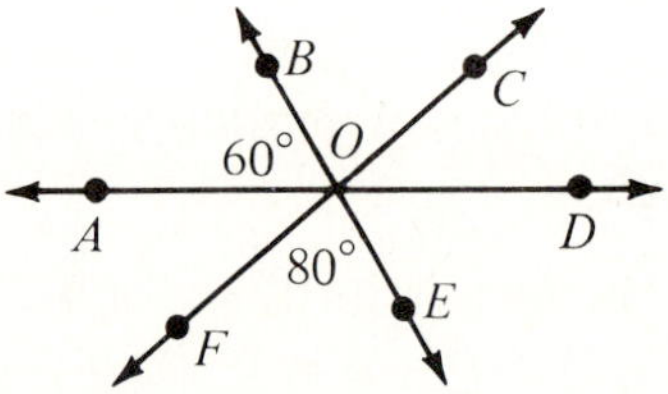

a. $\angle DOE$ and $\angle AOB$ are vertical angles. Therefore, both equal 60°.

b. $\angle BOC$ and $\angle FOE$ are vertical angles. Therefore, both equal 80°.

c.
$$\begin{aligned} \angle AOB + \angle BOC + \angle COD &= 180° \\ 60° + 80° + \angle COD &= 180° \\ 140° + \angle COD &= 180° \\ \angle COD &= 40° \end{aligned}$$

Classroom Practice

Find the measure of each angle.

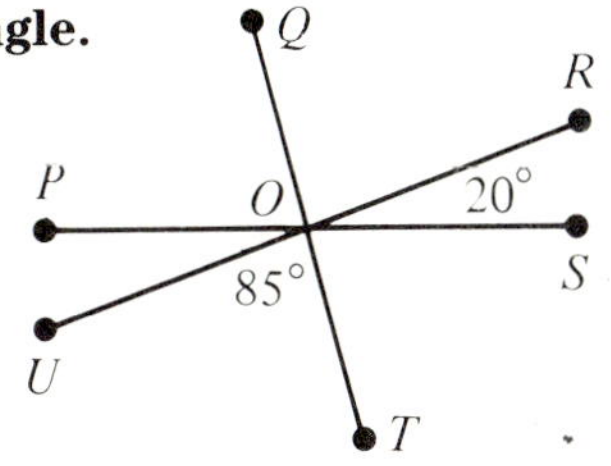

1. $\angle POU =$ __?__ °
2. $\angle QOR =$ __?__ °
3. $\angle POQ =$ __?__ °
4. $\angle SOT =$ __?__ °
5. There are six pairs of vertical angles in the diagram. Can you find all six pairs?

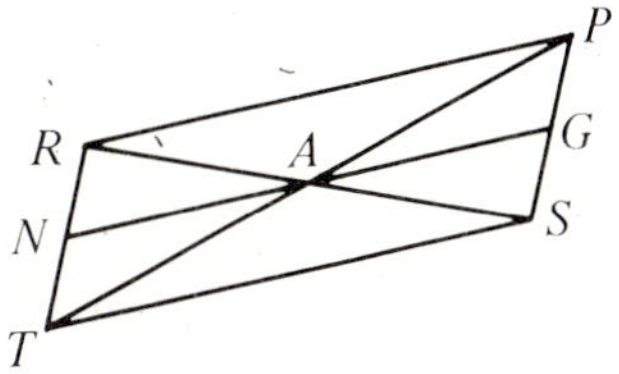

Written Exercises

Name an angle equal to the given angle.

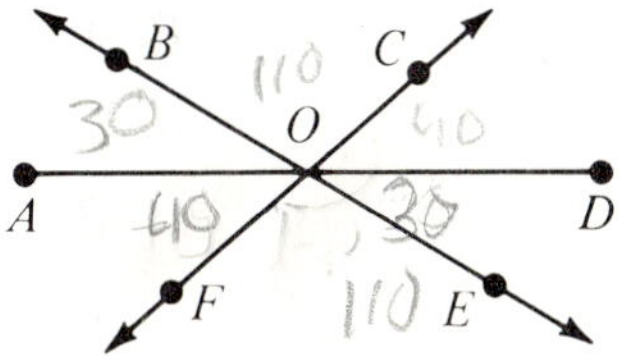

Exercises 1–8

A 1. $\angle BOA$ 2. $\angle BOC$ 3. $\angle AOC$ 4. $\angle DOC$

Suppose $\angle BOA = 30°$ and $\angle DOF = 140°$. Find the measure of each angle.

5. $\angle COD$ 6. $\angle BOC$ 7. $\angle COE$ 8. $\angle BOF$

In the diagram, $\overline{AE} \perp \overline{GC}$, $\angle GOH = 25°$, and $\angle AOB = 35°$. Find the measure of each angle.

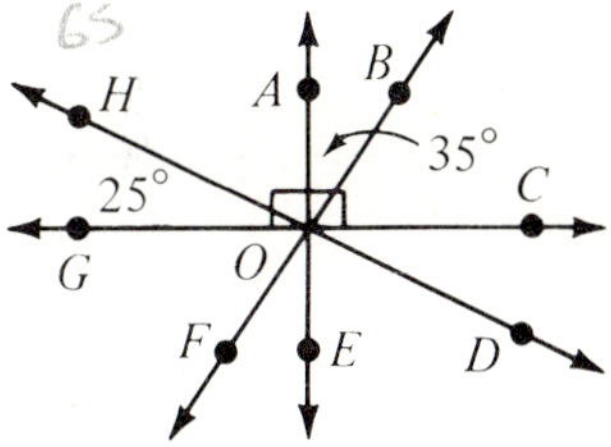

Exercises 9–14

9. $\angle AOH$ 10. $\angle BOC$ 11. $\angle COD$

12. $\angle DOE$ 13. $\angle EOF$ 14. $\angle FOG$

Find the values of x and y.

B 15.

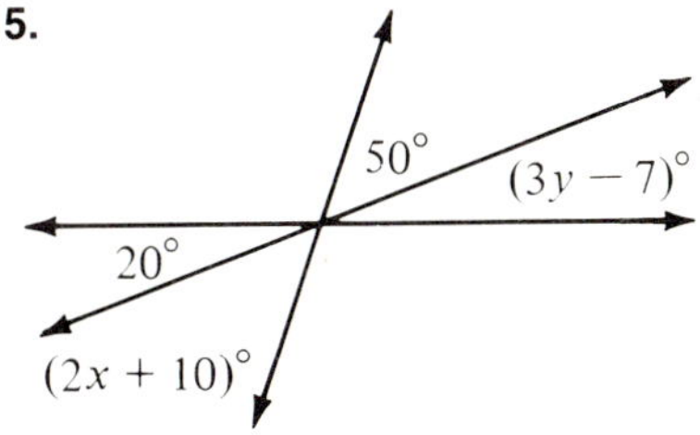

16.

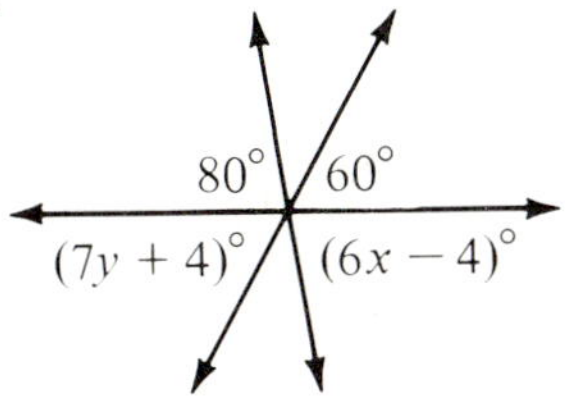

C 17. Draw a diagram showing $\overleftrightarrow{AB}$ and $\overleftrightarrow{CD}$ intersecting at point O so that $\angle AOD = \angle BOD$. Why must $\overleftrightarrow{AB}$ be perpendicular to $\overleftrightarrow{CD}$?

6 Angle Bisectors

A **bisector of an angle** is a ray or line which divides the angle into two equal angles. In the diagram below, $\overrightarrow{OX}$ *bisects* $\angle AOB$.

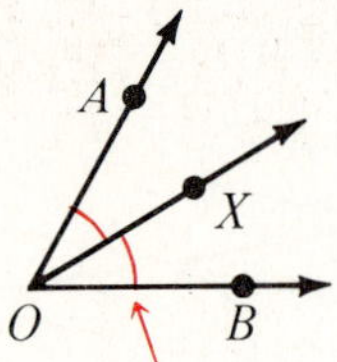

When $\overrightarrow{OX}$ bisects $\angle AOB$, $\angle AOX = \angle BOX$.

Thus $\angle AOX = \frac{1}{2} \angle AOB$

and $\angle BOX = \frac{1}{2} \angle AOB$.

The red marks show that $\angle AOX = \angle BOX$.

EXAMPLE

$\overrightarrow{OB}$ bisects $\angle AOC$.

Find the measure of $\angle AOB$.

Then find the number on the protractor that corresponds to $\overrightarrow{OB}$.

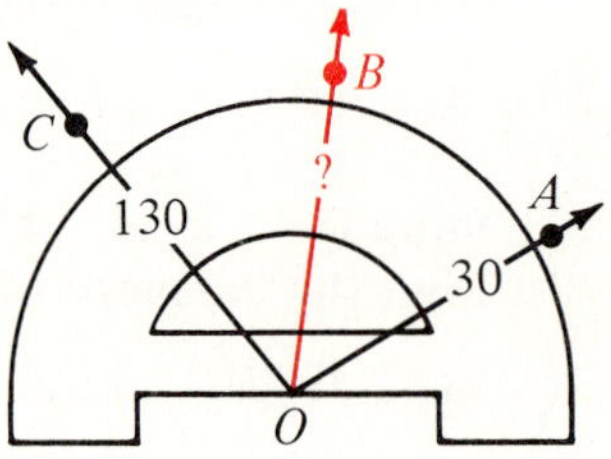

$\angle AOB = \frac{1}{2} \angle AOC$

$= \frac{1}{2}(130° - 30°)$

$= \frac{1}{2}(100°) = 50°$

$\overrightarrow{OB}$ corresponds to $30° + 50°$, or $80°$.

Classroom Practice

In each diagram, $\overrightarrow{SZ}$ bisects $\angle RST$.

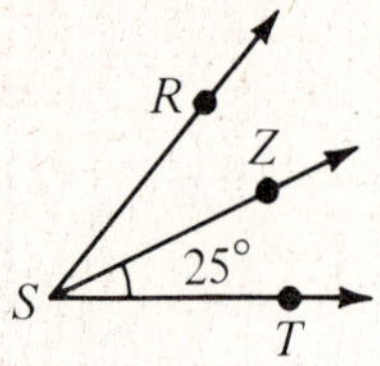

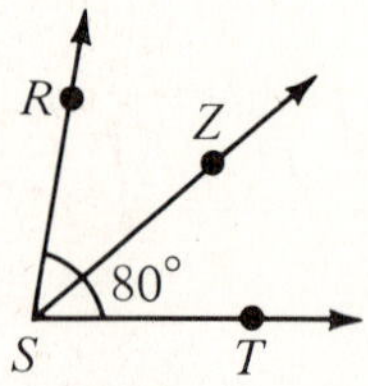

1. $\angle RSZ =$ __?__ °
2. $\angle RST =$ __?__ °
3. $\angle RSZ =$ __?__ °
4. $\angle ZST =$ __?__ °
5. How many lines can bisect a given angle?

Written Exercises

A

1. Draw an acute angle. Label it $\angle JOG$.
   Draw a bisector of $\angle JOG$ and call it $\overrightarrow{OP}$.
   Now draw $\overrightarrow{ON}$ bisecting $\angle JOP$.
   Suppose $\angle JOG = 40°$.
   Then $\angle JOP =$ __?__° and $\angle JON =$ __?__°.

In each exercise, $\overrightarrow{OB}$ bisects $\angle AOC$. Find
a. the measure of $\angle AOB$ and
b. the number on the protractor that corresponds to $\overrightarrow{OB}$.

2.

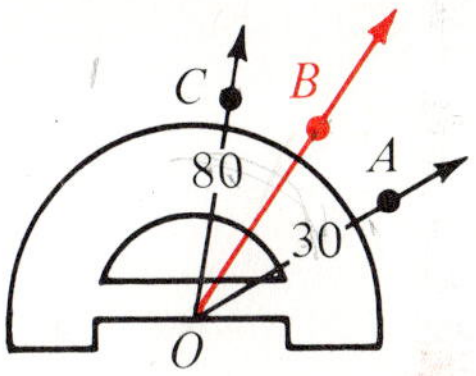

3.

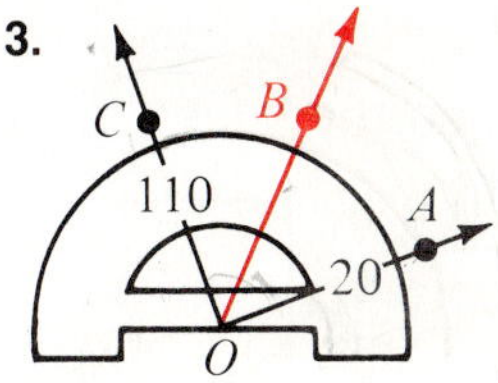

4.

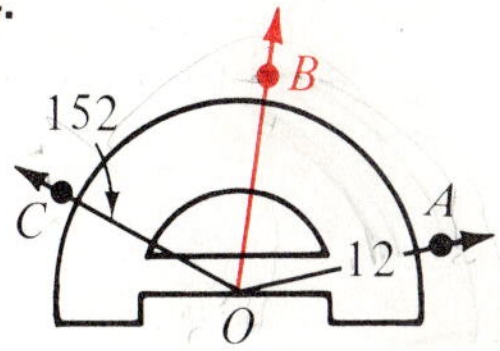

***LFAE* is a four-sided figure and $\overline{EF} \perp \overline{AL}$.**

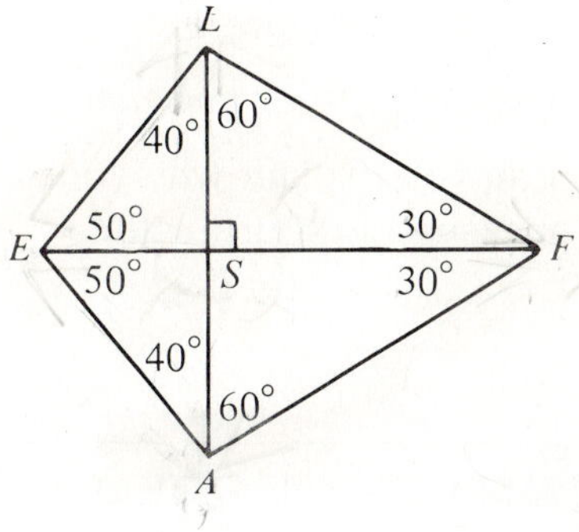

5. How many right angles are shown?
6. How many obtuse angles are shown?
7. How many acute angles are shown?
8. Name two pairs of vertical angles.
9. Name the bisector of $\angle LFA$.
10. Name the bisector of $\angle LEA$.

B

11. Make a paper airplane. Then unfold it and mark on the paper the measures of the angles between the folds.
12. Fold a rectangular sheet of paper to form a square.
13. a. Draw any angle and its bisector like that at the left below.

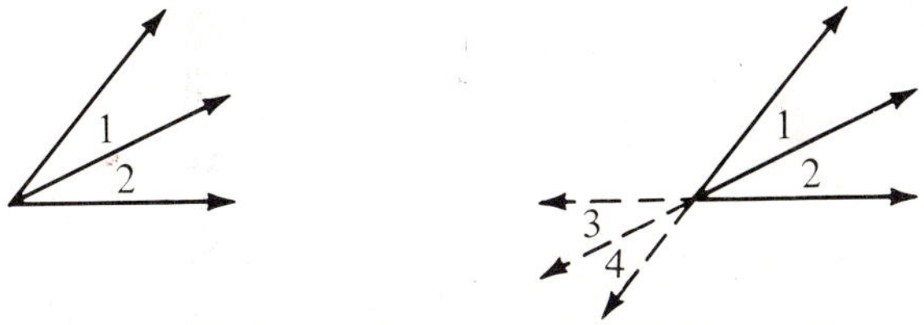

b. Extend each ray in the opposite direction as shown above.
c. What can you conclude about $\angle 3$ and $\angle 4$?
d. Explain why you think your conclusion is correct.

Experiments

Experiment 1

1. Fold a rectangular sheet of paper to form a 45° angle.
2. Can you discover how to form a $22\frac{1}{2}°$ angle?
3. Could an $11\frac{1}{4}°$ angle be formed by the folding process?

Experiment 2

1. Fold over a corner of a rectangular sheet of paper as in Figure 1 below.
2. Now fold the next corner so the edges touch as in Figure 2.

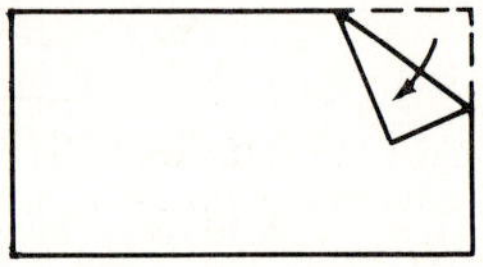

Figure 1

Figure 2

3. Open the paper and measure the angle between the creases.
4. Can you explain why your classmates' angles have the same measure as your angle?

SELF-TEST

Vocabulary

acute angle (p. 15)
right angle (p. 15)
perpendicular lines (p. 15)
obtuse angle (p. 15)
straight angle (p. 15)
triangle (p. 15)
vertices of a triangle (p. 15)
vertical angles (p. 18)
bisector of an angle (p. 20)

In the diagram, $\overline{AD} \perp \overline{BC}$. Classify each angle as acute, right, obtuse, or straight.

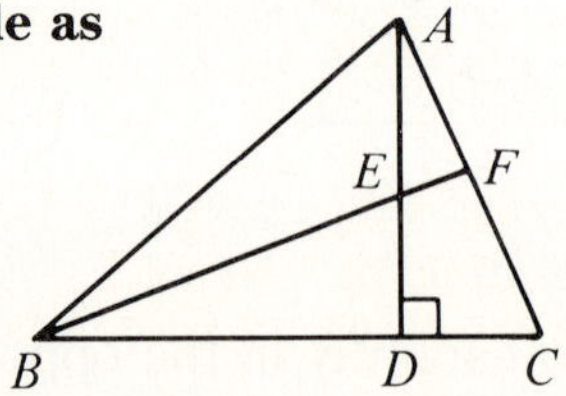

1. $\angle ADC$ **2.** $\angle ABC$ *(1-4)*

3. $\angle BEA$ **4.** $\angle AFC$

5. Name two pairs of vertical angles. *(1-5)*

6. If $\overrightarrow{BF}$ bisects $\angle ABC$, what two angles must be equal? *(1-6)*

APPLICATIONS

LATITUDE AND LONGITUDE

When traveling over land, you can sometimes tell your location by noticing landmarks that are near you. In the middle of the ocean, this would be very difficult to do. One of the ways that air and sea navigators describe their exact location is by using the system of *latitude* and *longitude*. This system consists of a grid of imaginary lines, east-west lines of latitude and north-south lines of longitude. Each point on the earth is at the crossing of a particular line of latitude and a particular line of longitude. Most globes and large maps show some of these lines.

The *equator*, a line that divides the earth in half, is used as the line of zero latitude. Other lines of latitude, also called parallels of latitude, are marked to indicate degrees north and south of the equator. We measure 90° of latitude north of the equator and 90° of latitude south.

Lines of longitude, also called meridians, run from the North Pole to the South Pole. These lines indicate degrees east or west of a zero line, called the prime meridian. We measure 180° of longitude west of the prime meridian and 180° of longitude east.

On navigational charts, points are drawn showing a ship's position at various times. Lines are drawn to connect these points. The woman shown is using parallel rulers to transfer such lines to another part of a chart. Navigators also need to know how to use a compass and dividers.

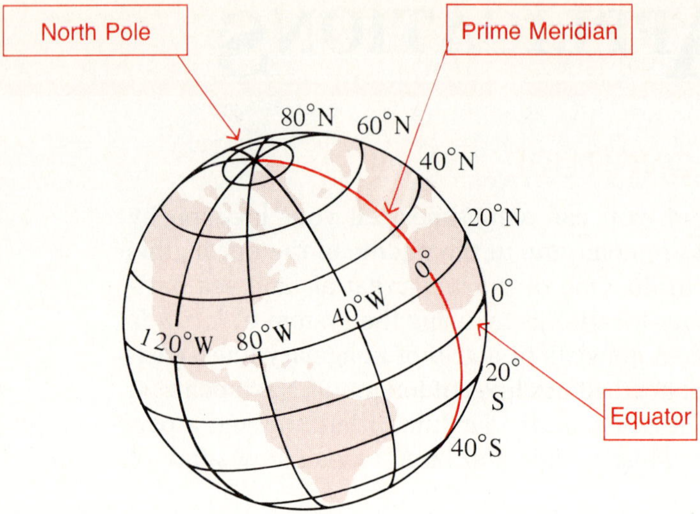

EXAMPLE

Look at the illustration of a globe shown above. Do you see that 40°N, 40°W is near the middle of the North Atlantic Ocean?

Find the location of 20°S, 80°W. It is near the west coast of South America.

There is no natural location for the prime meridian as there is for the equator. Originally each nation used the line of longitude passing through its own national observatory. This led to some confusion, and in 1884, most nations agreed to use the meridian passing through Greenwich, England, as the prime meridian.

EXERCISES

Use a map or globe showing latitude and longitude.

1. Name the ocean in which the following locations are found.
   a. 80°N, 160°W b. 5°S, 60°E c. 30°S, 10°W
   d. 40°N, 150°W e. 50°N, 20°W f. 50°S, 160°W
2. Find the approximate latitude and longitude for the place where you live.
3. The location 0° latitude, 0° longitude is near what continent?
4. Name the countries in Africa through which the equator passes.
5. Name the countries in Europe through which the prime meridian passes.

7 Three Constructions

Making a construction is a mathematical game in which figures are drawn using only a *straightedge* and *compass*. The use of a ruler or protractor for measuring purposes is not allowed.

CONSTRUCTION 1
Given: *An angle* Construct: *A bisector of the angle*

1.

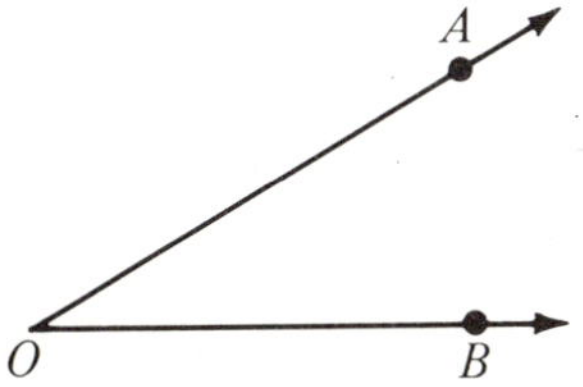

You are given $\angle AOB$.

2.

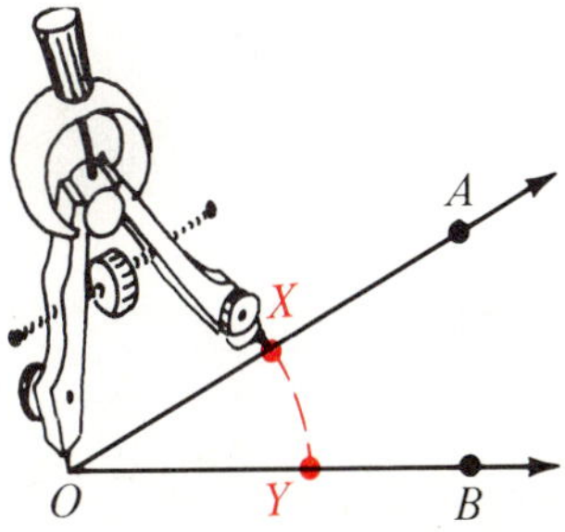

Put the point of your compass at O and swing an arc which cuts the sides of $\angle AOB$ at X and Y.

3.

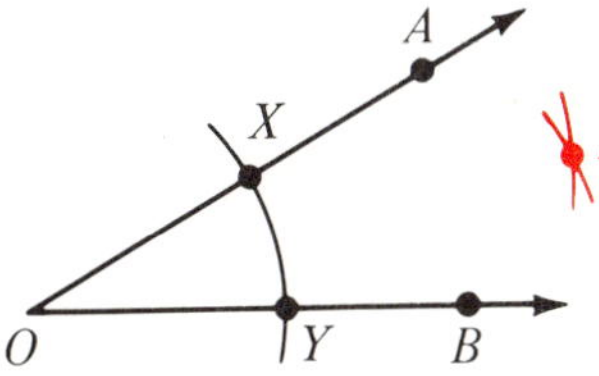

Now draw two arcs, one with the compass point at X and the other with the point at Y. Use the same compass setting for both arcs. The arcs meet at P.

4.

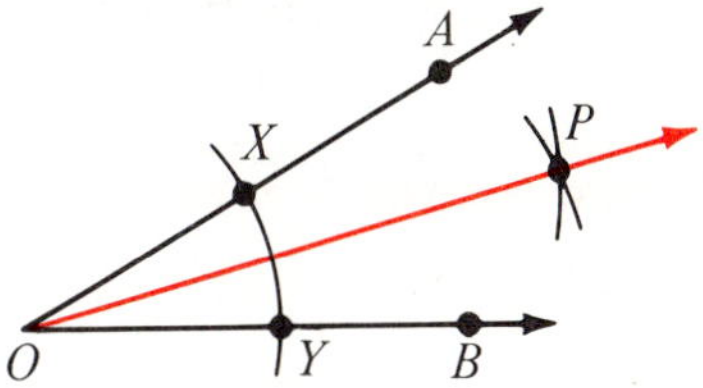

Join points P and O.
$\overrightarrow{OP}$ is a bisector of $\angle AOB$.

Suppose that $\angle AOB$ given in Construction 1 is a straight angle. When you bisect this straight angle with measure 180°, you form two angles with measure 90°. In other words, bisecting a straight angle produces perpendicular lines. This idea is used in the next construction.

CONSTRUCTION 2

Given: *A line and a point on the line*
Construct: *A perpendicular to the line through the point*

Turn back to Construction 1. You will notice that the steps in Construction 2 are almost the same as those of Construction 1.

1, 2.

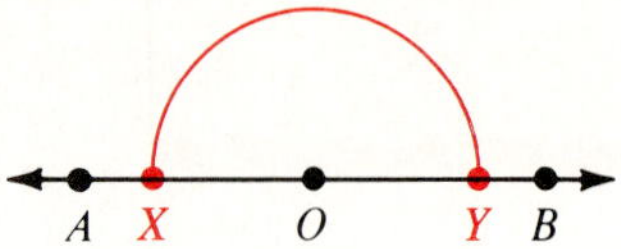

3.

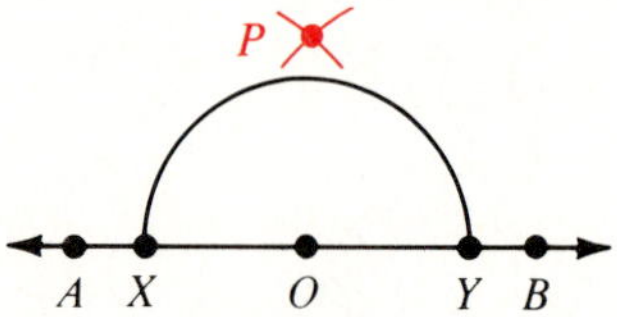

4.

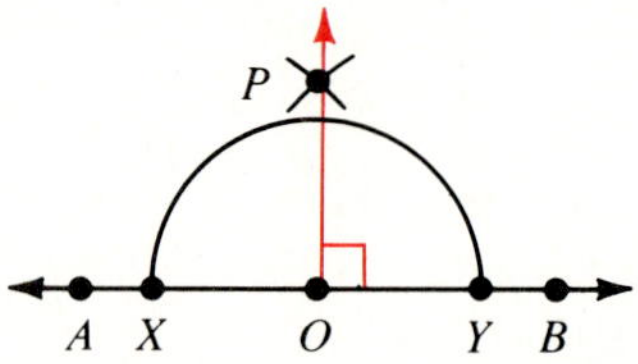

In Construction 2, the perpendicular passes through a point on the given line. In the next construction, the perpendicular passes through a point outside the given line.

CONSTRUCTION 3

Given: *A line and a point not on the line*
Construct: *A perpendicular to the line through the point*

Compare the steps in this construction with those of Constructions 1 and 2.

1.

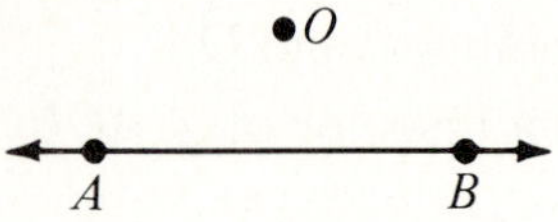

2.

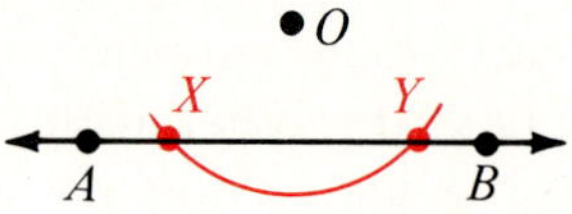

3.

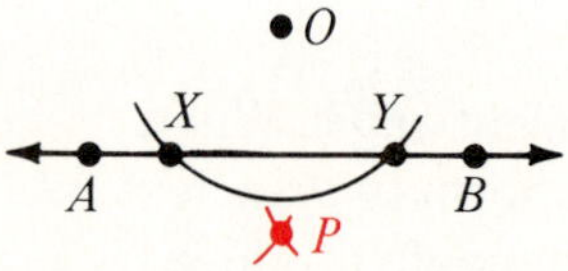

4.

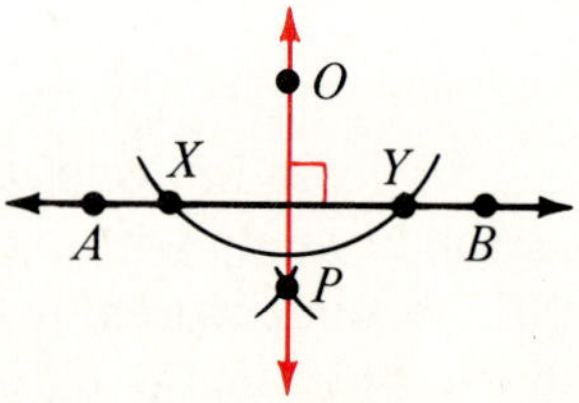

Classroom Practice

1. Draw an obtuse angle.
   Then bisect it, using a straightedge and compass.
2. Use Construction 2 to construct a $90°$ angle.
   Then construct a $45°$ angle by bisecting the $90°$ angle.
3. Draw a line l and choose a point O not on l.
   Use Construction 3 to construct a line, through O, which is perpendicular to line l.

Written Exercises

Draw an angle similar to, but larger than, the one shown. Then bisect it.

A

1.

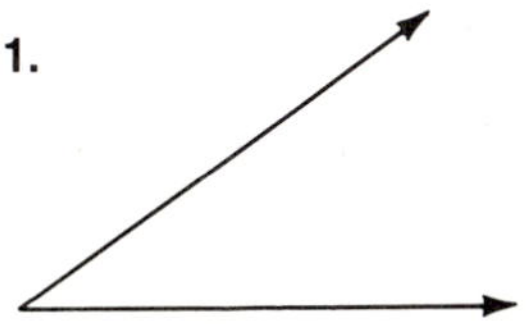

2.

3.

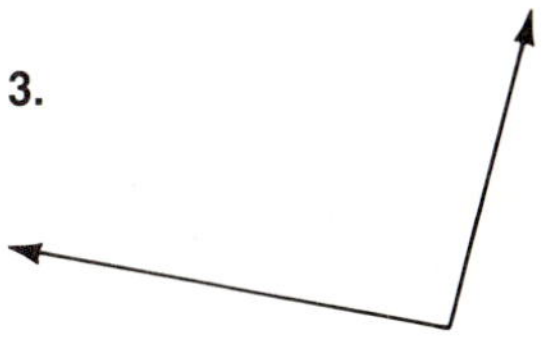

Draw a figure similar to, but larger than, the one shown. Then construct a perpendicular to $\overleftrightarrow{AB}$ through point O.

4.

5.

6.

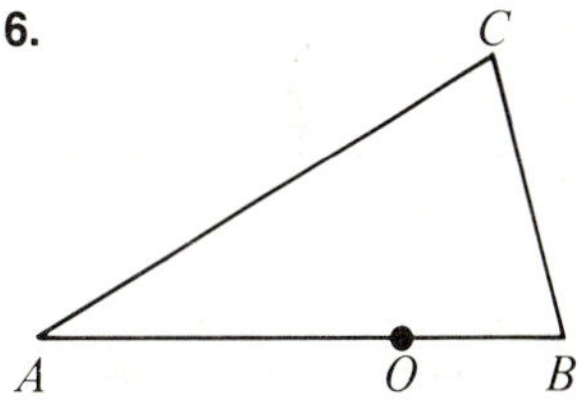

7.

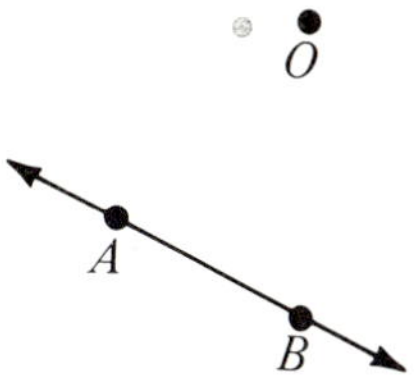

8.

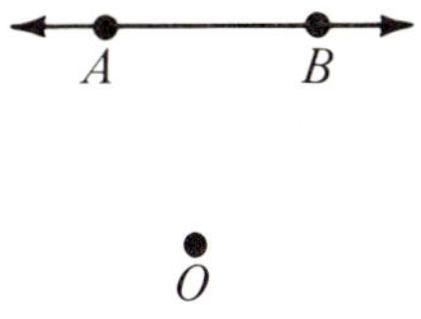

9.

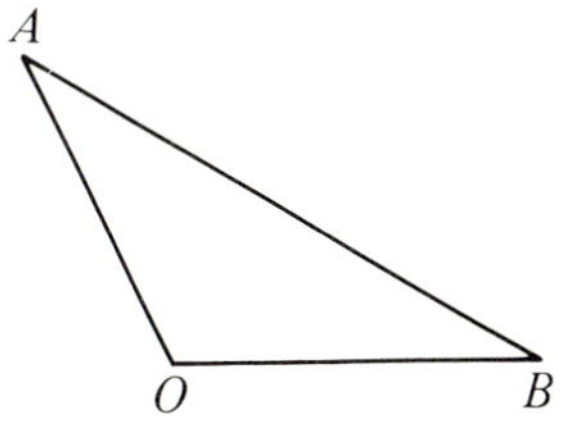

10. Draw an acute angle. Construct the bisector of the angle.
11. Draw an obtuse angle. Construct the bisector of the angle.
12. a. Draw a large triangle.
    b. Construct the bisectors of the three angles.
    The bisectors should meet in a point. Do they?

13. a. Draw a line m and choose a point K on the line.
    b. Construct four right angles, each of which has K as vertex.

14. Use your construction from Exercise 13. Construct a 45° angle.

15. Use your construction from Exercises 13 and 14. Construct a $22\frac{1}{2}°$ angle.

Draw a figure similar to, but larger than, the one shown. Then construct perpendiculars to $\overleftrightarrow{AB}$ from C and D.

B

16. 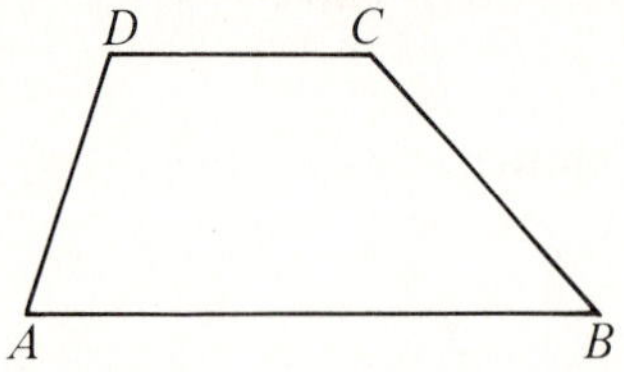

17. 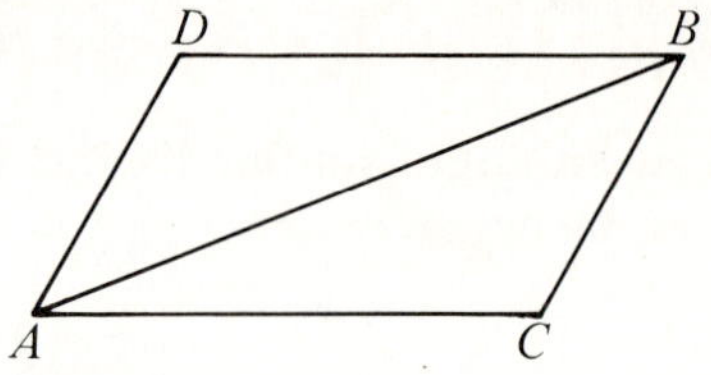

Draw a figure similar to, but larger than, the one shown. Construct a perpendicular to $\overleftrightarrow{RS}$ at S and a perpendicular to $\overleftrightarrow{RT}$ at T.

18. 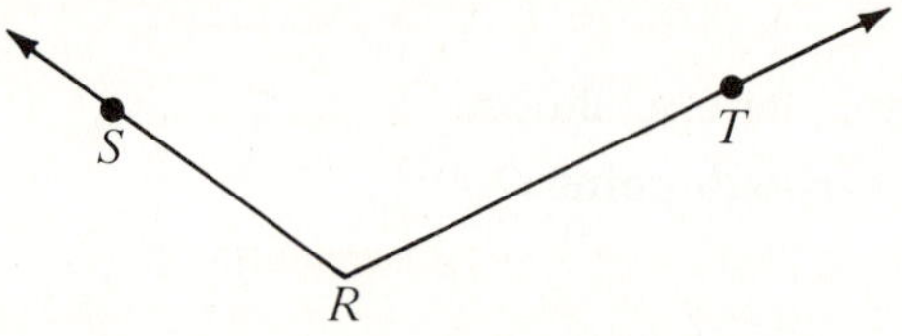

19. 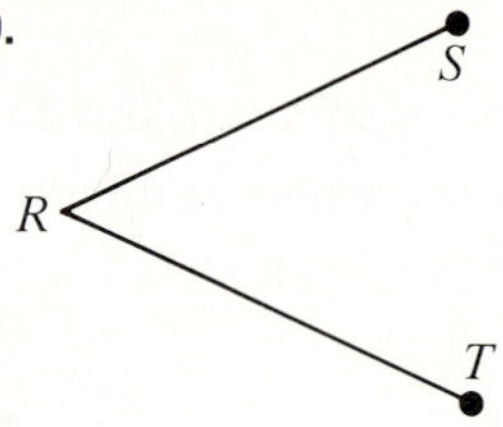

Draw a figure similar to, but larger than, the one shown. Then construct a bisector of $\angle UVW$ and a bisector of $\angle WVZ$.

20. 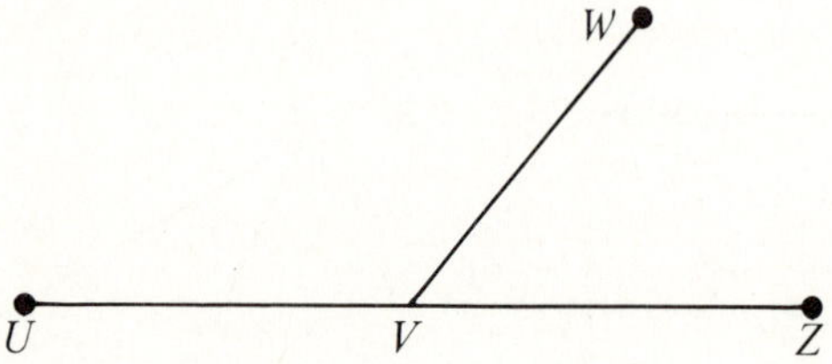

21. 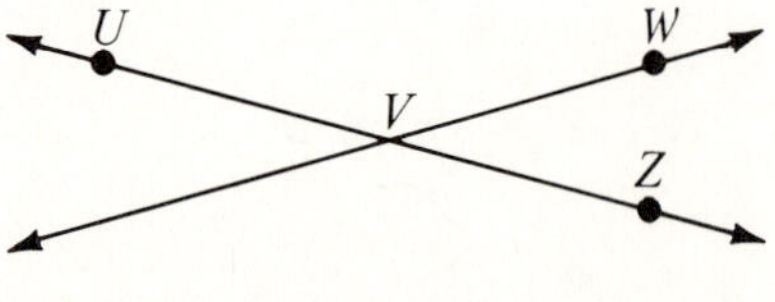

22. In Exercises 20 and 21, how do the two angle bisectors you constructed seem to be related?

C

23. Construct a 135° angle. (*Hint:* 135° = 180° − 45°)

24. Construct a square. Remember: You may not use your ruler to measure distances; you may use it only as a straightedge.

8 Postulates of Equality

The following facts about real numbers are called **postulates.** These are statements which are accepted without proof. Postulates are used to prove other statements called *theorems.*

THE ADDITION POSTULATE
If $a = b$ *and* $c = d$, *then* $a + c = b + d$.

EXAMPLE 1

In the diagram, $AB = CD$.

But we know that $BC = BC$.

Conclusion: $AB + BC = CD + BC$.

In other words, $AC = BD$.

THE SUBTRACTION POSTULATE
If $a = b$ *and* $c = d$, *then* $a - c = b - d$.

EXAMPLE 2

In the diagram, $\angle ABC = \angle ACB$

and also $\angle 1 = \angle 2$.

Conclusion: $\angle ABC - \angle 1 = \angle ACB - \angle 2$.

In other words, $\angle 3 = \angle 4$.

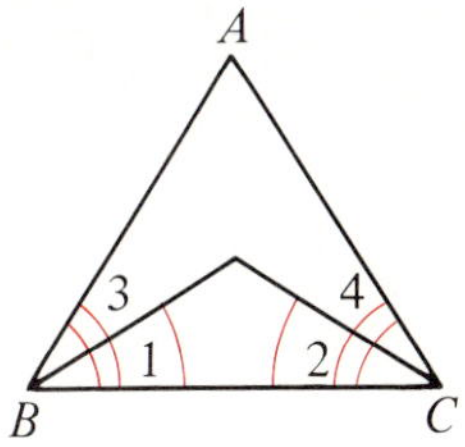

THE MULTIPLICATION POSTULATE
If $a = b$, *then* $ac = bc$.

EXAMPLE 3

In the diagram, M and N are midpoints of $\overline{AB}$ and $\overline{CD}$, and also $AM = CN$.

Conclusion: $2 \times AM = 2 \times CN$.

In other words, $AB = CD$.

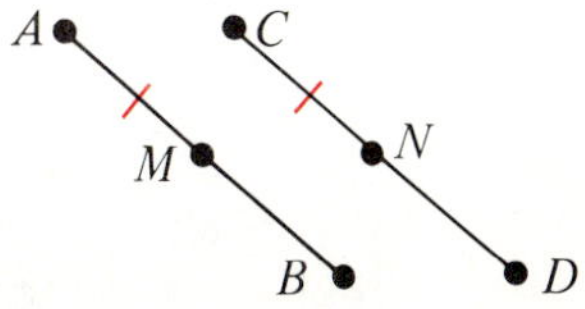

THE DIVISION POSTULATE
If $a = b$ and $c \neq 0$, then $\frac{a}{c} = \frac{b}{c}$.

THE SUBSTITUTION POSTULATE
If $a = b$, then a can be substituted for b in any equation or inequality.

EXAMPLE 4 In the diagram, $BC = BD$.

Since $AB + BC = AC$,
we can substitute ↓
and get $AB + BD = AC$.

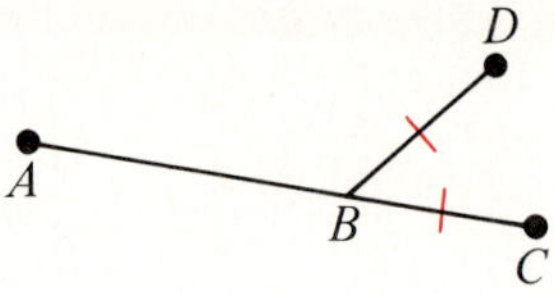

EXAMPLE 5 In the diagram, $\angle 1 = \angle 2$

and $\angle 3 = \angle 4$.

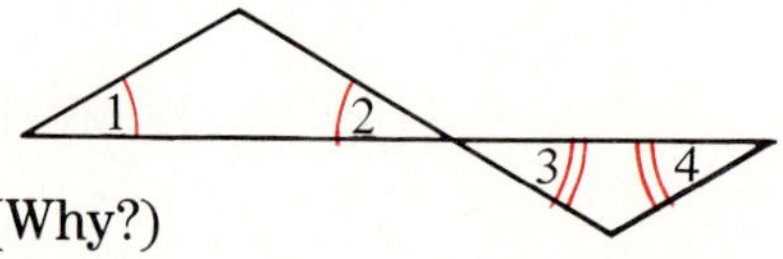

Since $\angle 2 = \angle 3$, (Why?)
we can substitute ↓ ↓
and get $\angle 1 = \angle 4$.

Classroom Practice

Name the postulate that can be used to justify each statement.

1. If $x - 5 = 21$, then $x = 26$.
2. If $\frac{1}{2}x = 5$, then $x = 10$.
3. If $3x + 4 = 19$, then $3x = 15$.
4. If $3x = 15$, then $x = 5$.
5. If $x + y = z$ and $z = 24$, then $x + y = 24$.
6. If $CL = AM$, then $CA = LM$.
7. If $CA = LM$, then $CL = AM$.

8. If $\angle 1 = \angle 3$ and $\angle 2 = \angle 4$, then $\angle ABC = \angle XYZ$.
9. If $\angle XYZ = \angle ABC$ and $\angle 1 = \angle 3$, then $\angle 2 = \angle 4$.

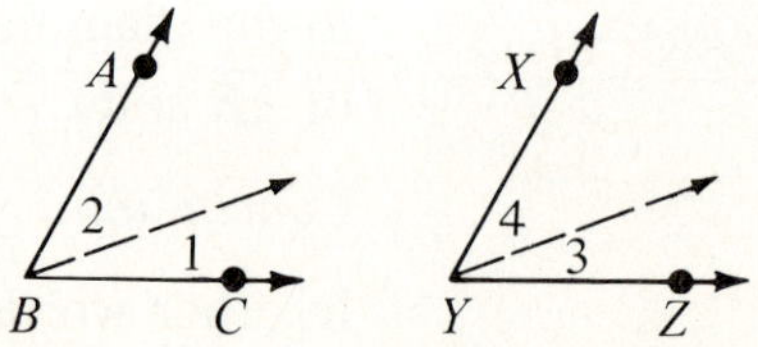

9 Postulates of Geometry

A **plane** is a flat surface that extends indefinitely in all directions. A floor suggests part of a plane. So does a wall.

Usually a plane is represented by a four-sided figure. The plane can be named by the four vertices of the figure or by a single capital letter.

plane *ABCD*

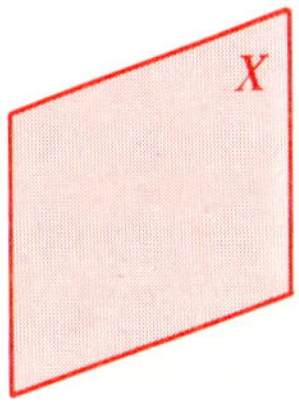

plane *X*

You should realize that drawings like those above only *suggest* planes. There is no good way of showing in a drawing that a plane extends indefinitely. An actual plane does *not* have sides or vertices.

Remember that points all on one line are called *collinear* points. Similarly, points all in one plane are called **coplanar** points.

P, *Q*, and *R* are *not* collinear. But they are coplanar.

This drawing suggests that there is exactly one plane containing three noncollinear points.

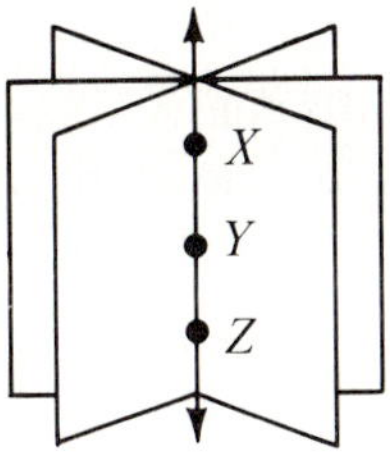

X, *Y*, and *Z* are collinear. They are also coplanar.

This drawing suggests that there are an unlimited number of planes containing three collinear points.

A summary of the relationships between points, lines, and planes is given by the postulates on the next page. Remember that postulates are statements which we assume to be true without proof.

POSTULATE 1

Through any two points there is exactly one line.

POSTULATE 2

Through any three noncollinear points there is exactly one plane.

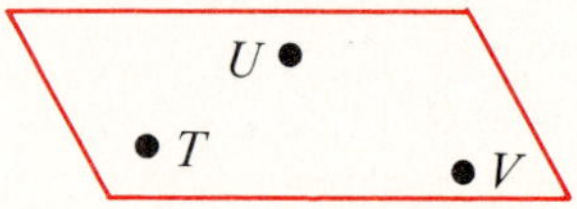

POSTULATE 3

If two points lie in a plane, then the line joining them lies in that plane.

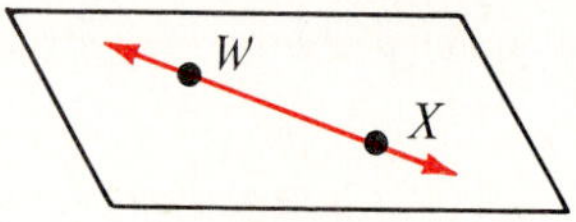

POSTULATE 4

If two planes intersect, then their intersection is a line.

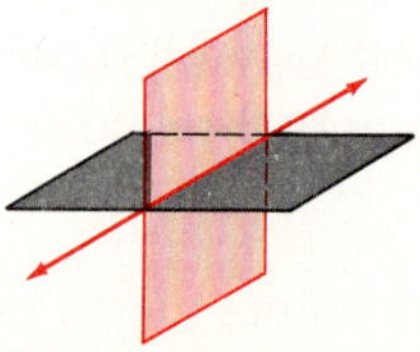

If two planes do not intersect, we say that they are parallel. Planes M and N, shown at the right, are parallel planes. We indicate this by writing "plane $M \parallel$ plane N."

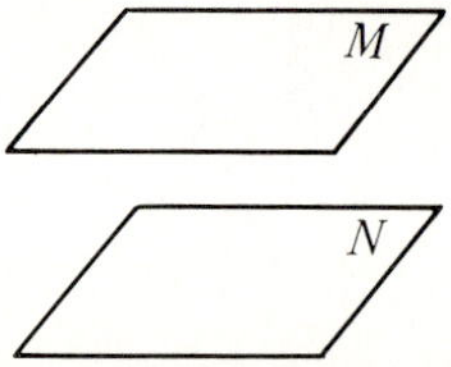

Postulates 1–4 deal with the geometric ideas of *points, lines,* and *planes.* Postulates 5 and 6 connect geometric and algebraic ideas. You do not need to memorize the statements of these two postulates. Just be sure that you understand them.

POSTULATE 5 (The Ruler Postulate)

Each point on a line can be paired with exactly one real number called its coordinate. The distance between two points is the positive difference of their coordinates.

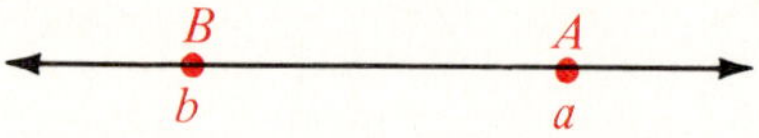

If $a > b$, *then* $AB = a - b$.

POSTULATE 6 (The Protractor Postulate)

Suppose O is a point of $\overleftrightarrow{XY}$. Consider all rays with endpoint O which lie on one side of $\overleftrightarrow{XY}$. Each ray can be paired with exactly one real number between 0 and 180, as shown.

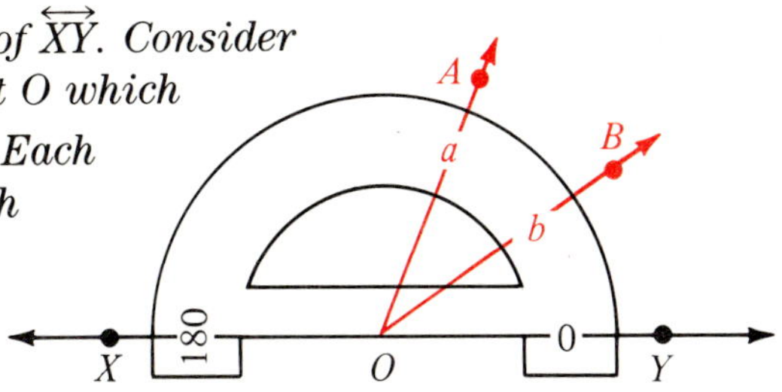

If $a > b$, then $\angle AOB = a - b$.

Do the ideas of the Ruler and Protractor Postulates seem familiar? When we found lengths of segments and measures of angles earlier, we were informally using the ideas expressed in these postulates.

Classroom Practice

Find each length.

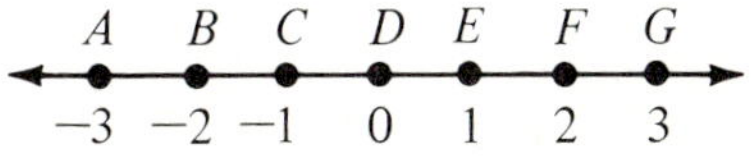

1. CF 2. EG 3. EA 4. BG

Find the measure of each angle.

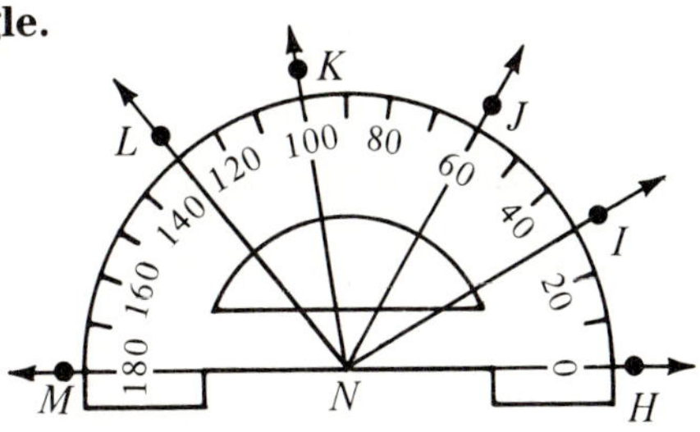

5. $\angle INL$
6. $\angle KNJ$
7. $\angle HNM$
8. $\angle JNM$

The diagram shows a figure called a *rectangular solid*. Decide if the given points are coplanar.

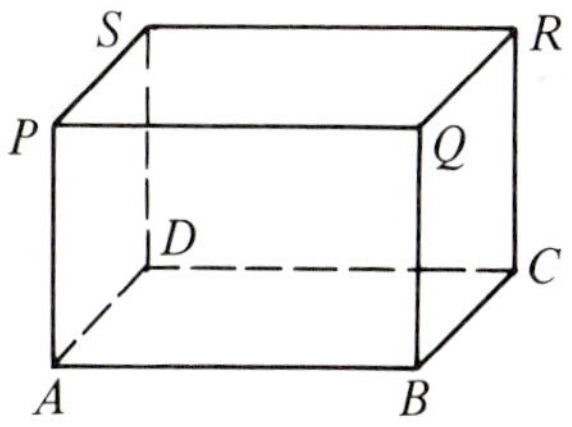

Exercises 9–18

9. P, Q, R, and S 10. P, Q, B, and A

11. B, C, R, and S 12. B, Q, S, and D

Name another point coplanar with the given points.

13. A, B, and D 14. A, D, and S

15. A, B, and R 16. A, P, and R

Complete each statement.

17. Plane $ABCD$ || plane _?_. 18. Plane $ABQP$ || plane _?_.

Written Exercises

The diagram shows a rectangular solid. Complete each statement.

A

1. Plane $ABCD \parallel$ plane _?_.
2. Plane $ADHE \parallel$ plane _?_.
3. Plane $ABFE \parallel$ plane _?_.
4. Plane $EFGH$ intersects plane $BCGF$ in the line _?_.
5. Plane $ABFE$ intersects plane $BCGF$ in the line _?_.
6. Points E, A, D, and _?_ are coplanar.
7. Points B, F, H, and _?_ are coplanar.
8. Points C, G, H, and _?_ are coplanar.

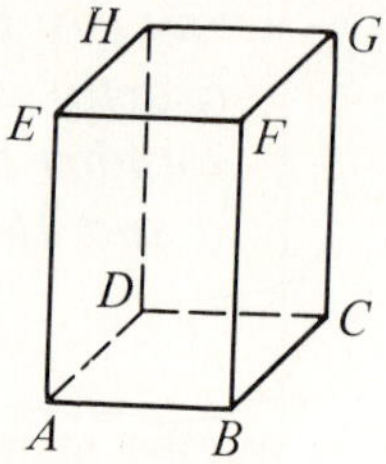

Draw each of the following.

9. Two parallel planes
10. Two intersecting planes
11. A corner of a room where two walls and the floor meet
12. Plane M and plane N both contain point P.
    a. Do the planes have any other points in common?
    b. State the postulate that answers this question.

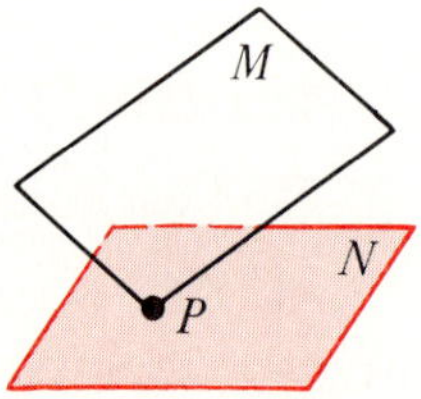

13. The diagram suggests what would happen if two different "lines" both contained points A and B. State the postulate that says this cannot happen.

14. A point is given. Is the point contained in just one line or in more than one line?
15. A line is given. Is the line contained in just one plane or in more than one plane?
16. A line and a point not on the line are given. Are the line and the point both contained in one plane or in more than one plane?
17. A rectangle is given. Is the rectangle contained in just one plane or in more than one plane?

Tell whether each statement is true or false.

B

18. If points P, Q, and R are in plane M and also in plane N, then the planes must be the same.
19. A triangle is contained in exactly one plane.
20. Any four points are contained in exactly one plane.

21. If points A and B are in plane X, then every point of $\overline{AB}$ is in X.

22. If points A and B are on a soup can, then every point of $\overline{AB}$ must be on the can.

C 23. When a three-dimensional figure is represented on a flat surface, it is usually drawn in perspective. For example, angles are drawn larger or smaller than they actually are. In the diagram, $\angle HJI$ is an acute angle but it represents a right angle in the actual building. Complete the table.

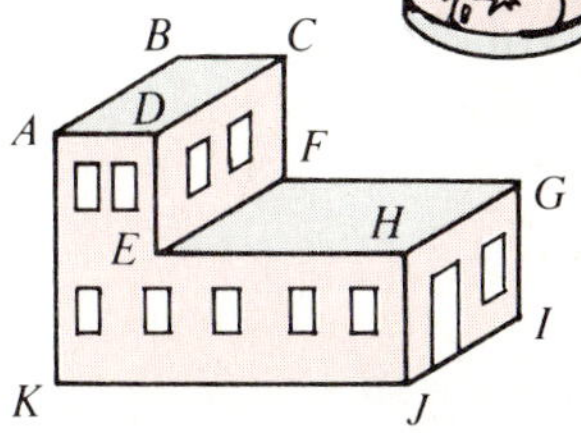

	$\angle HJI$	$\angle EDC$	$\angle BAD$	$\angle AKJ$	$\angle ABC$
In the diagram	acute	?	?	?	?
In the actual building	right	?	?	?	?

SELF-TEST

Vocabulary

straightedge (p. 25)
compass (p. 25)
postulate (p. 29)
plane (p. 33)
coplanar points (p. 33)
rectangular solid (p. 35)

1. Draw an acute angle. Bisect it using straightedge and compass. *(1-7)*

Which postulate can be used to justify each statement?

2. If $\angle 1 + \angle 2 = 180°$ and $\angle 2 = \angle 3$, then $\angle 1 + \angle 3 = 180°$. *(1-8)*

3. If $x - 7 = 3$, then $x = 10$.

4. Use the diagram at the right.
   Given information: $\overrightarrow{BD}$ bisects $\angle CBE$.
   Conclusion: $\angle 1 = \angle$ __?__ $= \angle$ __?__

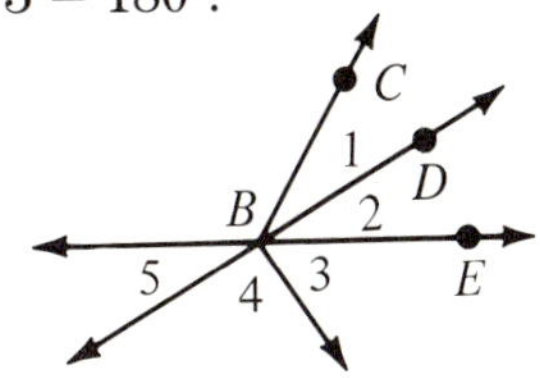

The diagram shows a rectangular solid. Complete each statement.

5. S, V, Z, and __?__ are coplanar points. *(1-9)*

6. Plane $STUV$ $\parallel$ plane __?__.

7. Plane $STXW$ intersects plane $WXYZ$ in the line __?__.

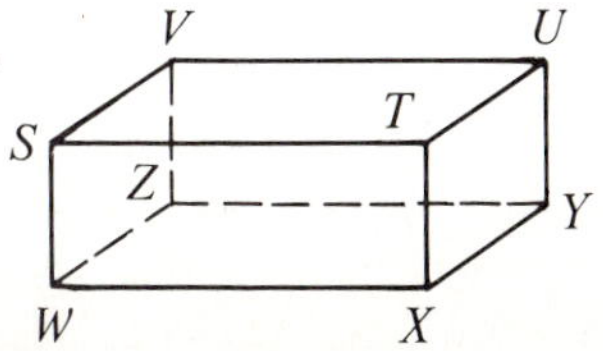

SKILLS REVIEW

ARITHMETIC OPERATIONS

Add. The samples will help you remember.

SAMPLES $7 + 4 = 11$ $\quad -7 + 4 = -3$
$7 + (-4) = 3$ $\quad -7 + (-4) = -11$

1. $-3 + 12$ **2.** $0 + (-5)$ **3.** $-4 + (-8)$ **4.** $9 + (-2)$
5. $-15 + (-6)$ **6.** $-11 + 19$ **7.** $3 + (-14)$ **8.** $-23 + 23$
9. $-5.7 + 13.8$ **10.** $6.2 + (-10.4)$ **11.** $-7.6 + (-8.5)$ **12.** $0 + (-9.5)$

Subtract.

SAMPLES $4 - 7 = 4 + (-7) = -3$ $\quad -4 - 7 = -4 + (-7) = -11$
$4 - (-7) = 4 + 7 = 11$ $\quad -4 - (-7) = -4 + 7 = 3$

13. $13 - 17$ **14.** $-2 - 8$ **15.** $9 - (-2)$ **16.** $10 - (-10)$
17. $-7 - 7$ **18.** $20 - 34$ **19.** $-16 - (-9)$ **20.** $-14 - 30$
21. $6.4 - 9.8$ **22.** $-5.7 - 4.9$ **23.** $7.2 - (-7.2)$ **24.** $-5.3 - (-8.6)$

Multiply.

SAMPLES $3(8) = 24$ $\quad 3(-8) = -24$
$-3(-8) = 24$ $\quad -3(8) = -24$

25. $5(12)$ **26.** $-4(20)$ **27.** $13(-7)$ **28.** $-8(-9)$
29. $-29(0)$ **30.** $-3(-50)$ **31.** $8(-11)$ **32.** $-6(25)$
33. $8(11.2)$ **34.** $-5.1(9)$ **35.** $7(-10.5)$ **36.** $-0.2(-0.6)$

Divide.

SAMPLES $40 \div 8 = 5$ $\quad -40 \div 8 = -5$
$-40 \div (-8) = 5$ $\quad 40 \div (-8) = -5$

37. $-63 \div 9$ **38.** $42 \div (-6)$ **39.** $-52 \div (-4)$ **40.** $-17 \div 17$
41. $-44 \div (-11)$ **42.** $-72 \div 8$ **43.** $39 \div (-13)$ **44.** $-125 \div 25$
45. $69.3 \div (-3)$ **46.** $-4.8 \div 12$ **47.** $-63 \div (-0.7)$ **48.** $8.6 \div -8.6$

Compare the numbers. Write <, =, or >.

49. $17 + (-10)$ __?__ 5 **50.** $(-5)^2$ __?__ 10 **51.** $-2(-9)$ __?__ $0 + 18$
52. $-(-3)^2$ __?__ 9 **53.** $-13 - (-4)$ __?__ -7 **54.** $0(-10)$ __?__ $5 - 8$

CHAPTER REVIEW

CHAPTER SUMMARY

1. The basic figures studied in this chapter are lines, rays, segments, and angles. Each of these is a set of points.
   Line AB is written $\overleftrightarrow{AB}$.
   Ray AB is written $\overrightarrow{AB}$. A is the endpoint.
   Segment AB is written $\overline{AB}$. A and B are endpoints.
   Angle AOB is written $\angle AOB$. O is the vertex.

2. Angles are classified as acute, right, obtuse, and straight according to their measures. Perpendicular lines form right angles.

3. Vertical angles are equal.

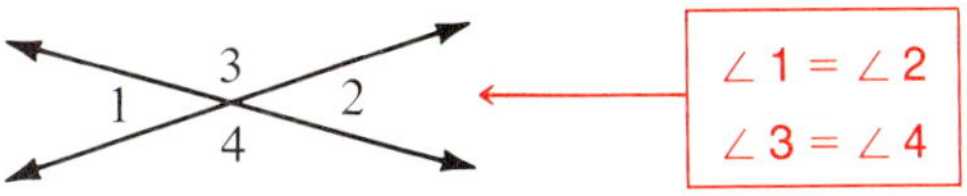

4. If M is the midpoint of $\overline{AB}$, then $AM = MB = \frac{1}{2}AB$.
   If $\overrightarrow{OM}$ bisects $\angle AOB$, then $\angle AOM = \angle MOB = \frac{1}{2}\angle AOB$.

5. In constructing a geometric figure, the only instruments which may be used are a compass and a straightedge. Three basic constructions are shown on pages 25–26:
   (1) a bisector of an angle;
   (2) a line $\perp$ to a given line at a point on the line;
   (3) a line $\perp$ to a given line through a point not on the line.

6. Postulates are statements which are accepted without proof. They are used to prove other statements and theorems. You should review the postulates of equality on pages 29–30 and the geometric postulates on pages 34–35.

REVIEW EXERCISES

Complete. ***(See pp. 6–9.)***

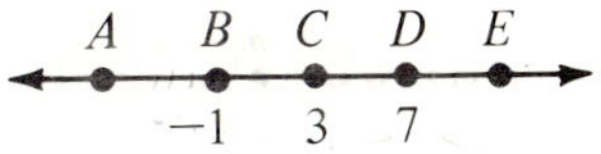

1. Another name for $\overleftrightarrow{AB}$ is __?__.
2. Another name for $\overrightarrow{BD}$ is __?__.
3. $BC =$ __?__
4. If $AB = 5$, then the coordinate of A is __?__.
5. If D is the midpoint of $\overline{CE}$, then the coordinate of E is __?__.

Complete. ***(See pp. 10–13.)***

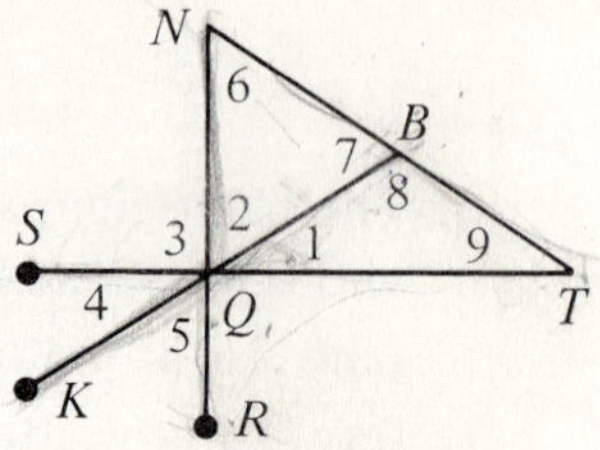

Exercises 6–16

6. The vertex of $\angle 1$ is ___?___.

7. $\angle 7 + \angle 8 =$ ___?___ °

8. The sides of $\angle 7$ are ___?___ and ___?___.

9. Another name for $\angle 5$ is ___?___. (Use three letters.)

10. $\angle SQB - \angle NQB = \angle$ ___?___

In the diagram above, $\overline{NQ} \perp \overline{ST}$. Classify each angle as acute, right, obtuse, or straight. ***(See pp. 15–17.)***

11. $\angle RQT$

12. $\angle QBN$

13. $\angle QBT$

14. $\angle KQB$

Complete. Use the diagram above. ***(See pp. 18–19.)***

15. $\angle 2$ and $\angle$ ___?___ are vertical.

16. If $\angle KQN = 122°$, $\angle$ ___?___ $= 122°$.

Solve. ***(See pp. 20–22.)***

17. When you bisect a right angle, you divide the angle into two smaller angles. What is the measure of each?

For each exercise, draw a large triangle like $\triangle RST$. Then construct the required figure. ***(See pp. 25–28.)***

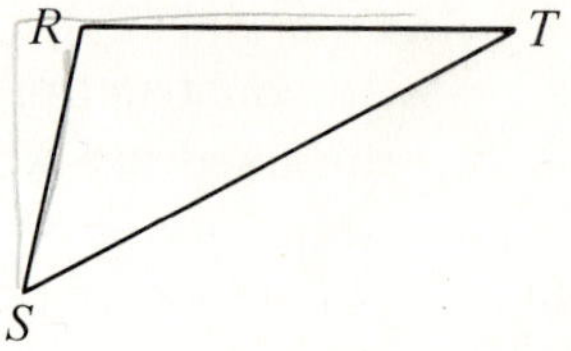

18. Construct a bisector of $\angle T$.

19. Construct a line perpendicular to $\overline{ST}$ through R.

Which postulate can be used to justify each statement? ***(See pp. 29–32.)***

20. If $5x = 35$, then $x = 7$.

21. If $\angle 1 = \angle 2$ and $\angle 4 = \angle 5$, then $\angle 1 + \angle 4 = \angle 2 + \angle 5$.

22. If $\angle 1 + \angle 2 = 180°$ and $\angle 2 = \angle 4$, then $\angle 1 + \angle 4 = 180°$.

Refer to the rectangular solid shown. Complete. ***(See pp. 33–37.)***

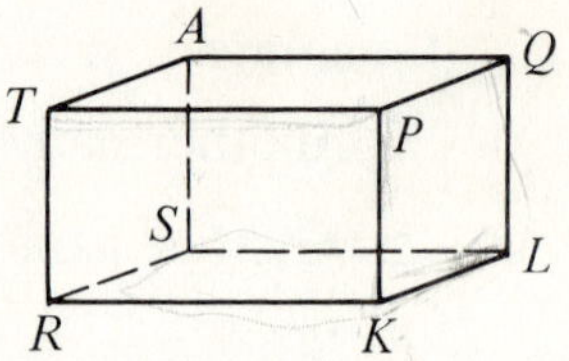

23. T, P, Q, and ___?___ are coplanar points.

24. A, T, R, and ___?___ are coplanar points.

25. Plane $RKPT \parallel$ plane ___?___.

26. Plane $PKLQ$ intersects plane $RKLS$ in the line ___?___.

27. How many planes contain three noncollinear points?

CHAPTER TEST

On a number line, point R has coordinate -7 and point S has coordinate -1.

1. $RS = \underline{\quad ? \quad}$ 2. Find the coordinate of the midpoint of $\overline{RS}$. *(1-2)*

3. If X lies on $\overrightarrow{RS}$, must X have a negative coordinate?

In Questions 4–10, refer to the diagram below.

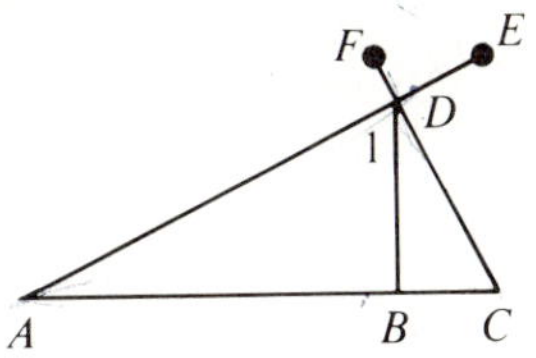

4. Name an angle with measure 180°. *(1-3)*

5. Use your protractor to find the approximate measure of $\angle A$.

6. $\angle ADC - \angle 1 = \angle \underline{\quad ? \quad}$

7. If $\overline{CD} \perp \overline{AE}$, name **(a)** a right angle with vertex D and **(b)** an obtuse angle with vertex D. *(1-4)*

8. If $\overline{DA} \perp \overline{DC}$ and $\angle 1 = 68°$, find the measure of $\angle BDC$.

9. Can a triangle contain two obtuse angles?

10. Explain why $\angle ADF$ and $\angle CDE$ must be equal. *(1-5)*

Exercises 11–12

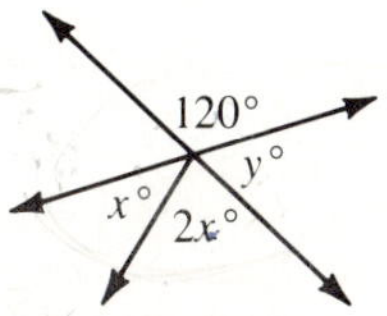

11. Find the value of x. 12. Find the value of y.

13. If you bisect a right angle forming $\angle 1$ and $\angle 2$, then $\angle 1 = \underline{\quad ? \quad}°$. *(1-6)*

14. If $\overrightarrow{OE}$ bisects $\angle DOF$, $\angle DOE = (y + 6)°$, and $\angle DOF = 40°$, then $y = \underline{\quad ? \quad}$.

15. Draw a large obtuse angle, $\angle XYZ$. Construct $\overrightarrow{YW}$, a bisector of $\angle XYZ$. *(1-7)*

16. Draw $\overleftrightarrow{QR}$. Construct a right angle, $\angle PQR$.

Which postulate can you use to justify each statement?

17. If $AB + BC = BC + CD$, then $AB = CD$. *(1-8)*

18. If $\angle 1 + \angle 2 = 90°$ and $\angle 3 = 90°$, then $\angle 1 + \angle 2 = \angle 3$.

19. If $x = y$ and $x = z$, then $2x = y + z$.

20. Two perpendicular lines are given. How many planes contain the lines?

21. Can three planes intersect in a line? in a point? *(1-9)*

22. If A and B lie in plane X, what can you conclude about $\overleftrightarrow{AB}$?

MIXED REVIEW

ALGEBRA PROBLEMS

Solve each equation. Show your work.

SAMPLES

$$\begin{aligned} 3x - 1 &= 17 \\ 3x - 1 + 1 &= 17 + 1 \\ 3x &= 18 \\ \frac{3x}{3} &= \frac{18}{3} \\ x &= 6 \end{aligned}$$

$$\begin{aligned} m + 4m &= 90 \\ 5m &= 90 \\ \frac{5m}{5} &= \frac{90}{5} \\ m &= 18 \end{aligned}$$

1. $x + 120 = 180$ **2.** $z - 10 = 25$ **3.** $7 - k = 1$

4. $5 - c = 8$ **5.** $3x = 60$ **6.** $7n = -14$

7. $2a - 1 = 7$ **8.** $5b + 2 = 2$ **9.** $8q + 12 = 4$

10. $4d - 15 = -3$ **11.** $19 - 10j = -11$ **12.** $-6s - 70 = 20$

13. $2x - 3x = -4$ **14.** $7y + 2y = 3$ **15.** $t - 3t = 8$

16. $6(z - 7) = 3$ **17.** $-(2 + w) = 2$ **18.** $4(9p + 2) = 0$

SAMPLES

$$\begin{aligned} 2(r - 1) &= 1 + 2r \\ 2r - 2 &= 1 + 2r \\ 2r &= 3 + 2r \\ 0 &= 3 \end{aligned}$$

No solution

$$\begin{aligned} 7r - 1 &= 5 - (6 - 7r) \\ 7r - 1 &= 5 - 6 + 7r \\ 7r - 1 &= 7r - 1 \\ -1 &= -1 \end{aligned}$$

All real numbers are solutions.

19. $11g + 7 = 5(3 - g)$ **20.** $2(6c - 1) = 3(4c + 7)$

21. $-(3 - 2v) = v + v$ **22.** $5(4 + x) = -2(2x - 1)$

23. $1 - 2(7 - h) = -4$ **24.** $7(y + 5) - 2y = 5(y + 7)$

25. $-8(2k - 1) = -3k - 31$ **26.** $2l + 2(4 - l) = 12$

27. $h + 2h + 4h = -7h$ **28.** $(9 - k) - (2 - k) = 7$

29. $5n - (2 + n) = 2(2n - 1)$ **30.** $\frac{1}{2}(4x - 8) = -3x + 11$

GEOMETRY PROBLEMS

Draw a diagram similar to, but larger than, the one shown.

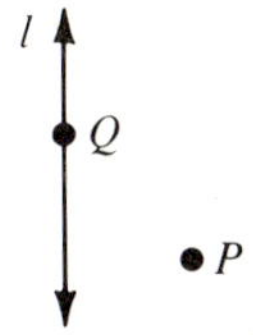

1. Construct a line perpendicular to l through P.
2. Construct a ray perpendicular to l through Q.
3. How many lines contain both P and Q? State the postulate that answers this question.

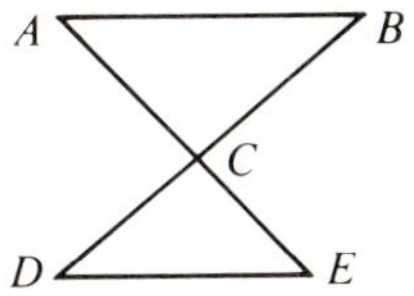

4. Name an obtuse angle shown.
5. $\angle ACB + \angle BCE = \angle$ _?_ = _?_ °
6. Given that $\angle ACD = 85°$, find the measures of $\angle ECB$ and $\angle DCE$.

Tell whether each statement is true or false.

7. R lies on $\overrightarrow{ST}$.
8. If A is the midpoint of $\overline{SR}$ and T is the midpoint of $\overline{SA}$, then $ST = \frac{1}{3} SR$.
9. The plane of the paper is the only plane containing $\overleftrightarrow{ST}$.
10. If $ST = TA$ and $SA = AR$, then $TA = AR - ST$.

11. Which segment is longer, $\overline{RS}$ or $\overline{XY}$?
12. Use your protractor to draw an angle with measure 130°. Construct the bisector of the angle.
13. Find the values of x and y.

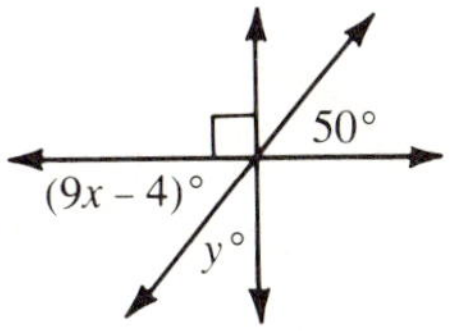

Exercise 13

Write the name of the postulate used.

14. If $5x - 1 = y$ and $y = 9$, then $5x - 1 = 9$.
15. If $5x - 1 = 9$, then $5x = 10$.
16. If $5x = 10$, then $x = 2$.

Complete each sentence with *always*, *sometimes*, or *never*.

17. Three different points are _?_ collinear.
18. Three different points are _?_ coplanar.

Here's what you'll learn in this chapter:

To use theorems about complementary, supplementary, and vertical angles.

To write "If . . . then" statements and their converses.

To understand and write basic geometric proofs.

To use the properties of parallel lines cut by a transversal.

To prove lines parallel.

To construct parallel lines.

This curving fence is used to hold sand dunes in place and prevent them from drifting. The upright stakes that make up the fence suggest parallel lines, discussed in this chapter.

Chapter 2

Introducing Proof

1 Three Theorems

Logical reasoning is an important part of geometry. You are given certain facts about a geometric figure and you use these facts to reach conclusions. So far, your conclusions have been based on definitions of figures and on the postulates of equality. Here is an example of logical reasoning.

EXAMPLE *Given information:* $\overline{AB} \perp \overline{BC}$
$\overline{PQ} \perp \overline{QR}$
$\angle 1 = \angle 2$

Conclusion: $\angle 3 = \angle 4$

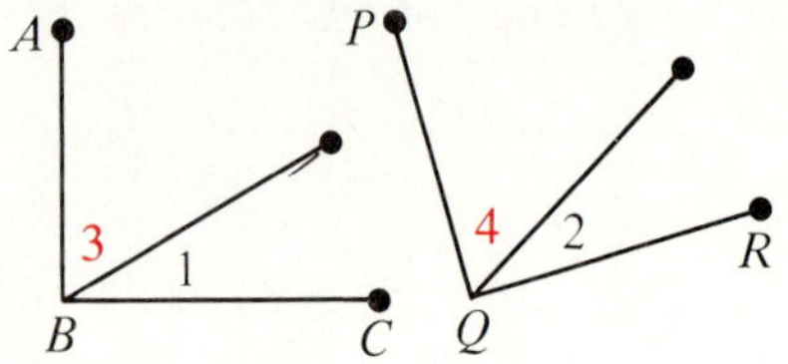

These are the steps used in reaching the conclusion:

Step 1 $\angle ABC = 90°$ because $\overline{AB} \perp \overline{BC}$, and perpendicular lines (by definition) form 90° angles.

Step 2 $\angle PQR = 90°$ because $\overline{PQ} \perp \overline{QR}$.

Step 3 $\angle 1 + \angle 3 = \angle 2 + \angle 4$ because both sums equal 90°.

Step 4 $\angle 1 = \angle 2$ because this is given information.

Step 5 $\angle 3 = \angle 4$ by using the Subtraction Postulate.

In the example above, $\angle 1$ and $\angle 3$ are called *complementary angles* or just *complements.* **Complementary angles** are two angles whose measures total 90°. In the diagram above, $\angle 2$ and $\angle 4$ are also complementary angles. So are $\angle A$ and $\angle B$ in the diagram at the right.

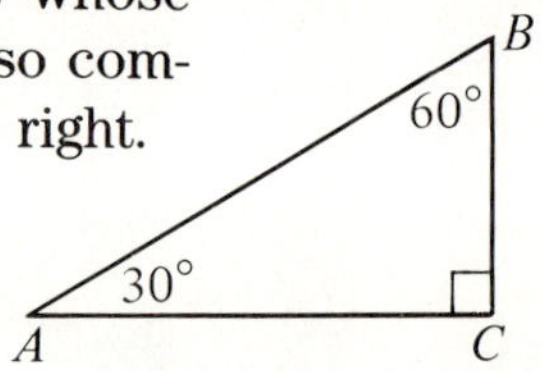

The conclusion reached in the example is really a statement about complementary angles:

If two angles are equal (such as $\angle 1 = \angle 2$), then their complements are equal ($\angle 3 = \angle 4$).

Let us list this statement as Theorem 1. Remember that a **theorem** is nothing more than a statement which has been proved. Of course, we won't list every proved statement as a theorem. We'll list only the ones which will be most helpful to us in our later work.

THEOREM 1

If two angles are complements of equal angles (or of the same angle), then the two angles are equal.

The next theorem refers to **supplementary angles** (sometimes called **supplements).** These are two angles whose measures total 180°.

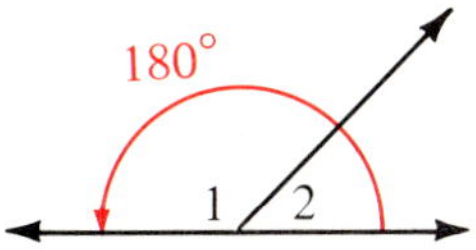

∠ 1 and ∠ 2 are supplementary angles.

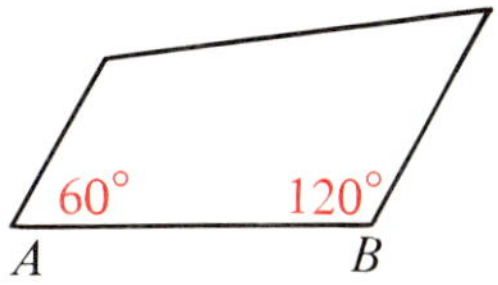

∠ *A* and ∠ *B* are supplements.

THEOREM 2

If two angles are supplements of equal angles (or of the same angle), then the two angles are equal.

Here is a diagram and proof of Theorem 2.

Given information: ∠ 1 and ∠ 2 are supplements.
∠ 3 and ∠ 4 are supplements.
∠ 1 = ∠ 3

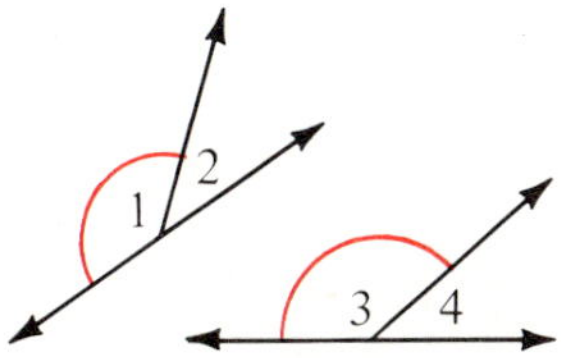

Conclusion: ∠ 2 = ∠ 4

Step 1 ∠ 1 + ∠ 2 = 180° because ∠ 1 and ∠ 2 are supplements, and supplements (by definition) total 180°.

Step 2 ∠ 3 + ∠ 4 = 180° because ∠ 3 and ∠ 4 are supplements.

Step 3 ∠ 1 + ∠ 2 = ∠ 3 + ∠ 4 because both sums are 180°.

Step 4 ∠ 1 = ∠ 3 because this is given information.

Step 5 ∠ 2 = ∠ 4 by using the Subtraction Postulate.

Our next theorem is the familiar statement that vertical angles are equal. Of course, we introduced this result on page 18.

THEOREM 3

Vertical angles are equal.

Step 1 ∠ 1 and ∠ 2 are supplements.

Step 2 ∠ 2 and ∠ 3 are supplements.

Step 3 ∠ 1 = ∠ 3 because supplements of the same angle are equal.

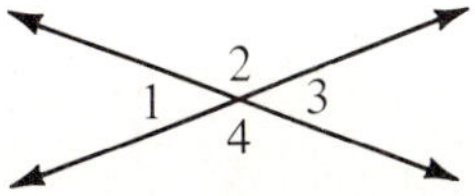

Classroom Practice

1. If $\angle A = 70°$, what is the measure of a complement of $\angle A$? a supplement of $\angle A$?
2. If $\angle B = 89°$, what is the measure of a complement of $\angle B$? a supplement of $\angle B$?
3. If $\angle C = x°$, what is the measure of a complement of $\angle C$? a supplement of $\angle C$?

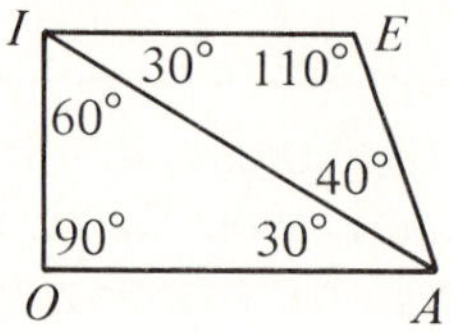

4. Name two pairs of complementary angles.
5. Name a supplement of $\angle OAE$.
6. Name another pair of supplementary angles.

7. Name three pairs of supplementary angles.

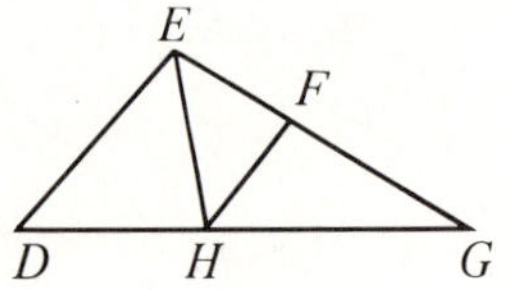

Exercise 7

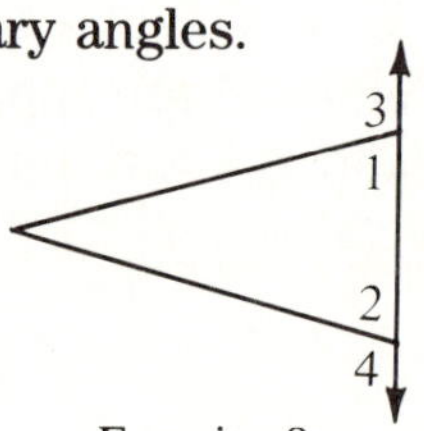

Exercise 8

8. Suppose that $\angle 1 = \angle 2$. State the theorem that allows you to conclude that $\angle 3 = \angle 4$.

Suppose $\overline{RY} \perp \overline{SY}$ and $\overline{UY} \perp \overline{TY}$.

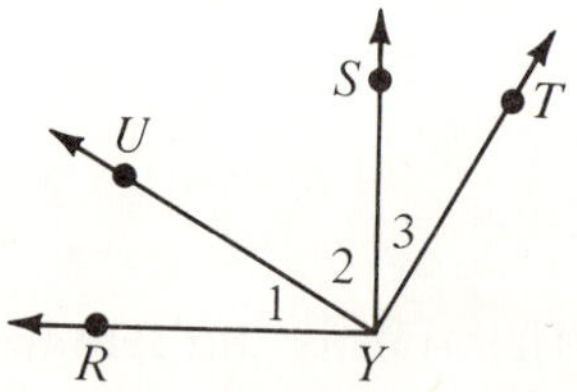

9. Name a complement of $\angle 1$.
10. Name a complement of $\angle 3$.
11. State the theorem that allows you to conclude that $\angle 1 = \angle 3$.

Written Exercises

Copy and complete the table.

A

	1.	2.	3.	4.	5.	6.	7.
$\angle A$	50°	62°	18°	?	?	?	?
Complement of $\angle A$	?	?	?	70°	48°	?	?
Supplement of $\angle A$	?	?	?	?	?	100°	135°

In the diagram, $\overline{AB} \perp \overline{BC}$ and $\overline{DC} \perp \overline{BC}$.

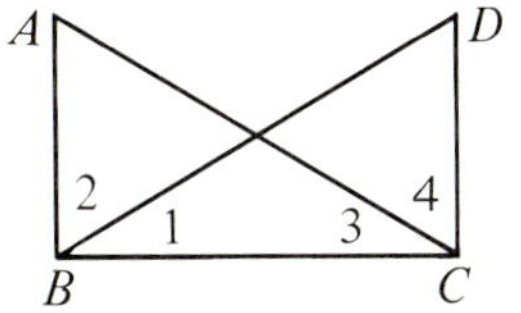

8. Name a complement of $\angle 1$.
9. Name a complement of $\angle 3$.
10. If $\angle 1 = \angle 3$, state the theorem that allows you to conclude that $\angle 2 = \angle 4$.

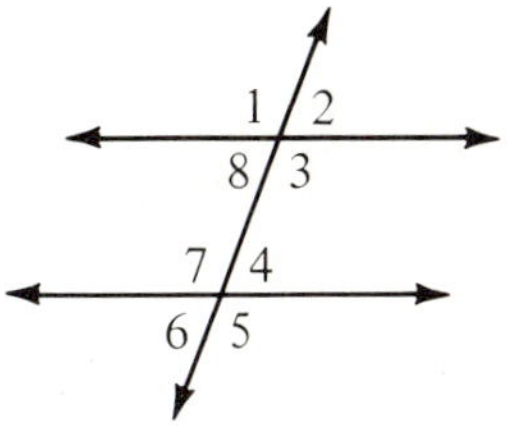

11. In the diagram, name two different supplements of $\angle 3$.
12. If $\angle 3$ and $\angle 4$ are supplements, state the theorem that allows you to conclude that $\angle 2 = \angle 4$.
13. From Exercise 12, you know that $\angle 2 = \angle 4$. State the theorem that allows you to conclude that $\angle 1 = \angle 7$.

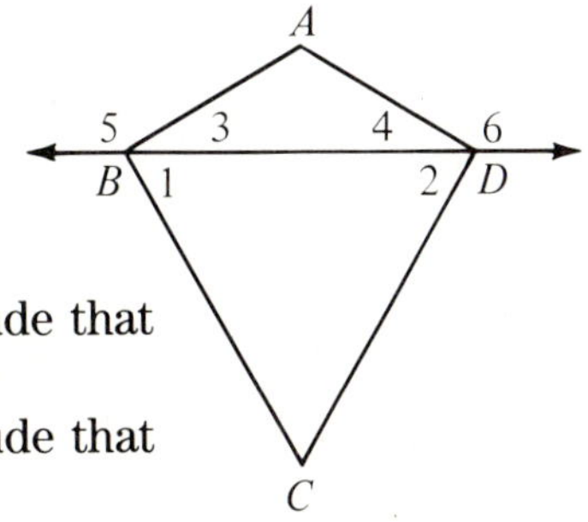

14. *Given information:* $\angle ABC = 90°$
$\angle ADC = 90°$
$\angle 1 = \angle 2$
   a. What theorem allows you to conclude that $\angle 3 = \angle 4$?
   b. What theorem allows you to conclude that $\angle 5 = \angle 6$?
15. Draw an acute angle, $\angle ABC$. Then use Construction 2, page 26, to construct the complement of $\angle ABC$.

Name the postulate of equality that justifies each statement.

16. If $\frac{5x + 3}{4} = 7$, then $5x + 3 = 28$.
17. If $5x + 3 = 28$, then $5x = 25$.
18. If $5x = 25$, then $x = 5$.

B 19. If $3x - 4y = 5$ and $2x + 4y = 10$, then $5x = 15$.

20. If $\frac{2x}{3} = \frac{7}{5}$, then $10x = 21$.
21. Two angles are complementary and equal. What is the measure of each?
22. Two angles are supplementary and equal. What is the measure of each?

2 "If . . . then" Statements

"If . . . then" statements are common in everyday speech and in mathematics. Here are some examples.

1. **If** Jolynda is 16, **then** she can apply for a driver's license.
2. **If** we win, **then** we'll be in first place.
3. **If** two angles are equal, **then** their supplements are equal.

These "if . . . then" statements all have this form:

If *A*, **then** *B*.

A is called the **hypothesis;** *B* is called the **conclusion.**

Sometimes you can express an "if . . . then" idea without even using the words *if* and *then.*

EXAMPLE 1 All owls are birds.

Restatement: **If** X is an owl,
then X is a bird.

This is a true statement.

A **true statement** is *always* true.

EXAMPLE 2 All birds are owls.

Restatement: **If** X is a bird,
then X is an owl.

This is a false statement.

A **false statement** is *not always* true.

The statements in Examples 1 and 2 are *converses,* and they say very different things. The **converse** of a statement is formed by exchanging the hypothesis and the conclusion of the statement.

Statement: If *A*, then *B*. Switch

Converse: If *B*, then *A*.

Some true statements have converses that are true. Other true statements have converses that are false.

1. True statement: If two lines are $\perp$, then they form 90° angles.
   True converse: If two lines meet at a 90° angle, then they are $\perp$.

2. True statement:
   If Kai lives in Texas,
   then he lives in North America.

   False converse:
   If Kai lives in North America,
   then he lives in Texas.

True

False!

Classroom Practice

State the hypothesis and the conclusion of each statement.

1. If the anchor gets loose, then the boat drifts.

2. If I am smart, then you are a genius.

3. If $3x - 7 = 23$, then $x = 10$.

4. I will go if you will. (*Warning:* This exercise shows that the hypothesis is not always mentioned first.)

5. We shall succeed if we try.

6. If you want to play lead guitar, then you must practice.

7–12. State the converse of each statement in Exercises 1–6.

Express each statement in "If . . . then" form.

13. The integer n is even when $n + 1$ is odd.

14. All Olympic athletes like competition.
(*Hint:* Begin with "If a person is . . .")

15. All record collectors are interested in music.

16. Too many sweets spoil the appetite.

Express each statement in "If . . . then" form.

17. $y = 14$ when $y - 8 = 6$.

18. Every positive number less than 100 has a square root less than 10.

In Exercises 19–22: **a. decide if the statement is true;**
b. state the converse;
c. decide if the converse is true.

19. If $t + 4 = 8$, then $8 = t + 4$.

20. If $a = 2$ and $b = 3$, then $ab = 6$.

21. If x is an even integer, then $2x$ is an even integer.

22. All right angles have the same measure.

Written Exercises

Write the hypothesis and the conclusion of each statement.

A **1.** If the sum of two angles is 90°, then the angles are complements.

2. If it rains, it pours.

3. If $5x + 7 = 27$, then $x = 4$.

4. They'll be here at noon if they take the train.

5. We can go to the beach Saturday, provided it's a nice day.

6. If a number is greater than 4, it is greater than 3.

7–12. Write the converse of each statement in Exercises 1–6.

Write each statement in "If . . . then" form.

13. All spiders are insects. (*Hint:* Begin with "If X is . . .")

14. All snakes are poisonous.

15. Vertical angles are equal.
(*Hint:* Begin with "If two angles . . .")

16. $x^2 = 4$ when $x = 2$ or $x = -2$.

17. Every turtle is a reptile.

18. Every history student knows about Paul Revere.

19. People who live in glass houses should not throw stones.

20. An obtuse angle has a measure that is greater than 90°.

In Exercises 21–28: **a. decide if the statement is true;**
b. write the converse;
c. decide if the converse is true.

B 21. If two angles are equal, then they are vertical angles.

22. If $x = -3$, then $x^2 = 9$.

23. If a figure is a square, then all its sides have the same length.

24. If $AM = MB$, then M is the midpoint of $\overline{AB}$.

C 25. Every rectangle is a square.

26. If $a < 0$, then $a^2 > 0$.

27. Every number divisible by five is also divisible by ten.

28. Whenever each of two numbers is negative, the sum of the numbers is negative.

SELF-TEST

Vocabulary

complementary angles (p. 46)
theorem (p. 46)
supplementary angles (p. 47)
"If . . . then" statement (p. 50)
hypothesis (p. 50)
conclusion (p. 50)
true statement (p. 50)
false statement (p. 50)
converse (p. 50)

1. If $\angle A = 20°$, what is the measure of a complement of $\angle A$? *(2-1)*

2. If $\angle B = 38°$, what is the measure of a supplement of $\angle B$?

In the diagram, $\overline{AL} \perp \overline{RK}$.

3. Name a complement of $\angle RAS$.

4. Name a supplement of $\angle RAS$.

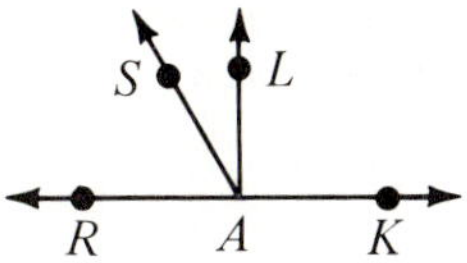

Consider the statement "If you read the newspaper every day, then you are well informed."

5. Write the hypothesis and the conclusion of the statement. *(2-2)*

6. Write the converse of the statement.

APPLICATIONS

ART AND GEOMETRY

In the period between 1400 and 1700, artists began to use geometric ideas to make their paintings more realistic. The sketch below illustrates some of their techniques.

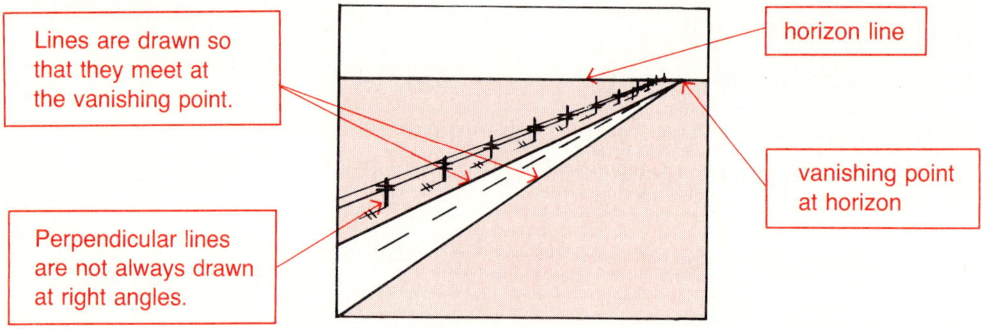

Baldassare Peruzzi, a sixteenth century Italian architect and painter, drew this picture of ancient Roman monuments. Notice his use of the vanishing point.

EXERCISES

1. Draw the tiled hallway shown by following these steps:

   (1) Draw the horizon and three vanishing points A, B, and C with $AB = BC$.

   (2) Draw $DEFG$ with $DE = EF = FG$. Draw $\overline{BD}$, $\overline{BE}$, $\overline{BF}$, and $\overline{BG}$.

   (3) Draw $\overline{DC}$ and $\overline{GA}$, marking the points in which they intersect $\overline{BD}$, $\overline{BE}$, $\overline{BF}$, and $\overline{BG}$.

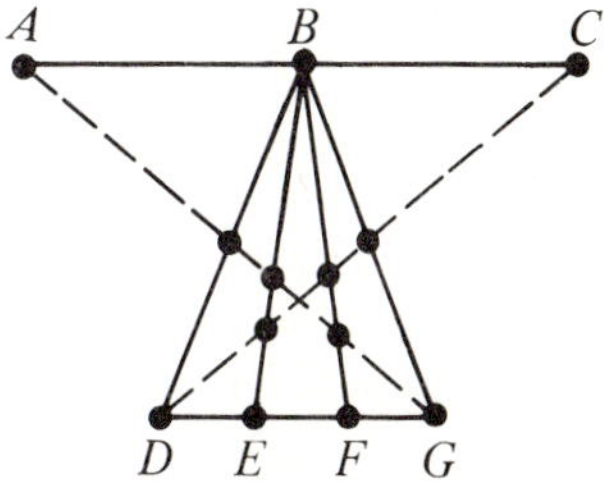

   (4) Draw horizontal lines through the intersection points.

   (5) Erase the construction lines and shade every other square.

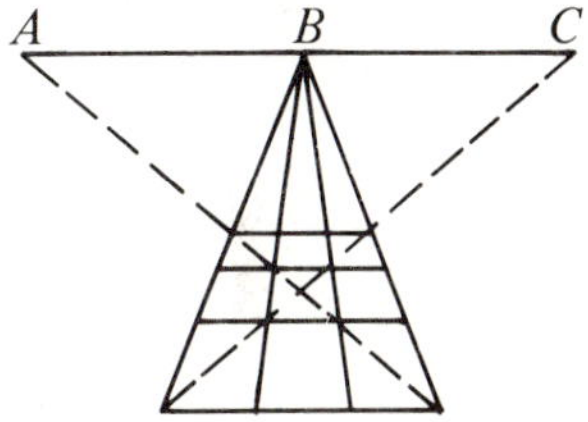

2. The humorous picture at the right is called "False Perspective." It was created in 1753 by the famous British artist William Hogarth. Describe some of the false perspectives in this work.

3. Cut a picture from a magazine or newspaper and enlarge it by the method described below.
   (1) Draw a rectangular grid on the picture.
   (2) Draw a larger rectangular grid.
   (3) Copy each square of the picture in the corresponding square of the larger grid.

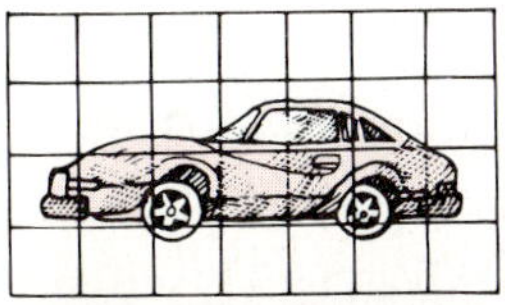

ORIGINAL PICTURE

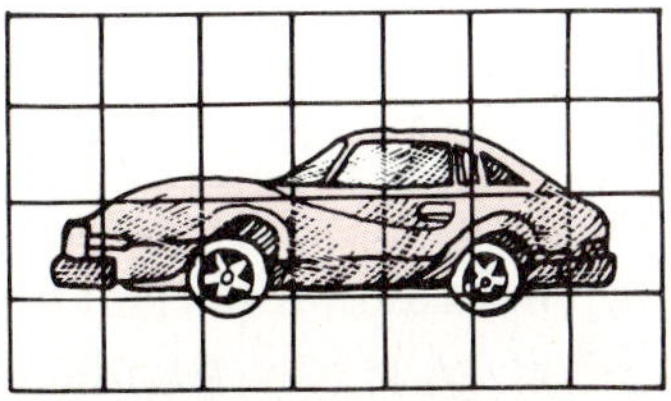

ENLARGEMENT

3 Reaching Conclusions (Optional)

When we try to prove something to others, we often use "If . . . then" statements. In this section, we shall use these statements to reach conclusions. Four kinds of simple reasoning are shown below. The first two are correct, and the last two are incorrect.

CORRECT REASONING

Type 1 *Given information:*

(1) If a figure is a square, then it has four right angles.

(2) *ABCD* is a square.

CORRECT CONCLUSION: *ABCD* has four right angles.

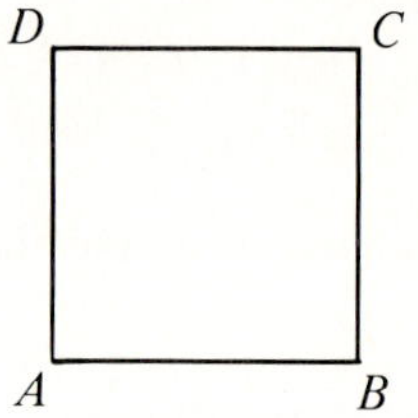

Type 2 *Given information:*

(1) If a figure is a square, then it has four right angles.

(2) *EFGH* does not have four right angles.

CORRECT CONCLUSION: *EFGH* is not a square.

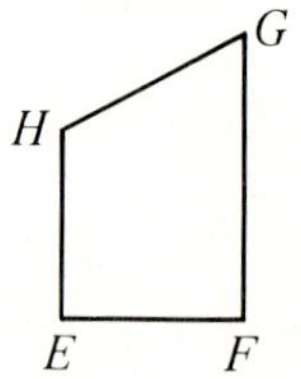

INCORRECT REASONING

Type 3 *Given information:*

(1) If a figure is a square, then it has four right angles.

(2) *PQRS* has four right angles.

INCORRECT CONCLUSION: *PQRS* is a square.

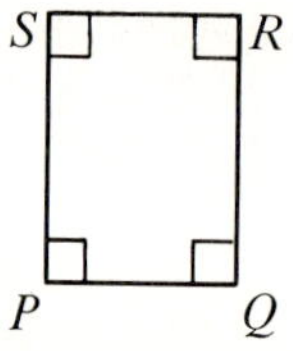

Type 4 *Given information:*

(1) If a figure is a square, then it has four right angles.

(2) *WXYZ* is not a square.

INCORRECT CONCLUSION: *WXYZ* does not have four right angles.

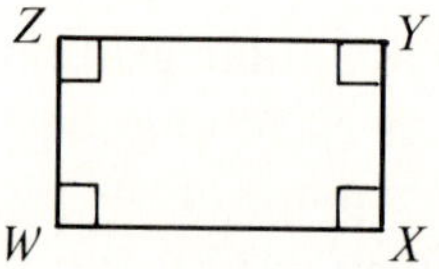

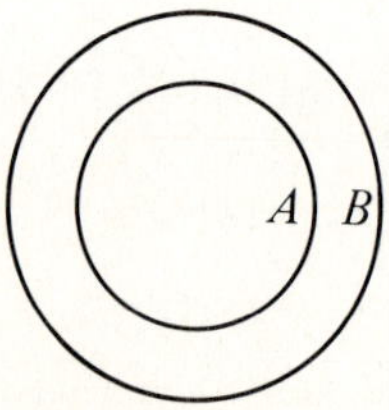

The basic form of each type of reasoning follows, along with a diagram. The diagram at the left illustrates the statement "If *A*, then *B*." When *A* is true, we place a point inside circle *A*. When *A* is not true, we place a point outside circle *A*.

Type 1
(1) If A, then B.
(2) A is true.
CORRECT CONCLUSION: B is true.

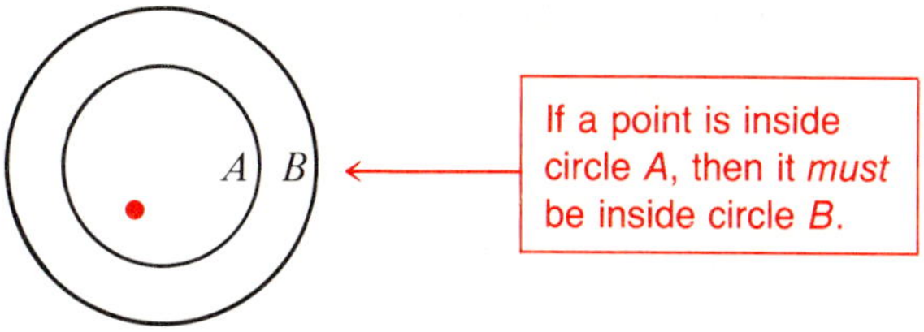

Type 2
(1) If A, then B.
(2) B is not true.
CORRECT CONCLUSION: A is not true.

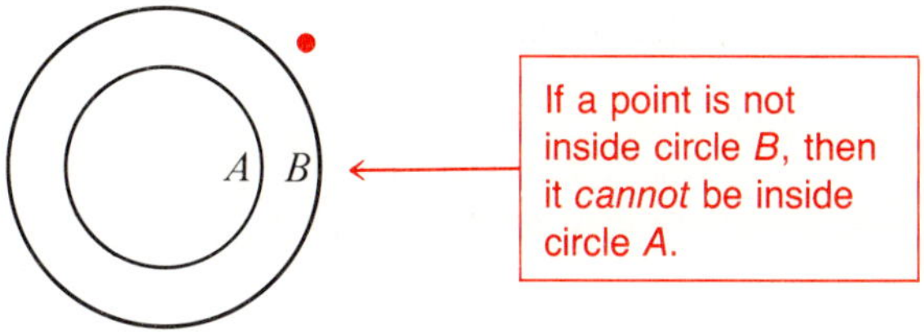

Type 3
(1) If A, then B.
(2) B is true.
NO CONCLUSION POSSIBLE.

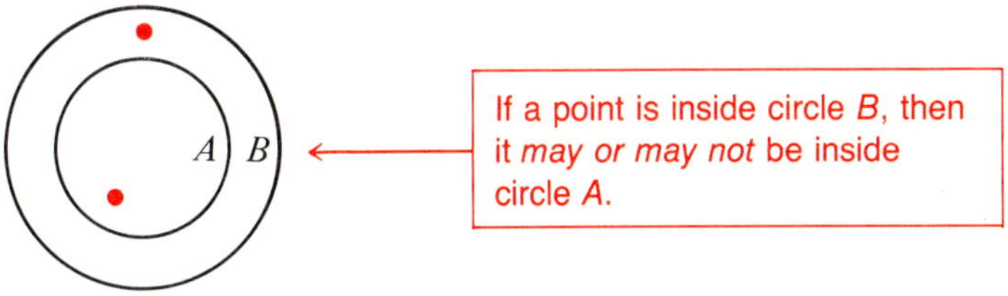

Type 4
(1) If A, then B.
(2) A is not true.
NO CONCLUSION POSSIBLE.

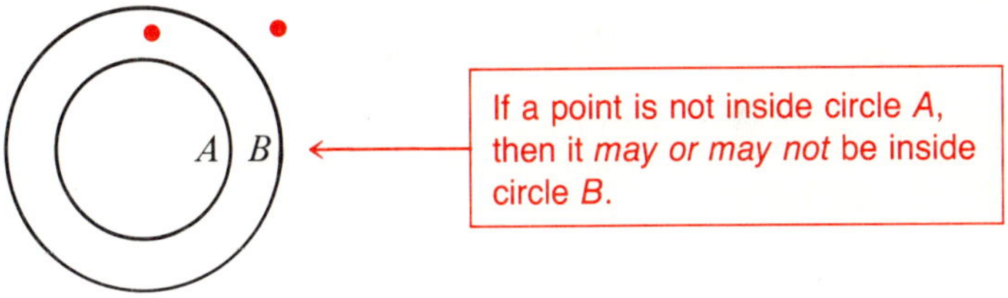

Classroom Practice

Accept this statement as true:

"All basketball players are tall."

1. Reword this statement in "If . . . then" form.
2. Make a circle diagram of this statement.
3. For each statement below, tell what you can conclude. If no conclusion is possible, say so.
   a. Julia is tall.
   b. A giraffe is not a basketball player.
   c. My cousins play basketball.
   d. Irving is short.

In Exercises 4–5: a. decide whether the reasoning is Type 1, 2, 3, or 4;
b. decide whether the reasoning is correct.
(*Hint:* Rewrite (1) in "If . . . then" form.)

4. (1) All comedians love laughter.
(2) Larry Larynx loves laughter.
Conclusion: Larry Larynx is a comedian.

5. (1) When the product of two numbers is zero, one of the numbers must be zero.
(2) $(x - 2) \cdot (x - 3) = 0$
Conclusion: $x - 2 = 0$ or $x - 3 = 0$.

6. *Given information:* (1) $\angle 1 = \angle 2$
(2) $\angle 3 = \angle 4$
What, if anything, can you conclude about $\angle 1$ and $\angle 3$?

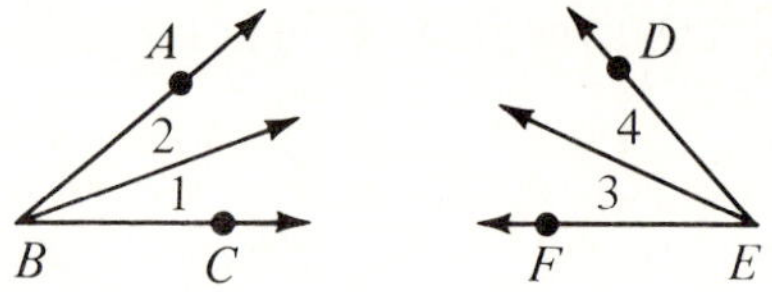

Written Exercises

Accept this statement as true: "All spiders have hairy legs."

A 1. Express the statement in "If . . . then" form.

2. Make a circle diagram of this statement.

3. For each statement, tell what you can conclude.
   a. Daddy Long-Legs is a spider.
   b. A tarantula has hairy legs.
   c. Charlotte is not a spider.
   d. An octopus does not have hairy legs.

4. Accept this statement as true: "If a figure is a rectangle, then the opposite sides have the same length." For each statement, tell what, if anything, you can conclude.
   a. Opposite sides of figure *ABCD* have the same length.
   b. Opposite sides of figure *PQRS* do not have the same length.
   c. Figure *WXYZ* is a rectangle.

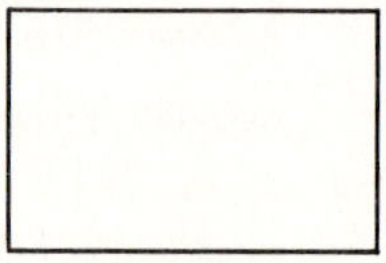
rectangle

Accept the numbered steps as true. For each exercise:
a. decide whether the reasoning is Type 1, 2, 3, or 4;
b. decide whether the reasoning is correct.

5. (1) If $x = 5$, then $y = 0$.
(2) $x = 5$
Conclusion: $y = 0$

6. (1) If $x = 5$, then $y = 0$.
(2) $y = 0$
Conclusion: $x = 5$

7. (1) Vertical angles are equal.
(2) $\angle 1$ and $\angle 2$ are not vertical angles.
Conclusion: $\angle 1 \neq \angle 2$

8. (1) Vertical angles are equal.
(2) $\angle 3 \neq \angle 4$
Conclusion: $\angle 3$ and $\angle 4$ are not vertical angles.

9. (1) If $\angle D$ is larger than $\angle E$, then $\overline{EF}$ is longer than $\overline{DF}$.
(2) $\overline{EF}$ is shorter than $\overline{DF}$.
Conclusion: $\angle D$ is not larger than $\angle E$.

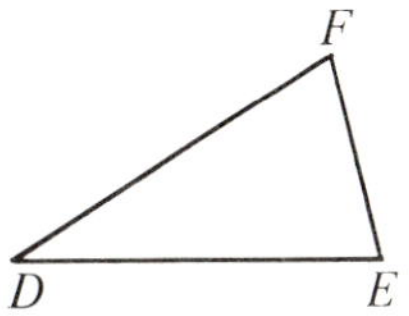

In Exercises 10–11, try to draw a conclusion from the given information. If none is possible, say so.

B 10. *Given information:* (1) $\overrightarrow{EN}$ bisects $\angle AEG$.
(2) $\overrightarrow{EG}$ bisects $\angle NEL$.

What, if anything, can you conclude about $\angle 1$ and $\angle 3$?

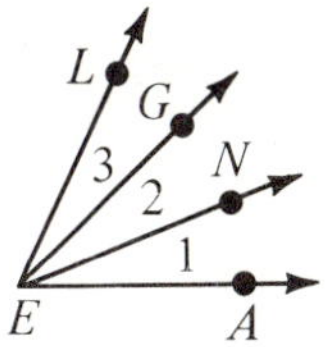

11. *Given information:* (1) $AE = EG = GH$
(2) $NO = OS = ST$

What, if anything, can you conclude about $\overline{AE}$ and $\overline{NO}$?

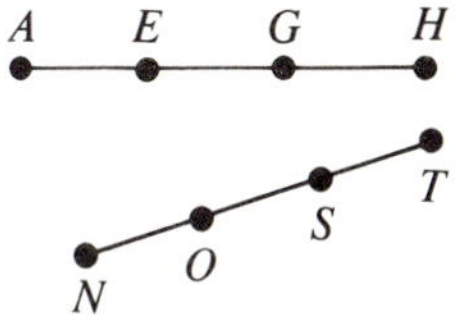

CAREER NOTES

MARINE DIETICIAN

Have you ever tried tuna hot dogs or fish luncheon rolls? These are two examples of food products that marine dieticians have developed. Marine dieticians are both nutrition experts and oceanographers. They analyze possible food sources in the oceans.

Most of the marine life that can be harvested is not considered desirable as human food. One of the most challenging problems facing marine dieticians is the conversion of such marine life into acceptable food products.

Science and math courses will help prepare you for this field. Marine dieticians often begin their careers as lab assistants.

4 Writing Proofs

A geometric proof consists of steps that show how a conclusion follows logically from other statements. As your proofs become longer than just a few steps, it helps to organize them in two columns as shown below.

Given: $\angle 1$ and $\angle 4$ are supplements.

Prove: $\angle 2 = \angle 3$

Here is the proof:

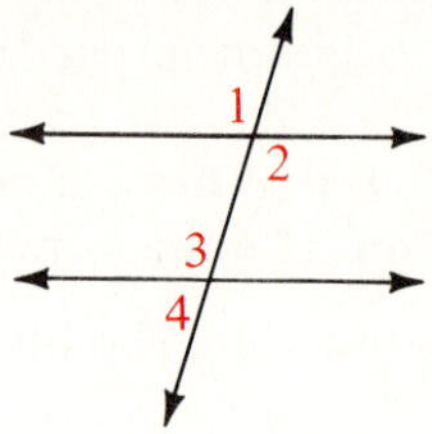

STATEMENTS	REASONS
1. $\angle 1$ and $\angle 4$ are supplements.	1. Given
2. $\angle 3$ and $\angle 4$ are supplements.	2. If the measures of two angles total 180°, the angles are supplements. (A definition)
3. $\angle 1 = \angle 3$	3. Supplements of the same angle are equal. (A theorem)
4. $\angle 1 = \angle 2$	4. Vertical angles are equal. (A theorem)
5. $\angle 2 = \angle 3$	5. Substitution Postulate (Using Steps 3 and 4)

The comments in parentheses are not really necessary. They are included to show you that there are four kinds of reasons which can be used to justify a step in a proof.

REASONS USED IN A PROOF

1. *Given information*
2. *Definitions*
3. *Postulates (These are statements accepted without proof.)*
4. *Theorems (These are statements which have been proved.)*

These four kinds of reasons are used in the following proof. Notice that Steps 5 and 6 use both the given information and the definition of angle bisector.

Given: $\angle 1 = \angle 2$
$\overrightarrow{BX}$ bisects $\angle ABC$.
$\overrightarrow{CX}$ bisects $\angle ACB$.

Prove: $\angle 3 = \angle 4$

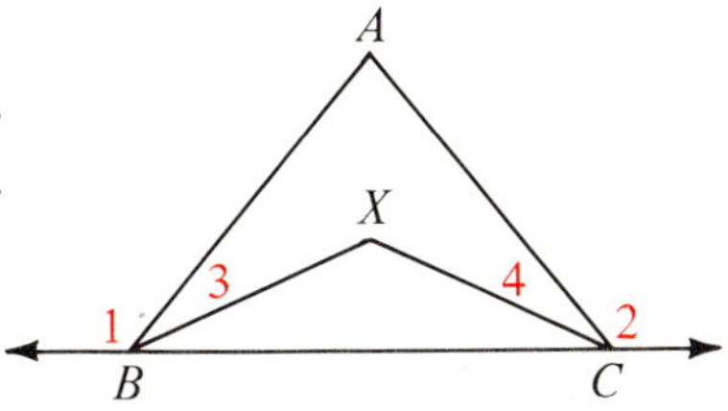

STATEMENTS	REASONS
1. $\angle 1$ and $\angle ABC$ are supplements. $\angle 2$ and $\angle ACB$ are supplements.	1. Definition of supplementary angles
2. $\angle 1 = \angle 2$	2. Given
3. $\angle ABC = \angle ACB$	3. Supplements of equal angles are equal.
4. $\frac{1}{2}\angle ABC = \frac{1}{2}\angle ACB$	4. Multiplication Postulate
5. $\frac{1}{2}\angle ABC = \angle 3$	5. *Given:* $\overrightarrow{BX}$ bisects $\angle ABC$.
6. $\frac{1}{2}\angle ACB = \angle 4$	6. *Given:* $\overrightarrow{CX}$ bisects $\angle ACB$.
7. $\angle 3 = \angle 4$	7. Substitution Postulate

Classroom Practice

Supply the missing reasons in the proofs.

1. *Given:* $\angle 1 = \angle 2$

Prove: $\overleftrightarrow{LI} \perp \overleftrightarrow{NE}$

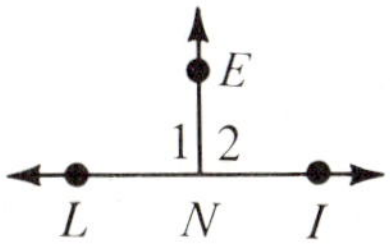

STATEMENTS	REASONS
1. $\angle 1 + \angle 2 = 180°$	1. Their sum is a ___?___ angle.
2. $\angle 1 = \angle 2$	2. ___?___
3. $\angle 1 + \angle 1 = 180°$, or $2 \cdot \angle 1 = 180°$	3. ___?___
4. $\angle 1 = 90°$	4. ___?___
5. $\overleftrightarrow{LI} \perp \overleftrightarrow{NE}$	5. ___?___

2. *Given:* $\angle 1$ and $\angle 3$ are supplementary.

Prove: $\angle 2 = \angle 4$

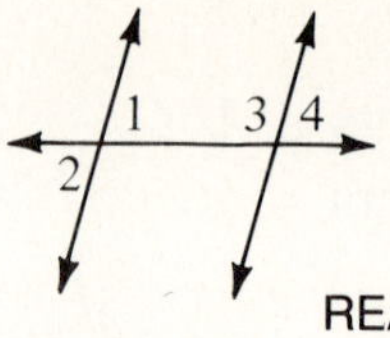

STATEMENTS	REASONS
1. $\angle 1$ and $\angle 3$ are supplementary.	1. ___?___
2. $\angle 4$ and $\angle 3$ are supplementary.	2. ___?___
3. $\angle 1 = \angle 4$	3. ___?___
4. $\angle 1 = \angle 2$	4. ___?___
5. $\angle 2 = \angle 4$	5. ___?___

Written Exercises

Supply the missing statements and reasons in the proofs.

A **1.** *Given:* $\angle 1 = \angle 2$
$\angle 3 = \angle 4$

Prove: $\angle 1 = \angle 4$

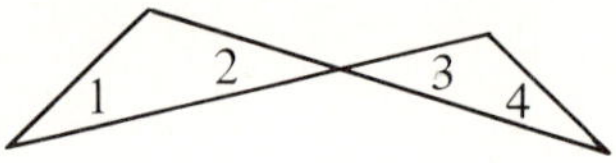

STATEMENTS	REASONS
1. $\angle 1 = \angle 2$ $\angle 3 = \angle 4$	1. ___?___
2. $\angle 2 = \angle 3$	2. ___?___
3. $\angle 1 = \angle 4$	3. ___?___

2. *Given:* $\angle 1$ and $\angle 2$ are complements.
$\angle 3$ and $\angle 4$ are complements.

Prove: $\angle 1 = \angle 4$

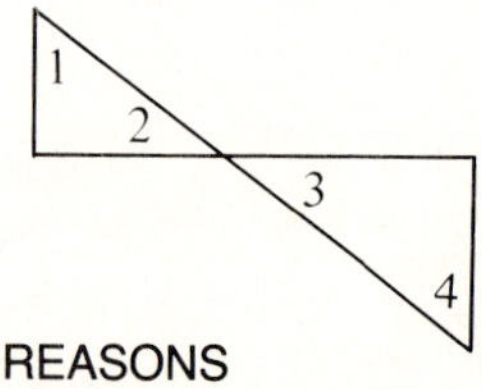

STATEMENTS	REASONS
1. $\angle 1$ and $\angle 2$ are complements. $\angle 3$ and $\angle 4$ are complements.	1. ___?___
2. $\angle 2 = \angle 3$	2. ___?___
3. $\angle 1 = \angle 4$	3. ___?___

3. *Given:* $\angle 1 = \angle 4$

Prove: $\angle 3$ and $\angle 2$ are supplements.

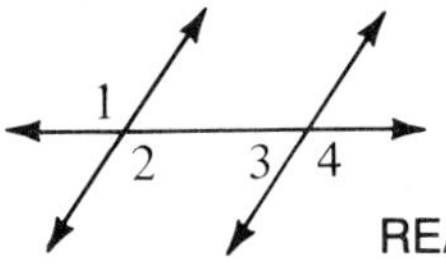

STATEMENTS	REASONS
1. $\angle 1 = \angle 4$	1. ___?___
2. $\angle 2 = \angle 1$	2. ___?___
3. $\angle 2 = \angle 4$	3. ___?___
4. $\angle 3 + \angle 4 = 180°$	4. Their sum is a ___?___ angle.
5. $\angle 3 + \angle 2 = 180°$	5. ___?___
6. ___?___	6. Def. of supplementary angles

4. *Given:* $\overline{AB} \perp \overline{BD}$
$\overrightarrow{BD}$ bisects $\angle EBC$.

Prove: $\angle 1$ and $\angle 3$ are complements.

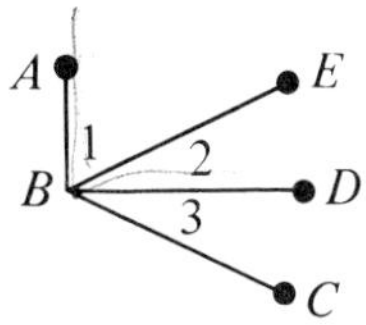

STATEMENTS	REASONS
1. $\overline{AB} \perp \overline{BD}$	1. ___?___
2. $\angle ABD = 90°$	2. ___?___
3. $\angle 1 + \angle$ ___?___ $= 90°$	3. Subst. Post.: $\angle ABD = \angle 1 + \angle 2$
4. $\angle 2 = \angle$ ___?___	4. *Given:* $\overrightarrow{BD}$ bisects $\angle EBC$.
5. $\angle 1 + \angle 3 = 90°$	5. ___?___
6. $\angle 1$ and $\angle 3$ are complements.	6. ___?___

5. *Given:* $\overleftrightarrow{XY}$ bisects $\angle AOC$.

Prove: $\overleftrightarrow{XY}$ bisects $\angle DOB$.

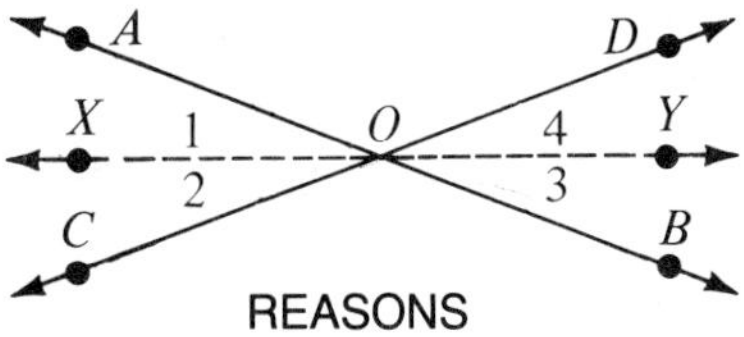

STATEMENTS	REASONS
1. $\angle 1 = \angle 2$	1. *Given:* ___?___
2. $\angle 1 = \angle 3$	2. ___?___
3. $\angle 2 = \angle$ ___?___	3. ___?___
4. $\angle 3 = \angle 4$	4. ___?___
5. ___?___	5. Definition of angle bisector

6. *Given:* $AC = BD$
O is the midpoint of $\overline{AC}$.
O is the midpoint of $\overline{BD}$.

Prove: $AO = DO$

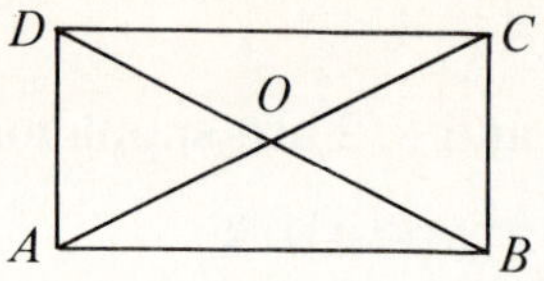

STATEMENTS	REASONS
1. $AC = BD$	1. ___?___
2. $\frac{1}{2}AC = \frac{1}{2}BD$	2. ___?___
3. $AO = \frac{1}{2}AC$	3. *Given:* ___?___
4. $DO = \frac{1}{2}BD$	4. ___?___
5. ___?___ = ___?___	5. Substitution Postulate (Steps 2, 3, and 4)

B 7. *Given:* $\overrightarrow{RO}$ bisects $\angle YRS$.
$\overrightarrow{SO}$ bisects $\angle YSR$.
$\angle 1 = \angle 2$

Prove: $\angle 5 = \angle 6$

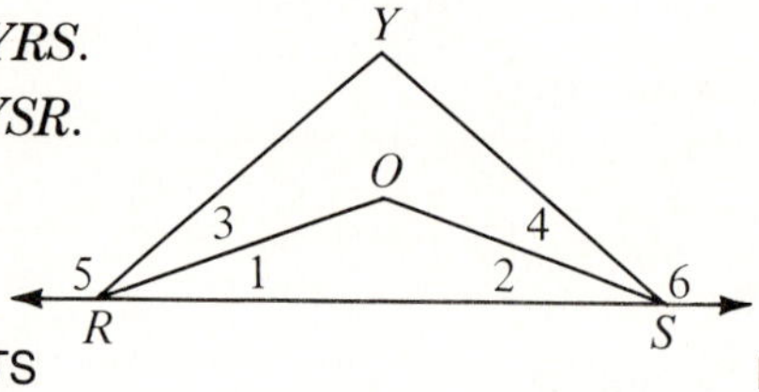

STATEMENTS	REASONS
1. $\angle 1 = \angle 3$	1. ___?___
2. $\angle 2 = \angle 4$	2. ___?___
3. $\angle 1 = \angle 2$	3. ___?___
4. $\angle 3 = \angle 4$	4. ___?___ (Steps 1, 2, and 3)
5. $\angle 1 + \angle 3 = \angle 2 + \angle 4$, or $\angle YRS = \angle YSR$	5. ___?___
6. $\angle YRS$ and $\angle 5$ are supplements. $\angle YSR$ and $\angle 6$ are supplements.	6. ___?___
7. $\angle 5 = \angle 6$	7. ___?___

8. *Given:* $\angle ABC = 90°$
$\angle ADC = 90°$
$\angle 1 = \angle 4$

Prove: $\angle 2 = \angle 5$

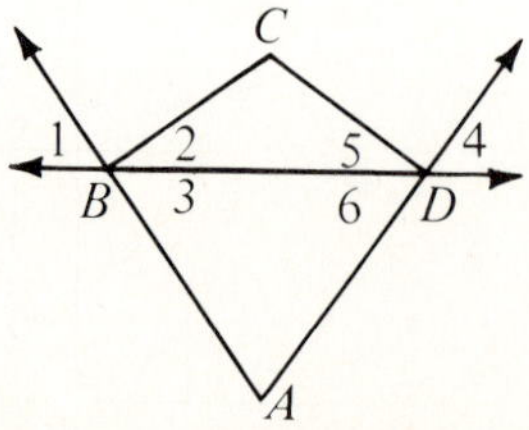

STATEMENTS	REASONS
1. $\angle 1 = \angle 3$	1. ___?___
2. $\angle 4 = \angle 6$	2. ___?___
3. $\angle 1 = \angle 4$	3. ___?___
4. $\angle 3 = \angle 6$	4. ___?___
5. $\angle 2$ and $\angle$ ___?___ are complements.	5. *Given:* $\angle ABC = 90°$
6. $\angle 5$ and $\angle$ ___?___ are complements.	6. ___?___
7. $\angle 2 = \angle 5$	7. ___?___

For each exercise, copy what is given and what is to be proved. Write a proof in two-column form. Refer to the diagrams below.

9. *Given:* $\angle 3 = \angle 5$

Prove: $\angle 4 = \angle 6$

10. *Given:* $\angle 2 = \angle 8$

Prove: $\angle 4 = \angle 6$

11. *Given:* $\angle 3$ and $\angle 6$ are supplements.

Prove: $\angle 2 = \angle 6$

12. *Given:* $\angle 1 = \angle 5$

Prove: $\angle 4$ and $\angle 5$ are supplements.

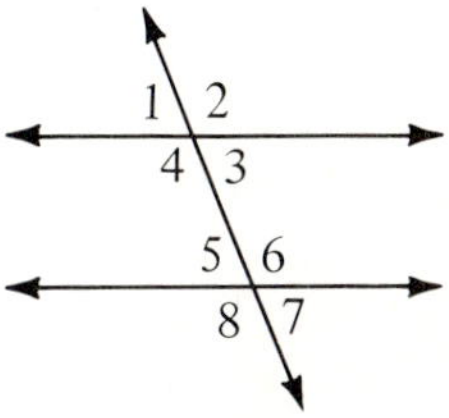

Exercises 9–12

C **13.** *Given:* $\angle 2 = \angle 3$
$\angle 3 = \angle 5$

Prove: $\angle 1 = \angle 6$

14. *Given:* $\angle 4$ and $\angle 5$ are supplements.
$\angle 2 = \angle 3$

Prove: $\angle 2 = \angle 5$

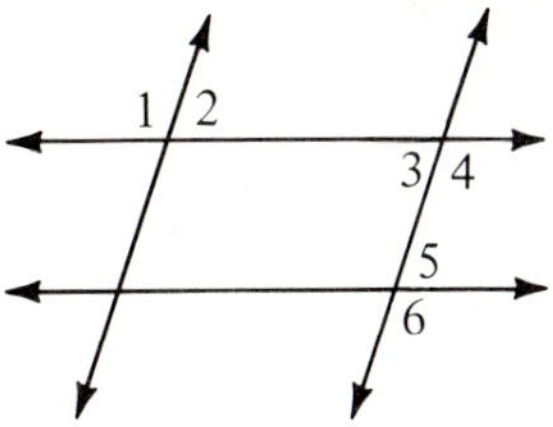

Exercises 13, 14

15. *Given:* $\angle 1 = \angle 2$
$\angle 3$ and $\angle 5$ are complements.
$\angle 4$ and $\angle 6$ are complements.

Prove: $\angle 5 = \angle 6$

16. *Given:* $\overline{AB} \perp \overline{BD}$
$\overline{CD} \perp \overline{BD}$
$\angle 5 = \angle 6$

Prove: $\angle 1 = \angle 2$

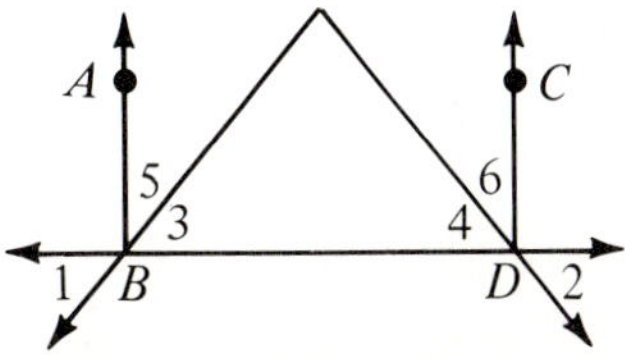

Exercises 15, 16

5 Parallel Lines

If two lines do not intersect, they are either *parallel* or *skew.*

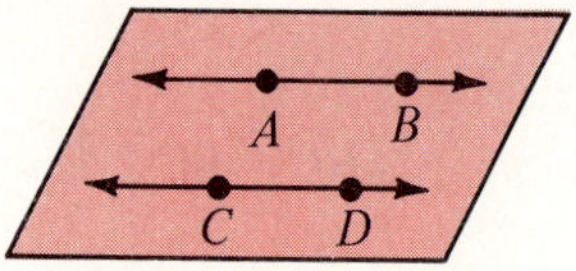

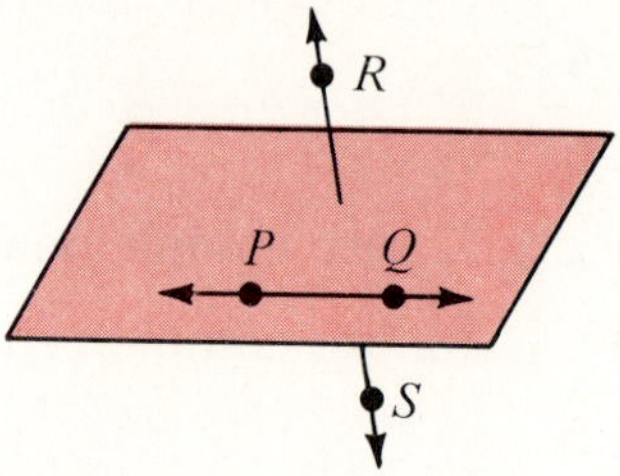

$\overleftrightarrow{AB}$ and $\overleftrightarrow{CD}$ are **parallel lines.**
(1) They are coplanar lines.
(2) They do not intersect.

$\overleftrightarrow{PQ}$ and $\overleftrightarrow{RS}$ are **skew lines.**
(1) They are *not* coplanar lines.
(2) They do not intersect.

We abbreviate the statement "$\overleftrightarrow{AB}$ is parallel to $\overleftrightarrow{CD}$" by writing $\overleftrightarrow{AB} \parallel \overleftrightarrow{CD}$. We shall also say that $\overline{AB} \parallel \overline{CD}$ because the segments are contained in parallel lines. In a drawing, we indicate parallel lines and segments by arrowheads and, if necessary, double arrowheads.

The single arrowheads show that $\overline{AD} \parallel \overline{BC}$.

The double arrowheads show that $\overline{AB} \parallel \overline{CD}$.

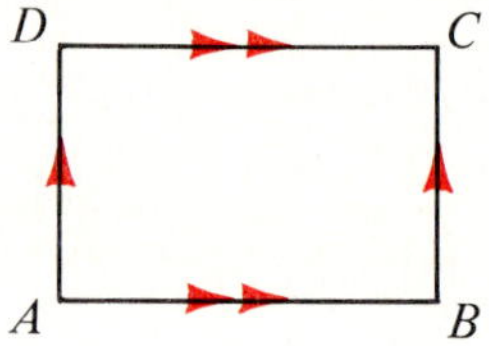

The diagram at the right shows two parallel lines cut by a third line. This cutting line is called a **transversal.** Angles 1, 2, 3, and 4 formed on the inside of the parallel lines are called **interior angles.** Angles formed by parallel lines are given special names as follows:

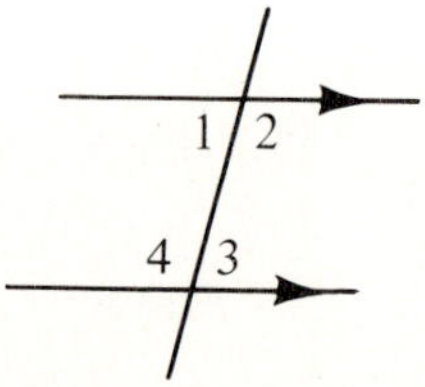

Alternate interior angles are interior angles on alternate sides of the transversal. (Examples above: $\angle 1$ and $\angle 3$; also $\angle 2$ and $\angle 4$) To spot alternate interior angles, look for a "Z-shaped" figure in various positions.

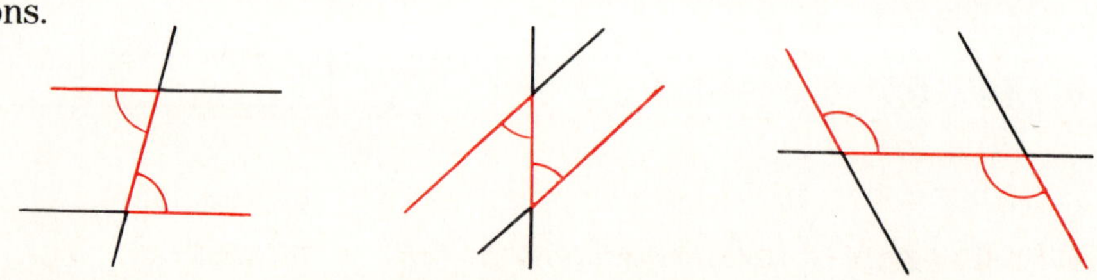

Same-side interior angles are interior angles on the same side of the transversal. (Examples on page 66: $\angle 1$ and $\angle 4$; also $\angle 2$ and $\angle 3$) To spot same-side interior angles, look for a "U-shaped" figure in various positions.

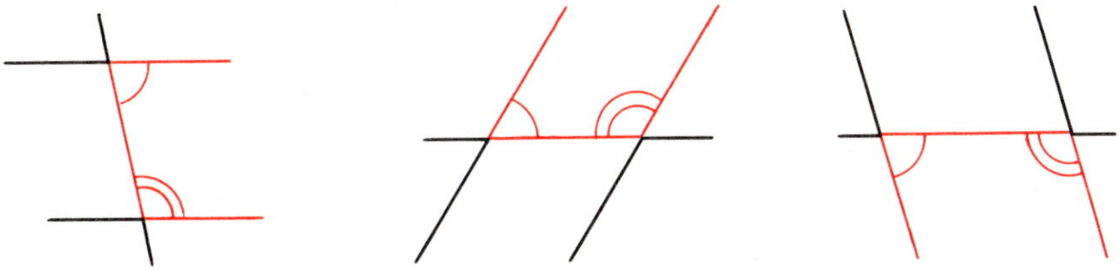

Corresponding angles are so named because they appear to be in corresponding positions in relation to the two lines. To spot corresponding angles, look for an "F-shaped" figure in various positions.

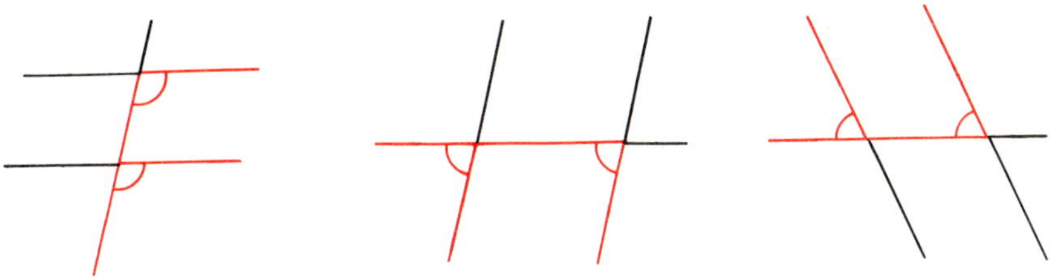

Even when lines cut by a transversal are not parallel, we still use the same vocabulary. For example:

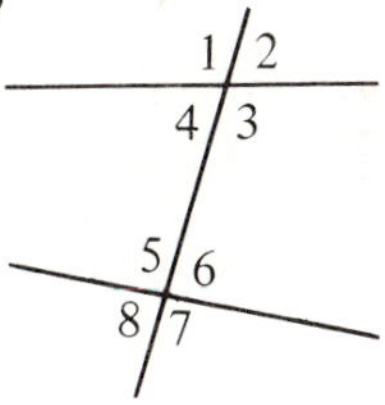

$\angle 4$ and $\angle 8$ are corresponding angles;

$\angle 4$ and $\angle 6$ are alternate interior angles;

$\angle 4$ and $\angle 5$ are same-side interior angles.

The diagram at the right suggests that when two parallel lines are cut by a transversal, corresponding angles are equal. You can convince yourself of this by drawing and measuring many pairs of corresponding angles formed by different transversals. You may suspect that we can prove this property. However, there is just not enough information in our previous postulates and theorems to deduce this result as a theorem. Let us therefore accept it without proof as a postulate.

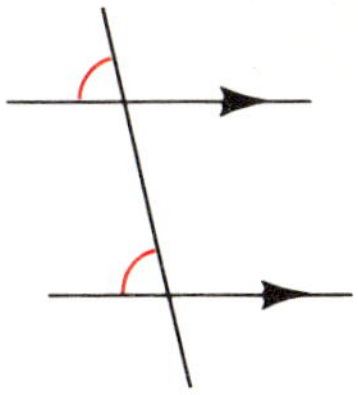

POSTULATE 7

If two parallel lines are cut by a transversal, then corresponding angles are equal.

Once we have this postulate, we can immediately deduce the following theorems. The proofs are left as Exercises 31 and 32 on page 71.

THEOREM 4
If two parallel lines are cut by a transversal, then alternate interior angles are equal.
THEOREM 5
If two parallel lines are cut by a transversal, then same-side interior angles are supplementary.

Classroom Practice

In the diagram, $l \parallel m$.

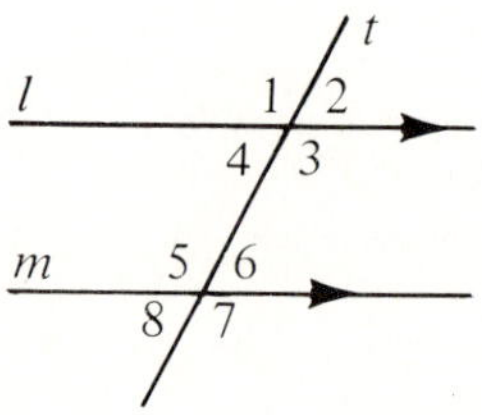

1. Name two pairs of alternate interior angles.
2. Name two pairs of same-side interior angles.
3. Name four pairs of corresponding angles.
4. What is the special name given to line t?
5. Suppose $\angle 4 = 60°$. Find the measures of the other seven angles.
6. Suppose $\angle 4 = x°$. Find the measures of $\angle 5$ and $\angle 6$.
7. Although we have not discussed *alternate exterior angles*, you may be able to guess what they are. Name two pairs of them.

State the special name for each pair of angles.

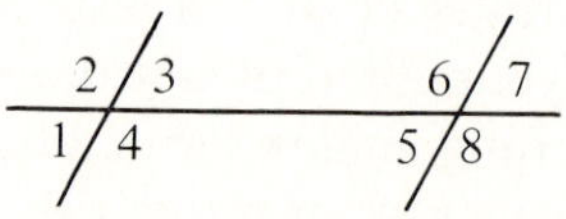

8. $\angle 3$ and $\angle 5$
9. $\angle 2$ and $\angle 6$
10. $\angle 3$ and $\angle 6$
11. $\angle 1$ and $\angle 5$

Find the values of x and y in each diagram.

12.

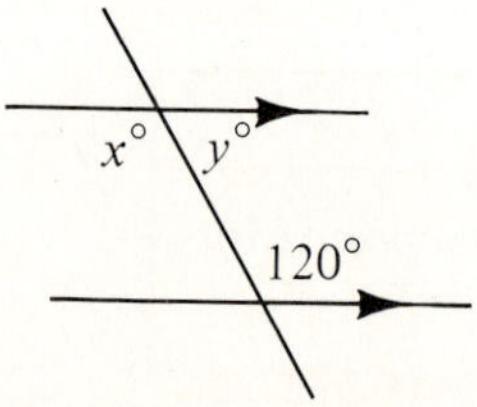

13.

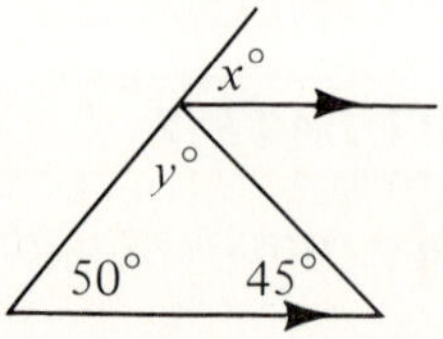

14.

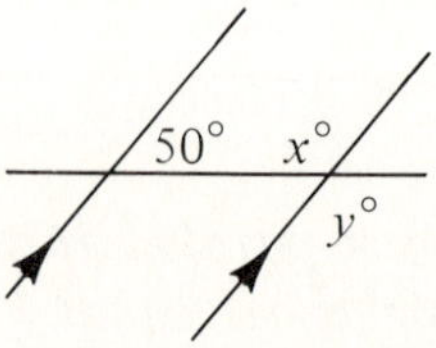

Written Exercises

Classify each pair of angles as (1) alternate interior angles, (2) same-side interior angles, (3) corresponding angles, or (4) none of these.

A

1. $\angle 1$ and $\angle 5$
2. $\angle 3$ and $\angle 8$
3. $\angle 2$ and $\angle 8$
4. $\angle 4$ and $\angle 8$
5. $\angle 2$ and $\angle 7$
6. $\angle 3$ and $\angle 5$
7. $\angle 2$ and $\angle 5$
8. $\angle 3$ and $\angle 7$

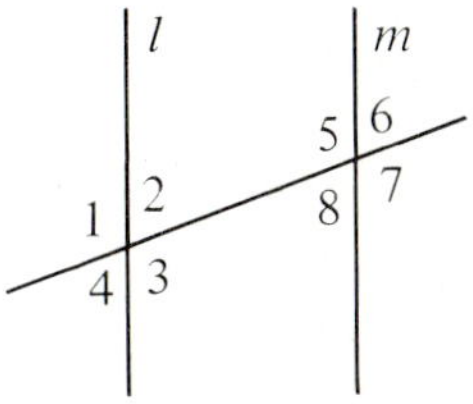

In the diagram above, suppose $l \parallel m$.

9. If $\angle 1 = 110°$, find the measures of the other seven angles.
10. If $\angle 2 = x°$, find the measures of $\angle 5$ and $\angle 8$.

Complete each statement.

11. $\angle 1$ and $\angle 2$ are formed by lines __?__ and __?__ and the transversal __?__.

    $\angle 3$ and $\angle 4$ are formed by lines __?__ and __?__ and the transversal __?__.

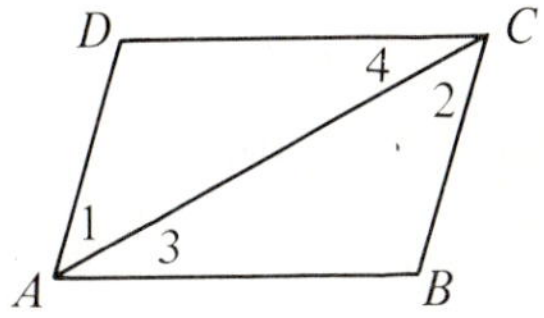

12. $\angle 5$ and $\angle 6$ are formed by lines __?__ and __?__ and the transversal __?__.

    $\angle 7$ and $\angle 8$ are formed by lines __?__ and __?__ and the transversal __?__.

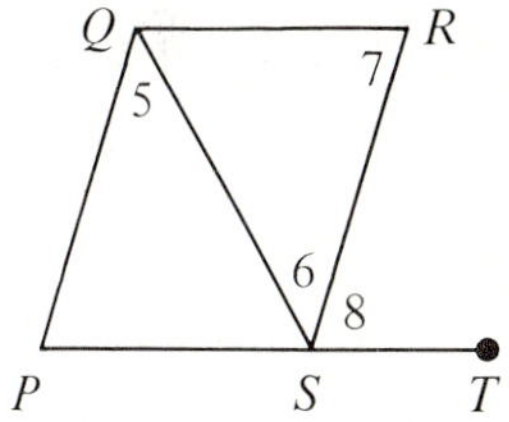

In the diagram, $a \parallel b$ and $c \parallel d$. State the special name for each pair of angles. Then tell if the angles are equal or supplementary.

13. $\angle 6$ and $\angle 10$
14. $\angle 7$ and $\angle 9$
15. $\angle 3$ and $\angle 14$
16. $\angle 2$ and $\angle 14$
17. $\angle 9$ and $\angle 16$
18. $\angle 8$ and $\angle 10$
19. $\angle 9$ and $\angle 13$
20. $\angle 8$ and $\angle 12$

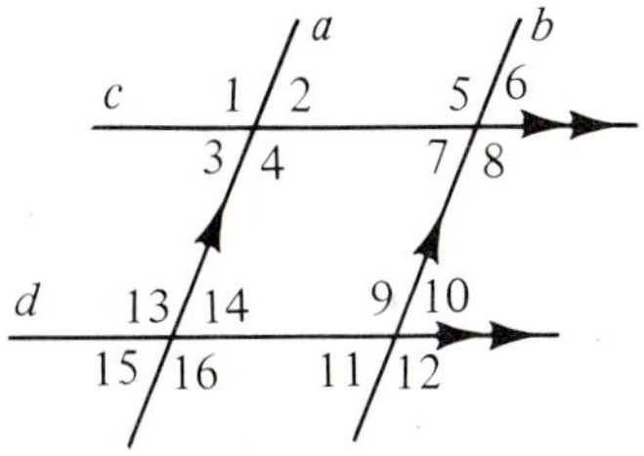

21. Refer to the diagram above. Name seven angles equal to $\angle 4$.
22. Name eight angles supplementary to $\angle 4$.

Find the values of x and y in each diagram.

23.

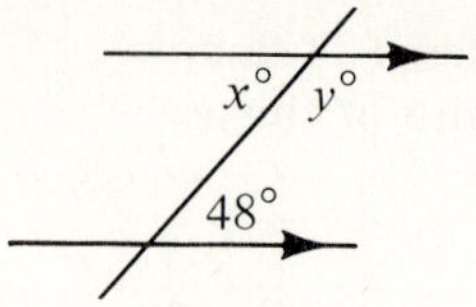

24.

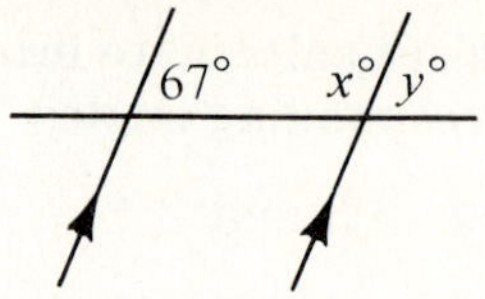

25.

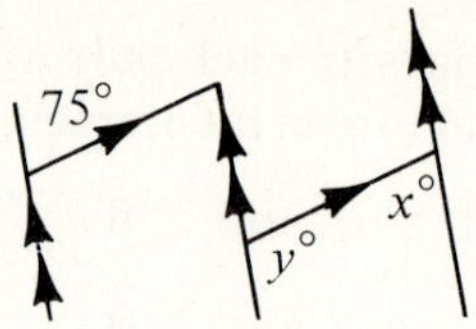

B **26.**

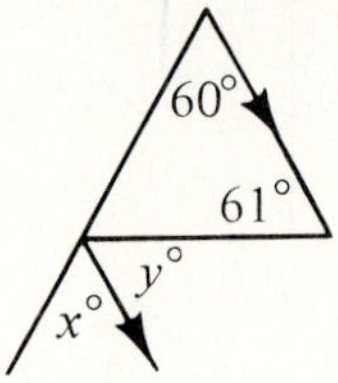

27.

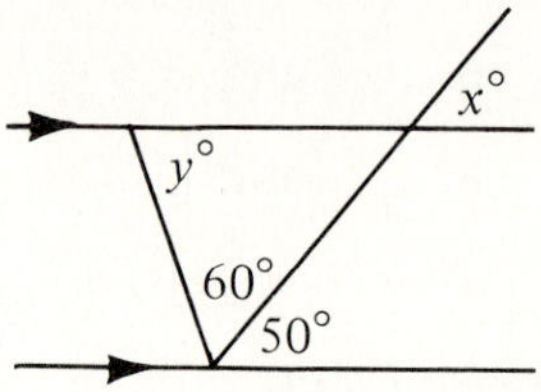

28.

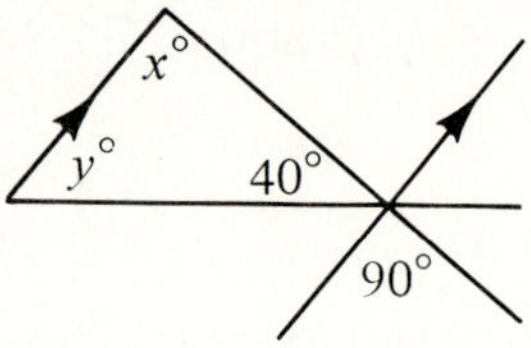

Supply the missing statements and reasons in each proof.

29. *Given:* $\overline{RU} \parallel \overline{ST}$
$\overline{RS} \parallel \overline{UT}$

Prove: $\angle R = \angle T$

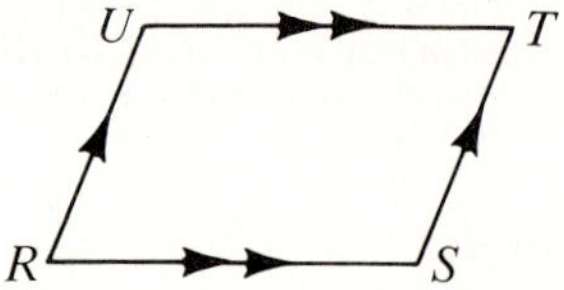

STATEMENTS	REASONS
1. $\overline{RU} \parallel \overline{ST}$	1. ___?___
2. $\angle R$ and $\angle S$ are supplements.	2. If 2 $\parallel$ lines are cut by a trans., then ___?___.
3. $\overline{RS} \parallel \overline{UT}$	3. ___?___
4. $\angle S$ and $\angle T$ are supplements.	4. ___?___
5. $\angle R = \angle T$	5. ___?___

30. *Given:* $\angle B = \angle C$
$\overline{AD} \parallel \overline{BC}$

Prove: $\angle 1 = \angle 2$

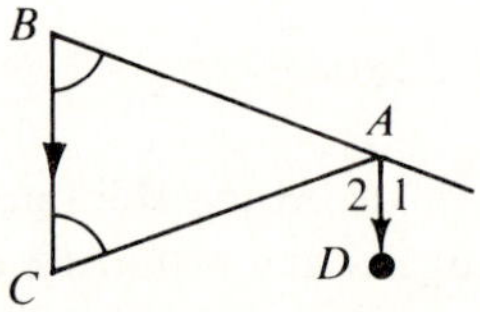

STATEMENTS	REASONS
1. $\overline{AD} \parallel \overline{BC}$	1. ___?___
2. $\angle 1 = \angle B$	2. If 2 $\parallel$ lines are cut by a trans., then ___?___.
3. $\angle 2 = \angle C$	3. ___?___
4. $\angle B = \angle C$	4. ___?___
5. $\angle 1 = \angle$ ___?___	5. Subst. Post. (Steps 2, 3, 4)

C **31.** Prove Theorem 4: If two parallel lines are cut by a transversal, then alternate interior angles are equal.

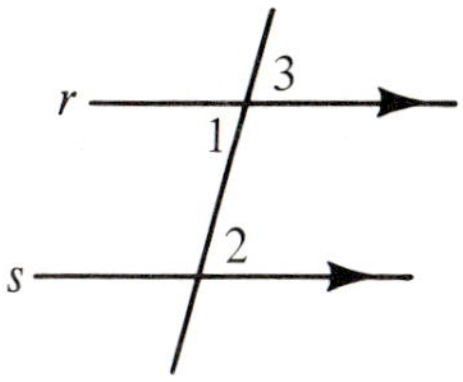

Given: $r \parallel s$

Prove: $\angle 1 = \angle 2$ (*Hint:* Use $\angle 3$.)

32. Prove Theorem 5: If two parallel lines are cut by a transversal, then same-side interior angles are supplementary.

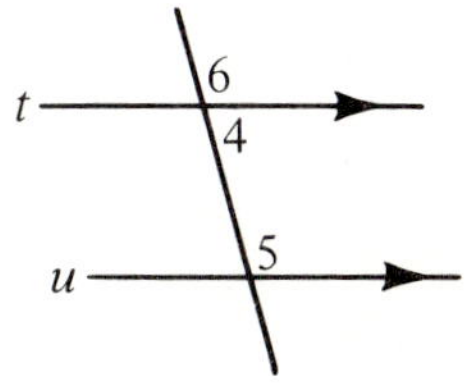

Given: $t \parallel u$

Prove: $\angle 4$ and $\angle 5$ are supplementary. (*Hint:* Use $\angle 6$.)

CALCULATOR ACTIVITIES

Perform the indicated operations using a calculator. Round your answers to three decimal places.

1. 42.31 + 13.26 + 8.79	**2.** 2.3 + 9.41 + 0.00048	**3.** 107.236 + 2.0007 + 44
4. 69.1 + 332 + 0.4445	**5.** 99 + 0.11 + 10.0909	**6.** 3.3333 + 77.7 + 0.555
7. 153.2 − 89.378	**8.** 1436 − 725.23	**9.** 5.698 − 0.7531
10. 0.74963 − 0.54296	**11.** 3647.02 − 897.001	**12.** 24.27 − 19.0005
13. 6.7 × 418	**14.** 0.38 × 0.57	**15.** 46.12 × 0.9
16. 0.763 × 97.2	**17.** 1.1111 × 26	**18.** 15.6 × 89.21
19. 437 ÷ 33	**20.** 56 ÷ 8.2	**21.** 115.3 ÷ 0.3
22. 8.426 ÷ 0.002	**23.** 0.567 ÷ 40	**24.** 215 ÷ 5.375

Use the division key to find the *terminating* decimal *equal* to each fraction. Write your answer exactly as it appears on the calculator.

25. $\frac{15}{75}$ **26.** $\frac{17}{16}$ **27.** $\frac{31}{32}$ **28.** $\frac{12}{25}$ **29.** $\frac{43}{64}$ **30.** $\frac{23}{92}$

The fractions below can be expressed as *repeating* decimals. Use the division key to find a decimal *approximation* for each fraction by rounding to four decimal places.

31. $\frac{5}{9}$ **32.** $\frac{2}{3}$ **33.** $\frac{4}{9}$ **34.** $\frac{6}{11}$ **35.** $\frac{5}{6}$ **36.** $\frac{2}{11}$

6 Proving Lines Parallel

In this section, we shall study ways of proving lines parallel. The first way is given in Postulate 8. Postulate 7 is listed with Postulate 8 so that you can see that these two postulates are converses.

POSTULATE 7

If two parallel lines are cut by a transversal, then corresponding angles are equal.

POSTULATE 8

If two lines and a transversal form equal corresponding angles, then the lines are parallel.

Postulates 7 and 8 can be written in convenient picture statements as follows:

Postulate 7: If 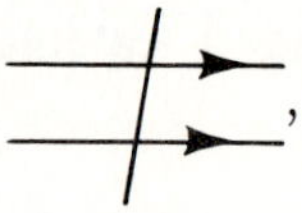, then 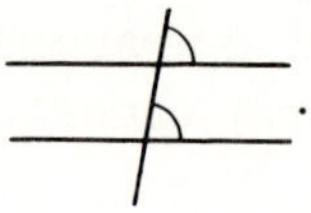.

Postulate 8: If 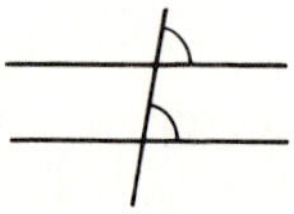, then 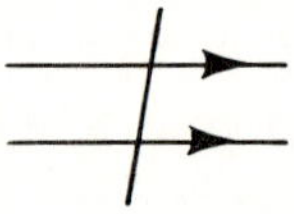.

From Postulate 8, we can easily determine other ways to prove lines parallel. These are given in Theorems 6, 7, and 8 below.

THEOREM 6

If two lines and a transversal form equal alternate interior angles, then the lines are parallel.

THEOREM 7

If two lines and a transversal form supplementary same-side interior angles, then the lines are parallel.

THEOREM 8

In a plane, if two lines are each perpendicular to a third line, then the two lines are parallel.

WAYS TO PROVE TWO LINES PARALLEL
1. *Show that corresponding angles are equal.*
2. *Show that alternate interior angles are equal.*
3. *Show that same-side interior angles are supplementary.*
4. *Show that the lines are both perpendicular to a third line.*

EXAMPLE

Which lines are parallel?

$\overleftrightarrow{BP}$ and $\overleftrightarrow{CR}$ are not parallel because two corresponding angles are not equal ($50 \neq 51$).

$\overleftrightarrow{BQ}$ and $\overleftrightarrow{CS}$ are parallel because two corresponding angles are equal ($50 + 30 = 80$ and $51 + 29 = 80$).

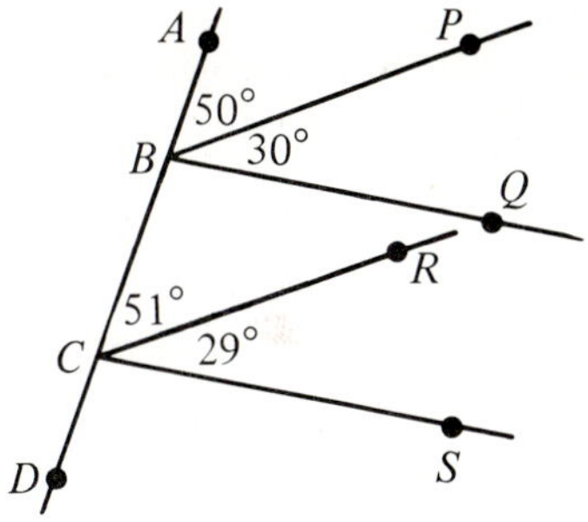

Classroom Practice

For each figure, tell whether you can correctly conclude that $l \parallel m$.

1.

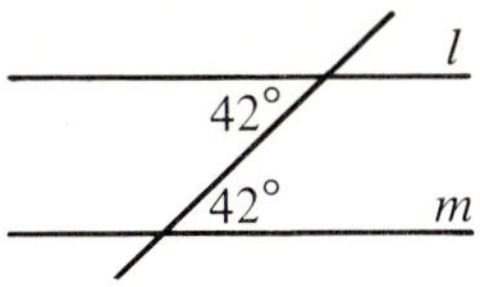

2.

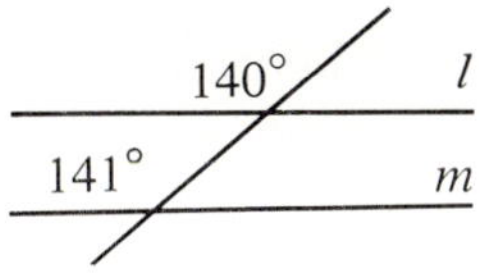

3.

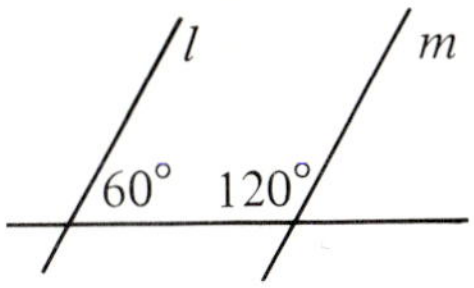

In each figure, tell which lines are parallel.

4.

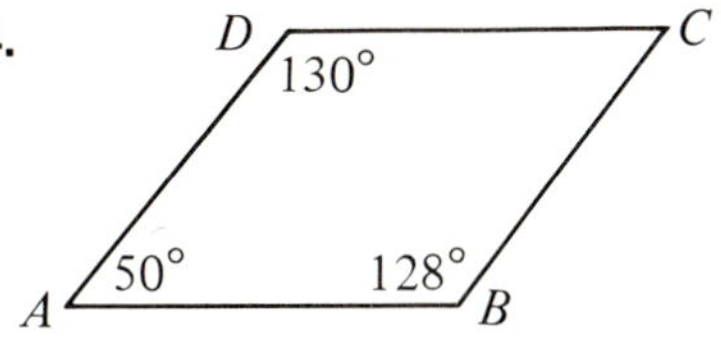

5.

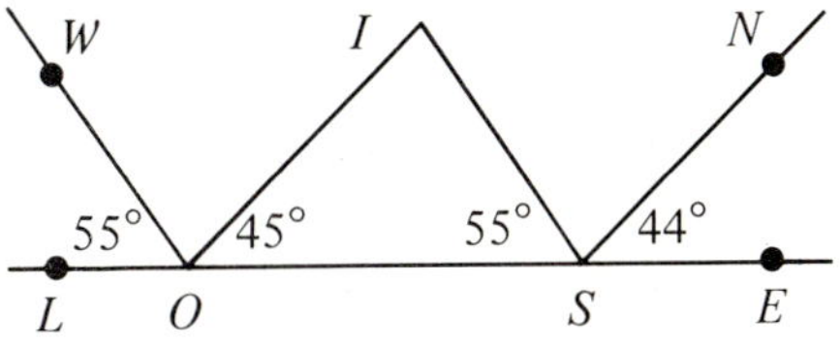

6. Suppose $\angle 1 = \angle 2$. Tell which lines are parallel in each figure.

a.

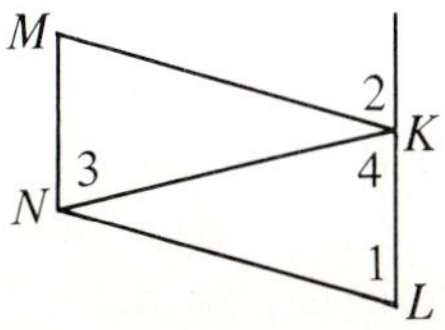

b.

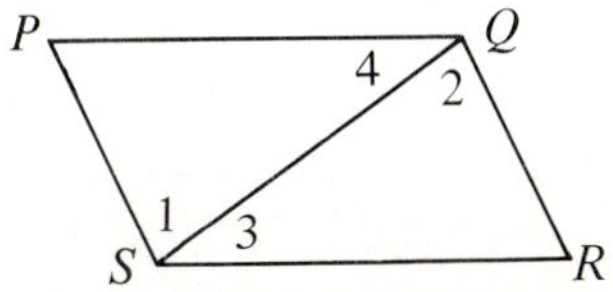

7. Draw a picture statement for Theorem 7 on page 72.

8. Suppose $\angle 1 = \angle 2$.
   a. Tell why $\overline{AB} \parallel \overline{CD}$.
   b. Tell why $\angle 3 = \angle 4$.

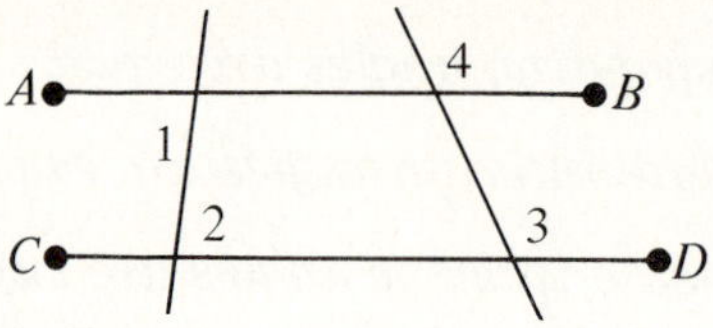

Written Exercises

For each figure, tell if you can correctly conclude that $u \parallel v$.

A

1.
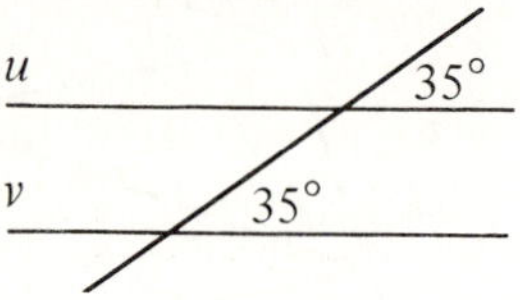

2.
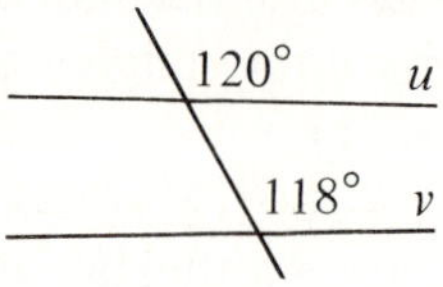

3.
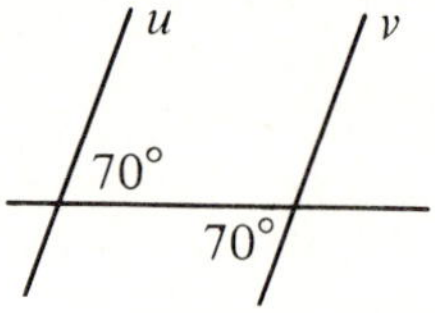

4.
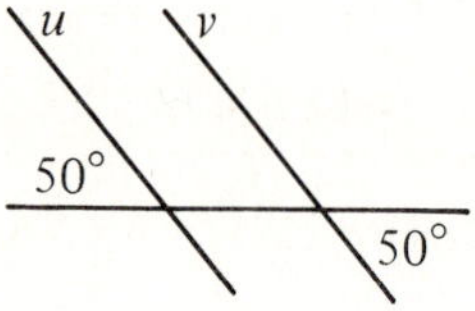

5.
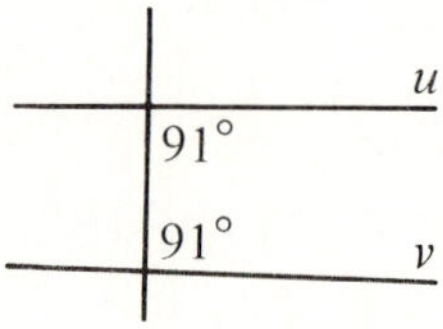

6.
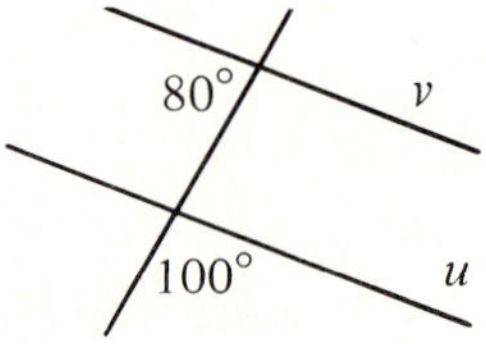

Draw a picture statement for each theorem.

7. Theorem 4 (page 68)

8. Theorem 6 (page 72)

9. Theorem 8 (page 72)

For each figure, name two pairs of parallel lines.

10.
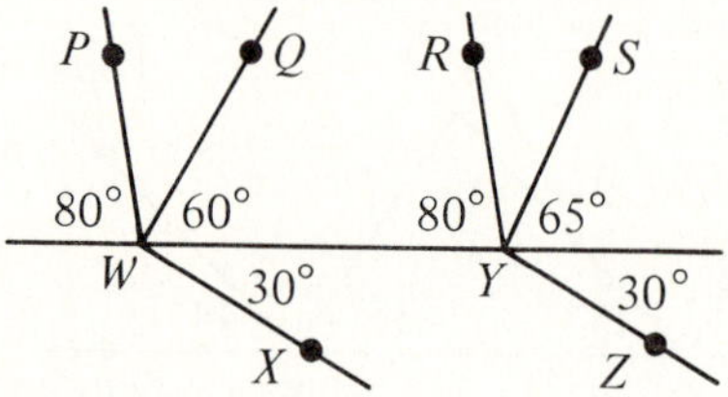

11.
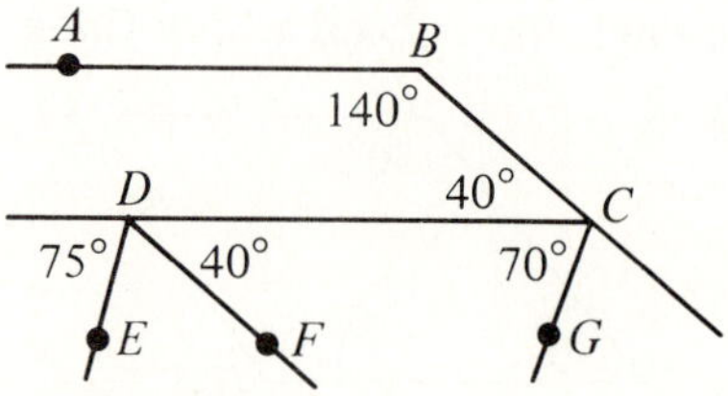

Find the value of x which will make lines a and b parallel.

12.
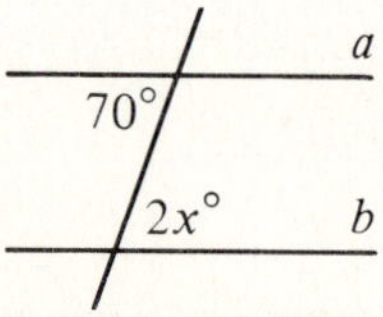

13.
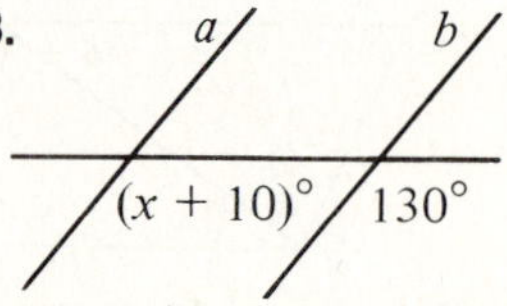

14.
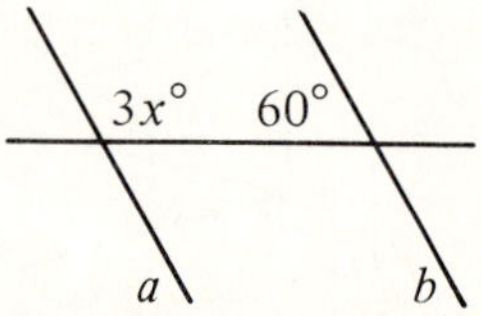

Use the given information to tell which lines, if any, must be parallel.

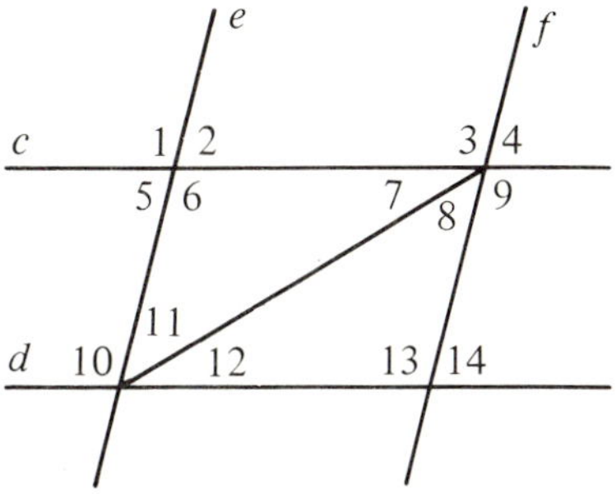

SAMPLE $\angle 1 = \angle 3$

Since $\angle 1$ and $\angle 3$ are equal corresponding angles, $e \parallel f$.

15. $\angle 2 = \angle 4$

16. $\angle 4 = \angle 14$

17. $\angle 8 = \angle 11$

18. $\angle 7 = \angle 13$

19. $\angle 5 + \angle 10 = 180°$

20. $\angle 2 + \angle 3 = 180°$

21. $\angle 1 = \angle 10$

22. $\angle 1 = \angle 9$

23. $\angle 9 + \angle 14 = 180°$

24. Complete the following proof of Theorem 6.

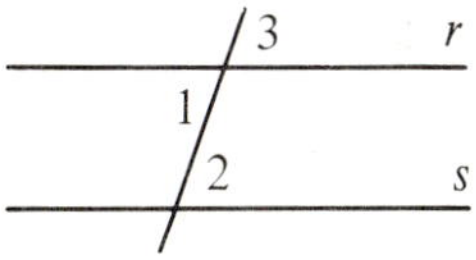

Given: $\angle 1 = \angle 2$

Prove: $r \parallel s$

STATEMENTS	REASONS
1. $\angle 1 = \angle 2$	1. ?
2. $\angle 1 = \angle 3$	2. ?
3. $\angle 2 = \angle 3$	3. ?
4. ?	4. If 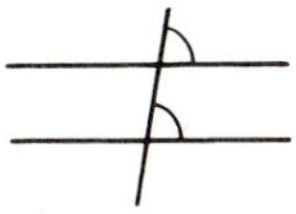, then ?.

25. Complete the following proof of Theorem 7.

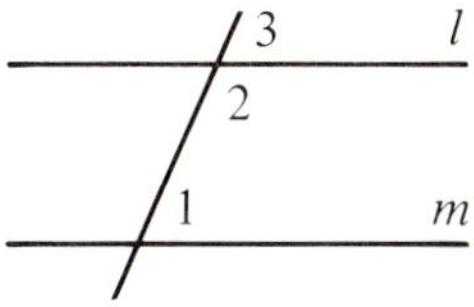

Given: $\angle 1$ and $\angle 2$ are supplements.

Prove: $l \parallel m$

STATEMENTS	REASONS
1. $\angle 1$ and $\angle 2$ are supplements.	1. ?
2. $\angle 3$ and $\angle 2$ are supplements.	2. Definition of ?
3. $\angle 1 = \angle 3$	3. Supplements of ?.
4. $l \parallel m$	4. ?

B **26.** A technical artist sets a plastic triangle in position ABC and draws $\overline{AB}$. Then the artist slides the triangle along the T-square to position XYZ and draws $\overline{XY}$. Why is $\overline{AB} \parallel \overline{XY}$?

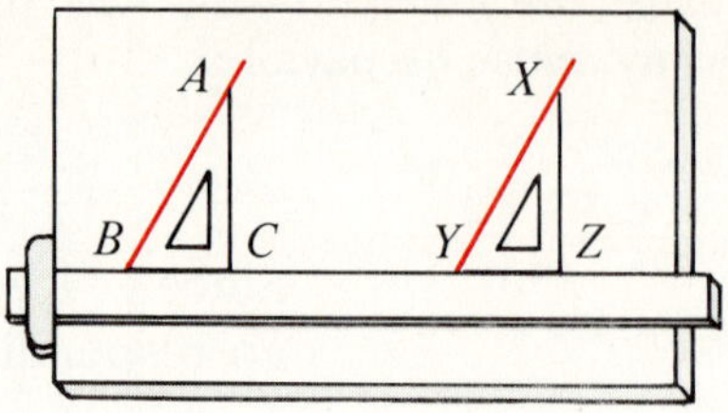

27. Explain why the phrase *In a plane* is needed in Theorem 8. Draw a diagram to illustrate your answer.

28. *Given:* $\angle 1 = \angle 2$

Prove: $\angle 3 = \angle 4$

(*Hint:* This proof requires just three steps.)

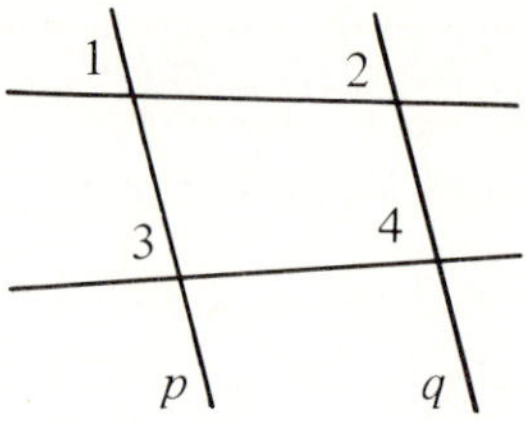

C **29.** *Given:* $\angle 5 = \angle 6$
$\angle 7 = \angle 8$

Prove: $\overline{AB} \parallel \overline{CD}$

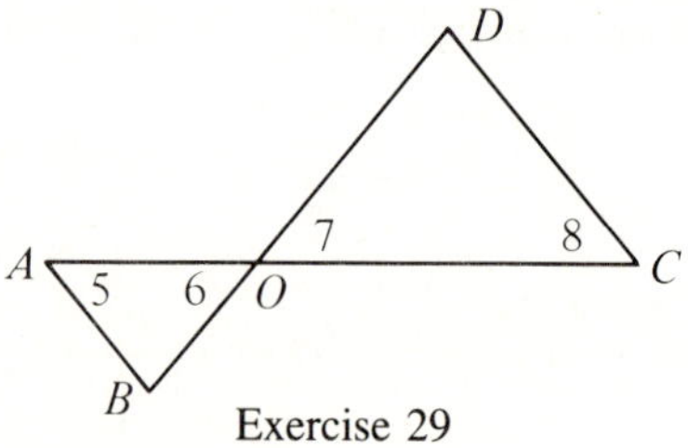

Exercise 29

30. Prove Theorem 8.

Given: $a \perp b$; $a \perp c$

Prove: $b \parallel c$ (*Hint:* Use Postulate 8.)

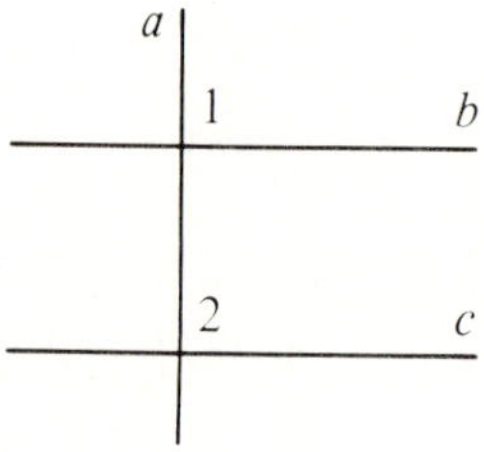

Exercise 30

CONSUMER APPLICATIONS

ENERGY USAGE

Every time you toast a piece of bread, use a hair dryer, or play the radio, you use energy. Some machines use more energy than others. The chart at the right lists this usage in terms of the length of time the machine would run if supplied with the amount of energy in a liter of gasoline.

electric clock	31 weeks
black & white television	$2\frac{3}{4}$ days
color television	$1\frac{1}{2}$ days
toaster	$8\frac{1}{2}$ hours
oven	3 hours

1. Which of the machines in the chart uses the most energy?
2. Estimate the cost of running an electric clock for one year. (Assume that gasoline costs 32¢ per liter.)

7 Constructing Parallel Lines

Our goal in this section is to construct a line that passes through a given point P and is parallel to a given line l. To do this, we must first be able to copy an angle with a compass and straightedge.

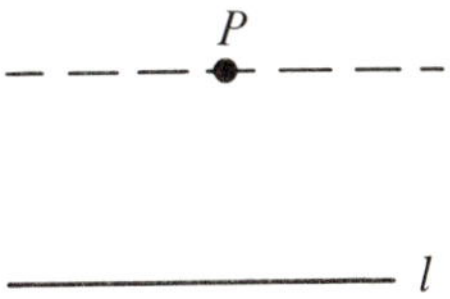

CONSTRUCTION 4
Given: *An angle*
Construct: *An angle equal to the given angle*

1. Draw a ray with endpoint O.

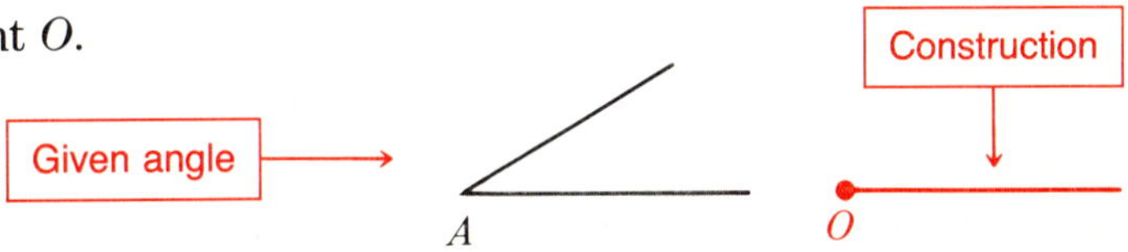

2. Construct an arc with center at A.
3. Without changing your compass, construct an arc with center at O.

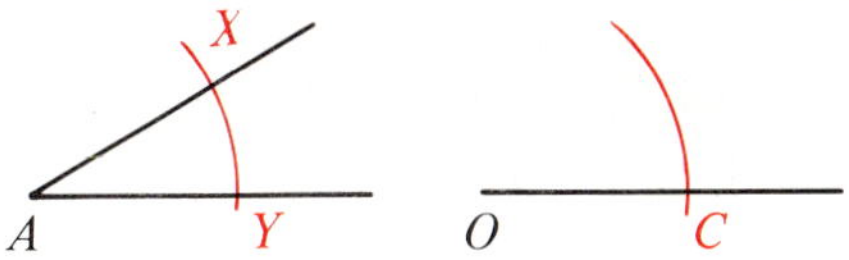

4. Put the compass point at Y. Then adjust your compass so that you can draw an arc passing through X.
5. Without changing your compass, construct an arc with center at C.

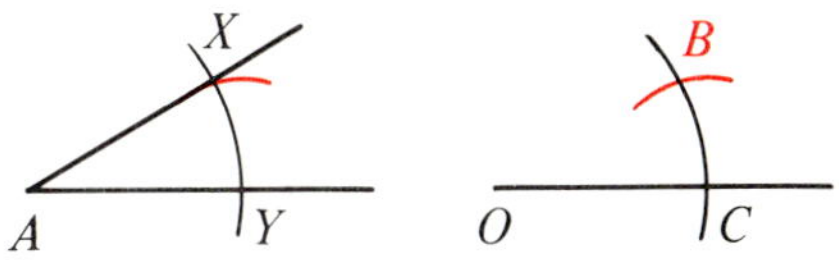

6. Draw $\overrightarrow{OB}$. $\angle BOC = \angle A$.

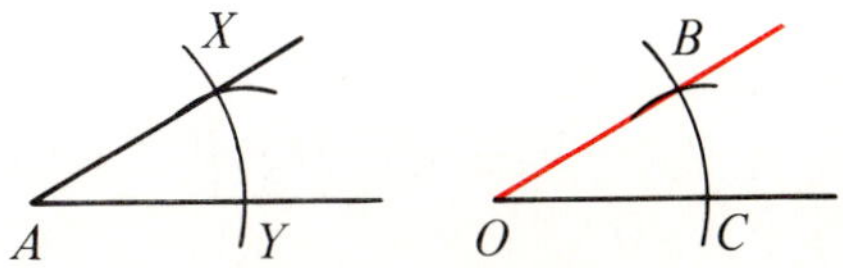

CONSTRUCTION 5

Given: *A line and a point not on the line*

Construct: *A line that passes through the point and is parallel to the given line*

Follow these steps:

1.

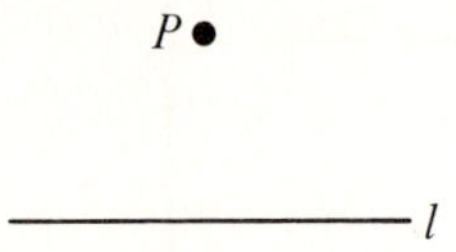

You are given l and P.

2.

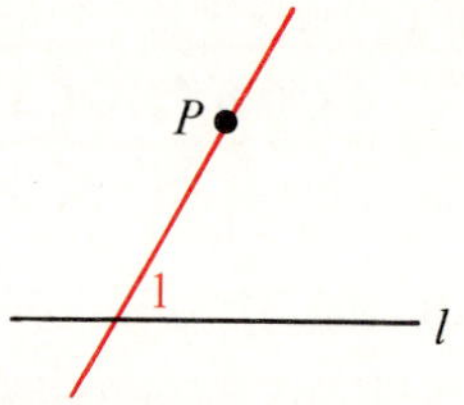

Draw any transversal through P. Label $\angle 1$.

3.

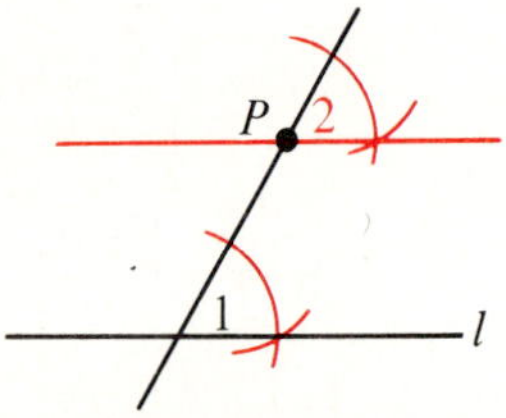

At P, construct $\angle 2$, a corresponding angle equal to $\angle 1$.

4.

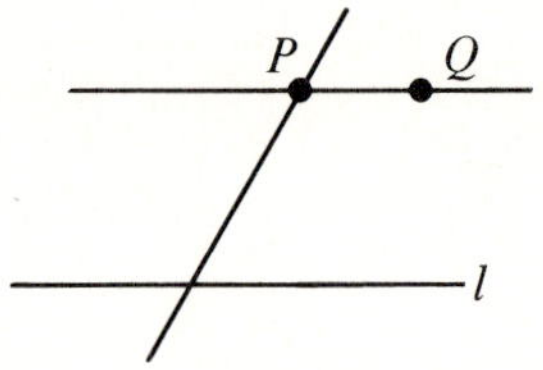

$\overleftrightarrow{PQ} \parallel l$. Do you know why?

Classroom Practice

1. Draw an acute angle.
   Then copy the angle using compass and straightedge.
2. Draw an obtuse angle.
   Then copy the angle using compass and straightedge.
3. Draw a line l and choose a point P not on l.
   Construct a line through P and parallel to l.
4. Draw a triangle similar to, but larger than, the one shown.
   Then construct a line through O parallel to $\overleftrightarrow{JY}$.

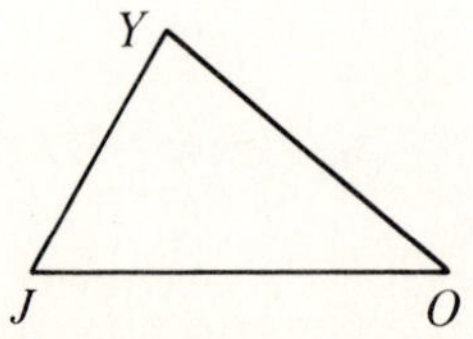

Written Exercises

Draw an angle similar to, but larger than, the one shown. Then copy the angle using compass and straightedge.

A **1.** 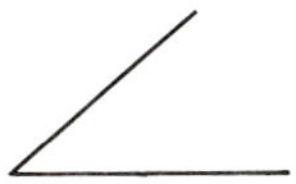 **2.**  **3.** 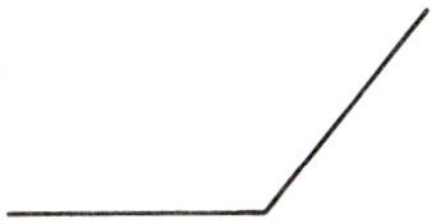

Draw a figure similar to, but larger than, the one shown. Then construct a parallel to $\overleftrightarrow{RS}$ through point P.

4. 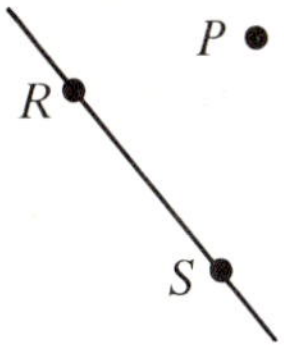

5. 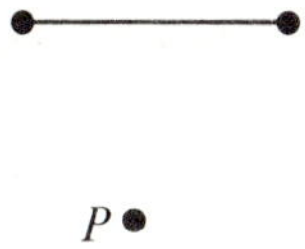

6. 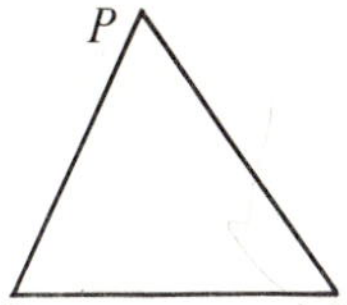

7. Draw a line l and a point P not on l.
Construct a line which passes through P and is parallel to l by constructing equal alternate interior angles.

8. a. Using lined paper, draw two parallel lines and a transversal as shown. Then construct the bisectors of two alternate interior angles.
b. What appears to be true about the bisectors?
c. Explain why your answer in part (b) is true.

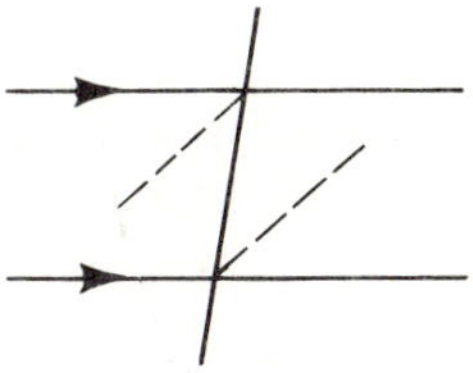

9. a. Using lined paper, draw two parallel lines and a transversal. Then construct the bisectors of two same-side interior angles.
b. Measure an angle formed by the bisectors.
c. Repeat the experiment with a different transversal.
d. What appears to be true about the angles formed by the bisectors?

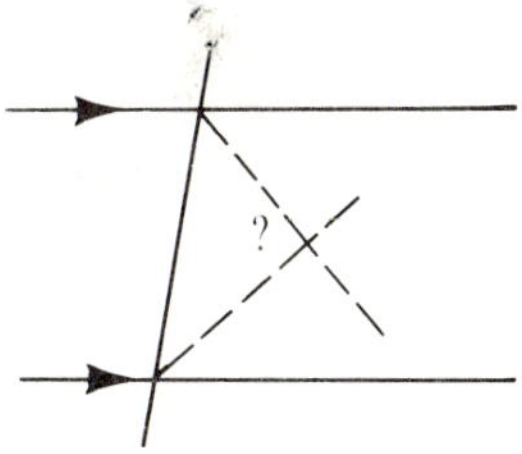

10. a. Construct a line through P and parallel to l by these steps:
(1) Use Construction 3 on page 26 to construct m, a line perpendicular to l through P.
(2) Use Construction 2 on page 26 to construct n, a line perpendicular to m through P.
b. Tell why $l \parallel n$.

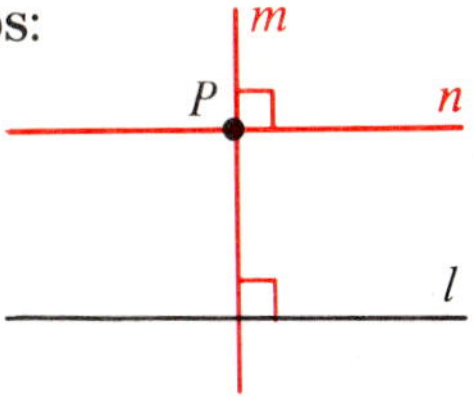

B **11.** Draw an acute angle. Call it $\angle B$.
Draw a line t and choose a point Q on t.
At Q, construct an angle that is supplementary to $\angle B$.

12. Repeat Exercise 11, using an obtuse $\angle B$.

SAMPLE Given $\angle A$ and $\angle B$, construct $\angle A + \angle B$.

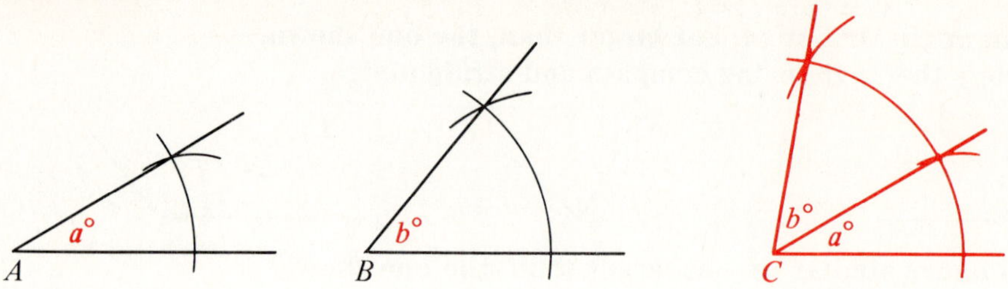

13. Draw two acute angles, $\angle X$ and $\angle Y$. Construct $\angle X + \angle Y$.

14. Draw an acute angle, $\angle Z$. Construct an angle with measure $2 \cdot \angle Z$.

SELF-TEST

Vocabulary

parallel lines (p. 66)
skew lines (p. 66)
transversal (p. 66)
interior angles (p. 66)
alternate interior angles (p. 66)
same-side interior angles (p. 67)
corresponding angles (p. 67)

1. What is the difference between a postulate and a theorem? *(2-4)*

2. Name two pairs of alternate interior angles. *(2-5)*

3. Name two pairs of same-side interior angles.

4. Suppose $\angle 6 = 130°$. Find the measures of the other numbered angles.

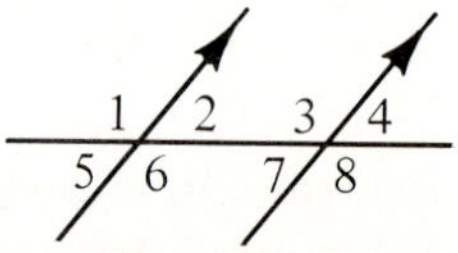

5. Find the values of x and y in the diagram.

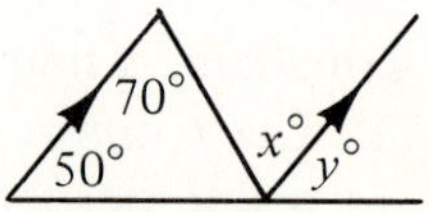

Exercise 5

6. List four ways to prove that two lines are parallel. *(2-6)*

7. *Given:* $l \parallel m$
   $\angle 1 = \angle 2$
   Prove: $\angle 1 = \angle 3$

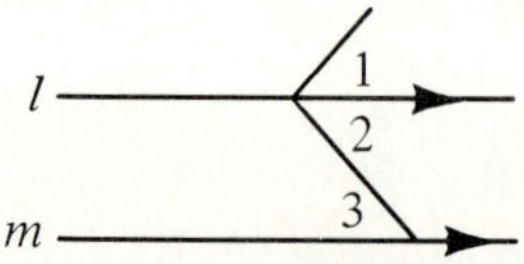

8. Draw a line m. Choose a point A that is several centimeters from m. Construct a line through A parallel to m. *(2-7)*

READING GEOMETRY

DIAGRAMS

Have you ever thought what this book would be like if everything had to be described in words, without the help of diagrams? Diagrams are very useful in geometry because they provide a great deal of information in a form that can be easily understood.

Most of the explanations in this book are illustrated by diagrams. You need to read the text and the diagrams *together* in order to get the most out of your reading. Next to a diagram you may sometimes see a label in a red box. Be sure to look at all the labels when you read the diagrams. Some have arrows that point to important parts of the figure, and some state facts that you need to know about the diagram.

Many of the exercises include diagrams. When you work an exercise, you are allowed to use *all* the information given in the diagram, but you are not allowed to take anything for granted that isn't actually given. For example, you may not say that two sides of a figure are parallel just because they *look* parallel to you.

Sometimes you may want to draw a diagram of your own to help you understand an explanation or a problem. In such a case, you don't need to make a careful construction. A reasonably accurate sketch is usually all you need. Be sure to letter the points correctly and to show only what is given in the description of the figure.

EXERCISES

Use the information given in the diagrams. Tell whether each statement is true or false.

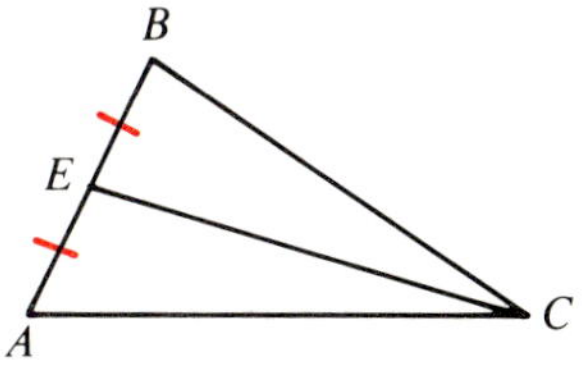

1. E is the midpoint of $\overline{AB}$.
2. $\angle BCE = \angle ACE$
3. $\angle BEC = \angle AEC$

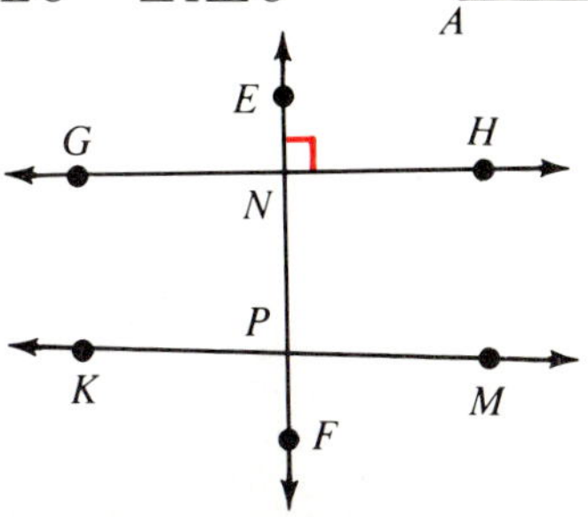

4. $\overleftrightarrow{EF} \perp \overleftrightarrow{GH}$
5. $\overleftrightarrow{KM} \parallel \overleftrightarrow{GH}$
6. $\angle GNE$ is a right angle.
7. $\angle MPE$ is a right angle.
8. $\angle KPF = \angle MPE$

PROBLEM SOLVING STRATEGIES

ASKING QUESTIONS

Here are some questions to ask when solving geometry problems.

1. DO I UNDERSTAND THE PROBLEM?
   Did I read the problem carefully?
   Do I know what all the words mean?

2. WHAT DO I KNOW?
   What is the given information?
   What conditions are part of the problem?
   Can I draw a picture of the information?
   Can I make a chart?

3. WHAT DON'T I KNOW?
   What information is unknown?
   What is the problem asking for?
   Do I have to find something else first?
   Do I have to prove something?

4. LOOK AT THE WHOLE PROBLEM.
   How can I use the information I have?
   How can I tie together the known and the unknown?

EXERCISES

Answer the questions that follow each problem.

1. Gail brings some herbs to sell at the Farmers' Market.
   In the morning, she sells them for \$1.50 each and receives \$16.50.
   In the afternoon, Gail lowers the price to \$1.00 each.
   She sells three times as many as in the morning.
   What is her gross income for the entire day?

a. You know Gail's gross income from morning sales is $16.50.
b. What is the problem asking for?
c. Find Gail's gross income from afternoon sales:

You need to know how many herbs Gail sold in the afternoon. To do this you need to find out how many herbs she sold in the morning. Use the information about her morning sales to find the number of herbs she sold then.

How many herbs did Gail sell in the afternoon?

What was her gross income for the afternoon?

d. What was Gail's gross income for the entire day?

2. Find the smallest integer that will be divisible by each of the following: 6, 20, 33
   a. What do you have to find in this problem?
   b. List two conditions this integer must satisfy.
   c. Write out the prime factors of 6, 20, and 33.
   d. How can you find an integer that is divisible by all of these factors?
   e. What is the smallest such integer?

3. Prove that if the same-side interior angles of two parallel lines and a transversal are bisected, then the two angle bisectors will form a 90° angle.
   a. Here is a diagram of the situation.
      Copy the diagram and mark in the information that is given.

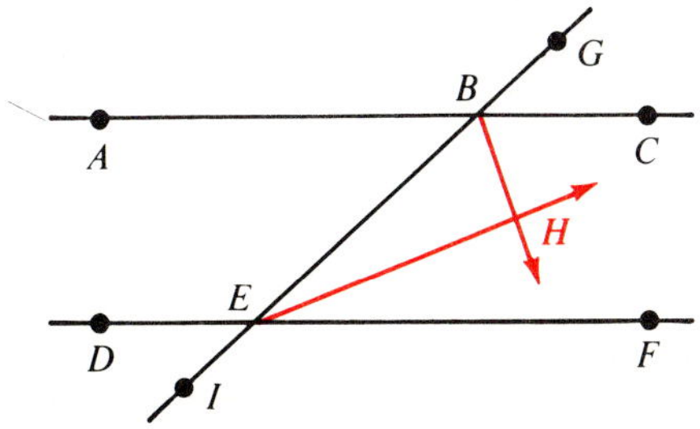

   b. Using the letters in the diagram, state what you are being asked to prove.

SKILLS REVIEW

COORDINATES IN THE PLANE

Give the coordinates of each point.

SAMPLE Point A, $(-3, 3)$

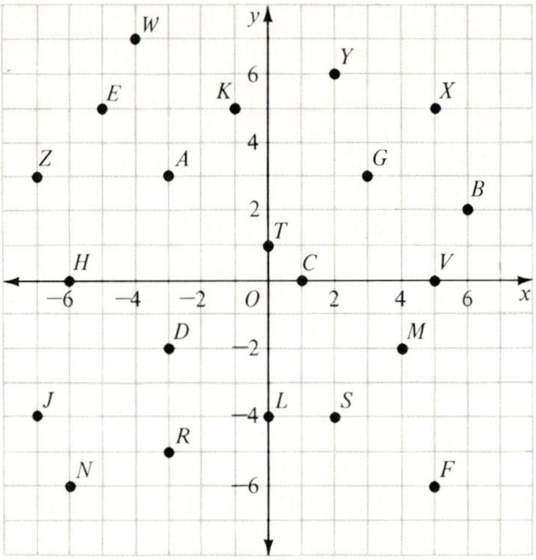

1. B **2.** C **3.** D

4. E **5.** F **6.** G

7. H **8.** J **9.** K

Name the point by its letter.

10. $(-6, -6)$ **11.** $(0, -4)$

12. $(-4, 7)$ **13.** $(5, 0)$

14. $(-3, -5)$ **15.** $(4, -2)$

16. $(2, 6)$ **17.** $(-7, 3)$

18. $(0, 1)$ **19.** $(2, -4)$

Plot the points in Exercises 20–31. Then draw line segments to connect the points in order. You should now have the outline of half a fir tree.

20. $(0, 6)$ **21.** $(-2, 4)$ **22.** $(-1, 4)$ **23.** $(-3, 2)$

24. $(-2, 2)$ **25.** $(-5, -1)$ **26.** $(-3, -1)$ **27.** $(-6, -4)$

28. $(-1, -4)$ **29.** $(-1, -6)$ **30.** $(-2, -7)$ **31.** $(0, -7)$

Complete the ordered pairs in Exercises 32–43 to tell how to draw the other half of the fir tree. Start at the top again.

32. (0, _?_) **33.** (2, _?_) **34.** (_?_, 4) **35.** (3, _?_)

36. (_?_, 2) **37.** (_?_, −1) **38.** (_?_, −1) **39.** (6, _?_)

40. (_?_, −4) **41.** (_?_, _?_) **42.** (_?_, −7) **43.** (_?_, _?_)

44. Plot and connect the points $(2, -1)$ and $(2, -5)$. Name the vertices of the two squares that can be drawn with this segment as a side.

CHAPTER REVIEW

CHAPTER SUMMARY

1. Complementary angles are two angles whose measures total 90°. Supplementary angles are two angles whose measures total 180°.
2. "If . . . then" statements are important in logical reasoning.
   Statement: If A, then B. Converse statement: If B, then A.
3. There are four kinds of reasons used in a geometric proof.
   Given information Definition Postulate Theorem
4. Parallel lines are coplanar lines which do not intersect. Skew lines are not coplanar lines, and they do not intersect.
5. When two parallel lines are cut by a transversal:
   a. corresponding angles are equal;
   b. alternate interior angles are equal;
   c. same-side interior angles are supplementary.
6. Ways to prove that two lines cut by a transversal are parallel:
   a. show that corresponding angles are equal;
   b. show that alternate interior angles are equal;
   c. show that same-side interior angles are supplementary;
   d. show that both lines are perpendicular to the transversal.
7. It is possible to construct an angle equal to a given angle. (Construction 4, page 77)
   Given a line and a point not on the line, you can construct a line that passes through the point and is parallel to the given line. (Construction 5, page 78)

REVIEW EXERCISES

In the diagram shown, $\overline{TR} \perp \overline{RO}$ and $\overline{SO} \perp \overline{RO}$. *(See pp. 46–49.)*

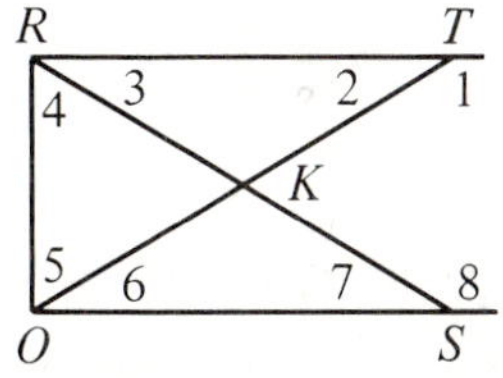

1. If $\angle 2 = 30°$, then a complement of $\angle 2$ has measure ___?___°.
2. If $\angle 2 = 30°$, then a supplement of $\angle 2$ has measure ___?___°.
3. Name a complement of $\angle 3$.
4. If $\angle 3 = \angle 6$, which theorem lets you conclude $\angle 4 = \angle 5$?

Use the statement, "Every acute angle has a measure less than 90°." *(See pp. 50–53.)*

5. Write the statement in "if . . . then" form.

6. State the hypothesis and the conclusion.

7. State the converse of the given statement.

(Optional) Accept as true: "All cats have sharp claws." *(See pp. 56–59.)*

8. For each statement, tell what you can conclude.
   a. Cyrus is not a cat. b. Catarina has sharp claws.
   c. Cinnamon is a cat. d. Cleo does not have sharp claws.

Write a proof in two-column form. *(See pp. 60–65.)*

9. *Given:* $\angle 1 = \angle 4$

   Prove: $\angle 2 = \angle 5$

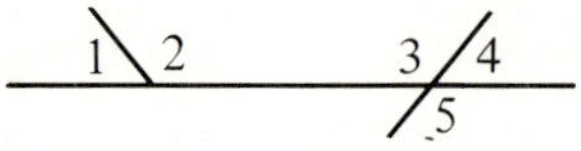

Classify each pair of angles as alternate interior angles, same-side interior angles, or corresponding angles. *(See pp. 66–71.)*

10. $\angle 1$ and $\angle 3$ 11. $\angle 2$ and $\angle 6$

12. $\angle 7$ and $\angle 6$ 13. $\angle 8$ and $\angle 6$

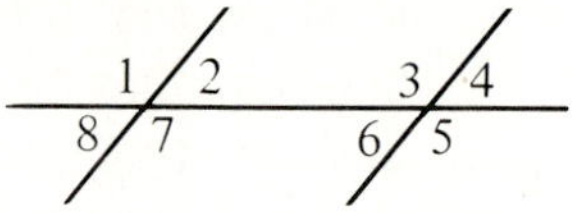

Find the values of x and y in each diagram. *(See pp. 66–71.)*

14.

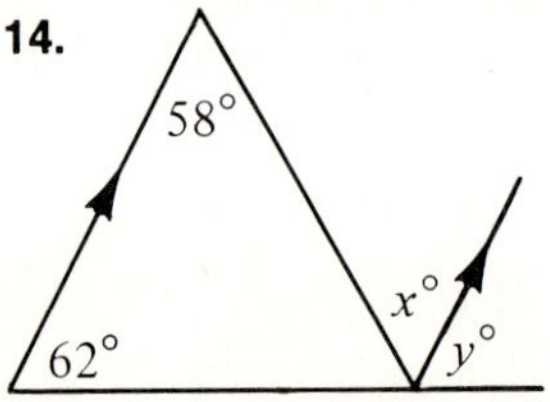

15.

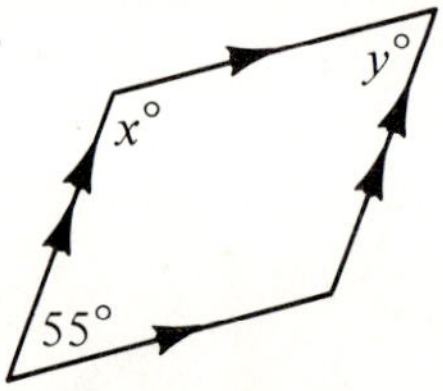

16.

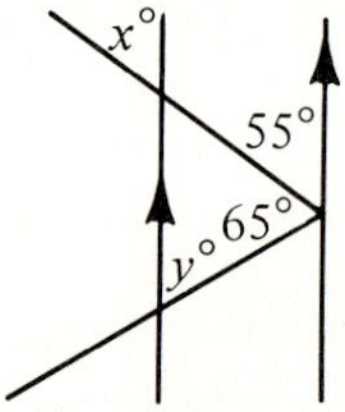

Use the given information to tell which lines must be parallel. *(See pp. 72–76.)*

17. $\angle 1 = \angle 10$ 18. $\angle 7 = \angle 12$

19. $\angle 8 = \angle 11$ 20. $\angle 2 + \angle 3 = 180°$

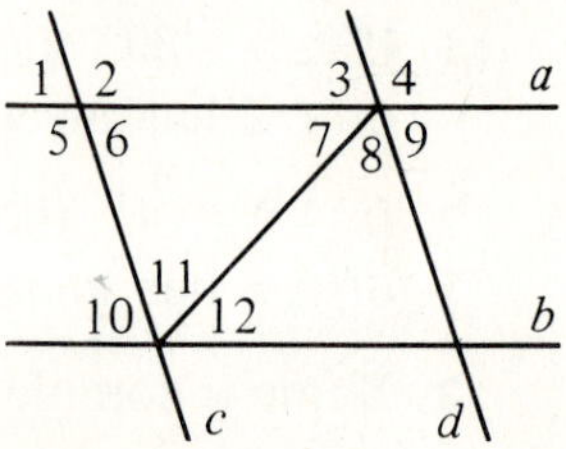

Construct the required figure. *(See pp. 77–80.)*

21. Draw a line l and choose a point P not on l. Construct a line through P and parallel to l.

CHAPTER TEST

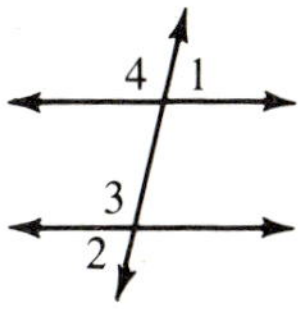

1. If $\angle A = 57°$, find the measures of (**a**) a supplement of $\angle A$ and (**b**) a complement of $\angle A$. *(2-1)*
2. If $\angle 1 = \angle 2$, what do you know about $\angle 3$ and $\angle 4$?
3. Write the theorem that justifies your conclusion in Exercise 2.

In Exercises 4–9 accept as true: "If $x = 0$, then $x + x = 0$."

4. Write the hypothesis and the conclusion. *(2-2)*
5. Write the converse.
6. Is the converse true?

(Optional) Tell what, if anything, you can conclude.

7. $x + x = 0$
8. $x + x \neq 0$
9. $x \neq 0$ *(2-3)*

10. *Given:* $\angle 1$ and $\angle 2$ are complements. *(2-4)*
$\angle 3$ and $\angle 4$ are complements.
$\overrightarrow{BD}$ bisects $\angle EBF$.

Prove: $\angle 1 = \angle 4$

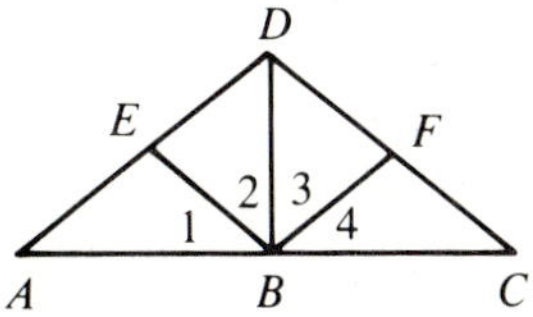

11. $\angle 3$ and $\angle$ __?__ are corresponding angles. *(2-5)*
12. $\angle 1$ is supplementary to which four angles?
13. $\angle 6$ and $\angle$ __?__ are same-side interior angles. Is it possible for these angles to be equal angles? to be complementary?
14. Name two pairs of alternate interior angles.

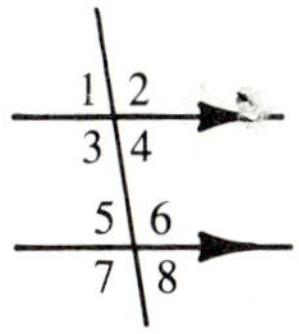

Use the given information to tell which segments must be parallel.

15. $\angle 1 = \angle 2$ *(2-6)*
16. $\overline{RZ} \perp \overline{RS}$; $\overline{RS} \perp \overline{SY}$
17. $\angle TRZ + \angle XZR = 180°$
18. $\angle RSY = \angle STX$

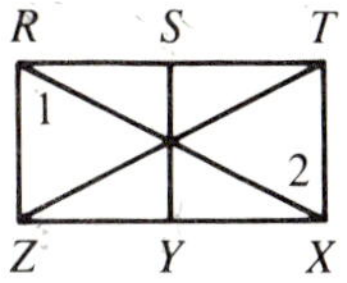

Draw a large acute angle, $\angle 1$.

19. Construct an angle $\angle RST$ equal to $\angle 1$. *(2-7)*
20. Construct a line through T and parallel to $\overleftrightarrow{RS}$.

UNIT A CUMULATIVE REVIEW

Find the following:

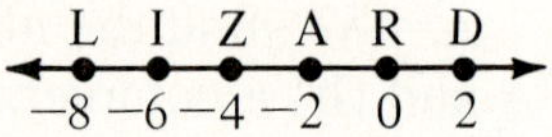

1. *AD*
2. *LR*
3. *ZR*
4. The coordinate of *Z*
5. The midpoint of $\overline{IA}$

Tell whether each statement is true or false.

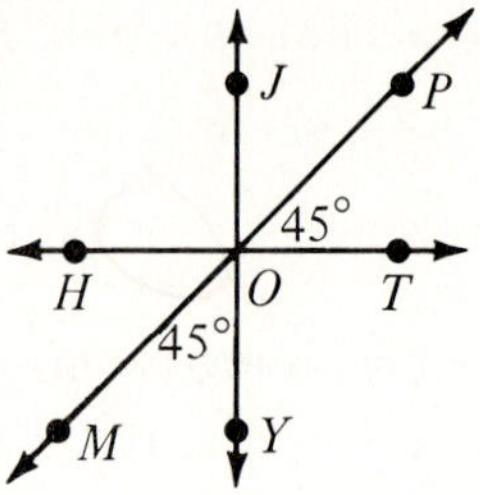

6. $\angle HOJ + \angle JOP = \angle HOP$
7. $\angle MOY$ is an obtuse angle.
8. $\angle HOM = 45°$
9. $\overrightarrow{OM}$ bisects $\angle HOY$.
10. $\angle HOY$ is an acute angle.
11. It is possible to draw two different lines containing both points *P* and *T*.
12. Exactly one plane contains points *P*, *O*, and *Y*.
13. $\angle TOP$ and $\angle HOM$ are complementary angles.
14. Complete the following proof.

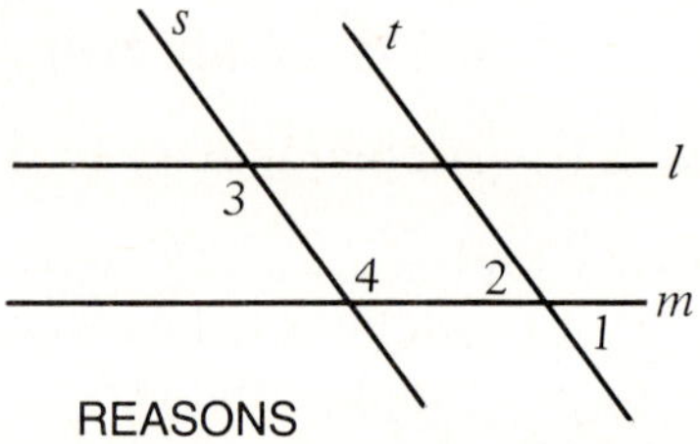

Given: $\angle 1$ and $\angle 3$ are supplements.
$s \parallel t$

Prove: $l \parallel m$

STATEMENTS	REASONS
1. $\angle 1$ and $\angle 3$ are supplements.	1. ?
2. $\angle 2$ and $\angle 4$ are supplements.	2. ?
3. $\angle 1 = \angle 2$	3. ?
4. $\angle 3 = \angle 4$	4. ?
5. ?	5. ?

15. Write the converse of the statement: "If two angles are vertical angles, then they are equal." Is the converse true?
16. Draw a line *l* and points *P* and *Q* not on *l*.
    a. Construct a perpendicular to *l* through *P*.
    b. Construct a line through *Q* parallel to *l*.

UNIT

B

Here's what you'll learn in this chapter:

To use the properties of the angles of a triangle.

To classify triangles according to their sides or angles.

To name congruent triangles and their corresponding parts.

To use postulates and theorems to prove that triangles are congruent.

Because triangles are rigid figures, they are often used in the construction of bridges and other buildings. The frames of the bridges shown above contain many congruent triangles. *(See the Application on pages 113–114.)*

Chapter 3

Triangles

1 The Angle Sum of a Triangle

Draw a large triangle on a sheet of paper and cut the triangle out. Then tear off the three corners of the triangle and arrange them as shown. What appears to be true about the angles of the triangle? Try this experiment with a few other triangles and see if your results are the same.

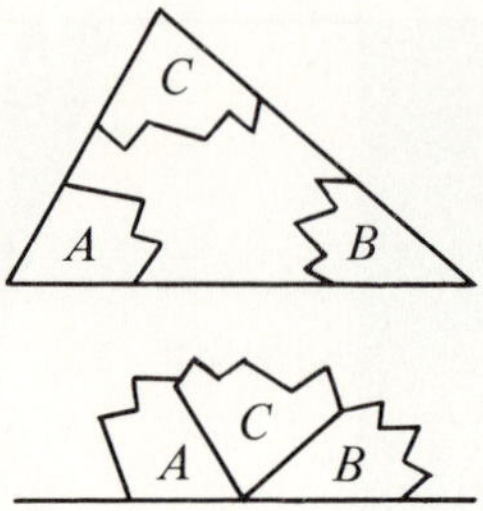

The property illustrated by this experiment is one of the best known and most important theorems in geometry.

THEOREM 1
The angle sum of a triangle is 180°.

Given: $\triangle ABC$

Prove: $\angle 1 + \angle 2 + \angle 3 = 180°$

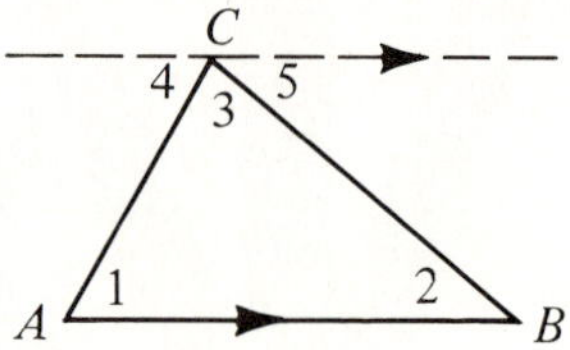

STATEMENTS	REASONS
1. Through C, draw a line parallel to $\overline{AB}$.	1. Through any point not on a line, a parallel can be drawn to the line. (Recall Construction 5, page 78.)
2. $\angle 4 = \angle 1$ and $\angle 5 = \angle 2$	2. If 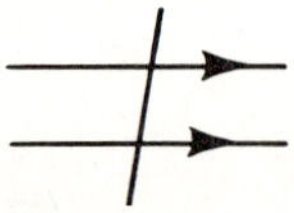, then 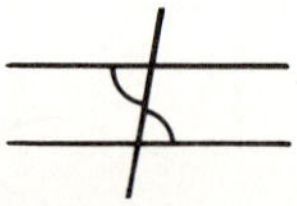
3. $\angle 4 + \angle 5 + \angle 3 = 180°$	3. Their sum is a straight angle.
4. $\angle 1 + \angle 2 + \angle 3 = 180°$	4. Substitution Postulate

EXAMPLE

Find the measures of $\angle 1$ and $\angle 2$.

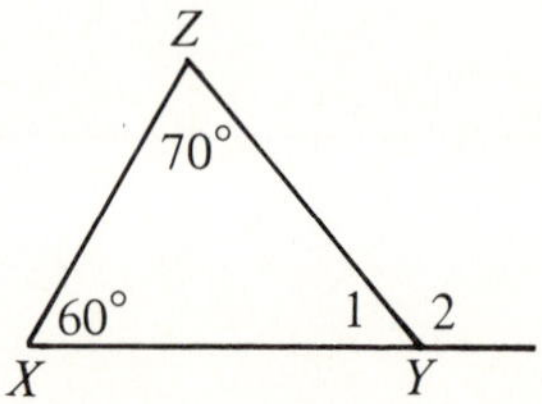

$\angle 1 + \underbrace{60° + 70°} = 180°$

$\angle 1 + \quad 130° \quad = 180°$

$\angle 1 = 50°$

Since $\angle 1$ and $\angle 2$ are supplements, $\angle 2 = 180° - 50° = 130°$.

In the preceding example, $\angle 2$ is called an **exterior angle** of $\triangle XYZ$. Notice that $\angle 2 = 130° = \angle X + \angle Z$.

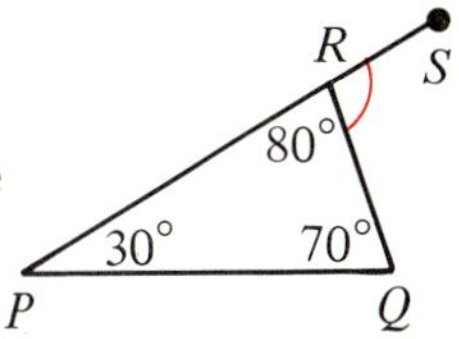

Likewise, in the diagram at the right, $\angle QRS$ is an exterior angle of $\triangle PQR$. Do you see that $\angle QRS = 100° = \angle P + \angle Q$?

These observations lead to an important *corollary* of Theorem 1. (A **corollary** of a theorem is a result which follows from the theorem with very little extra work.) The proof of the corollary is left as an exercise.

COROLLARY
An exterior angle of a triangle is equal to the sum of the two opposite angles of the triangle.

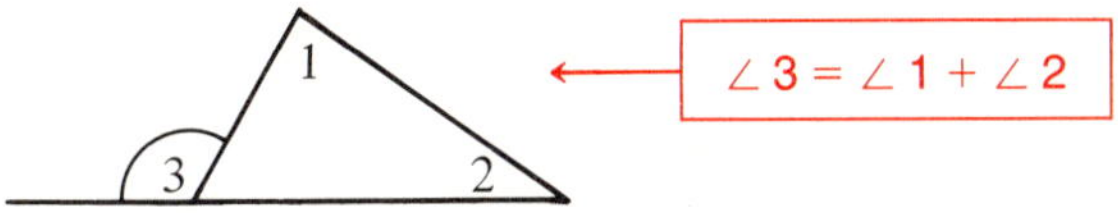

Classroom Practice

Find the measure of each numbered angle.

1.
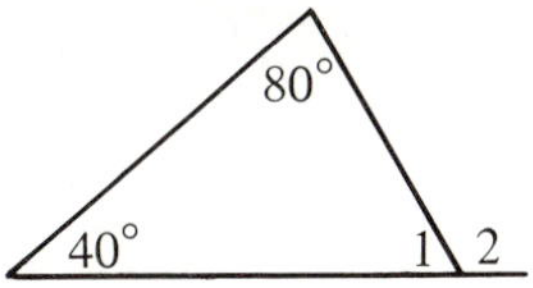

2.
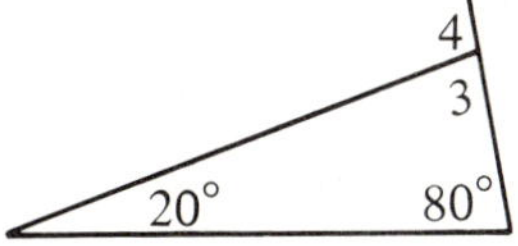

3.
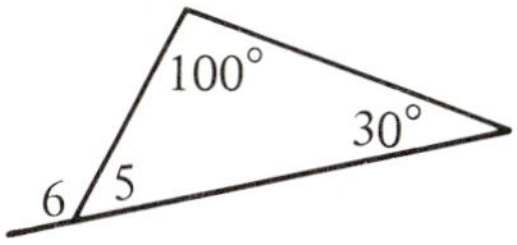

4.
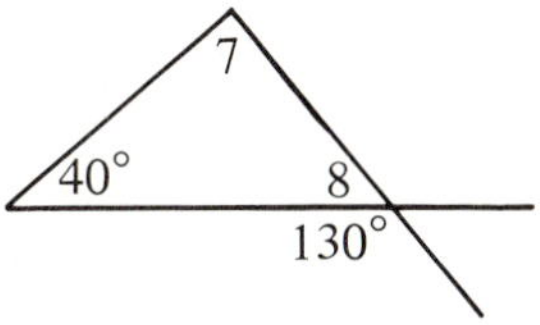

5.
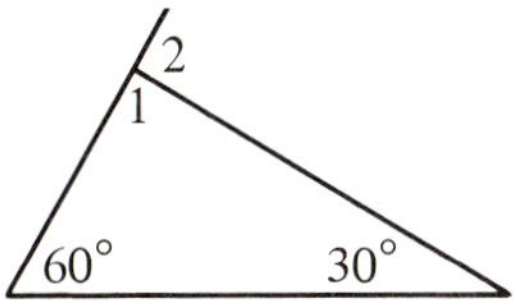

6.
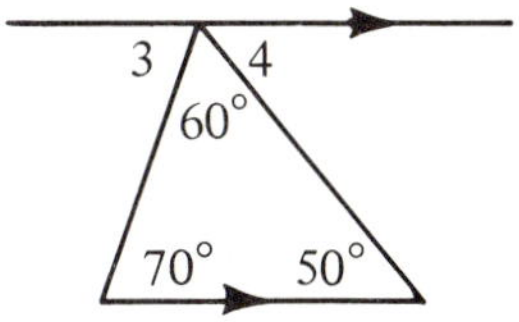

In $\triangle ABC$, $\angle C = 90°$.

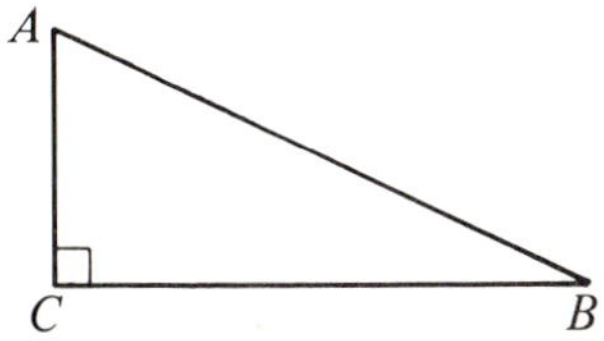

7. Find $\angle B$ if $\angle A = 60°$.

8. Find $\angle B$ if $\angle A = 70°$.

9. Find $\angle B$ if $\angle A = 65°$.

10. From Exercises 7–9, we see that if a triangle has a right angle, then the acute angles are __?__

11. In the diagram, $\angle 1 + \angle 2 =$ __?__ °;
$\angle 1 + \angle 3 =$ __?__ °.

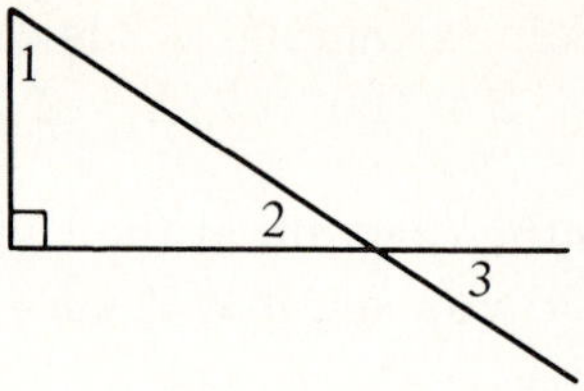

Find the value of x in each diagram.

12.

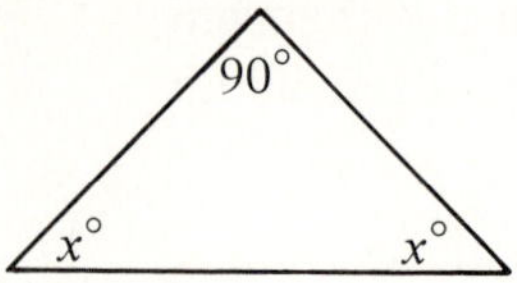

13.

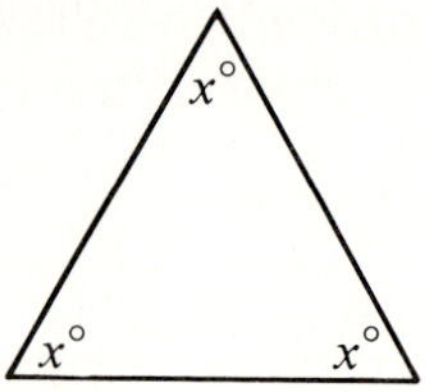

14.

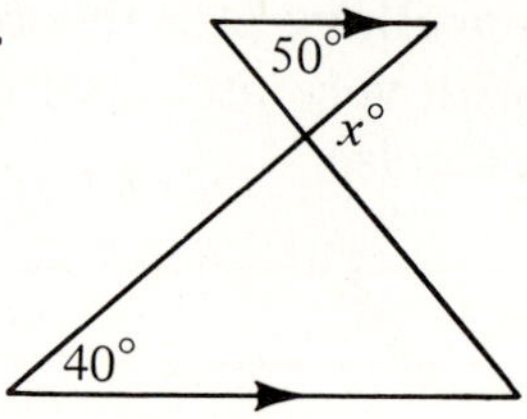

15. How many right angles can a triangle have? Explain your answer.

16. How many obtuse angles can a triangle have? Explain your answer.

Written Exercises

Find the measure of each numbered angle.

A 1.

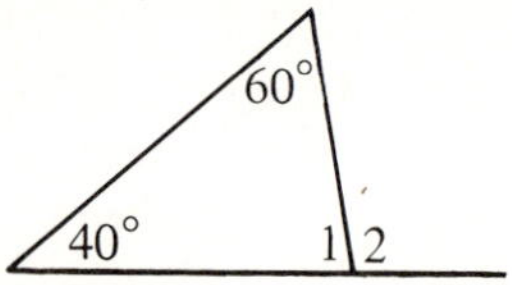

2.

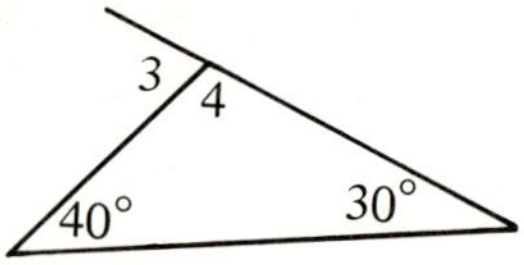

3.

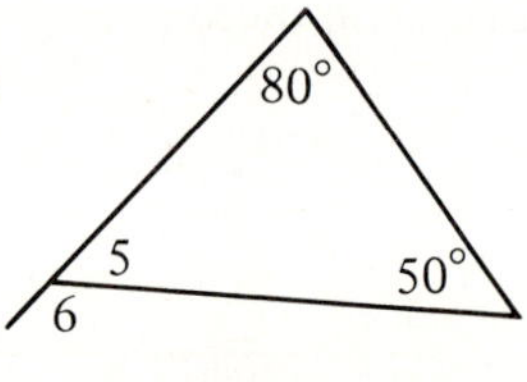

4.

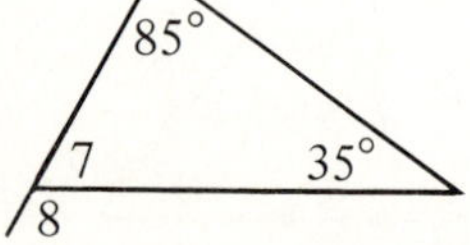

5.

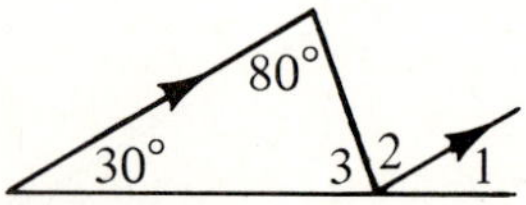

6.

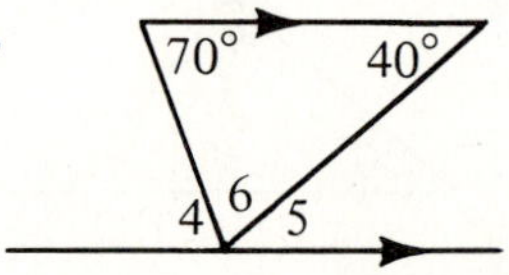

In Exercises 7–10, the measures of two angles are given.
Find the measure of the third angle using the Corollary of Theorem 1.

	7.	8.	9.	10.
$\angle 1$	25°	40°	30°	?
$\angle 2$	65°	?	70°	65°
$\angle 3$	?	110°	?	115°

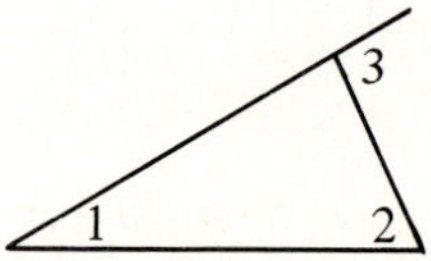

Find the measures of ∠ 1 and ∠ 2.

11.

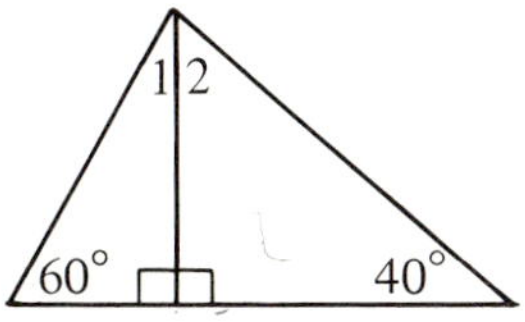

12.

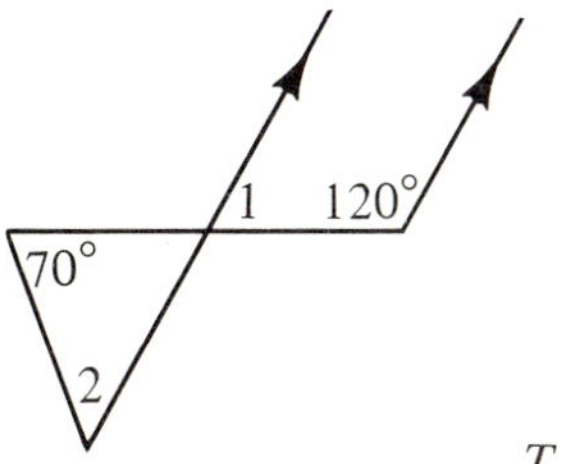

13.

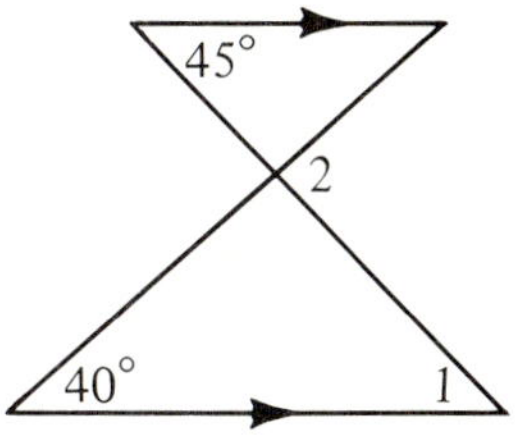

14. The three angles of a certain triangle are all equal. Find the measure of each angle.

15. Find the numerical measure of each angle of △ *RST*.

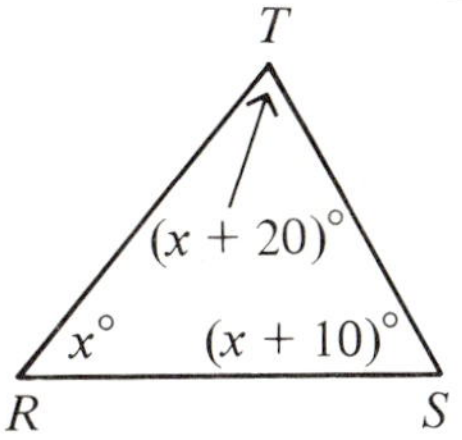

Find the value of *x* in each diagram.

16.

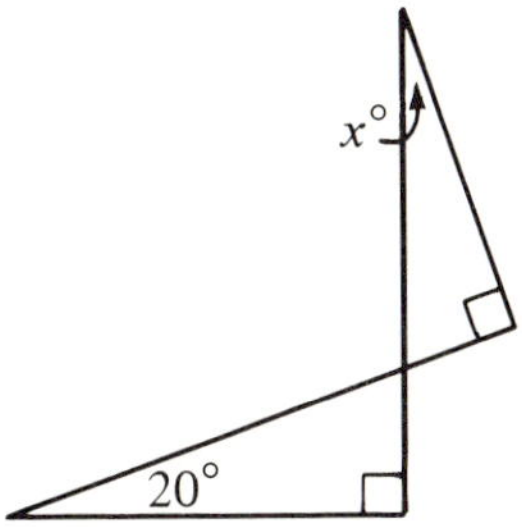

17.

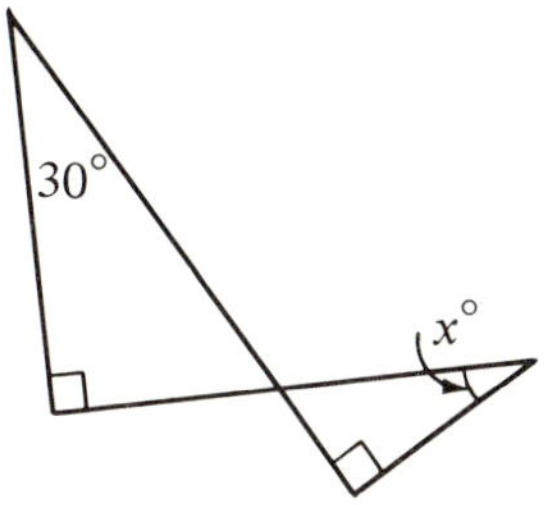

18.

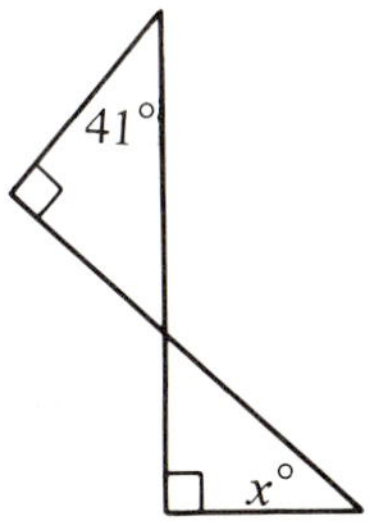

B **19.** If two angles of one triangle are equal to two angles of another triangle, what can you say about the third angles of the triangles? Explain your answer.

In each diagram, $\overrightarrow{RT}$ and $\overrightarrow{ST}$ are angle bisectors. Find the value of *x*.

20.

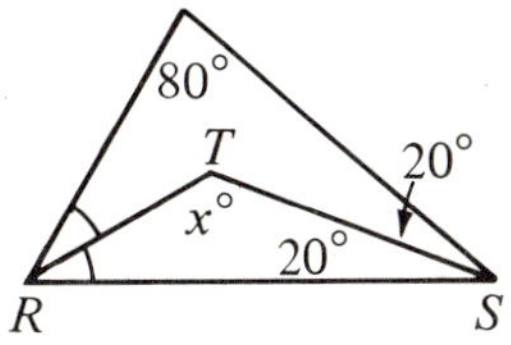

21.

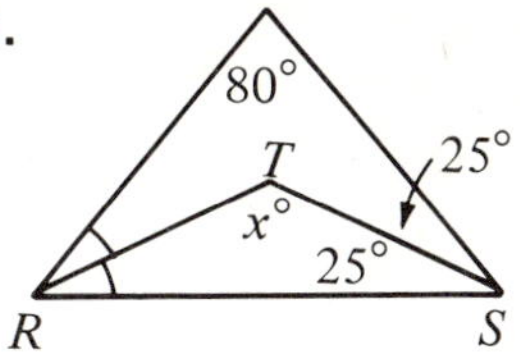

22.

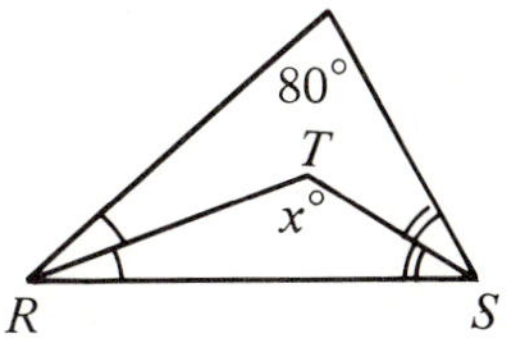

23. In △ *DEF*, ∠ *E* is 15° larger than ∠ *D*.
Also, ∠ *F* is 45° larger than ∠ *D*.
Find the measure of each angle.
(*Hint:* Let ∠ *D* = *x*°. Then ∠ *E* = __?__° and ∠ *F* = __?__°.)

24. In △ *ABC*, ∠ *B* is twice as large as ∠ *A*. Also, ∠ *B* = ∠ *C*. Find the measure of each angle.

25. Complete this proof of the Corollary of Theorem 1.

Given: $\triangle ABC$

Prove: $\angle 3 = \angle 1 + \angle 2$

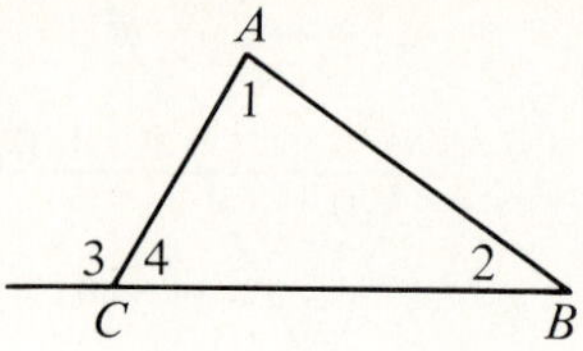

STATEMENTS	REASONS
1. $\angle 1 + \angle 2 + \angle 4 = 180°$	1. ?
2. $\angle 1 + \angle 2 = 180° - \angle 4$	2. ?
3. $\angle 4 + \angle 3 = 180°$	3. Their sum is a ? angle.
4. $\angle 3 = 180° - \angle 4$	4. ?
5. $\angle 3 = \angle 1 + \angle 2$	5. ?

In each exercise, find the indicated sum.

26.

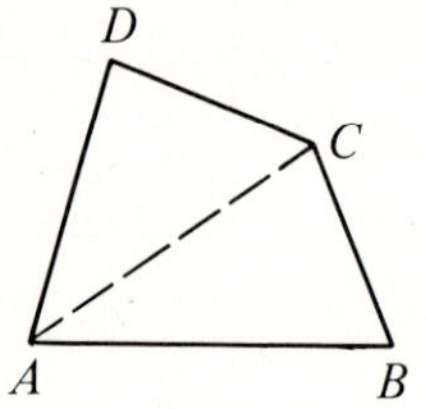

$\angle A + \angle B + \angle C + \angle D =$? °

27.

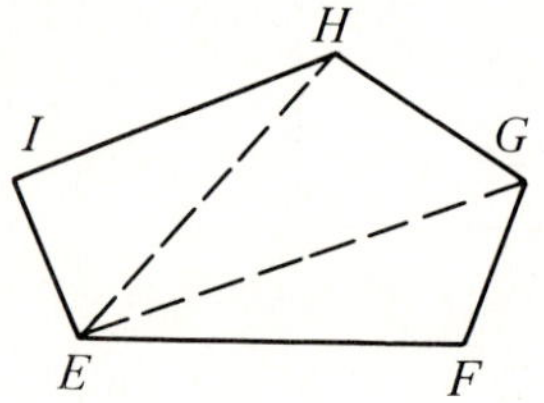

$\angle E + \angle F + \angle G + \angle H + \angle I =$? °

In each exercise, find the value of x.

C **28.**

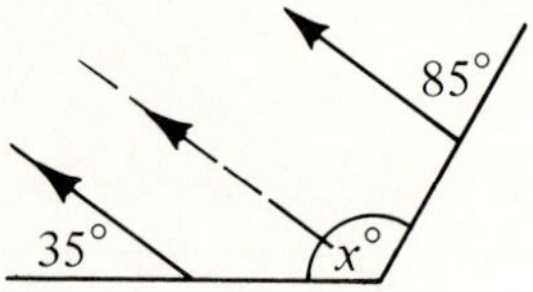

29.

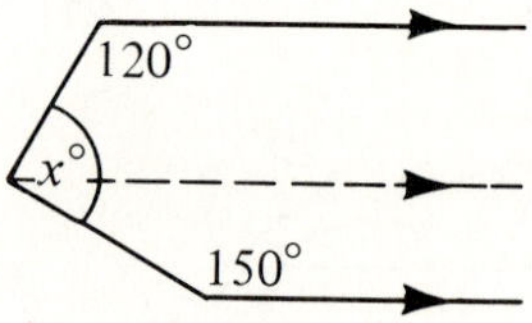

30. *Given:* $\angle B = \angle C$

$\overrightarrow{AE}$ bisects $\angle DAC$.

Prove: $\overline{AE} \parallel \overline{BC}$

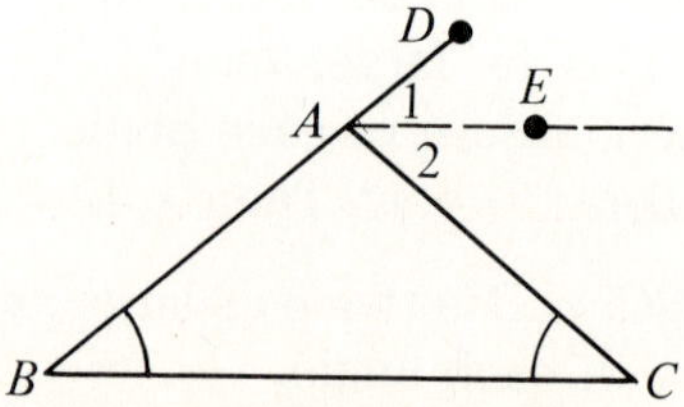

2 Classifying Triangles

$\triangle ABC$ is shown at the right. The segments, $\overline{AB}$, $\overline{BC}$, and $\overline{AC}$ are called the **sides** of the triangle. Notice that two of these sides have equal lengths. We say that the two sides are equal. (This is like saying that two angles are equal if they have equal measures.)

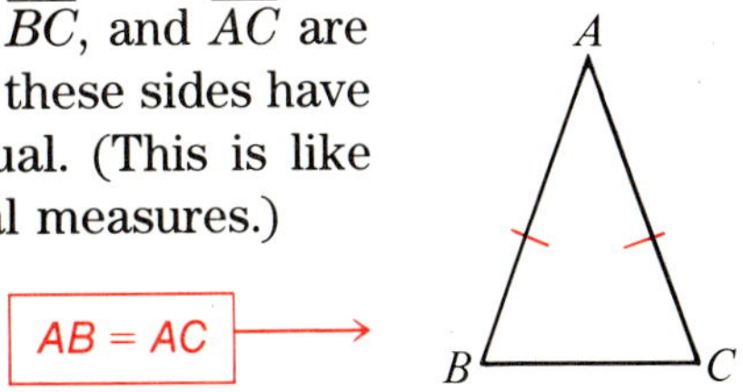

Classification of Triangles by Sides:

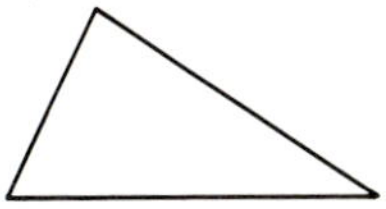

Scalene triangle

No sides are equal.

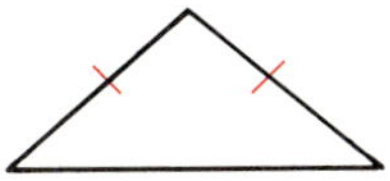

Isosceles triangle

At least *two* sides are equal.

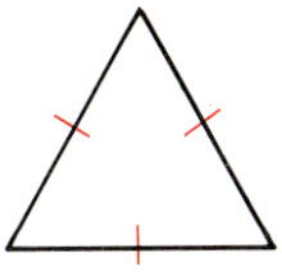

Equilateral triangle

All *three* sides are equal.

Notice that an equilateral triangle is also isosceles.

Classification of Triangles by Angles:

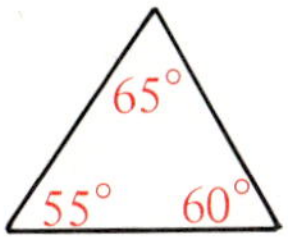

Acute triangle

All angles are acute.

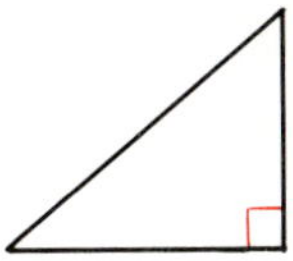

Right triangle

One angle is right.

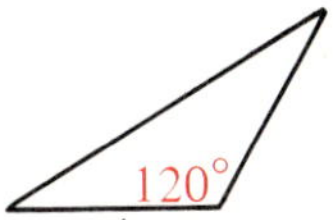

Obtuse triangle

One angle is obtuse.

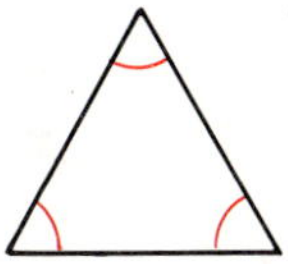

Equiangular triangle

All angles are equal.

Special Names for Sides:

The two equal sides of an isosceles triangle are called **legs.** The third side is the **base.**

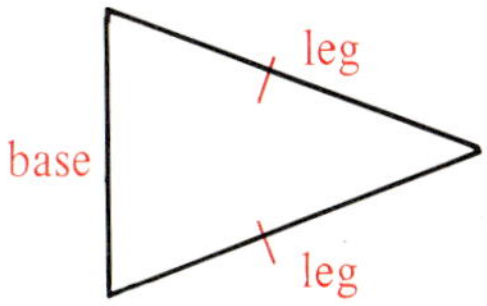

The side opposite the right angle of a right triangle is called the **hypotenuse.** The other sides are called **legs.**

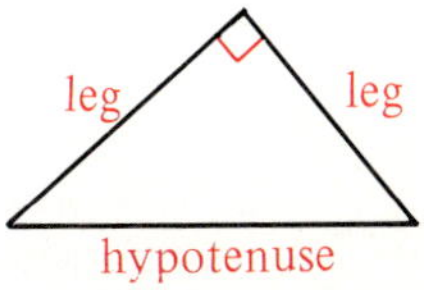

Classroom Practice

Classify each triangle shown as scalene, isosceles, or equilateral.

1. 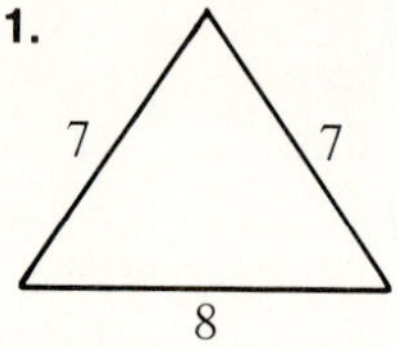

2. 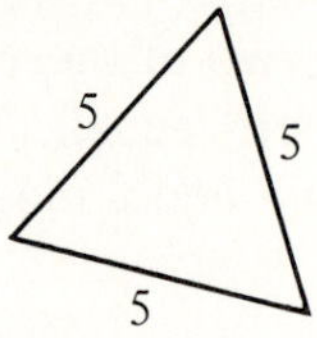

3. 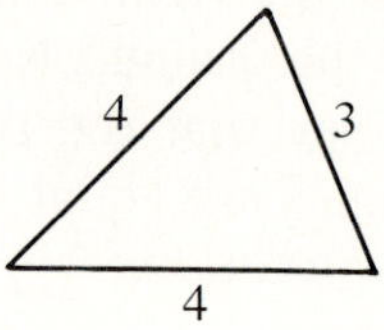

4. 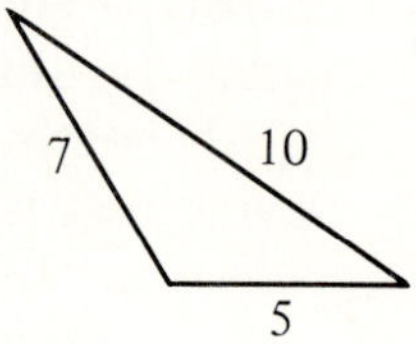

Classify each triangle shown as acute, right, obtuse, or equiangular.

5. 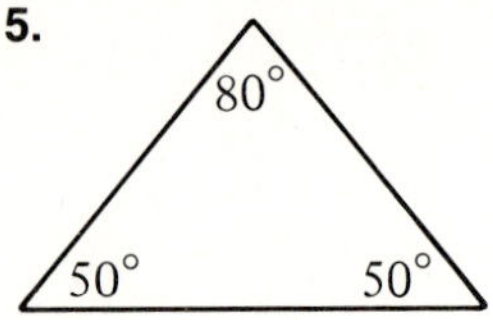

6. 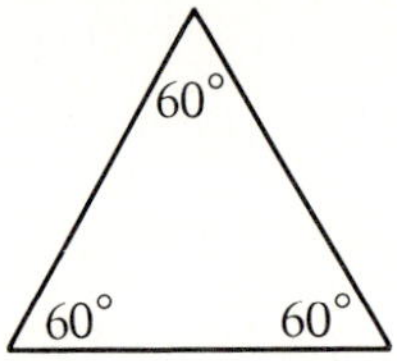

7. 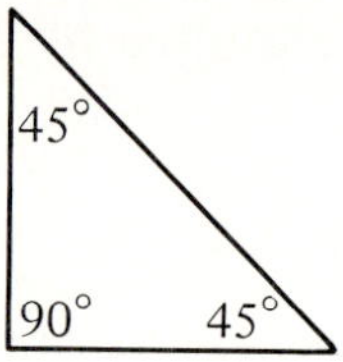

8. 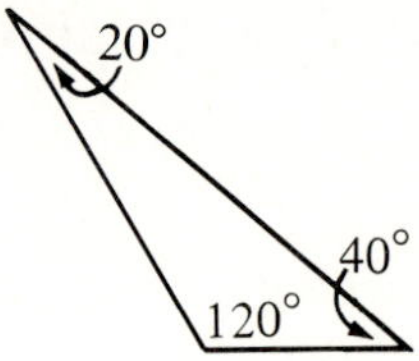

9. $\triangle ABC$ is a right triangle.
Sides $\overline{AC}$ and $\overline{BC}$ are called __?__.
Side $\overline{AB}$ is called the __?__.

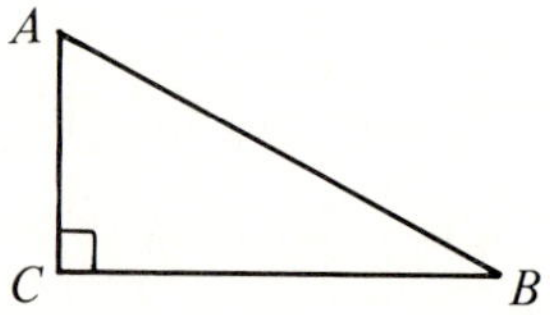

10. $\triangle PQR$ is an isosceles triangle.
Sides $\overline{PQ}$ and $\overline{PR}$ are called __?__.
Side $\overline{QR}$ is called the __?__.

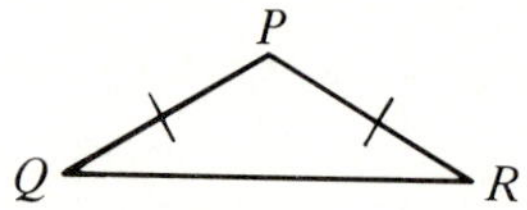

Given: $\overline{BE} \perp \overline{AC}$.

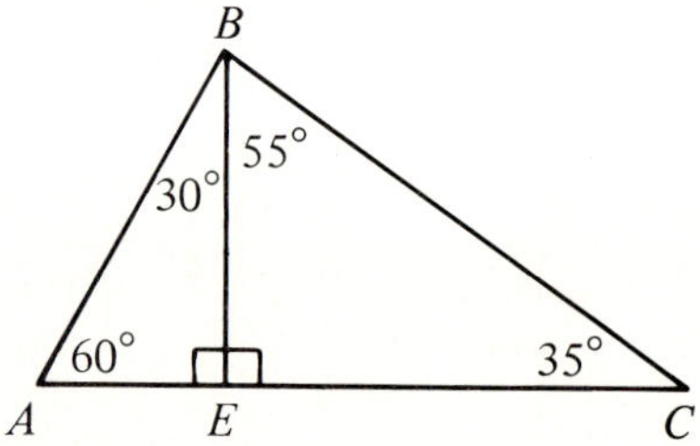

11. Name two right triangles.
12. Name an acute triangle.
13. Name an exterior angle of $\triangle BEC$.
14. Name a side common to $\triangle ABE$ and $\triangle BEC$.
15. Name a vertex common to $\triangle ABE$ and $\triangle BEC$.
16. Name an angle common to $\triangle ABE$ and $\triangle ABC$.

17. If a triangle is equilateral, is it also isosceles?
18. If a triangle is isosceles, is it also equilateral?
19. The *perimeter* of a triangle is the sum of the lengths of its sides. If the perimeter of an equilateral triangle is 30 cm, find the length of each side.

Written Exercises

Classify each triangle shown as scalene, isosceles, or equilateral.

A

1.
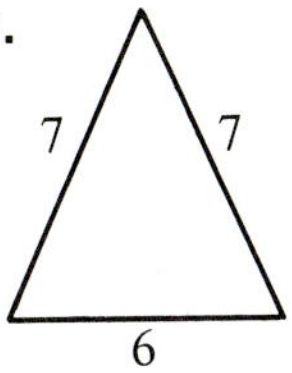

2.
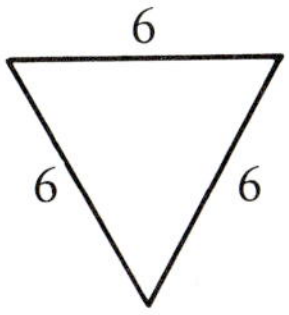

3.
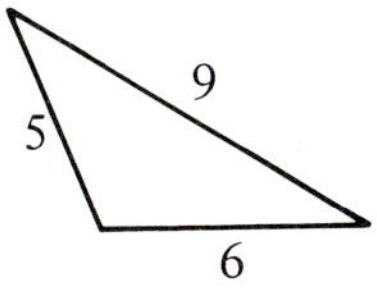

4.
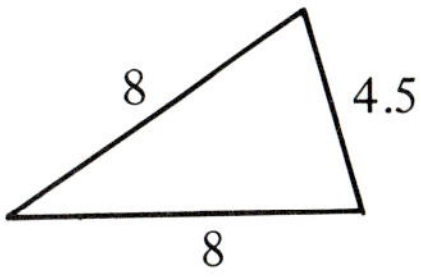

Classify each triangle shown as acute, right, obtuse, or equiangular.

5.
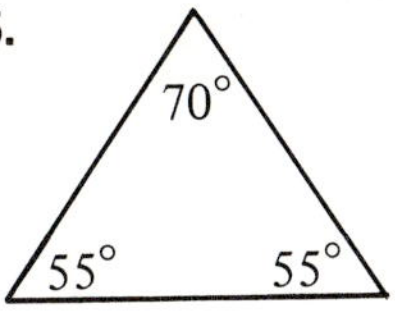

6.
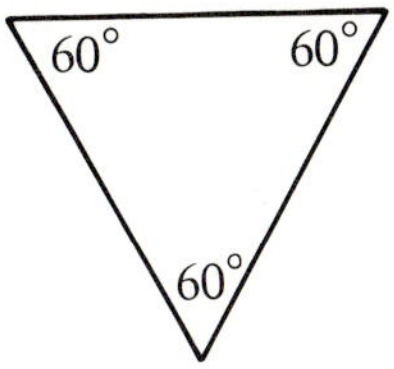

7.
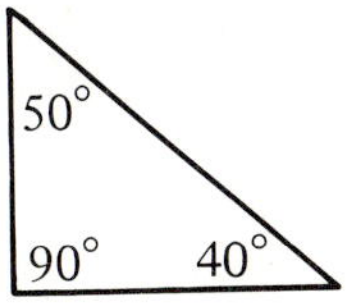

8.
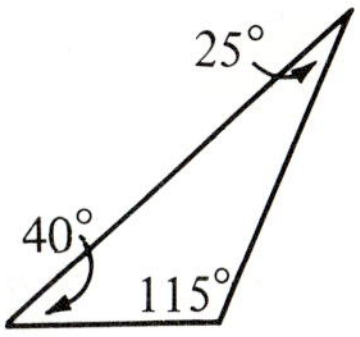

9. a. Draw an acute scalene triangle. **b.** Draw an acute isosceles triangle.

10. a. Draw an obtuse scalene triangle. **b.** Draw an obtuse isosceles triangle.

11. a. Draw a right scalene triangle. **b.** Draw a right isosceles triangle.

12. a. Fold a sheet of paper in half.

b. Use scissors to cut from any point on the fold to a corner point.

c. What kind of triangle is formed when you unfold the paper?

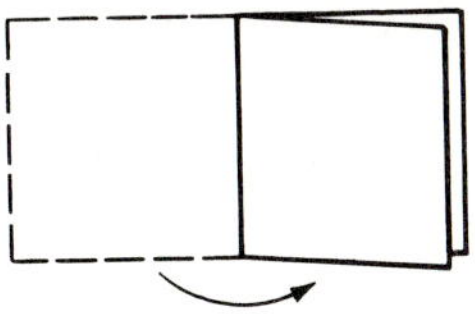

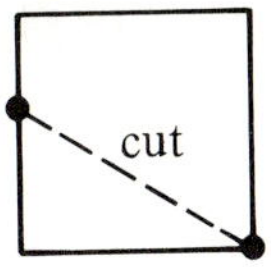

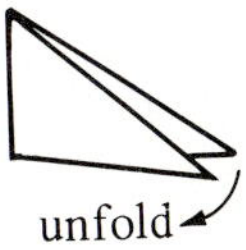

13. *Given:* $\triangle ABC$ is equiangular.
$\overrightarrow{AX}$ bisects $\angle BAC$.

Then: $\angle C = \underline{\ ?\ }°$
$\angle CAX = \underline{\ ?\ }°$
$\angle AXC = \underline{\ ?\ }°$

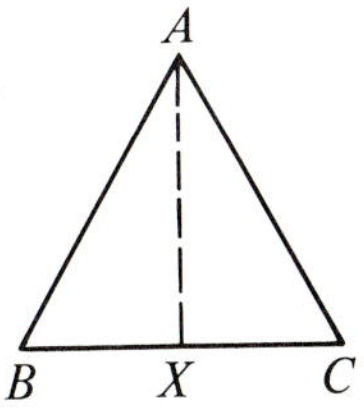

14. Figure *ABCD* has four right angles.
(In other words, it is a rectangle.)

a. Name two right triangles with hypotenuse $\overline{AC}$.

b. Name two right triangles with hypotenuse $\overline{BD}$.

In Exercises 15–17, find the lengths of the sides of the triangle.

B **15.**

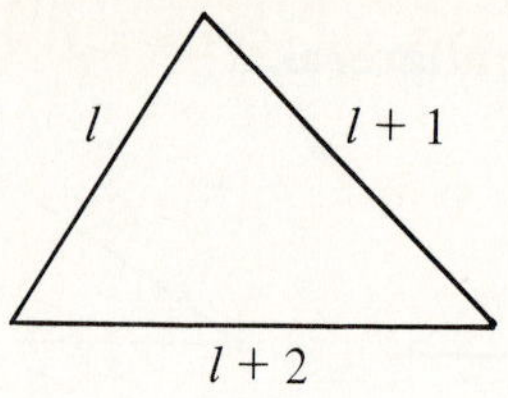

The perimeter is 18 cm.

16.

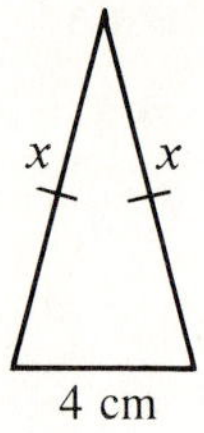

The perimeter is 20 cm.

17.

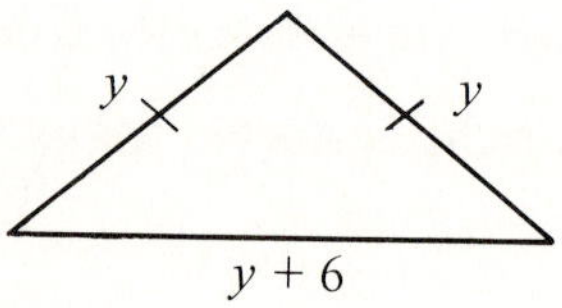

The perimeter is 36 cm.

Exercises 18–20 suggest some important properties of triangles. Proofs of these properties will be postponed until Chapter 4.

18. a. Construct a triangle with two equal sides. Using a protractor, carefully measure its three angles.

b. Repeat this process for other triangles with two equal sides.

c. What do you discover about the angles of such triangles?

19. a. Construct a triangle with two equal angles. Using a ruler, carefully measure its three sides.

b. Repeat this process for other triangles with two equal angles.

c. What do you discover about the sides of such triangles?

C **20. a.** Express your discoveries of Exercises 18 and 19 in "If . . . then" form.

b. As a pair, the statements can be called __?__.

SELF-TEST

Vocabulary

exterior angle (p. 93)
corollary (p. 93)
scalene triangle (p. 97)
isosceles triangle (p. 97)
equilateral triangle (p. 97)
acute triangle (p. 97)
right triangle (p. 97)
obtuse triangle (p. 97)
equiangular triangle (p. 97)
base (p. 97)
legs (p. 97)
hypotenuse (p. 97)

Find the value of x. *(3-1)*

1. 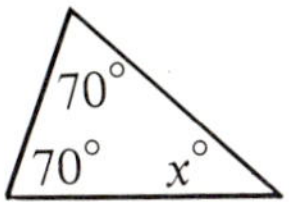

2. 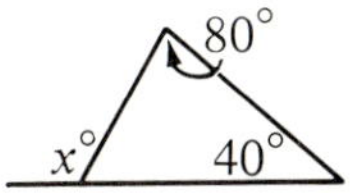

3. 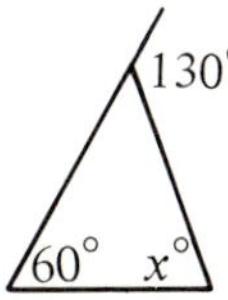

4. 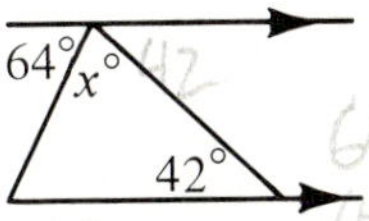

Write a, b, c, or d to identify a triangle of the type named. *(3-2)*

5. Scalene triangle

6. Obtuse triangle

7. Right triangle

8. Equilateral triangle

a. 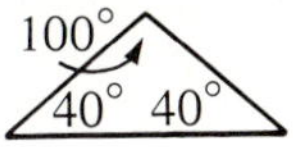

b. 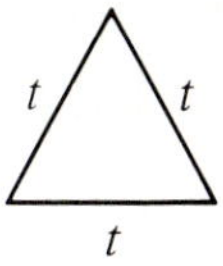

c. 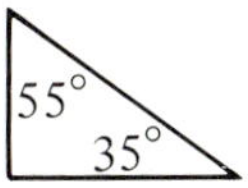

d. 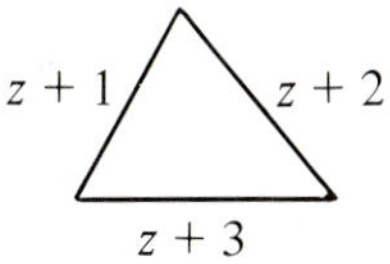

PUZZLE ◆ PROBLEMS

Suppose you don't like obtuse triangles. How can you divide the triangle shown into acute triangles, by using only straight line cuts?

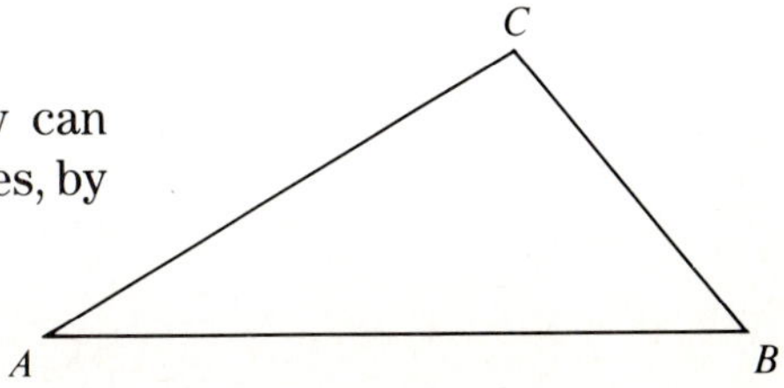

3 Defining Congruent Triangles

When you buy a package of notebook paper, you get sheets that are all alike. Any two sheets have the same size and the same shape. One sheet will fit directly over another sheet. In the language of geometry, one sheet is *congruent* to another sheet.

Likewise $\triangle EFG$ and $\triangle PQR$ are congruent. They have the same size and shape.

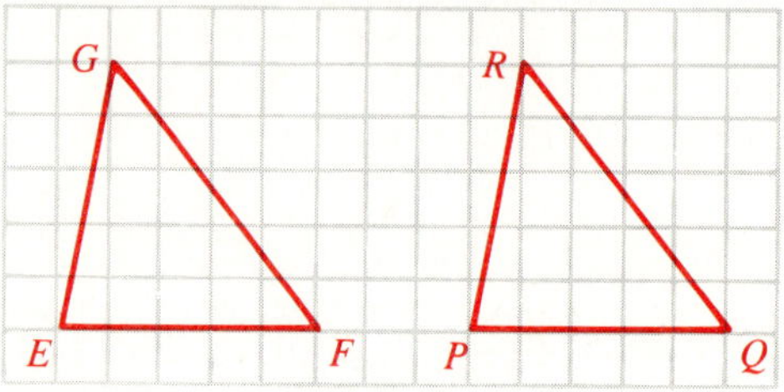

Imagine sliding one triangle over to fit on the other triangle. You put point E on point P; point F on point Q; point G on point R.

When the vertices are matched in this way:

$\angle E$ and $\angle P$ are called **corresponding angles;**
$\overline{EG}$ and $\overline{PR}$ are called **corresponding sides.**

We often refer to corresponding angles and corresponding sides as corresponding **parts** of the triangles. In congruent triangles, *corresponding parts are always equal.*

When $\triangle EFG$ **is congruent to** $\triangle PQR$ we write

$$\triangle EFG \cong \triangle PQR$$

$\triangle EFG \cong \triangle PQR$ means that all of the following are true:

$$\angle E = \angle P \qquad \angle F = \angle Q \qquad \angle G = \angle R$$
$$EF = PQ \qquad FG = QR \qquad EG = PR$$

Suppose that instead of writing $\triangle EFG \cong \triangle PQR$ you started to write:

$$\triangle GEF \cong$$

Having started with G to name one triangle, you must start with the corresponding letter, R, to name the other triangle. Corresponding parts are named in the same order. When you begin with $\triangle GEF$, the complete correct statement is:

$$\triangle GEF \cong \triangle RPQ$$

EXAMPLE The two triangles shown are congruent. Name the (**a**) corresponding angles and (**b**) corresponding sides.

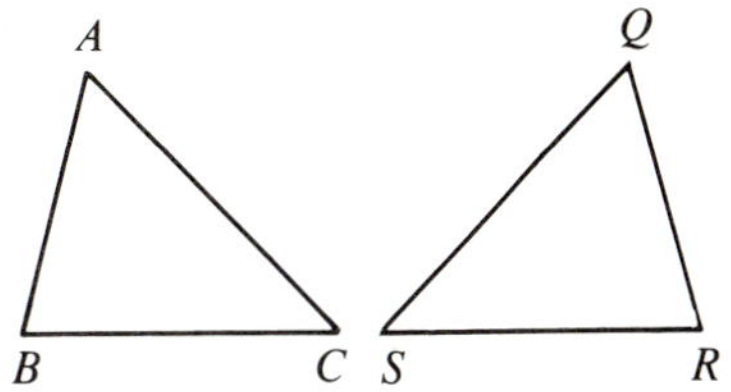

a. *Corresponding angles:*
$\angle A$ and $\angle Q$, $\angle B$ and $\angle R$, $\angle C$ and $\angle S$

b. *Corresponding sides:*
$\overline{AB}$ and $\overline{QR}$, $\overline{BC}$ and $\overline{RS}$, $\overline{CA}$ and $\overline{SQ}$

Classroom Practice

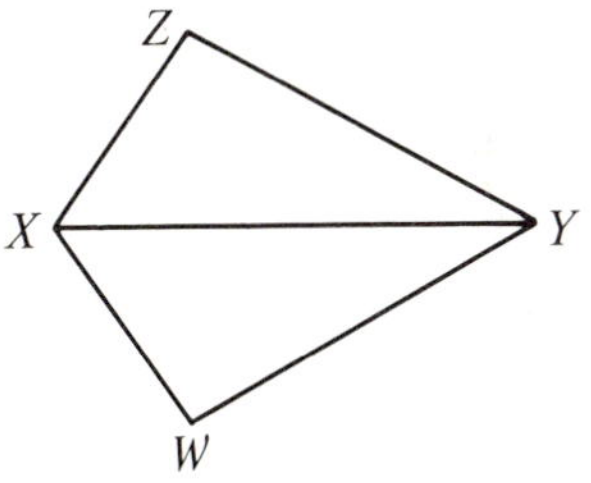

1. Name the six parts of $\triangle XYZ$.

2. Name the six parts of $\triangle XYW$.

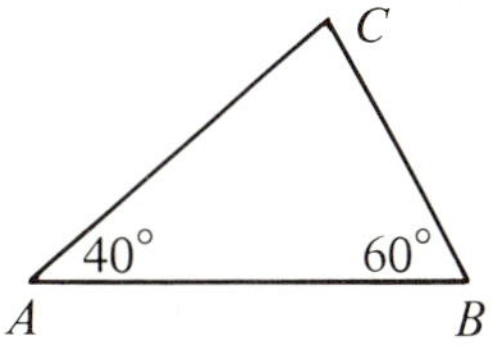

3. Kim drew a triangle that is congruent to $\triangle ABC$. Is Kim's triangle acute, right, or obtuse?

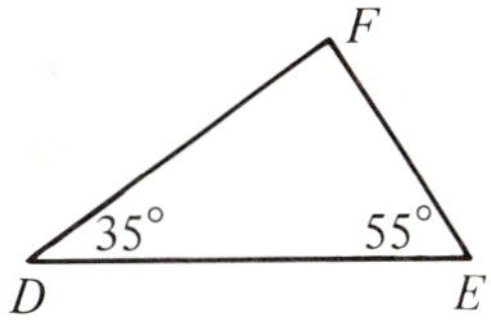

4. Erin drew a triangle that is congruent to $\triangle DEF$. Is Erin's triangle acute, right, or obtuse?

5. Do the triangles appear to be congruent? If you can't decide, trace $\triangle ABC$ and slide the tracing over $\triangle XYZ$. Does one triangle fit exactly over the other?

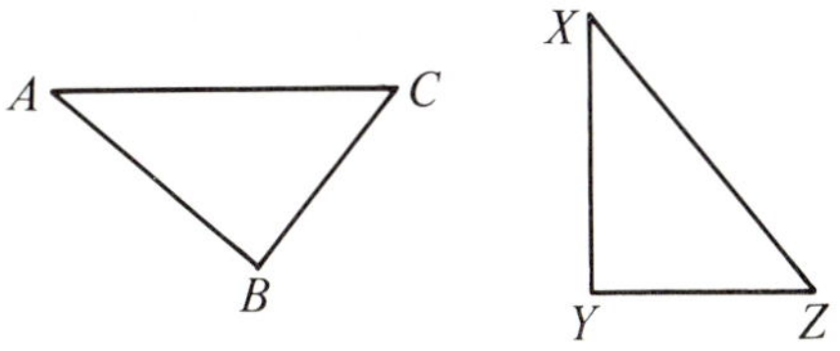

6. Do the triangles appear to be congruent? If you can't decide, trace $\triangle DEF$ and see if it will fit exactly over $\triangle RST$.

7. *Given:* $\triangle ABC \cong \triangle RST$.

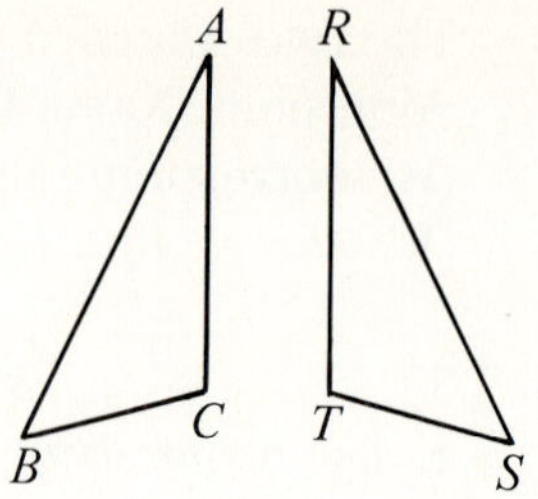

$\angle R$ corresponds to __?__.
$\angle T$ corresponds to __?__.
$\angle S$ corresponds to __?__.

$\overline{RS}$ corresponds to __?__.
$\overline{RT}$ corresponds to __?__.
$\overline{TS}$ corresponds to __?__.

8. *Given:* $\triangle DEF \cong \triangle VWZ$.
State the angle measures and the segment lengths.

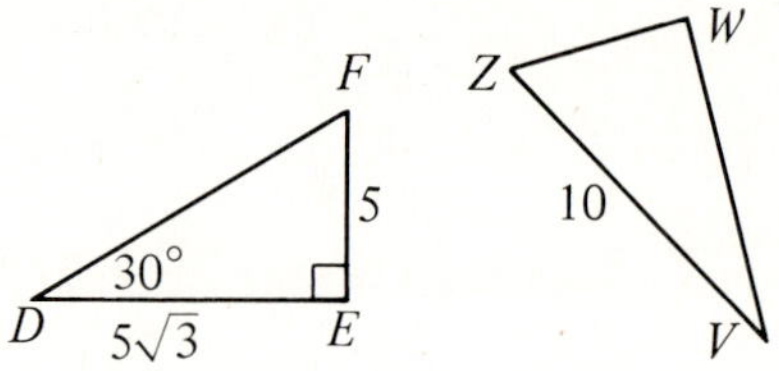

$\angle W =$ __?__ $\angle V =$ __?__ $\angle Z =$ __?__
$DF =$ __?__ $ZW =$ __?__ $WV =$ __?__

9. *Given:* $\triangle ABC \cong \triangle RST$.
Complete each statement.

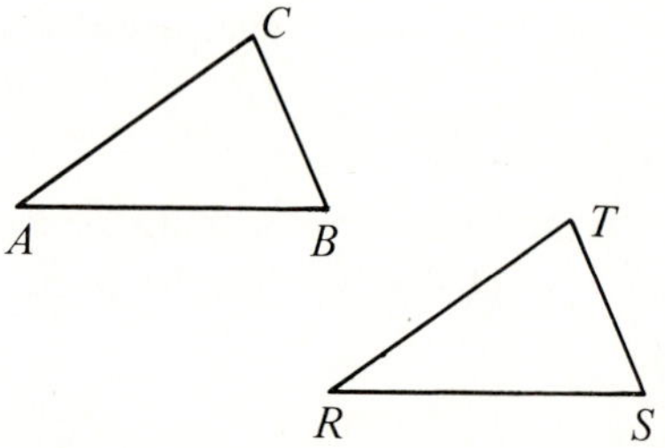

a. $\triangle ACB \cong \triangle$ __?__ b. $\triangle RST \cong \triangle$ __?__

c. $\triangle BAC \cong \triangle$ __?__ d. $\triangle RTS \cong \triangle$ __?__

e. $\triangle CAB \cong \triangle$ __?__ f. $\triangle TSR \cong \triangle$ __?__

g. $\triangle CBA \cong \triangle$ __?__ h. $\triangle STR \cong \triangle$ __?__

10. Sandy said: "Take any three points X, Y, and Z. Draw $\overline{XY}$, $\overline{YZ}$, and $\overline{XZ}$ and you form a triangle." Was Sandy correct?

11. Carlos said: "I know a way to pick out three points so you don't get a triangle. Take one point in the middle of the front wall, another point in the ceiling, and a third point on a side wall." Was Carlos correct?

Written Exercises

In Exercises 1 and 2, copy and complete each statement.

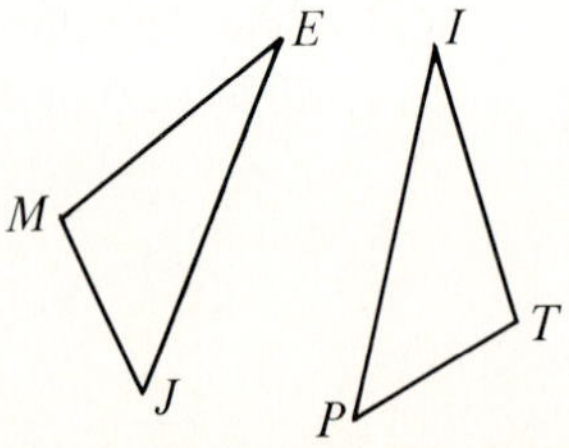

A

1. Suppose you know that $\triangle EJM \cong \triangle IPT$.

a. $\angle J =$ __?__ b. $\angle M =$ __?__

c. $\angle E =$ __?__ d. $JE =$ __?__

e. __?__ $= IP$ f. __?__ $= PT$

2. The statement $\triangle ABC \cong \triangle DEF$ can be a true statement only if:

a. $\angle A =$ ___?___ b. ___?___ $= \angle E$ c. ___?___ $= \angle F$

d. ___?___ $= DE$ e. ___?___ $= DF$ f. $BC =$ ___?___

3. *Given:* $\triangle RST \cong \triangle XYZ$. Write six statements, about equal parts, that follow from the definition of congruent triangles.

4. List the six requirements that must be met for the statement $\triangle JKM \cong \triangle PUG$ to be true.

In Exercises 5–8, do the following:

a. Plot points *A*, *B*, and *C* on graph paper. Draw $\triangle ABC$.

b. Plot points *R*, *E*, and *W*. Draw $\triangle REW$.

c. If possible, complete the statement $\triangle ABC \cong \triangle$ ___?___. Otherwise, write *not congruent*.

SAMPLE

$A(-3, -1)$
$B(2, -1)$
$C(2, 1)$

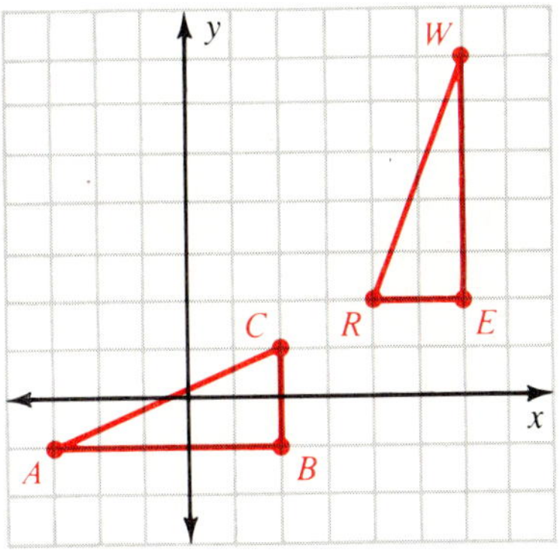

$R(4, 2)$
$E(6, 2)$
$W(6, 7)$

Answer: $\triangle ABC \cong \triangle WER$

	A	B	C	R	E	W
5.	(–3, 1)	(2, 1)	(0, 3)	(4, 2)	(9, 2)	(7, 4)
6.	(–1, 2)	(4, 2)	(2, 4)	(5, –1)	(7, 1)	(10, –1)
7.	(–6, –3)	(–2, –3)	(–4, 0)	(0, 1)	(4, 1)	(2, 2)
8.	(1, 1)	(8, 1)	(4, 3)	(3, –7)	(5, –3)	(3, 0)

Basing your decision on the appearance of the figure, judge whether the statement is correctly written.

9. $\triangle ABC \cong \triangle OND$
10. $\triangle ABC \cong \triangle DON$
11. $\triangle BAC \cong \triangle NOD$
12. $\triangle CAB \cong \triangle NDO$

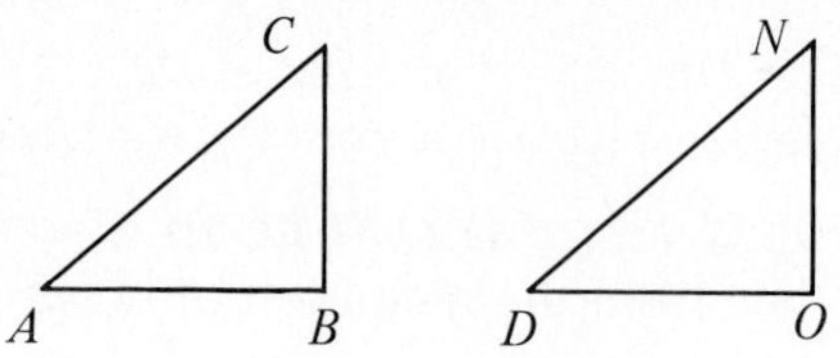

B 13. $\triangle RSJ \cong \triangle CNE$
14. $\triangle JRS \cong \triangle ECN$
15. $\triangle RJS \cong \triangle NEC$
16. $\triangle SRJ \cong \triangle ENC$

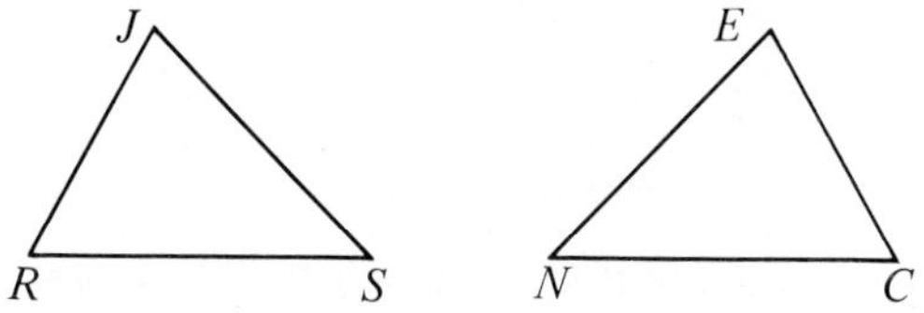

In each exercise it is given that two triangles are congruent. Judge, from the appearance of the figure, which triangles are congruent. List, in pairs, the corresponding parts.

17.

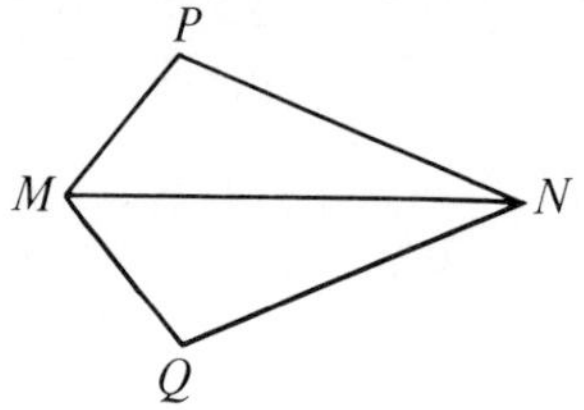

18.

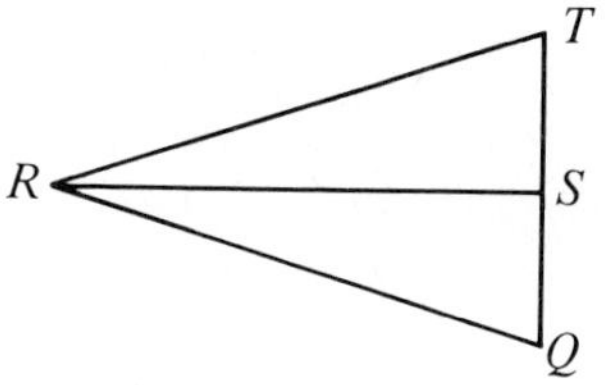

19.

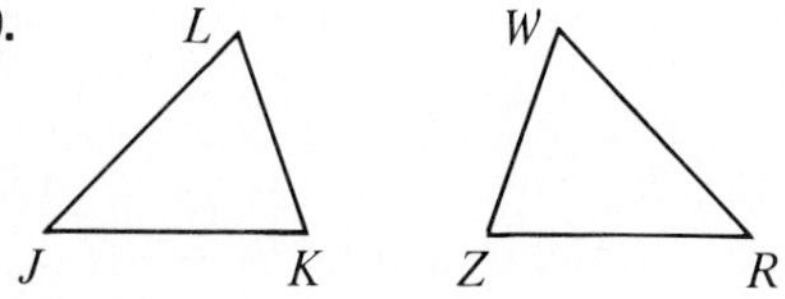

20.

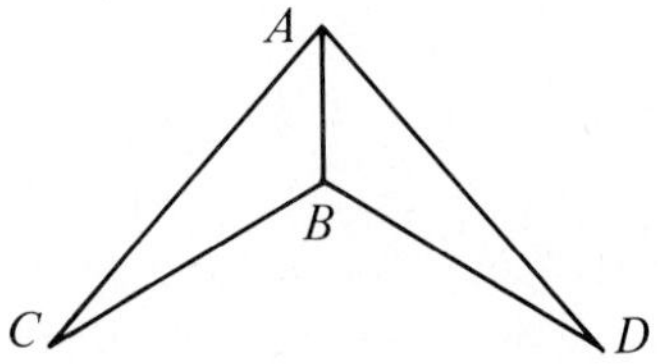

21.

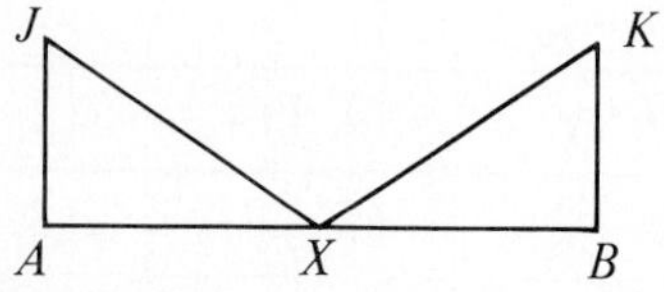

22.

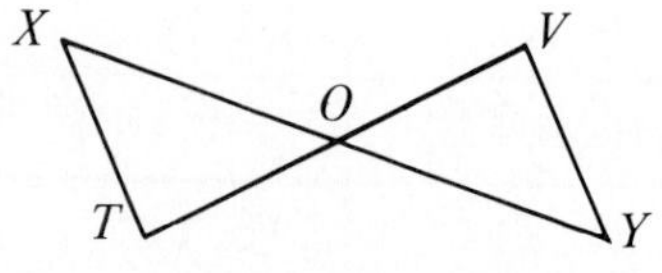

4 The SSS Postulate

Suppose that you are handed the three sticks shown. You fasten the sticks together at the ends to form a triangle.

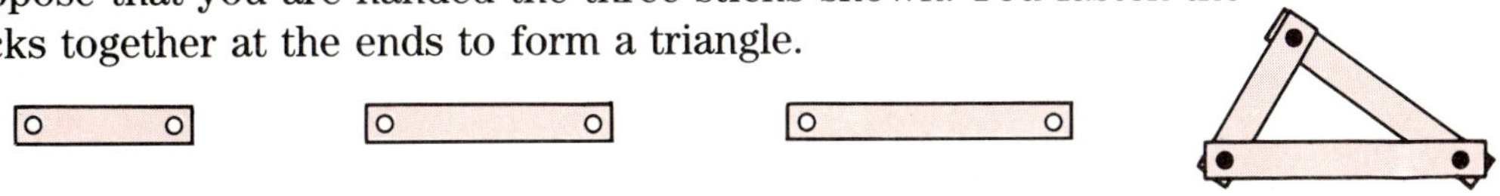

Now suppose someone has a duplicate set of sticks and tries to form a different-looking triangle by joining the sticks at their ends. It is impossible to do this. Only one kind of triangle can be formed from three particular sticks.

The idea suggested by the sticks is a postulate of geometry.

POSTULATE 9 (SSS Postulate)

If three sides of one triangle are equal to the corresponding parts of another triangle, the triangles are congruent.

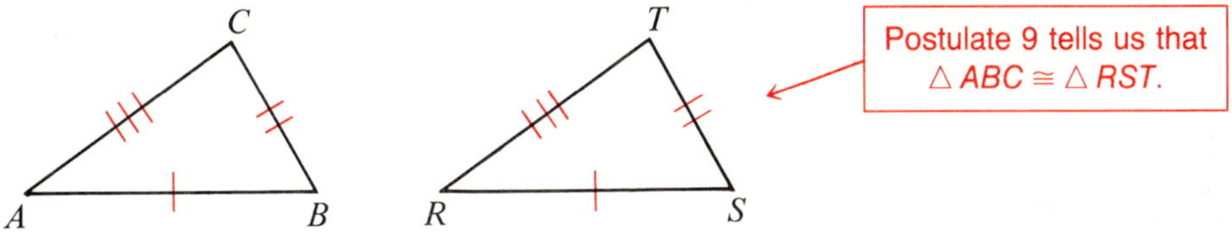

It follows from the definition of congruent triangles that all six parts of $\triangle ABC$ must be equal to the corresponding parts of $\triangle RST$. Besides the three given pairs of equal sides, we have three pairs of equal angles:

$$\angle A = \angle R \qquad \angle B = \angle S \qquad \angle C = \angle T.$$

Notice in the diagram above that equal sides of the triangles are indicated with marks. $\overline{AB}$ and $\overline{RS}$ each have one mark, $\overline{BC}$ and $\overline{ST}$ each have two, and so on. It is often convenient when writing a proof to mark corresponding sides and angles of congruent triangles.

Suppose someone cannot see that the triangles below are congruent. You can point out:

$AC = RT$ (each length is 10);
$AB = RS$ (each length is 12);
$BC = ST$ (each length is 15).

Then $\triangle ABC \cong \triangle RST$ by the SSS Postulate.

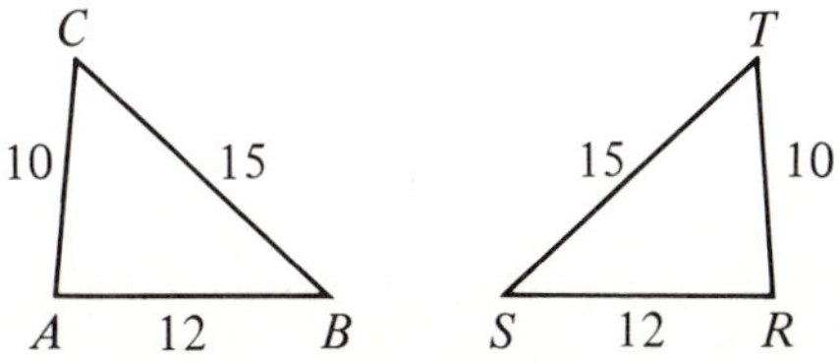

Can you prove that $\triangle XAZ \cong \triangle XBZ$? You can do so if you notice that $\overline{XZ}$ is a side of each of the triangles. Recall, from algebra, that $XZ = XZ$. The length of any segment is equal to itself. We use this idea in the following proof.

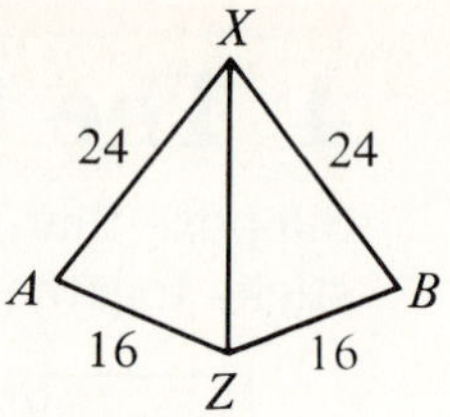

Given: $AX = 24$; $BX = 24$; $AZ = 16$; $BZ = 16$

Prove: $\triangle XAZ \cong \triangle XBZ$

STATEMENTS	REASONS
1. $AX = BX$	1. *Given:* each length is 24.
2. $AZ = BZ$	2. *Given:* each length is 16.
3. $XZ = XZ$	3. From algebra
4. $\triangle XAZ \cong \triangle XBZ$	4. SSS Postulate

Classroom Practice

1. To use the SSS Postulate to prove $\triangle JKV \cong \triangle RNC$, you must first show that:

$JK =$ __?__; $KV =$ __?__; $JV =$ __?__.

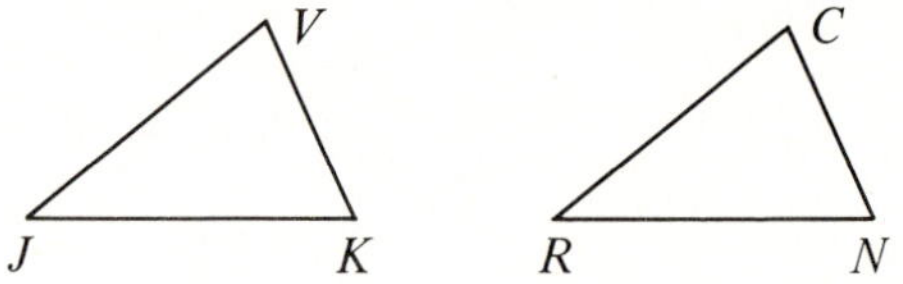

2. To use the SSS Postulate to prove $\triangle ABC \cong \triangle VTS$, you must first show that:

$AC =$ __?__; $AB =$ __?__; $BC =$ __?__.

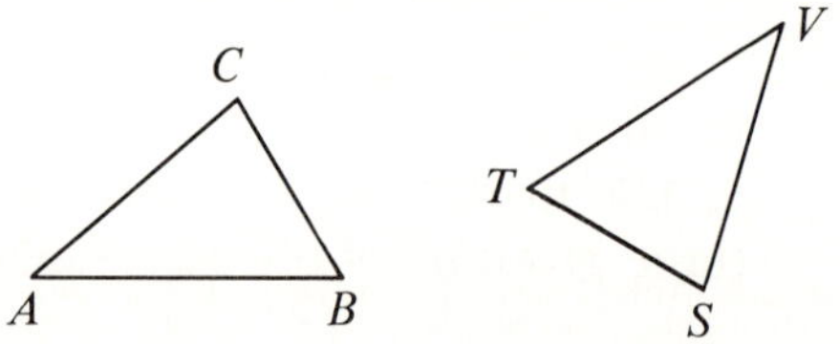

In Exercises 3–8, the goal is to prove that the two triangles are congruent.

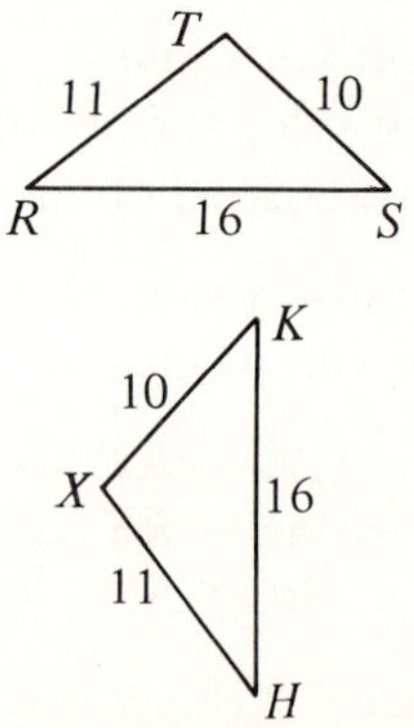

3. Vertex R should be matched with vertex __?__.

4. Vertex T should be matched with vertex __?__.

5. $\overline{TS}$ and __?__ are corresponding sides.

6. __?__ and $\overline{XH}$ are corresponding sides.

7. Is $\triangle RST \cong \triangle KHX$?

8. Is $\triangle RTS \cong \triangle HXK$?

Written Exercises

Copy what is shown. Complete the proofs by supplying the reasons.

A

1. *Given:* X is the midpoint of $\overline{AB}$ and $\overline{CD}$.
$AC = BD$

Prove: $\triangle AXC \cong \triangle BXD$

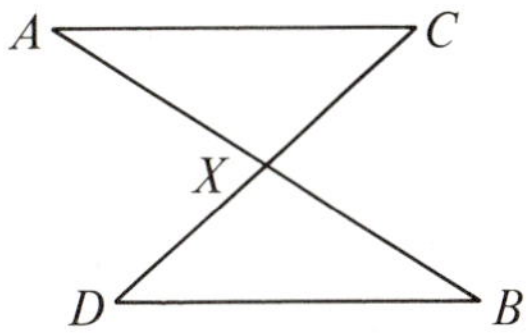

STATEMENTS	REASONS
1. $CX = DX$	1. *Given:* X is the midpoint of _?_.
2. $AX = BX$	2. _?_
3. $AC = BD$	3. _?_
4. $\triangle AXC \cong \triangle BXD$	4. _?_

2. *Given:* S is the midpoint of $\overline{TV}$.
$TR = VR$

Prove: $\triangle TSR \cong \triangle VSR$

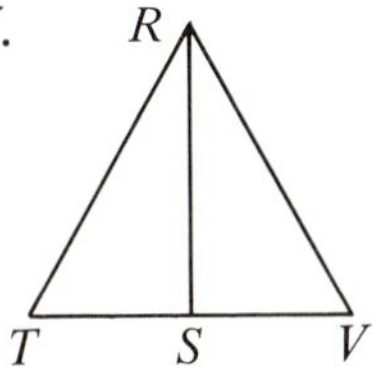

STATEMENTS	REASONS
1. $TS = VS$	1. _?_
2. $RS = RS$	2. _?_
3. $TR = VR$	3. _?_
4. $\triangle TSR \cong \triangle VSR$	4. _?_

3. *Given:* $ZY = WX$
$ZW = YX$

Prove: $\triangle WZY \cong \triangle YXW$

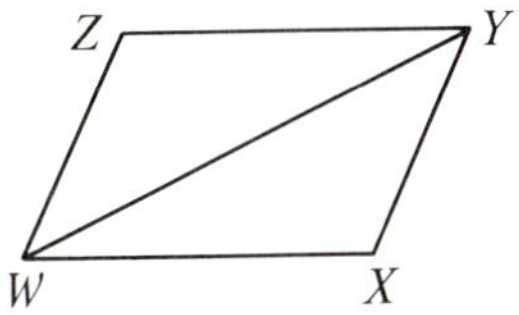

STATEMENTS	REASONS
1. $ZY = WX$; $ZW = YX$	1. _?_
2. $WY = WY$	2. _?_
3. $\triangle WZY \cong \triangle YXW$	3. _?_

Suppose you wish to use the SSS Postulate to prove that $\triangle DEF \cong \triangle JKM$. What value must x have?

4.

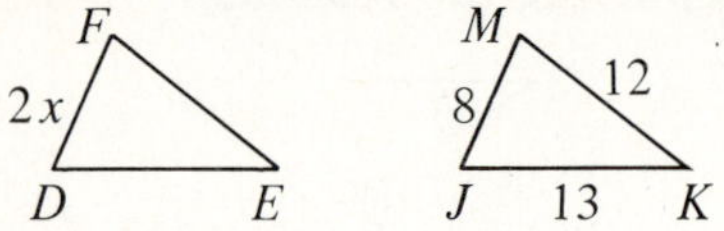

5.

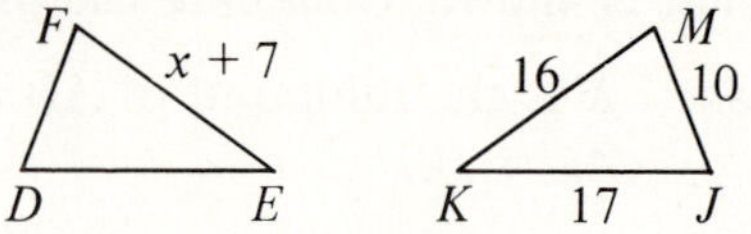

6.

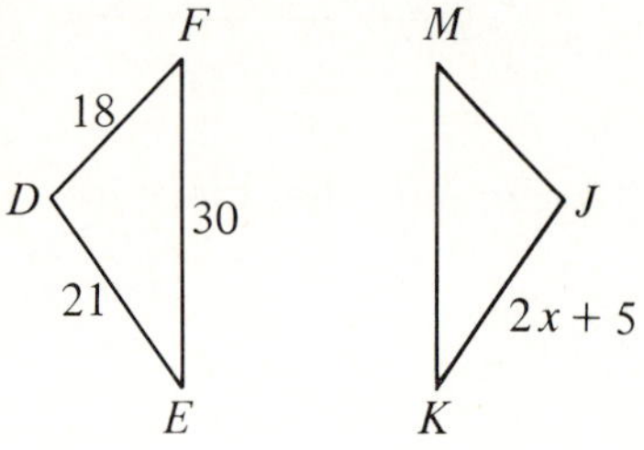

7.

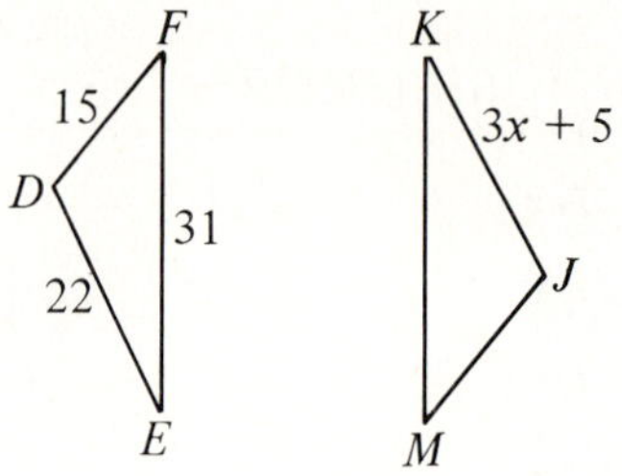

Given: **$\triangle ABC$ and $\triangle RST$, with $AB = RS$, $BC = ST$, and $AC = RT$. Copy and complete each row of angle measures.**

	$\angle A$	$\angle B$	$\angle C$	$\angle R$	$\angle S$	$\angle T$
8.	70°	80°	?	?	?	?
9.	100°	?	20°	?	?	?
10.	?	52°	?	?	?	63°
B **11.**	?	?	$x°$	$y°$	?	?

Draw on your paper three segments roughly like those shown.

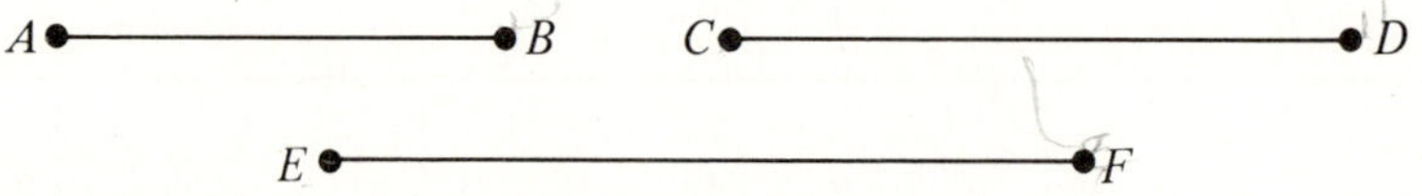

12. Construct a segment $\overline{XY}$ so that $XY = AB$.
With X as center, draw an arc with radius $= CD$.
With Y as center, draw an arc with radius $= EF$.
Use Z to label the point where the arcs intersect.
Draw $\overline{XZ}$ and $\overline{YZ}$.

13. Construct a segment $\overline{RU}$ so that $RU = AB$.
With R as center, draw an arc with radius $= EF$.
With U as center, draw an arc with radius $= CD$.
Use N to label the point where the arcs intersect.
Draw $\overline{RN}$ and $\overline{UN}$.

14. a. Refer to Exercises 12 and 13 to complete the statement:

$$\triangle XYZ \cong \triangle \underline{\ \ ?\ \ }.$$

b. What postulate supports your answer?

15. a. Plot points $A(-7, 1)$, $B(-3, 1)$, and $C(-6, 3)$ on a graph. Draw $\triangle ABC$.

b. Plot points $R(2, 0)$ and $S(6, 0)$ on the same graph. Draw $\overline{RS}$.

c. On your graph, locate a point T so that the statement $\triangle ABC \cong \triangle RST$ will be true. (*Note:* There is more than one correct point. How many can you find?)

16. a. Plot points $D(2, 0)$, $E(2, 7)$, and $F(5, 7)$ on a graph. Draw $\triangle DEF$.

b. Plot points $V(-6, 0)$ and $W(-6, 7)$ on the same graph. Draw $\overline{VW}$.

c. Locate a point K so that the statement $\triangle DEF \cong \triangle VWK$ will be true. (*Note:* There is more than one correct point.)

d. Locate a point L so that the statement $\triangle DEF \cong \triangle WVL$ will be true. (*Note:* There is more than one correct point.)

17. To strengthen a triangular frame, Myra fastened three additional sticks as shown. How many triangles were formed?

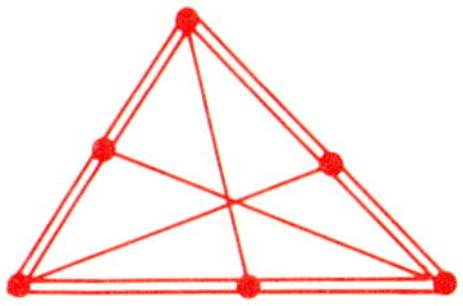

For each exercise, copy what is shown. Then write a complete proof in two-column form.

C **18.** *Given:* N is the midpoint of $\overline{AB}$.
$AX = BY$
$NX = NY$

Prove: $\triangle AXN \cong \triangle BYN$

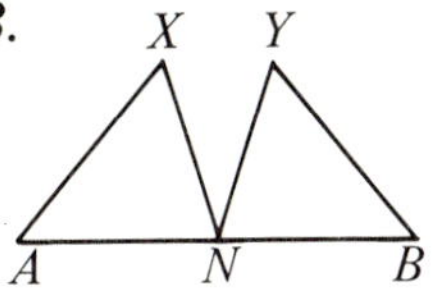

19. *Given:* $RT = RV$
$TS = VS$

Prove: $\triangle RST \cong \triangle RSV$

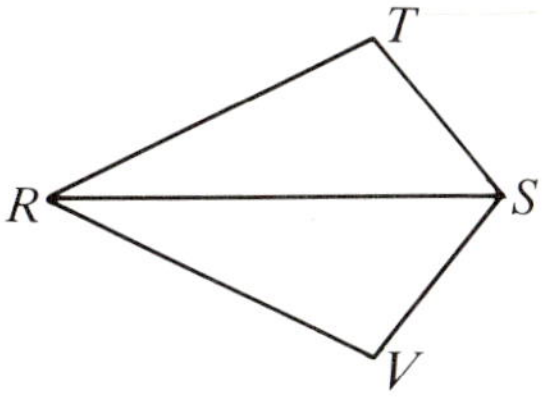

20. *Given:* $JK = MN$
$NJ = 9$
$KM = 9$

Prove: $\triangle NJK \cong \triangle KMN$

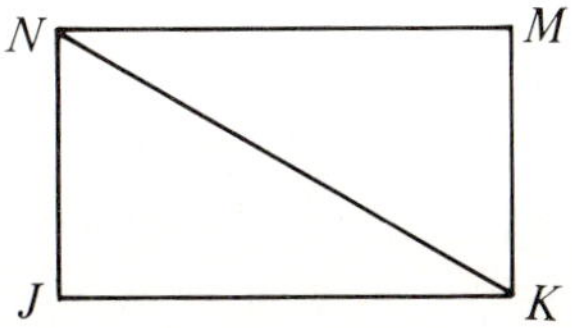

SELF-TEST

Vocabulary

corresponding angles (p. 102) congruent (p. 102)
corresponding sides (p. 102)

You are given: △ *ENP* ≅ △ *OBJ*. Complete the statements.

1. $\triangle NEP \cong \triangle$ __?__ **2.** $\angle E =$ __?__ **3.** $OJ =$ __?__ *(3-3)*

4. Supply reasons to complete the proof. *(3-4)*

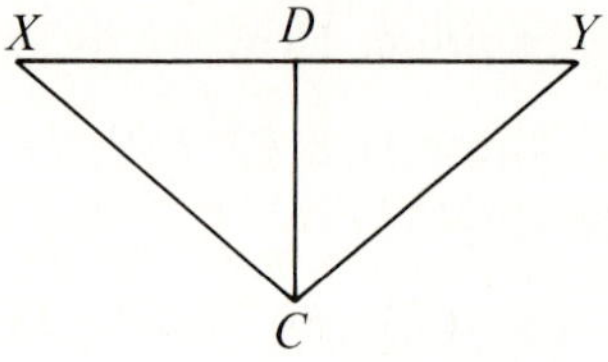

Given: $CX = CY$
D is the midpoint of $\overline{XY}$.

Prove: $\triangle CDX \cong \triangle CDY$

STATEMENTS	REASONS
1. $CX = CY$	1. __?__
2. $CD = CD$	2. __?__
3. $DX = DY$	3. __?__
4. $\triangle CDX \cong \triangle CDY$	4. __?__

PUZZLE ◆ PROBLEMS

1. Arrange twelve toothpicks as in Figure 1. Now can you rearrange three toothpicks to form three squares of the same size?

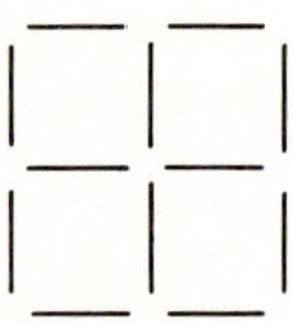

Figure 1

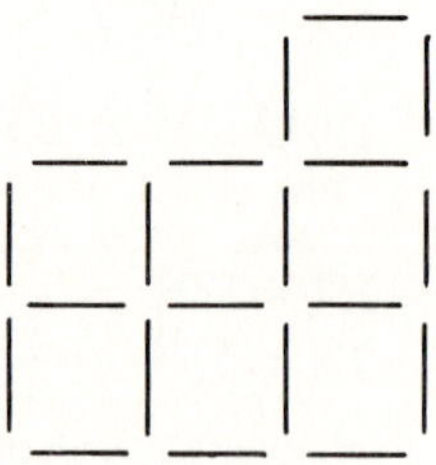

Figure 2

2. In Figure 2, rearrange three toothpicks to form five squares of the same size.

3. Form four equilateral triangles with six toothpicks. (This puzzle is solved in a way completely different from the last two puzzles. Can you see how?)

APPLICATIONS

RIGIDITY OF TRIANGLES

The SSS Postulate states that if three sides of one triangle are equal to the corresponding sides of another triangle, the triangles are congruent. We have seen that to change the shape of a triangle we must change the length of at least one of its sides. Thus, we say that a triangle is a *rigid* figure.

Architects and builders use the fact that triangles are rigid when they want to build a structure that is strong enough to hold its shape when acted upon by external and internal forces.

Many roofs have a triangular cross-section. In a properly designed triangular roof, all the force is exerted down. None of the force pushes out. As a result, the walls do not need to be reinforced to hold their shape.

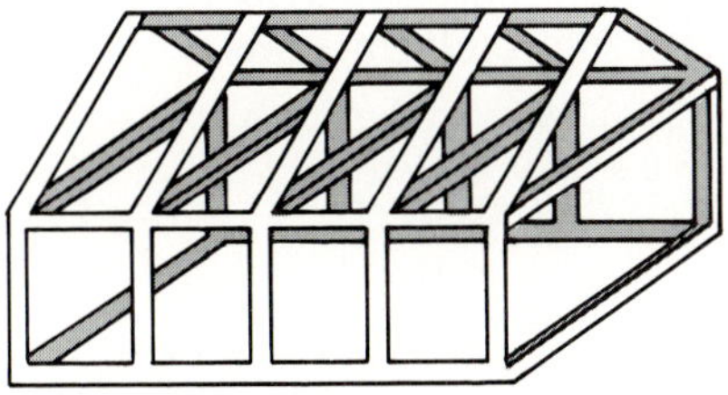

The roof of this building will have a triangular cross-section.

Another example of the use of triangles in construction is the truss bridge. Builders often combine several triangles to make a strong bridge with a wide span. Many covered bridges built in the late 18th century and early 19th century are of truss-type construction.

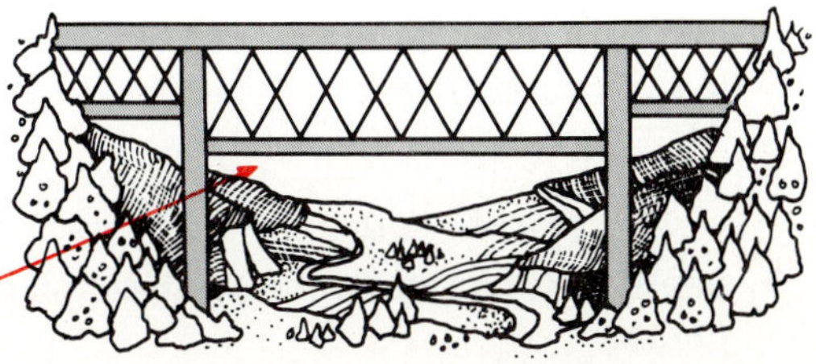

Notice the overlapping triangles in this truss bridge.

R. Buckminster Fuller used the rigidity of triangles in designing his famous geodesic dome. His design combines many triangular surfaces to form a dome that is strong, but lightweight. Large geodesic domes form pavilions for fairs, sporting events, and horticultural exhibits. Smaller domes make interesting houses.

EXERCISES

Cut some cardboard into 1 cm strips of various lengths. Punch holes in the strips 1 cm from each end. Use metal fasteners to attach the ends and form models like those pictured here.

Which of the models are rigid figures?

1.

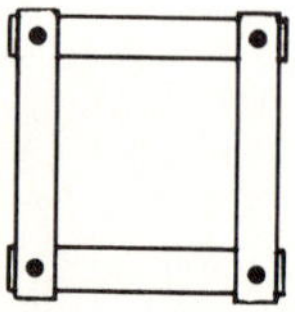

2.

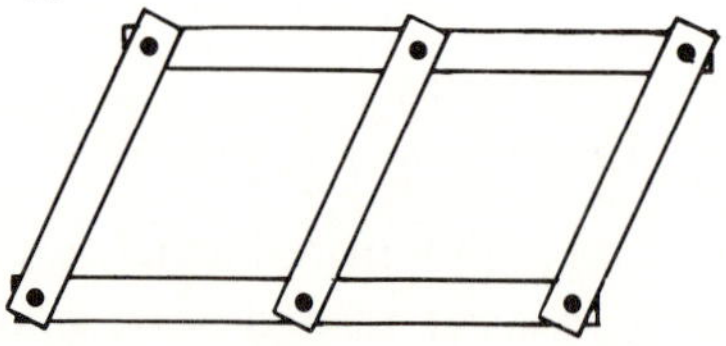

3.

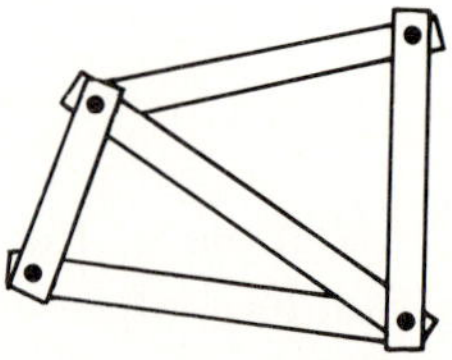

4.

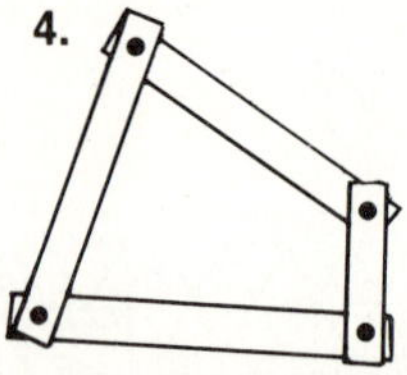

5.

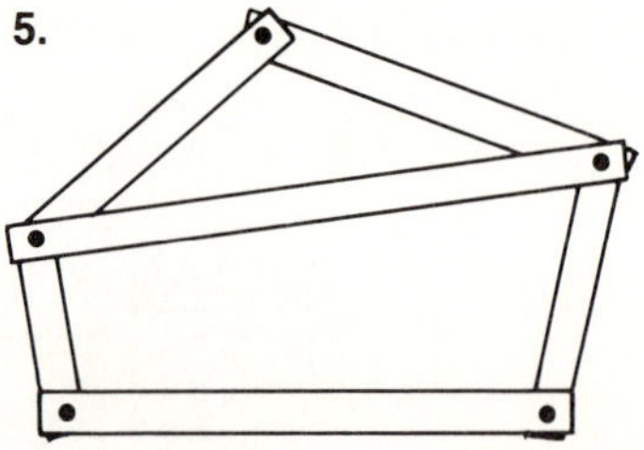

6.

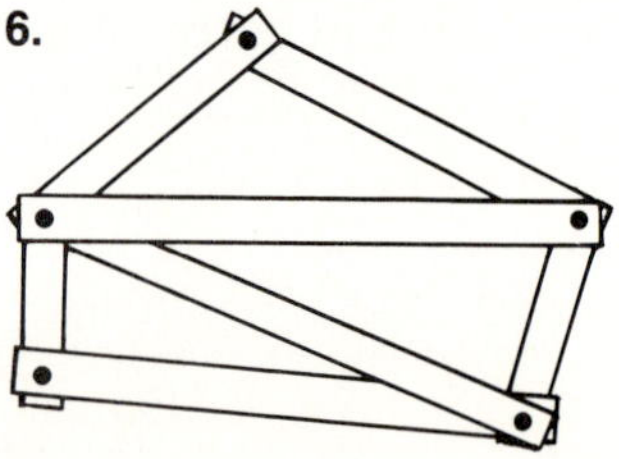

See if you can combine your strips into other models that form rigid figures.

5 The SAS Postulate

Suppose you have two sticks. You fasten them together at a 40° angle.

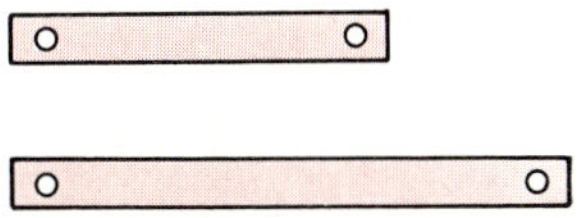

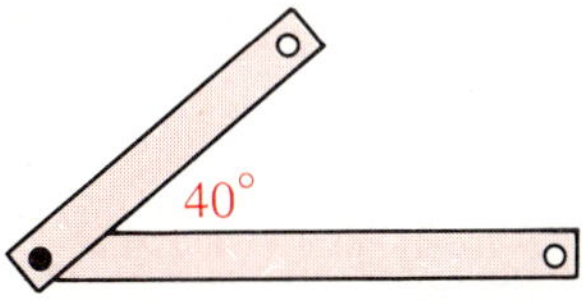

There is only one way to finish forming a triangle. A third stick must be of just the right length.

Think again of the two sticks. What would happen if the sticks were fastened like this?

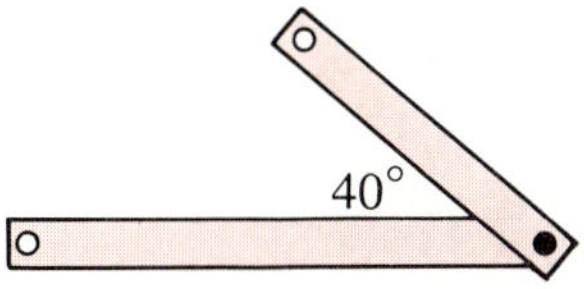

Would the triangle you could form be a different-looking triangle? To decide, you can perform an experiment in your mind.

Flip

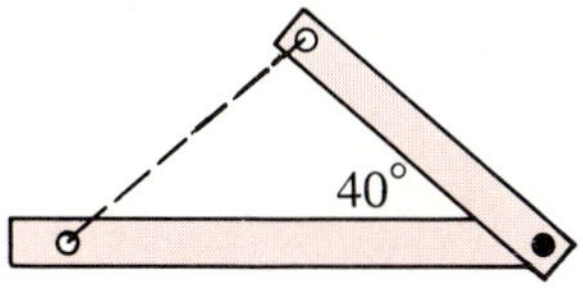

to get

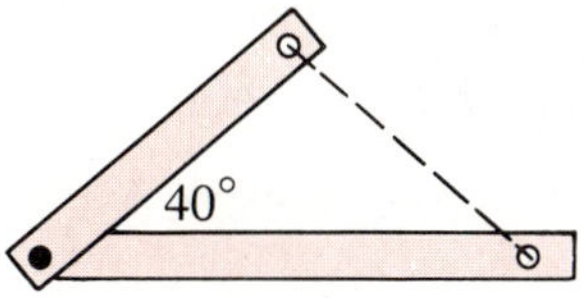

The models are identical.

When two sticks are joined at a certain angle, a third stick must be of just the right length to form a triangle. The triangle can have only one size and shape.

The 40° angle is included between the sides formed by the two sticks. Look at $\triangle DEF$.

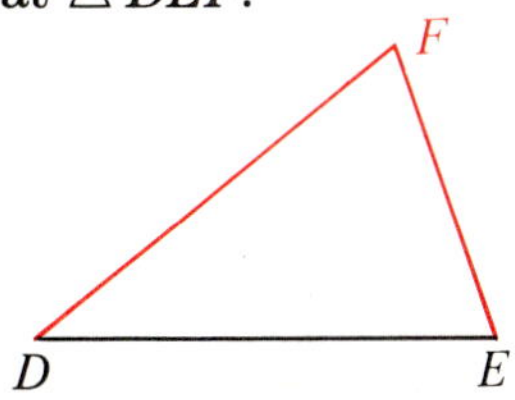

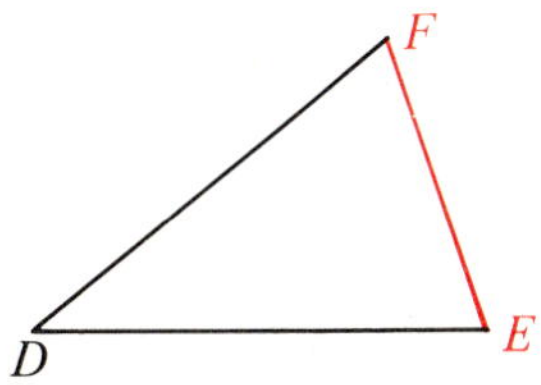

$\angle F$ is *included between* sides $\overline{FD}$ and $\overline{FE}$. $\angle F$ is *opposite* side $\overline{DE}$.

Side $\overline{EF}$ is *included between* $\angle E$ and $\angle F$. $\overline{EF}$ is *opposite* $\angle D$.

POSTULATE 10 (SAS Postulate)

If two sides and the included angle of one triangle are equal to the corresponding parts of another triangle, the triangles are congruent.

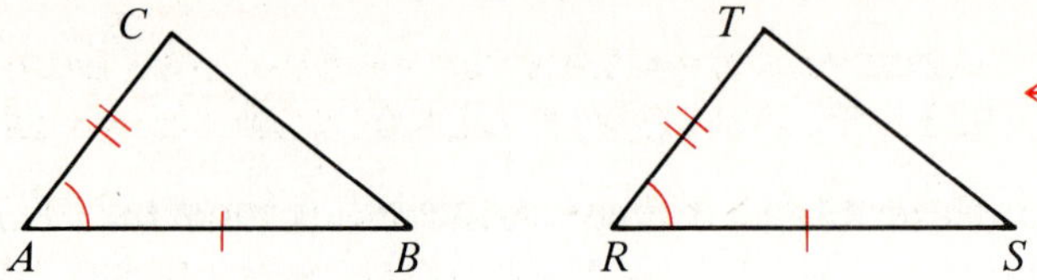

Postulate 10 tells us that $\triangle ABC \cong \triangle RST$.

Classroom Practice

Complete each statement.

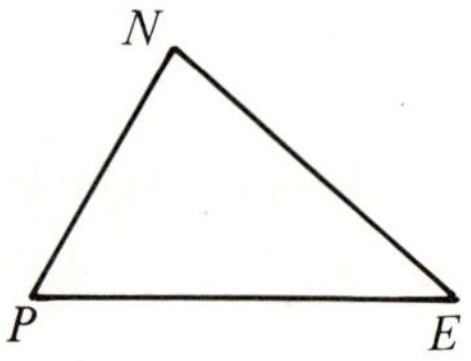

1. $\angle E$ lies opposite side ___?___.
2. $\angle E$ is included between sides ___?___ and ___?___.
3. $\angle$ ___?___ lies opposite side $\overline{PE}$.
4. $\angle$ ___?___ is included between sides $\overline{PN}$ and $\overline{PE}$.
5. Side $\overline{NE}$ is included between $\angle$ ___?___ and $\angle$ ___?___.
6. Side ___?___ is included between $\angle P$ and $\angle N$.

In Exercises 7–12, two triangles are to be proved congruent.

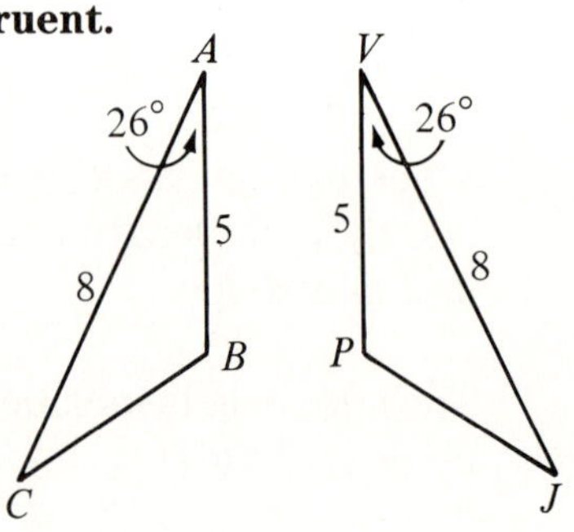

7. Vertex A should be paired with vertex ___?___.
8. Vertex ___?___ should be paired with vertex J.
9. $\overline{AB}$ and ___?___ are corresponding sides.
10. ___?___ and $\overline{VJ}$ are corresponding sides.
11. Is $\triangle ACB \cong \triangle VJP$?
12. Is $\triangle CAB \cong \triangle JVP$?

In each exercise, two triangles are congruent.
In order to fit one triangle over the other could you:
a. slide the first triangle over the second?
b. flip the first triangle and then slide it over the second?

13.

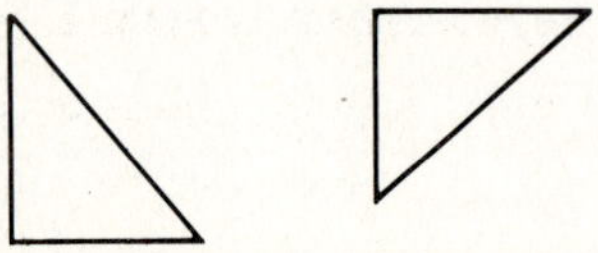

14.

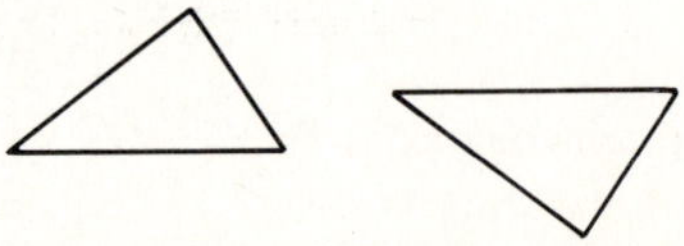

15.

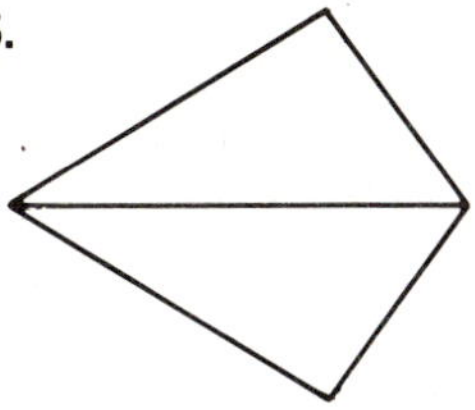

16.

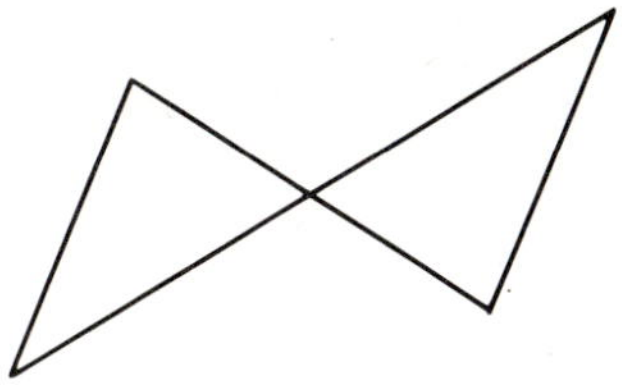

17.

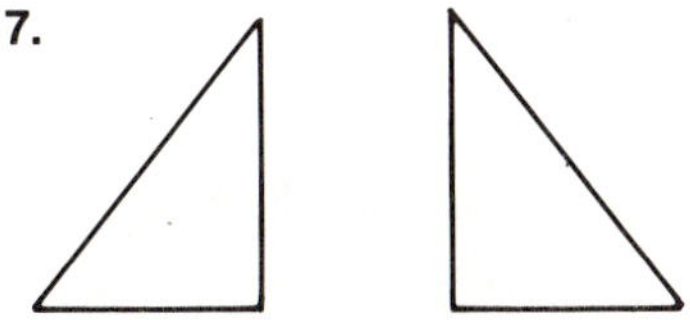

18.

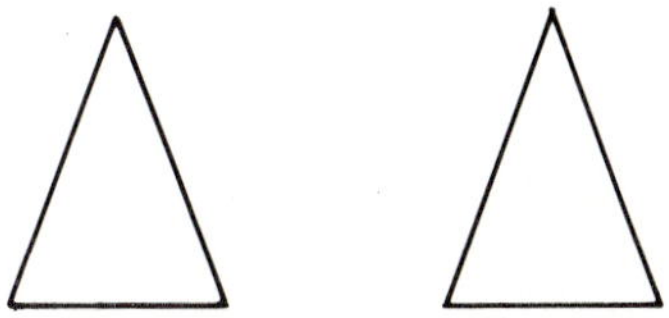

Written Exercises

In Exercises 1–4, we will prove that $\triangle ABC \cong \triangle JKN$.

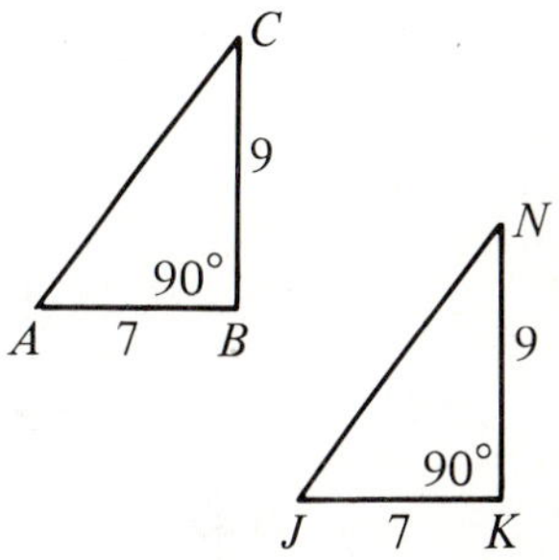

A 1. Is $AB = JK$? Is $BC = KN$? Why?

2. Is $\angle B = \angle K$? Why?

3. Do you know that two sides and the included angle of $\triangle ABC$ are equal to the corresponding parts of $\triangle JKN$?

4. State the postulate that supports the statement $\triangle ABC \cong \triangle JKN$.

In Exercises 5–8, we will prove that $\triangle PXQ \cong \triangle RXS$.

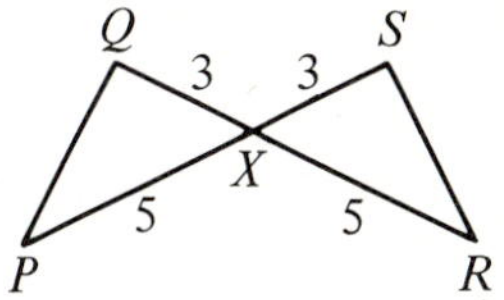

5. In $\triangle PXQ$, what angle is included between sides $\overline{XQ}$ and $\overline{XP}$? In $\triangle RXS$, what angle is included between sides $\overline{XS}$ and $\overline{XR}$?

6. Is $\angle PXQ = \angle RXS$? Why?

7. Is $QX = SX$? Is $PX = RX$?

8. To support the statement $\triangle PXQ \cong \triangle RXS$, would you use the SSS Postulate or the SAS Postulate?

State whether the SSS Postulate or the SAS Postulate could be used to prove the triangles congruent.

9.

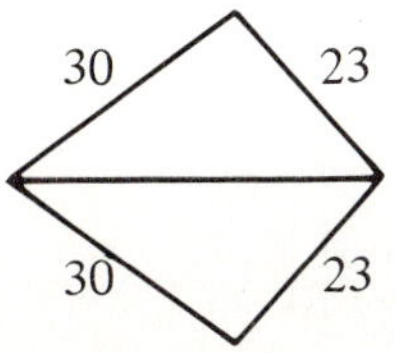

10.

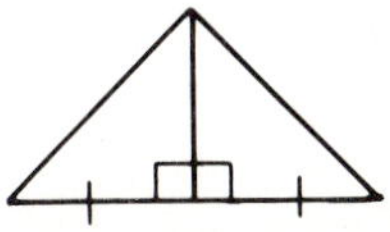

11.

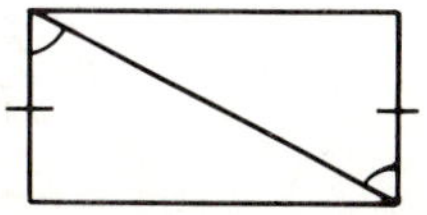

Complete the proofs by supplying the reasons.

12. *Given:* $\angle A = 40^\circ$
$\angle R = 40^\circ$
$AB = RS$
$AC = RT$

Prove: $\triangle ABC \cong \triangle RST$

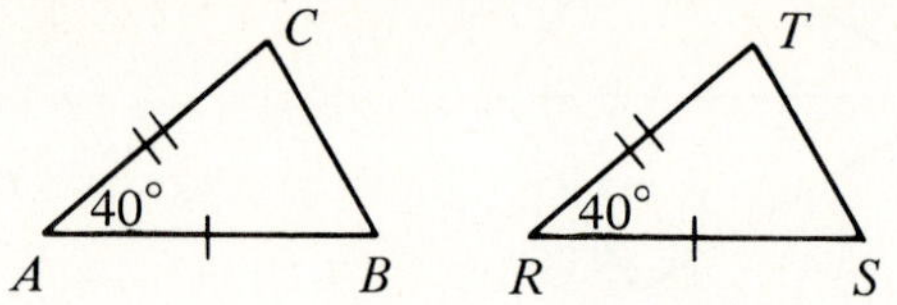

STATEMENTS	REASONS
1. $\angle A = \angle R$	1. *Given:* Each angle has measure __?__.
2. $AB = RS$; $AC = RT$	2. __?__
3. $\triangle ABC \cong \triangle RST$	3. __?__

13. *Given:* $\overrightarrow{ZW}$ bisects $\angle XZY$.
$XZ = YZ$

Prove: $\triangle XWZ \cong \triangle YWZ$

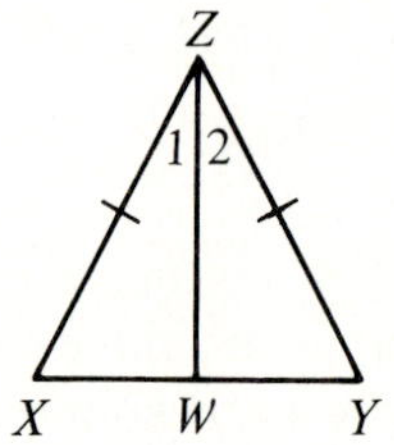

STATEMENTS	REASONS
1. $\angle 1 = \angle 2$	1. __?__
2. $ZW = ZW$	2. __?__
3. $XZ = YZ$	3. __?__
4. $\triangle XWZ \cong \triangle YWZ$	4. __?__

Draw, on your paper, an angle and two segments like those shown.

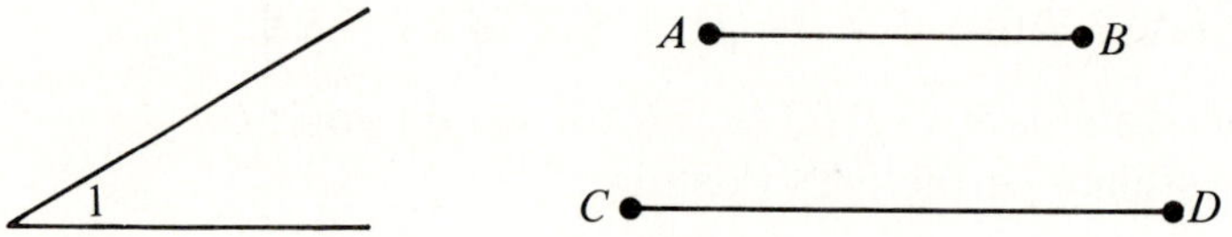

B **14.** Construct an angle equal to $\angle 1$. Call it $\angle R$.
On one side of $\angle R$, construct $\overline{RS}$ so that $RS = AB$.
On the other side of $\angle R$, construct $\overline{RT}$ so that $RT = CD$.
Draw $\overline{ST}$.

15. Repeat Exercise 14, using X, Y, and Z instead of R, S, and T.
What postulate supports the statement: $\triangle RST \cong \triangle XYZ$?

16. Complete the proof by supplying the reasons.

Given: X is the midpoint of $\overline{AB}$ and $\overline{CD}$.

Prove: $\triangle AXC \cong \triangle BXD$

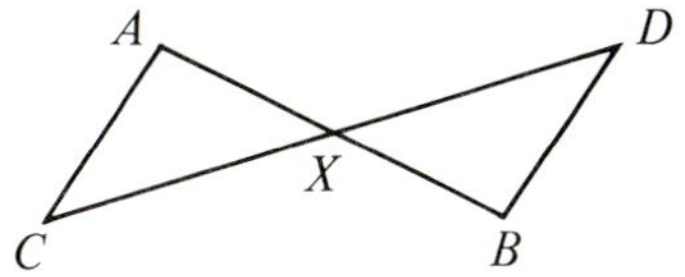

STATEMENTS	REASONS
1. $CX = DX$	1. *Given: X* is __?__.
2. $AX = BX$	2. __?__
3. $\angle AXC = \angle BXD$	3. __?__
4. $\triangle AXC \cong \triangle BXD$	4. __?__

17. *Given:* $RS = VT$
$\overline{RS} \parallel \overline{VT}$

Prove: $\triangle VRS \cong \triangle STV$

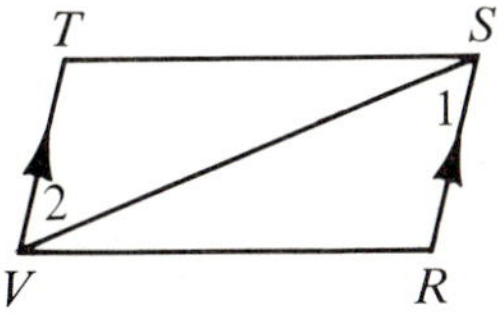

Copy what is shown.
Then write a proof in two-column form.

Suggested strategy:

A. Use the parallel lines to prove two angles equal.

B. Use the fact that $\overline{SV}$ is a side of each triangle.

C. List enough equal parts so that you can use the SSS Postulate or the SAS Postulate.

Copy what is shown. Then write a proof in two-column form.

C **18.** *Given:* $AC = AD$
$\overrightarrow{AB}$ bisects $\angle CAD$.

Prove: $\triangle ABC \cong \triangle ABD$

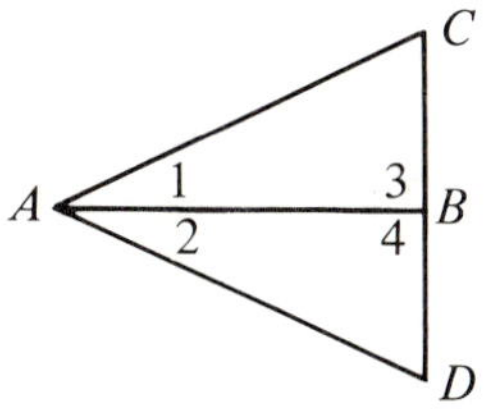

19. *Given:* $\overline{AB} \perp \overline{CD}$
B is the midpoint of $\overline{CD}$.

Prove: $\triangle ABC \cong \triangle ABD$

20. Refer to the figure shown. We wish to prove the following: If the two legs of one right triangle are equal to the two legs of another right triangle, then the triangles are congruent.

a. List what is given and what is to be proved, in terms of the diagram.

b. Write a proof in two-column form.

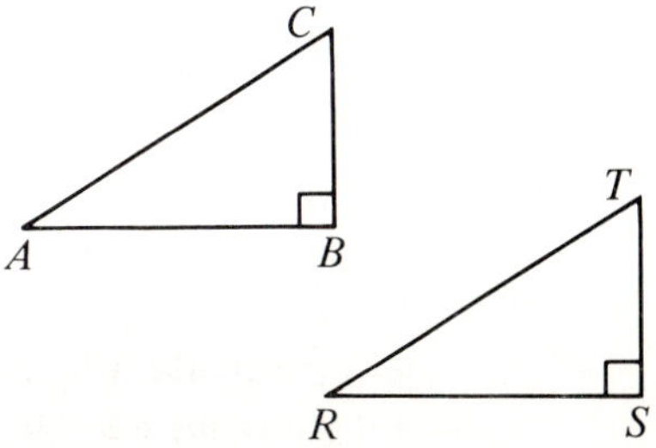

Experiments

Each student needs paper, scissors, a pencil, a ruler, and a protractor.

1. Draw a segment 20 cm long.
2. Choosing either endpoint of the segment, draw a ray so that a 35° angle is formed.
3. From the endpoint of that ray, mark off a segment 15 cm long.
4. Connect the two free endpoints to form a triangle.
5. Shade the triangular region and cut out the triangle.

Now students should compare their models until they agree on answers to the following questions.

a. Are there some pairs of triangles that fit exactly, one over the other, each with the shaded side up?

b. Are there some pairs that fit exactly when the shaded side of one triangle faces up and that of the other triangle faces down?

c. Are there any two models that don't fit at all? Should there be?

CONSUMER APPLICATIONS

DO YOU GET WHAT YOU PAY FOR?

Consumers often have to decide, "Which one should I buy?" For example, you may have to decide among two, three, or a dozen different radios. Several different winter coats may look attractive. You may find a bargain and be able to buy the best product at the lowest price. Often, however, you have to pay more for a better quality item.

Is the extra expense worth it? The answer depends on your own needs, tastes, and financial resources. The radio which costs an extra $15 may produce a much better sound. The camera which costs twice as much may last three times longer. Sometimes, however, extra dollars only buy extra features or frills which are of no use.

Here are some common-sense suggestions for wise purchasing:
(1) Decide exactly what you want and how long it must last.
(2) Talk with people who have bought a similar item.
(3) Consult a trade magazine or a consumer magazine.
(4) Compare prices in several different stores before you buy.

6 The ASA Postulate

The diagram shows three sticks. The fasteners at A and B allow sticks l and m to turn.

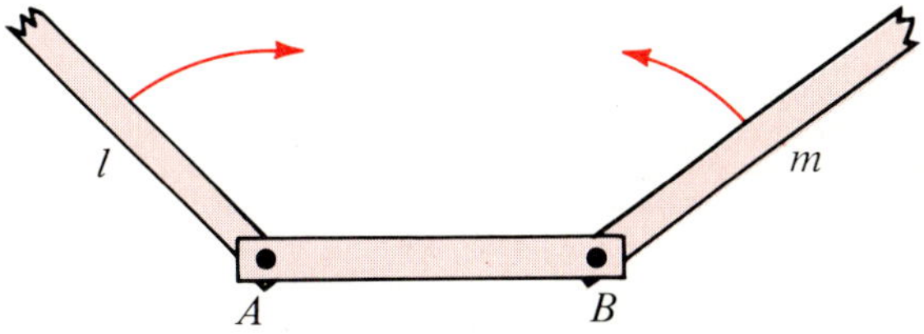

1. Stick l is rotated until $\angle A = 40°$.

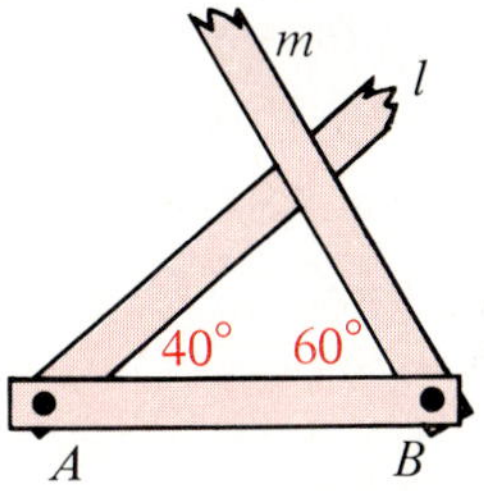

2. Stick m is rotated until $\angle B = 60°$.

3. The extra wood is sawed off and a fastener is attached at C.

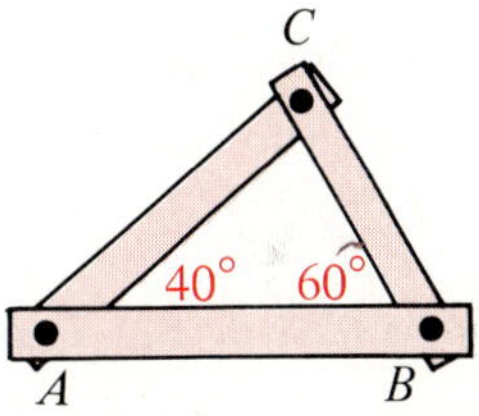

Do you see that for a particular length AB, a 40° angle at A, and a 60° angle at B, only one kind of triangle is possible?

POSTULATE 11 (ASA Postulate)

If two angles and the included side of one triangle are equal to the corresponding parts of another triangle, the triangles are congruent.

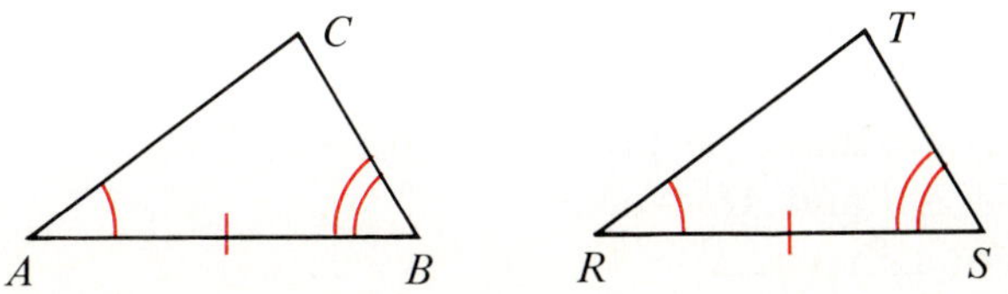

Postulate 11 tells us that $\triangle ABC \cong \triangle RST$.

Classroom Practice

Name the side included between the two angles.

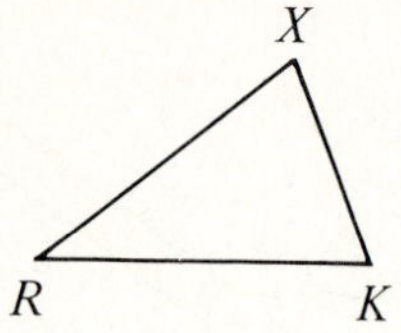

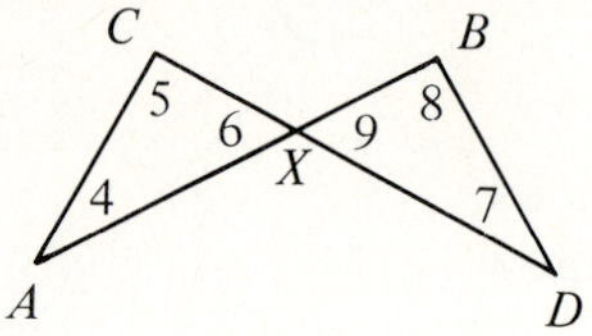

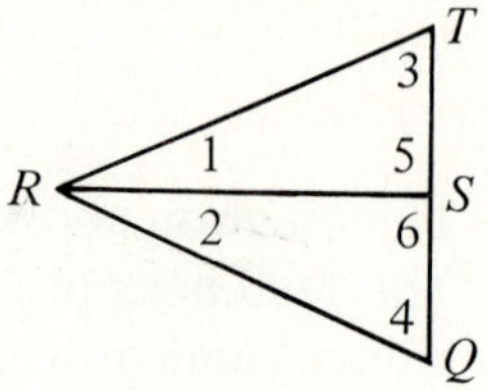

1. $\angle R$ and $\angle K$
2. $\angle X$ and $\angle R$
3. $\angle 5$ and $\angle 6$
4. $\angle 7$ and $\angle 8$
5. $\angle 1$ and $\angle 5$
6. $\angle 6$ and $\angle 4$

Suppose you know that $\angle K = \angle X$ and $KL = XO$.

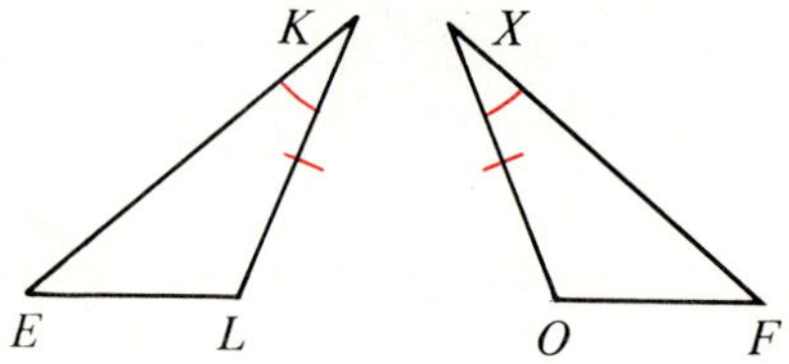

7. To use the ASA Postulate to prove that $\triangle KEL \cong \triangle XFO$, you must also show that __?__.
8. To use the SAS Postulate to prove that $\triangle KEL \cong \triangle XFO$, you must also show that __?__.
9. *Given:* $\overrightarrow{RS}$ bisects $\angle XRY$.
   S is the midpoint of $\overline{XY}$.
   $\overline{RS} \perp \overline{XY}$

   Could you prove that $\triangle RSX \cong \triangle RSY$

   a. by using the SSS Postulate?

   b. by using the SAS Postulate?

   c. by using the ASA Postulate?

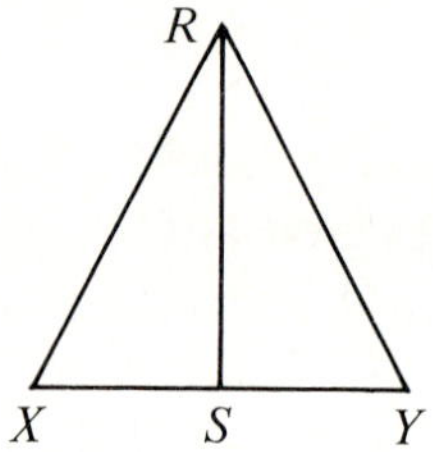

Written Exercises

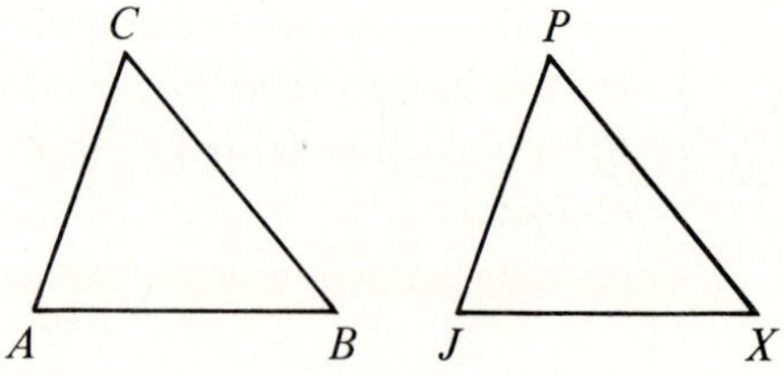

Exercises 1–4

A

1. If $\angle A = \angle J$, $\angle B = \angle X$, and $AB = JX$, then you can use the ASA Postulate to conclude $\triangle ABC \cong \triangle$ __?__.
2. You intend to use the ASA Postulate to prove $\triangle ABC \cong \triangle JXP$. You have stated that $\angle C = \angle P$ and $\angle B = \angle X$. You also need the statement __?__ = __?__.
3. You intend to use the SAS Postulate to prove $\triangle ABC \cong \triangle JXP$. You have stated that $\angle A = \angle J$ and $AB = JX$. You also need the statement __?__ = __?__.

4. You intend to use the ASA Postulate to prove $\triangle ABC \cong \triangle JXP$. You have stated that $AC = JP$ and $\angle C = \angle P$. You also need the statement __?__ = __?__.

In Exercises 5–6, state the reasons needed to complete the proofs.

5. *Given:* $\angle D = \angle R$; $\angle F = \angle T$
$DF = 8$; $RT = 8$

Prove: $\triangle DEF \cong \triangle RST$

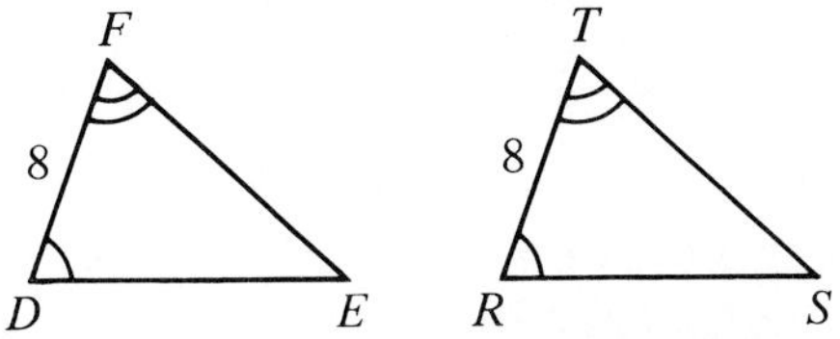

STATEMENTS	REASONS
1. $\angle D = \angle R$; $\angle F = \angle T$	1. __?__
2. $DF = RT$	2. *Given:* each length __?__.
3. $\triangle DEF \cong \triangle RST$	3. __?__

6. *Given:* M is the midpoint of $\overline{YZ}$.
$\overrightarrow{MY}$ bisects $\angle OMV$.
$\angle Y = \angle Z$

Prove: $\triangle YOM \cong \triangle ZUM$

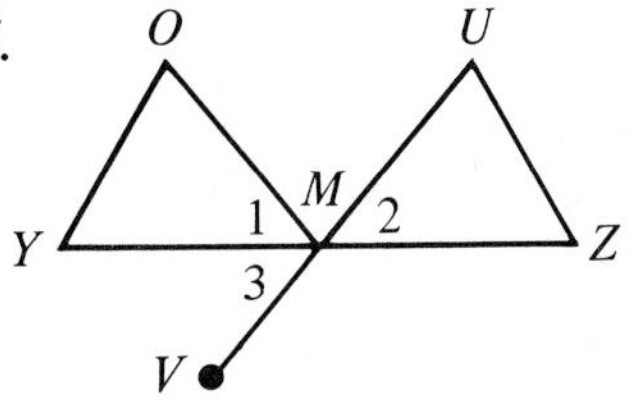

STATEMENTS	REASONS
1. $\angle 1 = \angle 3$	1. __?__
2. $\angle 2 = \angle 3$	2. __?__
3. $\angle 1 = \angle 2$	3. __?__
4. $YM = ZM$	4. __?__
5. $\angle Y = \angle Z$	5. __?__
6. $\triangle YOM \cong \triangle ZUM$	6. __?__

Draw, on your paper, two angles and a segment roughly like those shown. Use your drawings for Exercises 7–9 on page 124.

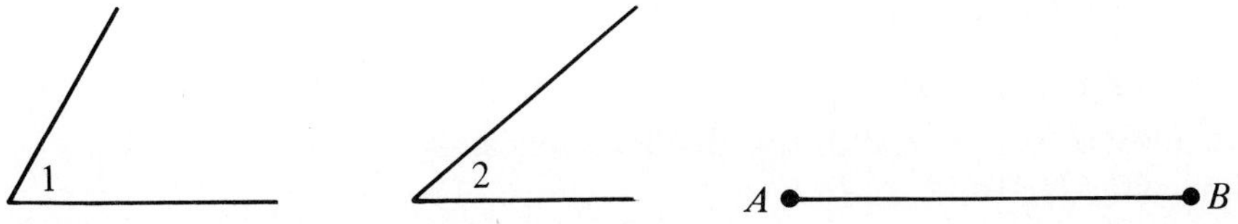

For Exercises 7–9, use your drawings from page 123.

B **7.** Construct a segment $\overline{RS}$ so that $RS = AB$.
Using $\overline{RS}$ as one side, construct $\angle R = \angle 1$.
Using $\overline{SR}$ as one side, construct $\angle S = \angle 2$.
Use T to label the point where two rays intersect.

8. Construct a segment $\overline{EZ}$ so that $EZ = AB$.
Using $\overline{EZ}$ as one side, construct $\angle E = \angle 2$.
Using $\overline{ZE}$ as one side, construct $\angle Z = \angle 1$.
Use N to label the point where two rays intersect.

9. a. Exercises 7 and 8 show that $\triangle RTS \cong \triangle$ __?__.
b. What postulate supports your answer?

Copy what is shown. Write complete proofs in two-column form.

10. *Given:* $AC = AD$
$BC = BD$
Prove: $\triangle ACB \cong \triangle ADB$

11. *Given:* $CB = DB$
$\angle ABC = \angle ABD$
Prove: $\triangle ACB \cong \triangle ADB$

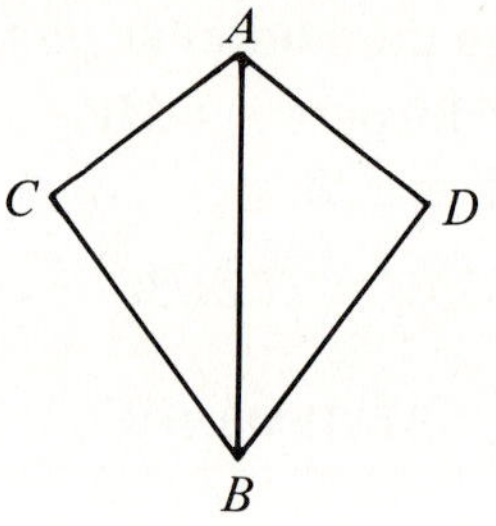

Exercises 10–13

C **12.** *Given:* $\overleftrightarrow{AB}$ bisects $\angle CAD$.
$\overleftrightarrow{AB}$ bisects $\angle CBD$.
Prove: $\triangle ACB \cong \triangle ADB$

13. *Given:* $AC = AD$
$\overrightarrow{AB}$ bisects $\angle CAD$.
Prove: $\triangle ACB \cong \triangle ADB$

14. *Given:* $\overline{RT} \parallel \overline{VS}$
$RT = VS$
Prove: $\triangle RMT \cong \triangle SMV$

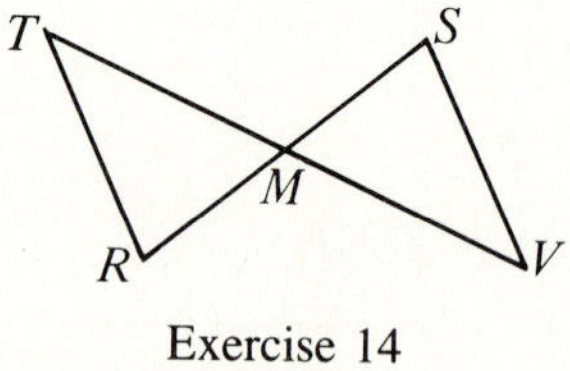

Exercise 14

15. a. Draw a large right triangle, $\triangle ABC$, with $\angle B = 90°$.
b. Use Construction 2, page 26, to construct a $90°$ angle. Label it $\angle K$.
c. Construct $KL = BC$.
d. Construct $\angle L = \angle C$. Label the third vertex as J.
e. Explain why $\triangle ABC \cong \triangle JKL$.

SELF-TEST

Write proofs in two-column form.

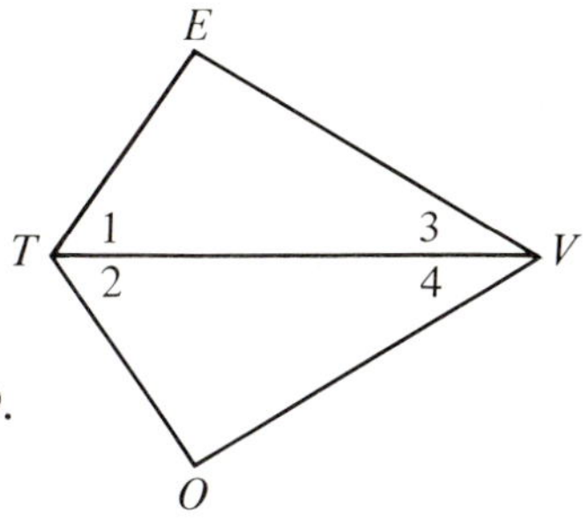

1. *Given:* $\overrightarrow{TV}$ bisects $\angle ETO$.
   $TE = TO$

   Prove: $\triangle TEV \cong \triangle TOV$ ***(3-5)***

2. *Given:* $\overleftrightarrow{TV}$ bisects both $\angle ETO$ and $\angle EVO$.

   Prove: $\triangle TEV \cong \triangle TOV$ ***(3-6)***

CAREER NOTES

REALTOR

MARSTON, 8 rm. older home in excel. cond., 2 full baths, porch, walking distance to shopping and bus, $108,900. Call 555-6789.

Do you ever read the newspaper advertisements for houses? Chances are that this ad was placed by a realtor.

Realtors examine the property to help determine the value of the real estate. They collect information about tax rates, insurance, and financing. A realtor often has to calculate percentages when working with any of these figures.

If the property is residential, realtors find the locations of schools, public transportation, and nearby shopping areas. For commercial property, the realtor may have to determine the price of office rental space. This is often based upon the number of square feet of floor space available. When his or her research is complete, the realtor uses advertisements to locate buyers or renters.

All realtors must obtain licenses by passing written tests. Then they may represent people who wish to sell or rent their property. Realtors may obtain part or all of their income from a commission based upon a percentage of their dollar sales.

7 The AAS and HL Theorems

In the diagram below, two angles and a *non*-included side of $\triangle RST$ are equal to the corresponding parts of $\triangle XYZ$. Can we prove that the triangles are congruent?

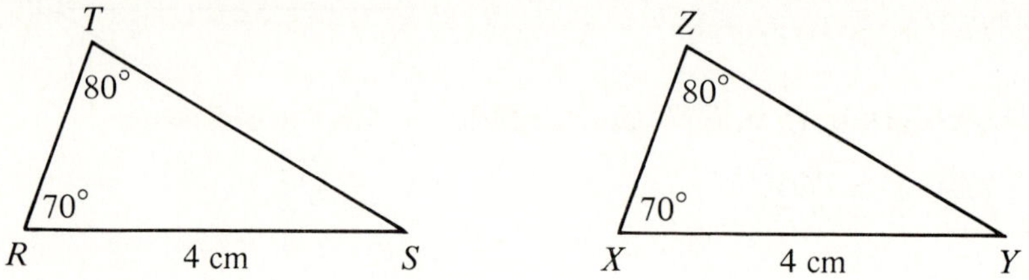

If we use the fact that $\angle R + \angle S + \angle T = 180°$, we can show that $\angle S = 30°$. We can likewise show that $\angle Y = 30°$. Then $\angle S = \angle Y$, and $\triangle RST \cong \triangle XYZ$ by the ASA Postulate.

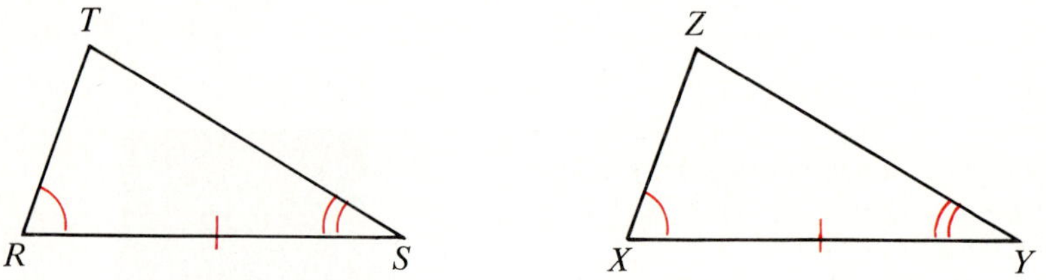

The discussion above suggests our next theorem. You will be asked to prove this theorem in Exercise 30, page 131.

THEOREM 2 (AAS Theorem)

If two angles and a non-included side of one triangle are equal to the corresponding parts of another triangle, the triangles are congruent.

A theorem about right triangles is stated here without proof.

THEOREM 3 (HL Theorem)

If the hypotenuse and a leg of one right triangle are equal to the corresponding parts of another right triangle, the triangles are congruent.

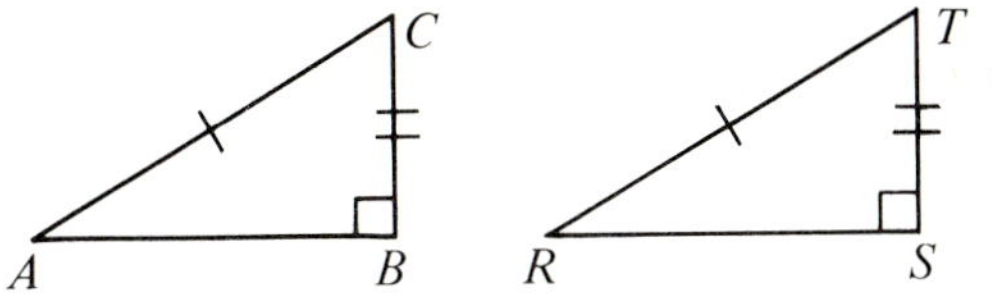

Theorem 3 tells us that $\triangle ABC \cong \triangle RST$.

WAYS TO PROVE TRIANGLES CONGRUENT				
SSS	*SAS*	*ASA*	*AAS*	*HL*

Classroom Practice

What is the measure of the angle marked in color?

SAMPLE

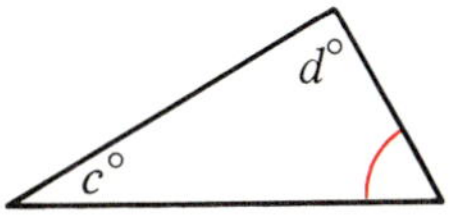

Answer: $180° - c° - d°$
Also correct: $180° - (c° + d°)$

1.

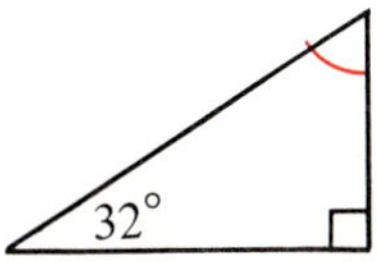

2.

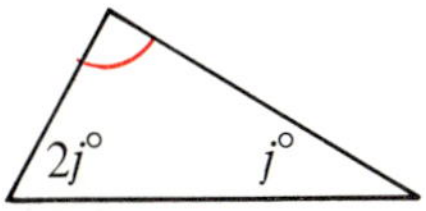

3.

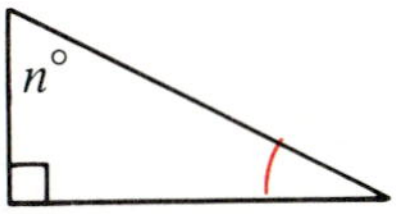

In Exercises 4–7, identify all right triangles shown in each figure. Name the hypotenuse of each right triangle.

4.

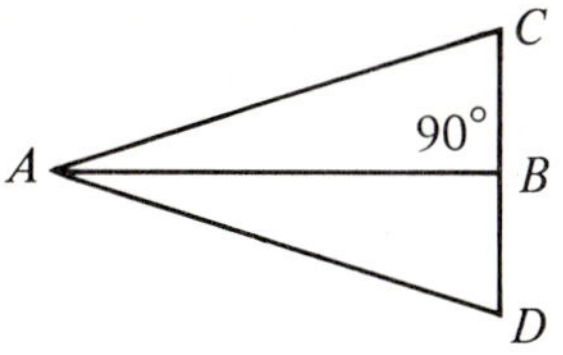

5.

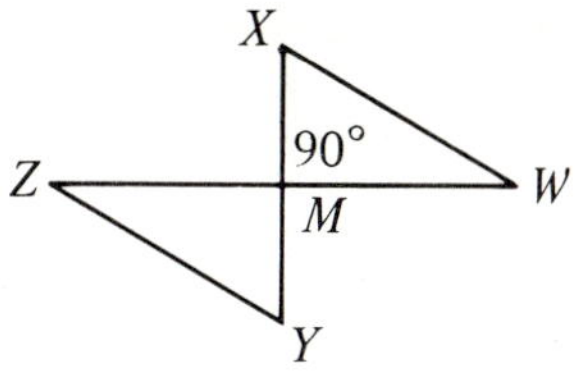

6.

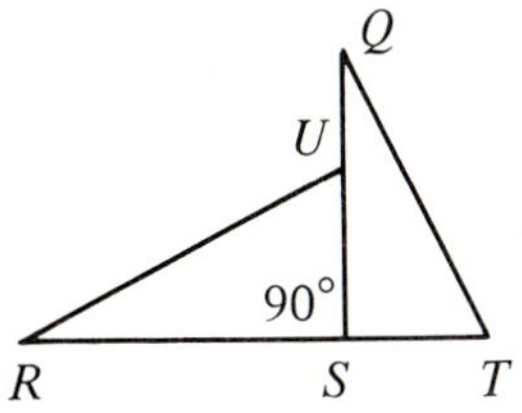

7.

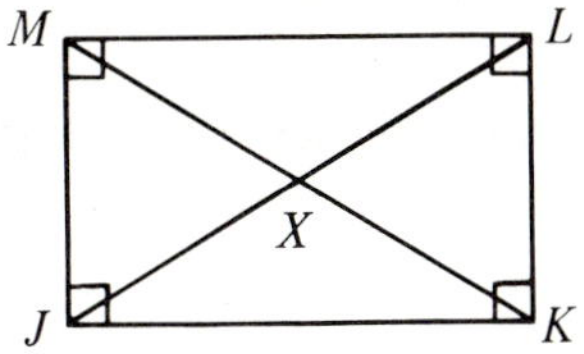

In Exercises 8–17, state which postulate or theorem you would use to prove the two triangles congruent. If no method applies, say so.

8.

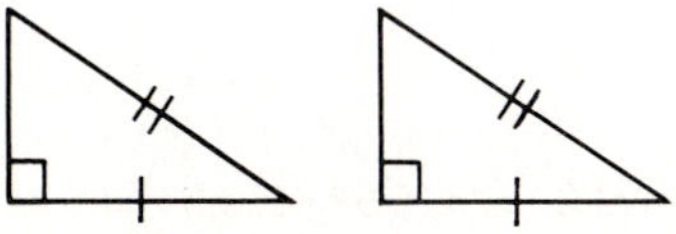

9.

State the postulate or theorem you would use to prove the two triangles congruent. If no method applies, say so.

10.
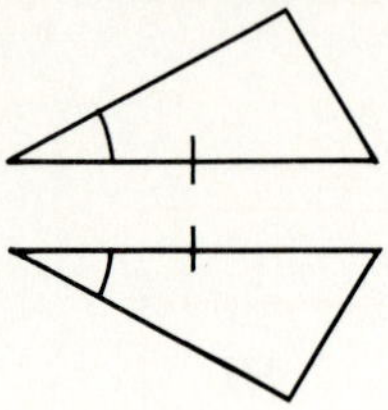

11.
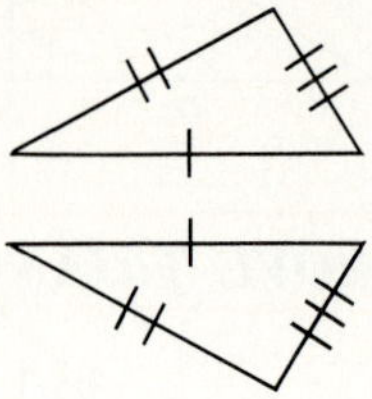

12.
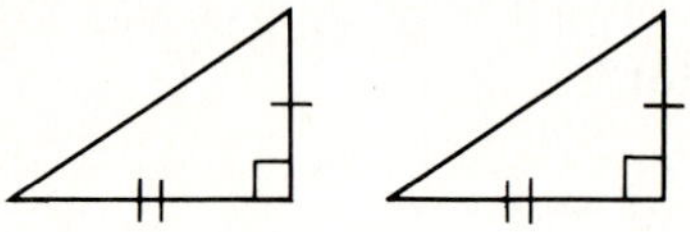

13.
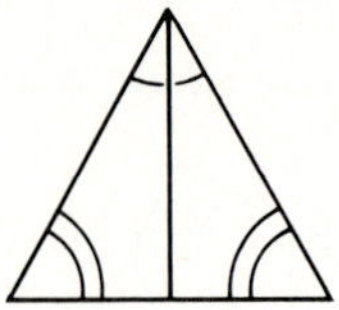

14.
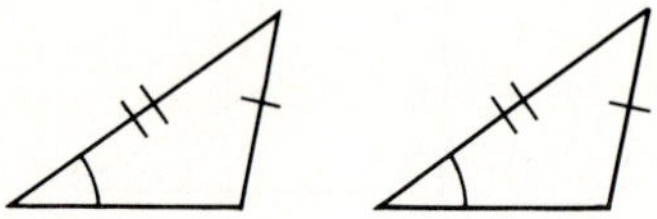

15.
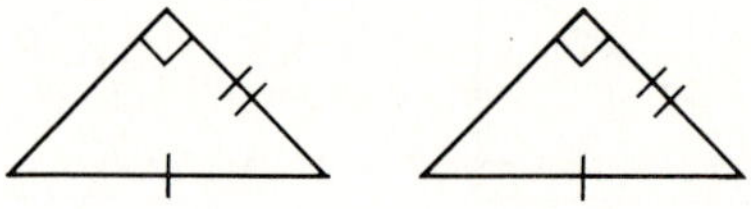

16.
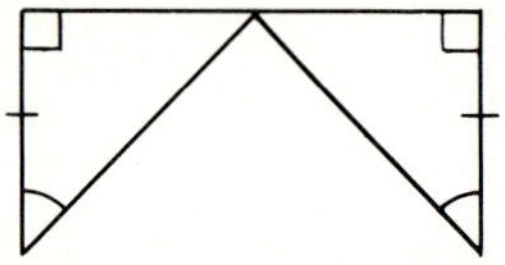

17.
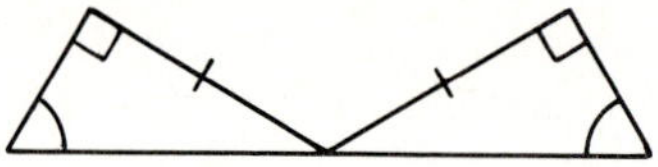

Written Exercises

Write SSS, SAS, ASA, AAS, or HL to state the method needed to prove the two triangles congruent. If no method applies, write *none*.

A

1.
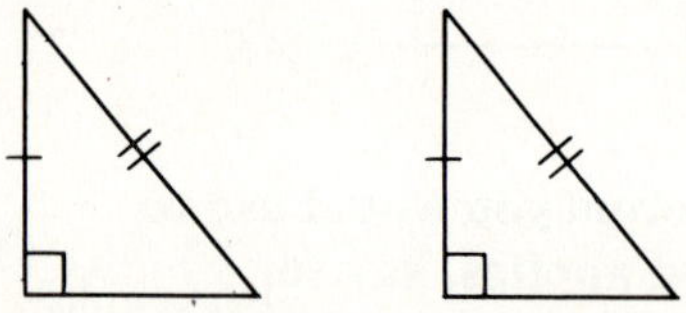

2.
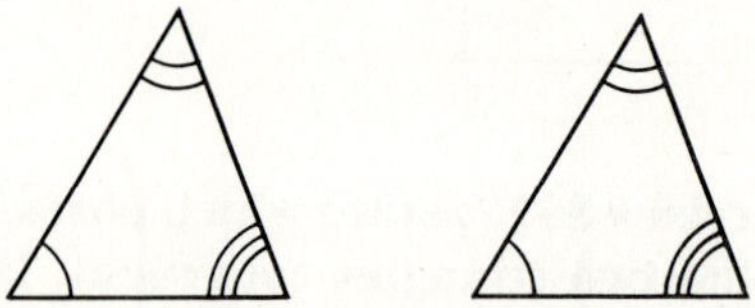

3.
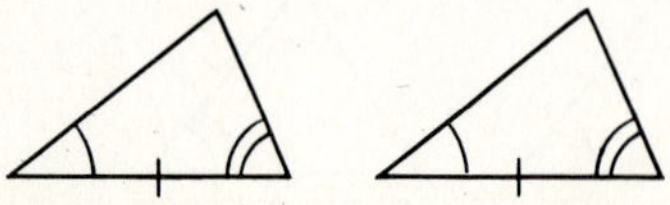

4.

5. 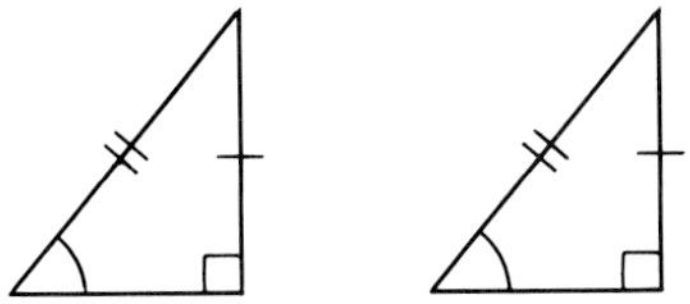

6. 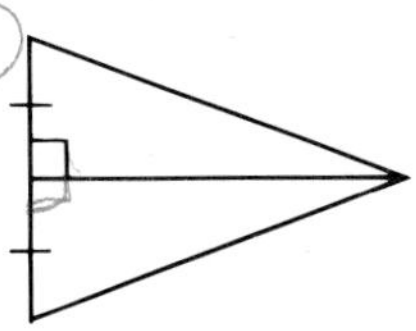

7. 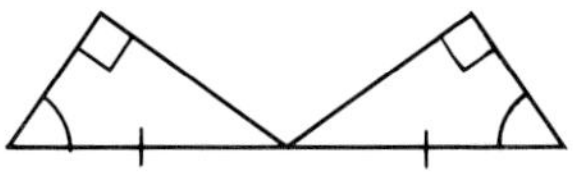

8. 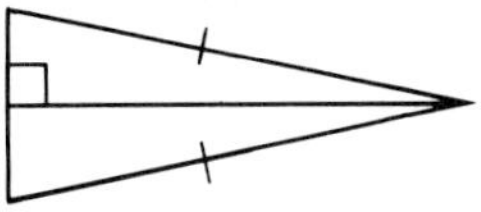

9. 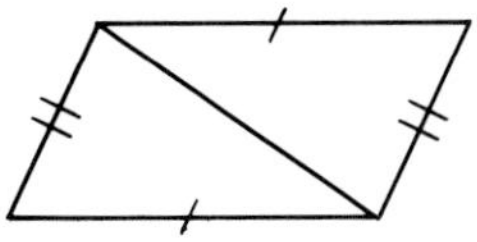

10. 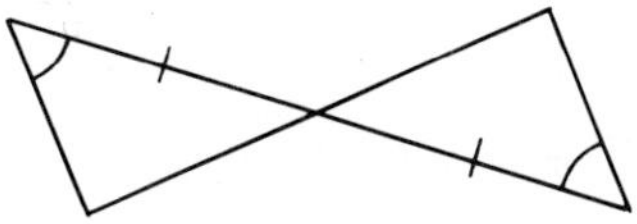

11. 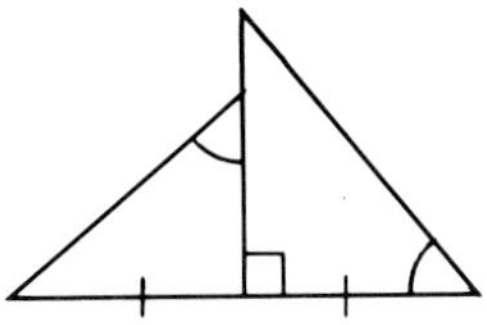

12. 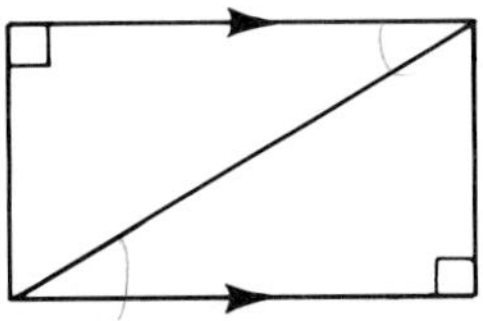

In Exercises 13 and 14, list *all* possible methods that could be used to prove the two triangles congruent.

13. 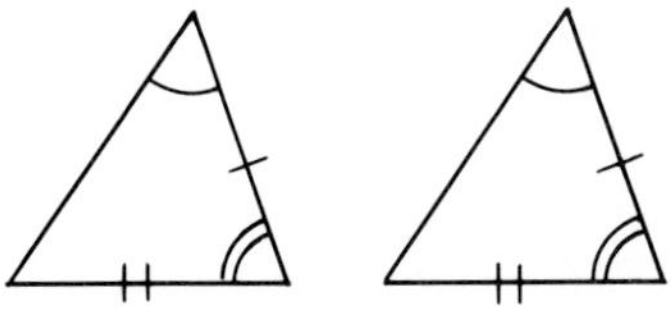

14. 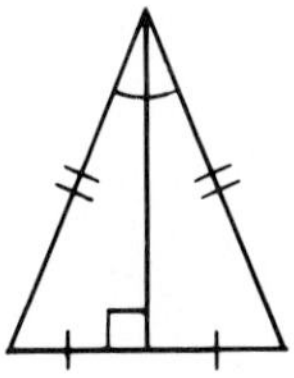

In Exercises 15–17, supply the reasons to complete the proofs.

15. *Given:* $\angle B$ and $\angle X$ are right angles.
$BY = AX$

Prove: $\triangle ABY \cong \triangle YXA$

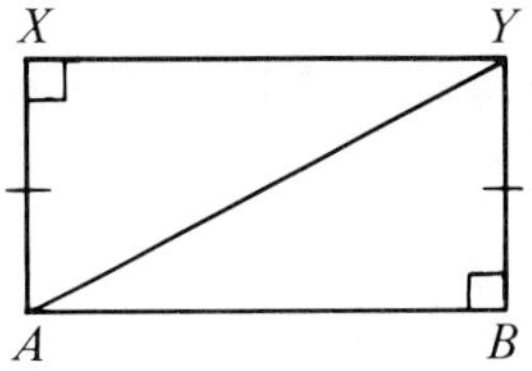

STATEMENTS	REASONS
1. $\angle B$ and $\angle X$ are right angles.	1. ?
2. $AY = AY$	2. ?
3. $BY = AX$	3. ?
4. $\triangle ABY \cong \triangle YXA$	4. ?

16. *Given:* $\overrightarrow{BD}$ bisects $\angle ABC$.
$\angle A = \angle C$

Prove: $\triangle ABD \cong \triangle CBD$

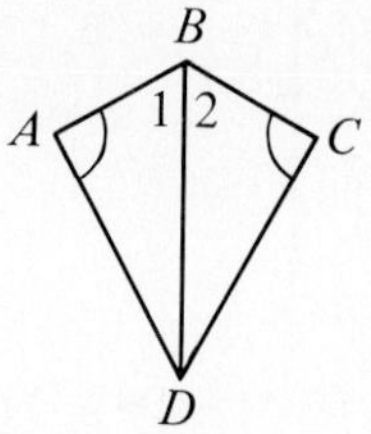

STATEMENTS	REASONS
1. $\angle 1 = \angle 2$	1. ___?___
2. $\angle A = \angle C$	2. ___?___
3. $BD = BD$	3. ___?___
4. $\triangle ABD \cong \triangle CBD$	4. ___?___

17. *Given:* $\angle 1 = \angle 2$
$\angle 5 = \angle 6$

Prove: $\triangle WXY \cong \triangle WXZ$

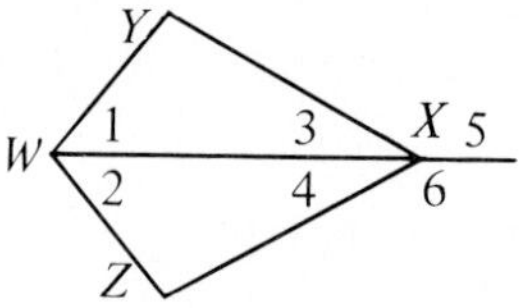

STATEMENTS	REASONS
1. $\angle 3$ and $\angle 5$ are supplements. $\angle 4$ and $\angle 6$ are supplements.	1. ___?___
2. $\angle 5 = \angle 6$	2. ___?___
3. $\angle 3 = \angle 4$	3. ___?___
4. $\angle 1 = \angle 2$	4. ___?___
5. $WX = WX$	5. ___?___
6. $\triangle WXY \cong \triangle WXZ$	6. ___?___

Each diagram shows two congruent triangles. Find the values of *x*, *y*, and *z*.

18.

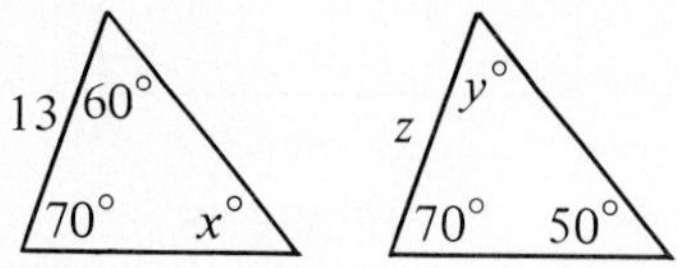

19.

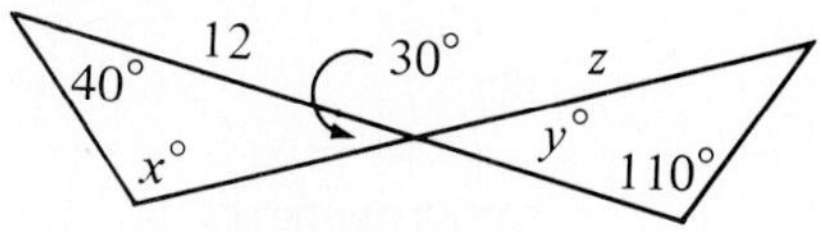

B **20.**

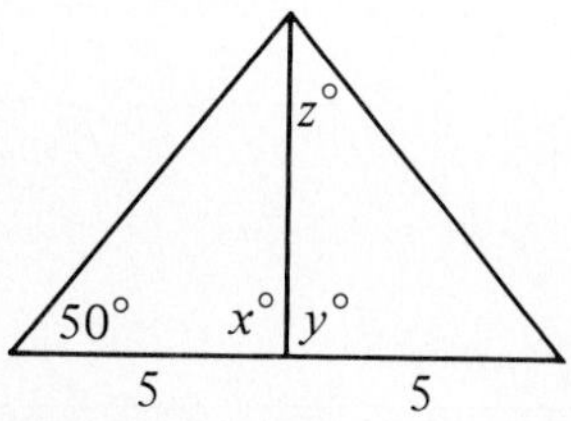

21.

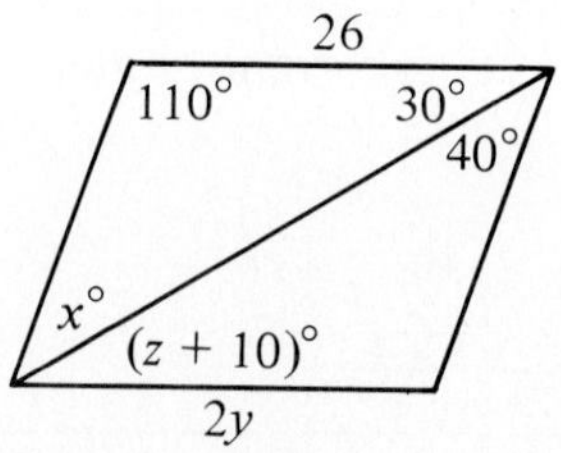

A geometry class is asked to write this proof:

Given: $\overline{AB} \parallel \overline{XY}$

M is the midpoint of $\overline{AY}$.

Prove: $\triangle AMB \cong \triangle YMX$

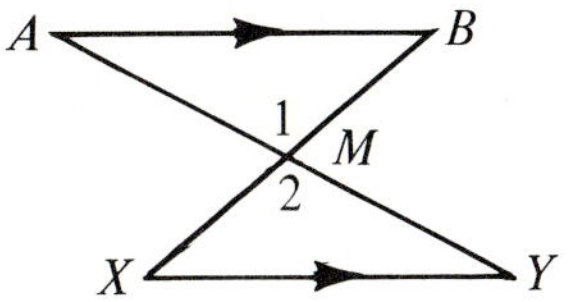

22. Jason uses the ASA Postulate. Name the three parts of $\triangle AMB$ that he uses in his proof.

23. Jasmine writes $\angle 1 = \angle 2$, and then decides to use the AAS Theorem. List the two parts of $\triangle YMX$, in addition to $\angle 2$, that she uses in her proof.

24. Miguel uses the AAS Theorem, but doesn't use $\angle 1$. List the three parts of $\triangle AMB$ that he uses.

25. Leora remarks: "If I knew that M was the midpoint of $\overline{BX}$, then I could use something other than the ASA Postulate or the AAS Theorem." What theorem or postulate could she use?

26. *Given:* M is the midpoint of $\overline{YZ}$.

M is the midpoint of $\overline{AB}$.

$\angle Y$ and $\angle Z$ are right angles.

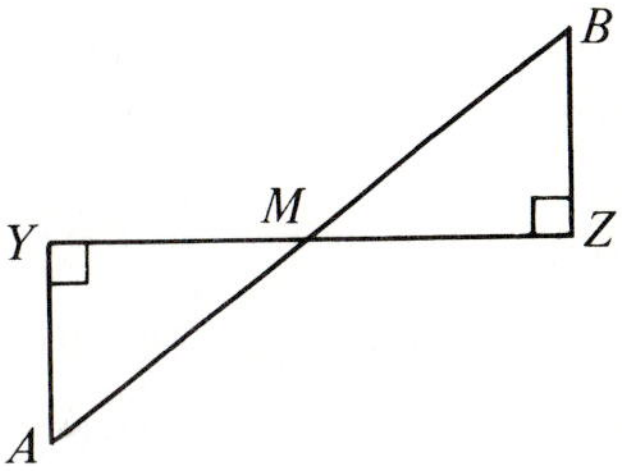

a. List four methods you could use to prove

$\triangle AYM \cong \triangle BZM$.

b. Choose one method and write a complete proof.

In each exercise, equal parts are indicated. Can the two triangles be proved congruent? If so, what postulate or theorem would be simplest to use?

27.

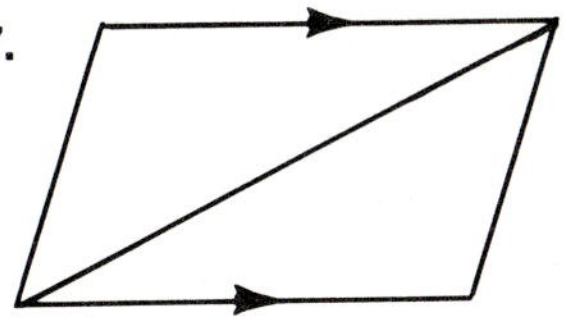

28.

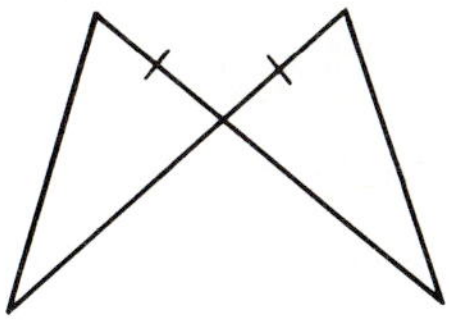

29.

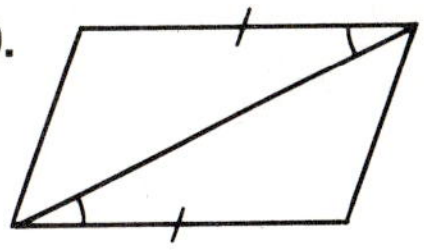

C **30.** Prove the AAS Theorem.

Given: $\angle R = \angle X$; $\angle T = \angle Z$

$RS = XY$

Prove: $\triangle RST \cong \triangle XYZ$

(*Hint:* Let $\angle R = j°$ and $\angle T = k°$. Show that $\angle S = \angle Y$.)

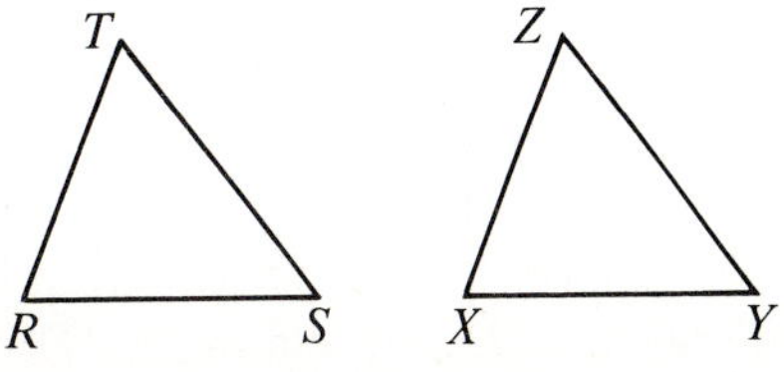

Experiments

Each student needs paper, scissors, a pencil, a ruler, and a compass.

1. Draw a segment 8 cm long.
2. Choosing either endpoint of the segment, construct a ray so that a 90° angle is formed.
3. Set your compass for a radius of 12 cm. Using the free endpoint of the original segment as center, draw an arc intersecting the ray.
4. Connect the two points to form a triangle.
5. Cut out your triangle and compare it with triangles that other students have made. If each student works carefully, all the triangles should be congruent.

SELF-TEST

1. *Given:* $\overline{AC} \parallel \overline{DB}$; $AX = BX$; $CX = DX$. (3-7)

Indicate, by abbreviation, three different postulates or theorems you could use to prove $\triangle AXC \cong \triangle BXD$.

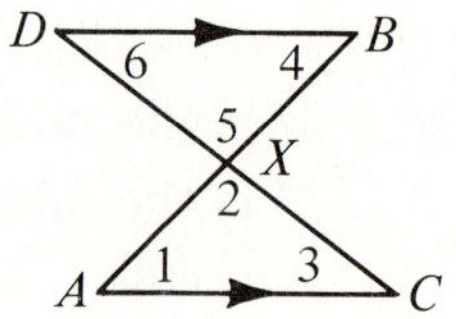

2. Write a proof in two-column form.

Given: $\overline{RT} \perp \overline{SV}$
$RS = TV$
$RM = TM$

Prove: $\triangle RMS \cong \triangle TMV$

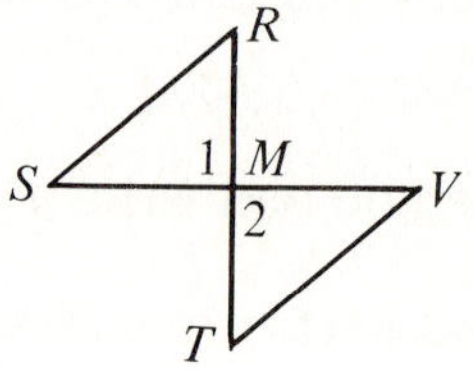

PUZZLE ◆ PROBLEMS

Sid and Gregory were hiking in the woods when they came to the Red Hawk River. Sid challenged Gregory to guess the width of the river. This is what Gregory did:

He stood at B and looked out across the river. He then adjusted his cap until the tip of his visor was in line with his eye and point X. Keeping his neck stiff, he turned and noted the point Z, on the ground, that was in line with his eye and the tip of the visor. By pacing, he found that the distance BZ was about 12 m. He claimed, "The river is about 12 meters wide." Explain why Gregory was right.

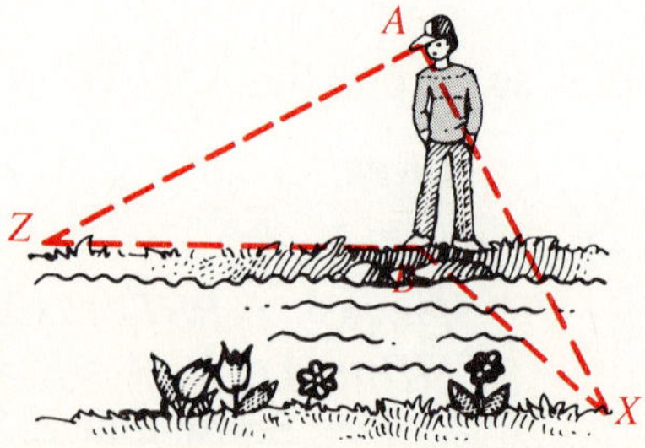

READING GEOMETRY

SYMBOLS

The symbols used in this book tell you a lot about geometry without using many words. You find symbols in the explanations, in theorems and proofs, in exercises, and even in diagrams. In order to get the message, you must pay just as much attention to the symbols as you do to the words of the text. See the list following the Table of Contents if you have forgotten the meaning of any symbol.

Look at the diagram at the right. The symbols $\overline{AB}$, $\overleftrightarrow{AB}$, $\overrightarrow{AB}$, $\overrightarrow{BA}$, and AB can all be used in referring to this simple diagram. Each symbol has a different meaning. Can you tell what each one means? (The answers are printed upside down at the bottom of the page.)

Symbols can be parts of diagrams. Examples are the right-angle mark ⌝, the little marks, often in color, that tell you lines or angles are equal, and the heavy arrowheads used to show that lines are parallel. Look for such symbols as you study the lessons.

EXERCISES

1. Do the symbols $\angle ABC$ and $\angle BAC$ stand for the same angle? Why or why not?
2. The triangles shown at the right are congruent. Is "$\triangle MNO \cong \triangle QRP$" a correct way of saying this in symbols? Why or why not?
3. How would you mark the diagram at the right to show that the triangles are congruent by the SAS Postulate?

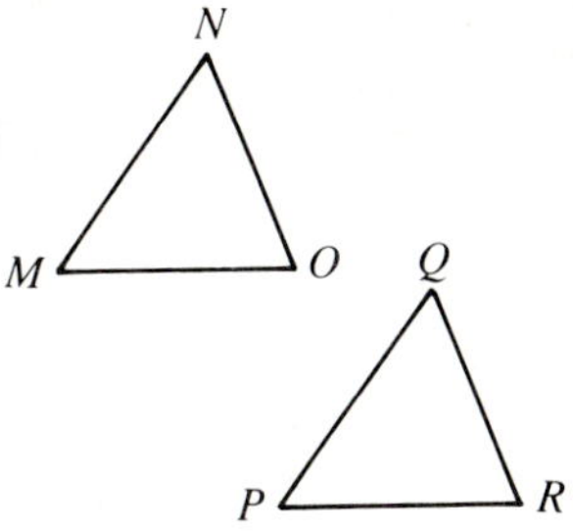

4. Using a different inequality symbol, write an expression with the same meaning as "$\angle FGH > \angle JKL$".
5. On page 72, Postulates 7 and 8 are stated in words and then written in a short form that uses diagrams and symbols. Do Exercises 7–9 on page 74 if you have not already done so.

In order as above: segment with endpoints A and B; line through points A and B; ray with endpoint A through point B; ray with endpoint B through point A; length of segment AB

COMPUTER ACTIVITIES

COMPLEMENTARY AND SUPPLEMENTARY ANGLES

In the exercises that follow, you must find the complement or supplement of many angles. A computer is an excellent tool for doing repeated or tiresome calculations. If you know what kinds of calculations are needed to solve a problem, you can write a computer program for it. The program is a plan of the procedure to be followed in solving the problem. It functions as a set of instructions for the computer. The programs in this book are written in BASIC, one of many computer programming languages.

You may already know some BASIC, or you may learn some in these computer activities. But you do not need to master BASIC or programming to be able to use a computer successfully. Every day millions of people who know little or nothing about computer languages or programming use computers at school, at home, or on the job.

The program below is a very simple one. You are asked to complete it in the Exercises. The purpose is to compute the measure of the complement of a given angle. Note that each line has a line number. This tells the computer the order in which its instructions are to be followed. The INPUT statement in line 40 allows the person using the computer to supply information. Statements beginning with REM are ignored by the computer. Their purpose is to explain the program itself.

```
10  REM***ANGLE PROBLEMS
20  PRINT "WHAT IS THE ANGLE"
30  PRINT "(TYPE 0 WHEN DONE)";
40  INPUT A
50  IF A = 0 THEN 200
60  REM***FIGURING COMPLEMENT
70  C = __?__
80  REM***CHECK FOR IMPOSSIBLE ANSWERS
90  REM***AND PRINT RESULTS
100  PRINT "ANGLE:   ";A
110  IF A > = 90 THEN 140
120  PRINT "COMPLEMENT:   ";C
130  GOTO 10
140  PRINT "NO COMPLEMENT EXISTS"
150  GOTO 10
200  END
```

EXERCISES

1. Line 70 is not complete. After the equals sign, what expression would you use to represent the complement of angle *A*? Use this expression to complete line 70 when you type it into the computer.

2. RUN the program to find the complement of each angle. All numbers must be entered into the computer as decimals. The computer will continue until you signal you've finished by giving 0, an impossible angle measure.

   a. 22° **b.** 46° **c.** 90° **d.** 78.9° **e.** $12\frac{1}{2}°$ **f.** 150°

3. Using trial and error (trying many values until you find the correct one), RUN the program to find an angle satisfying the given condition.
   a. The angle measures 20° more than its complement.
   b. The angle measures 15° less than its complement.
   c. The measure of the angle is twice the measure of its complement.
   d. The measure of the angle is the same as that of its complement.

If you want the program to figure the supplement also, you need to add to the program. Type in the following lines to expand the program. The computer will put these lines in order by line number, no matter what order you have typed them in.

```
74  REM***FIGURING SUPPLEMENT
78  S = __?__
105  IF A> =180 THEN 135
107  PRINT "SUPPLEMENT:  ";S
135  PRINT "NO SUPPLEMENT EXISTS"
```

4. Complete line 78 with an expression for the supplement of an angle whose measure is *A*.

5. RUN the expanded program. Find an angle satisfying the condition.
   a. The angle is equal to its supplement.
   b. The angle is triple its supplement.
   c. The supplement of the angle is triple the complement of the angle.
   d. The supplement of the angle is 20° more than twice the complement of the angle.

SKILLS REVIEW

WHOLE NUMBERS AND FRACTIONS

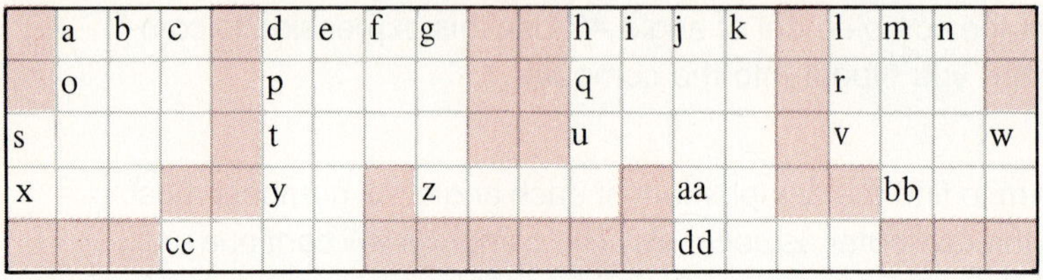

Copy the pattern above on squared paper. Then work the exercises and fill in the blanks just as you would in a crossword puzzle.

Across

a. $905 - 219$ **d.** 5×847 **h.** $6421 - 3804$ **l.** $2826 \div 9$

o. $(19)^2$ **p.** 96×74 **q.** $709 + 517$ **r.** $1172 \div 4$

s. 25×85 **t.** $995 + 7918$ **u.** $7309 - 719$ **v.** 21×301

x. $5992 \div 8$ **y.** 2^4 **z.** $2823 - 819$ **aa.** $2139 \div 69$

bb. 3×227 **cc.** $1833 + 2517$ **dd.** 200×47

Down

a. $5406 + 908$ **b.** $9303 - 674$ **c.** 15×41 **d.** $53{,}913 - 6098$

e. 72×305 **f.** $3913 \div 13$ **g.** 8×679 **h.** $968 + 1196$

i. $2500 \div 4$ **j.** $5347 + 7592$ **k.** 618×123 **l.** $49{,}552 \div 152$

m. $(44)^2$ **n.** $8001 - 3673$ **s.** 3^3 **w.** $528 \div 48$

Simplify.

1. $\frac{7}{8} + \frac{5}{8}$ **2.** $\frac{9}{16} - \frac{5}{16}$ **3.** $12 - \frac{5}{6}$ **4.** $\frac{7}{16} + \frac{1}{2}$

5. $\frac{11}{12} - \frac{3}{4}$ **6.** $8\frac{5}{9} - 2\frac{1}{3}$ **7.** $3\frac{2}{5} + 4\frac{3}{10}$ **8.** $4\frac{3}{8} - 2\frac{1}{2}$

9. $\frac{5}{6} \times \frac{2}{3}$ **10.** $8 \div \frac{2}{7}$ **11.** $\frac{3}{4} \div \frac{9}{10}$ **12.** $\frac{1}{3} \times \frac{4}{5} \times \frac{3}{8}$

13. $10\frac{2}{3} \div \frac{1}{3}$ **14.** $4 \times 2\frac{5}{16}$ **15.** $24 \div 1\frac{1}{5}$ **16.** $3\frac{3}{4} \times \frac{1}{10} \times \frac{2}{9}$

17. $6\frac{1}{4} - \frac{7}{8}$ **18.** $\frac{9}{16} \div 1\frac{1}{2}$ **19.** $3\frac{1}{7} \times 56$ **20.** $8\frac{1}{2} + \frac{9}{10}$

21. $\frac{5}{12} \div \frac{5}{8}$ **22.** $\frac{19}{20} + \frac{4}{5}$ **23.** $9\frac{5}{6} - 1\frac{2}{3}$ **24.** $\frac{1}{6} \times 1\frac{3}{5} \times \frac{3}{16}$

CHAPTER REVIEW

CHAPTER SUMMARY

1. a. In any triangle, the sum of the angles is 180°.
   b. In any triangle, an exterior angle is equal to the sum of the two opposite angles.

2. Corollaries, like theorems, are proved. A corollary follows directly from a theorem.

3. Triangles can be classified as scalene, isosceles, or equilateral according to their sides.

4. Triangles can be classified as acute, right, obtuse, or equiangular according to their angles.

5. $\triangle ABC \cong \triangle DEF$ means that each part of $\triangle ABC$ is equal to the corresponding part of $\triangle DEF$. Corresponding parts are named in the same order. That is, $\angle A = \angle D$, $\angle B = \angle E$, $\angle C = \angle F$, $AB = DE$, $BC = EF$, and $AC = DF$.

6. Postulates and theorems used to prove triangles congruent are abbreviated by: SSS SAS ASA AAS HL

REVIEW EXERCISES

Find the value of x in each diagram. ***(See pp. 92–96.)***

1.

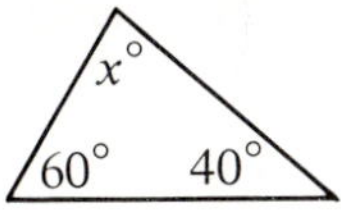

2.

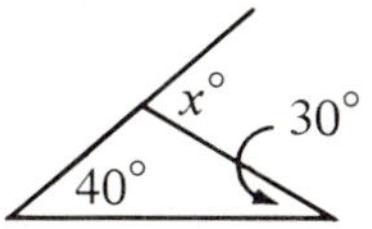

3.

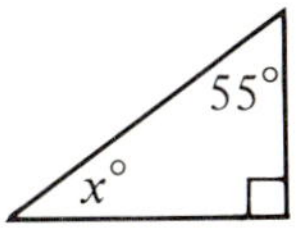

4.

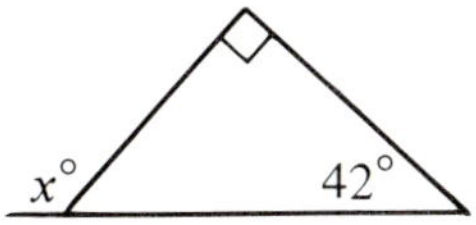

5.

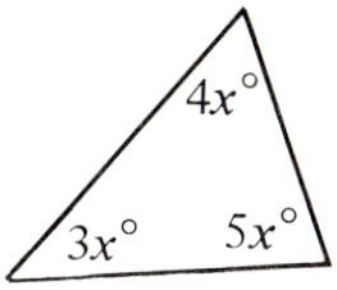

6.

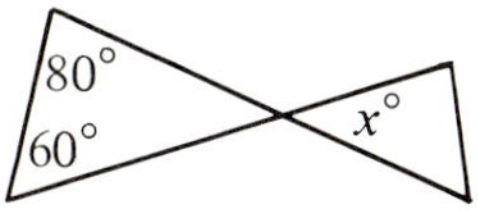

Complete. ***(See pp. 97–100.)***

7. If the perimeter of an equilateral triangle is 42 cm, the length of each side is __?__ cm.

8. The base of an isosceles triangle is 8 cm long, and a leg is 7 cm long. The perimeter of the triangle is __?__ cm.

9. An isosceles triangle cannot be ___?___.
scalene/equilateral

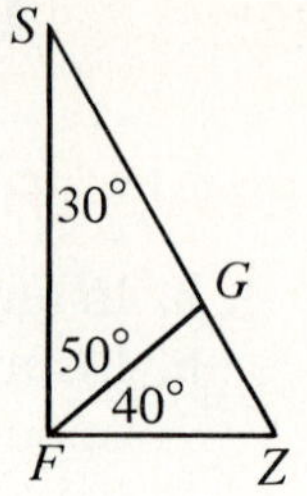

10. A right triangle found in the diagram is ?.

11. An acute triangle found in the diagram is ?.

12. An obtuse triangle found in the diagram is ?.

Complete. ***(See pp. 102–106.)***

If $\triangle CAP \cong \triangle DIR$, then:

13. $\angle P = \angle$?

14. $PC =$?

15. $\triangle PCA \cong \triangle$?

Write SSS, SAS, ASA, AAS, or HL to indicate the method you would use to prove the two triangles congruent. Do not write the proof. ***(See pp. 107–132.)***

16. *Given:* $AY = BY$ and $AX = BX$

Prove: $\triangle AXY \cong \triangle BXY$

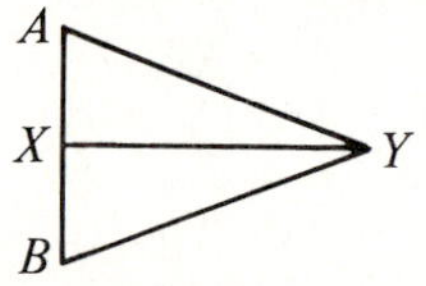

17. *Given:* $\angle C = \angle X$ and $CO = XO$

Prove: $\triangle BOC \cong \triangle YOX$

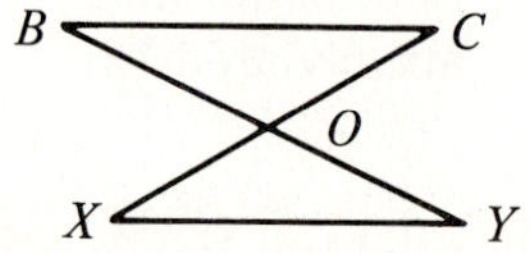

18. *Given:* $\overline{AB} \perp \overline{CD}$ and $AC = AD$

Prove: $\triangle ABC \cong \triangle ABD$

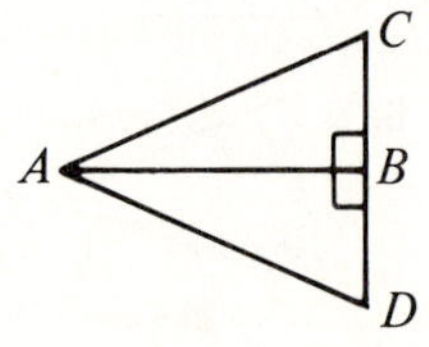

19. *Given:* $\angle S$ and $\angle V$ are rt. $\angle$s.
M is the midpoint of $\overline{SV}$.
$RS = TV$

Prove: $\triangle RSM \cong \triangle TVM$

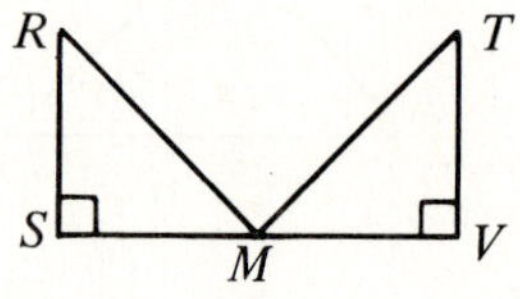

20. Write a complete proof in two-column form.

Given: $\angle R = \angle S$ and $\overrightarrow{GH}$ bisects $\angle RGS$.

Prove: $\triangle RHG \cong \triangle SHG$

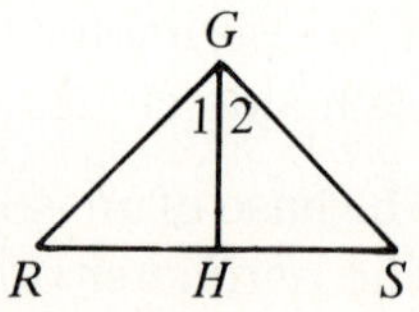

CHAPTER TEST

1. If $x + 3$, $2x - 6$, and $4x + 1$ are the measures of the angles of a triangle, find the numerical measure of the largest angle. *(3-1)*

2. Find the measures of $\angle 1$ and $\angle 2$.

3. Write the property that you used in Exercise 2 to find $\angle 1$.

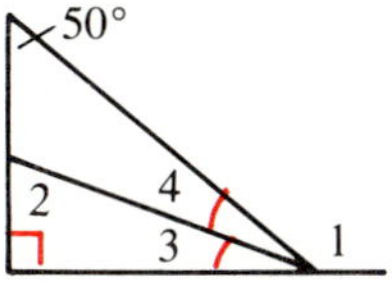

4. Name a scalene right triangle. *(3-2)*

5. Name an obtuse triangle.

6. Name the base of an isosceles triangle.

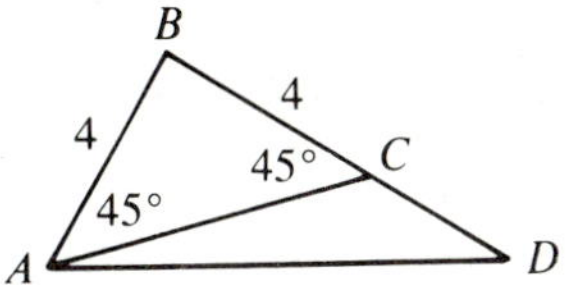

Given: $\triangle FLO \cong \triangle REW$

7. Write six statements about equal parts that must be true. *(3-3)*

8. Complete: $\triangle EWR \cong \triangle$ __?__.

9. Write the SSS Postulate. *(3-4)*

10. To prove that $\triangle WXY \cong \triangle WZY$ by the SSS Postulate, what given information is needed?

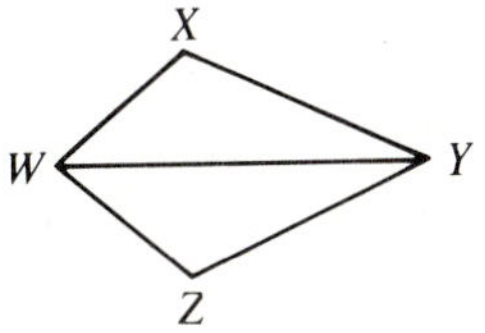

11. Name two pairs of equal sides. *(3-5)*

12. Name a pair of equal angles and explain why they must be equal.

13. Name two congruent triangles and tell why they must be congruent.

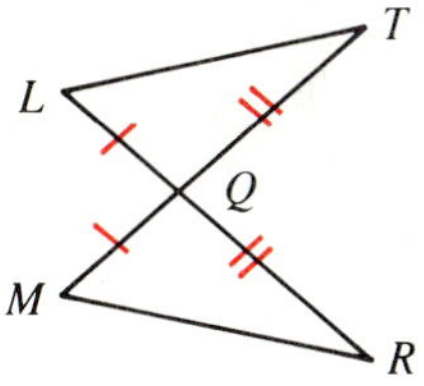

14. Since $\overline{EF} \parallel \overline{GH}$, $\angle$ __?__ $= \angle$ __?__ *(3-6)*

15. Since $\overline{FG} \parallel \overline{HE}$, $\angle$ __?__ $= \angle$ __?__

16. Write a reason to support your answers to Exercises 14 and 15.

17. By the __?__ Postulate, $\triangle EFG \cong \triangle$ __?__.

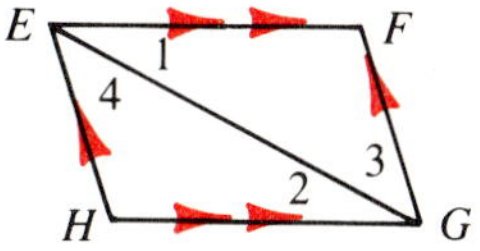

In Exercises 18–20, list *all* methods that could be used to prove the two triangles congruent. If no method exists, write "none."

18. 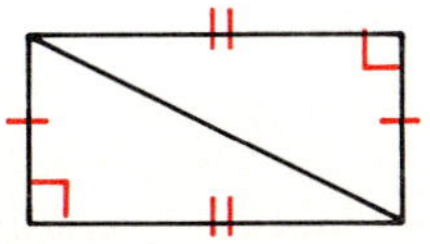 **19.** 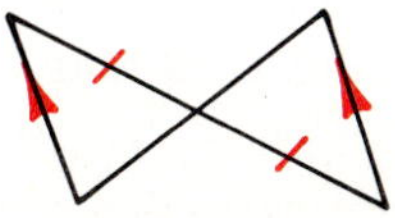 **20.** 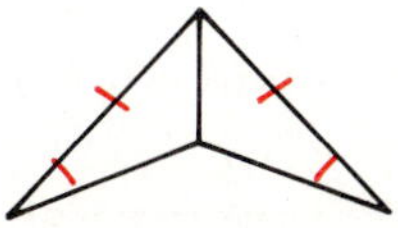 *(3-7)*

MIXED REVIEW

Consider the statement "Perpendicular lines form right angles."

1. Write the statement in "If . . . then" form.

2. Write the converse of the statement.

3. Supply the reasons to complete the proof.

Given: $AR = BS$; $\angle 1 = \angle 2$; $\angle A = \angle B$

Prove: $\triangle ACR \cong \triangle BCS$

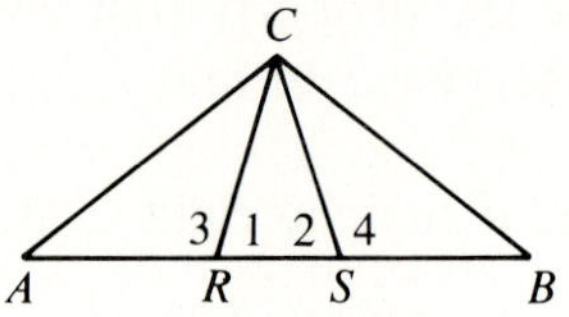

STATEMENTS	REASONS
1. $\angle 1 = \angle 2$	1. _?_
2. $\angle 3 = \angle 4$	2. _?_
3. $AR = BS$; $\angle A = \angle B$	3. _?_
4. $\triangle ACR \cong \triangle BCS$	4. _?_

4. If $\angle 1 = x°$, give the measure of a complement of $\angle 1$ in terms of x.

Suppose B is the midpoint of $\overline{AC}$ and C is the midpoint of $\overline{BD}$.

5. Explain why $AB = BC$ and $BC = CD$.

6. Which postulate justifies the statement $AB = CD$?

7. Which postulate justifies the statement $AC = BD$?

8. If $\overrightarrow{OJ}$ bisects $\angle AOB$, $\angle AOJ = (2x - 3)°$, and $\angle AOB = (3x + 10)°$, then $x =$ _?_ and the numerical measure of $\angle BOJ =$ _?_°.

9. If A has coordinate -5 and C has coordinate 1, find the coordinate of D.

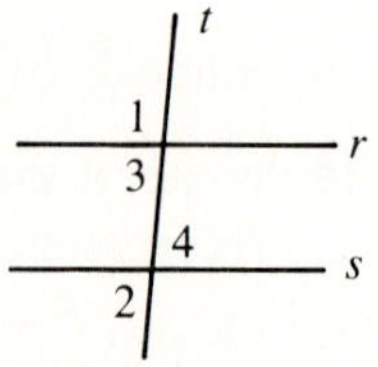

10. Does the given information tell you that $r \parallel s$?

a. $\angle 1 + \angle 2 = 180°$ **b.** $\angle 3 = \angle 4$
c. $\angle 1 = \angle 2$ **d.** $r \perp t$ and $s \perp t$

11. Is it possible for four points to have the stated relationships? Draw a diagram to support each answer.

a. The points are collinear. **b.** The points are not coplanar.
c. The points are coplanar but not collinear.

12. Draw a line l and a point A not on l. Construct a line m that is parallel to l and contains point A.

13. Refer to your construction in Exercise 12. Write the statement that supports the claim that $l \parallel m$.

Find the values of x and y.

14.

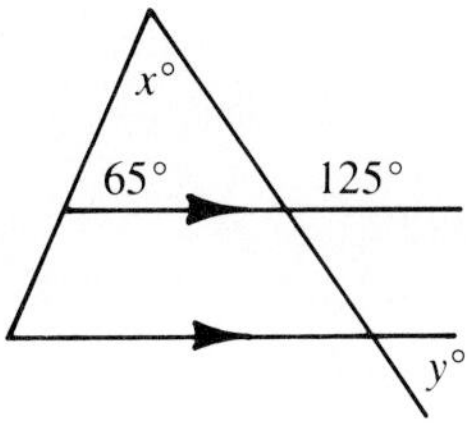

15.

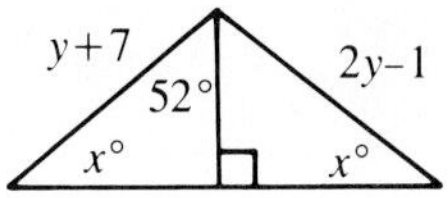

16.

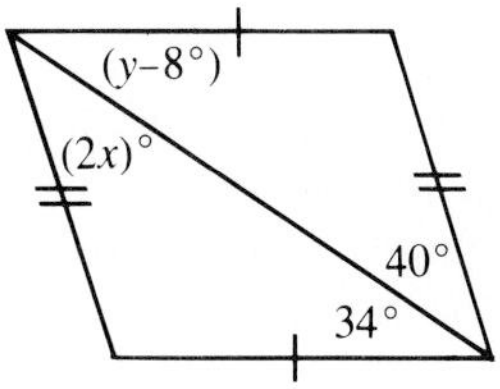

Suppose that $\triangle IJK \cong \triangle RST$. Which statements *must* be true?

17. $IK = TR$

18. $\triangle KJI \cong \triangle TSR$

19. $\angle K = \angle S$

Write SSS, SAS, ASA, AAS, or HL to indicate the method you would use to prove two triangles congruent.

20.

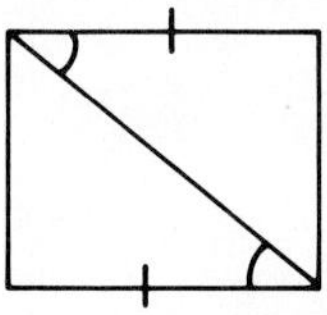

21.

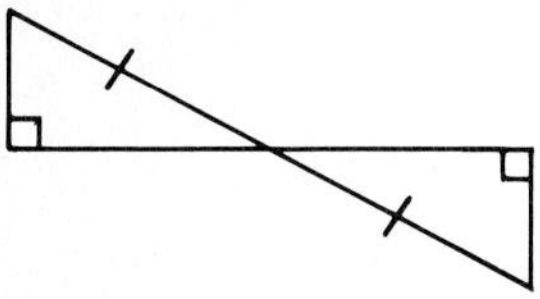

22.

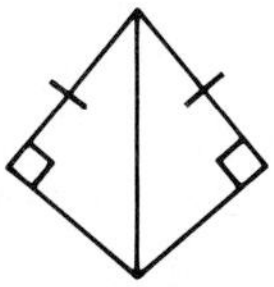

23. Draw a large $\triangle PQR$. Then use the SSS Postulate to construct a triangle congruent to $\triangle PQR$.

24. Construct a right angle and its bisector.

25. If two parallel lines are cut by a transversal and two same-side interior angles are equal, what is the measure of each?

26. If $\triangle ABC$ is an acute triangle, then each exterior angle of $\triangle ABC$ is a(n) __?__ angle.

27. True or false? If line j lies in plane P, line k lies in plane Q, and $j \parallel k$, then plane $P \parallel$ plane Q.

28. Write a two-column proof.

Given: X is the midpoint of $\overline{AC}$ and of $\overline{BD}$.

Prove: $\triangle AXB \cong \triangle CXD$

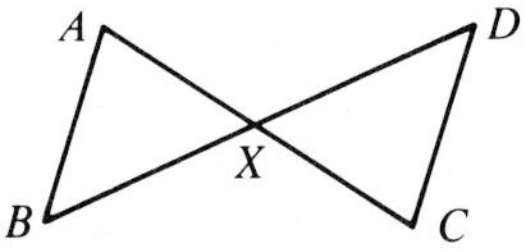

Here's what you'll learn in this chapter:

To use congruent triangles to prove that two segments or two angles are equal.

To use congruent triangles to justify constructions.

To name the altitudes and the medians of a triangle.

To inscribe a circle in a triangle and circumscribe a circle about a triangle.

To use the properties of isosceles triangles.

Patchwork is made from many small pieces of material sewed together to form patterns. Notice the congruent triangles in the patchwork quilt shown above.

Chapter 4

Using Congruent Triangles

1 Proving Corresponding Parts Equal

Suppose you are told that $\triangle ABC \cong \triangle RST$. Then you know that six things must be true. You should be able to complete the statements below:

$\angle A = \angle R$	$AB = RS$
$\angle B =$ ___?___	$BC =$ ___?___
$\angle C =$ ___?___	$AC =$ ___?___

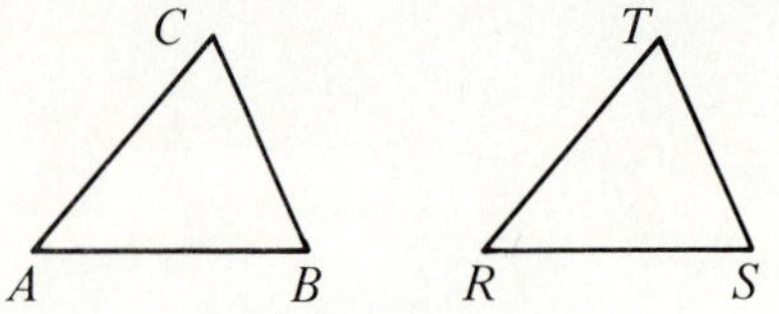

The six statements are true because of the definition of congruent triangles. *Corresponding parts of congruent triangles are equal.*

A STRATEGY FOR PROVING THAT TWO SEGMENTS OR TWO ANGLES ARE EQUAL

1. *Find two triangles in which the two sides or the two angles are corresponding parts.*
2. *Prove that the two triangles are congruent.*
3. *State that the two parts are equal, using as the reason, "Corr. parts of $\cong$ △ are $=$."*

This strategy will be used in the examples that follow.

EXAMPLE 1

Given: $JP = JQ$
$PK = QK$

Prove: $\angle P = \angle Q$

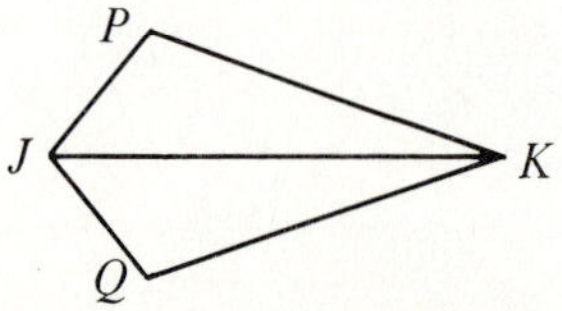

STATEMENTS	REASONS
1. $JP = JQ$	1. Given
2. $PK = QK$	2. Given
3. $JK = JK$	3. From algebra
4. $\triangle PJK \cong \triangle QJK$	4. SSS Postulate
5. $\angle P = \angle Q$	5. Corr. parts of $\cong$ △ are $=$.

EXAMPLE 2

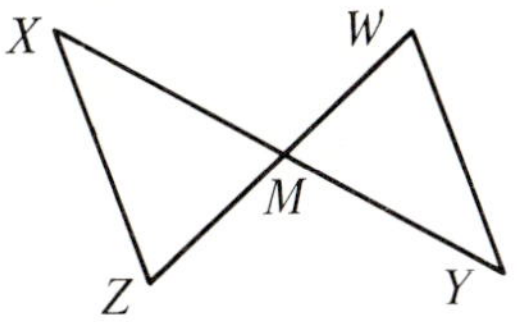

Given: M is the midpoint of $\overline{XY}$.
M is the midpoint of $\overline{ZW}$.

Prove: $XZ = YW$

STATEMENTS	REASONS
1. $XM = YM$	1. *Given:* M is the midpoint of $\overline{XY}$.
2. $ZM = WM$	2. *Given:* M is the midpoint of $\overline{ZW}$.
3. $\angle XMZ = \angle YMW$	3. ___?___
4. $\triangle XMZ \cong \triangle YMW$	4. SAS Postulate
5. $XZ = YW$	5. Corr. parts of $\cong$ $\triangle$s are $=$.

Classroom Practice

In Exercises 1–8, it is known that $\triangle ABC \cong \triangle JTN$.

1. $\angle A =$ ___?___ **2.** $\angle B =$ ___?___ **3.** $\angle C =$ ___?___

4. What reason supports each of statements 1, 2, and 3 above?

5. $AB =$ ___?___ **6.** $BC =$ ___?___ **7.** $AC =$ ___?___

8. What reason supports each of statements 5, 6, and 7 above?

Suppose, in Exercises 9–14, that you want to prove that $\angle 1 = \angle 2$. Does the figure suggest that congruent triangles might be used?

9.

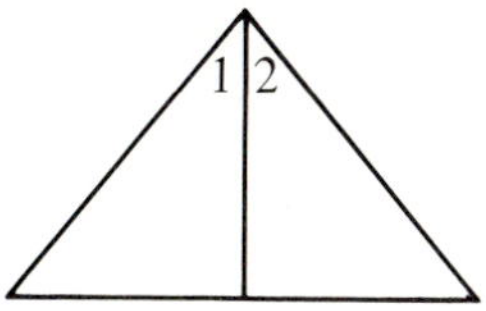

10.

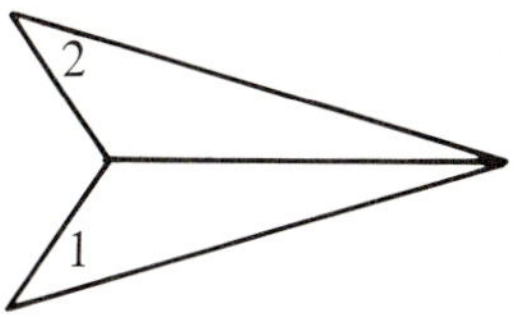

11.

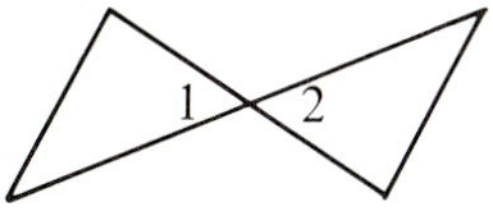

12.

13.

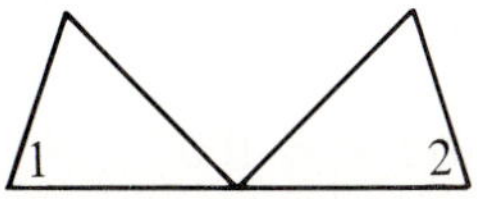

14.

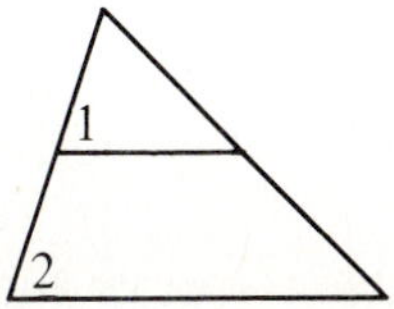

Written Exercises

A **1–6.** *Given:* $\triangle HIJ \cong \triangle UVW$
Write the six statements that can be supported by the reason:
Corr. parts of $\cong$ $\triangle$s are $=$.

Name the triangles you might try to prove congruent in order to prove each of the following.

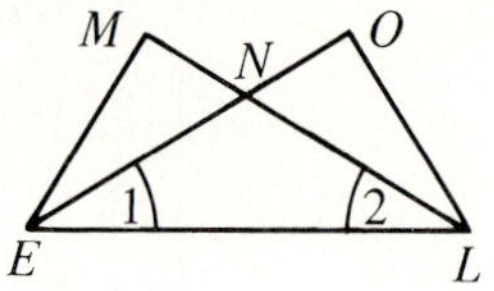

7. $NM = NO$

8. $EO = LM$

9. $\angle 1 = \angle 2$

10. $EN = LN$

In Exercises 11–14, supply the reasons to complete the proofs.

11. *Given:* $AC = BC$
$\overrightarrow{CD}$ bisects $\angle ACB$.

Prove: $AD = BD$

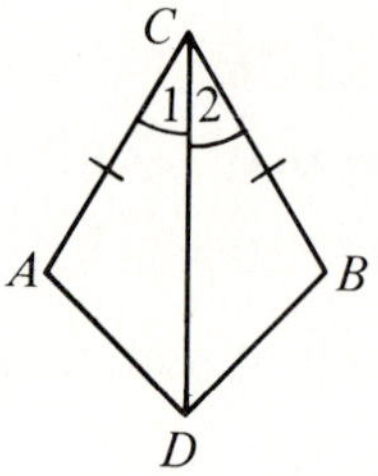

STATEMENTS	REASONS
1. $AC = BC$	1. ?
2. $\angle 1 = \angle 2$	2. ?
3. $CD = CD$	3. ?
4. $\triangle ACD \cong \triangle BCD$	4. ?
5. $AD = BD$	5. ?

12. *Given:* $\angle O$ and $\angle E$ are right angles.
$RO = ES$

Prove: $\overline{RO} \parallel \overline{ES}$

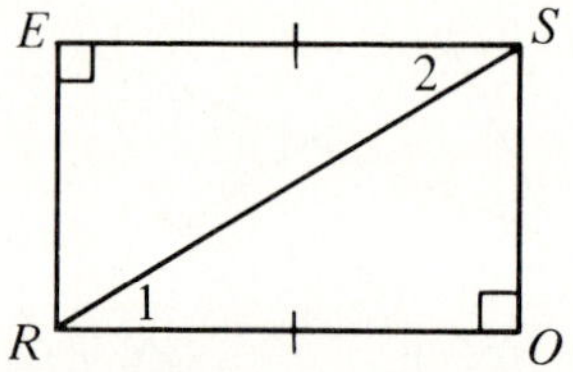

Strategy for proof:

A. Prove that two triangles are congruent.
B. Think of a transversal. Use corresponding parts to show that angles are equal.
C. Use the equal angles to prove that two lines are parallel.

STATEMENTS	REASONS
1. $\angle O$ and $\angle E$ are right angles.	1. ___?___
2. $RO = ES$	2. ___?___
3. $RS = RS$	3. ___?___
4. $\triangle ROS \cong \triangle SER$	4. ___?___
5. $\angle 1 = \angle 2$	5. ___?___
6. $\overline{RO} \parallel \overline{ES}$	6. ___?___

B **13.** *Given:* $\angle A = \angle B$
$\angle 1 = \angle 2$

Prove: $AX = BY$

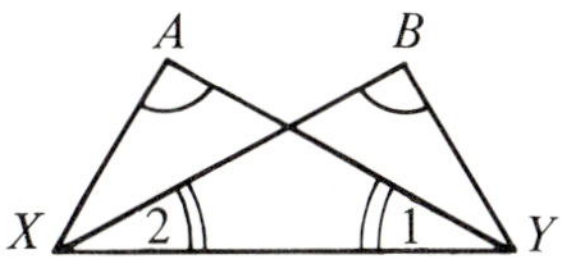

STATEMENTS	REASONS
1. $\angle A = \angle B$; $\angle 1 = \angle 2$	1. ___?___
2. $XY = XY$	2. ___?___
3. $\triangle AXY \cong \triangle BYX$	3. ___?___
4. $AX = BY$	4. ___?___

14. *Given:* $\overline{EB} \perp \overline{AC}$
$\angle E = \angle C$
$EB = BC$

Prove: $AB = BD$

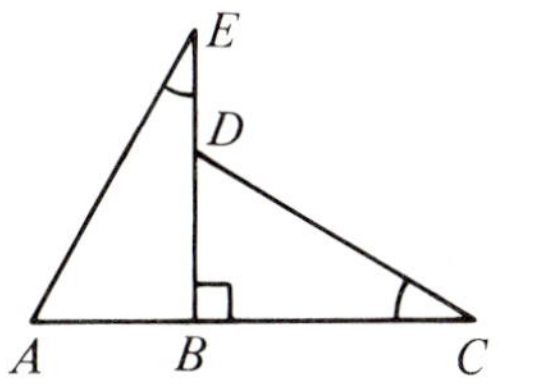

STATEMENTS	REASONS
1. $\angle EBA$ and $\angle DBC$ are right angles.	1. ___?___
2. $\angle EBA = \angle DBC$	2. ___?___
3. $\angle E = \angle C$; $EB = BC$	3. ___?___
4. $\triangle EBA \cong \triangle CBD$	4. ___?___
5. $AB = BD$	5. ___?___

Copy what is shown. Then write a proof in two-column form.

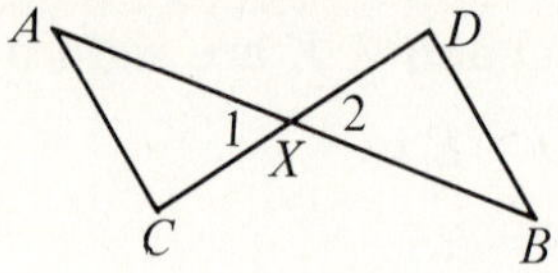

15. *Given:* $AX = BX$; $\angle A = \angle B$
Prove: $CX = DX$

16. *Given:* $AX = BX$; $\angle C = \angle D$
Prove: $AC = BD$

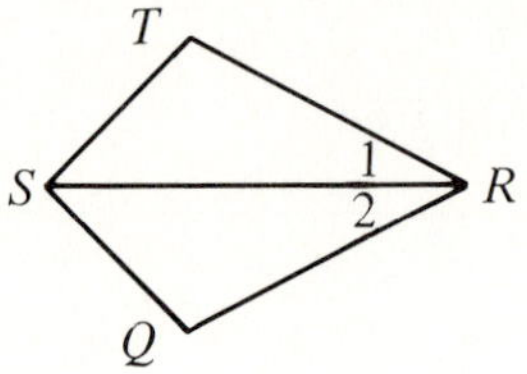

17. *Given:* $RT = RQ$; $ST = SQ$
Prove: $\angle T = \angle Q$

18. *Given:* $\angle 1 = \angle 2$; $RT = RQ$
Prove: $ST = SQ$

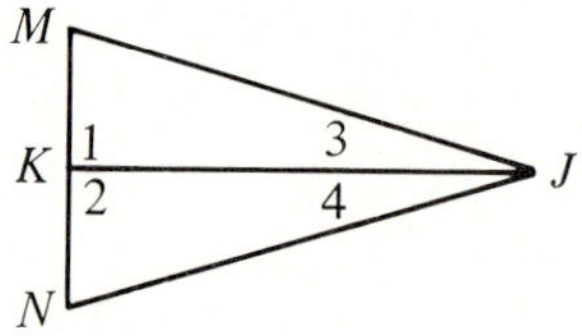

C 19. *Given:* $\overline{JK} \perp \overline{MN}$; $MJ = NJ$
Prove: $MK = NK$

20. *Given:* $\overline{JK} \perp \overline{MN}$; $\angle 3 = \angle 4$
Prove: $MJ = NJ$

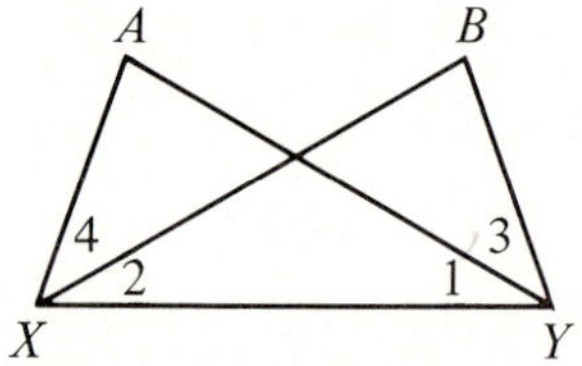

21. *Given:* $\angle 1 = \angle 2$; $AY = BX$
Prove: $\angle A = \angle B$

22. *Given:* $\angle 1 = \angle 2$; $\angle 3 = \angle 4$
Prove: $AX = BY$

CONSUMER APPLICATIONS

THE CONSUMER PRICE INDEX

The Consumer Price Index (CPI) is a number which keeps track of inflation. It takes into account the cost of food, clothing, housing, transportation, medical care, and other goods and services used by consumers. The CPI compares the *total price now* for a selection of typical consumer purchases to the *former price* during a base year.

If the CPI is 175, then the current total price of a selection of typical consumer purchases is 175% of the total price during the base year.

Consult a newspaper or your library to find out the current CPI. What year is used as the base year? Get a copy of your family's last grocery bill. How much would the bill have been during the base year?

2 Congruent Triangles and Constructions

In this section, we shall use congruent triangles to show that our construction methods are correct.

Bisecting an Angle (Construction 1, page 25)

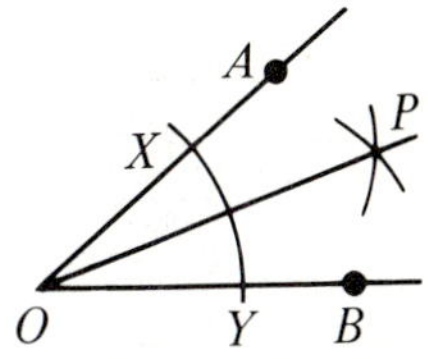

To show that $\angle AOB$ really is bisected, draw $\overline{XP}$ and $\overline{YP}$. Then:

1. $OX = OY$ — By construction
2. $XP = YP$ — By construction
3. $OP = OP$ — From algebra
4. $\triangle XOP \cong \triangle YOP$ — SSS Postulate
5. $\angle XOP = \angle YOP$ — Corr. parts of $\cong$ ⚠ are =.

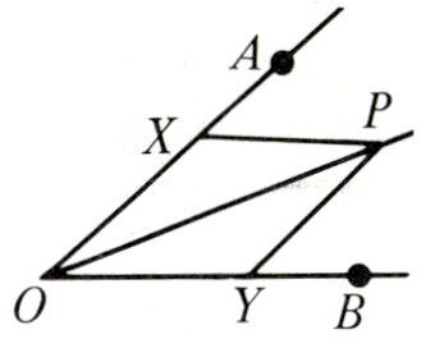

Therefore, $\angle AOB$ is bisected.

Constructing a Perpendicular at a Point on a Line
(Construction 2, page 26)

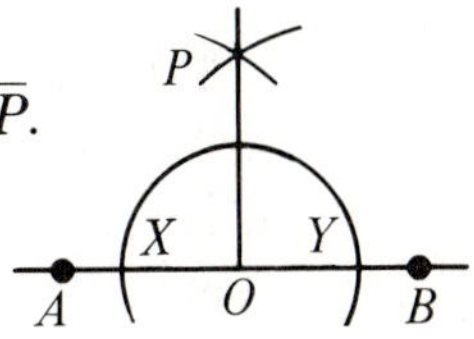

To show that $\overrightarrow{OP}$ really is perpendicular to $\overleftrightarrow{AB}$, draw $\overline{XP}$ and $\overline{YP}$.
Use the SSS Postulate to show that $\triangle XOP \cong \triangle YOP$.
Then you know that $\angle XOP = \angle YOP$.
You can show that $\angle XOP$ and $\angle YOP$ are right angles (see Written Exercise 8). Therefore, $\overrightarrow{OP} \perp \overleftrightarrow{AB}$.

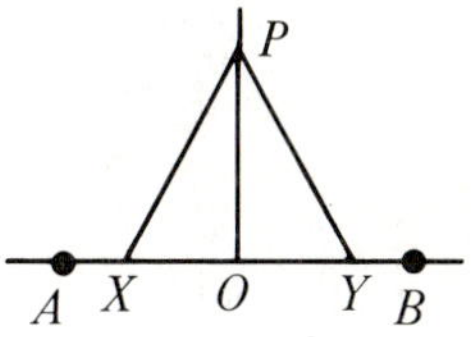

After a construction has been proved to be correct, the construction may be used in proofs. See, for example, Step 1 of Theorem 3 on page 167.

Classroom Practice

Elka started out, in the usual way, to bisect $\angle AOB$. But Elka likes to experiment. So she set her compass for a good-sized radius, used X and Y as centers, and drew arcs to get point P as shown. Finally she drew $\overline{OP}$ and claimed: $\overline{OP}$ bisects $\angle AOB$.

1. Was Elka's statement correct? Explain.

2. How could Elka add to her diagram, without using a compass, so that she would clearly show a bisector of $\angle AOB$?

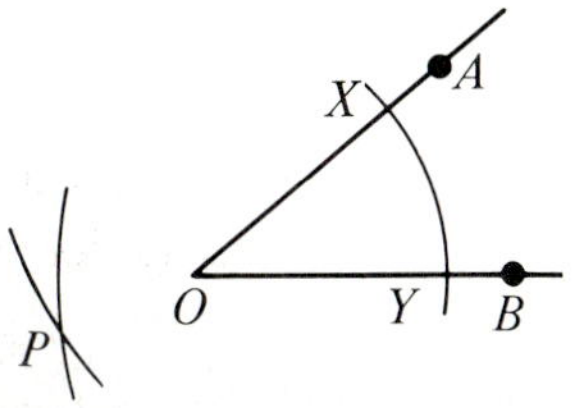

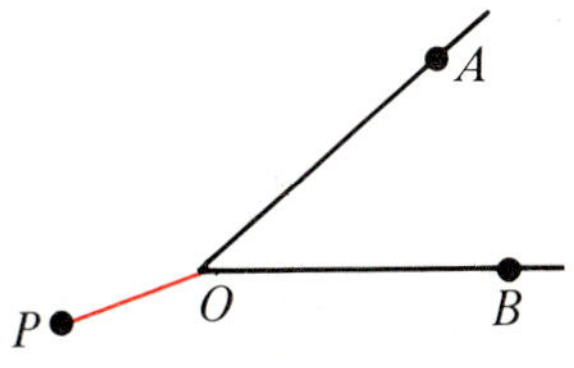

Written Exercises

On your paper, draw an acute $\triangle XYZ$ roughly like, but much larger than, the one shown. Use your triangle in Exercises 1–3.

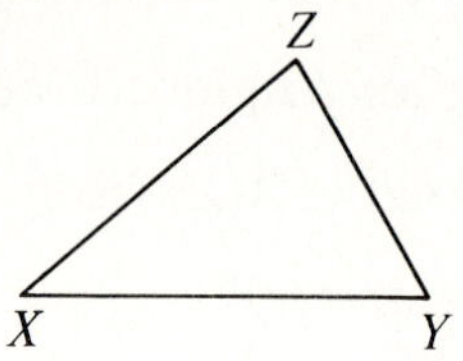

A **1. a.** Draw a line. On it choose a point D.

b. Set your compass for a radius equal to XY. Then with D as center, draw an arc that intersects the line at a point E.

c. At D, copy $\angle X$.

d. At E, copy $\angle Y$.

e. Use F to label the point at which the new sides of $\angle D$ and $\angle E$ intersect.

f. Explain why $\triangle DEF \cong \triangle XYZ$.

2. a. Draw a line. On it choose a point G.

b. Set your compass for a radius equal to XY. Then with G as center, draw an arc that intersects the line at a point H.

c. At G, copy $\angle X$.

d. Set your compass for a radius equal to XZ. On the new side of $\angle G$, draw an arc.

e. Use J to label the point at which the arc intersects the side of $\angle G$. Draw $\overline{JH}$.

f. Explain why $\triangle GHJ \cong \triangle XYZ$.

3. a. Draw a line. On it choose a point K.

b. Set your compass for a radius equal to XY. Then with K as center, draw an arc that intersects the line at a point M.

c. Set your compass for a radius equal to XZ. Use K as center and draw an arc.

d. Set your compass for a radius equal to YZ. Use M as center and draw an arc.

e. Use N to label the point at which the arcs intersect.

f. Draw $\overline{KN}$ and $\overline{MN}$.

g. Explain why $\triangle KMN \cong \triangle XYZ$.

4–6. On your paper, draw an obtuse $\triangle XYZ$ roughly like, but much larger than, the one shown. Using $\triangle XYZ$, repeat Exercises 1–3.

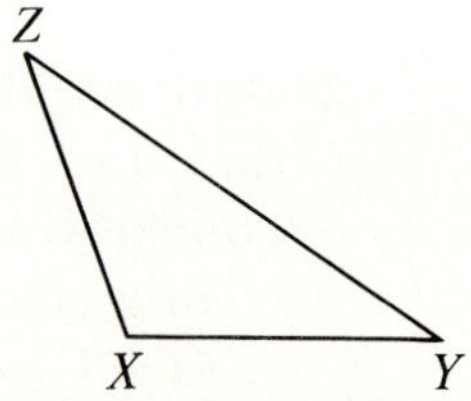

B

7. The diagram at the left, below, illustrates Construction 3 on page 26. When $\overline{RS}$, $\overline{RT}$, $\overline{US}$, and $\overline{UT}$ are drawn, we have the diagram at the right, below.

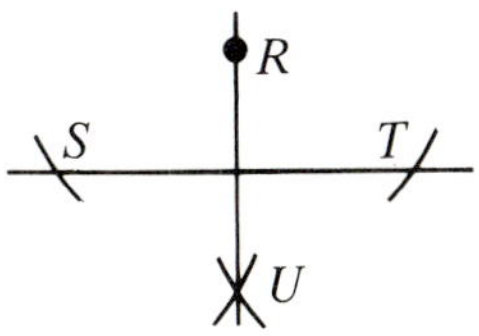

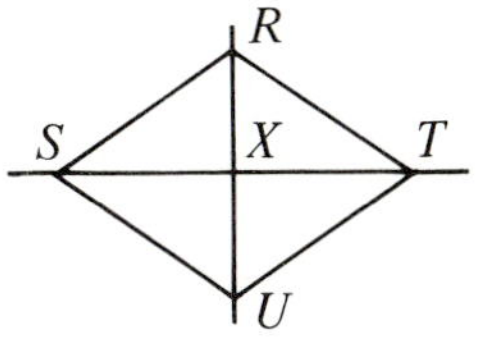

The steps below are the main ones in the proof that $\overleftrightarrow{RU} \perp \overleftrightarrow{ST}$. Complete each statement.

a. $\triangle RSU \cong \triangle RTU$ by __?__ (SSS/SAS/ASA/AAS).

b. Then $\angle SRU = \angle TRU$ because __?__.

c. Using $\triangle RXS$ and $\triangle RXT$:
$RS = RT$ by __?__ (algebra/construction);
$\angle SRX = \angle TRX$ because $\angle SRU = \angle TRU$ (see step **b**);
$RX = RX$ by __?__; $\triangle RXS \cong \triangle RXT$ by __?__.

d. $\angle RXS = \angle RXT$ because __?__.

e. Since $\angle RXS = \angle RXT$ and $\angle RXS + \angle RXT = 180°$, both $\angle RXS$ and $\angle RXT$ have measure __?__°.
Then $\overleftrightarrow{RU} \perp \overleftrightarrow{ST}$.

8. Refer to the construction at the right, which illustrates Construction 2 on page 26. Explain why $\overrightarrow{OP} \perp \overleftrightarrow{XY}$. (*Hint:* See step **e**, above.)

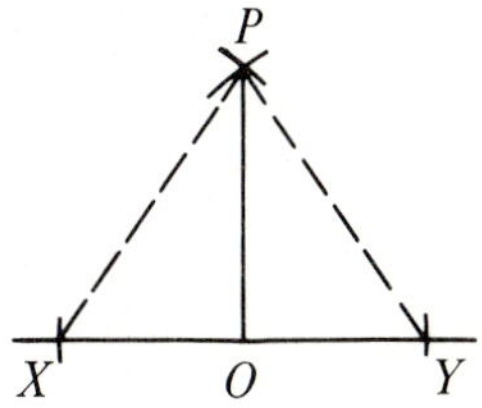

9. In the diagram at the right, $\angle A$ was given. $\angle O$ was constructed using Construction 4 on page 77. Explain why $\angle O = \angle A$.

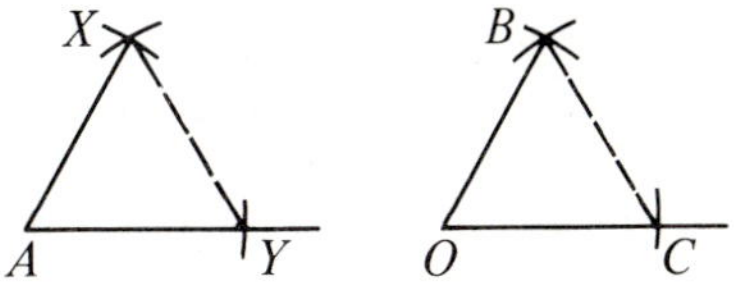

10. On your paper, draw a $\triangle DEZ$ with DZ clearly shorter than EZ. Construct the bisector of $\angle Z$. Use V to label the point where the bisector meets $\overline{DE}$. Is $\triangle DVZ \cong \triangle EVZ$? Explain.

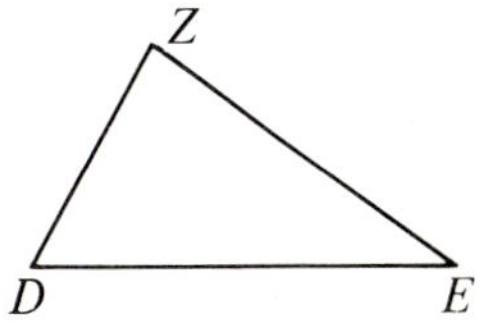

11. Draw a segment $\overline{DE}$. Using one setting of your compass, construct two arcs, one with center at D and the other with center at E. Z is the point where the arcs meet.

a. What kind of triangle is $\triangle DEZ$? Explain.

b. Construct the bisector of $\angle Z$. Use V to label the point where the bisector meets $\overline{DE}$.

c. Is $\triangle DVZ \cong \triangle EVZ$? Explain.

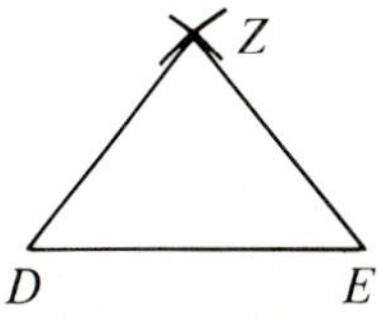

C **12.** Using a straightedge and lined paper, draw (you need not construct) a rectangle $ABCD$ that is clearly longer than it is wide. Construct the bisector of $\angle A$ and the bisector of $\angle B$. Use M to label the point where the bisectors intersect. Explain why $\triangle AMB$ is a right triangle. (*Hint:* Each angle of a rectangle is a right angle.)

13. As in Exercise 12, draw a rectangle $ABCD$ that is clearly longer than it is wide. Construct the bisector of $\angle A$ and the bisector of $\angle C$. Explain why the bisectors are parallel. (*Hint:* Label the angle measures in your diagram. If necessary, glance back at Section 6 of Chapter 2.)

14. Draw a quadrilateral $ABCD$ roughly like the one shown, but much larger. Draw $\overline{AC}$. Construct a $\triangle A'B'C'$ that is congruent to $\triangle ABC$. Building on $\overline{A'C'}$, construct a $\triangle A'C'D'$ that is congruent to $\triangle ACD$. Does each side and each angle of quadrilateral $A'B'C'D'$ equal the corresponding part of quadrilateral $ABCD$?

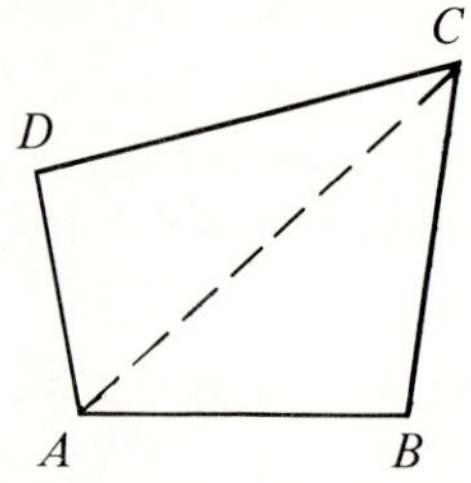

15. Draw a pentagon $ABCDE$. Draw diagonals $\overline{AC}$ and $\overline{AD}$. Construct a pentagon $A'B'C'D'E'$ that is congruent to pentagon $ABCDE$.

16. Draw a pentagon $RSTVW$. Choose any point P inside the pentagon. Draw $\overline{PR}$, $\overline{PS}$, $\overline{PT}$, $\overline{PV}$, and $\overline{PW}$. Construct a pentagon $R'S'T'V'W'$ that is congruent to pentagon $RSTVW$.

SELF-TEST

Write proofs in two-column form.

1. *Given:* $RM = TM$
$UM = SM$

Prove: $RU = TS$ **(4-1)**

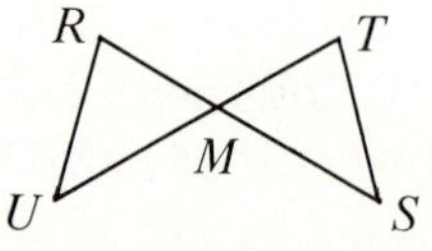

2. *Given:* $AX = AY$
$\angle X = \angle Y = 90°$

Prove: $\angle 1 = \angle 2$

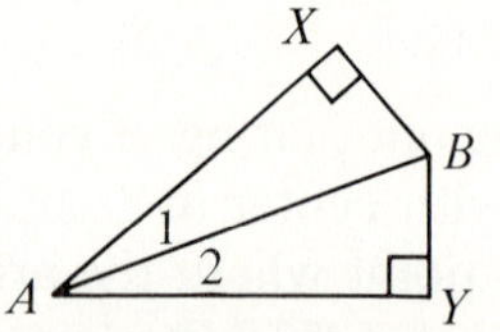

3. Suppose you want to prove that your construction of a perpendicular from P to m is correct. What postulate supports the statement: $\triangle TEP \cong \triangle TIP$? **(4-2)**

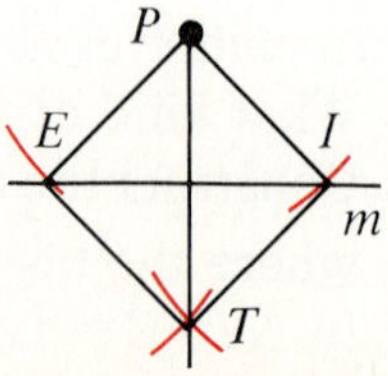

READING GEOMETRY

INDEPENDENT STUDY

When traveling, you usually have a destination in mind. Similarly, when studying geometry, you want to know where you are headed. At the beginning of every chapter in this book, you will find a list of goals to guide you through the chapter. Keeping these goals in mind, you can take a quick look through the chapter to get a preview of it before you begin to study the first lesson. Pay attention to the lesson titles and to statements that are highlighted by red frames or pointers.

As you study each lesson, pay special attention to new words and phrases, in heavy type, and to any constructions, postulates, or theorems that may be included. Read the explanations slowly and think as you read. If the lesson describes a construction, follow the directions and do the construction yourself. If the lesson includes a proof, follow it step by step. Study any worked-out examples in the same way. Try solving some examples before you look at the answers in the book.

After carefully reading a lesson, check your understanding by trying a few of the exercises. You may find that you need to reread some parts of the lesson. If there is anything that you still don't understand, make sure to discuss it with your teacher.

The Self-Tests generally cover several lessons. Before taking one of them, look back through the lessons that it covers and review important words and ideas. Answers to the Self-Tests are printed at the back of the book so that you can use them to check your progress. The Chapter Reviews and Cumulative Reviews will also help you.

EXERCISES

Look at pages 144–164 to find answers to the following questions.

1. Which lesson shows how to use congruent triangles to justify constructions?
2. What new words appear on page 154? What is a segment bisector?
3. Any point on the perpendicular bisector of a segment is __?__ from the endpoints of the segment.
4. Is a median of a triangle perpendicular to one of the sides?
5. A circle that fits exactly inside a triangle is a(n) __?__ circle.

3 Segment Bisectors

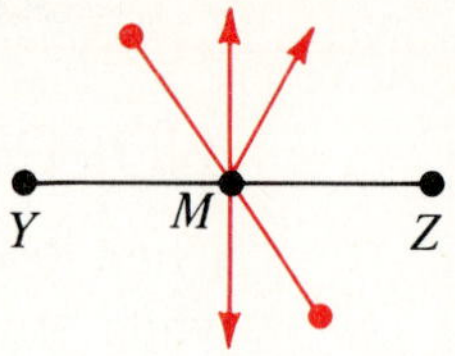

Suppose that M is the midpoint of $\overline{YZ}$. Any segment, ray, or line that passes through M *bisects* $\overline{YZ}$ and is a **bisector** of $\overline{YZ}$. The diagram suggests that a segment has an unlimited number of bisectors.

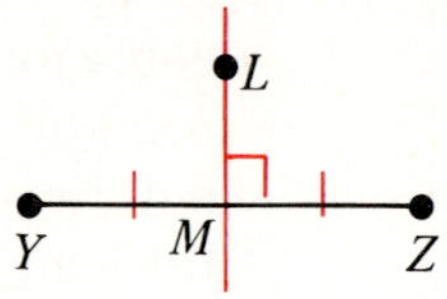

A bisector of a segment that is perpendicular to the segment is called a **perpendicular bisector** of the segment. As shown at the right, $YM = MZ$ and $\overline{LM} \perp \overline{YZ}$, so $\overline{LM}$ is a perpendicular bisector of $\overline{YZ}$.

CONSTRUCTION 6
Given: *A segment* Construct: *A perpendicular bisector of the segment*

Follow these steps:

1.

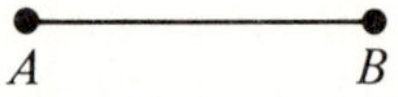

You are given $\overline{AB}$.

2.

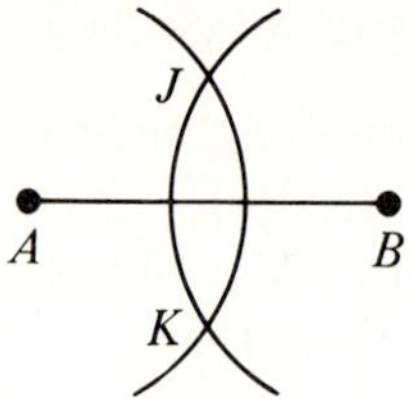

Set your compass for a convenient radius. Using A and B as centers, draw arcs that meet at J and K.

3.

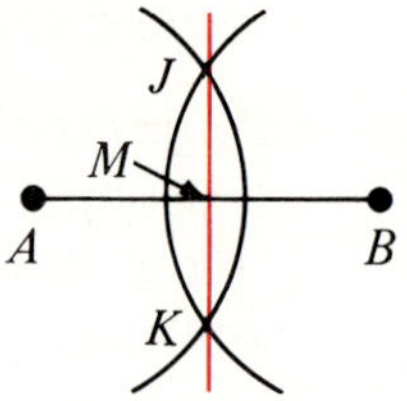

Draw $\overleftrightarrow{JK}$. $\overleftrightarrow{JK}$ is a perpendicular bisector of $\overline{AB}$. M is the midpoint of $\overline{AB}$.

A segment, $\overline{CD}$, and its perpendicular bisector are shown at the right. Is point R closer to point C or to point D? Neither one; $RC = RD$, and we say that R is *equidistant* from C and D. Point S is also equidistant from C and D because $SC = SD$.

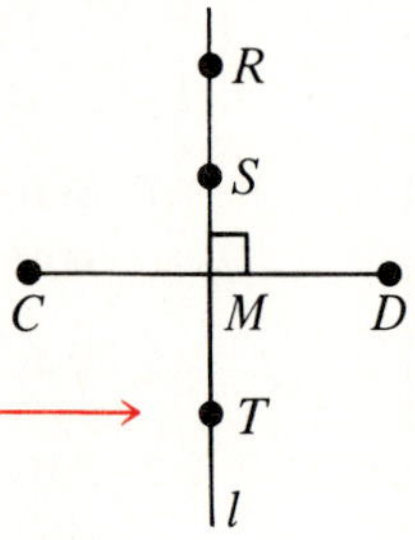

Any point on ℓ is equidistant from C and D.

THEOREM 1
Any point on the perpendicular bisector of a segment is equidistant from the endpoints of the segment.

Given: $\overleftrightarrow{RM}$ is the perpendicular bisector of $\overline{CD}$.

Prove: $RC = RD$

See if you can supply the reasons for the proof.

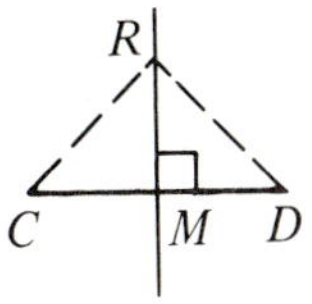

STATEMENTS	REASONS
1. $\overleftrightarrow{RM} \perp \overline{CD}$	1. ?
2. $\angle RMC$ and $\angle RMD$ are right angles.	2. ?
3. $\angle RMC = \angle RMD$	3. ?
4. $CM = DM$	4. ?
5. $RM = RM$	5. ?
6. $\triangle RMC \cong \triangle RMD$	6. ?
7. $RC = RD$	7. ?

Theorem 2 is the converse of Theorem 1.
The proof is left as Exercise 16 on page 157.

THEOREM 2

Any point that is equidistant from the endpoints of a segment is on the perpendicular bisector of the segment.

Classroom Practice

In Exercises 1–7, line l is a perpendicular bisector of $\overline{AB}$. Classify each statement as true or false.

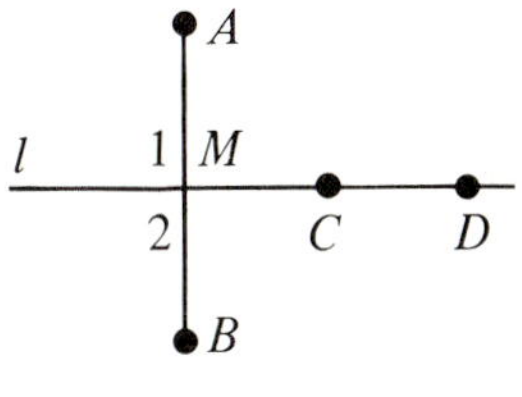

1. M is the midpoint of $\overline{AB}$.

2. $\angle 1$ is an acute angle.

3. $l \perp \overline{AB}$

4. $AD = BD$

5. $AB = CD$

6. $\angle 1 = \angle 2$

7. Any point on $\overline{CD}$ is equidistant from A and B.

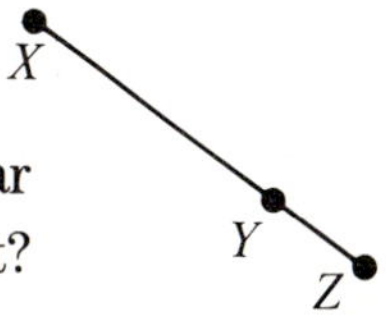

8. Think of the perpendicular bisector of $\overline{XY}$ and the perpendicular bisector of $\overline{YZ}$. Do the two perpendicular bisectors intersect? Explain.

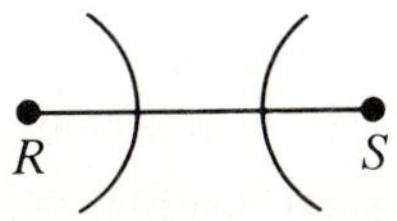

9. A student tried to construct a perpendicular bisector as shown. What went wrong?

Written Exercises

In Exercises 1–5, $JK = KL$. Complete each statement.

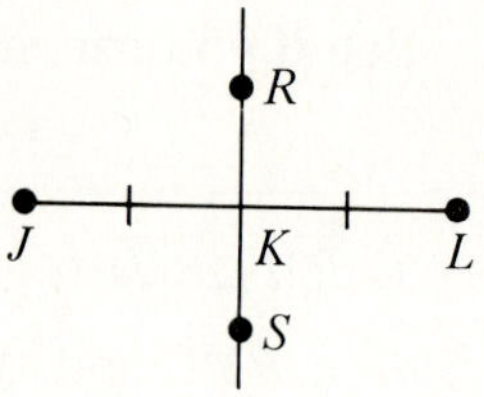

A 1. K is the _?_ of $\overline{JL}$.

2. Since $\overleftrightarrow{RS}$ passes through K, $\overleftrightarrow{RS}$ is a _?_ of $\overline{JL}$.

3. If $\overline{RK} \perp \overline{JL}$, then $\overline{RK}$ is a _?_ of $\overline{JL}$.

4. If $SJ = SL$, then S is _?_ from J and L.

5. Suppose $\overleftrightarrow{RS}$ is the perpendicular bisector of $\overline{JL}$. State the theorem that allows you to conclude that $RJ = RL$.

In Exercises 6–11, draw a figure roughly like the one shown. Then construct the indicated perpendicular.

6. The perpendicular bisector of $\overline{AB}$

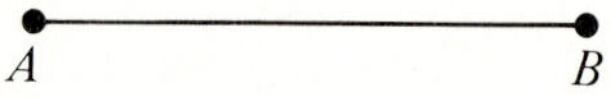

7. The perpendicular bisector of $\overline{XY}$

8. The perpendicular to $\overline{CD}$ at P

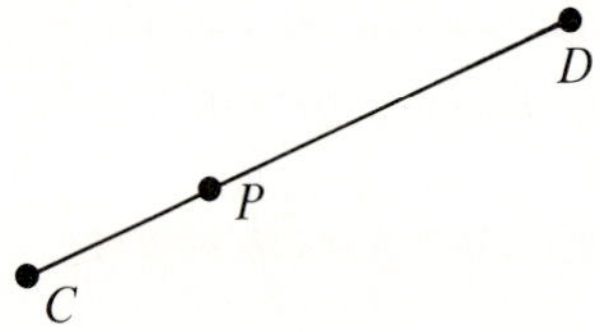

9. The perpendicular to $\overline{JK}$ through R

10. The perpendicular bisector of $\overline{PQ}$

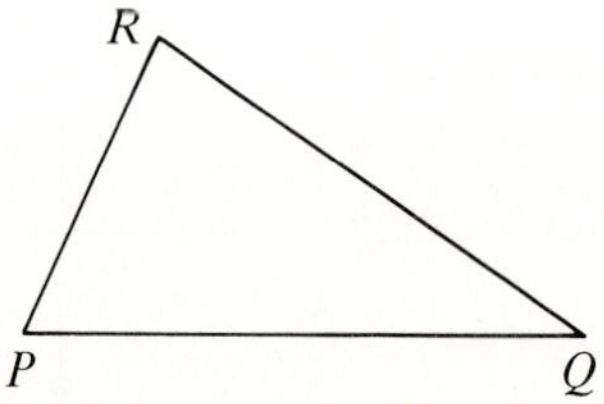

11. The perpendicular bisector of $\overline{BD}$

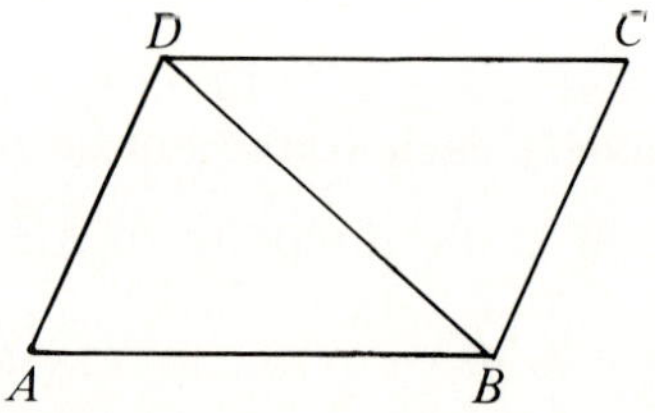

B 12. Construct a 90° angle. Then draw a segment to form a right triangle. Construct the perpendicular bisector of each leg of the right triangle. Where do the perpendicular bisectors appear to meet?

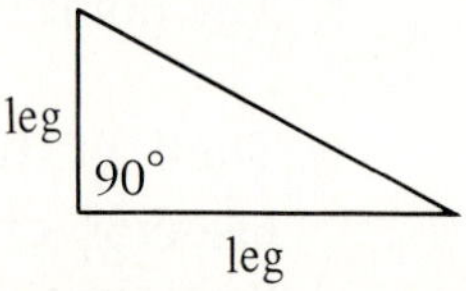

13. Repeat Exercise 12, using a different-looking right triangle. The perpendicular bisectors of the legs of a right triangle appear to meet at _?_.

14. Construct a large isosceles triangle, using the method outlined in Exercise 11 on page 151. Then construct the perpendicular bisector of the base. Through what point does the perpendicular bisector seem to pass?

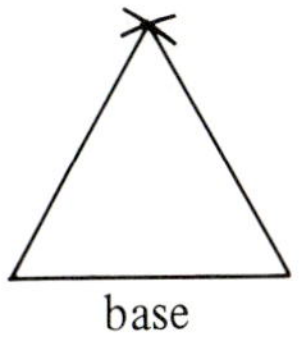

15. Repeat Exercise 14, using a different-looking isosceles triangle. The perpendicular bisector of the base of an isosceles triangle appears to pass through __?__.

C 16. One way to prove Theorem 2 is suggested below. Write the proof.

Given: $RC = RD$
M is the midpoint of $\overline{CD}$.

Prove: $\overleftrightarrow{RM}$ is the perpendicular bisector of $\overline{CD}$.

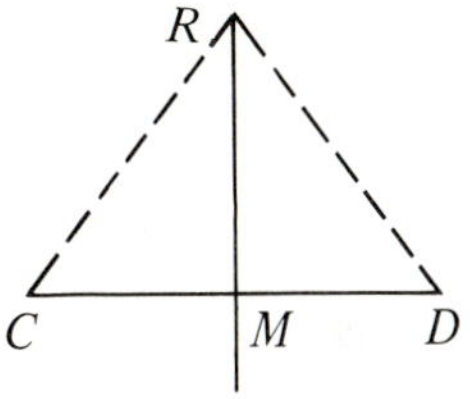

Strategy for proof:

A. Prove that $\triangle CRM \cong \triangle DRM$.
B. Prove that $\angle RMC$ and $\angle RMD$ are equal and are right angles.
C. Use the right angles to prove that $\overline{RM} \perp \overline{CD}$.

17. Use congruent triangles to show that Construction 6 does produce the perpendicular bisector of a segment.

Given: $JA = JB$; $KA = KB$

Prove: $AM = MB$; $\overleftrightarrow{JK} \perp \overline{AB}$

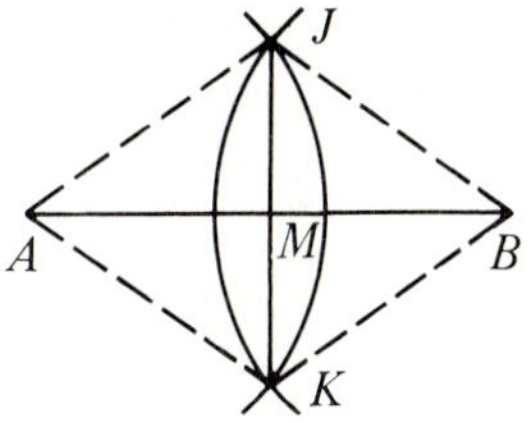

Strategy for proof:

A. Prove that $\triangle AJK \cong \triangle BJK$.
B. Prove that $\angle AJM = \angle BJM$.
C. Prove that $\triangle AJM \cong \triangle BJM$.
D. Prove that $AM = MB$ and $\angle AMJ = \angle BMJ$.
E. Use right angles to prove that $\overleftrightarrow{JK} \perp \overline{AB}$.

PUZZLE ◆ PROBLEMS

Draw a large triangle on cardboard. Construct the midpoints of the three sides. Draw the medians (see page 158) and let P be the intersection point. Cut the triangle out.

Hold the model horizontally and place your finger tip under it at P. The model should balance. Point P is called the *center of gravity* of the triangle.

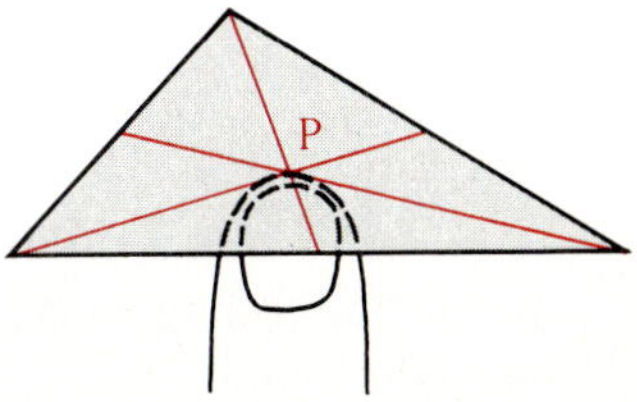

Try this experiment with other triangles.

4 Altitudes and Medians of a Triangle

The diagram at the right shows how Construction 3 can be used to construct a perpendicular to side $\overline{AB}$ through vertex C. We call $\overline{CR}$ the *altitude* of $\triangle ABC$ drawn to side $\overline{AB}$. In general, an **altitude** of a triangle is a segment, drawn from any vertex, perpendicular to the line that contains the opposite side.

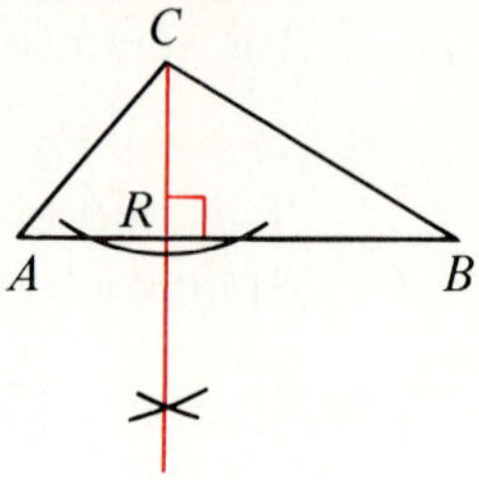

Every triangle has three altitudes, one from each vertex.

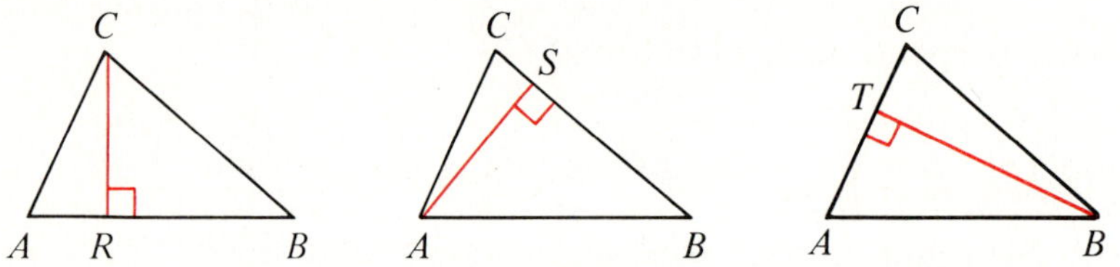

Altitude $\overline{CR}$ | Altitude $\overline{AS}$ | Altitude $\overline{BT}$

Sometimes it isn't so easy to draw an altitude.
Here the altitude from F doesn't intersect *segment DE*.
But it does intersect *line DE*.

$\overline{FH}$ is an altitude. →

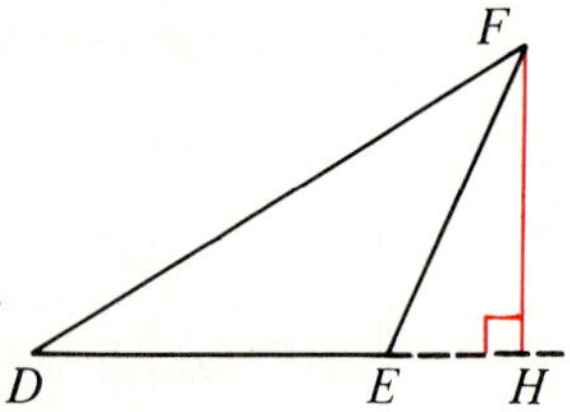

In any triangle, the three lines containing the altitudes meet in one point.

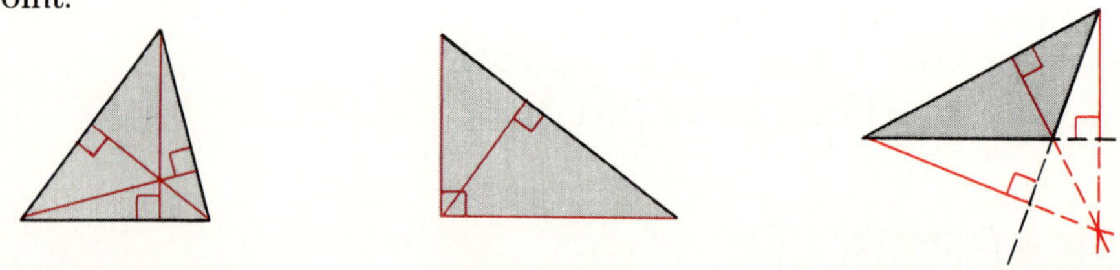

Acute triangle

The lines meet inside the triangle.

Right triangle

The lines meet at the vertex of the right angle.

Obtuse triangle

The lines meet outside the triangle.

Let's look at $\triangle ABC$ once again. Since M is the midpoint of side $\overline{AB}$, $\overline{CM}$ is a *median* of the triangle. A **median** is a segment that joins a vertex of the triangle to the midpoint of the opposite side.

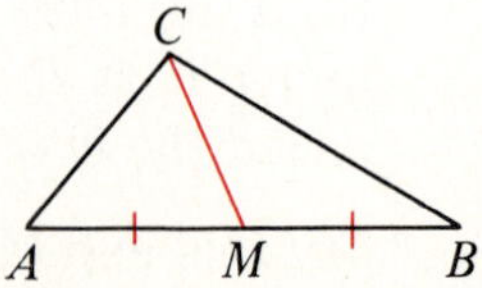

Every triangle has three medians, and they always meet inside the triangle.

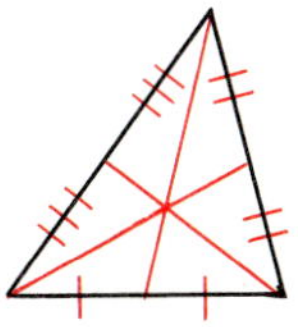

Acute triangle

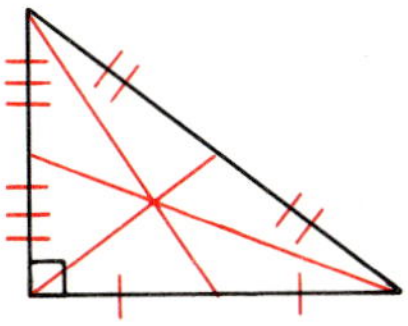

Right triangle

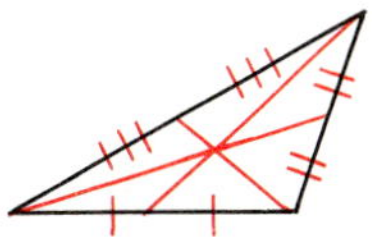

Obtuse triangle

Compare the altitudes and the medians of the triangles shown. Notice that, in general, the altitude and the median to a side of a triangle are *different* segments.

Classroom Practice

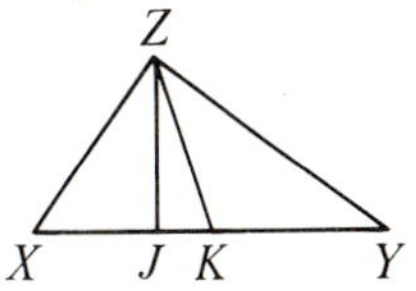

1. Point K is the midpoint of $\overline{XY}$, and $\overline{ZJ} \perp \overline{XY}$.

a. The altitude shown is segment __?__.

b. The median shown is segment __?__.

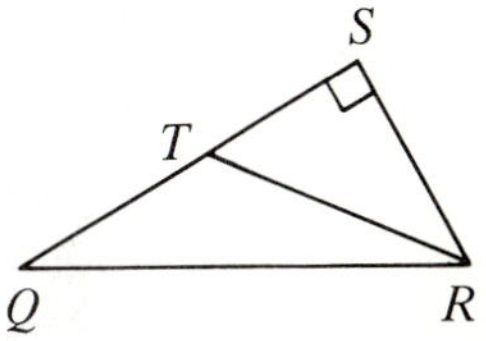

2. Point T is the midpoint of leg $\overline{SQ}$ in right $\triangle QRS$.

a. The altitude to side $\overline{SQ}$ is segment __?__.

b. The median to side $\overline{SQ}$ is segment __?__.

For each exercise, think of the three altitudes of the triangle shown. (Do not draw in your book.) Tell whether each altitude lies inside, on, or outside the triangle.

3.

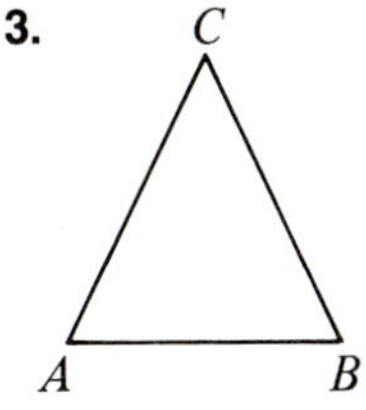

4.

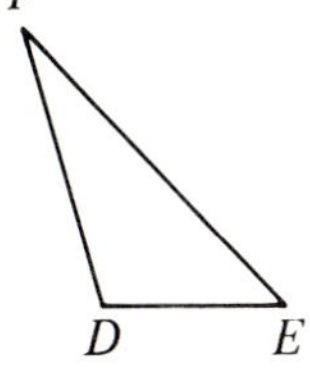

5.

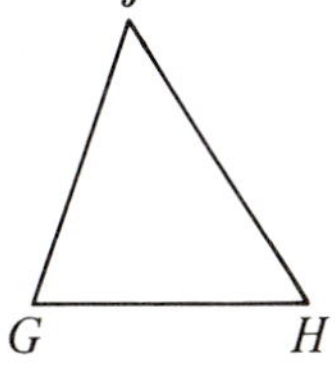

6.

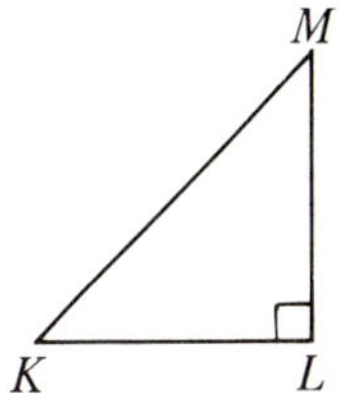

7–10. Use the figures for Exercises 3–6. Think of the three medians of the triangle shown. Tell whether each median lies inside, on, or outside the triangle.

11. You know that, in general, the altitude and the median to a side of a triangle are different segments. Describe a triangle in which an altitude and a median are the very same segment.

Written Exercises

Draw a triangle roughly like the one shown, but larger. Draw and label the altitude to side $\overline{AB}$.

A

1.

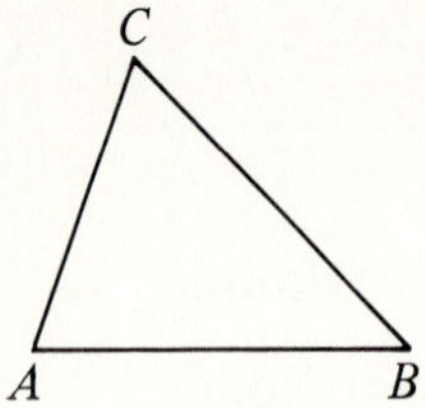

2.

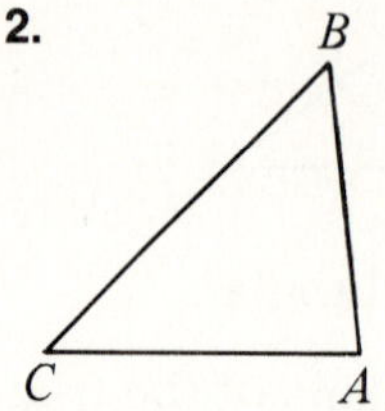

3.

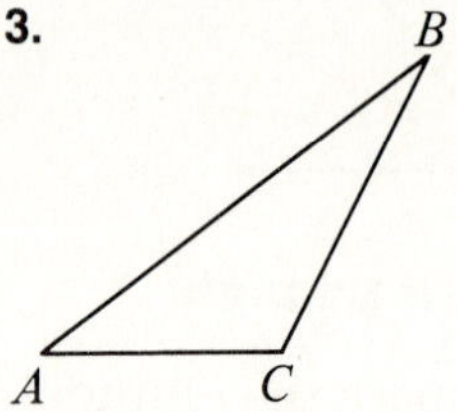

Draw a triangle roughly like the one shown, but larger. Construct and label the midpoint of side $\overline{YZ}$. Then draw the median to $\overline{YZ}$.

4.

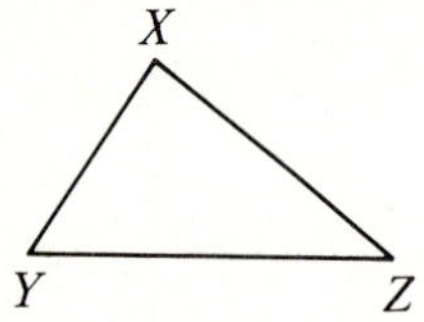

5.

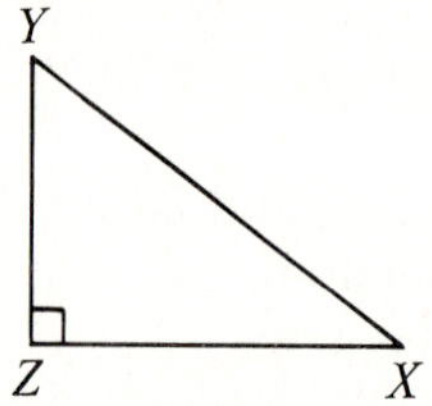

6.

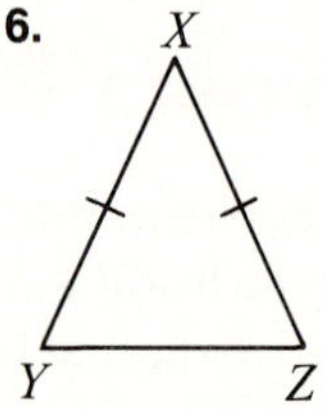

7. Draw a triangle roughly like $\triangle RST$, but larger. Then draw each of the following (you need not construct):

a. The altitude from R

b. The median from R

c. The perpendicular bisector of $\overline{ST}$

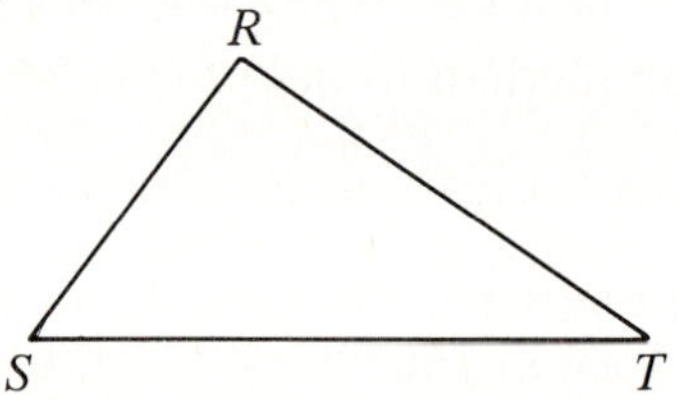

8. Repeat Exercise 7, using an isosceles $\triangle RST$ with $RS = RT$. What do you notice?

Draw a triangle roughly like the one shown, but larger. Construct the three altitudes of the triangle.

B

9.

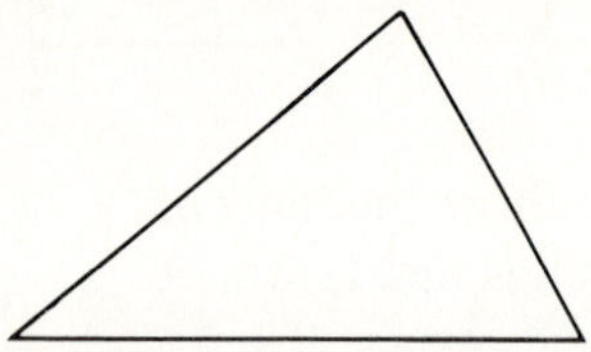

10.

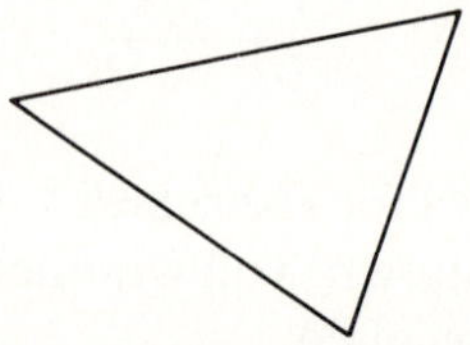

11. a. Draw an acute triangle.

b. Use Construction 6 to locate the midpoint of each side.

c. Draw the three medians of the triangle.

12. a. Construct a right angle.
b. Draw a segment to form a right triangle.
c. Use Construction 6 to locate the midpoint of each side.
d. Draw the three medians of the triangle.

13. a. Draw a large acute $\triangle ABC$.
b. Construct: the midpoint M of $\overline{BC}$;
the midpoint N of $\overline{AC}$;
the midpoint K of $\overline{AB}$.
c. Draw the three medians.
Use P to label the point of intersection.

d.

By measurement		*By arithmetic*
$AP = \underline{\ ?\ }$	$AM = \underline{\ ?\ }$	$\frac{2}{3}(AM) = \underline{\ ?\ }$
$BP = \underline{\ ?\ }$	$BN = \underline{\ ?\ }$	$\frac{2}{3}(BN) = \underline{\ ?\ }$
$CP = \underline{\ ?\ }$	$CK = \underline{\ ?\ }$	$\frac{2}{3}(CK) = \underline{\ ?\ }$

14. Repeat Exercise 13 using an obtuse triangle.

15. Examine part **d** of Exercises 13 and 14.
What do you observe about the first and third columns?
Complete: The medians of a triangle meet in a point which is $\underline{\ ?\ }$ of the way from any vertex to the opposite side.

C

16. a. Draw a large obtuse triangle.
b. Construct the three altitudes of the triangle.
c. If you extend the altitudes, will they meet in a point?

17. Repeat Exercise 16, using a different-looking obtuse triangle.

18. The goal is to prove that *corresponding altitudes of congruent triangles are equal.* Copy what is shown and write a proof in two-column form.

Given: $\triangle RST \cong \triangle XYZ$
$\overline{TV} \perp \overline{RS}$; $\overline{ZW} \perp \overline{XY}$

Prove: $TV = ZW$

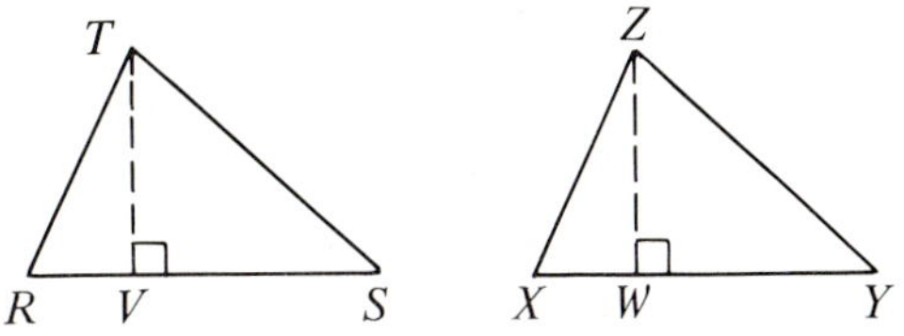

Strategy for proof:
A. Use the congruent triangles to prove $RT = XZ$ and $\angle R = \angle X$.
B. Use perpendicular lines to prove $\angle RVT = \angle XWZ$.
C. Prove $\triangle RVT \cong \triangle XWZ$.
D. Use the fact that corr. parts of $\cong$ △s are $=$.

Experiments

Experiment 1

1. Take a large sheet of paper and fold it down the middle. Cut from the corner to any point on the fold as shown. Unfold.

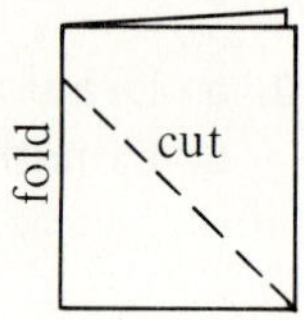

2. What kind of triangle did you form?
3. Explain why the fold is both an altitude and a median.
4. Complete: The median and the altitude to the base of an __?__ triangle are the same segment.

Experiment 2

1. Plot points $A(0, 0)$, $B(6, 12)$, and $C(12, 0)$ on graph paper. Draw $\triangle ABC$.
2. Draw dots at the midpoints of the sides of $\triangle ABC$.
   Draw the three medians.
   Use P to label the point where the medians meet.
   Use M to label the midpoint of $\overline{BC}$.
3. M is point (__?__, __?__) and P is point (__?__, __?__).
4. Measure $\overline{AP}$ and $\overline{AM}$ with your ruler.
   If you worked carefully, $AP = \frac{2}{3}(AM)$.

Experiment 3

Repeat Experiment 2, using points $A(0, 0)$, $B(0, 6)$, and $C(12, -12)$.

Experiments 2 and 3 suggest this property: *The medians of a triangle intersect in a point which is two-thirds of the way along each median.*

PUZZLE ♦ PROBLEMS

Here is another way to show that the sum of the angles of a triangle equals 180°. Draw a triangle ABC and construct the inscribed circle, as shown on page 164. Cut out the triangle and fold $\angle A$, $\angle B$, and $\angle C$ to the center of the circle. There are six angles at the center, the three original angles and three vertical angles.

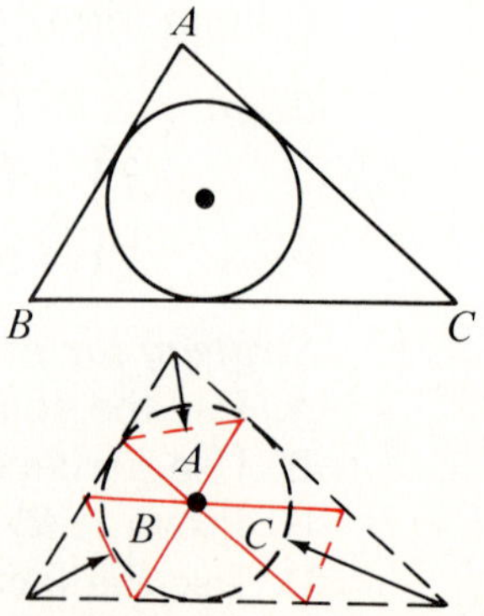

$$2\angle A + 2\angle B + 2\angle C = 360°,$$
$$\text{so } \angle A + \angle B + \angle C = 180°.$$

5 Inscribed and Circumscribed Circles

We know that the medians of a triangle meet in a point. So do the lines containing the altitudes. What about the perpendicular bisectors of the sides?

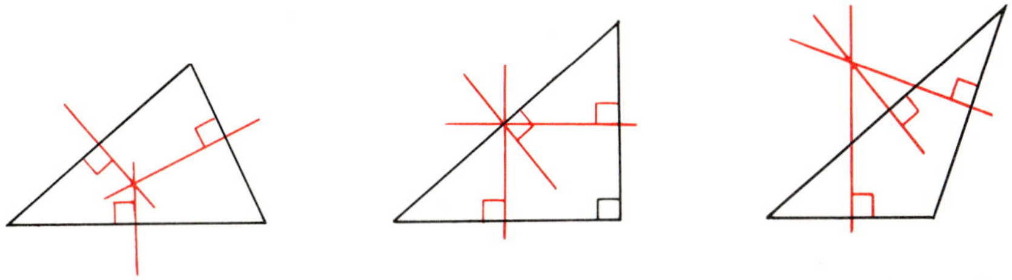

Acute triangle | **Right triangle** | **Obtuse triangle**

⊥ bisectors meet inside triangle.

⊥ bisectors meet on triangle.

⊥ bisectors meet outside triangle.

This special point in which the perpendicular bisectors meet is used in the following construction.

CONSTRUCTION 7

Given: *A triangle*
Construct: *A circle passing through the vertices of the triangle.*

1.

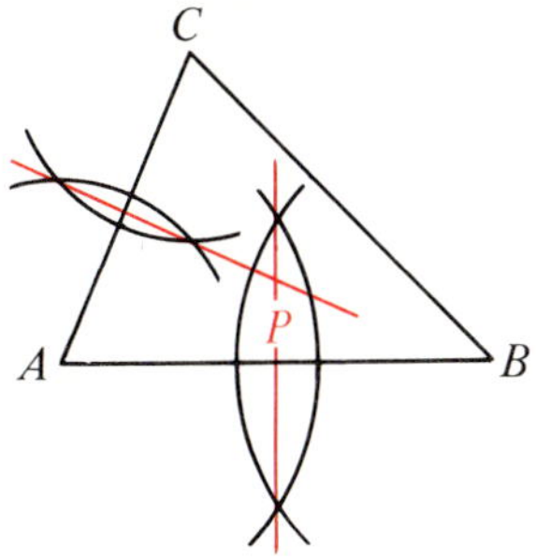

Given $\triangle ABC$, construct the perpendicular bisectors of any two sides. Label the meeting point P.

2.

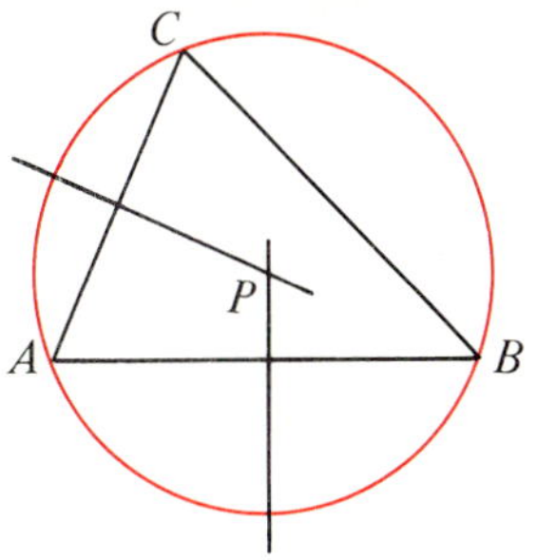

Using P as center and PA as radius, draw a circle. The circle should pass through A, B, and C.

When each vertex of a triangle is a point on a circle, we say that the circle is **circumscribed about** the triangle. In the construction above, for example, the circle shown is circumscribed about $\triangle ABC$. The figure at the right shows another **circumscribed circle.**

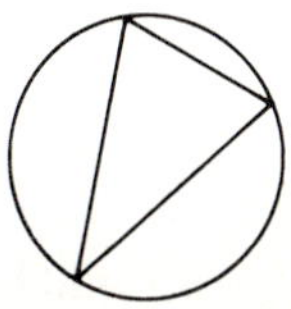

A circumscribed circle fits exactly *around* a triangle. On the other hand, an **inscribed circle** (see page 361) fits exactly *inside* a triangle. Here are some inscribed circles.

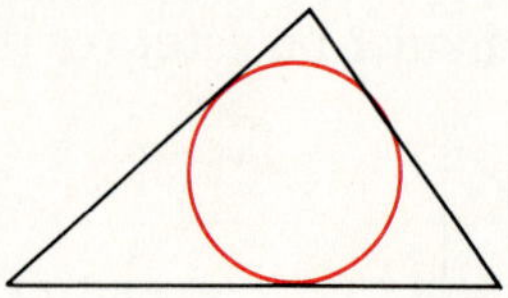

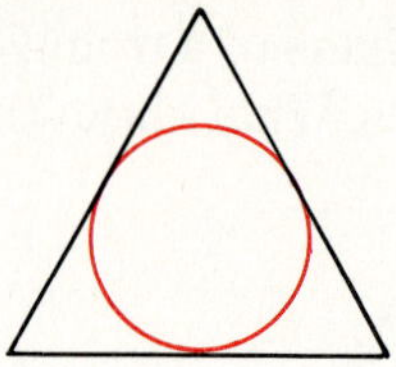

Suppose you are given a triangle, $\triangle DEF$. To find the center of the circle which can be inscribed in $\triangle DEF$, we use this fact: The angle bisectors of a triangle meet in a point which is also the center of the inscribed circle.

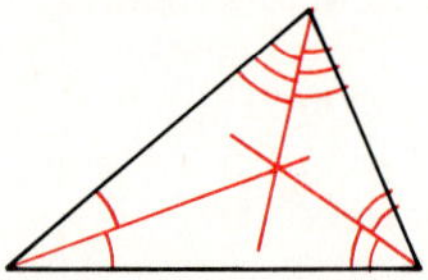

Acute triangle

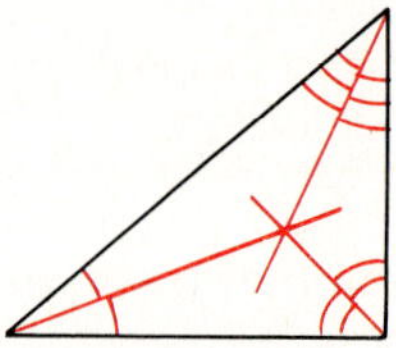

Right triangle

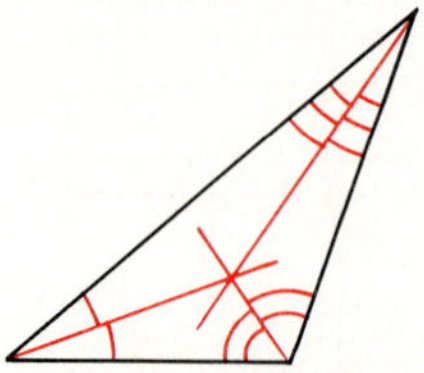

Obtuse triangle

CONSTRUCTION 8

Given: *A triangle*
Construct: *An inscribed circle*

1.

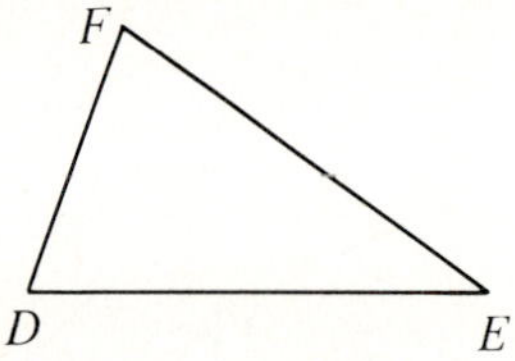

You are given $\triangle DEF$.

2.

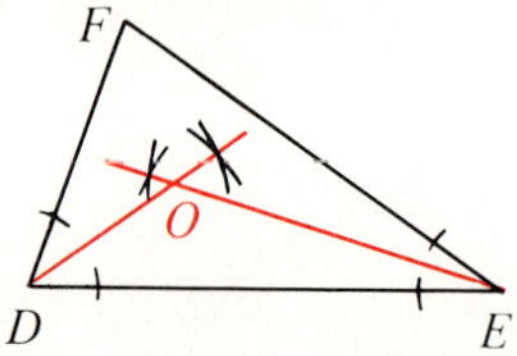

Construct the angle bisectors of any two angles. Label their meeting point O.

3.

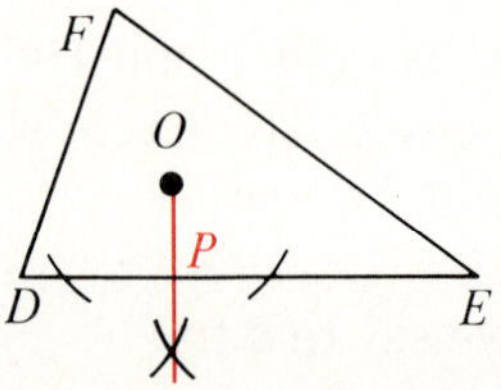

Construct a perpendicular from O to $\overline{DE}$. Label the meeting point P.

4.

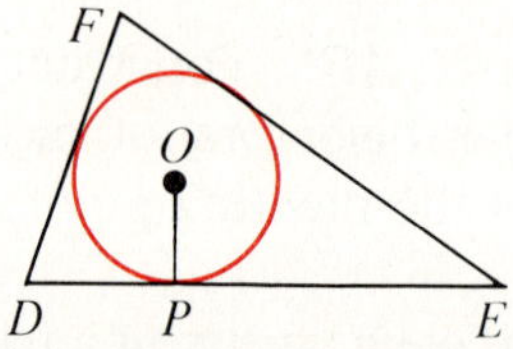

With O as center and OP as radius, draw a circle. The circle should just touch each side of $\triangle DEF$.

Classroom Practice

1. Which of the diagrams below show a circle circumscribed about a triangle?

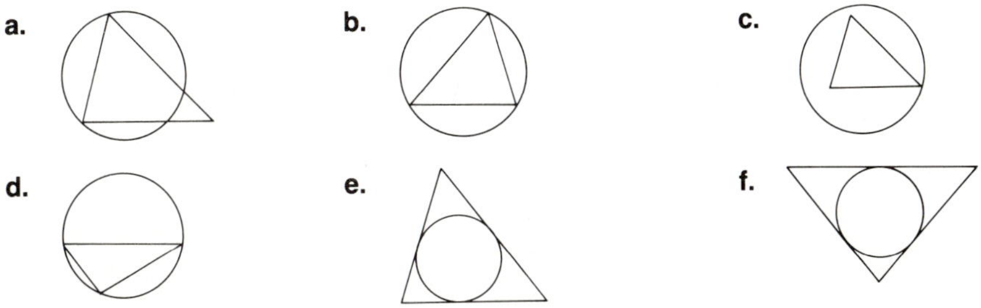

2. Which diagrams above show a circle inscribed in a triangle?

3. The center of a circle inscribed in a triangle is the point at which __?__.

4. The center of a circle circumscribed about a triangle is the point at which __?__.

5. The perpendicular bisector of $\overline{AC}$ and the perpendicular bisector of $\overline{BC}$ intersect at point X. We wish to show that X lies on the perpendicular bisector of $\overline{AB}$. Supply the required reasons.

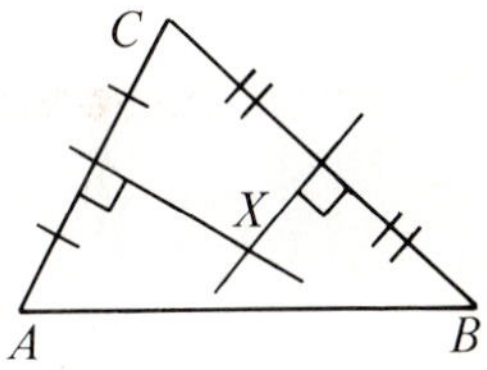

   a. Draw $\overline{AX}$, $\overline{BX}$, and $\overline{CX}$.
   b. $AX = CX$ (Why?)
   c. $BX = CX$ (Why?)
   d. Then $AX = BX$. (Why?)
   e. X is equidistant from A and B. (Why?)
   f. X lies on the perpendicular bisector of $\overline{AB}$. (Why?)

   Thus the perpendicular bisectors of the three sides of a triangle intersect in one point.

Written Exercises

Sketch each of the following. You need not construct.

A 1. A circle inscribed in an isosceles triangle

2. A circle inscribed in a scalene triangle

3. A circle circumscribed about an acute triangle

4. A circle circumscribed about an isosceles right triangle

In Exercises 5–7, draw a triangle roughly like the one shown, but much larger. Use Construction 7 to circumscribe a circle about the triangle.

5. 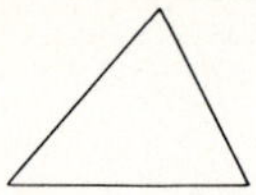

6. 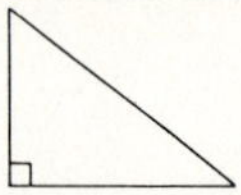

7. 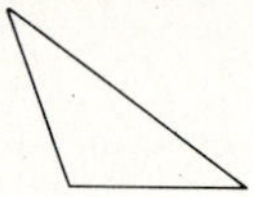

In Exercises 8–10, draw a triangle roughly like the one shown, but much larger. Use Construction 8 to inscribe a circle in the triangle.

8. 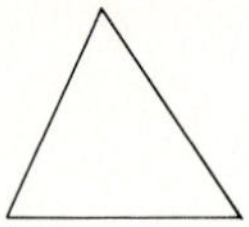

9. 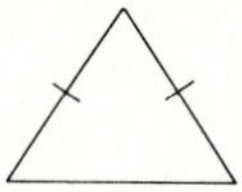

10. 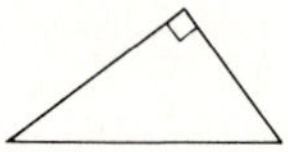

B **11.** Draw three noncollinear points on your paper and label them X, Y, and Z. Construct a circle that passes through the three points.

12. Repeat Exercise 11, using three different points.

13. Draw a scalene $\triangle ABC$. Circumscribe a circle about $\triangle ABC$ and inscribe a circle in $\triangle ABC$.

14. Repeat Exercise 13, using an isosceles $\triangle ABC$.

C **15.** Draw an obtuse $\triangle XYZ$. Construct a circle that passes through Y, Z, and the midpoint of $\overline{XZ}$.

SELF-TEST

Vocabulary

bisector (p. 154)
perpendicular bisector (p. 154)
altitude (p. 158)
median (p. 158)
circumscribed (p. 163)
inscribed (p. 164)

Draw three triangles roughly like, but much larger than, the ones shown.

1. In $\triangle ABC$, construct the altitude from C. *(4-4)*

2. In $\triangle ABC$, construct the median from A.

3. Circumscribe a circle about $\triangle DEF$. *(4-5)*

4. Find, by construction, the point which would be the center of the circle inscribed in $\triangle GHJ$.

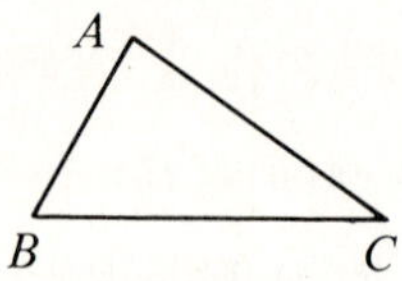

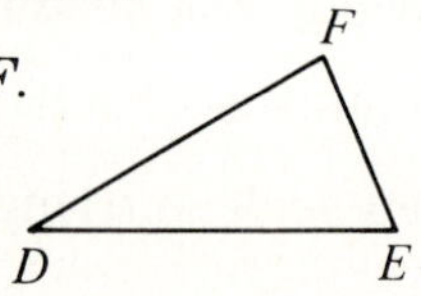

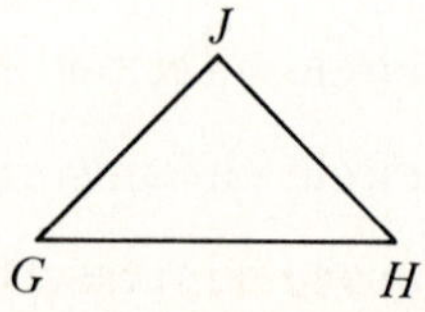

6 Triangles with Two Equal Sides

EXPLORATIONS

1. Draw a large isosceles $\triangle ABC$, with $AC = BC$. Cut out the triangle.
2. Make a fold that bisects $\angle C$.
3. Does vertex A fold onto vertex B?
4. Does $\angle A$ have the same measure as $\angle B$?

If you worked carefully, your answers to parts 3 and 4 were *yes*. A picture statement of the theorem suggested by the exploration is shown at the right. The exploration suggests a proof involving a "line down the middle."

If , then .

THEOREM 3

If two sides of a triangle are equal, then the angles opposite those sides are equal.

Given: $AC = BC$

Prove: $\angle A = \angle B$

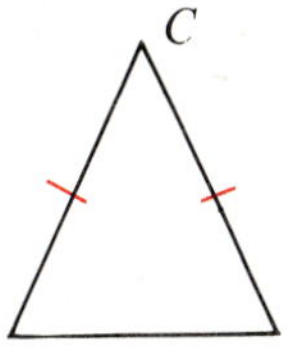

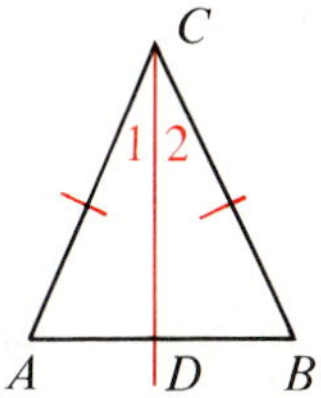

STATEMENTS	REASONS
1. Draw the bisector of $\angle C$.	1. Every angle has a bisector. (Recall Const. 1)
2. $\angle 1 = \angle 2$	2. Definition of angle bisector
3. $AC = BC$	3. Given
4. $CD = CD$	4. From algebra
5. $\triangle ADC \cong \triangle BDC$	5. SAS Postulate
6. $\angle A = \angle B$	6. Corr. parts of $\cong$ ⚠ are $=$.

COROLLARY

An equilateral triangle is also equiangular, and each angle has measure 60°.

See Exercise 21 on page 171 for an outline of a proof of the corollary on page 167.

We often use special terms to refer to parts of isosceles triangles.

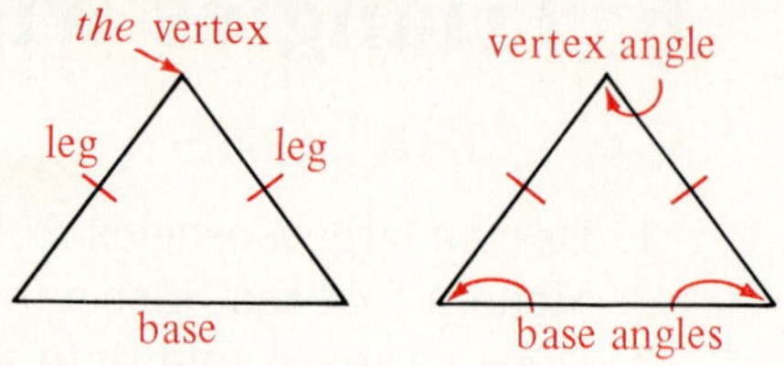

Some people like to state Theorem 3 in this form:

Base angles of an isosceles triangle are equal.

WAYS TO PROVE TWO ANGLES ARE EQUAL
1. *Show that they are corresponding parts of congruent triangles.*
2. *Show that they are opposite two equal sides of a triangle.*
3. *Show that they are corresponding angles or alternate interior angles of parallel lines.*

Classroom Practice

Exercises 1–6 refer to isosceles △*ABC*.

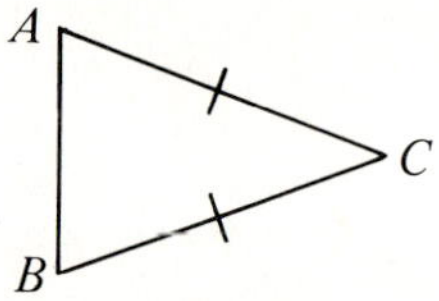

1. Name the base.
2. Name the vertex angle.
3. Name the legs.
4. Name the base angles.
5. Name the vertices.
6. Name *the* vertex.

In Exercises 7–12, name the angles that must be equal.

7. 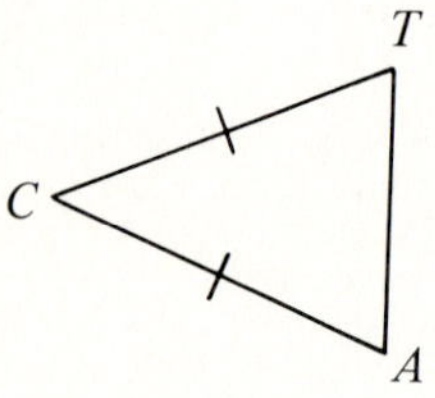

8. 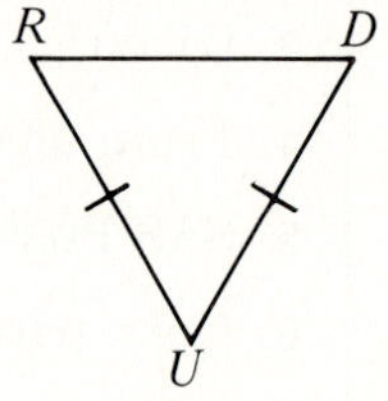

9. 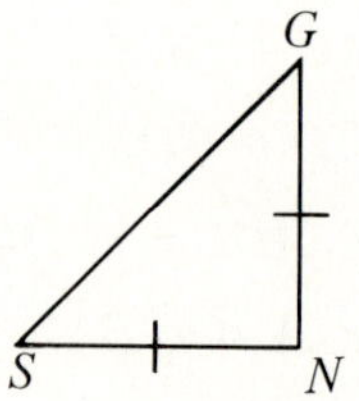

10. 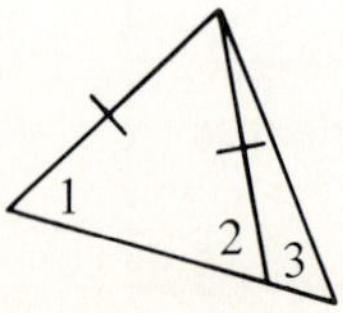

11. 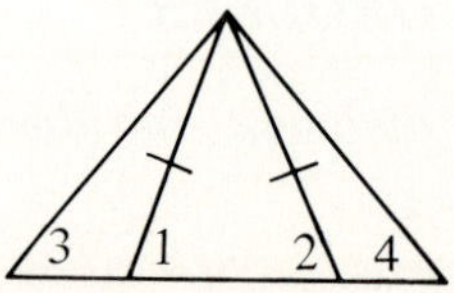

12. 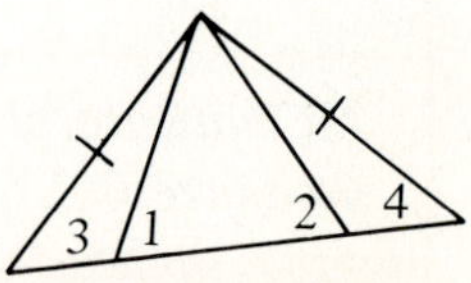

13. In the diagram for Exercise 7, you could say $CA = CT$, given.
Then you could conclude: $\angle A = \angle T$.
State the theorem that supports the conclusion.

14. In the diagram for Exercise 9, suppose $\angle N = 90°$.
Then $\angle S =$ _?_°.

15. In the diagram for Exercise 8, suppose that $DU = DR = RU$.
Then $\angle D =$ _?_°.

In Exercises 16–18, $XZ = YZ$ and $\overline{ZE}$ is drawn in the way described. State the method you would use to prove that $\triangle XEZ \cong \triangle YEZ$.

16. *Given:* $XZ = YZ$
$\overrightarrow{ZE}$ bisects $\angle XZY$.
Prove: $\triangle XEZ \cong \triangle YEZ$

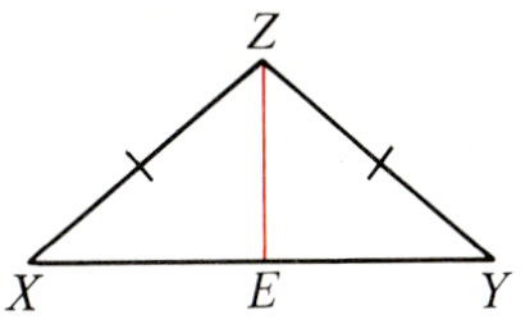

17. *Given:* $XZ = YZ$
$\overline{ZE}$ is a median.
Prove: $\triangle XEZ \cong \triangle YEZ$

18. *Given:* $XZ = YZ$
$\overline{ZE}$ is an altitude.
Prove: $\triangle XEZ \cong \triangle YEZ$

Written Exercises

In Exercises 1–10, use the diagram. Complete each row of the table.

A

	$\angle 1$	$\angle 2$	$\angle 3$	$\angle 4$	$\angle 5$
1.	65°	?	?	?	?
2.	?	?	120°	?	?
3.	?	?	?	116°	?
4.	?	70°	?	?	?
5.	68°	?	?	?	?
6.	?	?	112°	?	?
7.	?	?	?	?	42°
8.	?	?	?	?	50°
9.	$j°$	?	?	?	?
10.	$k°$	?	?	?	?

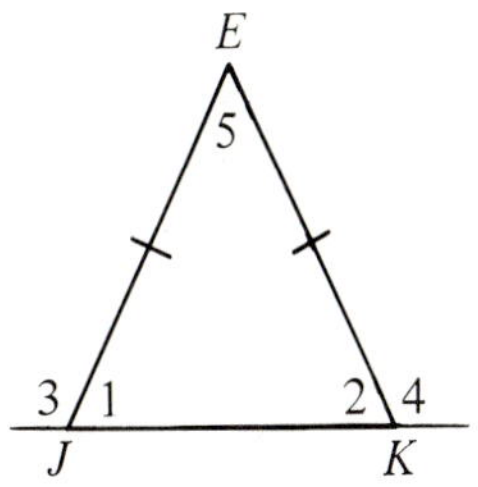

Hint: $\angle 1 + \angle 2 =$ _?_°

Your answers will involve j.

Your answers will involve k.

In Exercises 11–19, equal sides are marked. Find the value of x.

SAMPLE

$$\angle A + \angle B + \angle C = 180^\circ$$
$$2x^\circ + 2x^\circ + 92^\circ = 180^\circ$$
$$4x^\circ + 92^\circ = 180^\circ$$
$$4x^\circ = 88^\circ$$
$$x^\circ = 22^\circ$$

$\angle A = \angle B = 2x^\circ$

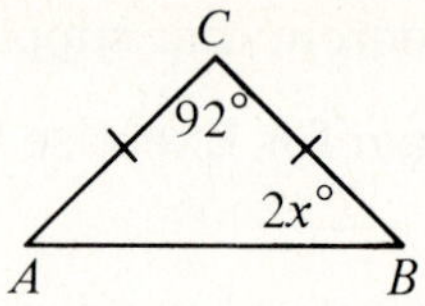

11.

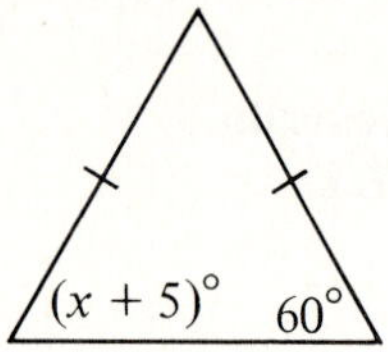

12.

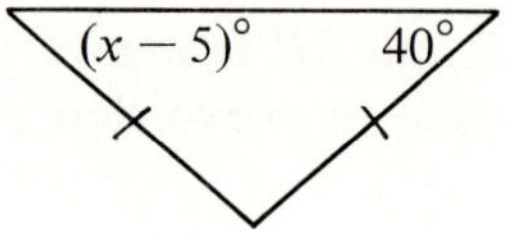

13.

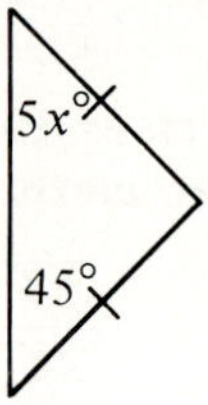

14.

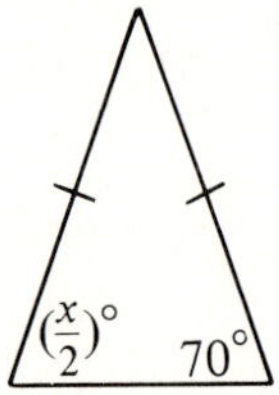

15.

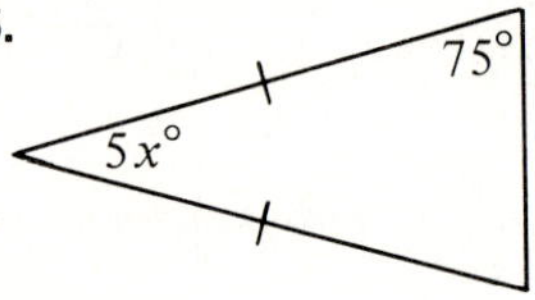

16.

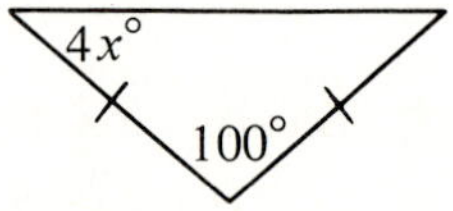

17.

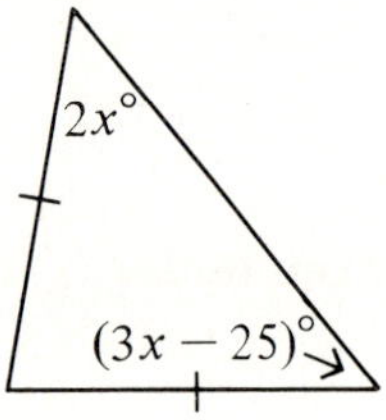

18.

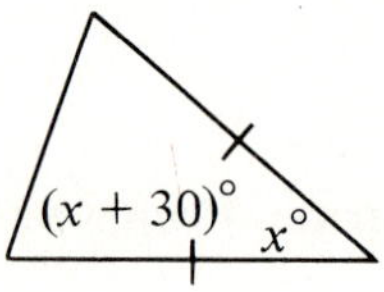

19.

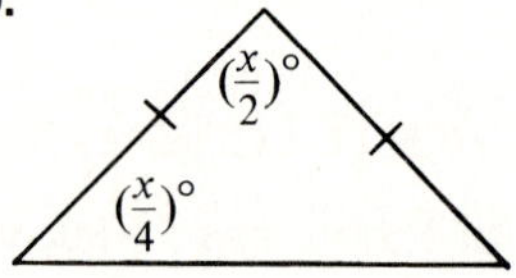

In Exercises 20–22, complete the proofs by supplying the reasons.

20. *Given:* $AC = BC$

Prove: $\angle 1 = \angle 3$

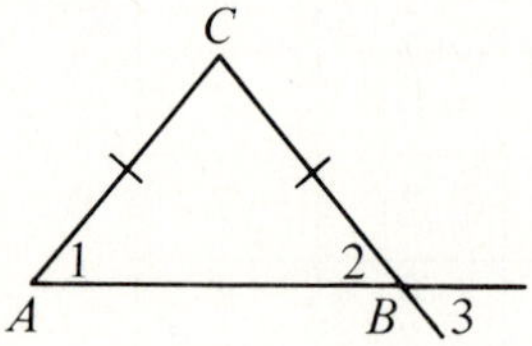

STATEMENTS	REASONS
1. $AC = BC$	1. ?
2. $\angle 1 = \angle 2$	2. ?
3. $\angle 3 = \angle 2$	3. ?
4. $\angle 1 = \angle 3$	4. ?

B **21.** Complete this proof of the corollary to Theorem 3:

An equilateral triangle is also equiangular, and each angle has measure 60°.

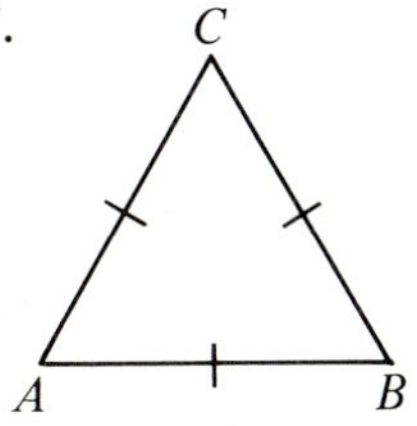

Given: $AB = BC = AC$

Prove: $\angle A = \angle B = \angle C = 60°$

STATEMENTS	REASONS
1. $AC = BC; AB = AC; AB = BC$	1. ?
2. $\angle A = \angle B; \angle B = \angle C; \angle C = \angle A$	2. ?
3. $\angle A = \angle B = \angle C$	3. ?
4. $\angle A + \angle B + \angle C = 180°$	4. ?
5. $\angle A + \angle A + \angle A = 180°$, or $3 \cdot \angle A = 180°$	5. ?
6. $\angle A = 60°$	6. ?
7. $\angle A = \angle B = \angle C = 60°$	7. ?

22. *Given:* $AX = DX; AB = DC$

Prove: $\angle 1 = \angle 2$

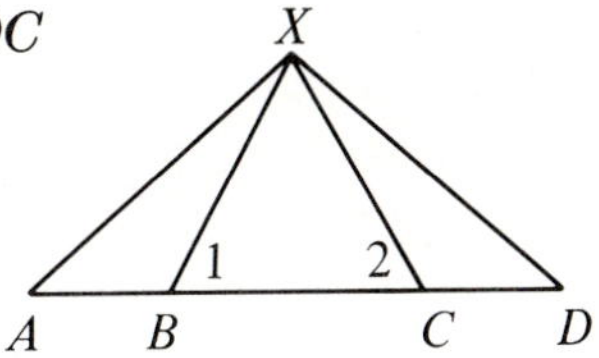

Strategy for proof:

A. Using isosceles $\triangle AXD$, prove that $\angle A = \angle D$.
B. Prove that $\triangle AXB \cong \triangle DXC$, and conclude that $BX = CX$.
C. Using isosceles $\triangle BXC$, prove that $\angle 1 = \angle 2$.

STATEMENTS	REASONS
1. $AX = DX$	1. ?
2. $\angle A = \angle D$	2. ?
3. $AB = DC$	3. ?
4. $\triangle AXB \cong \triangle DXC$	4. ?
5. $BX = CX$	5. ?
6. $\angle 1 = \angle 2$	6. ?

CAREER NOTES

PLUMBER

Have you ever taken something apart just because you were curious about how it worked? Can you figure out why the vacuum cleaner or lawnmower won't run, and then fix it? A plumber must be able to do similar types of problem solving.

Plumbers install and repair fixtures and appliances that are connected to water and sewer lines. They also work with gas pipes that supply energy to ranges, furnaces, and water heaters. Skilled plumbers design complex pipe systems for large commercial and industrial buildings.

It is important for plumbers to be able to visualize three-dimensional objects when looking at building plans and blueprints. Careful measurement is also necessary to avoid wasting materials. For example, a plumber may decide that drilling a hole through a wall or floor will save many lengths of pipe. Distances can also be shortened by joining pipes with 45° elbows.

A plumber uses mathematics often while on the job. He or she may have to calculate water pressure in different locations or determine the weight of water in a tank with a given volume. City or industrial codes may regulate such things as the grade (or slope) of sewer pipes and the rate of absorption of septic systems.

To become a plumber, you must usually complete a 4-year apprenticeship program after high school or vocational school. In many places you must then pass an exam on plumbing skills and local plumbing codes in order to get a license. Most plumbers work for contractors in new construction. Many others are self-employed, specializing in repairs and modernization.

7 Triangles with Two Equal Angles

In Section 6 you saw that when a triangle has two equal sides, it also has two equal angles. Experience suggests that when a triangle has two equal angles, it must also have two equal sides. Expressing this by diagram, we write:

Theorem 3 If , then .

Converse of Theorem 3 If , then .

We shall treat the converse as a theorem.

THEOREM 4

If two angles of a triangle are equal, then the sides opposite those angles are equal.

Given: $\angle A = \angle B$

Prove: $AC = BC$

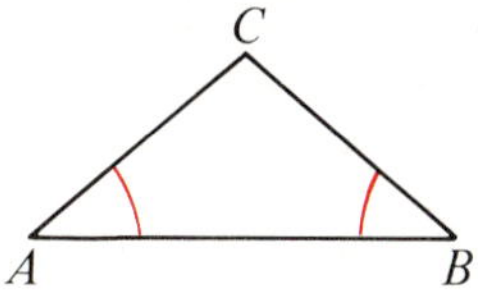

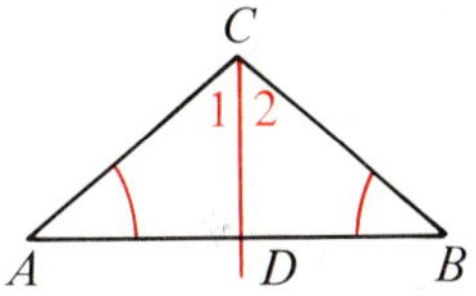

See if you can supply the reasons in the proof below.

STATEMENTS	REASONS
1. Draw the bisector of $\angle C$.	1. Every angle has a bisector. (Recall Construction 1.)
2. $\angle 1 = \angle 2$	2. ___?___
3. $\angle A = \angle B$	3. ___?___
4. $CD = CD$	4. ___?___
5. $\triangle ACD \cong \triangle BCD$	5. ___?___
6. $AC = BC$	6. ___?___

COROLLARY

If a triangle is equiangular, it is also equilateral.

See Exercise 11 on page 176 for an outline of a proof of the corollary on page 173. The corollary on page 167 states that an equilateral triangle is also equiangular, and that each angle has measure 60°. We can use this fact to construct a 60° angle.

CONSTRUCTION 9
Construct: *A 60° angle*

1.

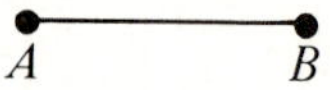

Draw a segment, $\overline{AB}$.

2.

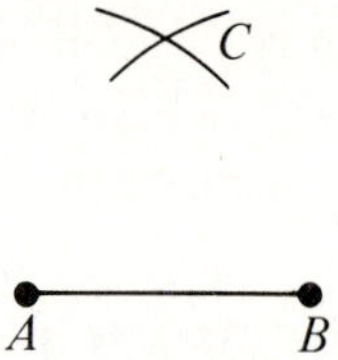

Set your compass for a radius of AB. Draw two arcs, one with A as center, and one with B as center.

3.

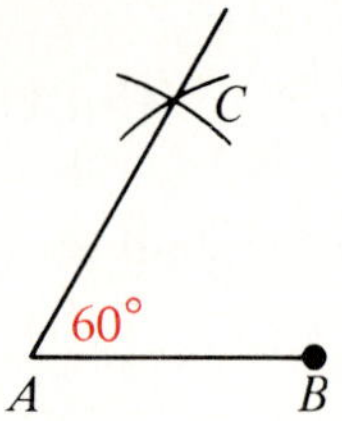

Draw $\overrightarrow{AC}$.
$\angle A$ is a 60° angle.

Can you explain why $\angle A$ must have measure 60°?

Try this:
1. Draw $\overline{BC}$.
2. $AB = BC = AC$ by construction.
3. $\angle A = \angle B = \angle C = 60°$ (Why?)

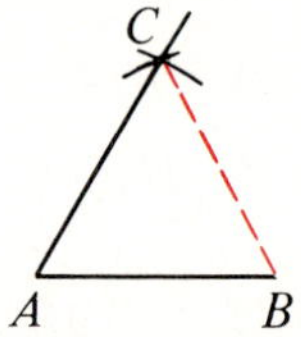

WAYS TO PROVE TWO SEGMENTS ARE EQUAL
1. *Show that they are corresponding parts of congruent triangles.*   2. *Show that they are opposite two equal angles of a triangle.*

Classroom Practice

In Exercises 1–6, name the sides that must be equal.

1.

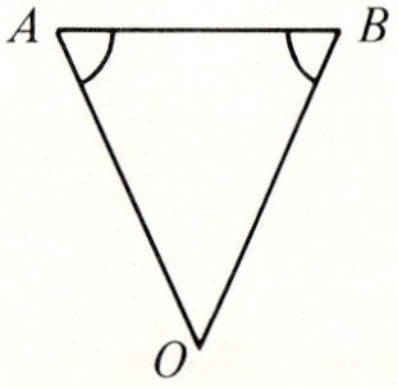

2.

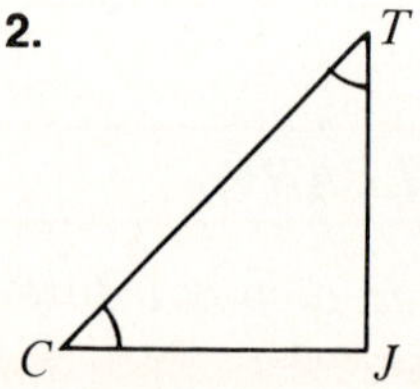

3.

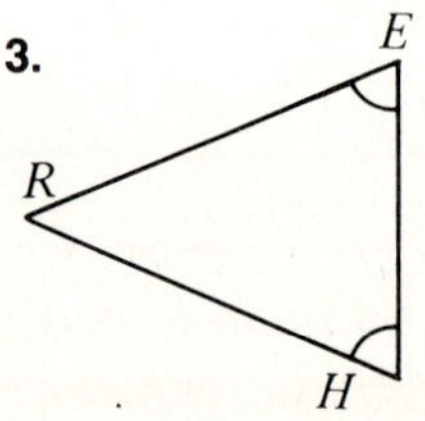

4.

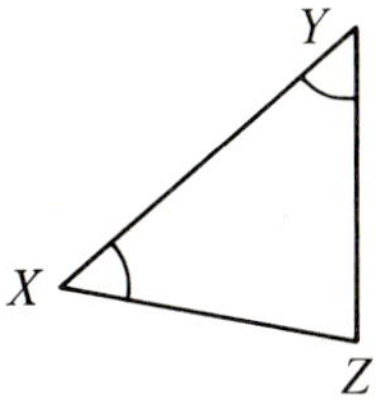

5.

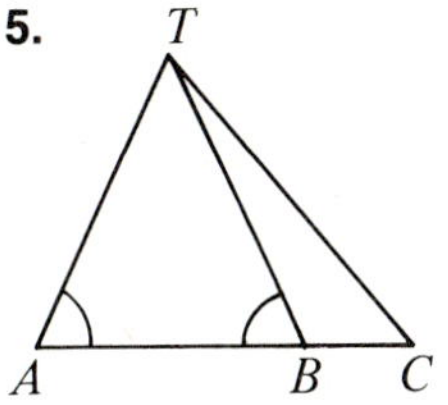

6.

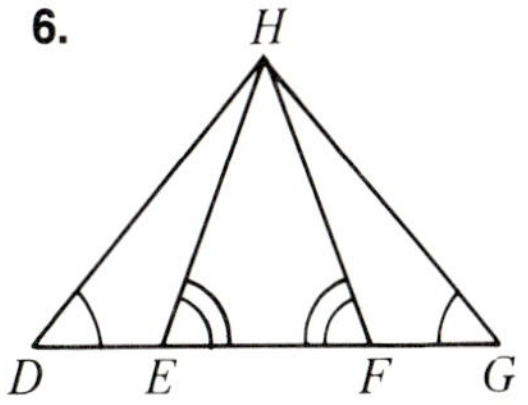

7. If you know that $AX = AY$, then you can conclude that $\angle X = \angle Y$. State the theorem that supports the conclusion.

8. If you know that $\angle X = \angle Y$, then you can conclude that $AX = AY$. State the theorem that supports the conclusion.

9. One student decided to try to prove $RT = ST$ as follows:
- **a.** Bisect $\angle RTS$ with a segment that is perpendicular to $\overline{RS}$.
- **b.** $\angle 1 = \angle 2$
- **c.** $\angle 3 = \angle 4$ (Each has measure 90°.)
- **d.** $TV = TV$
- **e.** $\triangle RVT \cong \triangle SVT$
- **f.** $RT = ST$

What is wrong with this reasoning?

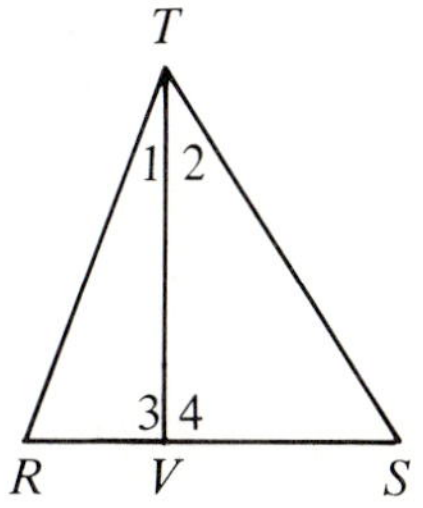

10. **a.** Construct a 60° angle.
b. Use your construction from part **a** to construct a 120° angle.

Written Exercises

Equal angles are marked. Find the value of *x*.

SAMPLE

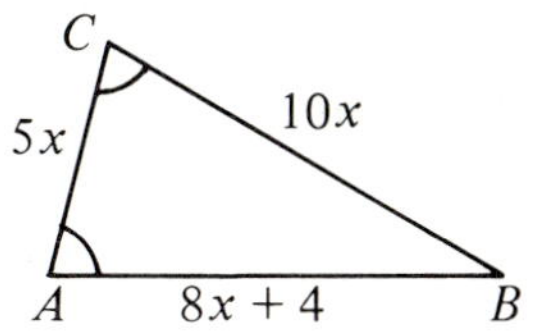

$BC = BA$
$10x = 8x + 4$
$2x = 4$
$x = 2$

A **1.**

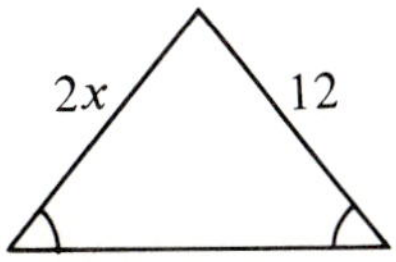

2.

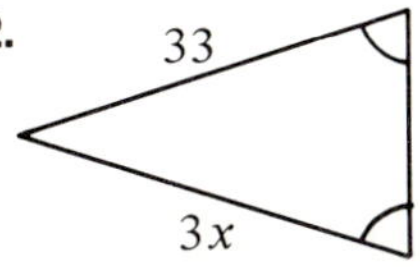

3.

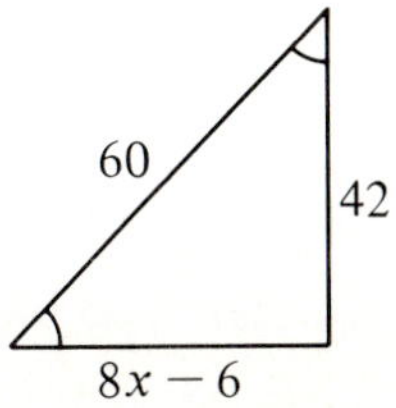

Equal angles are marked. Find the value of *x*.

4.

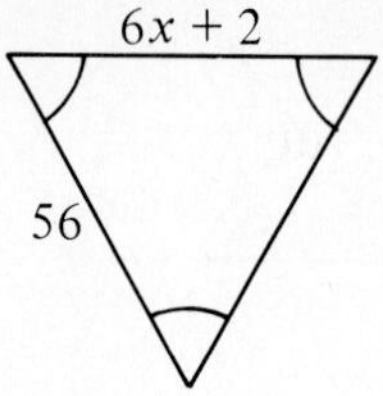

5.

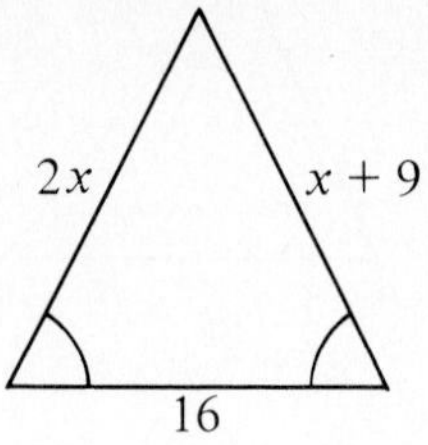

6.

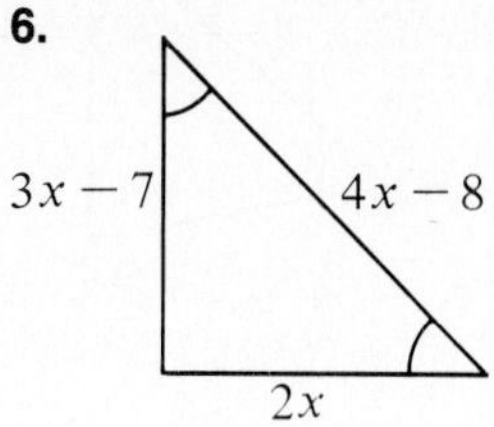

7.

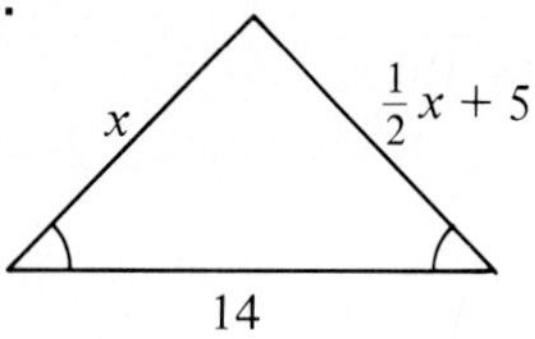

8.

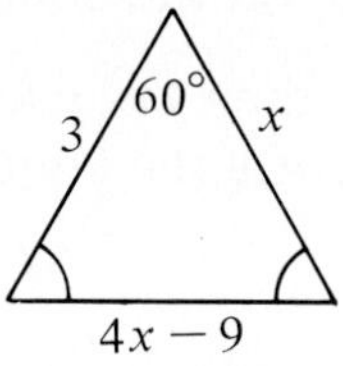

9.

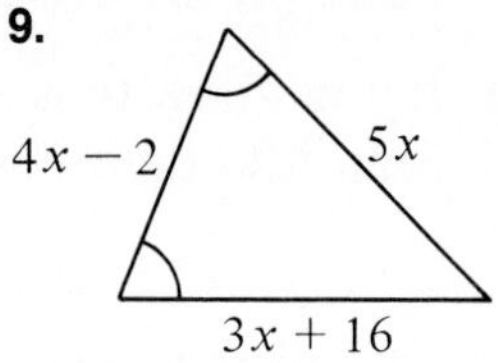

10. Complete the proof by supplying the reasons.

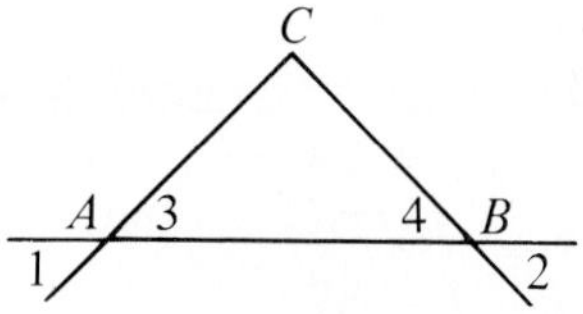

Given: $\angle 1 = \angle 2$

Prove: $AC = BC$

STATEMENTS	REASONS
1. $\angle 3 = \angle 1$; $\angle 4 = \angle 2$	1. ?
2. $\angle 1 = \angle 2$	2. ?
3. $\angle 3 = \angle 4$	3. ?
4. $AC = BC$	4. ?

11. Complete this proof of the corollary of Theorem 4:

If a triangle is equiangular, it is also equilateral.

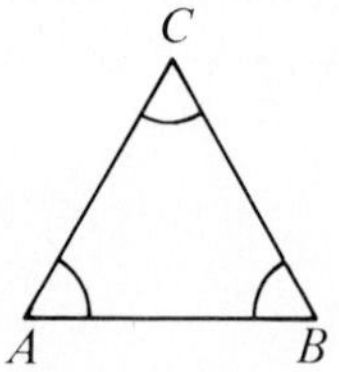

Given: $\angle A = \angle B = \angle C$

Prove: $AB = BC = CA$

STATEMENTS	REASONS
1. $\angle A = \angle C$; $\angle A = \angle B$; $\angle B = \angle C$	1. ?
2. $AB = BC$; $BC = CA$; $CA = AB$	2. ?
3. $AB = BC = CA$	3. ?

12. Construct a 60° angle.

13. Use your construction from Exercise 12. Construct a 30° angle.

14. Use your construction from Exercises 12 and 13. Construct a 15° angle.

In Exercises 15–17, refer to the sample on page 80.

B 15. Construct a 150° angle. Select a strategy.
One choice: Use the fact that 150° = 180° − 30°.
Another choice: Use the fact that 150° = 90° + 60°.

16. Construct a 105° angle.

17. Construct a 75° angle.

Equal angles are marked. Find the value of x.

18.

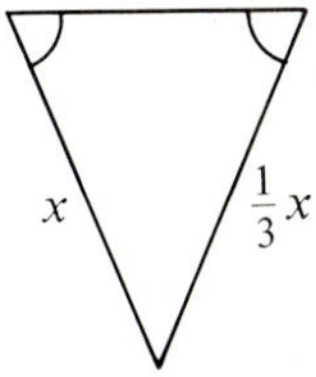

19.

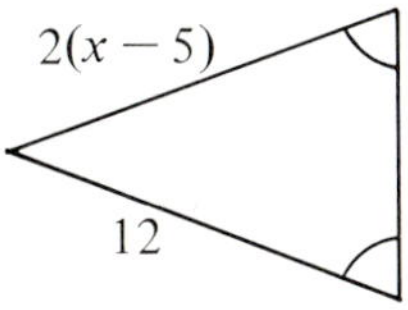

20.

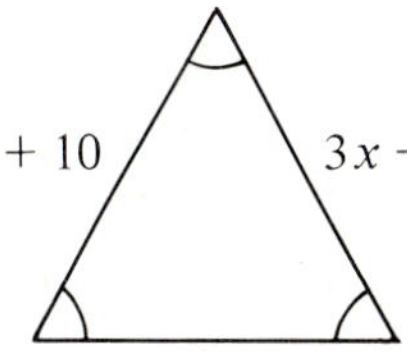

Copy what is shown. Then write a proof in two-column form.

21. *Given:* $JK = NM$; $\angle J = \angle N$

Prove: $\angle 5 = \angle 6$

Strategy for proof:

A. Think of $\angle J$ and $\angle N$ as angles of $\triangle JPN$. Prove that $PJ = PN$.

B. Prove that $\triangle JPK \cong \triangle NPM$ and conclude that $\angle 5 = \angle 6$.

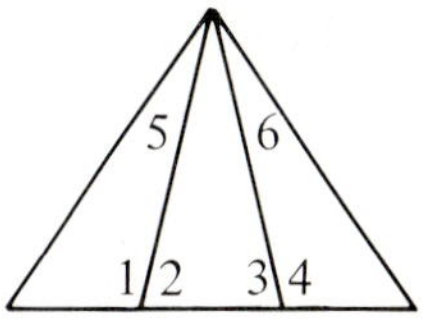

Exercises 21, 22

22. *Given:* $\angle 1 = \angle 4$

Prove: $\triangle PKM$ is an isosceles triangle.

C 23. The goal of this exercise is to prove the statement: In an isosceles triangle, the bisector of the vertex angle bisects the base and is perpendicular to the base.
- a. Draw and label an appropriate diagram.
- b. List what you are given and what you are to prove in terms of your diagram.
- c. Write a proof in two-column form.

Experiments

Counting Triangles You Cannot Draw

Begin with a very large equilateral triangle.

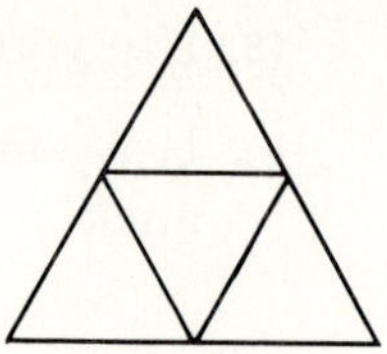

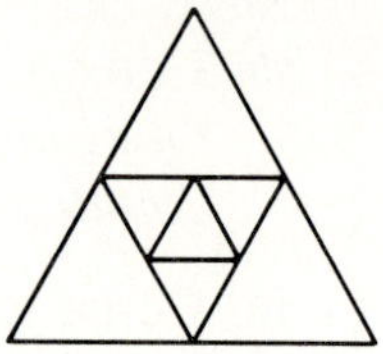

1. Join the midpoints of the three sides. How many equilateral triangles are shown now?
2. Using the "inside" triangle, again join the midpoints of the sides. How many equilateral triangles are there now?
3. Repeat this process and complete the table.

Step		1	2	3	4	5
No. of △	1	5	9	?	?	?

4. Look at your completed table. Each time you join the midpoints of a triangle, how many new triangles are formed?
5. How many triangles would be formed if you joined midpoints eight times?

SELF-TEST

1. Suppose you know that $TA = TE$. Write the theorem that supports the statement: $\angle A = \angle E$. *(4-6)*

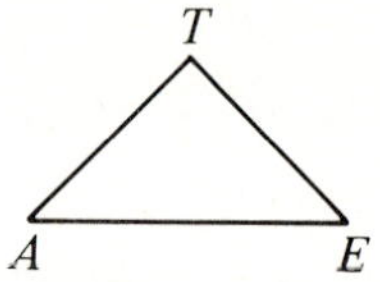

In each diagram, equal sides are marked. Find the value of y.

2. 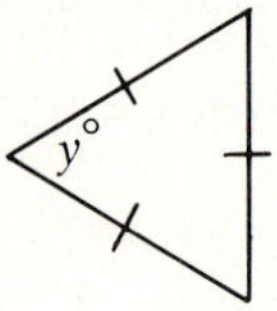

3. 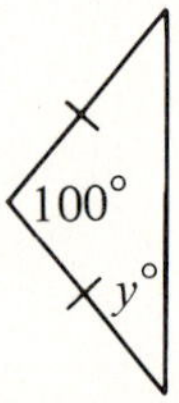

4. 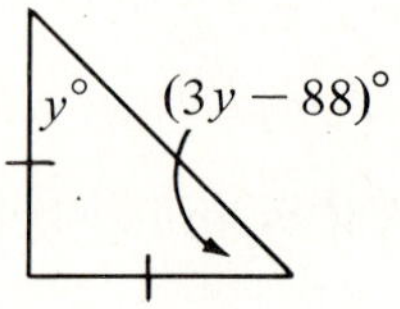

In each diagram, equal angles are marked. Find the value of x. *(4-7)*

5. 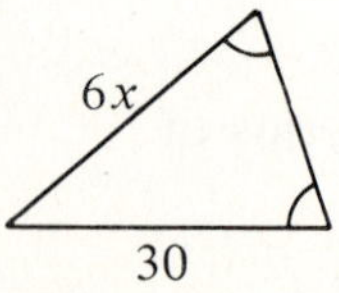

6. 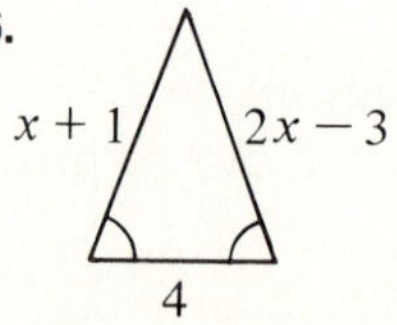

7. 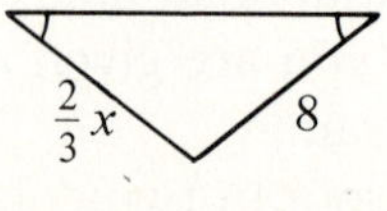

APPLICATIONS

THE TRIANGLE INEQUALITY

Suppose we are given the lengths of three line segments. Can we build a triangle with these segments? The SSS Postulate tells us that if we can, there is only one triangle possible.

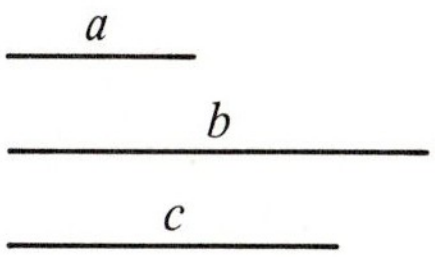

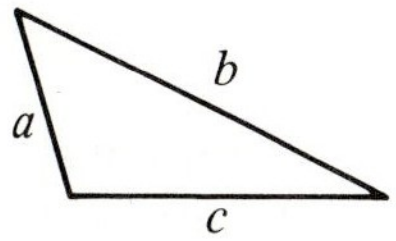

It may not be possible to make any triangle at all. Can you see that there is no triangle with sides of lengths 2, 4, and 10?

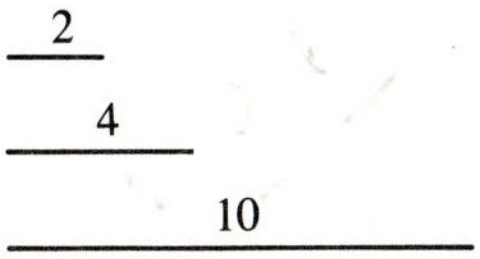

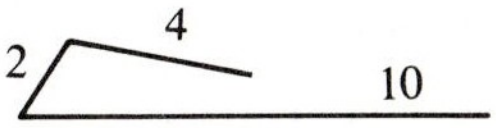

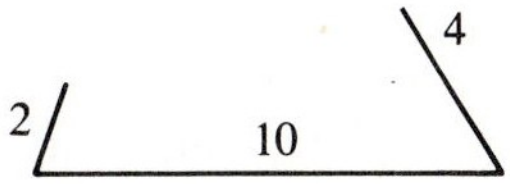

The triangle inequality is a rule which tells us whether three lengths can form the sides of a triangle.

THE TRIANGLE INEQUALITY
In a triangle, the sum of the lengths of any two sides must be greater than the length of the third side.

In the example above, the sum of 2 and 4 is not greater than 10. Therefore, we cannot make a triangle from these lengths.

Some people think of the triangle inequality as a special way of saying "the shortest path between two points is a straight line." You have probably used this idea yourself if you have ever taken a short cut across a corner. Look at $\triangle ABC$. If the triangle inequality were not true, the "short cut" from A to B would be longer than the "detour" through C!

EXERCISES

Can you draw a triangle using these lengths for the sides?

1. 3, 4, 5 **2.** 3, 4, 1 **3.** 3, 4, 20

4. 7, 7, 7 **5.** 7, 7, 14 **6.** 7, 7, 0.001

PROBLEM SOLVING STRATEGIES

DIAGRAMS AND CHARTS

If a problem contains a lot of information, it is often helpful to draw a picture or use a chart to organize the data. The following exercises will illustrate the use of a picture and a chart.

EXERCISES

1. Six people are in a room and each one shakes hands with each of the others exactly once. How many handshakes take place?

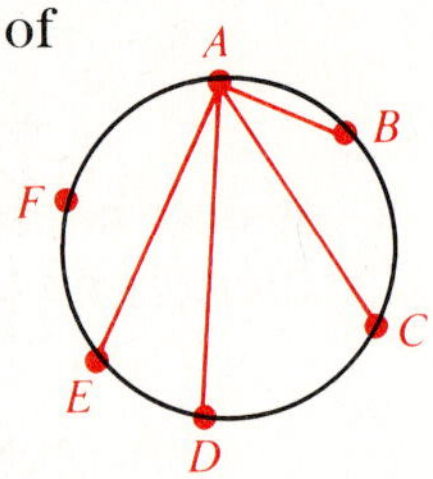

A possible picture that represents the conditions given in the problem could be a circle with six points around it. A handshake would then be a line drawn from one point to another.

The picture shows that A has shaken hands with B, C, D, and E. Complete the diagram to find the number of handshakes that take place.

2. Digger, Moose, Bigfoot, Goalie, and Fang are all dogs in the same neighborhood. Two are short-haired and three are long-haired. Two have short tails and three have long tails.

Digger and Moose have the same length hair;
Fang and Goalie have different length hair.

Bigfoot and Goalie have the same length tails;
Moose and Fang have different length tails.

What is the name of the dog with short hair and a short tail?

	Short hair	Long hair	Short tail	Long tail
Digger	✓		✓	
Moose	✓			
Bigfoot				
Goalie				
Fang				

Notice the chart shown above. It lists the names of the dogs at the left and their characteristics at the top.

a. Let's assume that the first dog, Digger, is the one with the short hair and a short tail. Check marks have been placed in the appropriate boxes by Digger's name.

One of the clues says that Digger and Moose have the same length hair, so a check mark is placed showing that Moose also has short hair.

Explain why it is impossible to satisfy the condition that Fang and Goalie have different length hair.

Now explain why it is impossible for Digger to have short hair and a short tail.

b. Now assume that the second dog, Moose, is the one with short hair and a short tail. Copy the chart without any check marks. Then fill in the boxes until you come to an impossible situation. You then know that Moose cannot be the dog with the short hair and a short tail.

c. Do the same thing for the rest of the dogs until you find the only dog that can have short hair and a short tail.

d. Find the type of hair and tail each dog has. Start by making check marks for the dog with short hair and a short tail that you found in part (c). Copy the chart without any checkmarks. Then use the given information and parts (a) and (b) to complete the chart.

SKILLS REVIEW

DECIMALS AND PERCENTS

Add or subtract. Be sure to keep the decimal points in line.

1. $5.372 + 2.197$ **2.** $91.08 - 16.37$ **3.** $7.65 + 0.829$ **4.** $56.7 - 29.54$

5. $26.03 + 62.9$ **6.** $5.003 - 1.315$ **7.** $9.666 - 7.209$ **8.** $1.62 + 4.7$

Multiply.

SAMPLE

$39.6 \times 2.04 = 80.784$

1 decimal place | 2 decimal places | 3 decimal places

9. 4×39.48 **10.** 100×61.2 **11.** 0.2×9.3 **12.** 5.9×0.766

13. 0.16×365 **14.** 35.27×7.75 **15.** 1.3×8.909 **16.** 81×3.14

Divide. If the division does not "come out even," round the answer to two decimal places.

SAMPLE

$0.51\overline{)4.7} \rightarrow 0.51\overline{)4.70\ 000}$ (quotient 9.215)

Round up

Answer: 9.22

17. $1.5\overline{)135}$ **18.** $0.7\overline{)2.93}$ **19.** $0.34\overline{)69.7}$ **20.** $0.8\overline{)11}$

21. $6.1\overline{)0.437}$ **22.** $0.03\overline{)12.8}$ **23.** $0.075\overline{)5}$ **24.** $9.4\overline{)6.18}$

Write as a percent.

SAMPLES

$0.9 = 0.90 = 90\%$

$\frac{9}{25} = \frac{36}{100} = 36\%$

$\frac{3}{8} = 3 \div 8 = 0.375 = 37.5\%$

25. $\frac{3}{4}$ **26.** 0.27 **27.** 0.6 **28.** $\frac{4}{5}$

29. $\frac{7}{8}$ **30.** $0.33\frac{1}{3}$ **31.** $\frac{1}{2}$ **32.** $\frac{7}{10}$

Complete.

33. 80% of 1125 is __?__.

34. 17 is __?__% of 51.

35. 25% of __?__ is 96.

36. 30% of 15.8 is __?__.

37. 13 is __?__% of 65.

38. __?__ is 97% of 482.

CHAPTER REVIEW

CHAPTER SUMMARY

1. One way to prove that two segments or two angles are equal is to show that they are corresponding parts of congruent triangles.

2. Congruent triangles can sometimes be used to show that constructions are correct.

3. Any segment, ray, or line that passes through the midpoint of a line segment is a bisector of the segment. Any point on the perpendicular bisector of $\overline{AB}$ is equidistant from A and B. Any point equidistant from A and B must lie on the perpendicular bisector of $\overline{AB}$.

4. Every triangle has three altitudes. Every triangle has three medians.

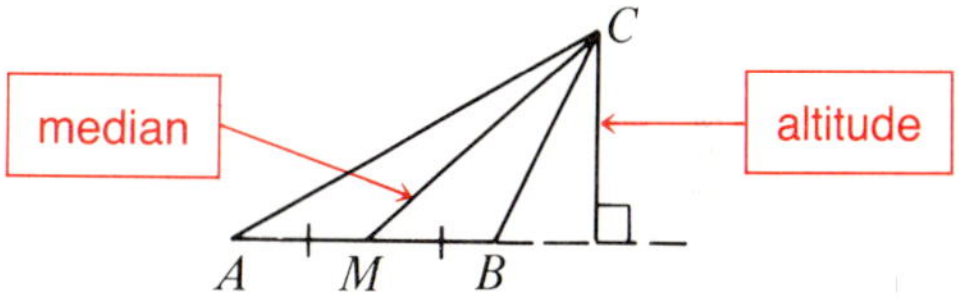

5. A circle can be circumscribed about a triangle (Construction 7). A circle can be inscribed in a triangle (Construction 8).

6. If two sides of a triangle are equal, then the angles opposite those sides are equal. An equilateral triangle is also equiangular, and each angle has measure 60°.

7. If two angles of a triangle are equal, then the sides opposite those angles are equal. If a triangle is equiangular, it is also equilateral.

8. Construction 9 tells how to construct a 60° angle.

REVIEW EXERCISES

Supply the reasons to complete the proof. *(See pp. 144–148.)*

1. *Given:* $CM = DM$; $EM = FM$

   Prove: $\overline{EC} \parallel \overline{DF}$

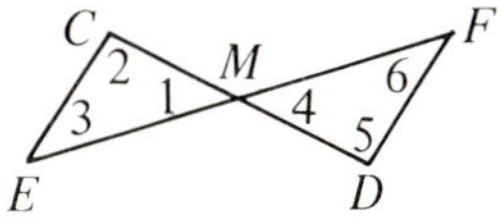

STATEMENTS	REASONS
1. $CM = DM$; $EM = FM$	1. ?
2. $\angle 1 = \angle 4$	2. ?
3. $\triangle CME \cong \triangle DMF$	3. ?
4. $\angle 3 = \angle 6$	4. ?
5. $\overline{EC} \parallel \overline{DF}$	5. ?

Supply reasons to complete a proof that the following construction is correct. ***(See pp. 149–152.)***

2. Angle A was given, and angle B was constructed to be equal to angle A.
   1. $BT = AR$ and $BU = AS$
   2. $TU = RS$
   3. $\triangle UBT \cong \triangle SAR$
   4. $\angle B = \angle A$

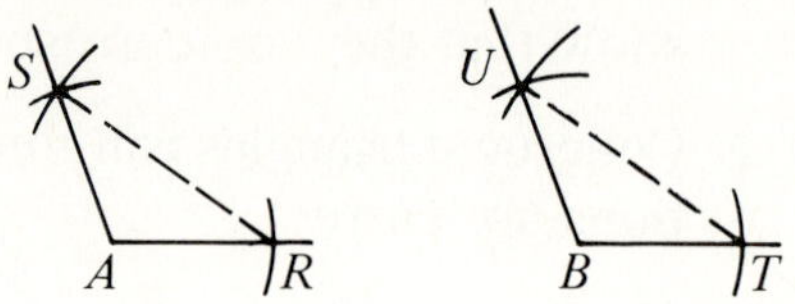

Draw a segment $\overline{CD}$. ***(See pp. 154–157.)***

3. Construct the perpendicular bisector of $\overline{CD}$, and label it k.

4. Refer to Exercise 3. Suppose you know, for some point X, that $CX = DX$. Where must X lie?

5. Suppose Z is a point on k. What can you say about Z with respect to C and D?

Refer to the figure shown. ***(See pp. 158–162.)***

6. Name an altitude shown.

7. Name a median shown.

8. Draw a large acute triangle, $\triangle MNO$. Construct the altitude from M.

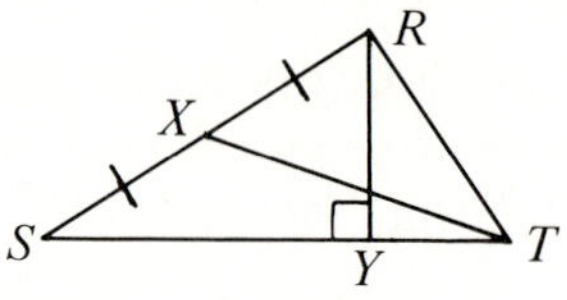

Draw two large figures, an obtuse $\triangle ABC$ and an acute $\triangle DEF$. ***(See pp. 163–166.)***

9. Circumscribe a circle about $\triangle ABC$.

10. Inscribe a circle in $\triangle DEF$.

Write a proof of the theorem. ***(See pp. 167–171.)***

11. If two sides of a triangle are equal, then the angles opposite those sides are equal.

    Given: $KX = KY$

    Prove: $\angle X = \angle Y$

    (*Hint:* Begin your proof by drawing the bisector of $\angle K$.)

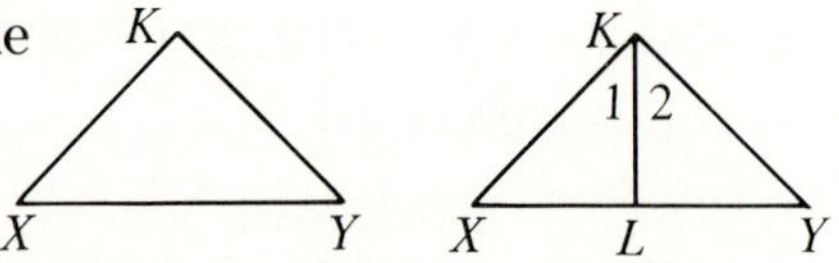

Equal angles are marked. Find x. ***(See pp. 173–178.)***

12.

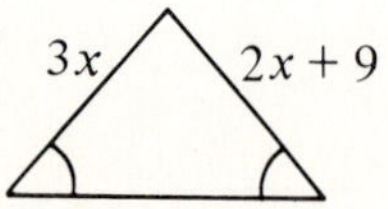

13.

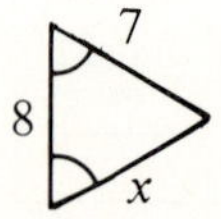

14.

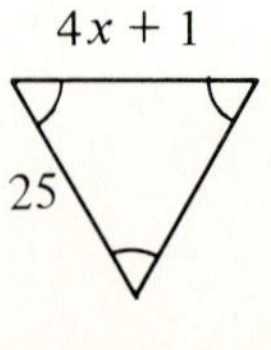

CHAPTER TEST

***Given:* M is the midpoint of $\overline{AB}$ and $\overline{CM} \perp \overline{AB}$.**

1. What method could be used to prove $\triangle ACM \cong \triangle BCM$? *(4-1)*

2. After proving the triangles congruent, you could conclude that $\triangle ABC$ is isosceles. What information and supporting reason could you use?

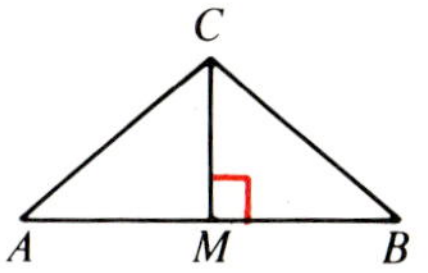

3. Tony was given $\triangle PQR$. By what method did he construct $\triangle XYZ$ congruent to $\triangle PQR$? *(4-2)*

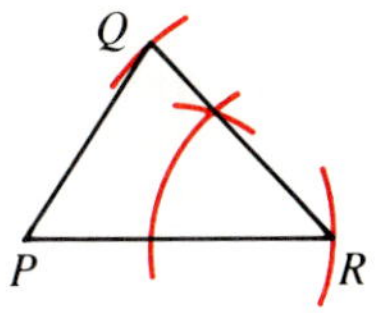

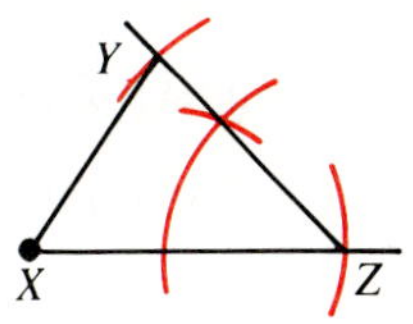

4. Draw a large triangle like $\triangle PQR$. By construction, locate the midpoint M of $\overline{QR}$. *(4-3)*

***Given:* $\overleftrightarrow{RS}$ is the perpendicular bisector of $\overline{AB}$.**

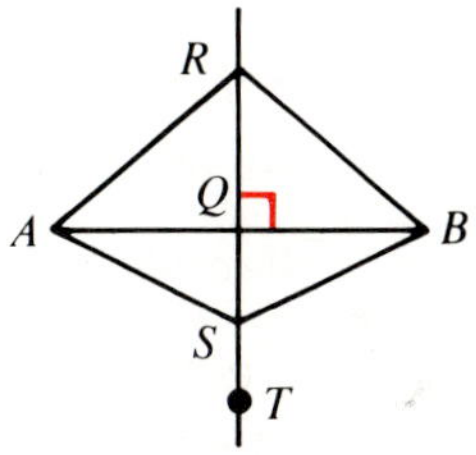

5. Point T is __?__ from A and B.

6. Why must $RA = RB$ and $SA = SB$?

7. Why is $\triangle RAS \cong \triangle RBS$?

8. Is $\overline{BQ}$ an altitude of $\triangle RBT$? a median? *(4-4)*

9. Draw a very large obtuse triangle. Draw and label the altitude and the median to the longest side. Which segment is longer?

10. Describe a method you could use to construct a circle that passes through the vertices of a given triangle. *(4-5)*

11. Draw a large acute triangle. Construct an inscribed circle in it.

***Given:* RS = TS and QS = US. Write the reason that supports each statement.**

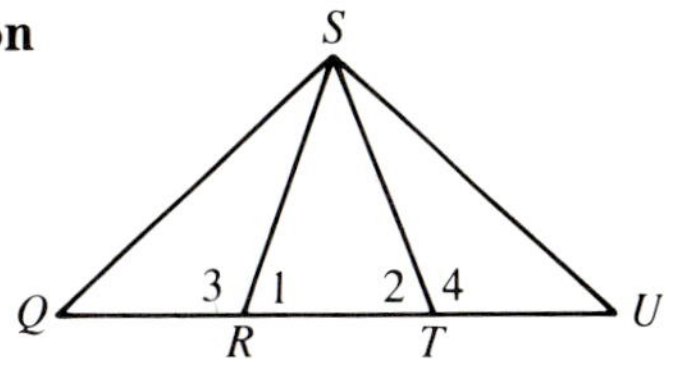

12. $\angle 1 = \angle 2$ 13. $\angle Q = \angle U$ *(4-6)*

14. $\angle 3 = \angle 4$ 15. $\triangle QRS \cong \triangle UTS$

16. In $\triangle ABC$, $\angle A = \angle B$. If $AB = 4x - 5$, $BC = 3x + 2$, and $AC = 4x - 3$, find x. *(4-7)*

UNIT B CUMULATIVE REVIEW

Complete each statement.

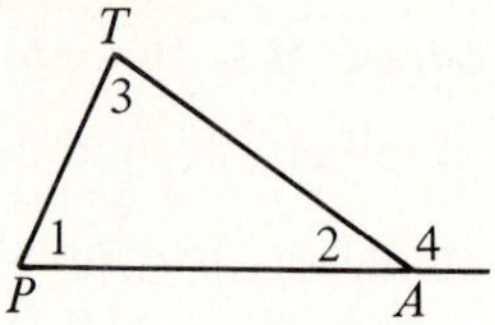

1. In $\triangle PAT$ at the right, $\angle 4 = \angle$ __?__ + $\angle$ __?__.
2. A triangle with no equal sides is called __?__.
3. A triangle with two equal sides is called __?__.
4. If a triangle has two 40° angles, the third angle equals __?__°.
5. If a triangle has a 110° angle, then the triangle is called __?__.
6. In a right triangle, the side opposite the right angle is called the __?__.
7. If $\triangle AUK \cong \triangle JIE$, then $\angle A = \angle$ __?__.
8. If $\triangle RMZ \cong \triangle AKC$, then $RM =$ __?__.

Using the given information, write SSS, SAS, ASA, AAS, or HL to tell which method you could use to prove $\triangle LBU \cong \triangle LBI$.

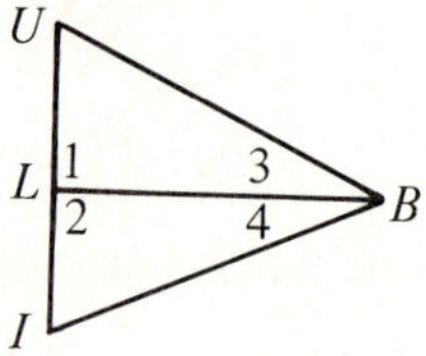

9. $\overline{BL} \perp \overline{UI}$; $BU = BI$
10. $BU = BI$; $UL = IL$
11. $\angle 1 = \angle 2$; $\angle 3 = \angle 4$
12. $\angle U = \angle I$; $\angle 3 = \angle 4$
13. $\angle 1 = 90°$; $\angle U = \angle I$
14. $BU = BI$; $\angle 3 = \angle 4$

Tell whether each statement is true or false.

15. A segment has exactly one bisector.
16. If point A is on a perpendicular bisector of $\overline{CD}$, then $AC = AD$.
17. The three lines containing the altitudes of a triangle intersect in a point.
18. Medians and altitudes are always the same segments.
19. In $\triangle ABC$, if $AB = AC$, then $\angle B = \angle C$.
20. An equilateral triangle is also equiangular.
21. Construct an equiangular triangle.
22. Draw an acute $\triangle DEF$. Circumscribe a circle about $\triangle DEF$.

UNIT C

Here's what you'll learn in this chapter:

To classify polygons.

To find the interior and exterior angle sums of a convex polygon.

To use the properties of parallelograms, rectangles, rhombuses, and squares.

To prove that a quadrilateral is a parallelogram.

To use properties of trapezoids.

To apply the Midpoints Theorem.

Have you ever noticed how pentagons and hexagons are combined on the surface of a soccer ball? Can you name other polygons that are suggested by the cords of the net shown in the photograph?

Chapter 5

Polygons

1 Introducing Polygons

The word *polygon* comes from the Greek words meaning *many angles.* The polygons shown in the first row below are called **convex polygons.** Do you see how they are different from the polygons in the second row?

One way to tell whether or not a polygon is convex is to imagine fitting a rubber band along the edges of the figure. If the rubber band fits snugly, then the polygon is convex. If it doesn't, the polygon is not convex.

Convex Polygons

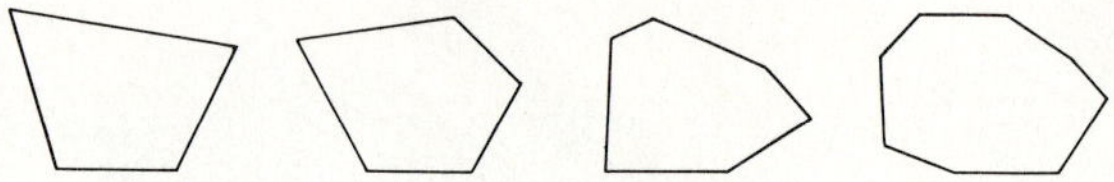

Not Convex Polygons

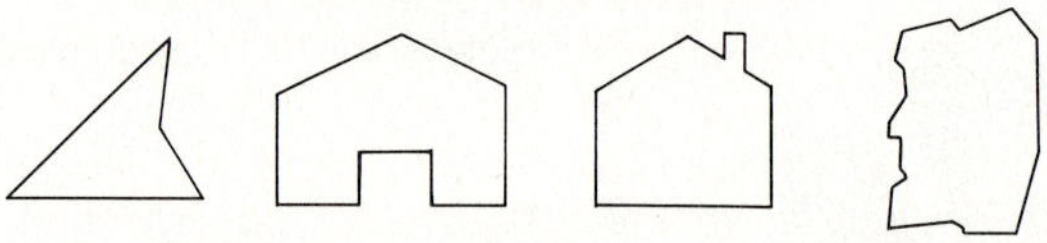

The figures should suggest what a polygon is. Even so, it is difficult to write a definition of *polygon.* Perhaps you and your class would like to try to write one. The figures below are *not* polygons and should not satisfy your definition.

Not Polygons

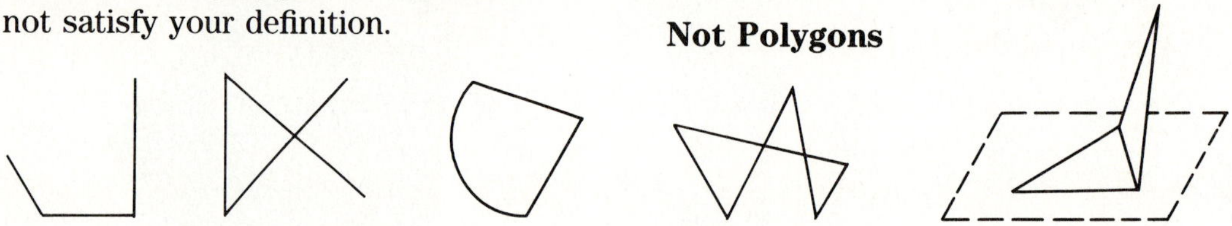

When referring to a polygon, we name its consecutive vertices in order. The *quadrilateral* shown at the top of the next page can be called, for example, *ABCD*, *DCBA*, or *BADC.* A polygon is classified according to the number of sides it has. The table at the right lists some special names.

Number of Sides	Name of Polygon
3	triangle
4	quadrilateral
5	pentagon
6	hexagon
8	octagon
10	decagon

The simplest polygon is a triangle. The terms which we defined for triangles (such as *sides*, *vertices*, and *angles*) also apply to other polygons.

Two sides which intersect are called **consecutive sides.** The endpoints of a side are called **consecutive vertices.** A segment which joins nonconsecutive vertices is called a **diagonal** of the polygon.

$\overline{AB}$ and $\overline{BC}$ are consecutive sides.
A and B are consecutive vertices.
$\overline{AC}$ and $\overline{BD}$ are diagonals.

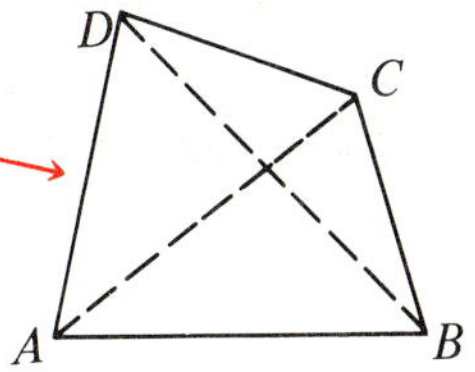

Polygons can be equiangular, equilateral, or both as shown below.

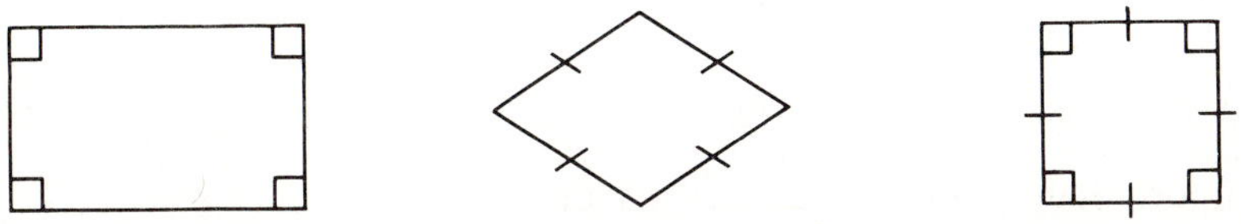

Equiangular (not equilateral)

Equilateral (not equiangular)

Both equiangular and equilateral

If a polygon is both equiangular and equilateral, it is called a **regular polygon.** A regular pentagon is shown at the right. The angle measures for regular polygons are explained in the next section.

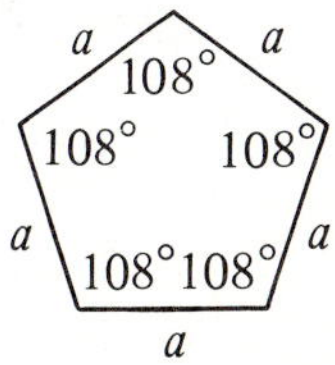

Classroom Practice

For each figure shown, decide (a) if the figure is a polygon and (b) if the figure is a convex polygon.

1. 
2. 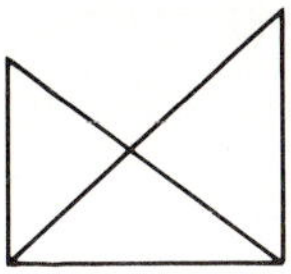
3. 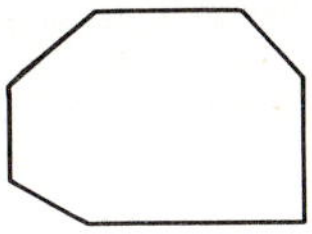
4. 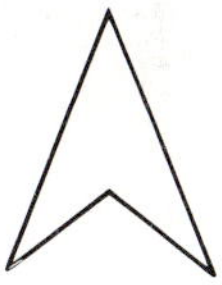
5. 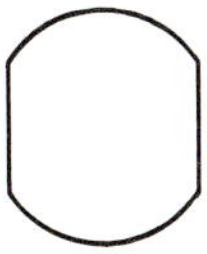
6. 
7. 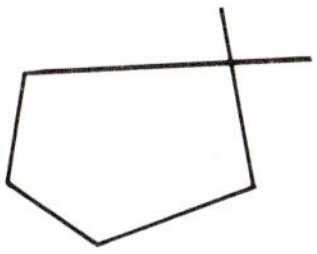
8. 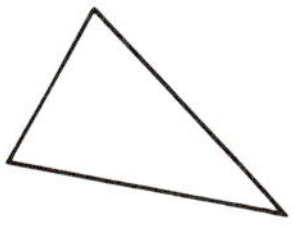

Which of the following are acceptable names for the pentagon shown?

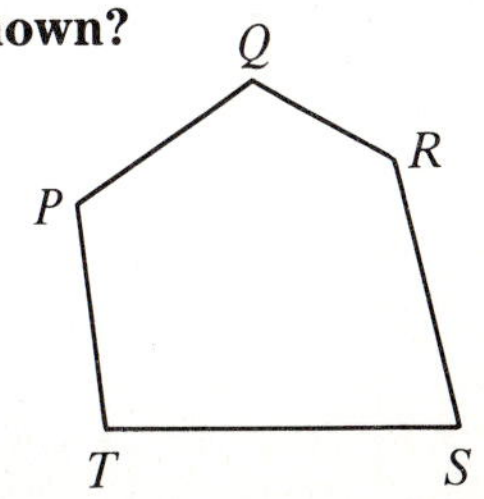

9. *PQRST*
10. *PTQRS*
11. *SRQPT*
12. Name any two consecutive sides of the pentagon.
13. Name any two consecutive vertices of the pentagon.
14. Name as many diagonals as you can.
15. A regular polygon with four sides is usually called a __?__.

Written Exercises

Is the polygon shown a convex polygon?

A

1. 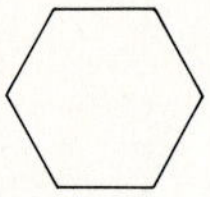

2. 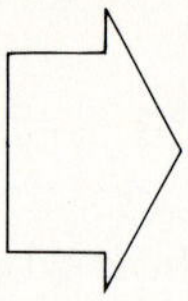

3. 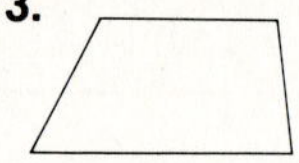

4. 

5. 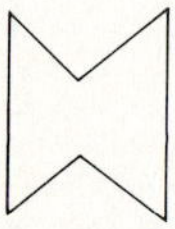

6. 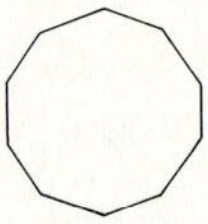

Which of the names below are acceptable for the hexagon shown?

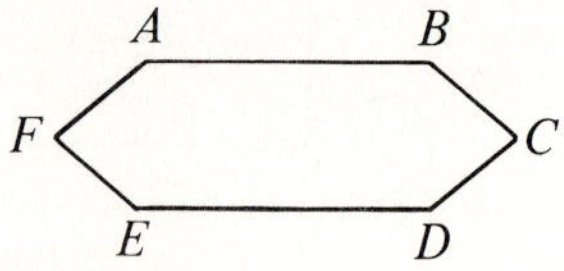

7. *ABCDEF* 8. *DCBAFE* 9. *CDFEAB*

10. How many diagonals can you draw from vertex *A*?

11. How many diagonals can you draw from vertex *F*?

12. $\overline{DE}$ and __?__ are consecutive sides. (Two answers are possible.)

Classify each polygon as equiangular, equilateral, or regular.

13. 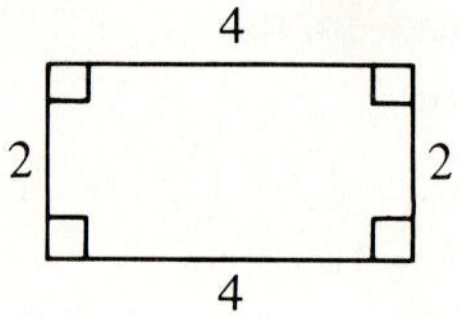

14. 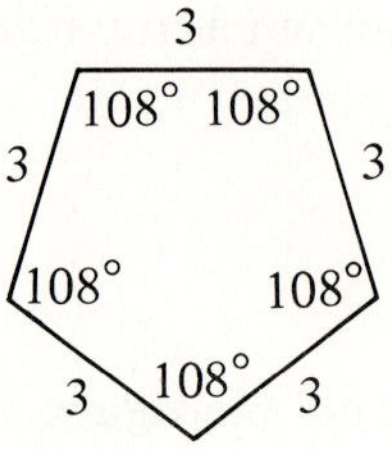

15. 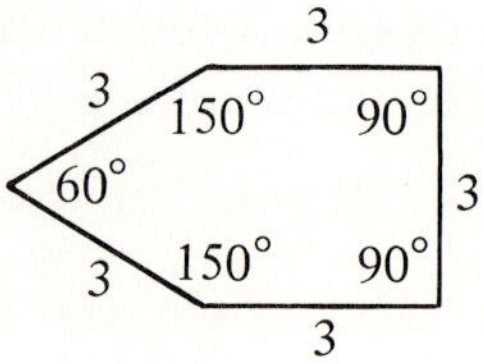

Draw a convex polygon which satisfies the conditions.

16. A quadrilateral which is equiangular but not equilateral

17. A quadrilateral which is equilateral but not equiangular

18. A regular triangle 19. A regular hexagon

B

20. A hexagon which is equilateral but not equiangular

21. A hexagon which is equiangular but not equilateral

In a convex pentagon, you can draw two diagonals from each vertex. This information is given in the table below. Copy and complete the table.

	22.	Sample	23.	24.	25.	26.	27.
Number of sides of a convex polygon	4	5	6	7	8	20	n
Number of diagonals from each vertex	?	2	?	?	?	?	?

Copy and complete the table below.
(*Hint:* Draw a diagram for each exercise.)

C

	28.	29.	30.	31.	32.
Number of sides of a convex polygon	4	5	6	7	8
Total number of diagonals	?	?	?	?	?

Experiments

Experiment 1

1. Cut out a strip of paper. A strip two or three centimeters wide cut from the long side of a sheet of notebook paper is convenient. Shade one side of your paper strip.

Figure 1

2. Begin to "tie" the paper as you would to form an ordinary overhand knot (see Figure 1).
3. Carefully press the folds flat. Trim off the excess paper labeled L and R in Figure 2. The polygon formed is a regular pentagon!

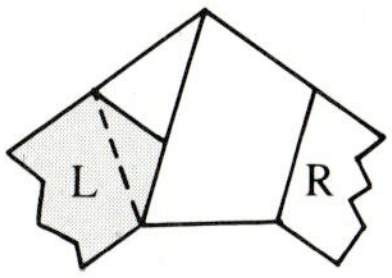

Figure 2

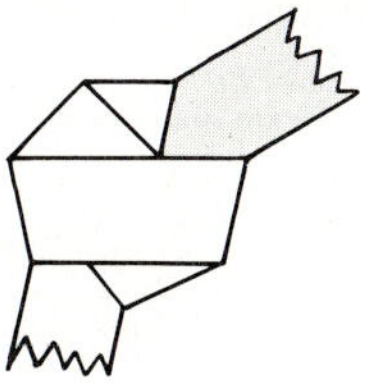

Figure 3

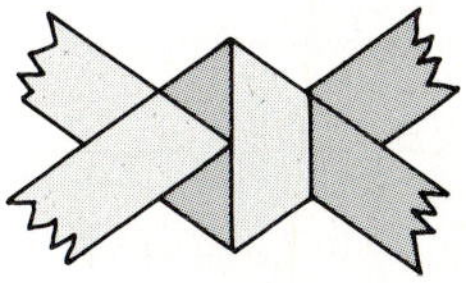

Figure 4

Experiment 2

1. Cut and shade one side of another strip of paper.
2. Begin an overhand knot. Then bring a free end around to the other side and tuck it under itself (see Figure 3).
3. Press the folds flat and trim off the excess paper. You now have a septagon!

Experiment 3

1. Cut two strips from different colored pieces of paper.
2. Make a hexagon by tying the two strips together (see Figure 4).

2 Angle Sums of Polygons

In Section 1 of Chapter 3 we proved that the angle sum of any triangle is 180°. In fact, we can find the angle sum of any convex polygon if we know how many sides the polygon has. The diagrams below suggest a way to find this sum.

Angle Sum of a Quadrilateral

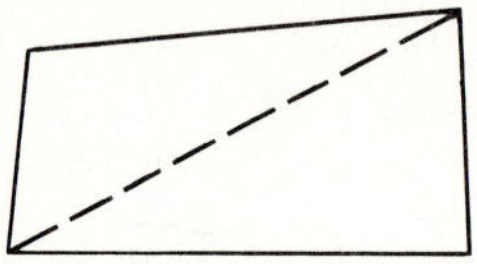

Angle sum of each triangle = 180°
Angle sum of quadrilateral = 2 × 180°
= 360°

Angle Sum of a Pentagon

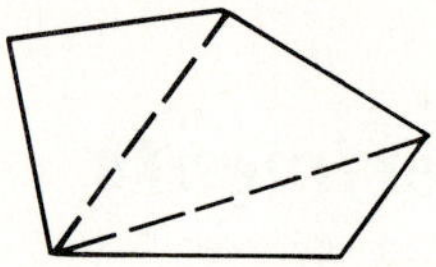

Angle sum of each triangle = 180°
Angle sum of pentagon = 3 × 180°
= 540°

Before reading further, try to find the angle sums of convex polygons with six and seven sides. Check your answers in the table below.

Number of sides of polygon	Number of triangles formed by drawing all the diagonals from one vertex	Angle sum of polygon
4	2	2 × 180° or 360°
5	3	3 × 180° or 540°
6	4	4 × 180° or 720°
7	5	5 × 180° or 900°
8	6	6 × 180° or 1080°

Do you see that if a convex polygon has n sides, you can form $n - 2$ triangles by drawing all the diagonals from one vertex? Therefore, the angle sum of the polygon is the same as the angle sum for $n - 2$ triangles. We state this as our next theorem.

THEOREM 1

If a convex polygon has n sides, then its angle sum is given by the formula

$$S = (n - 2) \times 180°.$$

EXAMPLE

A regular polygon has 12 sides.

a. Find its angle sum.

b. Find the measure of each angle of the polygon.

a. $S = (n - 2) \times 180° = (12 - 2) \times 180° = 1800°$

b. Since the polygon is regular, all 12 angles are equal.
The measure of each angle is $\frac{1}{12} \times 1800° = 150°$.

The angle sum of a polygon is sometimes called the *interior angle sum* to distinguish it from the *exterior angle sum* of the polygon. The exterior angle sum for the hexagon shown is the sum of the measures of the six indicated angles. (Notice that we consider only one of the two exterior angles at each vertex.) To find this sum, imagine yourself walking along the sides of the polygon. As you reach each vertex, you turn through the number of degrees in an exterior angle. When you return to your starting point, you will have turned through 360°.

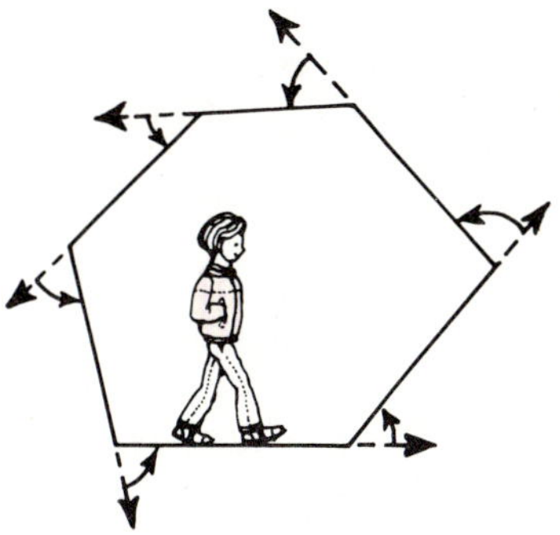

Do you see that given *any* convex polygon with any number of sides, you will turn through 360°? This fact is stated in Theorem 2. A more formal proof is suggested by Exercise 21, page 197.

THEOREM 2

The exterior angle sum of any convex polygon, one angle at each vertex, is 360°.

Classroom Practice

Copy and complete the table for convex polygons.

	Sample	1.	2.	3.	4.	5.	6.
Number of sides	5	6	8	10	12	?	?
Interior angle sum	3 × 180°	?	?	?	?	11 × 180°	20 × 180°
Exterior angle sum	360°	?	?	?	?	?	?

Written Exercises

Copy and complete the table for convex polygons.

A

	1.	2.	3.	4.	5.	6.
Number of sides	4	9	11	22	?	?
Interior angle sum	?	?	?	?	$10 \times 180°$	$15 \times 180°$
Exterior angle sum	?	?	?	?	?	?

7. In a regular triangle each interior angle has measure __?__, and each exterior angle has measure __?__.

8. a. In a regular quadrilateral each interior angle has measure __?__, and each exterior angle has measure __?__.

b. A regular polygon with four sides is called a __?__.

In Exercises 9–11, the polygon shown is a regular polygon.
a. Name the polygon.
b. Find the interior angle sum and the measure of each interior angle.
c. Find the exterior angle sum and the measure of each exterior angle.

9.

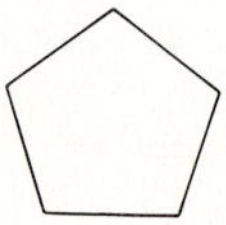

10.

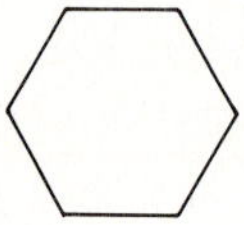

11.

Copy and complete the table for regular convex polygons.

B

	12.	13.	14.	15.	16.
Number of sides	12	?	?	?	?
Each interior angle	?	120°	140°	?	?
Each exterior angle	?	?	?	20°	10°

17. The neighboring atoms in a sheet of graphite consist of interlocking regular hexagons. Can you interlock STOP signs as you can hexagons? Use the results of Exercise 11 to explain your answer.

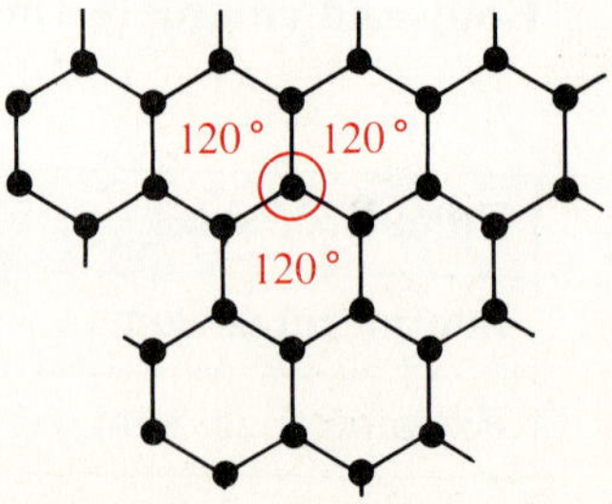

18. Could you tile an entrance way with regular pentagons? Use the results of Exercise 9 to explain your answer.

19. *Given:* Quadrilateral $ABCD$
$\angle A = \angle C = x°$
$\angle B = \angle D = y°$

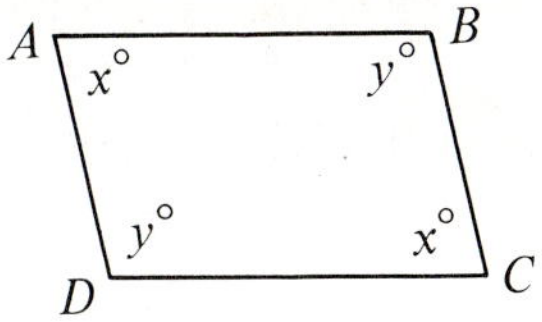

a. $2x° + 2y° = \underline{\ ?\ }°$ **b.** $x° + y° = \underline{\ ?\ }°$

c. $\overline{AD} \parallel \overline{BC}$ because same-side interior angles are ___?___.

d. $\overline{AB} \parallel \overline{CD}$ because ___?___.

C **20.** A regular pentagon and two diagonals are shown. Find the measure of $\angle 1$.

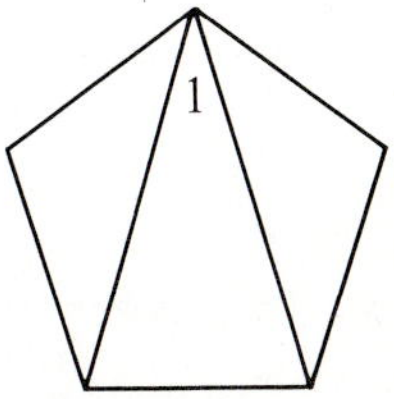

21. The diagram at the right below shows part of a polygon with n sides. Complete each statement.

a. At vertex A, $\angle 1 + \angle 2 = \underline{\ ?\ }°$.

b. At vertex B, $\angle 3 + \angle 4 = \underline{\ ?\ }°$.

c. At any vertex, the sum of an interior and an exterior angle is ___?___°.

d. If the polygon has n sides, then it has ___?___ vertices. The sum of the interior and the exterior angles at all the vertices is (___?___ × ___?___)°.

e. By Theorem 1, the sum of the interior angles is ___?___°.

f. By subtracting your answer in part (e) from your answer in (d), show that the sum of the exterior angles, one angle at each vertex, is 360°.

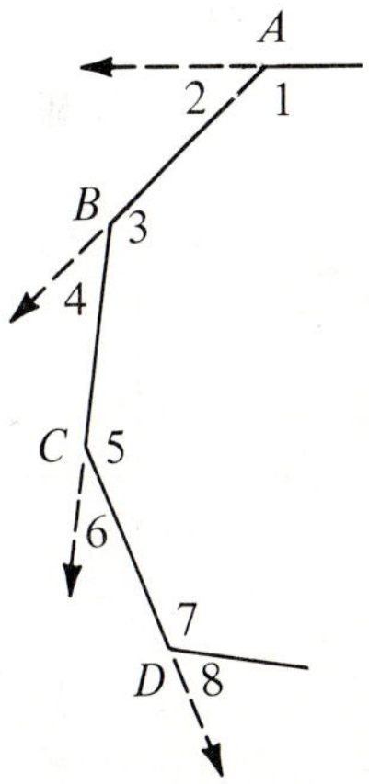

SELF-TEST

Vocabulary

convex polygon (p. 190)
consecutive sides (p. 190)
consecutive vertices (p. 190)
diagonal (p. 190)
regular polygon (p. 191)
interior angle sum (p. 195)
exterior angle sum (p. 195)

1. Draw a quadrilateral which is not convex. *(5-1)*
2. Draw a pentagon which is equilateral but not equiangular.
3. Draw a regular hexagon.
4. Find the measure of each interior angle of the hexagon. *(5-2)*
5. Find the measure of each exterior angle of the hexagon.

3 Special Quadrilaterals

Some special quadrilaterals are defined below.

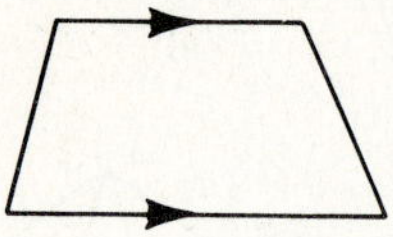

A **trapezoid** is a quadrilateral with just one pair of opposite sides parallel.

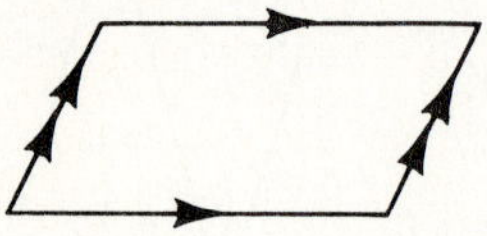

A **parallelogram** is a quadrilateral with both pairs of opposite sides parallel.

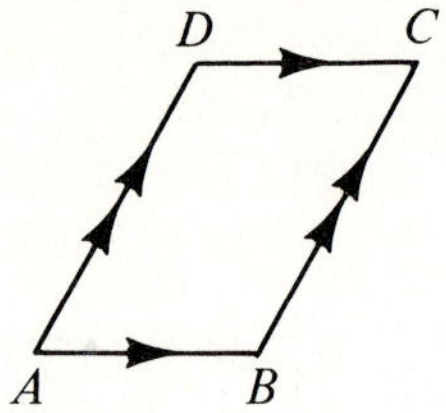

The symbol ▱ stands for parallelogram. In ▱$ABCD$, $\overline{AD} \parallel \overline{BC}$. Thus $\angle A$ and $\angle B$ must be supplementary angles. Also, $\angle B$ and $\angle C$ are supplementary because $\overline{AB} \parallel \overline{DC}$.

Three other special quadrilaterals are shown below with their definitions.

Rectangle

It has four right angles.

Rhombus

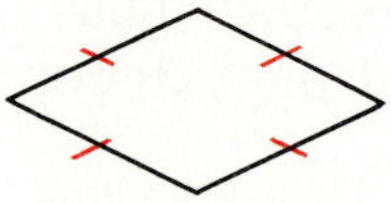

It has four equal sides.

Square

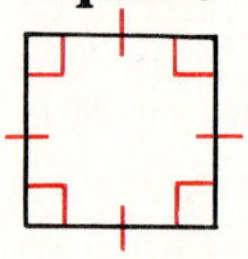

It has four right angles *and* four equal sides.

You can show that rectangles, rhombuses, and squares are all types of parallelograms (see Exercises 13–15). The diagram below illustrates how the special quadrilaterals relate to one another. Notice that every square is both a rectangle and a rhombus.

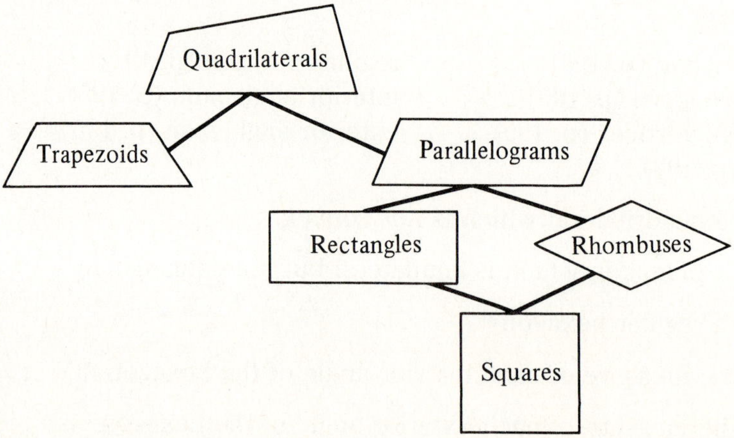

Classroom Practice

Classify each statement as true or false. Refer to the diagram at the bottom of page 198.

1. Every trapezoid is a quadrilateral.
2. Every rectangle is a quadrilateral.
3. Every parallelogram is a square.
4. Every square is a rhombus.
5. Every trapezoid is a parallelogram.
6. Every rectangle is a parallelogram.
7. List each quadrilateral below which appears to be:
   a. a parallelogram.
   b. a trapezoid.
   c. a rectangle.
   d. a rhombus.
   e. a square.

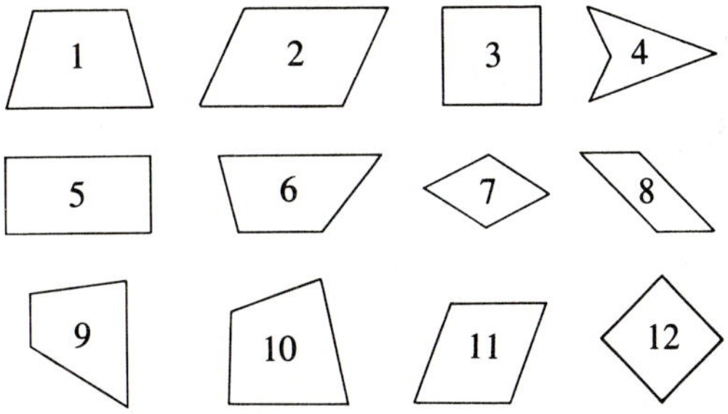

8. List each figure above which is *both* a rectangle *and* a rhombus.
9. List each figure above which is a rectangle but not a square.
10. List each figure above which is a rhombus but not a square.

In Exercises 11–14, assume that lines that appear to be parallel *are* parallel. Name all quadrilaterals that are parallelograms.

11.

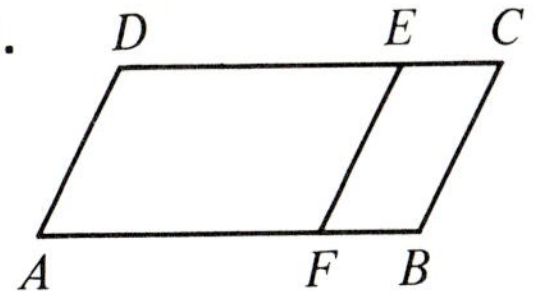

12.

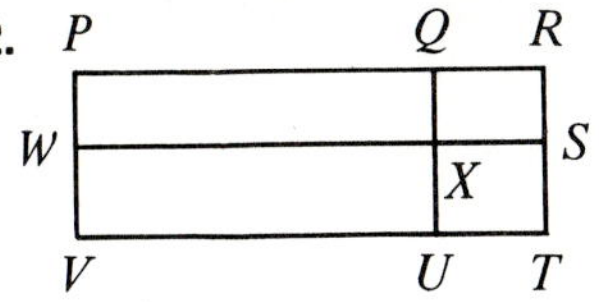

13.

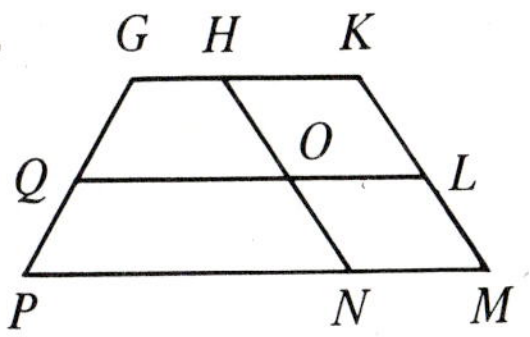

14.

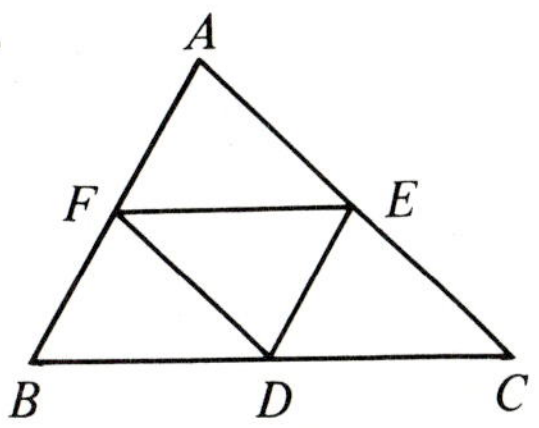

15. Refer to the figure for Exercise 13. Name all trapezoids shown.
16. Refer to the figure for Exercise 14. Name all trapezoids shown.

Written Exercises

A **1.** List each quadrilateral which appears to be:

a. a parallelogram.

b. a trapezoid.

c. a rectangle.

d. a rhombus.

e. a square.

f. both a rectangle and a rhombus.

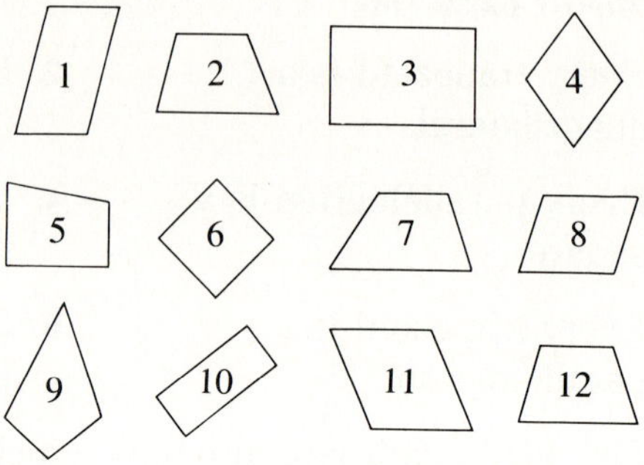

Tell whether each statement is true or false.

2. All squares are rectangles.

3. All rectangles are squares.

4. All squares are rhombuses.

5. All rhombuses are squares.

6. Imagine pushing against a rectangle with a fixed base to form a parallelogram. If you push against a square with a fixed base, what figure is formed?

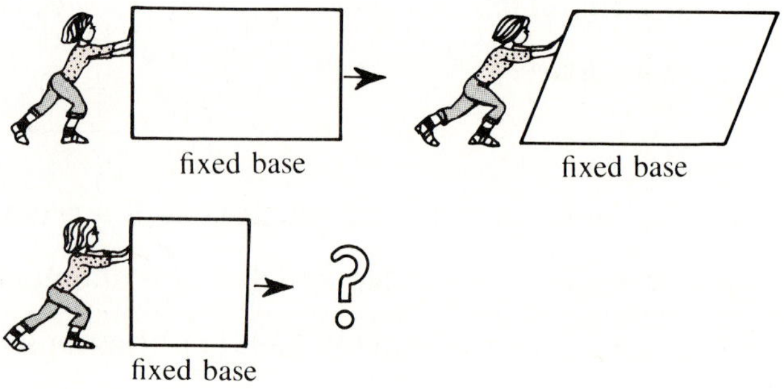

7. There are seven parallelograms in the figure below. Name as many as you can.

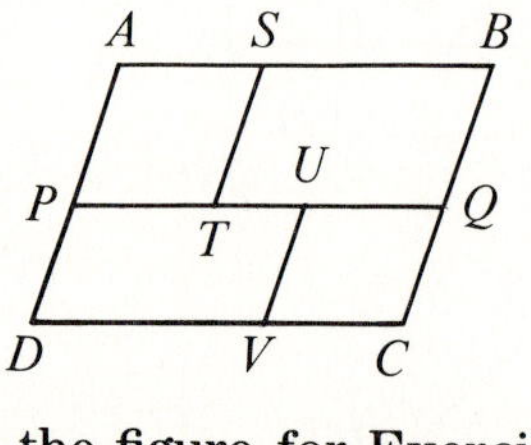

8. Name three parallelograms in the figure below.

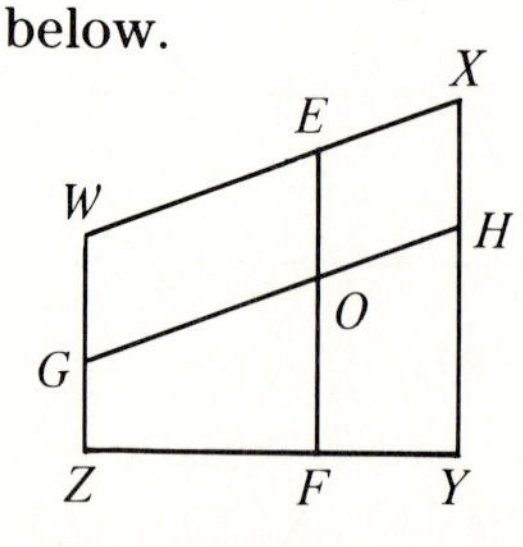

9. Refer to the figure for Exercise 7. Name two parallelograms which appear to be rhombuses.

10. Refer to the figure for Exercise 8. Name six trapezoids.

11. In the figure for Exercise 7, $\angle D = 70°$. Name four other angles with the same measure.

12. In the figure for Exercise 8, $\angle W = 110°$. Find the measure of:
a. $\angle X$ **b.** $\angle XHO$ **c.** $\angle HOE$

B

13. Our goal is to show that a rhombus must be a parallelogram. Supply the reasons to complete the proof.

Given: $ABCD$ is a rhombus.

Prove: $ABCD$ is a ▱.

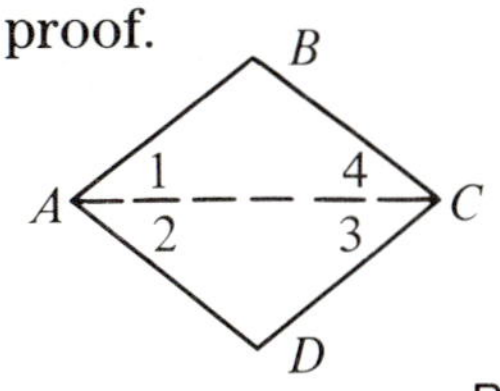

STATEMENTS	REASONS
1. $AB = CD$; $BC = DA$	1. ?
2. $AC = AC$	2. ?
3. $\triangle ABC \cong \triangle CDA$	3. ?
4. $\angle 1 = \angle 3$; $\angle 2 = \angle 4$	4. ?
5. $\overline{AB} \parallel \overline{DC}$; $\overline{AD} \parallel \overline{BC}$	5. ?
6. $ABCD$ is a ▱.	6. ?

14. Our goal is to show that a rectangle must be a parallelogram. Supply the reasons to complete the proof.

Given: $ABCD$ is a rectangle.

Prove: $ABCD$ is a ▱.

STATEMENTS	REASONS
1. $\angle A = 90°$; $\angle B = 90°$	1. ?
2. $\overline{AD} \perp \overline{AB}$; $\overline{BC} \perp \overline{AB}$	2. ?
3. $\overline{AD} \parallel \overline{BC}$	3. In a plane, if 2 lines are each $\perp$ to a third line, then ? .
4. $\angle B = 90°$; $\angle C = 90°$	4. ?
5. $\overline{AB} \perp \overline{BC}$; $\overline{DC} \perp \overline{BC}$	5. ?
6. $\overline{AB} \parallel \overline{DC}$	6. ?
7. $ABCD$ is a ▱.	7. ?

15. Explain why Exercise 14 lets you conclude that a square must be a parallelogram.

16. Construct a rectangle, using straightedge and compass.

17. Construct a trapezoid, using straightedge and compass.

18. Draw a quadrilateral $ABCD$ and locate the midpoint of each side. Join the midpoints to form a quadrilateral $EFGH$. What special kind of quadrilateral does $EFGH$ appear to be?

19. Repeat Exercise 18, beginning with a rectangle $ABCD$.

20. Repeat Exercise 18, beginning with a rhombus $ABCD$.

21. Repeat Exercise 18, beginning with a square $ABCD$.

Using straightedge and compass, construct:

C **22.** A rhombus that has a 60° angle

23. A parallelogram that has a 45° angle

CAREER NOTES

PHYSICAL THERAPIST

Did you ever squeeze a tennis ball, blow up a balloon, or assemble a jigsaw puzzle? There are millions of disabled people of all ages who find these simple tasks next to impossible. Physical therapists assist people who have nerve, joint, bone, and muscle diseases or injuries.

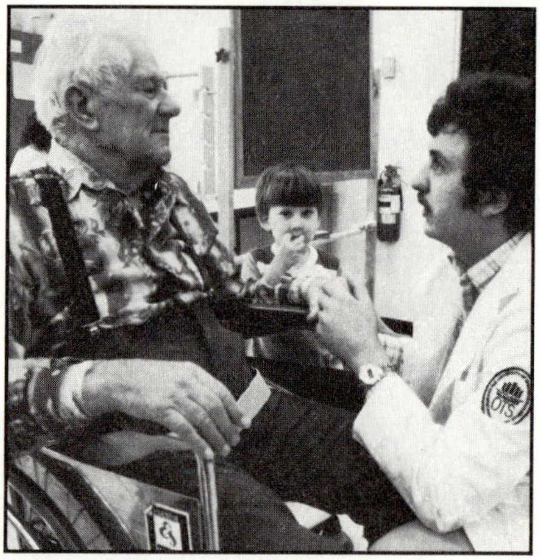

A significant part of the physical therapist's job is to develop and teach programs for treatment. Therapists use a wide range of tests to chart their patients' conditions. They test for muscular strength, muscular development, respiration, and circulation.

They then develop exercise programs for their patients. For example, a therapist may prescribe a series of breathing exercises for someone with asthma. Often, when a patient's legs are immobile, a therapist will encourage projects that require hand coordination, such as assembling a jigsaw puzzle.

Physical therapy involves a great deal of knowledge and responsibility. High school courses in physical education, science, and mathematics will help you prepare for this career. To obtain a license to practice, applicants must have earned a degree or certificate from a school of physical therapy and have passed an examination.

4 Properties of Parallelograms

Parallelograms have several special properties. Try the explorations below, and see how many properties you can find.

EXPLORATIONS

Carefully draw a very large parallelogram $ABCD$ which is neither a rectangle nor a rhombus. (Side $\overline{AB}$ should be at least 12 cm long.) Use your parallelogram for the exercises that follow.

A. The Angles of a Parallelogram

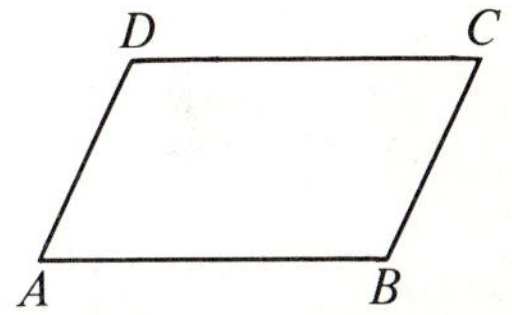

1. Carefully measure the four angles of $\square ABCD$.
2. Compare the measures of consecutive angles: $\angle A$ and $\angle B$, $\angle B$ and $\angle C$, and so on.
3. Compare the measures of opposite angles: $\angle A$ and $\angle C$, $\angle B$ and $\angle D$.
4. How can you tell, without measuring, that $\angle A + \angle B = 180°$? How can you tell, without measuring, that $\angle B + \angle C = 180°$?
5. How can you tell that $\angle A = \angle C$, using the two equations from Exercise 4?

B. The Sides of a Parallelogram

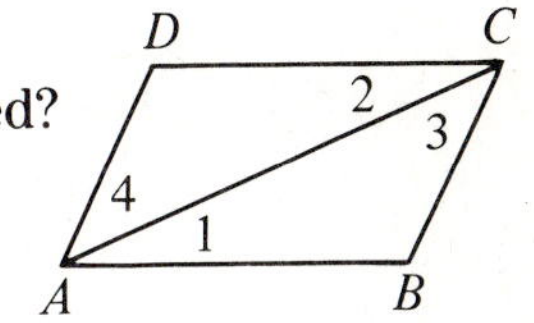

1. Carefully measure the four sides of $\square ABCD$.
2. Compare the lengths of opposite sides: $\overline{AB}$ and $\overline{CD}$, $\overline{BC}$ and $\overline{DA}$.
3. Now draw diagonal $\overline{AC}$.
4. What appears to be true about the two triangles formed?
5. How do you know that $\triangle ABC \cong \triangle CDA$?
6. Use the fact that $\triangle ABC \cong \triangle CDA$ to show that the opposite sides of $\square ABCD$ are equal.

C. The Diagonals of a Parallelogram

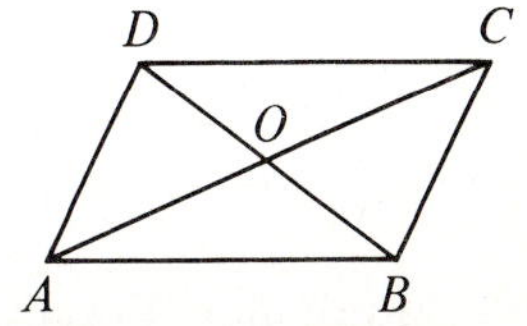

1. Draw the other diagonal, $\overline{BD}$. Let O be the intersection point of the diagonals.
2. Carefully measure both diagonals. Is $AC = BD$?
3. Carefully measure $\overline{AO}$ and $\overline{OC}$. Is $AO = OC$?
4. Carefully measure $\overline{BO}$ and $\overline{OD}$. Is $BO = OD$?
5. Would you say that the diagonals bisect each other?

By now, you have probably discovered many properties of parallelograms. Perhaps you have even proved some of these properties. A summary of them is given below. Proofs of the theorems are left as exercises.

PROPERTIES OF A PARALLELOGRAM
By definition: Opposite sides of a parallelogram are parallel.
THEOREM 3
Opposite sides of a parallelogram are equal.
THEOREM 4
Opposite angles of a parallelogram are equal.
THEOREM 5
Consecutive angles of a parallelogram are supplementary.
THEOREM 6
Diagonals of a parallelogram bisect each other.

Warning: The diagonals of a parallelogram are not usually equal.

Classroom Practice

$PQRS$ is a ▱. State the definition or theorem that justifies each statement.

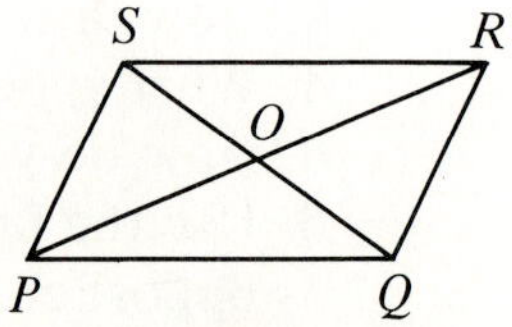

Exercises 1–9

1. $\overline{PQ} \parallel \overline{SR}$
2. $PQ = SR$
3. $SO = OQ$
4. $\angle PSR = \angle PQR$
5. $\angle SPQ + \angle PQR = 180°$
6. $SP = RQ$
7. $\overline{SP} \parallel \overline{RQ}$
8. $\overline{SQ}$ bisects $\overline{PR}$.
9. Name the four pairs of congruent triangles in the figure above.

In Exercises 10–17: $ACEF$ is a ▱; $BCDG$ is a ▱. Find the measure of each angle.

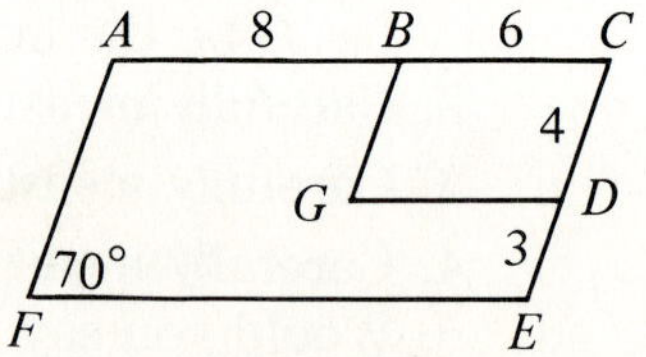

Exercises 10–17

10. $\angle C$
11. $\angle G$
12. $\angle GDC$
13. $\angle GBC$

Find each length.

14. FE
15. AF
16. BG
17. GD

Written Exercises

***Given:* PQRS is a ▱.**
Find each length and angle measure.

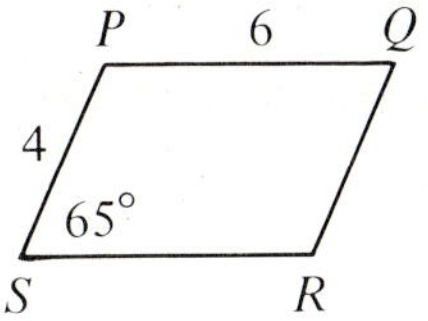

A **1.** SR **2.** RQ **3.** $\angle R$ **4.** $\angle Q$

***Given:* ▱MECO and rectangle EXIC lie in the same plane.**
Find each length and angle measure.

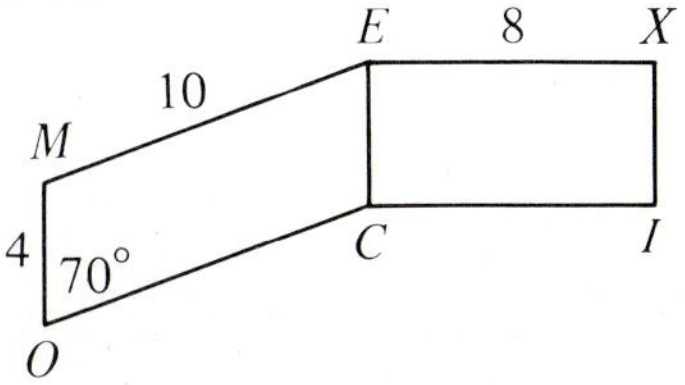

5. OC **6.** CI **7.** XI

8. $\angle I$ **9.** $\angle M$ **10.** $\angle XEM$

11. Is $\overline{MO} \parallel \overline{XI}$? Is $\overline{ME} \parallel \overline{CI}$?

***Given:* ▱ABCD; ▱AEFG**
Find each length and angle measure.

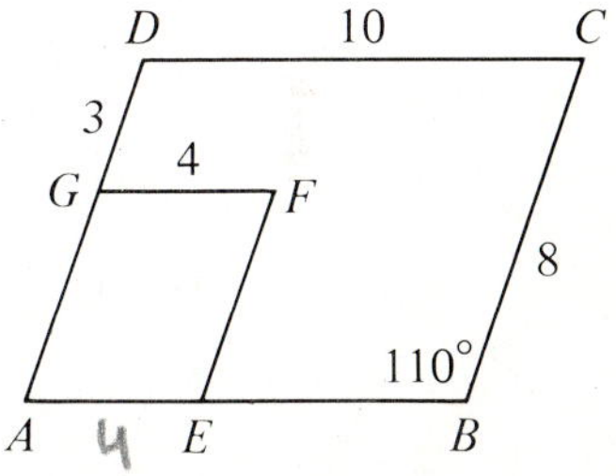

12. AG **13.** AE **14.** BE

15. $\angle C$ **16.** $\angle A$ **17.** $\angle F$

18. Is $\overline{GF} \parallel \overline{DC}$? Is $\overline{FE} \parallel \overline{CB}$?

***Given:* ▱TLAS**
Find each angle measure.

19. $\angle ALS$ **20.** $\angle TLS$ **21.** $\angle LAS$

Suppose that $LO = 9$ and $AO = 16$.
Find each length.

22. SO **23.** OT **24.** SL

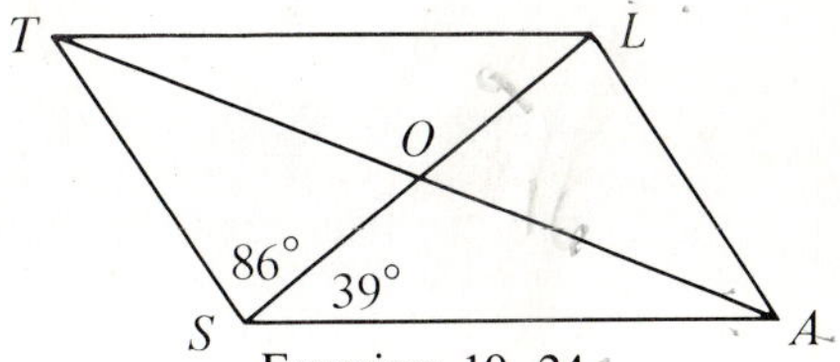

Exercises 19–24

25. $ABCD$ is a parallelogram. $AB = 8$ and $BC = 10$.
Find the perimeter of the parallelogram.
(*Hint:* The perimeter is the sum of the lengths of the sides.)

26. The length and width of a rectangle are 7 and 5.
Find the perimeter of the rectangle.

***Given:* ▱QRST**

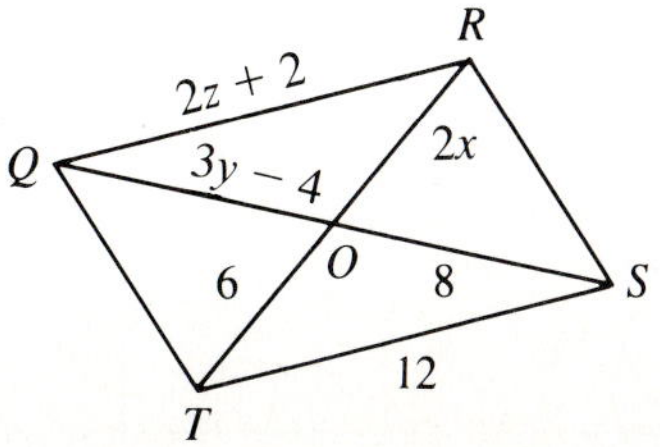

B **27.** Find the value of x.

28. Find the value of y.

29. Find the value of z.

The purpose of Exercises 30–32 is to prove Theorems 3 and 4.

30. Supply the reasons to complete the proof.

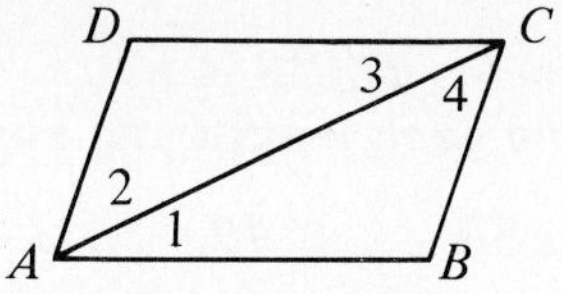

Given: $\square ABCD$

Prove: $\triangle ABC \cong \triangle CDA$

STATEMENTS	REASONS
1. $ABCD$ is a $\square$.	1. __?__
2. $\overline{AB} \parallel \overline{DC}$; $\overline{AD} \parallel \overline{BC}$	2. __?__
3. $\angle 1 = \angle 3$; $\angle 2 = \angle 4$	3. __?__
4. $AC = AC$	4. __?__
5. $\triangle ABC \cong \triangle CDA$	5. __?__

31. *Given:* $\square ABCD$

Prove: $AB = CD$; $BC = DA$

(*Hint:* Draw $\overline{AC}$. Then use the result of Exercise 30.)

Exercises 31, 32

32. *Given:* $\square ABCD$

Prove: $\angle B = \angle D$; $\angle A = \angle C$

(*Hint:* To prove that $\angle B = \angle D$, draw $\overline{AC}$ and use the result of Exercise 30. To prove that $\angle A = \angle C$, draw $\overline{BD}$.)

33. Complete this proof of Theorem 6 by supplying the reasons.

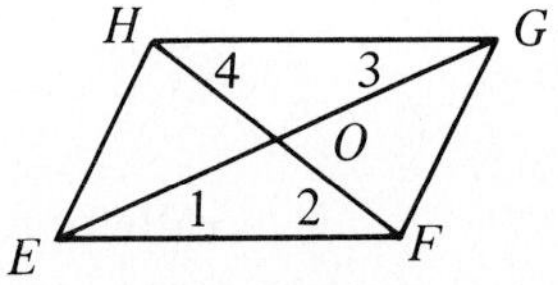

Given: $\square EFGH$ with diagonals meeting at O

Prove: $EO = GO$; $FO = HO$

STATEMENTS	REASONS
1. $EFGH$ is a $\square$.	1. __?__
2. $\overline{EF} \parallel \overline{GH}$	2. __?__
3. $\angle 1 = \angle 3$; $\angle 2 = \angle 4$	3. __?__
4. $EF = GH$	4. __?__
5. $\triangle EOF \cong \triangle GOH$	5. __?__
6. $EO = GO$; $FO = HO$	6. __?__

34. Write a proof of Theorem 5.

Given: ▱$WXYZ$

Prove: $\angle W$ and $\angle X$ are supplements.
$\angle W$ and $\angle Z$ are supplements.

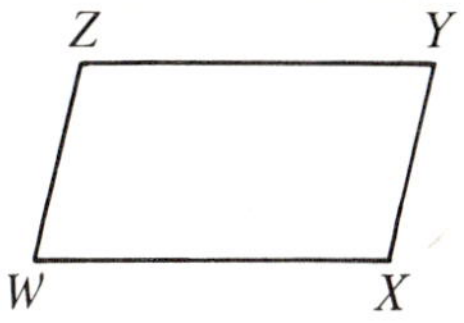

C **35.** *Given:* ▱$ABCD$

Prove: $RE = SE$

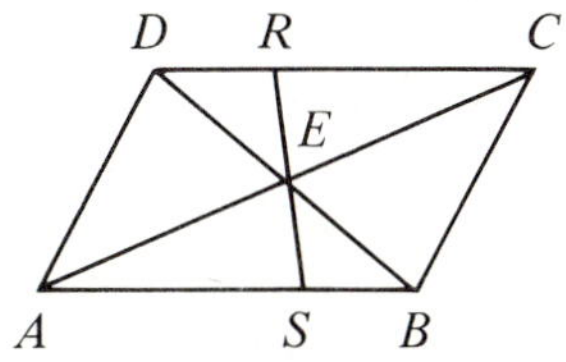

36. Our goal is to show that the distance between two parallel lines is always the same. To do this, we choose any two points on one line and prove that the points are the same distance from the other line. (The distance from a point to a line is defined to be the length of the *perpendicular* segment from the point to the line.)

Given: $l \parallel m$; $\overline{AX} \perp m$; $\overline{BY} \perp m$

Prove: $AX = BY$

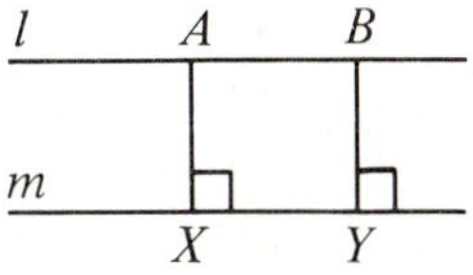

SELF-TEST

Vocabulary

trapezoid (p. 198)
parallelogram (p. 198)
rectangle (p. 198)
rhombus (p. 198)
square (p. 198)

Define each of the following special quadrilaterals.

1. Parallelogram **2.** Trapezoid **3.** Rhombus *(5-3)*

4. Rectangle **5.** Square

***ABCD* is a parallelogram. Complete each statement.** *(5-4)*

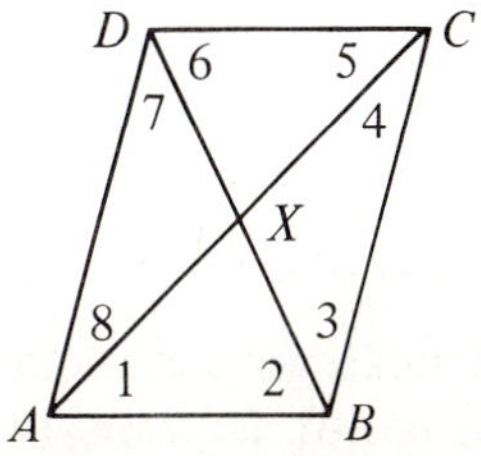

6. $AB =$ _?_ **7.** $\overline{BC} \parallel$ _?_

8. $DX =$ _?_ **9.** $\angle 1 = \angle$ _?_

10. $\angle DAB = \angle$ _?_ **11.** $\angle 7 = \angle$ _?_

12. $\angle ABC$ and $\angle$ _?_ are supplementary angles.

APPLICATIONS

GOLDEN RECTANGLE

Even though all rectangles have four right angles, not all rectangles are similar. Some are much longer than they are wide, while others (squares) are exactly as long as they are wide. In one type of rectangle, the golden rectangle, the length divided by the width will always equal $\frac{1+\sqrt{5}}{2}$, about 1.6. Ancient Greek geometers and architects thought this rectangular shape was the most beautiful.

This photo of the Parthenon shows how closely the dimensions fit within a golden rectangle. The dashed line indicates the original outline of this Greek temple.

You can construct a golden rectangle by following the steps below.

Construct square *ABCD*.

Find the midpoint *M* of $\overline{AB}$.

Using *M* as center and *MC* as radius, draw an arc intersecting $\overleftrightarrow{AB}$ at *E*.

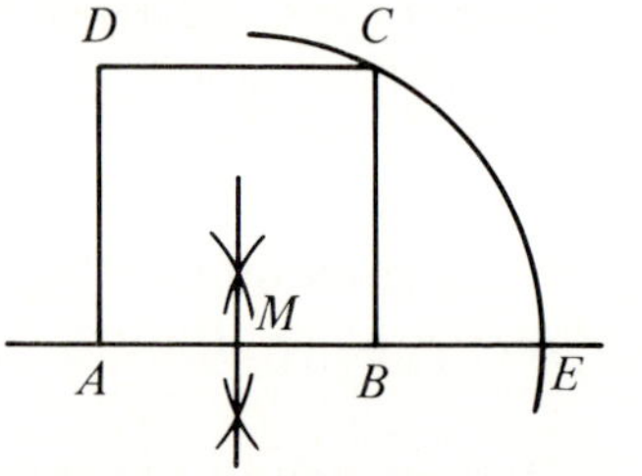

Construct a perpendicular to $\overleftrightarrow{AB}$ at point *E*.

Extend $\overline{DC}$ to meet the perpendicular at *F*.

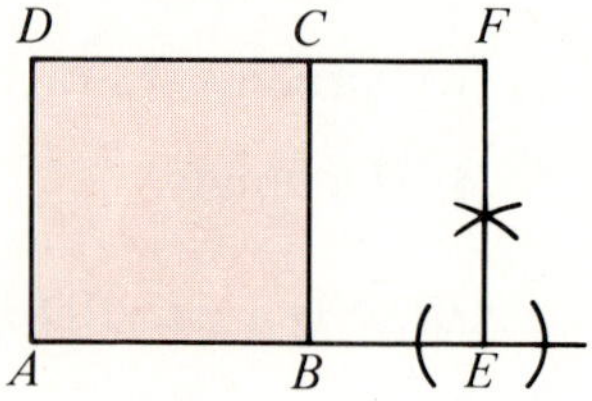

AEFD is a golden rectangle. If we remove the original square *ABCD*, the remaining rectangle *BEFC* is also a golden rectangle.

Construct a golden rectangle, measure the sides, and divide the length by the width. Is your result about 1.6?

5 Properties of Special Parallelograms

Since rectangles, rhombuses, and squares are parallelograms, they have all the properties listed on page 204. But because they are *special* parallelograms, they have other properties as well. Try the explorations below and see if you can discover what these properties are.

EXPLORATIONS

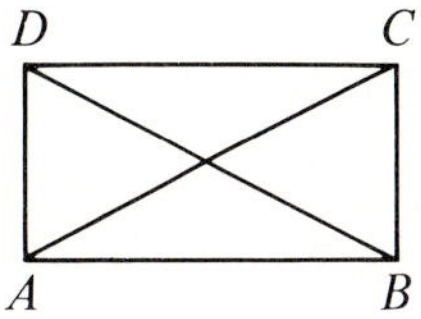

A. Carefully draw a very large rectangle that is not a square. Draw the two diagonals.
   1. Measure the diagonals. Are they equal?
   2. Are the diagonals perpendicular?
   3. Does each diagonal bisect two angles of the quadrilateral? (For example, in rectangle *ABCD*, does $\overline{BD}$ bisect $\angle ABC$ and $\angle ADC$?)

   Draw a different-looking rectangle and see whether your answers to questions 1–3 are the same.

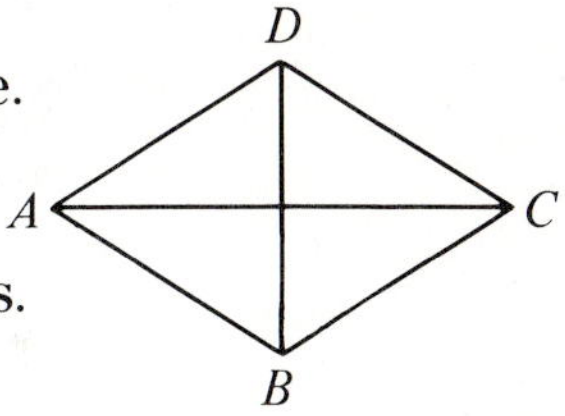

B. Carefully draw a very large rhombus that is not a square. Draw the two diagonals.
   Answer questions 1–3 in part A.
   Repeat this experiment with a different-looking rhombus.

C. Carefully draw a large square and its two diagonals.
   Answer questions 1–3 in part A.
   Repeat this experiment with a different square.

D. Now copy and complete the table.
   In each space write *Y* for *yes* or *N* for *not necessarily.*

	Rectangle	Rhombus	Square
Diagonals are equal.	?	?	?
Diagonals are perpendicular.	?	?	?
Each diagonal bisects two angles.	?	?	?

If you did the exploration exercises carefully, then you will not be surprised by the following theorems.

THEOREM 7
The diagonals of a rectangle are equal.

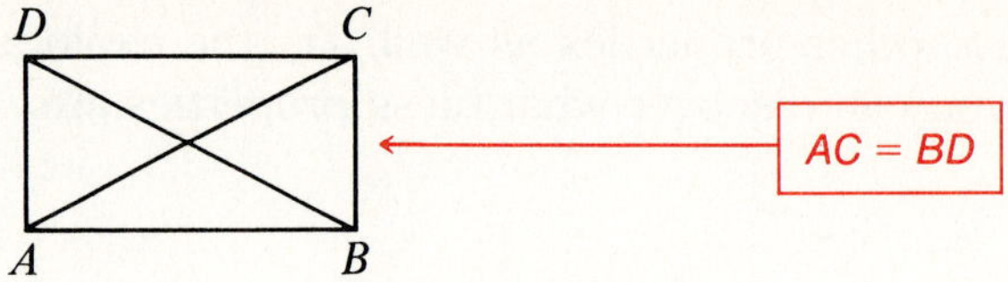

THEOREM 8
The diagonals of a rhombus are perpendicular, and they bisect the angles of the rhombus.

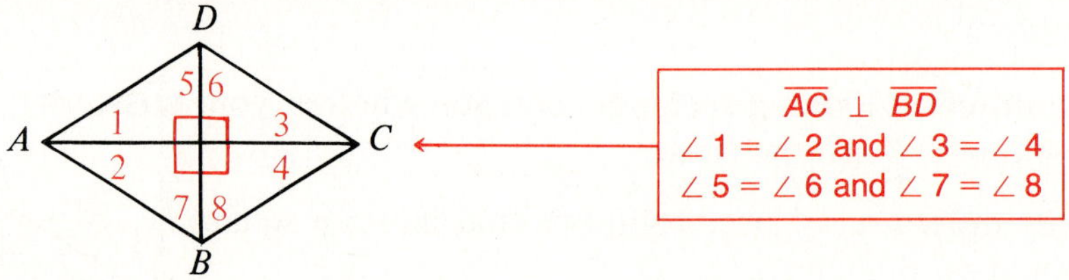

Both of these theorems apply to a square because a square is both a rectangle and a rhombus.

EXAMPLE In rectangle $WXYZ$, $WY = 8$. Find WO, XO, YO, and ZO.

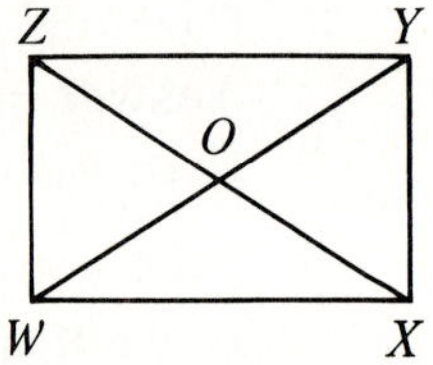

Since the diagonals of a rectangle are equal, WY and XZ both equal 8.
Since the diagonals of a parallelogram bisect each other, $WO = \frac{1}{2} \times 8 = 4$.
Likewise, XO, YO, and ZO are equal to 4.

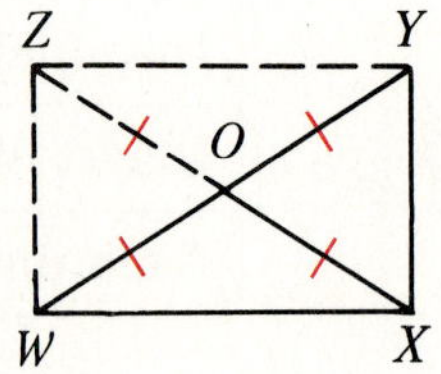

If half of rectangle $WXYZ$ is removed, we are left with a right triangle. Notice that O is the midpoint of hypotenuse $\overline{WY}$. The fact that $WO = XO = YO$ suggests the following theorem.

THEOREM 9
The midpoint of the hypotenuse of a right triangle is equidistant from the three vertices.

Classroom Practice

***ABCD* is a rhombus.**

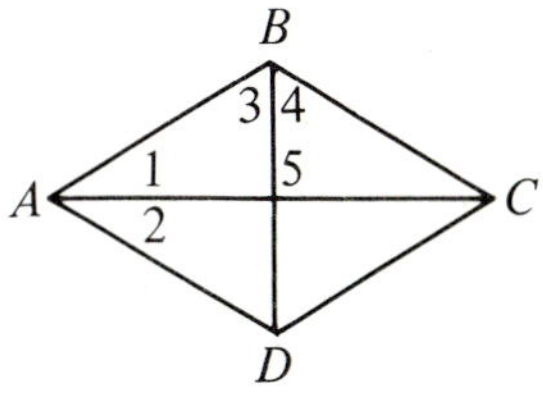

1. Suppose $\angle 1 = 20°$.
   Find the measure of each numbered angle.
2. Suppose $\angle 1 = 30°$.
   Find the measure of each numbered angle.

***PQRS* is a rectangle.**

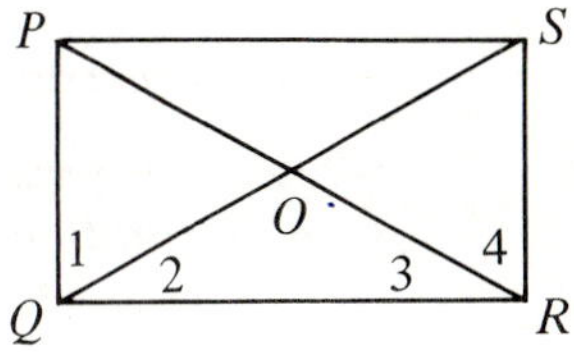

3. Suppose $SQ = 10$.
   Find SO, OQ, PO, and OR.
4. Suppose $\angle 1 = 62°$.
   Find the measure of each numbered angle.

$\triangle PQR$ is a right triangle. *O* is the midpoint of the hypotenuse.

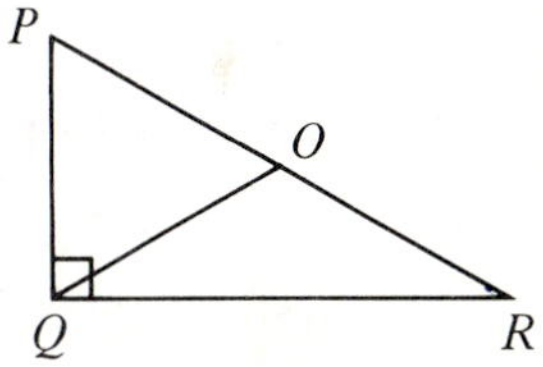

5. $\overline{QO}$ is called a(n) $\underline{\quad ? \quad}$.
   median/angle bisector/altitude
6. Suppose $PR = 10$. Find PO, OR, and OQ.
7. Suppose $QO = 6$. Find PO and PR.

Written Exercises

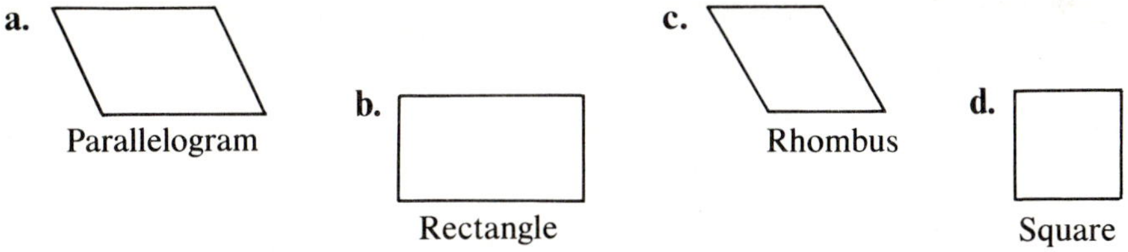

In Exercises 1–10, refer to the quadrilaterals shown above. For each exercise, state which quadrilaterals satisfy the given condition.

A

1. All angles are right angles.
2. Opposite sides are equal.
3. Diagonals are equal.
4. Diagonals bisect each other.
5. Opposite angles are equal.
6. Diagonals are perpendicular.
7. Each diagonal bisects two angles.
8. All sides are equal.
9. Diagonals are perpendicular bisectors of each other.
10. When a diagonal is drawn, two congruent triangles are formed.

***STAR* is a rhombus.**

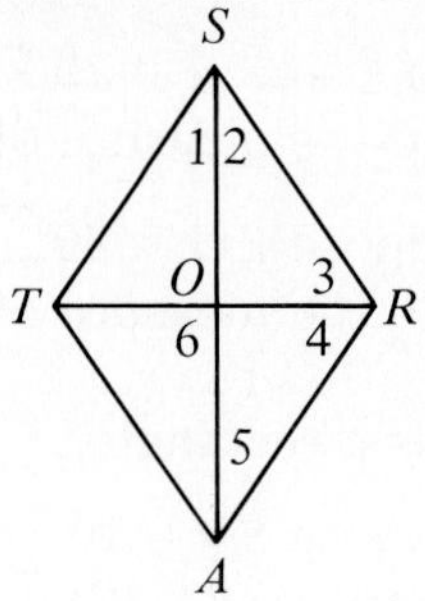

11. Suppose $\angle 1 = 25°$.
Find the measure of each numbered angle.

12. Suppose $\angle 5 = 32°$.
Find the measure of each numbered angle.

13. Suppose $SA = 8$ and $TR = 6$.
Find SO and TO.

14. Suppose $RA = 10$.
Find the perimeter of *STAR*.
(*Note:* The perimeter of a polygon is the sum of the lengths of its sides.)

***FLAT* is a rectangle.**

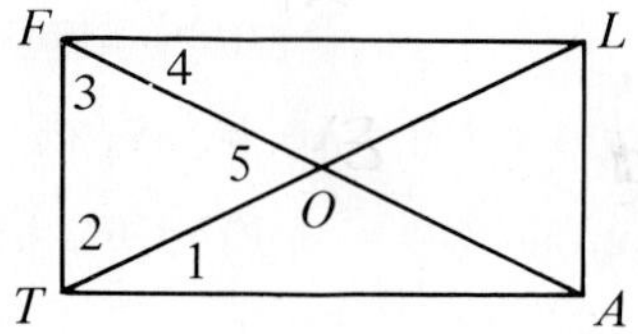

15. Suppose $FL = 12$ and $LA = 5$.
Find the perimeter of *FLAT*.

16. Suppose $FA = 13$. Find TL.

17. Suppose $FA = 14$. Find FO and TO.

18. Suppose $\angle 1 = 20°$.
Find the measure of each numbered angle.

19. Suppose $\angle 2 = 75°$.
Find the measure of each numbered angle.

20. Suppose $\angle 5 = 24°$.
Find the measure of each numbered angle.

***DEFG* is a square.**

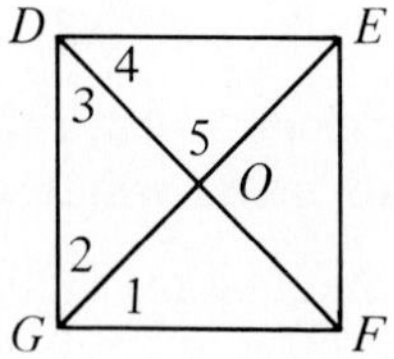

21. Find the measure of each numbered angle.

22. If $DO = 8$, find GE.

***M* is the midpoint of the hypotenuse of right △ *SEN*.**

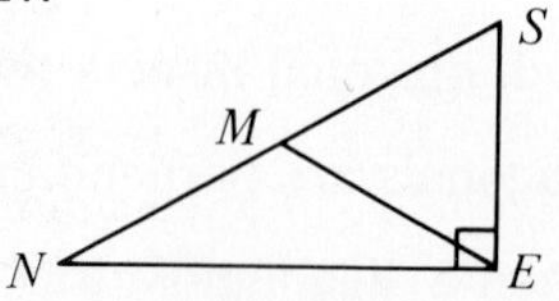

23. If $SN = 10$, find ME.

24. If $ME = 7$, find SM and SN.

B **25.** The purpose of this exercise is to prove Theorem 7. Supply the reasons needed to complete the proof.

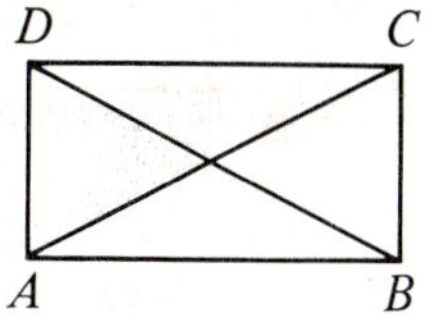

Given: Rectangle $ABCD$ with diagonals $\overline{AC}$ and $\overline{BD}$

Prove: $AC = BD$

STATEMENTS	REASONS
1. $ABCD$ is a rectangle.	1. _?_
2. $AD = BC$	2. _?_
3. $\angle DAB = \angle ABC = 90°$	3. _?_
4. $AB = AB$	4. _?_
5. $\triangle DAB \cong \triangle CBA$	5. _?_
6. $AC = BD$	6. _?_

The purpose of Exercises 26 and 27 is to prove Theorem 8.

26. Supply the reasons needed to complete the proof.

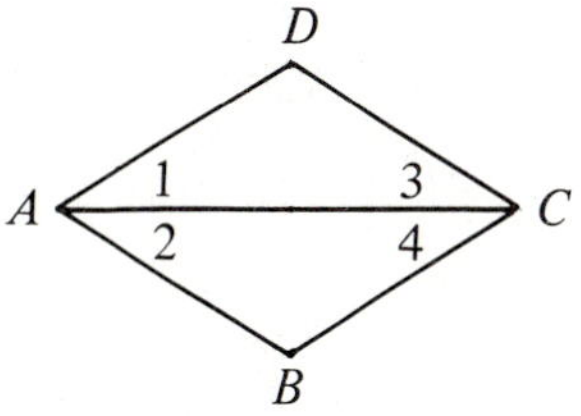

Given: Rhombus $ABCD$ with diagonal $\overline{AC}$

Prove: $\angle 1 = \angle 2$ and $\angle 3 = \angle 4$

STATEMENTS	REASONS
1. $ABCD$ is a rhombus.	1. _?_
2. $AB = AD$; $BC = DC$	2. _?_
3. $AC = AC$	3. _?_
4. $\triangle ABC \cong \triangle ADC$	4. _?_
5. $\angle 1 = \angle 2$ and $\angle 3 = \angle 4$	5. _?_

C **27.** Copy what is shown and write a two-column proof.

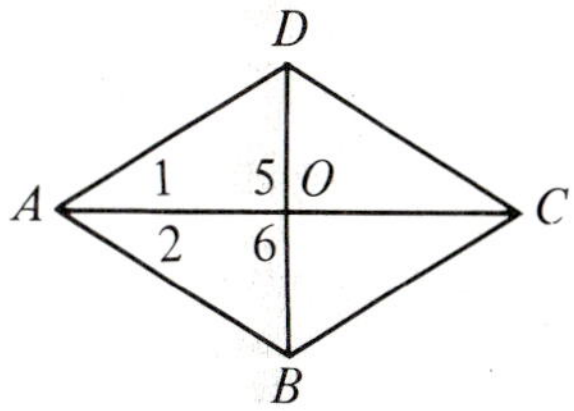

Given: Rhombus $ABCD$
Diagonals $\overline{AC}$ and $\overline{BD}$ meet at O.

Prove: $\overline{AC} \perp \overline{BD}$

(*Hint:* You know that $\angle 1 = \angle 2$ from Exercise 26. Show that $\triangle AOD \cong \triangle AOB$.)

28. Construct a rhombus and draw its diagonals. Then use a protractor to check that the diagonals are perpendicular.

6 Proving Figures Are Parallelograms

How can you show that quadrilateral *ABCD* is a parallelogram? Of course, one way is to show that its opposite sides are parallel. But there are other ways of showing that *ABCD* is a parallelogram. See if you can decide what these are by trying the explorations below.

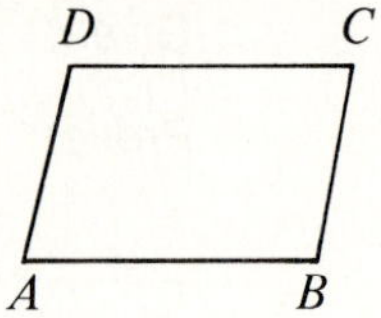

EXPLORATIONS

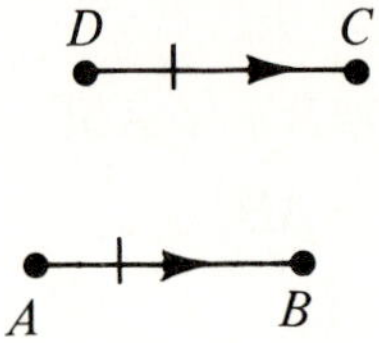

1. Use lined paper to draw two segments, $\overline{AB}$ and $\overline{DC}$, which are both parallel and equal.
2. Draw $\overline{AD}$ and $\overline{BC}$.
   Do $\overline{AD}$ and $\overline{BC}$ appear to be parallel?
3. What kind of quadrilateral is *ABCD*?

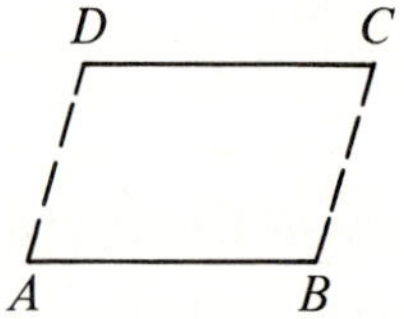

Repeat steps 1–3 using other segments which are both parallel and equal.

THEOREM 10

If a quadrilateral has one pair of opposite sides that are both parallel and equal, then the quadrilateral is a parallelogram.

Given: $\overline{AB} \parallel \overline{DC}$

$AB = DC$

Prove: *ABCD* is a ▱.

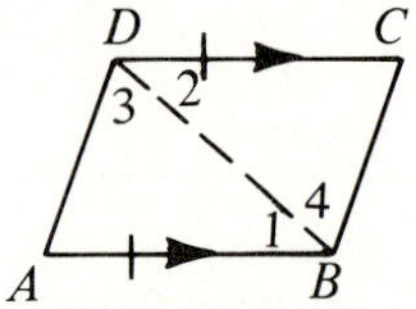

Key steps of proof:

1. Draw $\overline{DB}$. Then $\angle 1 = \angle 2$. (Why?)
2. $\triangle ABD \cong \triangle CDB$ (Why?)
3. $\angle 3 = \angle 4$ (Why?)
4. $\overline{AD} \parallel \overline{BC}$ (Why?)
5. *ABCD* is a ▱. (Why?)

EXPLORATIONS

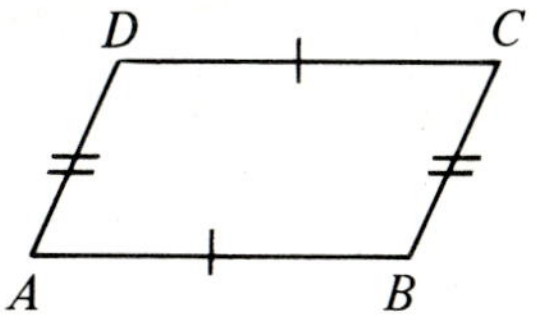

If *ABCD* is a parallelogram, we know that $AB = DC$ and $AD = BC$. Now let's look at the converse situation.

Suppose $AB = DC$ and $AD = BC$.
Can we conclude that *ABCD* is a ▱?

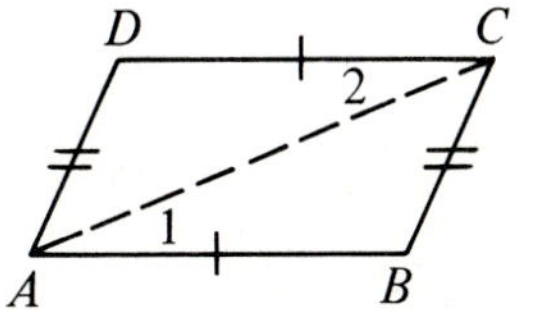

Follow the steps below to see how.

1. Draw $\overline{AC}$.
2. $\triangle ABC \cong \triangle CDA$ (Why?)
3. $\angle 1 = \angle 2$ (Why?)
4. $\overline{DC} \parallel \overline{AB}$ (Why?)
5. *ABCD* is a ▱. (Why?)

Hint: The answer is not *By definition.*

These five steps are the key ones in the proof of Theorem 11.

THEOREM 11

If a quadrilateral has both pairs of opposite sides equal, then the quadrilateral is a parallelogram.

EXPLORATIONS

We know that when *ABCD* is a parallelogram, diagonals $\overline{AC}$ and $\overline{BD}$ bisect each other. Is the converse true? Follow these steps.

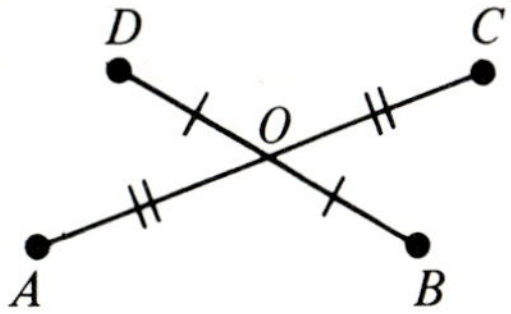

1. Draw two unequal segments, $\overline{AC}$ and $\overline{BD}$, which bisect each other at point *O*.
2. Draw quadrilateral *ABCD*.
3. $\triangle AOB \cong \triangle$ __?__ and $\triangle AOD \cong \triangle$ __?__
4. $AB = DC$ (Why?) $AD = BC$ (Why?)
5. *ABCD* is a ▱. (Why?)

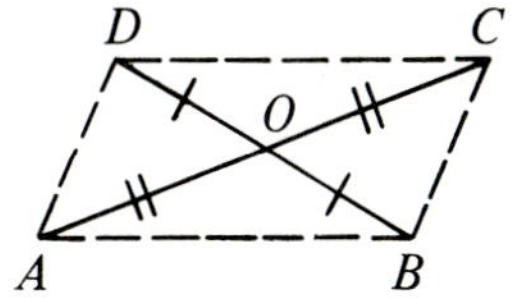

These steps are the key ones in the proof of Theorem 12.

THEOREM 12

If a quadrilateral has diagonals that bisect each other, then the quadrilateral is a parallelogram.

WAYS TO SHOW THAT A QUADRILATERAL IS A PARALLELOGRAM

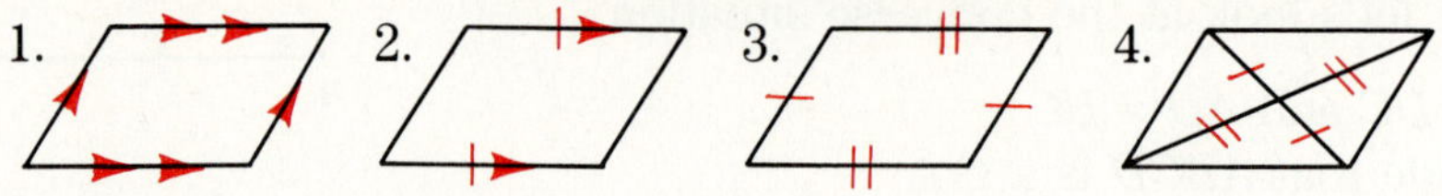

1. *Show that both pairs of opposite sides are parallel. (Definition of ▱)*
2. *Show that one pair of opposite sides are parallel and equal.*
3. *Show that both pairs of opposite sides are equal.*
4. *Show that the diagonals bisect each other.*

Classroom Practice

State the definition or theorem that allows you to conclude that *CDEF* is a parallelogram.

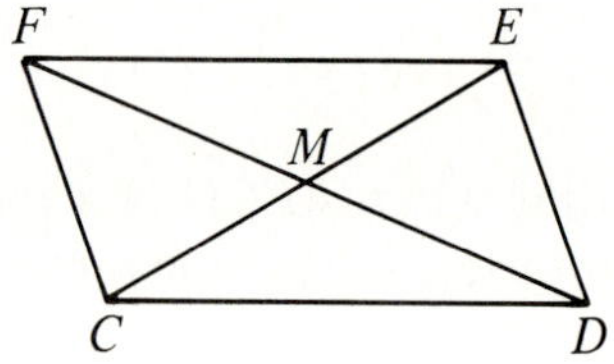

1. $CD = EF$; $DE = CF$
2. $\overline{CD} \parallel \overline{EF}$; $\overline{DE} \parallel \overline{CF}$
3. $CM = ME$; $DM = MF$
4. $CF = DE$; $\overline{CF} \parallel \overline{DE}$
5. M is the midpoint of $\overline{CE}$ and $\overline{DF}$.
6. $CD = EF$; $\angle CDE = 70°$; $\angle DEF = 110°$

Written Exercises

In Exercises 1–4, write the definition or theorem which supports the statement "*ABCD* is a ▱."

A **1.**

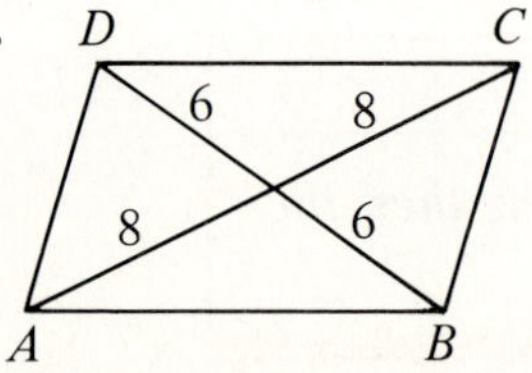

2.

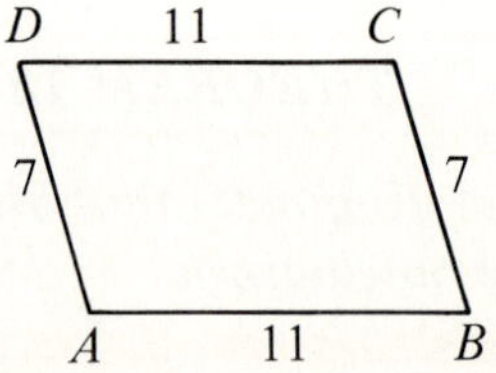

3.

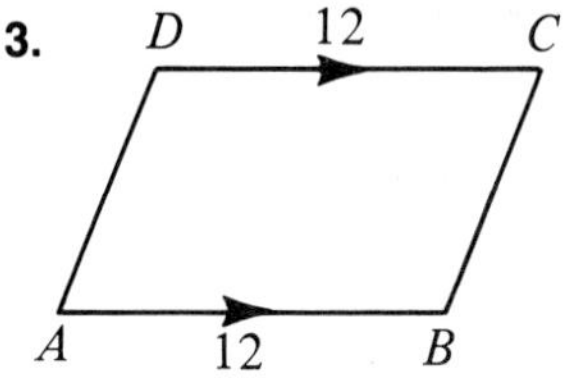

4.

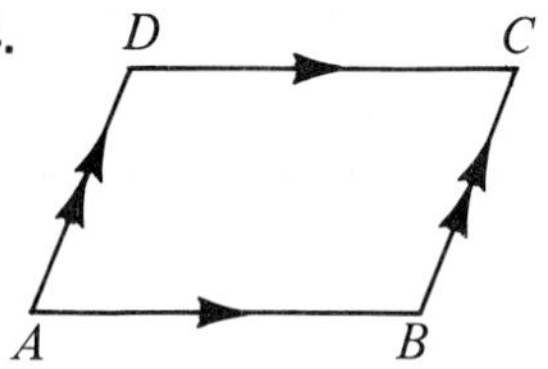

In Exercises 5–7, supply the reasons to complete the proofs.

5. *Given:* $PQ = ST$
$PS = QR$
$\angle 1 = \angle 2$

Prove: $PQTS$ is a ▱.

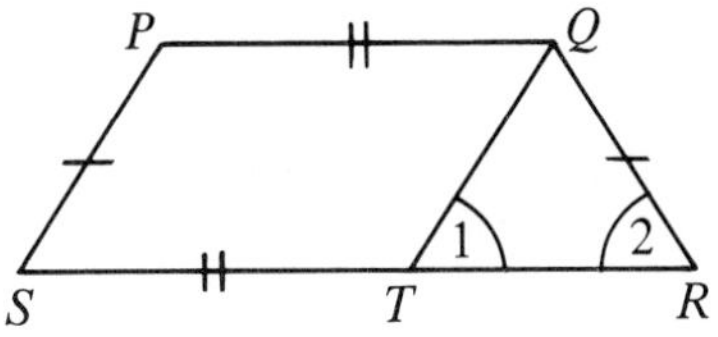

STATEMENTS	REASONS
1. $\angle 1 = \angle 2$	1. ?
2. $QT = QR$	2. ?
3. $PS = QR$	3. ?
4. $QT = PS$	4. ?
5. $PQ = ST$	5. ?
6. $PQTS$ is a ▱.	6. ?

6. *Given:* $WXYZ$ is a ▱.
M is the midpoint of $\overline{ZY}$.
N is the midpoint of $\overline{WX}$.

Prove: $WNYM$ is a ▱.

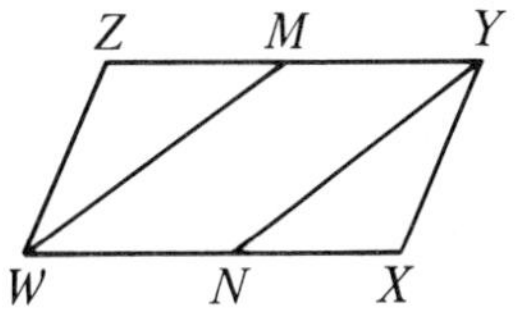

STATEMENTS	REASONS
1. $WX = ZY$	1. ?
2. $\frac{1}{2}WX = \frac{1}{2}ZY$	2. ?
3. $WN = \frac{1}{2}WX$; $MY = \frac{1}{2}ZY$	3. ?
4. $WN = MY$	4. ?
5. $\overline{WX} \parallel \overline{ZY}$ (and thus $\overline{WN} \parallel \overline{MY}$)	5. ?
6. $WNYM$ is a ▱.	6. ?

B **7.** *Given:* ABCD is a ▱.
P is the midpoint of $\overline{AO}$.
Q is the midpoint of $\overline{CO}$.

Prove: PBQD is a ▱.

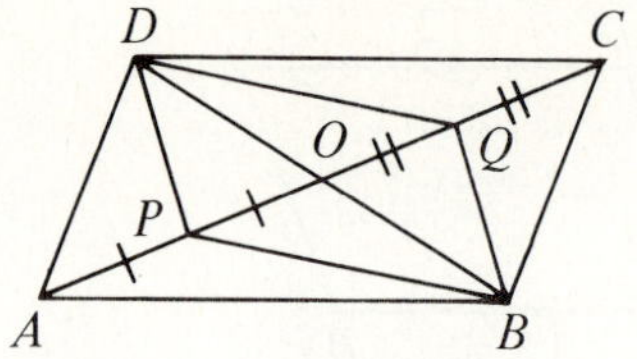

STATEMENTS	REASONS
1. $AO = CO$	1. ___?___
2. $\frac{1}{2}AO = \frac{1}{2}CO$	2. ___?___
3. $PO = \frac{1}{2}AO$; $QO = \frac{1}{2}CO$	3. ___?___
4. $PO = QO$	4. ___?___
5. $BO = DO$	5. ___?___
6. PBQD is a ▱.	6. ___?___

8. The quadrilaterals numbered 1, 2, 3, and 4 in the diagram are parallelograms. If you wanted to show that quadrilateral 5 is also a parallelogram, which method would you use? (*Hint:* If two lines are parallel to a third line, then they are parallel to each other.)

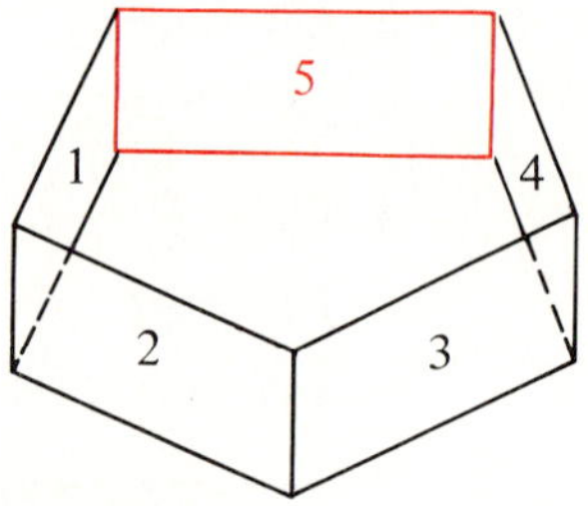

9. Draw a segment $\overline{AB}$ and its midpoint M. Draw a nonperpendicular segment $\overline{XY}$ that has M as its midpoint and is equal in length to $\overline{AB}$. Draw quadrilateral AXBY. What kind of quadrilateral is it? Explain.

10. Draw a segment $\overline{AB}$ and its midpoint M. Draw a perpendicular segment $\overline{XY}$ that has M as its midpoint but is unequal to $\overline{AB}$. Draw quadrilateral AXBY. What kind of quadrilateral is it? Explain.

11. Draw a segment $\overline{AB}$ and its midpoint M. Draw a perpendicular segment $\overline{XY}$ that has M as its midpoint and is equal to $\overline{AB}$. Draw quadrilateral AXBY. What kind of quadrilateral is it? Explain.

12. A kitchen utensils rack is built so that $PO = QO = RO = SO$. No matter how the rack is adjusted, the rows of pegs remain parallel to each other. Explain why this is so.

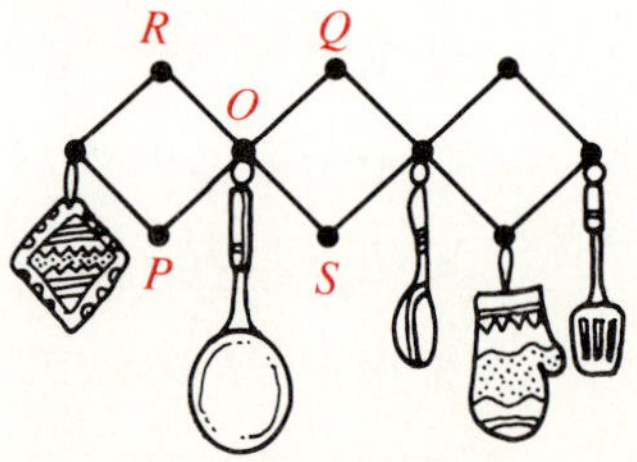

C **13.** Explain why the jaws of the pliers shown are always parallel.

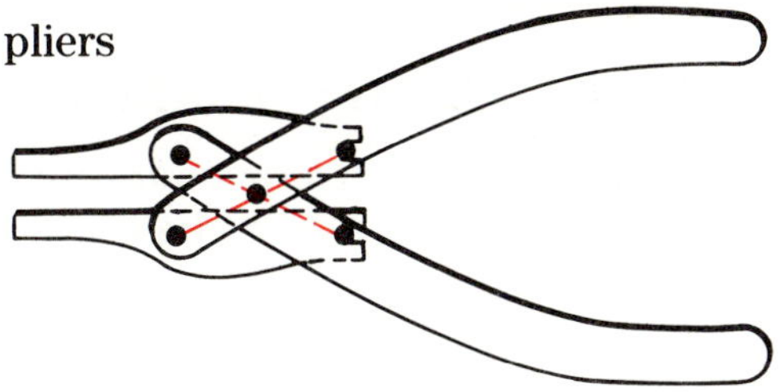

14. *Given:* $ABCD$ is a ▱.
$\overline{DE} \perp \overline{AC}$
$\overline{BF} \perp \overline{AC}$

Prove: $BEDF$ is a ▱.

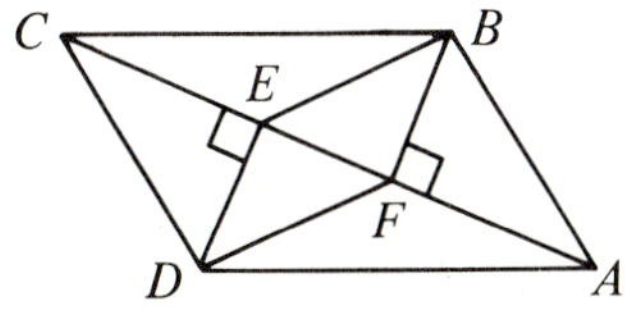

15. Prove that if a quadrilateral has equal opposite angles, then the quadrilateral is a parallelogram. (*Note:* This exercise provides a fifth way to show that a quadrilateral is a parallelogram.)

16. Prove that a parallelogram with perpendicular diagonals must be a rhombus. (*Hint:* Show that all four sides are equal.)

17. Prove that a parallelogram with equal diagonals must be a rectangle. (*Hint:* Show that all four angles are right angles.)

SELF-TEST

(5-5)

1. The diagonals of rectangle $ABCD$ meet at E. If $AC = 12$, find BD and BE.

2. $CROK$ is a rhombus with $\angle RCO = 25°$. Find the measures of $\angle 1$, $\angle 2$, $\angle 3$, and $\angle 4$.

3. If $CR = 7$, find the perimeter of $CROK$.

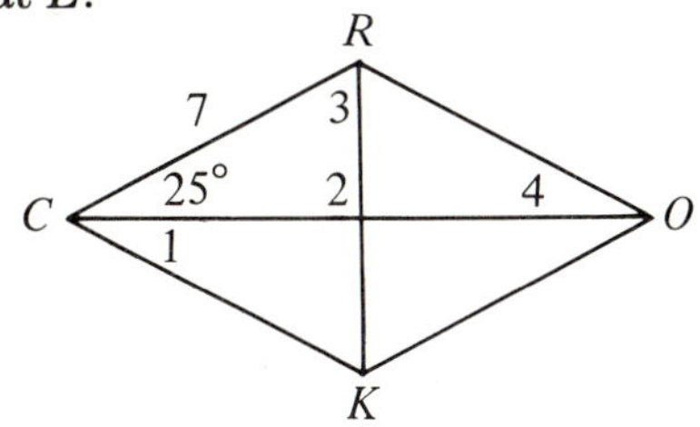

(5-6)

4. *Given:* $WZ = XY$
$\angle 1 = \angle 2$

Prove: $WXYZ$ is a ▱.

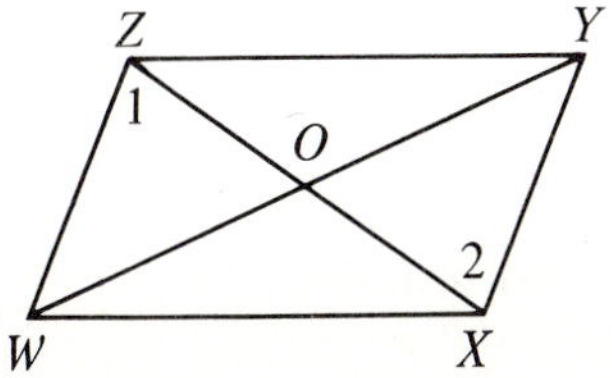

7 Properties of Trapezoids

Recall that a trapezoid is a quadrilateral with just one pair of parallel sides. These parallel sides are called **bases.** The other sides are called **legs.** If the legs are equal, the trapezoid is an **isosceles trapezoid.**

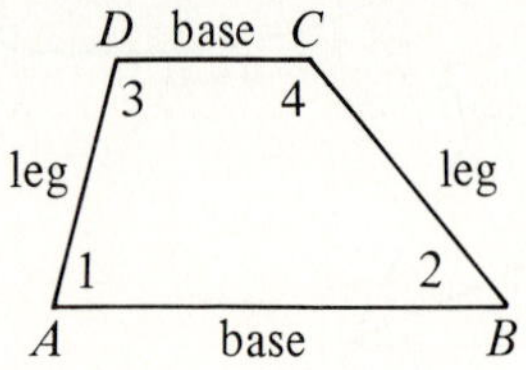

In trapezoid *ABCD*,
∠ 1 and ∠ 2 are base angles,
and so are ∠ 3 and ∠ 4.

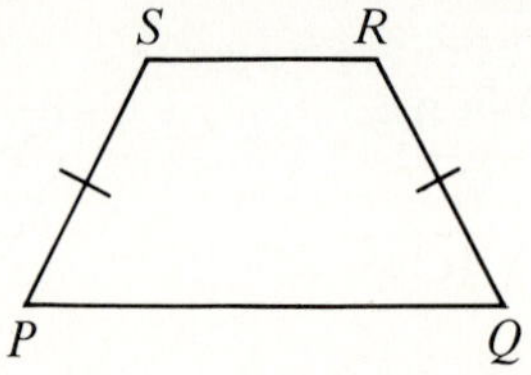

Isosceles trapezoid *PQRS* has equal legs, $\overline{PS}$ and $\overline{QR}$.

The two angles that include a base are called **base angles.** Every trapezoid has two pairs of base angles.

Try the following explorations and see if you can discover the special properties of trapezoids and isosceles trapezoids.

EXPLORATIONS

Draw a large non-isosceles trapezoid *ABCD* with $\overline{AB}$ as one base. Draw a large isosceles trapezoid *PQRS* with $\overline{PQ}$ as one base.

A. The Angles of a Trapezoid

1. Measure base angles, ∠ *A* and ∠ *B*, of trapezoid *ABCD*. Also measure ∠ *C* and ∠ *D*.
2. Measure base angles, ∠ *P* and ∠ *Q*, of trapezoid *PQRS*. Also measure ∠ *R* and ∠ *S*.
3. How are the base angles of an isosceles trapezoid related?

B. The Median of a Trapezoid

The segment joining the midpoints of the legs of a trapezoid is called the **median** of the trapezoid.

1. Draw the median, $\overline{EF}$, of trapezoid *ABCD*.
2. Does it appear that the median is parallel to the bases?
3. Measure $\overline{AB}$, $\overline{DC}$, and the median $\overline{EF}$. How does *EF* compare with *AB* + *DC*?

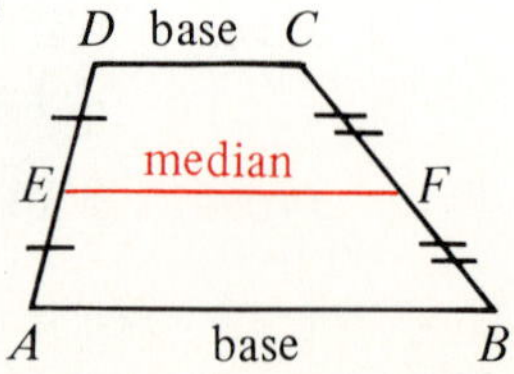

Repeat steps 1–3, using trapezoid *PQRS*.

By now, you may have discovered the following two theorems.

THEOREM 13

The base angles of an isosceles trapezoid are equal.

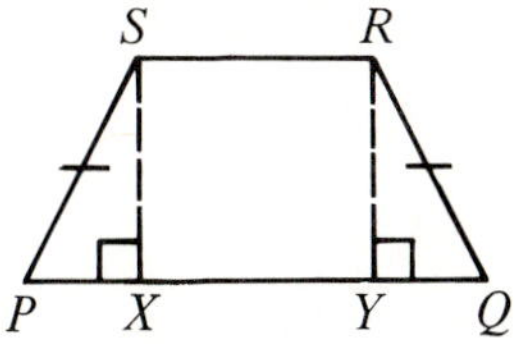

Given: $PQRS$ is a trapezoid with $SP = RQ$.

Prove: $\angle P = \angle Q$ and $\angle PSR = \angle QRS$

Key ideas of proof:

1. Draw $\overline{SX} \perp \overline{PQ}$ and $\overline{RY} \perp \overline{PQ}$.
2. $SX = RY$ because the distance between two parallel lines is always the same. (See Exercise 36 on page 207.)
3. $\triangle SPX \cong \triangle RQY$ (Why?)
4. Then $\angle P = \angle Q$.
5. $\angle PSR = \angle QRS$ (Supplements of equal angles are equal.)

THEOREM 14

The median of a trapezoid has two properties:

(1) It is parallel to the bases.
(2) Its length equals half the sum of the base lengths.

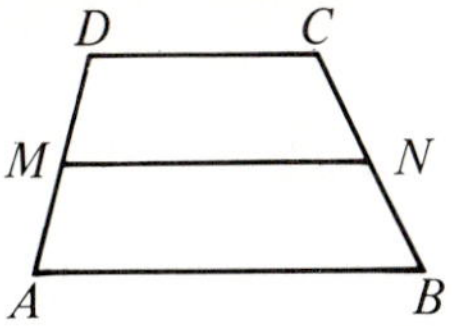

Given: Trapezoid $ABCD$ with median $\overline{MN}$

Prove: (1) $\overline{MN} \parallel \overline{AB}$ and $\overline{MN} \parallel \overline{DC}$

(2) $MN = \frac{1}{2}(AB + DC)$

At this stage of our work, a proof of Theorem 14 would be difficult. A proof using coordinate geometry is included in Chapter 12.

EXAMPLE A trapezoid and median are shown. Find the value of x.

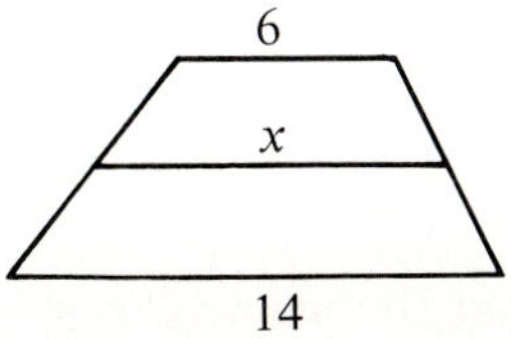

$x = \frac{1}{2}(14 + 6)$

$x = \frac{1}{2}(20)$

$x = 10$

EXAMPLE A trapezoid and its median are shown. Find the value of x.

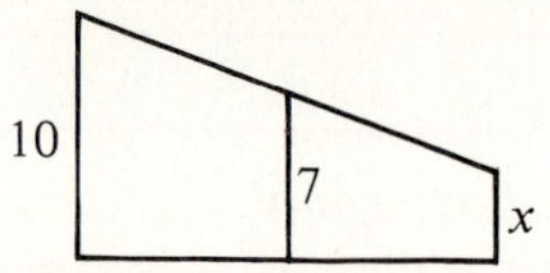

$$7 = \frac{1}{2}(10 + x)$$
$$14 = 10 + x$$
$$4 = x$$

Classroom Practice

1. *ABCD* is a trapezoid.

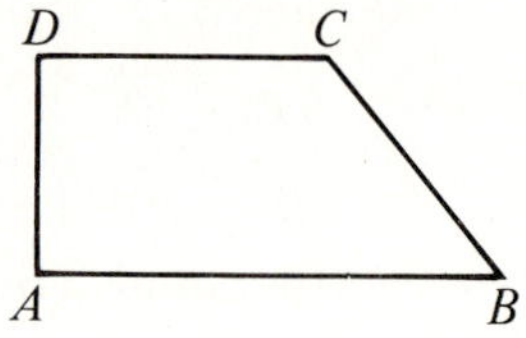

   a. Name its two bases. b. Name its legs.

   c. Name two pairs of base angles.

   d. Does the trapezoid appear to be isosceles?

   e. The median of the trapezoid joins the midpoint of __?__ and the midpoint of __?__.

   f. The median of the trapezoid is parallel to __?__ and __?__.

2. *WXYZ* is a trapezoid. Answer the questions given in Exercise 1.

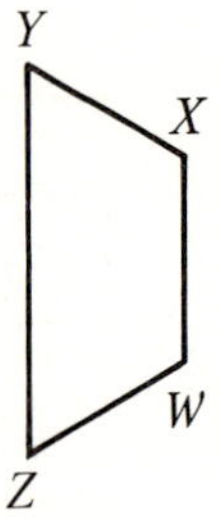

3. Suppose that in Exercise 1, $\angle A = 90°$ and $\angle B = 50°$. Find the measures of $\angle D$ and $\angle C$.

4. Suppose that in Exercise 2, *WXYZ* is an isosceles trapezoid. If $\angle Z = 60°$, find the measures of $\angle Y$ and $\angle X$.

Find the length of the median of each trapezoid.

5.

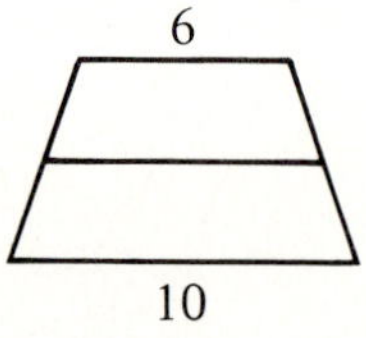

6.

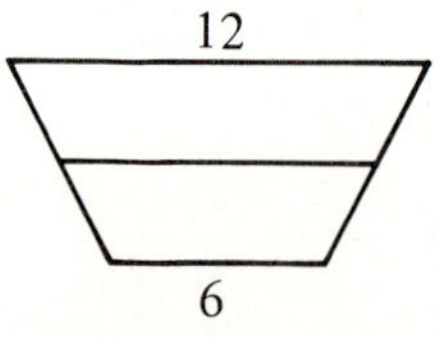

7.

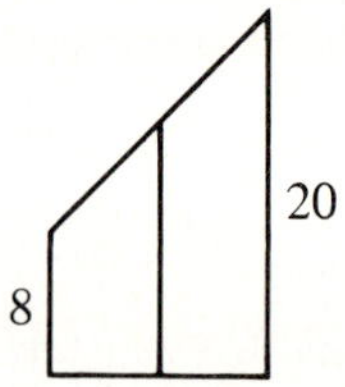

8. You can cut two congruent trapezoids out of cardboard and slide them together as shown. What kind of figure appears to be formed?

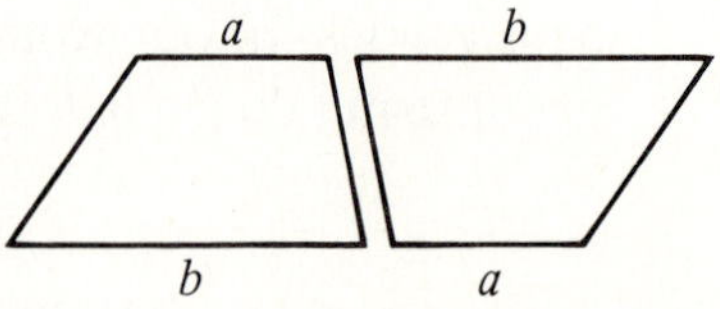

9. *P* and *R* are the midpoints of the opposite sides of the parallelogram shown.

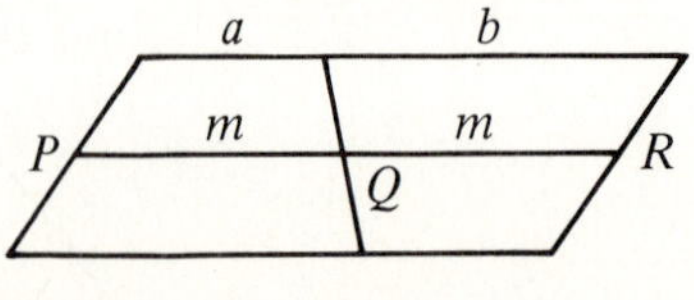

   a. Do $\overline{PQ}$ and $\overline{QR}$ appear to be the medians of the two trapezoids?

   b. Does the diagram suggest that $2m = a + b$?

   c. Why is $m = \frac{1}{2}(a + b)$? d. To what theorem is this result related?

Written Exercises

***PQRS* is an isosceles trapezoid.**

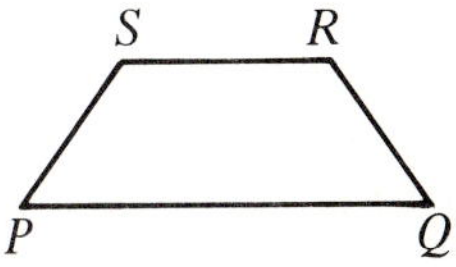

A 1. If $\angle P = 50°$, find the measures of $\angle Q$, $\angle R$, and $\angle S$.

2. If $\angle P = 60°$, find the measures of $\angle Q$, $\angle R$, and $\angle S$.

3. If $\angle S = 100°$, find the measures of $\angle P$, $\angle Q$, and $\angle R$.

4. If $\angle R = 110°$, find the measures of $\angle P$, $\angle Q$, and $\angle S$.

5. The median of the trapezoid joins the midpoints of __?__ and __?__.

6. If $SR = 8$ and $PQ = 16$, find the length of the median.

Find the length of the median of each trapezoid.

7.
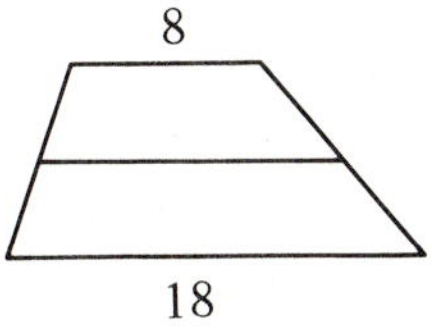

8.
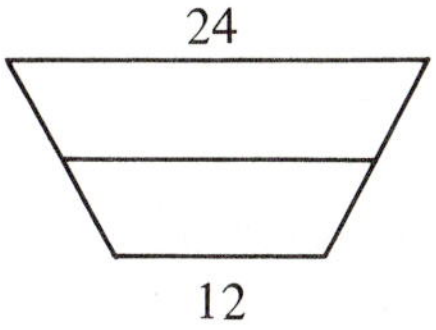

9.
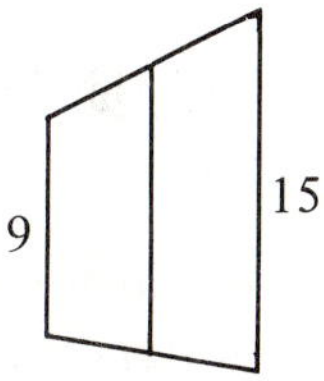

10.
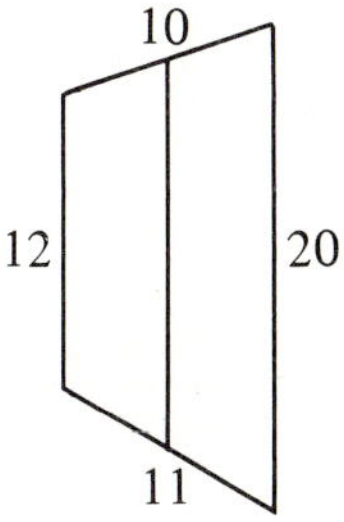

11.
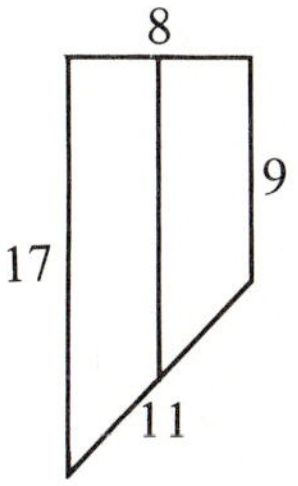

12.
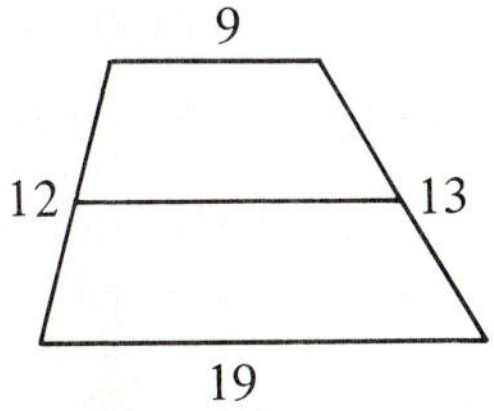

In Exercises 13–21, a trapezoid and its median are shown. Find the value of *x*.

13.
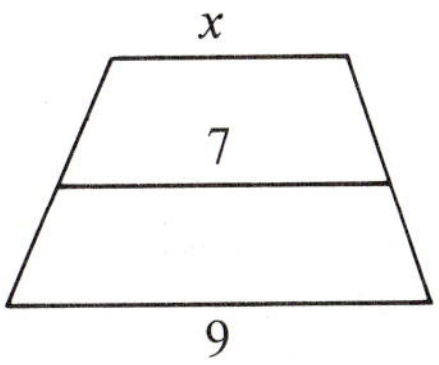

14.
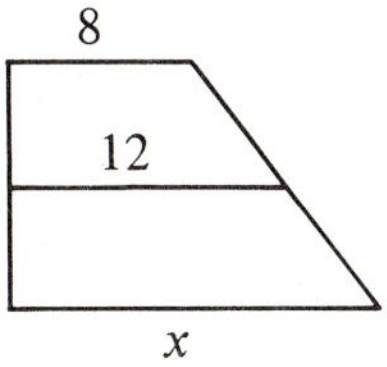

15.
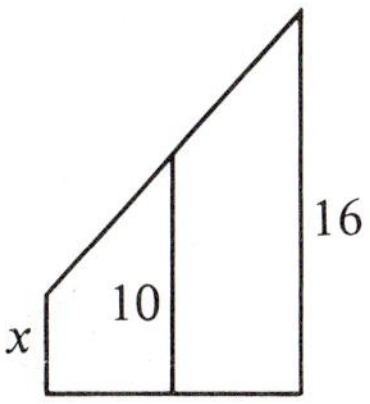

B 16.
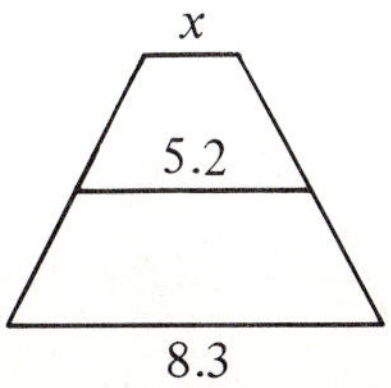

17.
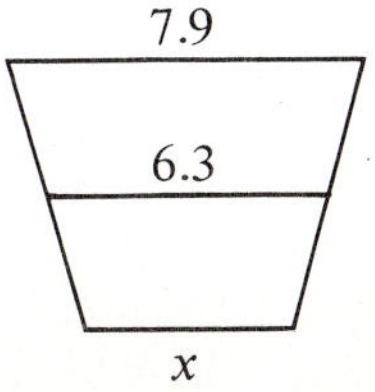

18.
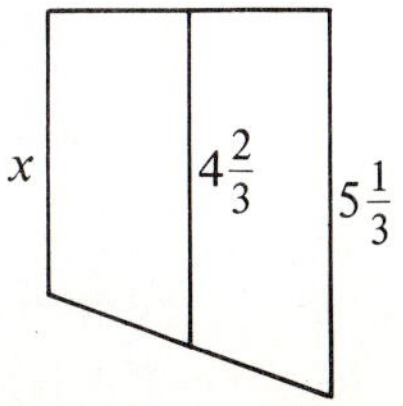

19.
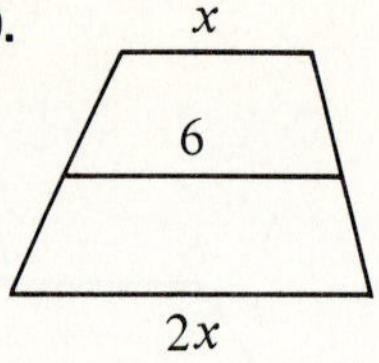

20.
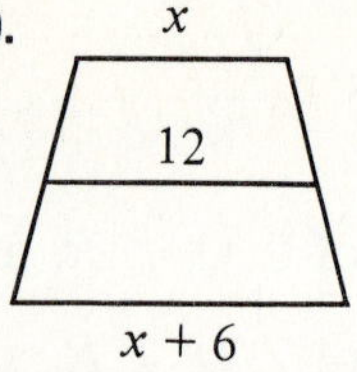

21.
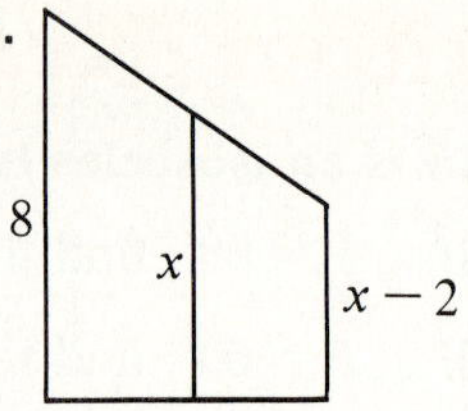

22. a. Theorem 13 can be expressed this way:
If a trapezoid is isosceles, then __?__.
b. What is the converse of this statement?
c. Do you think that the converse is a true statement?

C 23. Is it possible for the bases of a trapezoid to be equal? Explain.

24. Is it possible for the diagonals of a trapezoid to bisect each other? Explain.

25. Construct an isosceles trapezoid.
Measure the base angles with a protractor.

26. Our goal is to prove that the diagonals of an isosceles trapezoid are equal.

Given: Trapezoid $ABCD$ with $AD = BC$

Prove: $AC = BD$

(*Hint:* Use Theorem 13.)

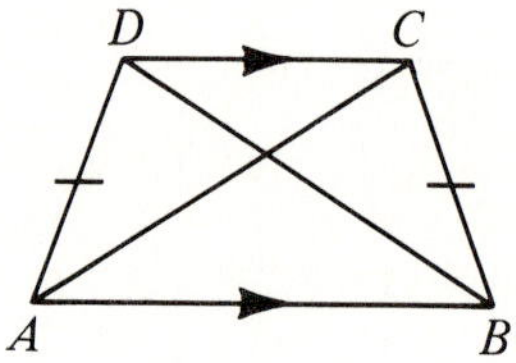

PUZZLE ◆ PROBLEMS

Given information:
(1) Three horses named Steady, Speedy, and Slowpoke ran a race.
(2) Tic, Tac, and Toe each owned one of the horses.
(3) Tac's horse nearly won.
(4) This was Speedy's fifth race.
(5) Toe owned the black horse.
(6) The horse owned by Tic had not previously raced.
(7) Steady broke an ankle after the start of the race.
(8) The horse that won was brown.

The winning horse was __?__.
The winning horse was owned by __?__.

8 The Midpoints Theorem

As you study the figures below from left to right, what do you notice? The base $\overline{DC}$ of trapezoid $ABCD$ becomes smaller and smaller. Finally, in figure (d), $\overline{DC}$ has shrunk to a single point and trapezoid $ABCD$ has become $\triangle ABC$.

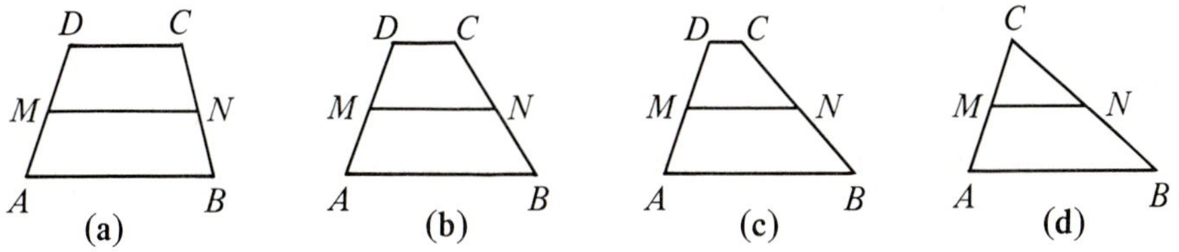

If we apply Theorem 14 to figures (a), (b), and (c) we have the following results for each figure:

$$(1)\ \overline{MN} \parallel \overline{AB}$$

$$(2)\ MN = \frac{1}{2}(AB + DC)$$

Theorem 14 can also suggest some information about figure (d). If we think of $\triangle ABC$ as a trapezoid $ABCD$ with $DC = 0$, we have

$$(1)\ \overline{MN} \parallel \overline{AB}$$

$$(2)\ MN = \frac{1}{2}(AB + 0) = \frac{1}{2}AB$$

These results are stated in the following theorem.

THEOREM 15 (The Midpoints Theorem)

The segment joining the midpoints of two sides of a triangle is parallel to the third side and half as long.

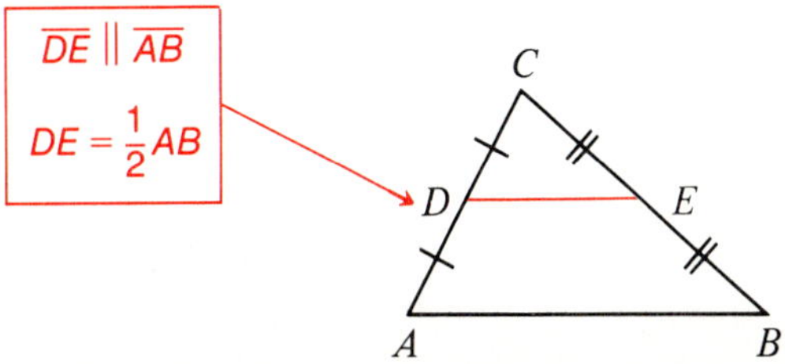

EXAMPLE M, N, and O are the midpoints of the sides of $\triangle ABC$. Find MN, NO, and MO.

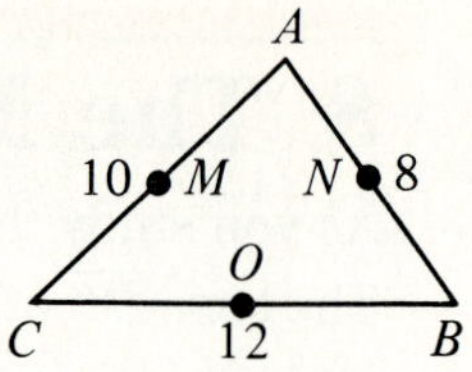

Use the Midpoints Theorem three times.

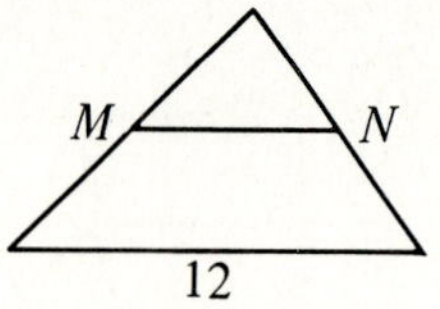

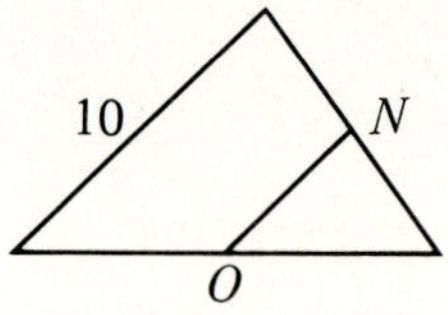

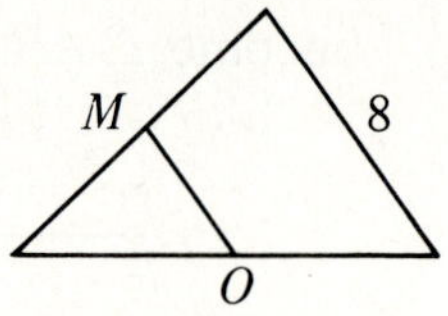

$MN = \frac{1}{2} \times 12 = 6$ $NO = \frac{1}{2} \times 10 = 5$ $MO = \frac{1}{2} \times 8 = 4$

Classroom Practice

***M*, *N*, and *O* are the midpoints of the sides of $\triangle ABC$. Complete each statement.**

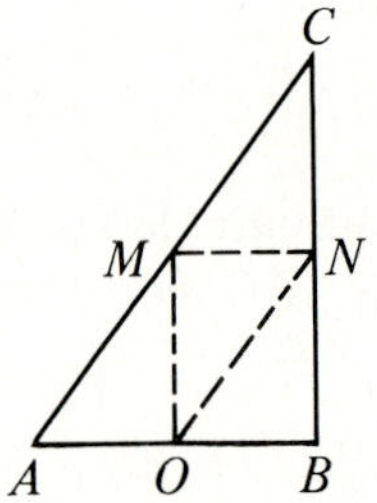

1. If $AB = 8$, then $MN =$ __?__.
2. If $AC = 14$, then $NO =$ __?__.
3. If $BC = 10$, then $MO =$ __?__.
4. If $MN = 11$, then $AB =$ __?__.
5. If $NO = 15$, then $AC =$ __?__.
6. If $MO = x$, then $BC =$ __?__.
7. Explain why $AMNO$ must be a parallelogram.
8. Name parallelograms, other than $AMNO$, shown in the figure.

Written Exercises

***M* and *N* are the midpoints of $\overline{XZ}$ and $\overline{YZ}$. Complete each statement.**

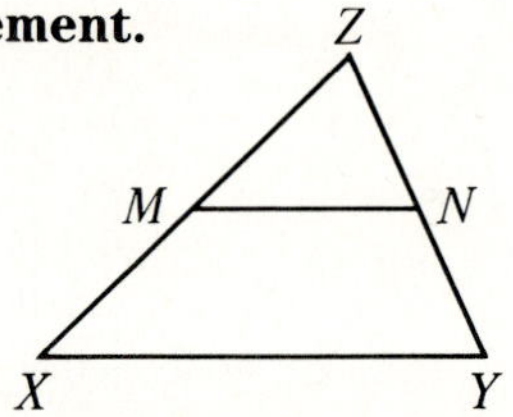

A

1. $\overline{MN} \parallel$ __?__.
2. If $XY = 12$, then $MN =$ __?__.
3. If $XY = 20$, then $MN =$ __?__.
4. If $MN = 7$, then $XY =$ __?__.
5. If $ZM = 10$, $MN = 12$, and $ZN = 7$, find the perimeter of $\triangle XYZ$.
6. If $XM = 6$, $MN = 7$, and $NY = 4$, find the perimeter of $\triangle XYZ$.

R, S, and T are midpoints of the sides of $\triangle ABC$. Copy and complete the table.

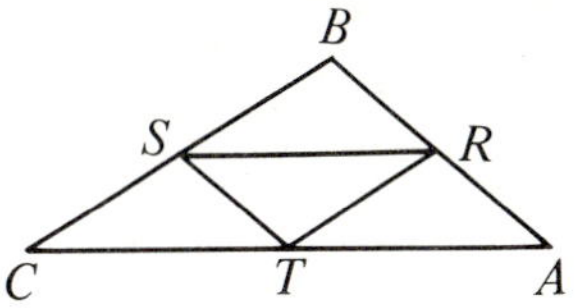

	AB	BC	AC	ST	RT	RS	Perimeter of $\triangle ABC$	Perimeter of $\triangle RST$
7.	8	10	12	?	?	?	?	?
8.	10	10	14	?	?	?	?	?
9.	?	?	?	4	6	5	?	?
10.	?	?	?	6	5	7	?	?
11.	14	?	?	?	6	4	?	?
12.	?	10	16	6	?	?	?	?

13. If the perimeter of $\triangle ABC$ is 32, the perimeter of $\triangle RST$ is __?__.

14. If the perimeter of $\triangle RST$ is 25, the perimeter of $\triangle ABC$ is __?__.

In Exercises 15–18, exactly one of the lengths represented by x, y, and z can be found. Find that length.

B **15.**

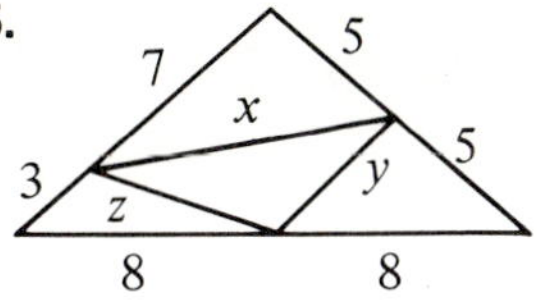

16.

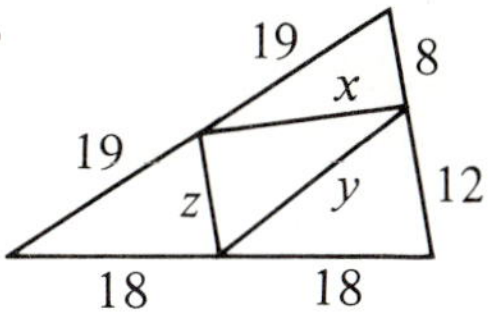

17.

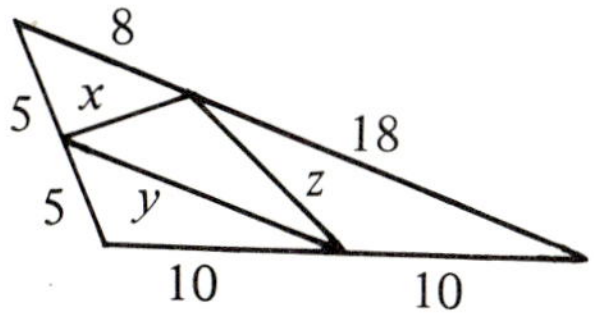

18.

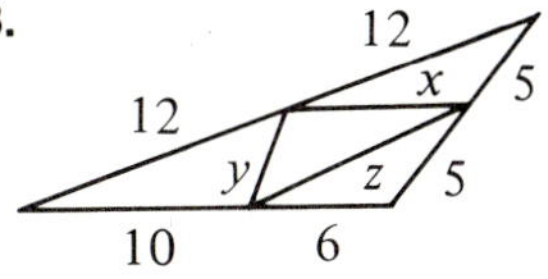

P, Q, and R are the midpoints of the sides of $\triangle XYZ$.

19. Find the perimeter of $\square PQRY$.

20. Find the perimeter of $\square XQRP$.

21. a. Name a parallelogram in the figure, other than $PQRY$ or $XQRP$.

b. Find the perimeter of this parallelogram.

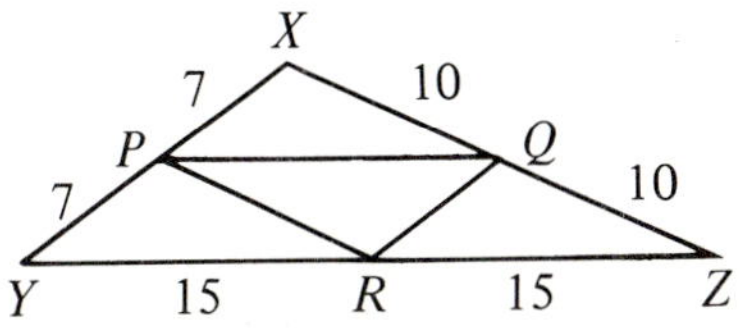

22. Draw a right $\triangle ABC$ with $\angle C = 90°$. Let X be the midpoint of $\overline{AC}$. Let Y be the midpoint of $\overline{BC}$. Let Z be the midpoint of $\overline{AB}$.
a. What kind of quadrilateral is $CXZY$? **b.** If $CZ = 10$, find XY and AB.

C **23.** Points M, N, and O are the midpoints of the sides of a triangle, $\triangle DEF$. Copy the figure shown and construct $\triangle DEF$.

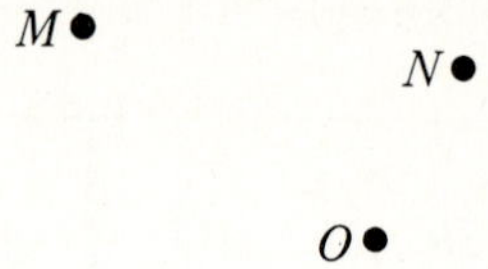

24. $ABCD$ is a trapezoid with median $\overline{MN}$. Find the values of x, y, and z.

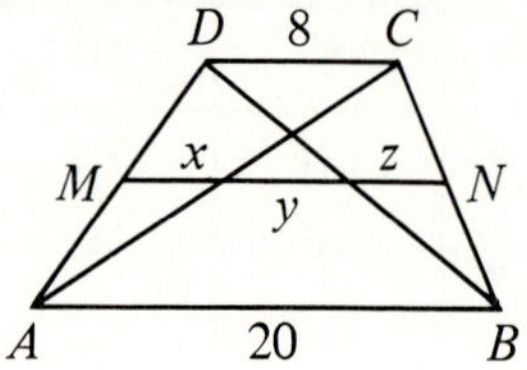

25. Let W, X, Y, and Z be the midpoints of the sides of any quadrilateral $EFGH$.

Prove: $WXYZ$ is a ▱.

(*Hint:* Draw $\overline{EG}$.)

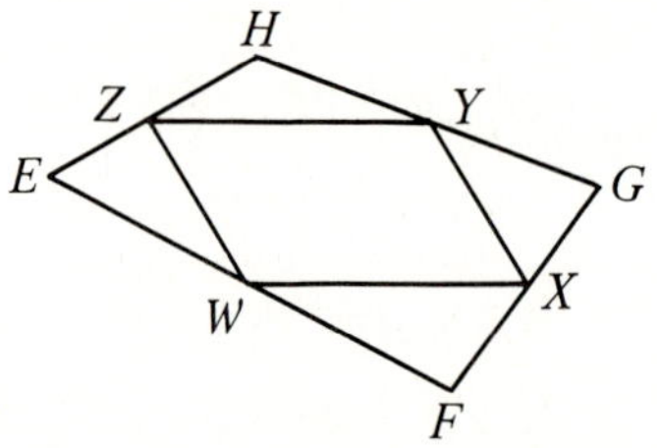

SELF-TEST

Vocabulary

bases of a trapezoid (p. 220)
legs of a trapezoid (p. 220)
isosceles trapezoid (p. 220)
base angles of a trapezoid (p. 220)
median of a trapezoid (p. 220)

***TGOA* is an isosceles trapezoid.**

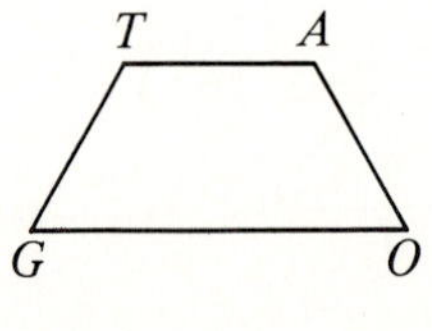

1. If $GO = 12$ and $TA = 6$, find the length of the median. *(5-7)*

2. If $\angle T = 120°$, find the measures of $\angle G$, $\angle O$, and $\angle A$.

In the diagram, *P*, *K*, and *T* are the midpoints of the sides of $\triangle SAL$.

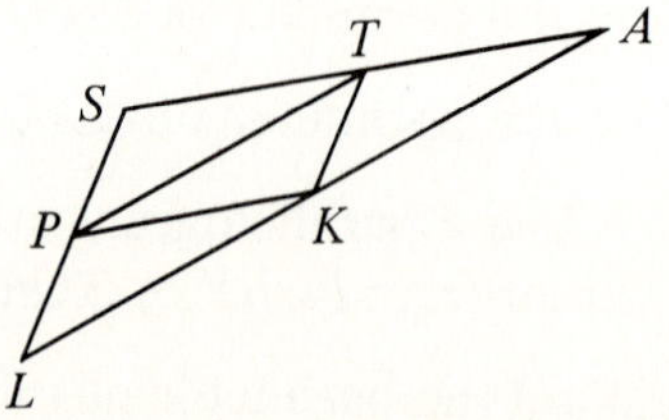

3. Name three parallelograms shown. *(5-8)*

4. If $PT = 5$, then $LA =$ __?__.

5. If the perimeter of $\triangle SAL = 18$, then the perimeter of $\triangle PKT =$ __?__.

READING GEOMETRY

WORDS AND MEANINGS

Theorem 8 on page 210 says, "The diagonals of a rhombus are perpendicular, and they bisect the angles of the rhombus." This is not a complicated sentence, but it won't make sense to you unless you know the meaning of *diagonals*, *rhombus*, *perpendicular*, *bisect*, and *angle*. Every one of these words stands for a geometric idea. When you read one of these words, you should have a clear picture or mental image to go with it. For example, perpendicular might suggest this:

Just learning words isn't very useful unless you understand the ideas that go with them. You also need to know how the ideas are related. Very often, in geometry, one statement that is known to be true may suggest several other related facts. If you can recall these related facts, you can use them in writing proofs and solving problems. For example, if you know that two given lines are parallel and that they are cut by a transversal, you can safely make statements about the angles that are formed. "Corresponding angles are equal" is one of them.

Sometimes a diagram or chart can help you organize your thoughts around related ideas. A quick look at the diagram at the bottom of page 198 helps you understand how special quadrilaterals are related. The chart on page 346 shows many important relationships that you will need to know when studying circles.

EXERCISES

1. Make a chart to show how triangles can be classified by their angles. Add two columns to indicate where the altitudes and perpendicular bisectors meet.
2. If a triangle is isosceles, do you know something special about its angles? Can you state a theorem that answers this question?
3. In what ways are a square and a rhombus alike? In what ways are they different?
4. If you know that two angles are opposite angles of a parallelogram, what other fact can you state about these angles?
5. If it is given that two sides of a quadrilateral are parallel, can you safely state that the figure is a parallelogram? Explain.

SKILLS REVIEW

SQUARES AND SQUARE ROOTS

$5^2 = 25$ ← square of 5

$\sqrt{25} = 5$ ← positive square root of 25

$(-5)^2 = 25$ ← square of -5

$-\sqrt{25} = -5$ ← negative square root of 25

Complete the table.

	1.	2.	3.	4.	5.	6.	7.	8.
Number	64	121	49	81	169	36	100	144
Positive Square Root	8	?	?	?	13	?	?	?
Negative Square Root	?	-11	?	?	?	?	?	?

Find the value. Leave your answer in simplest radical form if the number is not a perfect square. Assume that variables represent numbers greater than zero.

SAMPLES

$$\sqrt{900} = \sqrt{9 \cdot 100} = \sqrt{9} \cdot \sqrt{100} = 3 \cdot 10 = 30$$

$$\sqrt{72x^2} = \sqrt{36 \cdot 2 \cdot x^2} = \sqrt{36} \cdot \sqrt{2} \cdot \sqrt{x^2} = 6 \cdot \sqrt{2} \cdot x = 6x\sqrt{2}$$

9. $\sqrt{225}$ **10.** $\sqrt{18}$ **11.** $\sqrt{49n^2}$ **12.** $\sqrt{75}$ **13.** $\sqrt{63}$

14. $\sqrt{16y^2}$ **15.** $\sqrt{400}$ **16.** $\sqrt{125}$ **17.** $\sqrt{128}$ **18.** $\sqrt{50a^2}$

19. $\sqrt{98}$ **20.** $\sqrt{20k^2}$ **21.** $\sqrt{196}$ **22.** $\sqrt{162}$ **23.** $\sqrt{700}$

24. $\sqrt{32b^2}$ **25.** $\sqrt{324}$ **26.** $\sqrt{45}$ **27.** $\sqrt{256}$ **28.** $\sqrt{147}$

Simplify. Leave no perfect-square factor under the radical sign.

SAMPLES

$$\sqrt{2} \cdot \sqrt{8} = \sqrt{2 \cdot 8} = \sqrt{16} = 4 \qquad \frac{\sqrt{18}}{\sqrt{2}} = \sqrt{\frac{18}{9}} = \sqrt{9} = 3$$

29. $\sqrt{12} \cdot \sqrt{3}$ **30.** $\sqrt{10} \cdot \sqrt{2}$ **31.** $\sqrt{5} \cdot \sqrt{15}$ **32.** $\sqrt{14} \cdot \sqrt{7}$

33. $\frac{\sqrt{48}}{\sqrt{3}}$ **34.** $\frac{\sqrt{42}}{\sqrt{7}}$ **35.** $\frac{\sqrt{32}}{\sqrt{8}}$ **36.** $\frac{\sqrt{54}}{\sqrt{3}}$

CHAPTER REVIEW

CHAPTER SUMMARY

1. A polygon is classified according to the number of sides it has. A regular polygon is both equilateral and equiangular.
2. In a convex polygon with n sides, the interior angle sum is $(n - 2) \times 180°$. The exterior angle sum of any convex polygon, one angle at each vertex, is $360°$.
3. Properties of some special quadrilaterals are listed below.

	parallelogram	rectangle	rhombus	square
opposite sides parallel	✓	✓	✓	✓
opposite sides equal	✓	✓	✓	✓
opposite angles equal	✓	✓	✓	✓
consec. angles supplementary	✓	✓	✓	✓
diagonals bisect each other	✓	✓	✓	✓
diagonals equal		✓		✓
diagonals perpendicular			✓	✓
diagonals bisect angles			✓	✓

4. Four ways for showing that a quadrilateral is a parallelogram are given on page 216.
5. A trapezoid is a quadrilateral with just one pair of parallel sides. The median of a trapezoid is parallel to the bases and its length equals half the sum of the base lengths.
6. The Midpoints Theorem states that the segment joining the midpoints of two sides of a triangle is parallel to the third side and half as long.

REVIEW EXERCISES

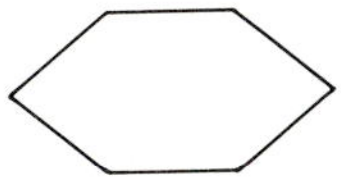

Refer to the figure at the right. *(See pp. 190–193.)*

1. Is the polygon convex?
2. Does the polygon appear to be equilateral?
3. Does the polygon appear to be equiangular?
4. Is the polygon regular?

Solve. *(See pp. 194–197.)*

5. Find the interior angle sum of a pentagon.
6. Find the exterior angle sum of a pentagon.
7. Find the measure of each angle of a regular octagon.

Tell whether each statement is true or false. *(See pp. 198–202.)*

8. Every rhombus is a square.
9. Every square is a parallelogram.
10. Every parallelogram is a rectangle.

Given ▱*PQRS*. *(See pp. 203–207.)*

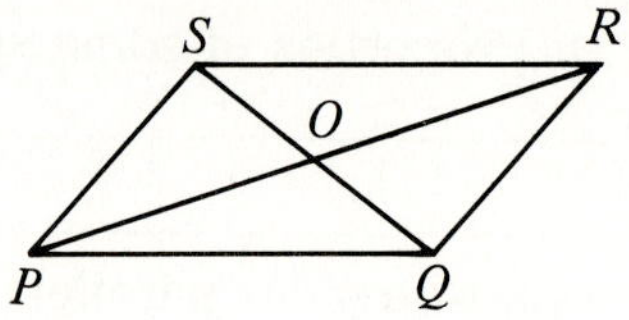

11. If $\angle RPQ = 18°$ and $\angle SPR = 30°$, find the measures of $\angle PRQ$ and $\angle PQR$.
12. If $PO = 12$ and $QO = 7$, find the lengths of $\overline{OS}$ and $\overline{PR}$.

Given rectangle *KNRX* and rhombus *SLFG*. *(See pp. 209–213.)*

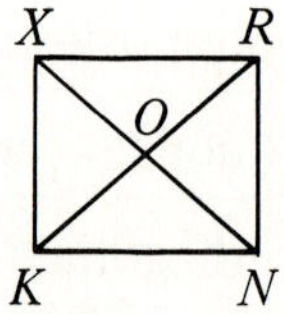

13. If $RK = 20$, then $ON =$ __?__.
14. If $\angle RKN = 40°$, find the measures of $\angle RKX$ and $\angle KXN$.
15. If $\angle LSG = 60°$, find the measures of $\angle FSG$ and $\angle LOF$.

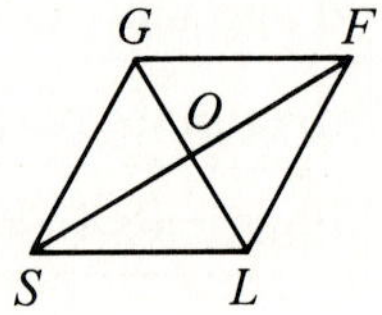

Write the definition or theorem that supports the statement "*ABCD* is a parallelogram."
(See pp. 214–219.)

16.

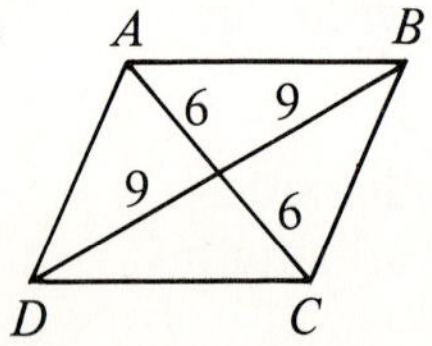

17.

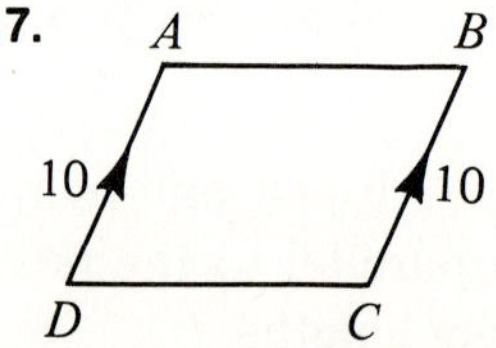

18.

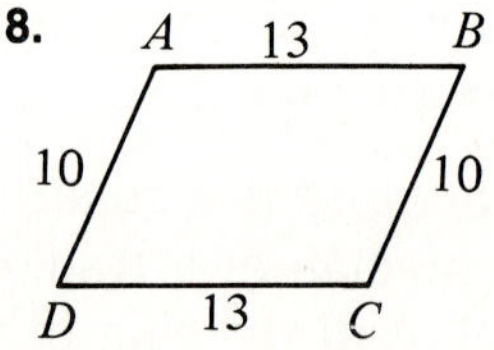

***ABXY* is an isosceles trapezoid with bases $\overline{AB}$ and $\overline{XY}$. $AY = BX = 6$, $XY = 8$, and $AB = 14$.** *(See pp. 220–224.)*

19. Find the length of the median of *ABXY*.
20. If $\angle A = 60°$, find the measures of $\angle B$, $\angle X$, and $\angle Y$.

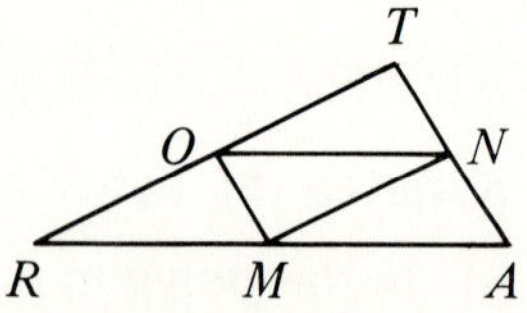

***M*, *N*, and *O* are the midpoints of the sides of △*TRA*. $RA = 14$, $AT = 6$, and $TR = 12$.** *(See pp. 225–228.)*

21. $MN =$ __?__
22. The perimeter of △*NOM* is __?__.

CHAPTER TEST

1. Sketch an octagon that is equiangular but not equilateral. *(5-1)*

2. A regular polygon is
   a. equiangular b. equilateral c. both (a) and (b)

A regular polygon has ten sides.

3. Find the measure of each exterior angle. *(5-2)*

4. Find the interior angle sum.

5. If $ABCDEF$ is a regular hexagon, find the measure of $\angle AFC$.

Tell whether each statement is true or false.

6. Some trapezoids are parallelograms. *(5-3)*

7. Some rectangles are squares. 8. Every rhombus is a quadrilateral.

The diagonals of ▱ WXYZ intersect at point P.
Tell whether the conclusion is true.

9. $WZ = XY$ 10. $\angle WXY + \angle XYZ = 180°$ 11. $\angle ZYW = \angle XWY$ *(5-4)*

***DAEC* is a rectangle.**

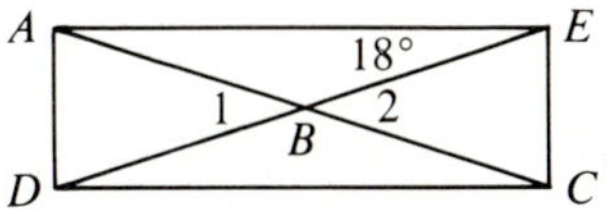

12. a. $\angle 1 =$ _?_ ° b. $\angle 2 =$ _?_ ° *(5-5)*

13. If $AC = 4.5$, then $ED =$ _?_.

Write the definition or theorem that allows you to conclude that the quadrilateral is a parallelogram. *(5-6)*

14. 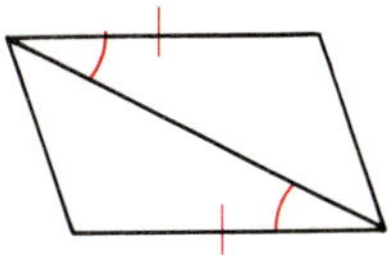

15. 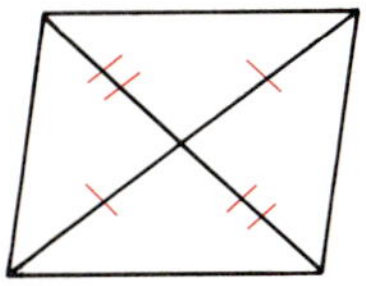

16. 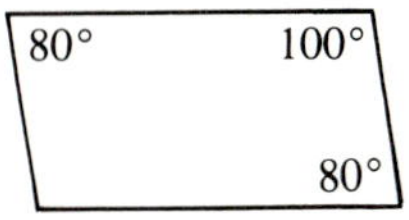

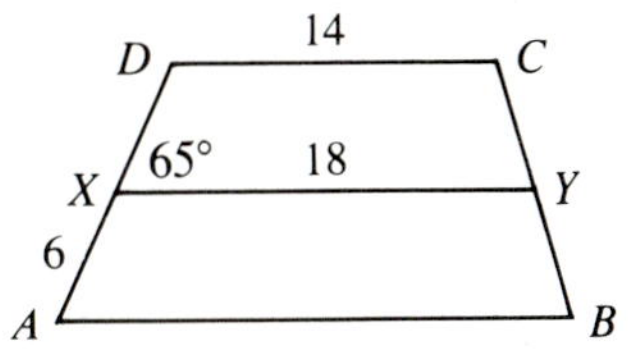

$\overline{XY}$ is the median of isosceles trapezoid *ABCD*. *(5-7)*

17. Find the measures of $\angle B$ and $\angle C$.

18. Find the lengths of $\overline{AB}$ and $\overline{BC}$.

In $\triangle DEF$, *M* is the midpoint of $\overline{DE}$ and *N* is the midpoint of $\overline{EF}$.

19. Explain why $\angle EMN = \angle D$. *(5-8)*

20. If $DF = 5x - 1$ and $MN = x + 4$, $x =$ _?_.

MIXED REVIEW

1. In $\triangle XYZ$, $\angle X = 53°$, $\angle Y = d°$, and $\angle Z = (2d - 5)°$. Find the value of d.

2. Write a two-column proof.

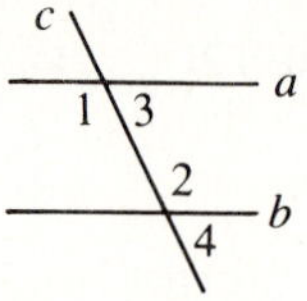

Given: $\angle 1 = \angle 2$

Prove: $\angle 3 = \angle 4$

3. In quadrilateral $ABCD$, $\angle A = 90°$, $\angle B = 50°$, and $\angle C = 130°$. Choose the best name for the quadrilateral.
   a. parallelogram **b.** trapezoid **c.** rectangle **d.** rhombus

In quadrilateral *EFGH*, $\overline{EF} \parallel \overline{GH}$ and $EF = GH$.

4. Which theorem supports the conclusion that *EFGH* is a parallelogram?

5. If *EFGH* is a parallelogram, must $EH = FG$? Explain.

Write *yes* or *no* to indicate whether the given method could be used to prove the two triangles congruent.

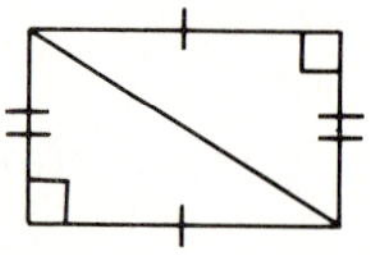

6. SSS **7.** SAS **8.** ASA **9.** AAS **10.** HL

11. Suppose many isosceles triangles have the same base. Then what must be true of the altitudes of the triangles drawn to the base?

12. Write a two-column proof.

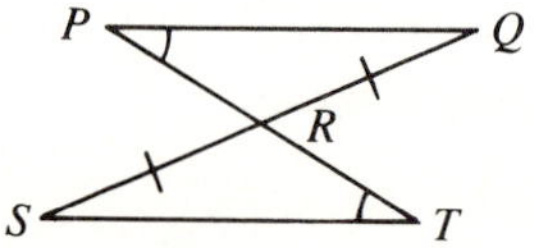

Given: $\angle P = \angle T$; $QR = RS$

Prove: $PQ = ST$

13. Construct a 45° angle.

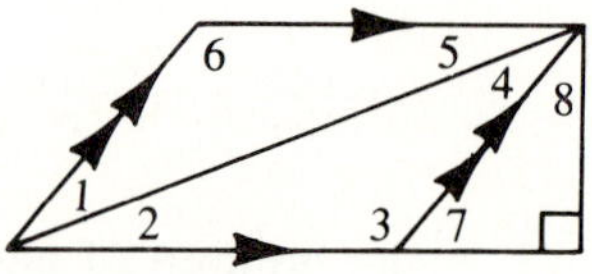

14. In the diagram, $\angle 1 = 30°$ and $\angle 2 = 20°$. Find the measures of the other numbered angles.

15. Write the converse of the statement "If a quadrilateral is a rectangle, then it has equal diagonals." Is the converse true or false?

16. Find the measure of each interior angle of a regular octagon.

17. Use one of the words *always, sometimes,* or *never* to complete: An obtuse triangle is __?__ equilateral.

18. Draw a very large triangle. Construct the inscribed circle.

19. The diagonals of $\square ABCD$ intersect at O. If $AO = 3x - 4$ and $AC = 4x + 6$, find the numerical length of $\overline{OC}$.

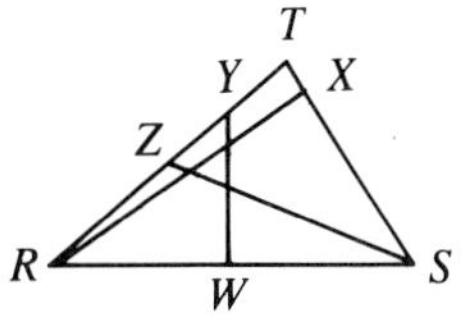

20. In $\triangle RST$, $RZ = ZT$, $RW = WS$, $\overline{YW} \perp \overline{RS}$, and $\overline{RX} \perp \overline{TS}$. __?__ is a median, __?__ is a perpendicular bisector, and __?__ is an altitude.

21. Points J and K have coordinates -17 and -5, respectively. Find the coordinate of the midpoint of $\overline{JK}$.

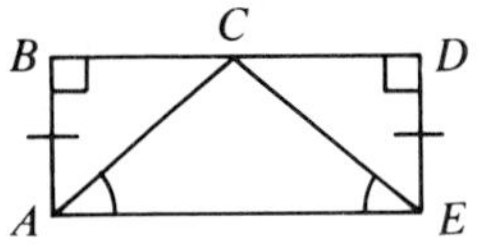

22. Can $\triangle ABC$ and $\triangle EDC$ be proved congruent? If so, what postulate or theorem would be simplest to use?

23. Sketch a rhombus that is not a square.

24. Name the four ways to prove that two lines are parallel.

25. In $\triangle JMW$, $JM = 9$, $MW = 7$, and $JW = 10$. Find the perimeter of the triangle formed by joining the midpoints of the sides of $\triangle JMW$.

26. *Must* the diagonals of a rhombus have the stated property?
    a. They are perpendicular. b. They are equal.
    c. They divide the rhombus into four congruent triangles.
    d. They intersect at the midpoint of each diagonal.

27. Use a protractor to draw an angle of 125°. Then construct an angle equal to the one you drew.

28. A trapezoid has bases of lengths 9 and 24 and a median of length $3x + \frac{3}{2}$. Find the value of x.

29. Supply the reasons to complete the proof.

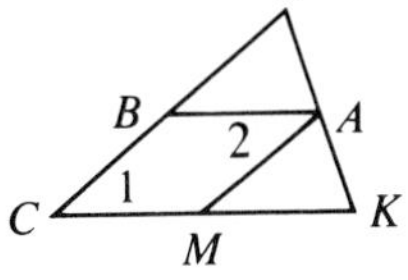

Given: A, B, and M are the midpoints of the sides of $\triangle CLK$.

Prove: $\angle 1 = \angle 2$

STATEMENTS	REASONS
1. $\overline{AB} \parallel \overline{CM}$; $\overline{AM} \parallel \overline{BC}$	1. __?__
2. $ABCM$ is a parallelogram.	2. __?__
3. $\angle 1 = \angle 2$	3. __?__

Here's what you'll learn in this chapter:

- To find the area of a rectangle, a square, and a parallelogram.
- To find the area of a triangle.
- To find the area of a trapezoid.
- To use the Pythagorean Theorem and its converse.
- To find the circumference and the area of a circle.

The fields on a large farm are often laid out in many different shapes. In order to plan for the best use of the land, a farmer needs to know the area of each field.

Chapter 6

Areas

1 Areas of Rectangles

Which of the rectangles shown at the right covers more surface? Count the squares to find out.

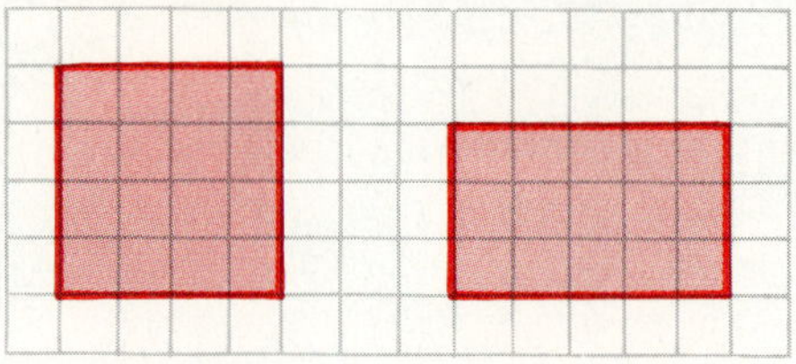

The 4 × 4 (read "4 by 4") rectangle is made up of 16 squares. The 5 × 3 rectangle is made up of 15 squares. We say that the 4 × 4 rectangle has the greater *area.* The amount of surface in a region is its **area.**

A square with sides one unit long has an area of *one square unit.* A common unit of area is the square centimeter.

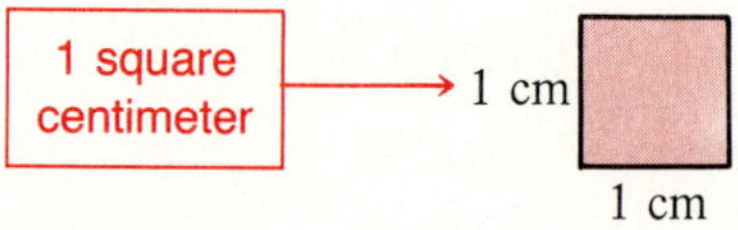

Rectangle *ABCD* has a *base* 5 cm long. Its *height* is 2 cm. Since the rectangle is made up of 10 square centimeters, its area is 10 square centimeters (written 10 cm^2). Notice that the area is the product of the lengths of the sides: 10 = 5 × 2.

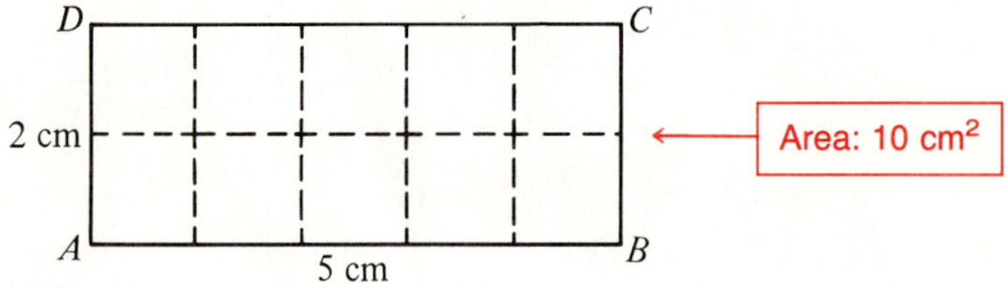

This example suggests the following postulate.

POSTULATE 12
The area of a rectangle is given by the formula: *Area = base × height*

EXAMPLE 1 Find the area of a rectangle with base length 12 and height 6.

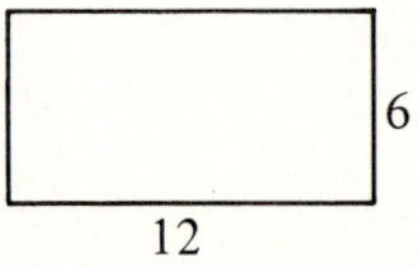

$A = bh$
$= 12 \times 6$
$= 72$
Answer: 72 square units

EXAMPLE 2 What is the area of a rectangle 2 m long and 40 cm wide?

When the unit of length is 1 centimeter:
Instead of using 2 m, use $2 \times 100 = 200$ cm.

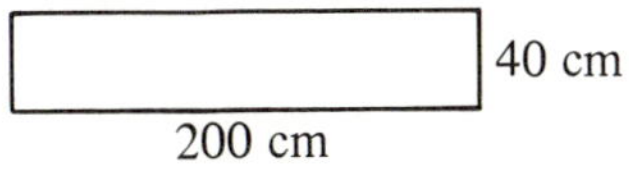

$A = bh = 200 \times 40 = 8000$

Answer: 8000 cm^2

When the unit of length is 1 meter:
Instead of using 40 cm, use $\frac{40}{100} = 0.4$ m

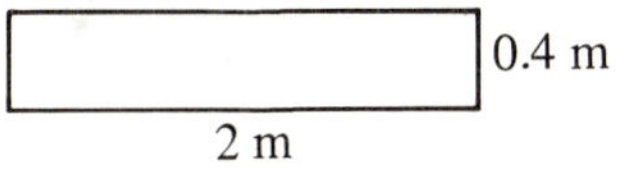

$A = bh = 2 \times 0.4 = 0.8$

Answer: 0.8 m^2

Both answers are correct.

You should see that 8000 cm^2 is the same amount of area as 0.8 m^2.

Recall: Since 1 m = 100 cm,
1 m^2 = 100 cm × 100 cm = 10,000 cm^2.

Every square is a rectangle, so the formula $A = bh$ applies to squares. In a square, b and h are equal. By substituting s for both b and h in the formula of Postulate 12, we obtain Theorem 1.

THEOREM 1

The area of a square is given by the formula:
Area = side squared

EXAMPLE 3 Find the area of a square with sides of length 8.

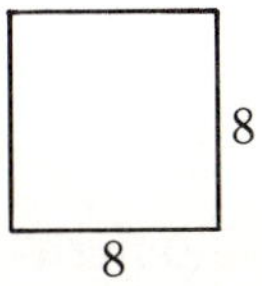

$A = s^2$
$= 8^2$
$= 8 \times 8 = 64$

Answer: 64 square units

Remember: The distance around a region is its **perimeter.** To find the perimeter of a rectangle, use the formula $\boldsymbol{p = 2b + 2h}$.

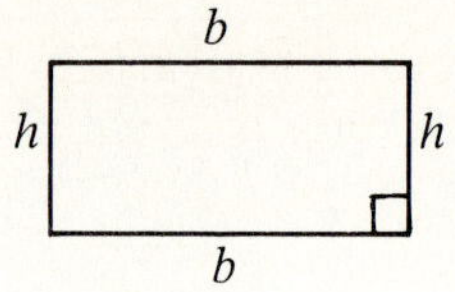

EXAMPLE 4

The perimeter of each rectangle shown below is 100. Which rectangle has the greatest area?

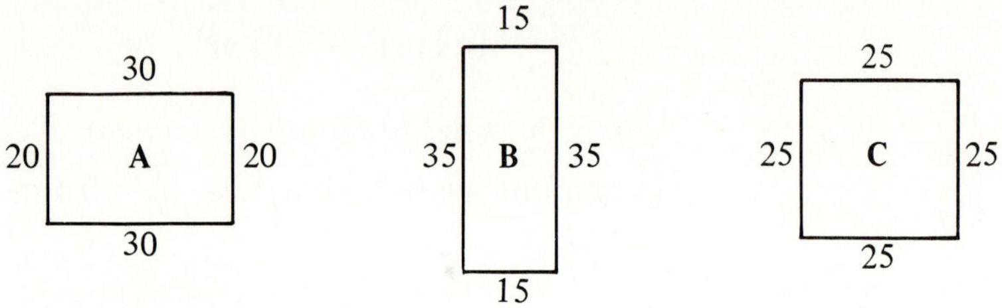

You should be able to show that rectangle C has the greatest area.

Classroom Practice

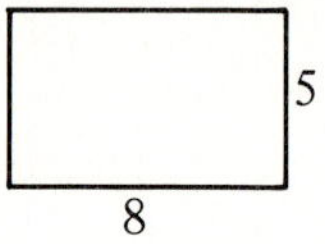

1. Find the perimeter of the rectangle shown.
2. Find the area of the rectangle shown.
3. Each rectangle in Example 4 has perimeter 100.
   a. Draw two different rectangles, each with perimeter 100.
   b. Find the area of each rectangle you drew.
4. Do you believe that you could draw a rectangle with perimeter 100 and an area as small as you please?
5. Draw a square with sides 1 m long on the chalkboard.
   The area = __?__ m^2.
6. Use the figure of Exercise 5. Each side is __?__ cm long.
   The area = __?__ cm^2.
7. Large regions are often measured in square kilometers (km^2). A large parking lot is in the form of a square, 1.2 km on each side.
   a. The area of the park = __?__ km^2.
   b. The area of the park = __?__ m^2. (1 km = 1000 m)
8. a. Suppose two rectangles have equal areas.
   Must the bases have the same length?
   b. Suppose two squares have equal areas.
   Must the sides of the squares have the same length?

Written Exercises

Find the area of each rectangle.

A

1. 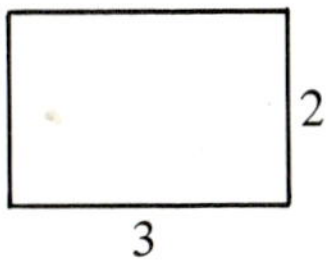

2. 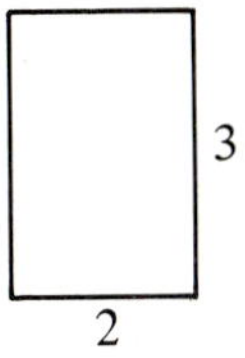

3. 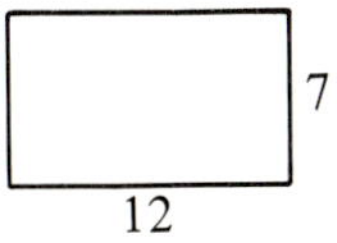

4. 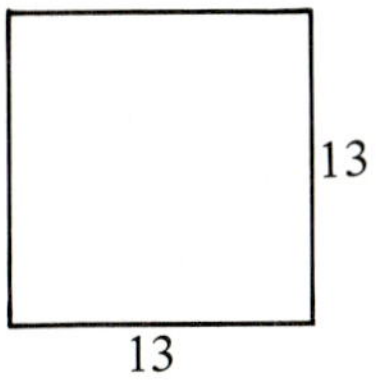

5. 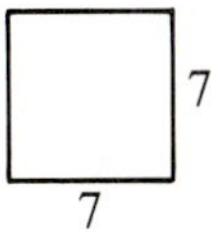

6. 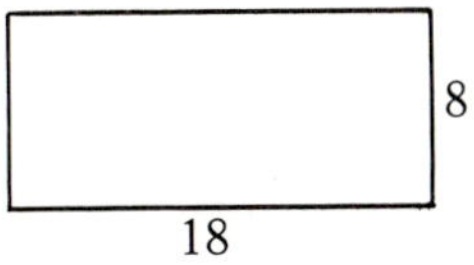

In Exercises 7–12, find the perimeter and the area of a rectangle with base length *b* and height *h*.

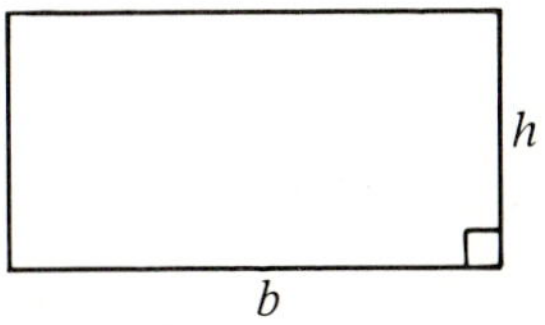

7. $b = 6, h = 2$
8. $b = 10, h = 7$
9. $b = 4, h = 3.2$
10. $b = 9.6, h = 4$
11. $b = 7\frac{2}{5}, h = 6$
12. $b = j$ cm, $h = k$ cm

In Exercises 13–18, find the perimeter and the area of a square with sides of length *s*.

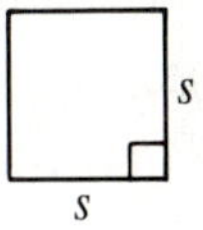

13. $s = 1$
14. $s = 2$
15. $s = 3$
16. $s = 4$
17. $s = 5$
18. $s = j$ km

Copy and complete the table about squares. In Exercises 23 and 24, express your answers in terms of *k*.

B

	19.	20.	21.	22.	23.	24.
side s	?	?	?	?	$2k$	$k + 3$
perimeter p	36	20	?	?	?	?
area A	?	?	49	64	?	?

Find the area (a) in square centimeters and (b) in square meters.

25. A square with sides 60 cm long
26. A square with sides 0.9 m long
27. A rectangle 3 m long and 50 cm wide
28. A rectangle 5 m long and 70 cm wide

29. An artist plans to design a series of posters. Each poster will have an area of 1855 cm^2 and a length of 53 cm. How wide will the posters be?

30. The area of a rectangular photo album page is 476 cm^2. The width is 17 cm. Find the length.

31. The Nickersons want to plant grass in their new backyard. How many square meters of lawn can they have if they also have a flower garden and a patio?

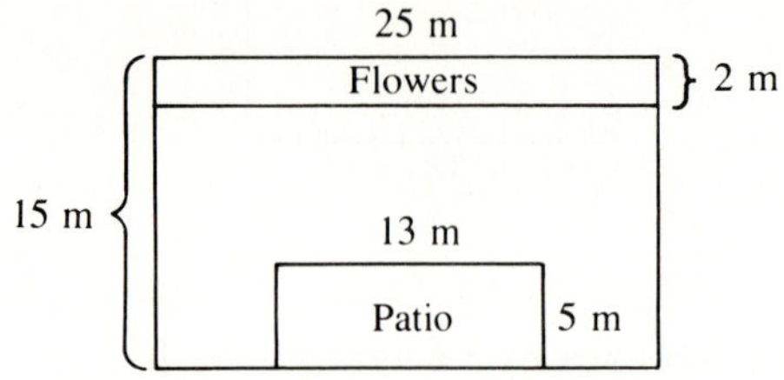

Exercise 31

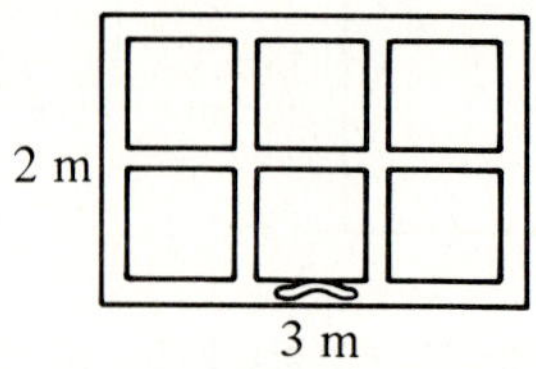

Exercise 32

C 32. Some pleated draperies must be twice as wide as the window that they cover.

a. How many square meters of material are needed to cover the window shown?

b. The drapery material is sold from a bolt of cloth 1.5 m wide. What length of material should be cut from the bolt?

33. A piece of sheet metal is cut and bent to form the box shown. The box has no top. Find the area of:

a. the bottom b. the two shaded sides c. the other two sides

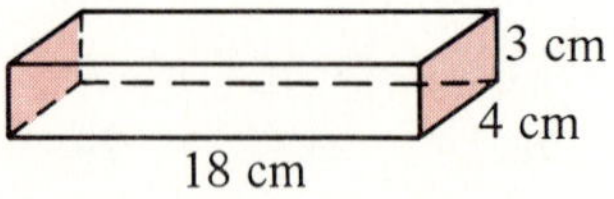

Exercise 33

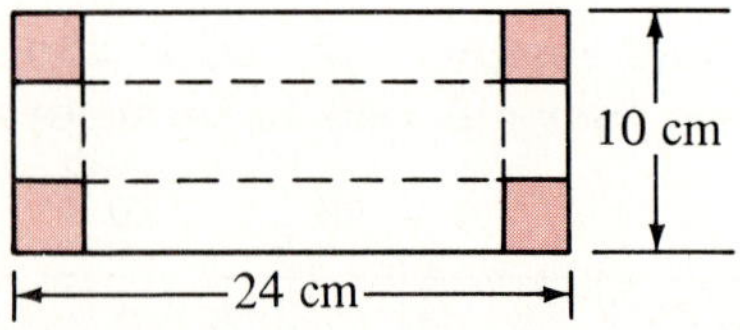

Exercise 34

34. The metal used for the box in Exercise 33 is 24 cm long and 10 cm wide. The dashed lines indicate folds. The shaded regions indicate wasted sheet metal. How much metal is wasted?

35. The roof of an A-frame cabin is to be shingled at a cost of \$90 a square. (A *square*, in shingling, is a region with an area of 100 square feet.) Find the cost of shingling the roof of the cabin shown.

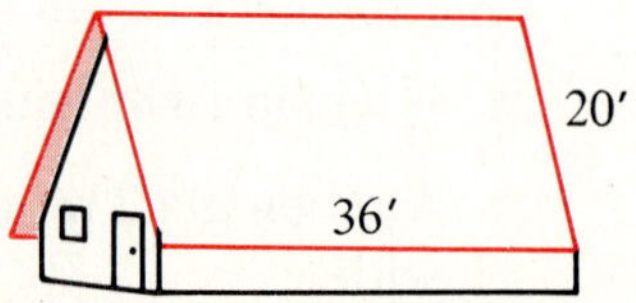

2 Areas of Parallelograms

Suppose you are given the four sticks shown.

You can join the sticks, with equal sides opposite each other, to form a quadrilateral. The quadrilateral has to be a parallelogram. It can be, but does not have to be, a rectangle.

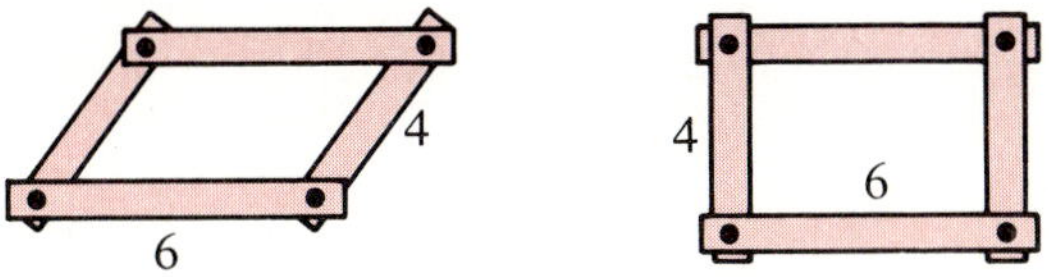

The area of the rectangle is 24, but the area of the parallelogram at the left is less than 24.

The area of a parallelogram is ***not*** *equal to the product of two consecutive sides, unless the parallelogram happens to be a rectangle.*

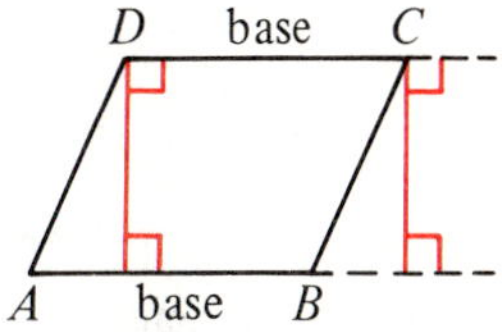

The area of the stick parallelogram depends on its *height*. The **height** of a parallelogram is the length of an *altitude*. This is a segment between, and perpendicular to, the lines containing the bases. Two altitudes are drawn to bases $\overline{AB}$ and $\overline{CD}$ in $\square ABCD$. Notice that these altitudes are equal.

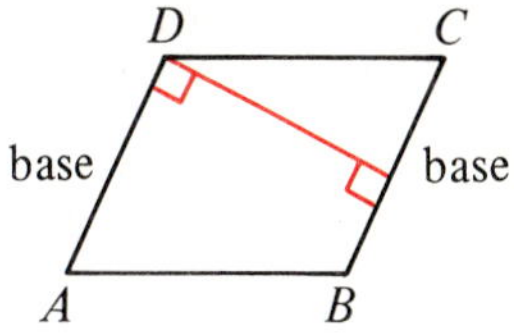

In any parallelogram, either pair of opposite sides may be considered the bases. If $\overline{AD}$ and $\overline{BC}$ are taken as bases, then the red segment is an altitude to those bases.

EXPLORATIONS

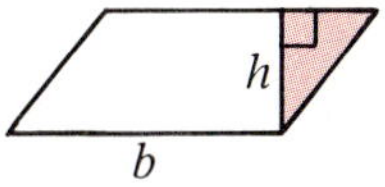

1. Cut a parallelogram out of cardboard. Cut along the indicated altitude, h.
2. Move the triangle as shown to form a rectangle. The area of the rectangle equals bh.

   Therefore the area of the original parallelogram is also equal to bh.

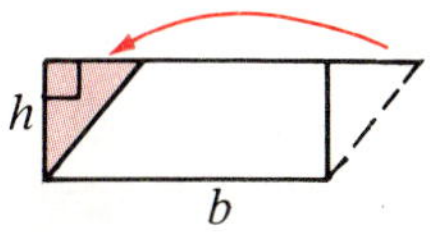

   This exploration suggests the theorem stated on page 244.

THEOREM 2
The area of a parallelogram is given by the formula: *Area = base × height*

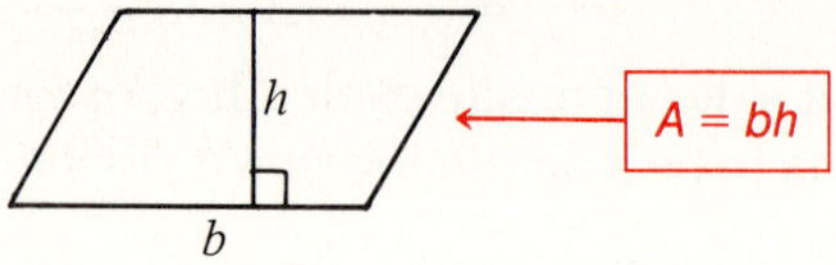

Classroom Practice

Find the area of each parallelogram.

1.

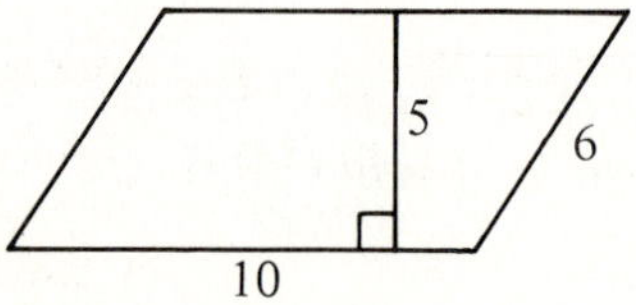

2.

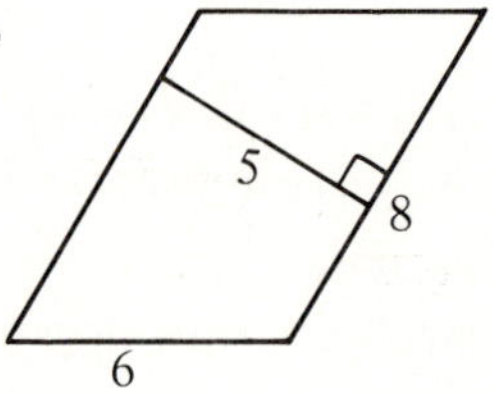

3. Take $\overline{QU}$ as a base of ▱*QUAD*.
   Area of ▱*QUAD* = 15 × _?_ = _?_.

4. Take $\overline{UA}$ as a base of ▱*QUAD*.
   Area of ▱*QUAD* = 10 × _?_ = _?_.

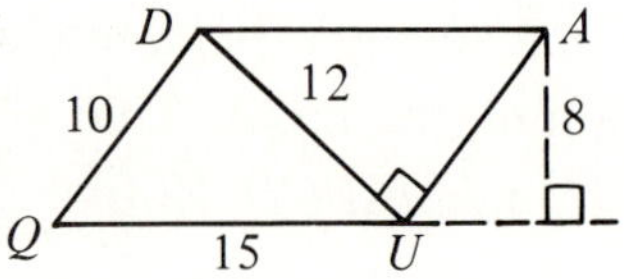

Exercises 5 and 6 refer to ▱*ABCD* shown below. The parallelogram is constructed so that its shape may be changed.

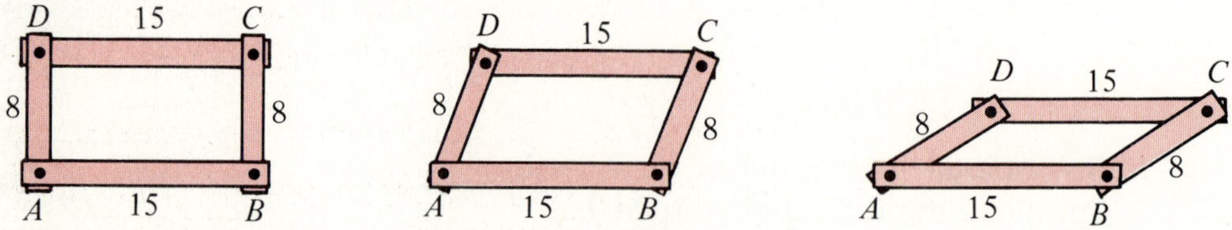

5. Find the greatest possible area that ▱*ABCD* can have.

6. As the height of ▱*ABCD* decreases, what happens to the area of the parallelogram?

7. For the parallelogram at the right:
   a. Find the perimeter in meters;
   b. Find the area in square meters.

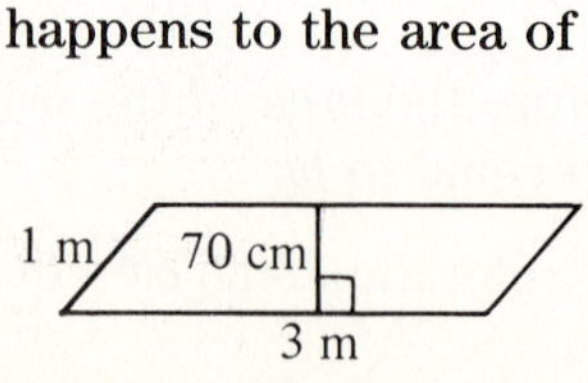

Written Exercises

Find the area of each parallelogram.

A **1.**
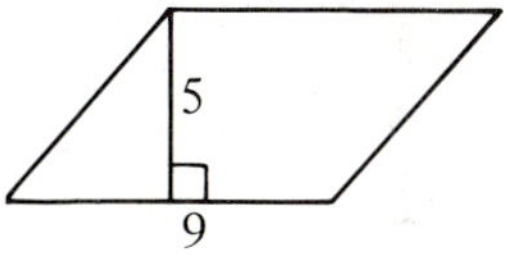

2.
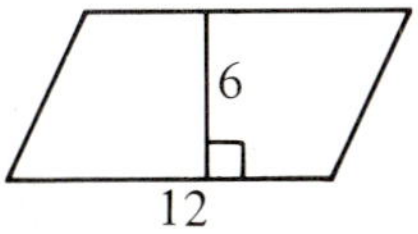

3.
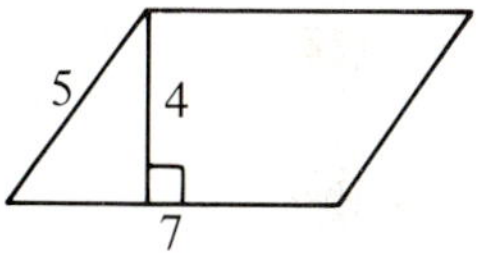

4.
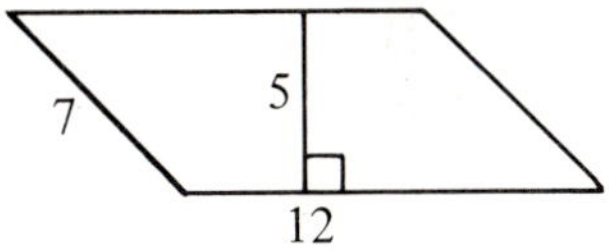

5.
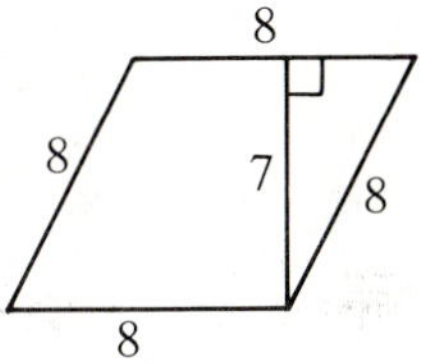

6.
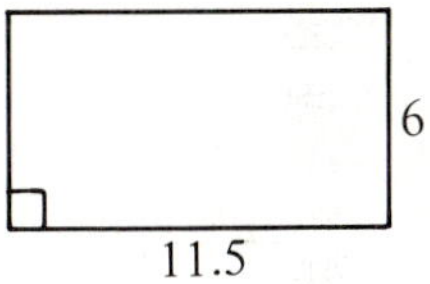

7.
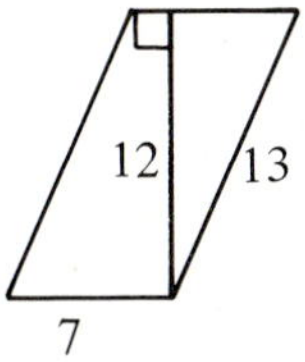

8.
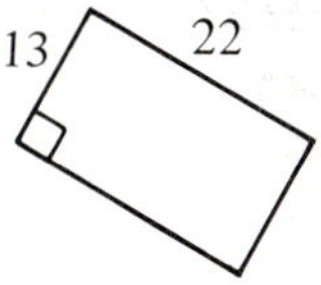

9.
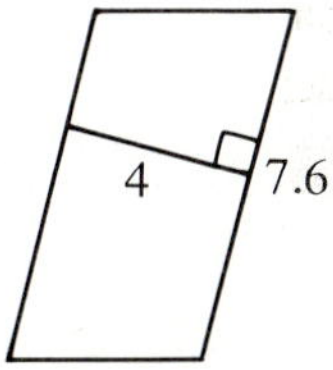

Find the perimeter and the area of each parallelogram. Express your answers in terms of *k*.

B **10.**
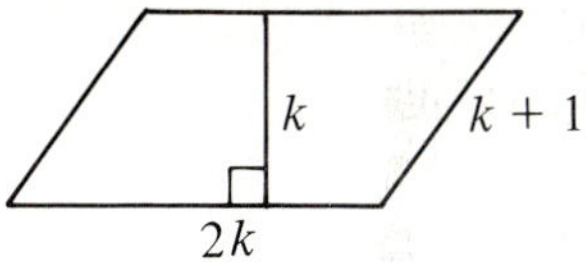

11.
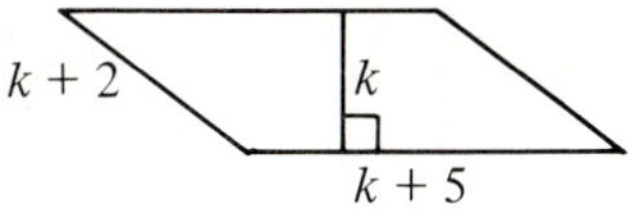

12.
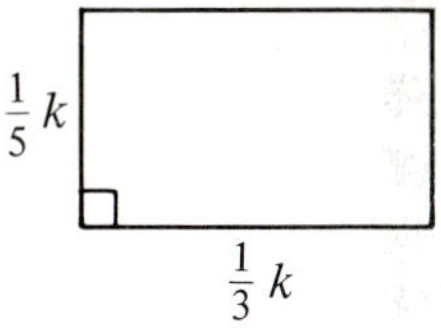

13.
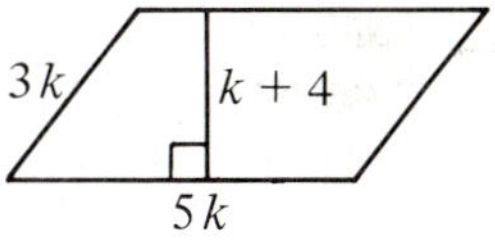

C **14.** The area of rectangle *CEFD* is 96.
Find the area of ▱*ABCD*.

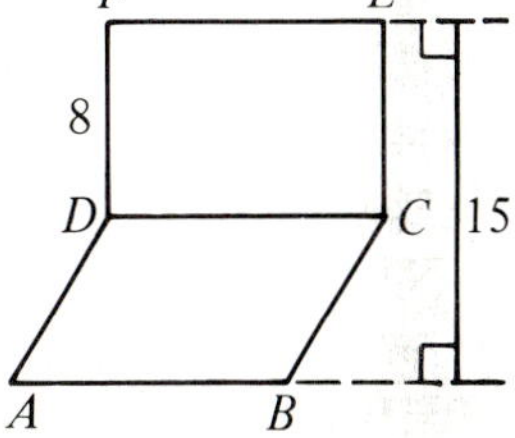

15. Construct a parallelogram *PQRS*.
Then construct a rectangle *WXYZ* so that the area of the rectangle equals the area of the parallelogram.

SELF-TEST

Vocabulary

area (p. 238) height of a parallelogram (p. 243)
perimeter (p. 240)

1. The sides of a rectangle have lengths 10 and 3. *(6-1)*
   a. The perimeter of the rectangle is __?__.
   b. The area of the rectangle is __?__.
2. The area of a square is 25 cm^2.
   a. The length of a side of the square is __?__.
   b. The perimeter of the square is __?__.

Find the area of each parallelogram.

3. 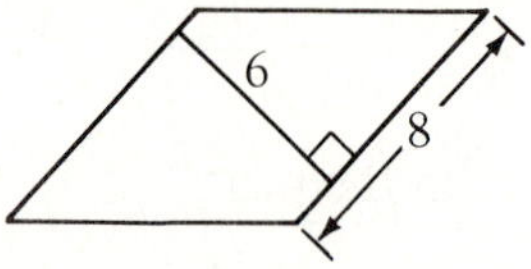

4. 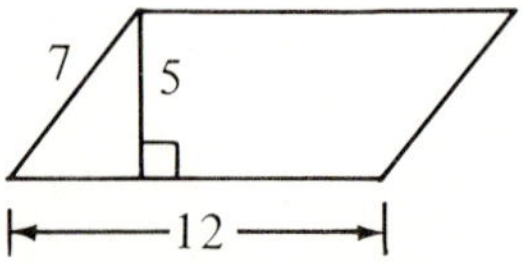

(6-2)

CONSUMER APPLICATIONS

THE COST OF DRIVING

Buying a car is a major investment. It takes money to operate a car, too. The owner must pay for gas, oil, repairs, parking fees, tolls, taxes, and insurance. A car also loses some of its resale value every day due to normal wear and tear. This loss is called *depreciation.* Taking all of these factors into account, we can estimate the average cost of 1 km of driving:

standard-size car: 17.3¢ compact car: 14.5¢ sub-compact car: 14.1¢

For many trips, a car is the most convenient and the least expensive method of transportation. This is particularly true if several people are traveling together. For some short trips, however, public transportation might cost less money. Don't forget that walking is free!

3 Areas of Triangles

Given any parallelogram, you can separate it into two congruent triangles. You simply draw a diagonal. Since congruent triangles have equal areas:

Area of each triangle $= \frac{1}{2}$(Area of parallelogram) $= \frac{1}{2}bh$

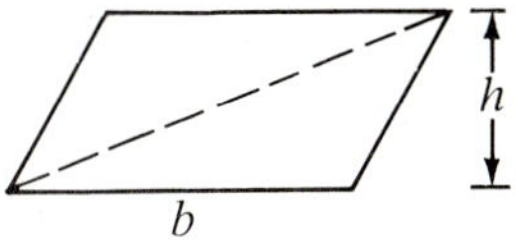

The diagrams below suggest that any triangle is "half" of a parallelogram that can be built on the triangle. The parallelogram has the same base length and height as the triangle.

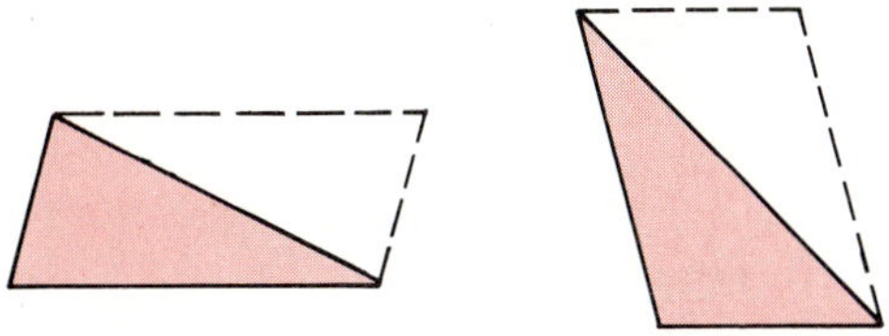

The discussion above contains the key ideas that are used in the proof of Theorem 3.

THEOREM 3

The area of a triangle is given by the formula:

$$Area = \frac{1}{2} \times base \times height$$

EXAMPLE Find the area of each triangle.

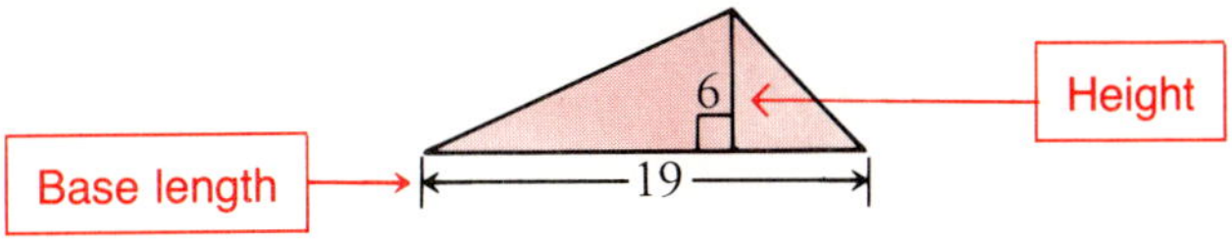

$$A = \frac{1}{2}bh = \frac{1}{2} \times 19 \times 6 = 57$$

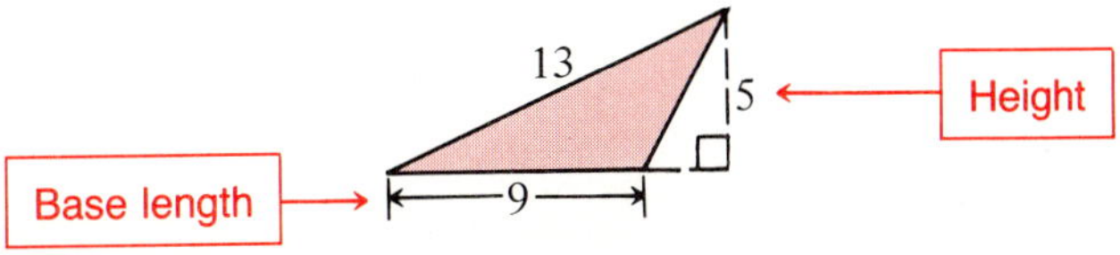

$$A = \frac{1}{2}bh = \frac{1}{2} \times 9 \times 5 = \frac{45}{2}, \text{ or } 22.5$$

Classroom Practice

1. $\overline{DE}$ is taken as the base of $\triangle DEF$. Name the altitude to $\overline{DE}$.

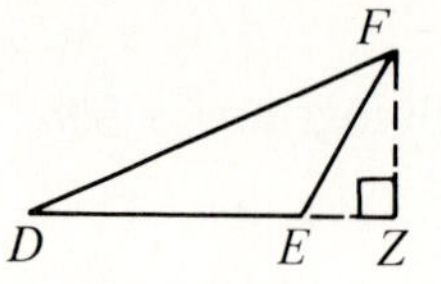

Exercise 1

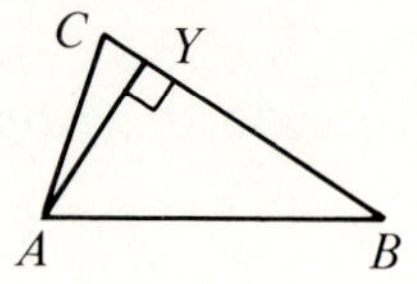

Exercise 2

2. $\overline{BC}$ is taken as the base of $\triangle ABC$. Name the altitude to $\overline{BC}$.

Exercises 3–5 refer to right $\triangle JKM$.

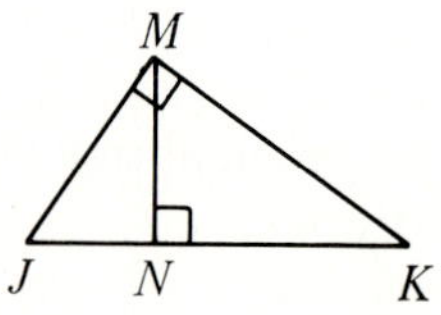

3. Name the altitude to $\overline{MK}$.
4. Name the altitude to $\overline{MJ}$.
5. Name the altitude to $\overline{JK}$.

In Exercises 6–10, use the diagram shown.

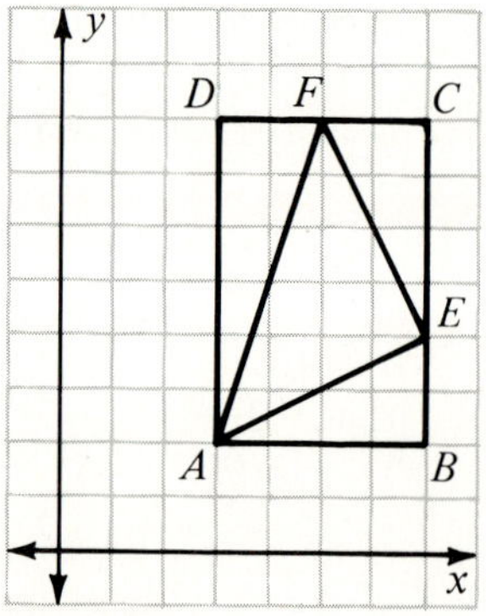

6. The area of rectangle $ABCD$ is __?__.
7. The area of $\triangle AEB$ is __?__.
8. The area of $\triangle ECF$ is __?__.
9. The area of $\triangle FDA$ is __?__.
10. Subtract the areas of these triangles from the area of the rectangle. The area of $\triangle AEF$ is __?__.

Exercises 11–13 refer to rhombus $ABCD$ with $AC = 12$ and $BD = 20$. Recall that the diagonals of a rhombus are perpendicular bisectors of each other.

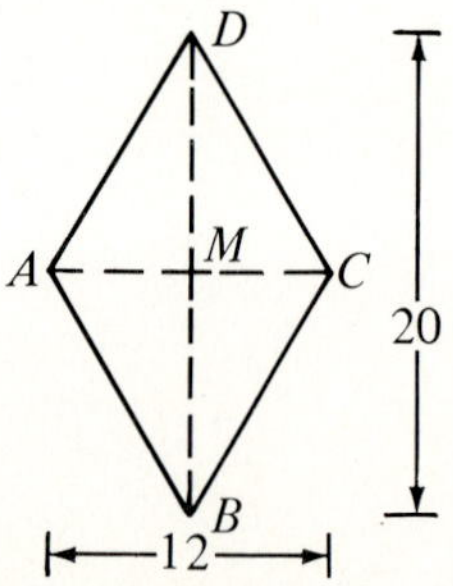

11. $MC =$ __?__ and $MD =$ __?__
12. Area of $\triangle DMC =$ __?__
13. Notice that the rhombus is separated into four congruent triangles. The area of rhombus $ABCD$ is __?__.

Written Exercises

Find the area of the shaded triangle.

A **1.**

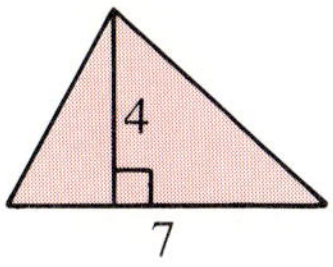

2.

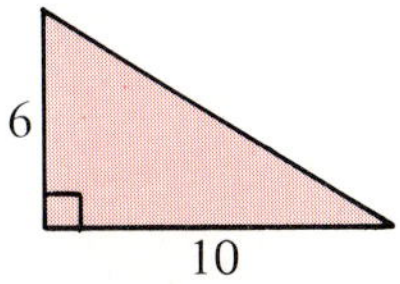

3.

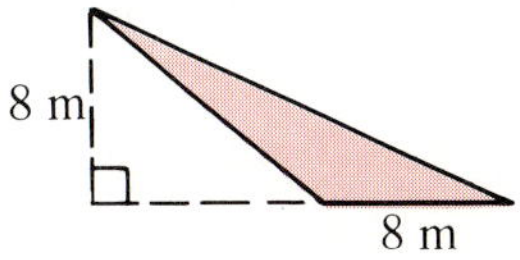

4.

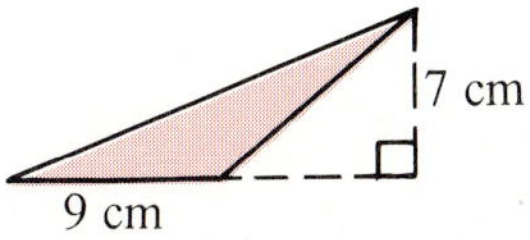

5.

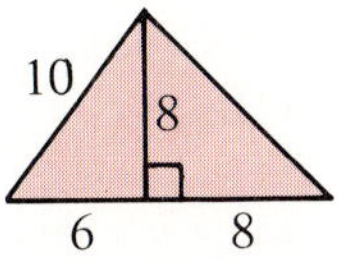

6.

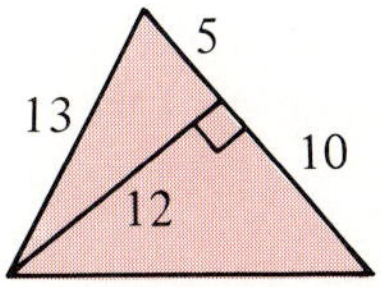

7.

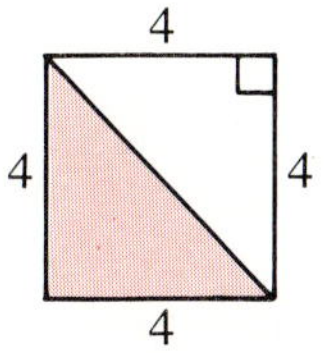

8.

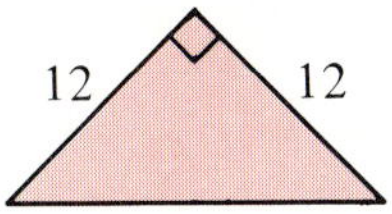

9.

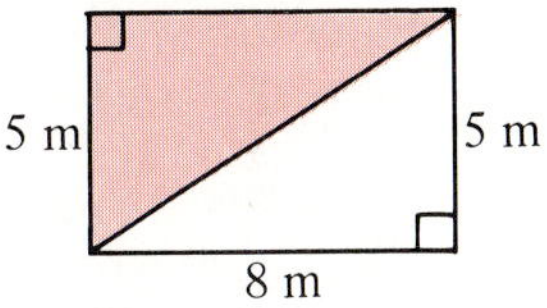

B **10.**

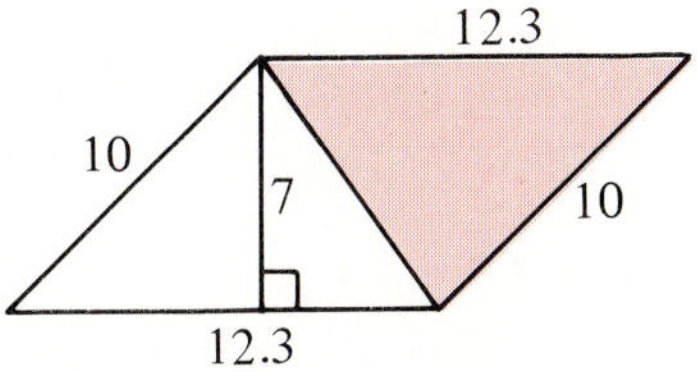

11.

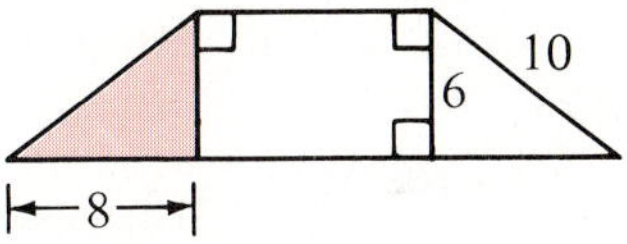

Express the area in terms of t.

12.

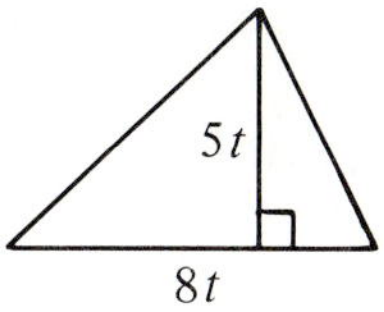

13.

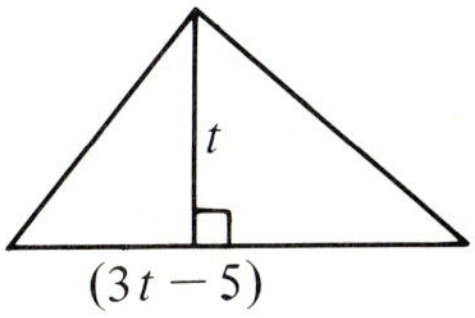

14.

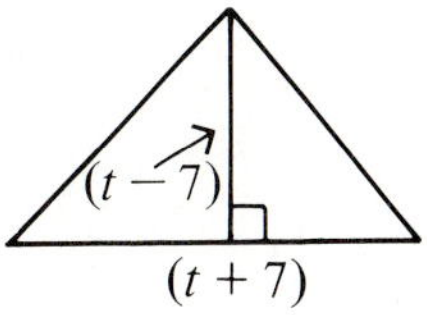

C **15.** Draw a triangle ABC. Then construct a rectangle $WXYZ$ so that the area of the rectangle equals the area of the triangle.

Strategy:

A. Construct $\overline{WX}$ so that $WX = \frac{1}{2}AB$.

B. Construct a line perpendicular to $\overline{WX}$ at W.

C. Mark off Z on the perpendicular so that WZ equals the length of the altitude from C in $\triangle ABC$.

D. Complete rectangle $WXYZ$.

4 Areas of Trapezoids

These explorations lead to a theorem about the area of a trapezoid.

EXPLORATIONS

Cut out three congruent trapezoids shaped roughly like trapezoid *RSTV*. The bases of your trapezoids should be 11 cm and 21 cm long. The altitudes should be 8 cm long.

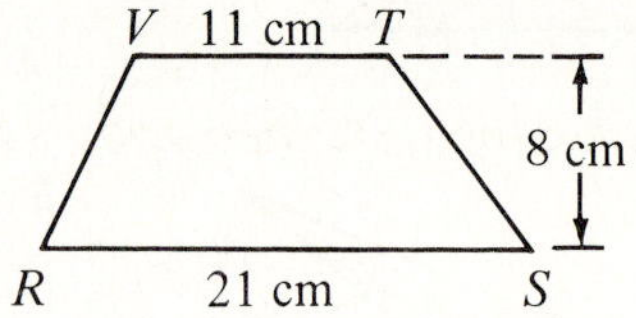

A. Cut along $\overline{VX}$ and $\overline{TY}$.
Since $RS = 21$ and $XY = 11$,
$RX + YS = 21 - 11 = 10$.

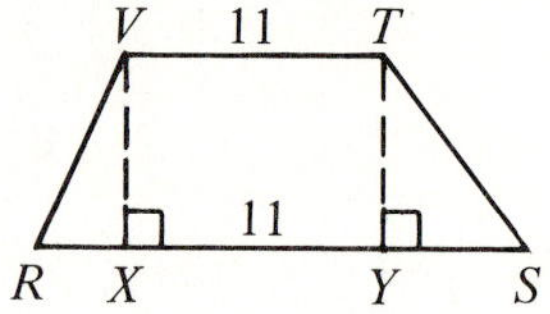

Place the two triangles alongside each other. Notice that rectangle *XYTV* and $\triangle RSZ$ both have altitudes 8 cm long.

Add the area of rectangle *XYTV* and the area of $\triangle RSZ$ to find the area of trapezoid *RSTV*.

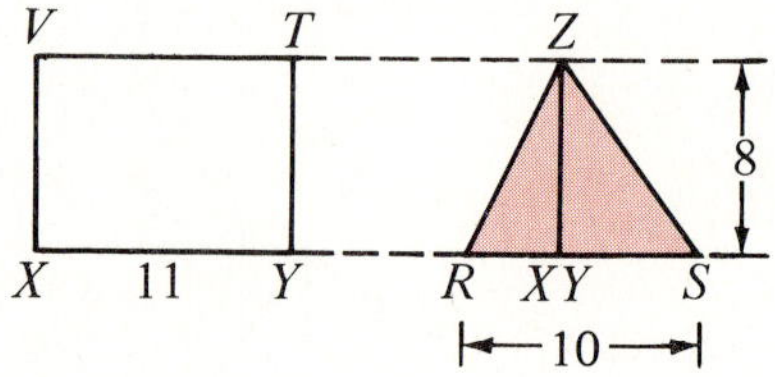

B. Take another copy of trapezoid *RSTV*. Cut along $\overline{VS}$.

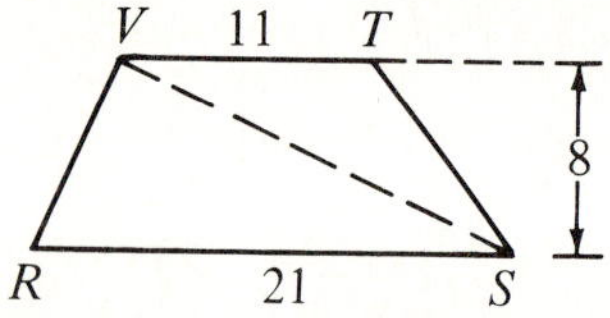

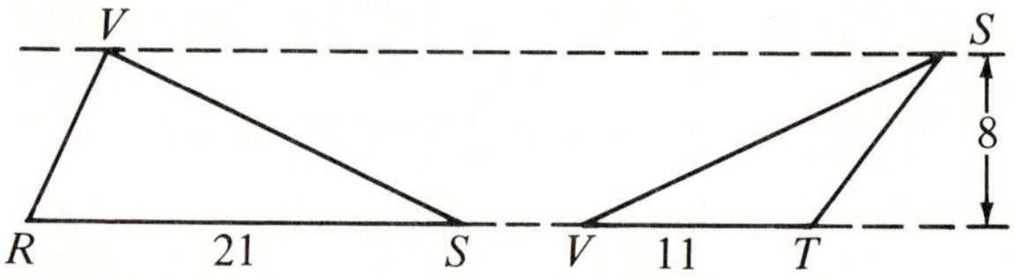

Place the triangles as shown above. Then add the areas of the triangles to find the area of the trapezoid.

C. Take the third copy of the trapezoid.
Draw median $\overline{MN}$.
Using Theorem 14 on page 221:

$$MN = \frac{1}{2}(21 + 11) = \frac{1}{2}(32) = 16.$$

Cut along $\overline{MX}$ and $\overline{NY}$.

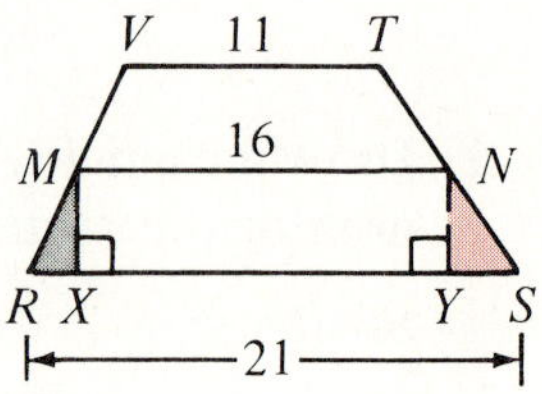

Move the triangles as shown. Find the area of trapezoid *RSTV* by computing the area of the rectangle formed.

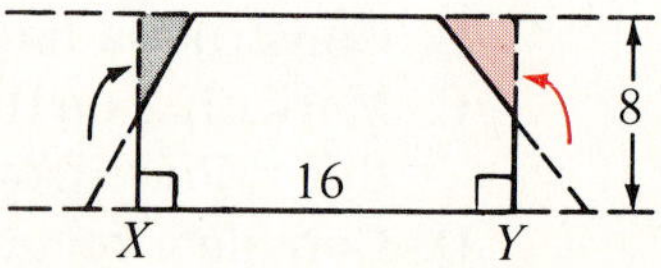

Each of the explorations suggests a way to prove Theorem 4.

THEOREM 4
The area of a trapezoid is given by the formula:
Area $= \frac{1}{2} \times$ *height* $\times$ *sum of the bases*

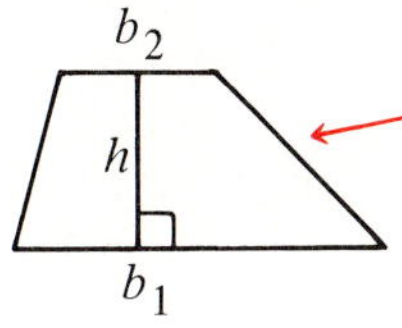

$A = \frac{1}{2}h(b_1 + b_2)$

EXAMPLE

Find the area.

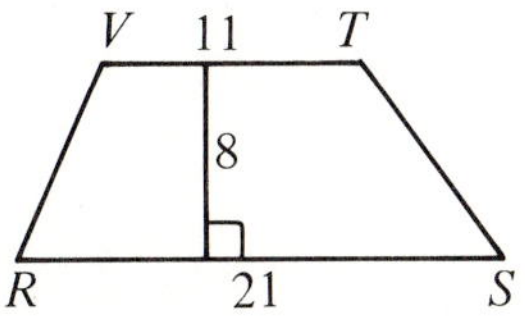

$$A = \frac{1}{2}h(b_1 + b_2)$$
$$= \frac{1}{2} \times 8 \times (21 + 11)$$
$$= 4 \times 32$$
$$= 128$$

Did you get this answer in Explorations A, B, and C?

Classroom Practice

Find the area of each trapezoid.

1.

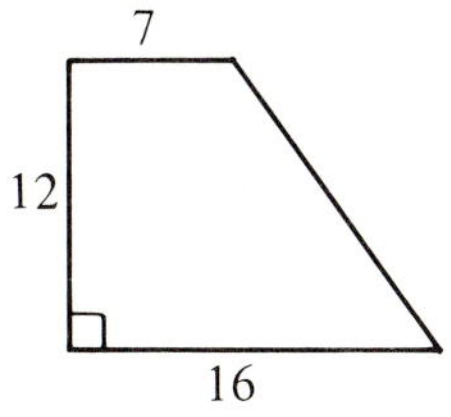

2.

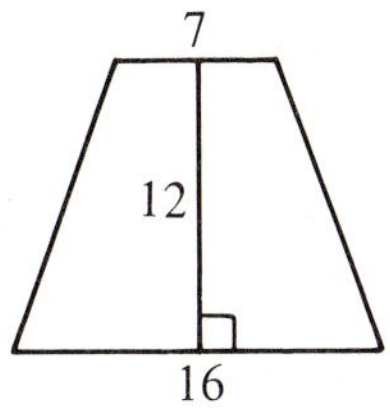

3.

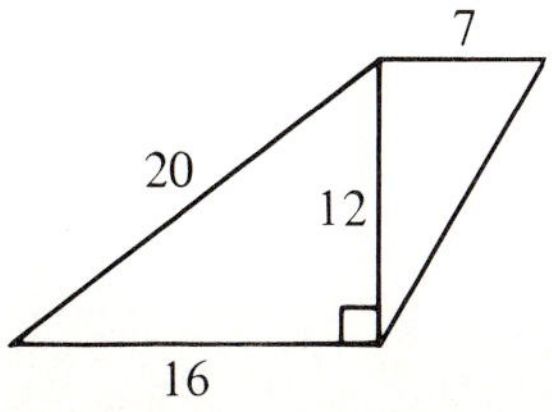

4.

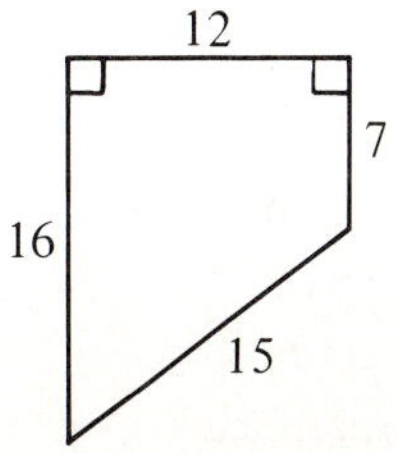

The figure shows congruent trapezoids *ABEF* and *DEBC*.

5. Why must quadrilateral *ACDF* be a parallelogram?

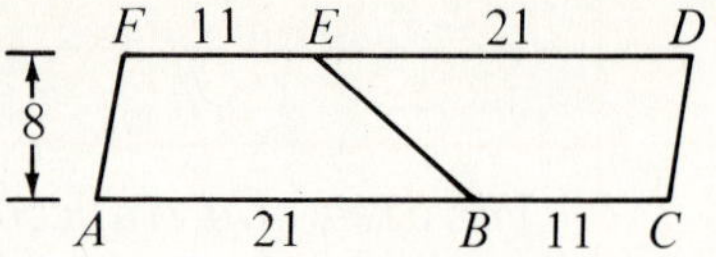

6. What is the area of parallelogram *ACDF*?

7. What is the area of trapezoid *ABEF*?

8. Refer to Explorations A–C. Which suggests that the area of a trapezoid equals the product of the median and the altitude?

Written Exercises

Select the correct area formula from the column at the right.

A

1. Triangle — a. $A = s^2$

2. Parallelogram — b. $A = \frac{1}{2}h(b_1 + b_2)$

3. Square — c. $A = bh$

4. Trapezoid — d. $A = \frac{1}{2}bh$

Find the area of each trapezoid.

5.

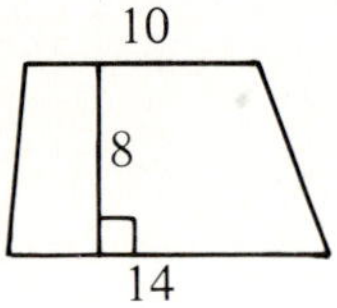

6.

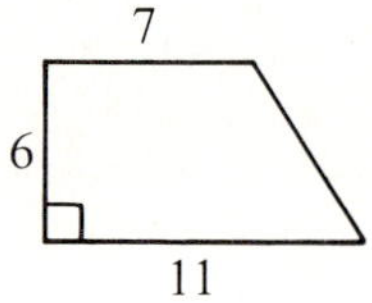

7.

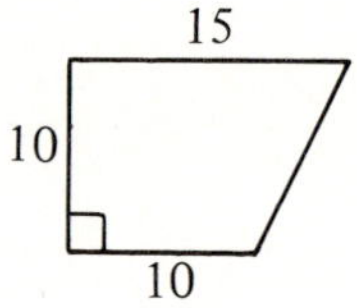

8.

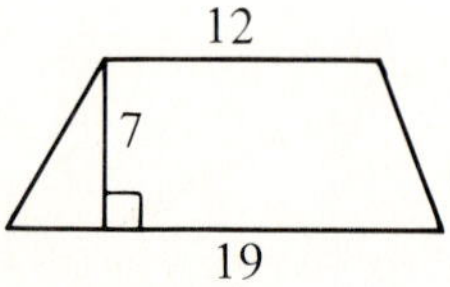

9.

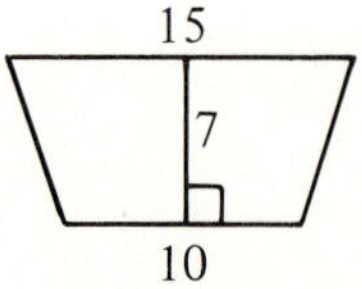

10.

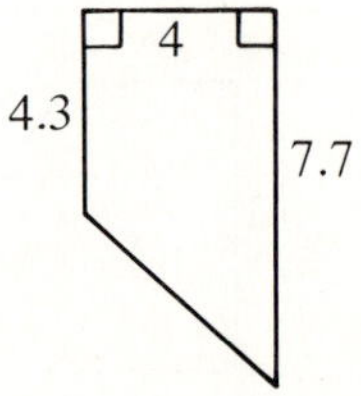

In Exercises 11–16, find the area of each polygon.

11.

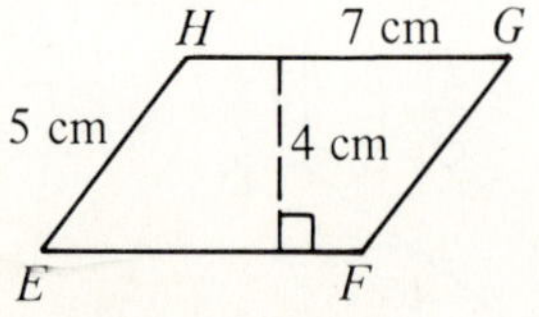

$\overline{HG} \parallel \overline{EF}$; $\overline{EH} \parallel \overline{FG}$

12.

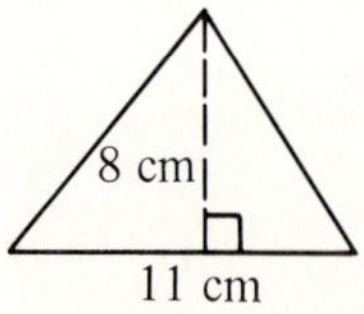

13.

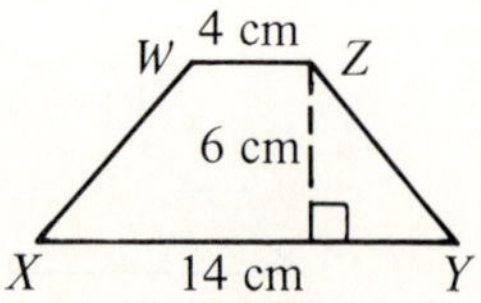

$\overline{WZ} \parallel \overline{XY}$

Find the area of each polygon.

14. 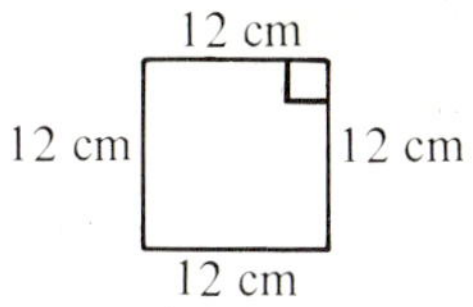

15. 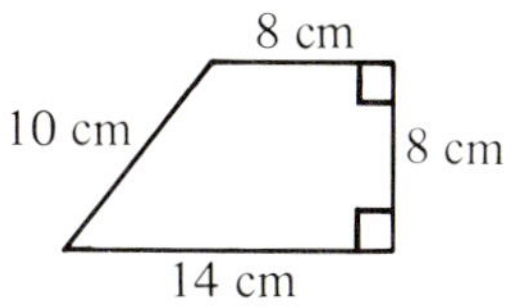

16. 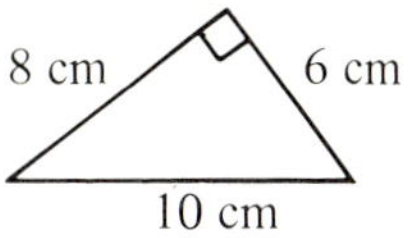

Copy and complete the table about trapezoids.

SAMPLE $A = 30$, $b_1 = 7$, $b_2 = 3$. Find h.

$$A = \frac{1}{2}h(b_1 + b_2)$$
$$30 = \frac{1}{2}h(7 + 3)$$
$$60 = h(10)$$
$$6 = h$$

B

	17.	18.	19.	20.	21.	22.	23.	24.
height h	7	8	7	j	?	?	8	7
base length b_1	10	10	12	k	14	18	5	?
base length b_2	4	5	7	n	6	12	?	3
area A	?	?	?	?	80	150	32	70

C **25.** A *hip roof* consists of two isosceles trapezoids and two isosceles triangles. How many squares of shingles are needed for the roof shown? (Recall that in talk about roofs, one square means 100 square feet.)

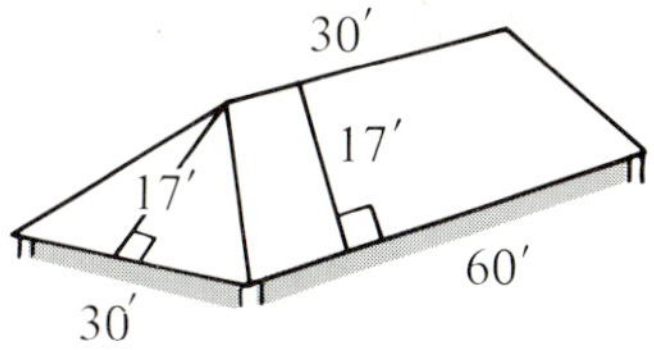

26. The Brock family used to farm one section of land. (A *section* is one square mile.) They sold a strip of land to permit construction of a highway. Then they sold a small triangular part that was awkward to reach to their neighbors.

To the nearest ten thousand square feet, find the area:

a. sold for the highway;
b. sold to neighbors;
c. in the Brock's new farm.

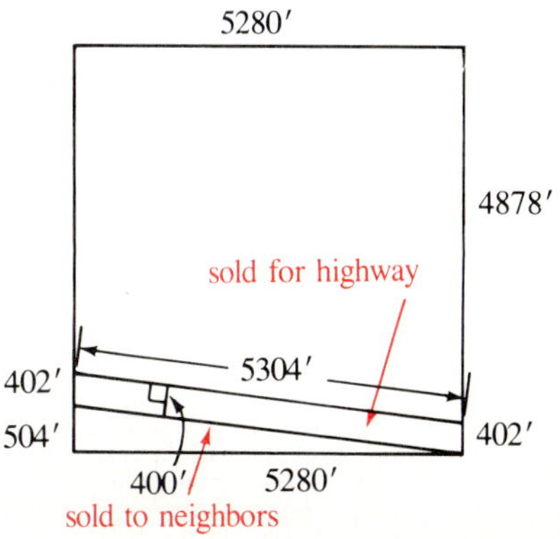

27. Carefully draw a trapezoid *EFGH*. Then construct a rectangle *WXYZ* so that the area of the rectangle equals twice the area of the trapezoid.

Strategy:

A. Construct $\overline{WX}$ so that *WX* equals the sum of the bases of trapezoid *EFGH*.

B. Construct a line perpendicular to $\overline{WX}$ at *W*.

C. Mark off *Z* on the perpendicular so that *WZ* equals the height of trapezoid *ABCD*.

D. Complete rectangle *WXYZ*.

SELF-TEST

Find the area of each triangle. *(6-3)*

1. 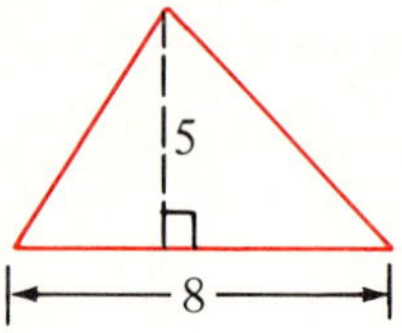

2. 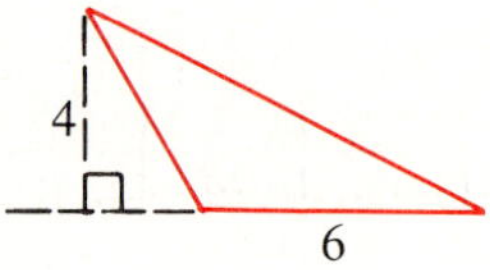

3. 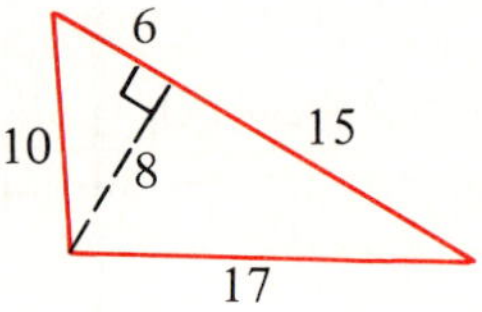

Find the area of each trapezoid. *(6-4)*

4. 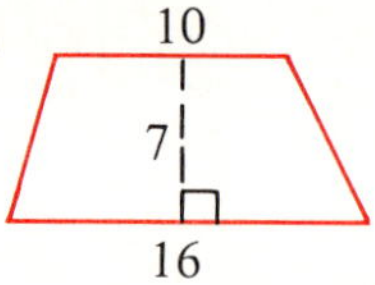

5. 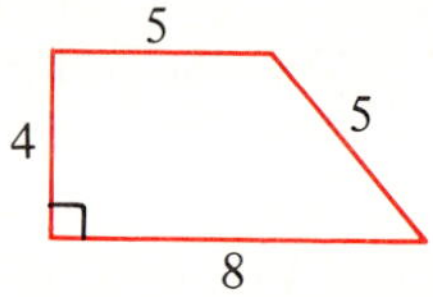

PUZZLE ◆ PROBLEMS

It is easy to join these sticks to form a rectangle with area 24.

How can you join the sticks to form a *nonrectangular* quadrilateral with area 24?

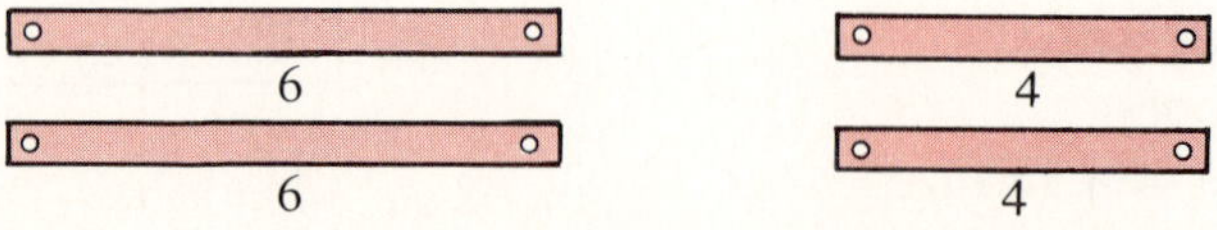

5 The Pythagorean Theorem

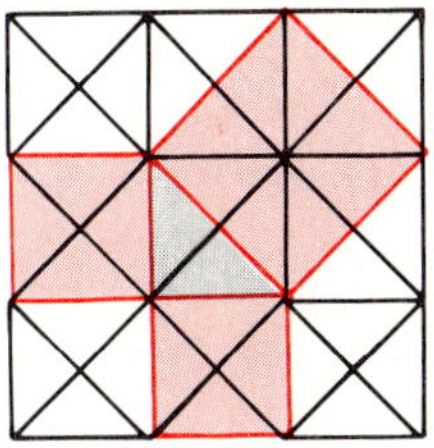

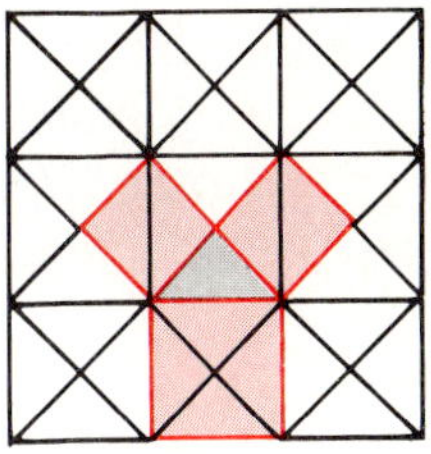

EXPLORATIONS

1. Part of a tiled floor is shown.
   Note the right triangle shaded in gray.
   Each side of that triangle is also the side of a square outlined and shaded in red.
   Count tiles to compare the area of the largest square with the sum of the areas of the smaller squares.
2. Another right triangle and some squares are shown.
   Compare the area of the largest square with the sum of the areas of the smaller squares.

The idea suggested above has often been expressed in this way:
The square on the hypotenuse of a right triangle is equal to the sum of the squares on the legs.

The explorations above deal with special right triangles.
The discussion that follows deals with *all* right triangles.

Begin with *any* right triangle.

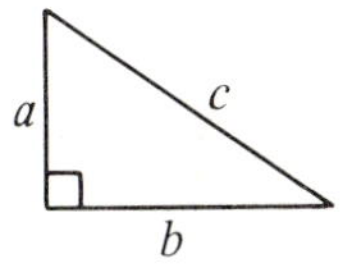

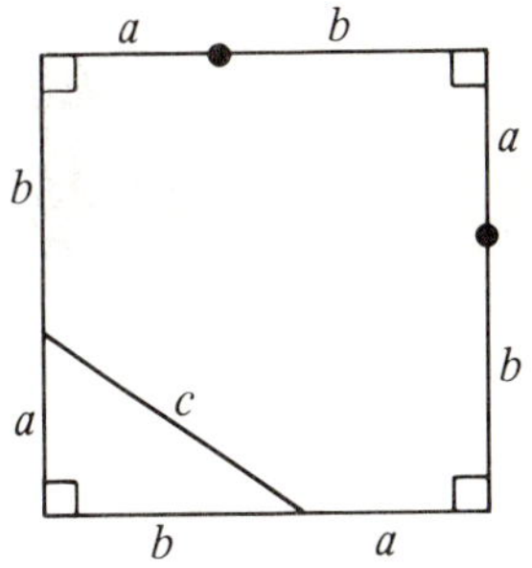

Draw a square, as shown, with sides of length $a + b$.

Locate two new points as shown.

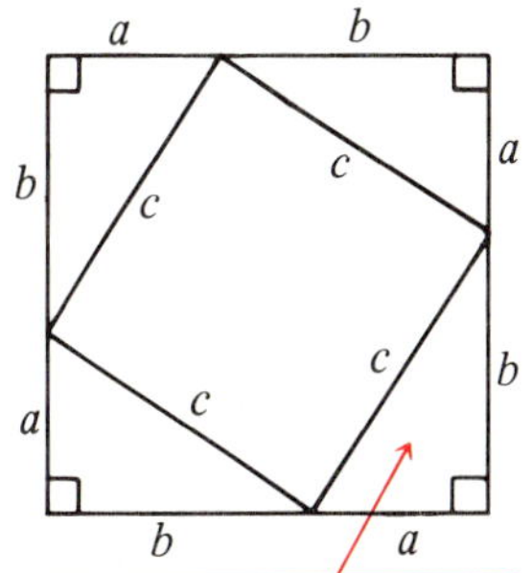

Each right triangle has area $\frac{1}{2}ab$.

Draw the rest of the inner quadrilateral.

The four right triangles are congruent by the SAS Postulate. Therefore, each hypotenuse has length c.

Each angle of the inner quadrilateral is a right angle (see Exercise 7 at the top of page 257).

The inner quadrilateral is a square with area c^2.

The areas of the two squares are compared on the next page.

Area of larger square	=	Area of smaller square	+	Areas of four triangles
$(a + b)^2$	=	c^2	+	$4(\frac{1}{2}ab)$

$$a^2 + 2ab + b^2 = c^2 + 2ab$$
$$a^2 + b^2 = c^2$$

THEOREM 5 (The Pythagorean Theorem)

In a right triangle, the square of the hypotenuse is equal to the sum of the squares of the legs.

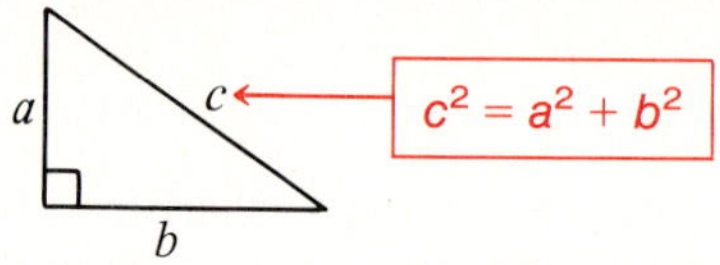

Theorem 5 is often called the **Pythagorean** (pi-**thag**-uh-**ree′**-an) **Theorem,** in honor of the Greek philosopher and mathematician Pythagoras. He is said to have written a proof of the theorem in the sixth century B.C.

EXAMPLE Find the value of x for each right triangle. Remember that the hypotenuse is the longest side of a right triangle.

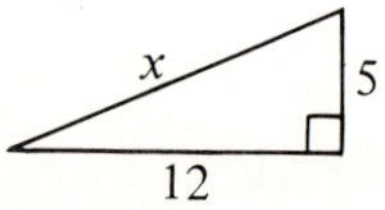

$$x^2 = 5^2 + 12^2$$
$$x^2 = 25 + 144$$
$$x^2 = 169$$
$$x = \sqrt{169}$$
$$x = 13$$

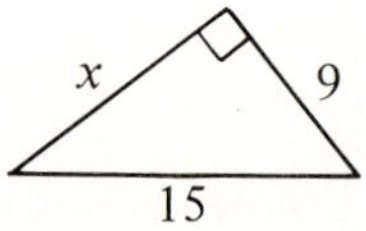

$$15^2 = x^2 + 9^2$$
$$225 = x^2 + 81$$
$$144 = x^2$$
$$\sqrt{144} = x$$
$$12 = x$$

Classroom Practice

In Exercises 1–6, state whether each equation is written correctly.

1. $a^2 + b^2 = c^2$

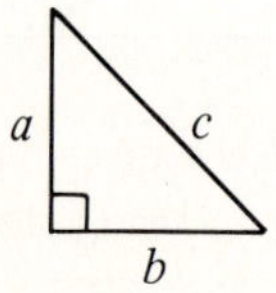

2. $c^2 = a^2 + b^2$

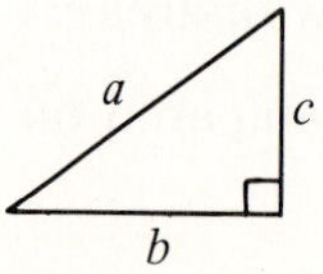

3. $r^2 = s^2 + t^2$

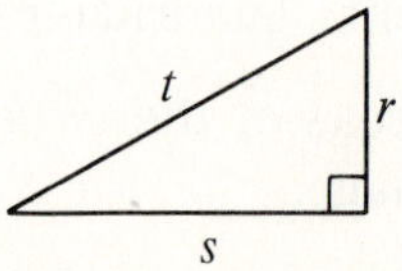

4. $k^2 = l^2 - m^2$

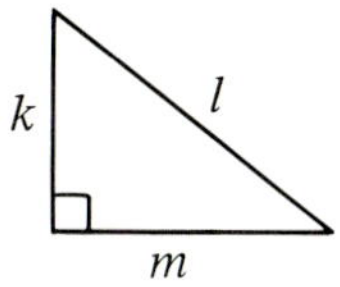

5. $d^2 = e^2 - f^2$

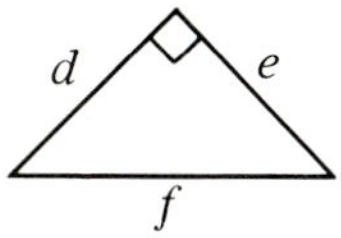

6. $p^2 = r^2 + q^2$

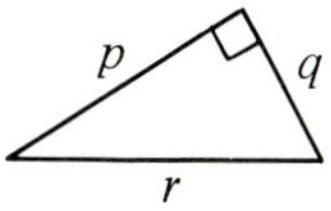

7. *Given:* $\angle 1$ and $\angle 5$ are right angles. Explain why $\angle 3$ must be a right angle. (*Hint:* Use the fact that $\triangle RST \cong \triangle TUV$.)

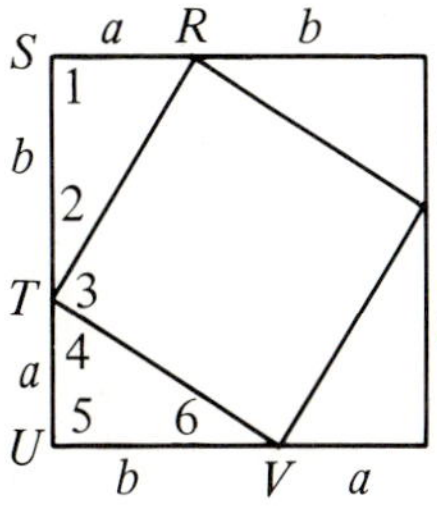

Written Exercises

In Exercises 1–6, write an equation for finding the value of *x*. Then solve the equation.

A

1.

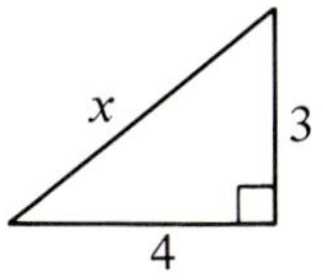

2.

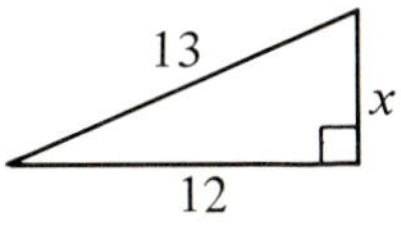

3.

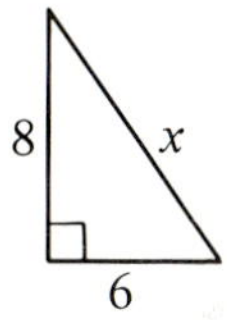

4.

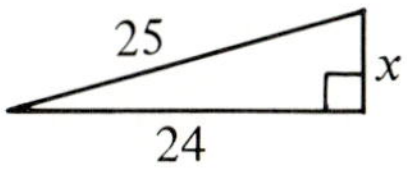

5.

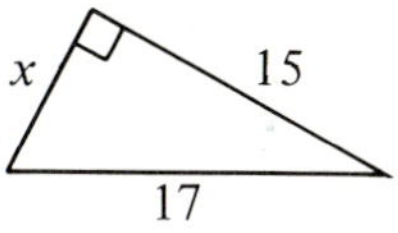

6.

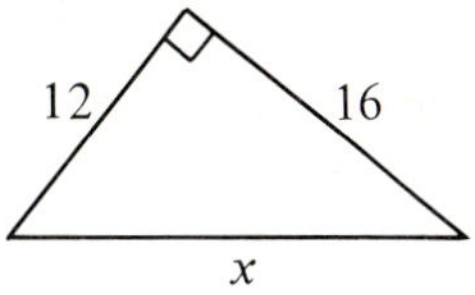

In Exercises 7–12, the lengths of two sides of a right triangle are given. Find the length of the third side.

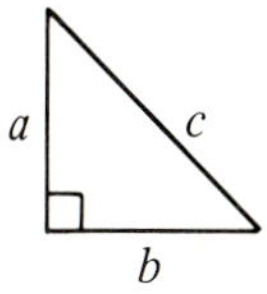

7. $a = 6, b = 8$

8. $a = 5, b = 12$

9. $a = 12, c = 15$

10. $b = 4, c = 5$

11. $a = 24, c = 26$

12. $b = 16, c = 20$

In Exercises 13–21, use the square root table on page 516 to write an approximation for *x* to the nearest tenth.

SAMPLE

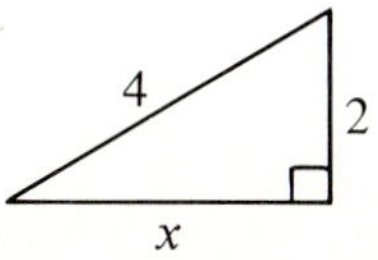

$$4^2 = 2^2 + x^2$$
$$16 = 4 + x^2$$
$$12 = x^2$$
$$x = \sqrt{12} \approx 3.5$$

B

13.

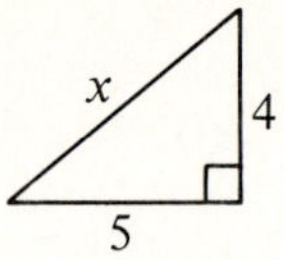

14.

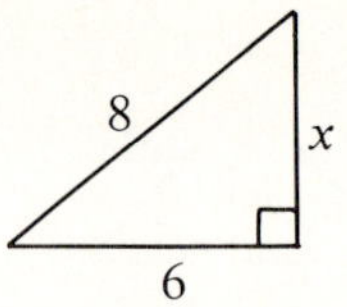

15.

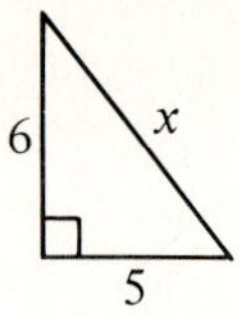

16.

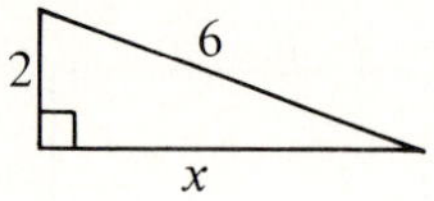

17.

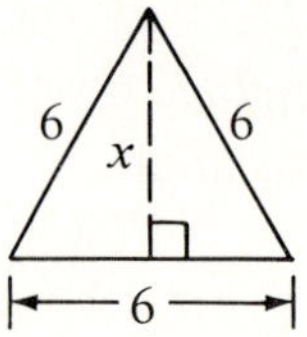

18.

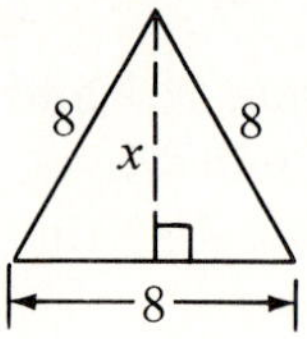

19.

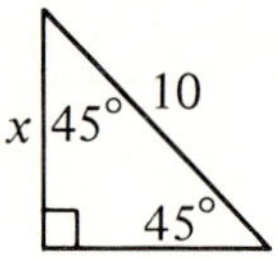

20.

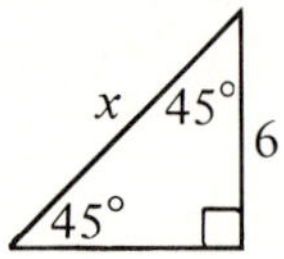

21.

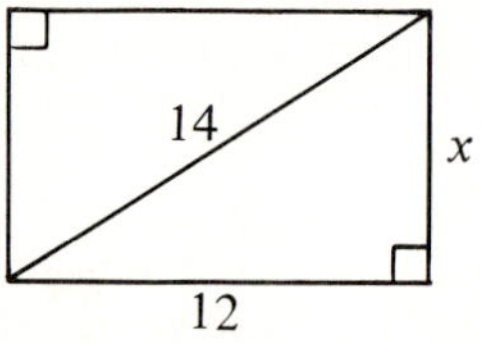

In Exercises 22–27, use the square root table on page 516 to find an approximation for x that is correct to the nearest tenth. Then find the area of each region.

C

22.

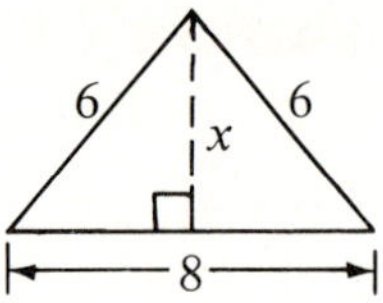

23.

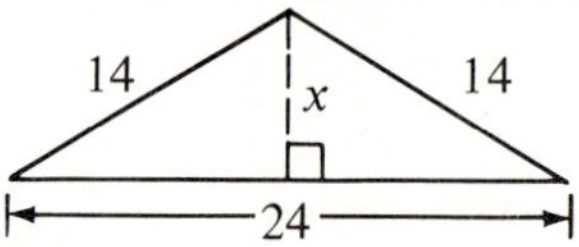

24.

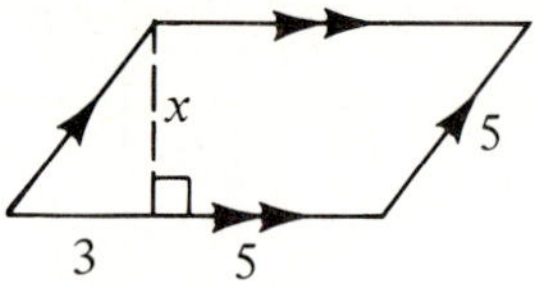

25.

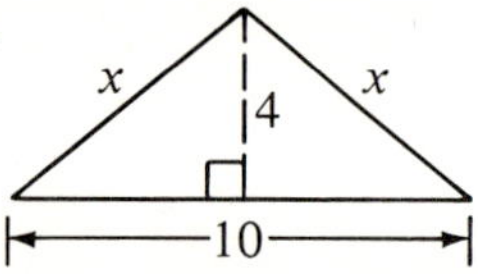

26.

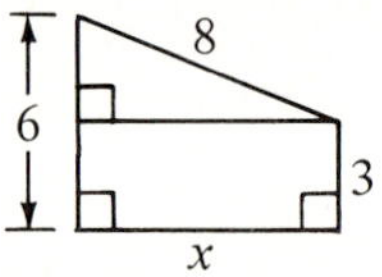

27.

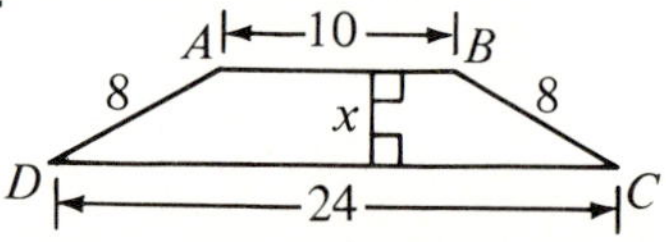

28. James A. Garfield (1831–1881), twentieth president of the United States, wrote a proof of the Pythagorean Theorem based on the diagram shown.

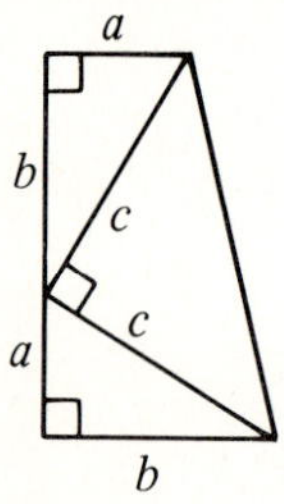

Area of the trapezoid	=	Total area of 2 ≅ $\triangle$s	+	Area of large rt. $\triangle$
?	=	?	+	?

Supply the missing measures in the equation above. Then simplify to show that $c^2 = a^2 + b^2$.

6 Converse of Pythagorean Theorem

Suppose you are asked if $\triangle RST$, shown below, is a right triangle. You notice:

$5^2 = 3^2 + 4^2$ because
$25 = 9 + 16.$

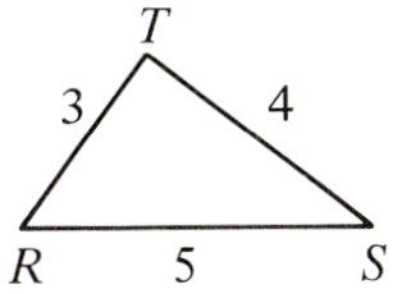

In order to conclude that $\triangle RST$ is a right triangle, you need to know whether the converse of the Pythagorean Theorem is true. Although some theorems have converses that are not true, the Pythagorean Theorem has a converse that *is* true. We omit the proof.

THEOREM 6

If the square of one side of a triangle is equal to the sum of the squares of the other two sides, then the triangle is a right triangle.

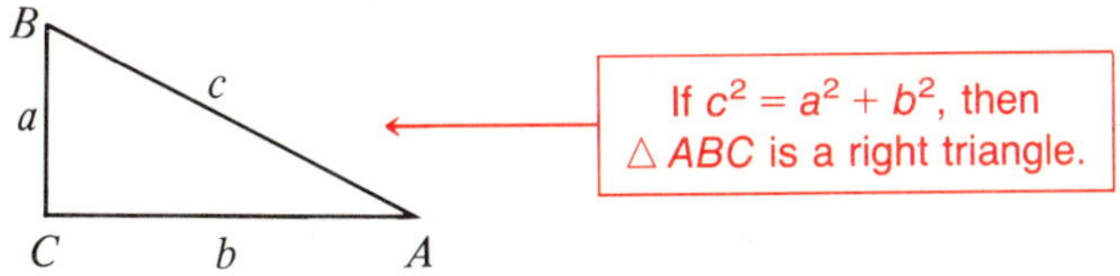

EXAMPLE Decide if each triangle shown is a right triangle.

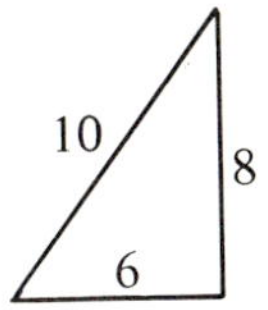

$6^2 + 8^2 \stackrel{?}{=} 10^2$

36 + 64	100
100	✓

The triangle *is* a right triangle.

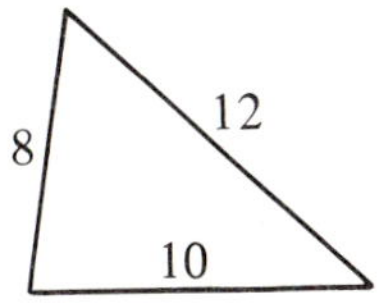

$8^2 + 10^2 \stackrel{?}{=} 12^2$

64 + 100	144
164	NO

The triangle *is not* a right triangle.

When you know the lengths of the three sides of a triangle, you can easily decide whether the triangle is acute, right, or obtuse. Look at the strategy outlined on the next page.

STRATEGY
1. *Let c be the longest side. Let a and b be the other two sides.* 2. *Compare c^2 to $a^2 + b^2$.* 3. *If $c^2 < a^2 + b^2$, then the triangle is* ***acute.*** *If $c^2 = a^2 + b^2$, then the triangle is a* ***right*** *triangle.* *If $c^2 > a^2 + b^2$, then the triangle is* ***obtuse.***

Classroom Practice

In each exercise, decide if the triangle is a right triangle. Explain your answers.

1.

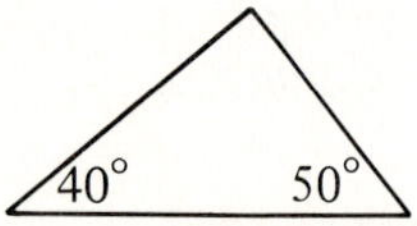

2.

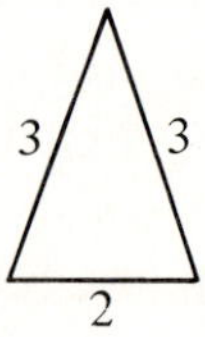

3.

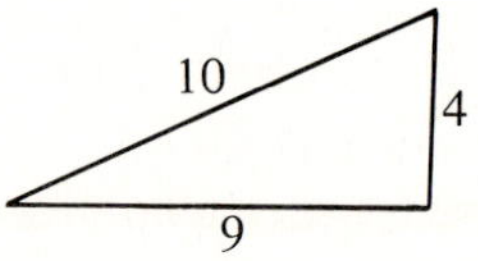

4.

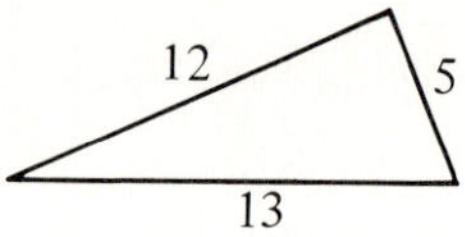

5.

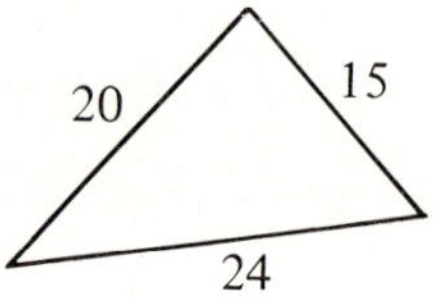

6.

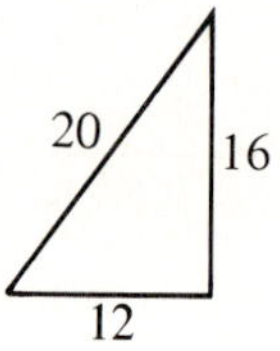

In each exercise, decide if $\triangle ABC$ is a right triangle.

7. $AB = 5$, $BC = 8$, $CA = 9$

8. $AB = 2$, $BC = 2$, $CA = 2$

9. $AB = 30$ cm, $BC = 40$ cm, $CA = 50$ cm

10. To check the accuracy of a corner in a foundation you can measure 7 m out from C to A, and 7 m out from C to B. A string stretched from A to B should be about 10 cm less than 10 m long.

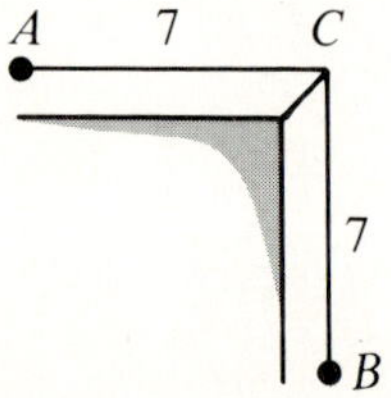

a. Suppose that $\overline{AB}$ is 10.1 m long.
Is $\angle ACB$ an acute angle, a right angle, or an obtuse angle?

b. Suppose that $\overline{AB}$ is 9.8 m long.
Is $\angle ACB$ an acute angle, a right angle, or an obtuse angle?

Written Exercises

Three sides of a triangle are given. Is the triangle a right triangle?

A

1. 3, 4, 5 **2.** 30, 40, 50 **3.** 2, 3, 4

4. 4, 5, 6 **5.** 8, 15, 17 **6.** 4, 6, 8

7. 10, 15, 20 **8.** 7, 24, 25 **9.** 0.9, 1.2, 1.5

10. 0.5, 1.2, 1.3 **11.** 16, 21, 27 **12.** 14, 14, 20

Three sides of a triangle are given. Classify the triangle as acute, right, or obtuse. (*Hint:* See the strategy on page 260.)

B

13.	14.	15.	16.	17.	18.	19.
6	7	10	7	11	15	12
8	7	10	11	60	16	21
10	10	14	16	61	21	25

C

20. Suppose that a, b, and c are the sides of a right triangle. Is a triangle with sides $7a$, $7b$, and $7c$ also a right triangle? Use algebra to show that your answer is correct.

21. Repeat Exercise 20 for a triangle with sides $a + 7$, $b + 7$, and $c + 7$.

22. Find the area of the quadrilateral shown.

23. Sketch $\square ABCD$ with $AB = 60$, $BC = 22$, and $AC = 64$. Which diagonal is longer: $\overline{AC}$ or $\overline{BD}$?

24. Sketch $\square WXYZ$ with $WX = 5$, $WY = 6$, and $XZ = 8$. What special kind of parallelogram is $WXYZ$?

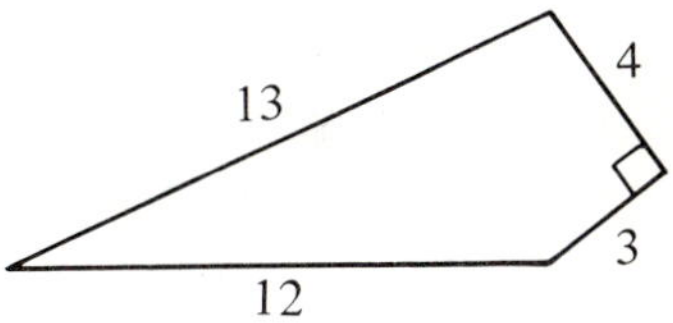

Exercise 22

PUZZLE ◆ PROBLEMS

Estelle, George, Jim, Keiko, Laura, and Luis are running a race.

Estelle is 20 m ahead of Jim.
Jim is 50 m ahead of George.
George is 10 m behind Laura.
Laura is 50 m behind Keiko.
Luis is 30 m ahead of Keiko.

Who is winning the race?
Who is second? Third? Fourth? Fifth? Last?

SELF-TEST

Find the value of x.

1.

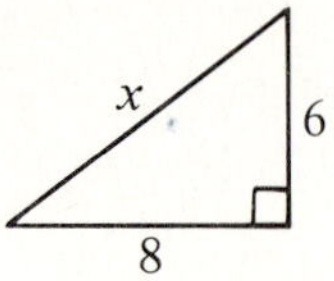

2.

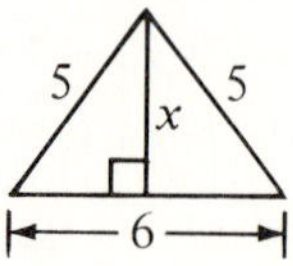

3.

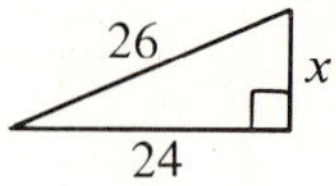

4.

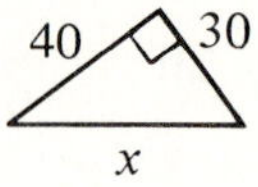

(6-5)

The lengths of three sides of a triangle are given.
Is the triangle right, acute, or obtuse?

5. $a = 2, b = 3, c = 4$

6. $a = 3, b = 4, c = 5$ *(6-6)*

7. $a = 6, b = 6, c = 6$

8. $a = 2.1, b = 2.8, c = 3.5$

CAREER NOTES

AUTO MECHANIC

In 1980, the number of passenger cars in use around the world was about 321,000,000 and in 1981 the number rose to about 331,000,000. Guess how many are in use today. Can you imagine the amount of work involved in keeping these cars running well?

Auto mechanics maintain and repair many different kinds of automobiles. They periodically examine, adjust, and replace car parts. They clean carburetors, balance wheels, replace spark plugs and distributor points, and grind valves.

Auto mechanics may work with sophisticated testing equipment or simple hand tools. Since parts for many cars are measured in metric units, mechanics must be familiar with both metric and nonmetric tools. High school or vocational school courses in machine shop, automobile repair, mathematics, and science help you learn these skills.

It takes three or four years of formal apprenticeship to become a senior auto mechanic. Some senior mechanics work for companies or repair garages. Others operate their own businesses. Mechanics may specialize in repairing particular car parts or particular types of cars.

7 Circumferences of Circles

A **circle** is a figure, in a plane, whose points are all the same distance from a particular point in the plane. That point is called the **center** of the circle. Circle P (written $\odot P$) is shown. The distance around $\odot P$ is called its **circumference.**

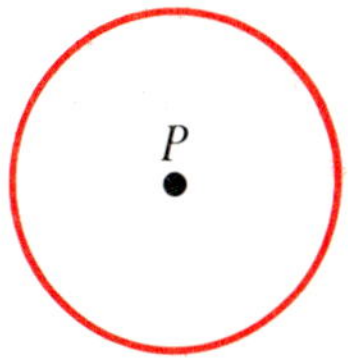

$\overline{PA}$, $\overline{PB}$, and $\overline{PC}$ are *radii* of $\odot P$. (The plural of radius is radii.) It follows from the definition of circle that *all radii of a circle are equal.*

$\overline{AC}$ is a *diameter* of $\odot P$. Can you see that any diameter of a circle is twice as long as a radius?

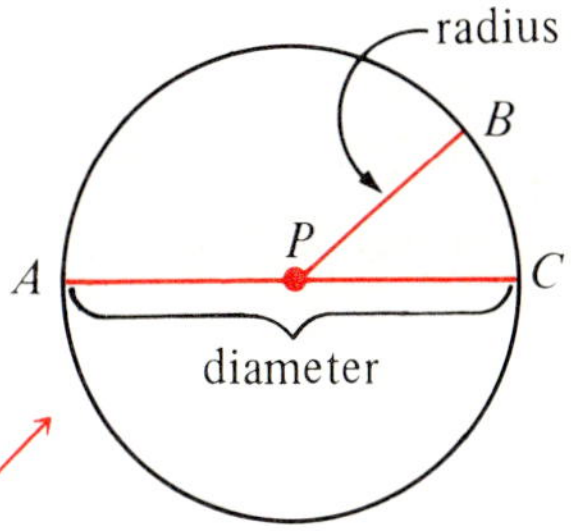

The words *radius* and *diameter* will sometimes refer to lengths of segments. We could say, for instance, that a circle has a radius of 5 and a diameter of 10.

$PA = PB = PC$;

$AC = 2(PA)$

A student measured the diameter of a half-dollar coin and reported: The diameter, d, is about 3.0 cm. Then she rolled the coin along a straight line until the coin had gone through one revolution.

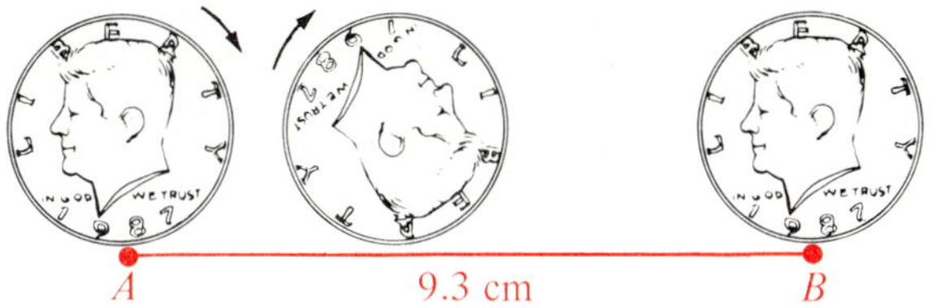

After measuring $\overline{AB}$ she said: The circumference, C, is about 9.3 cm. Next she divided the circumference by the diameter, getting

$$\frac{9.3}{3.0}, \text{ or } 3.1.$$

$\frac{C}{d} = 3.1$

Another student measured the diameter of a tin can. He found that the diameter was about 7.5 cm. The student then measured the distance around the rim of the can and reported: The circumference is about 24.0 cm. In this case,

$$\frac{\text{circumference}}{\text{diameter}} = \frac{24.0}{7.5}, \text{ or } 3.2.$$

$\frac{C}{d} = 3.2$

Notice that in the two cases on page 263 the values of $\frac{C}{d}$ are almost equal. Try the following experiment yourself.

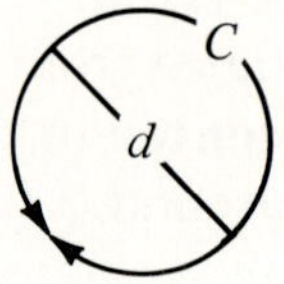

Select some circular objects such as coins, plates, and wheels. For each object:

1. measure the diameter and the circumference;
2. divide the circumference by the diameter.

If you work carefully, you should find that $\frac{C}{d}$ is usually a little greater than 3. Mathematicians have shown that the quotient $\frac{C}{d}$ is *exactly* the same for all circles. This quotient is denoted by π, a Greek letter pronounced *pie*. By definition,

$$\pi = \frac{C}{d}$$

The value of π has been computed to thousands of decimal places. Some approximations in common use are the following:

$\pi \approx 3.14 \qquad \pi \approx 3.1416 \qquad \pi \approx \frac{22}{7}$

is approximately equal to

None of these is exact. Unless specified otherwise, you may use any approximation that is convenient.

In a circle with radius r, diameter d, and circumference C:

$$\frac{C}{d} = \pi$$

$$C = \pi d$$

$$C = \pi \times 2r$$ ← $d = 2r$

$$C = 2\pi r$$

The circumference of a circle is given by the formula:

$$C = 2\pi r$$

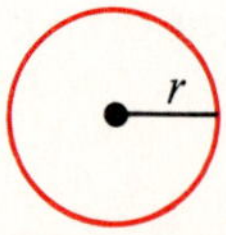

EXAMPLE

Find the circumference of a circle with radius 4 cm. Round your answer to the nearest tenth.

$C = 2\pi r$
$C = 2\pi \times 4 = 8\pi$ ← Exact answer: 8π cm
$C = 8\pi \approx 8 \times 3.14 = 25.12$ ← Approximate answer: 25.1 cm

Note that the exact circumference of a circle is expressed in terms of π. To find an approximate value, use an approximation of π.

Classroom Practice

Exercises 1–8 refer to the diagram below.

1. Name the circle shown.
2. Name the center of the circle.
3. Name all the radii shown.
4. Name all the diameters shown.
5. If $OC = 8$, then $OB =$ __?__.
6. If $OC = 3$, then $AC =$ __?__.
7. If $AC = 2x$, then $OB =$ __?__.
8. If a radius of the circle is 2 cm long, then the circumference of the circle is __?__.

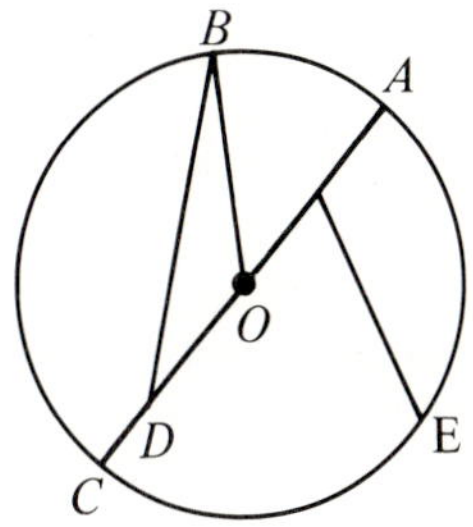

Find the circumference of a circle with the given radius. Express your answers in terms of π.

9.

r	3	6
C	6π	?

10.

r	11	22
C	?	?

11.

r	j	$2j$
C	?	?

12. Complete the statement suggested by Exercises 9–11: If the radius of one circle is twice as. . . .

Find the circumference of a circle with the given radius. Express your answers in terms of π.

13.

r	1	8
C	?	?

14.

r	1	13
C	?	?

15.

r	5	$5k$
C	?	?

16. Complete the statement suggested by Exercises 13–15: If the radius of one circle is k times as. . . .

Written Exercises

In Exercises 1–4, the radius of a circle is given. Find the diameter.

A

1. 7
2. 12 m
3. 5.1 cm
4. x

In Exercises 5–8, the diameter of a circle is given. Find the radius.

5. 16 cm
6. 26
7. 21.6
8. $2y$

In Exercises 9–17, find the circumference of a circle with the given radius r or diameter d. Express your answers in terms of π.

SAMPLE 1 $d = 4$ cm

$$C = \pi d$$
$$C = \pi \times 4 = 4\pi$$

9. $d = 10$
10. $d = 13$
11. $r = 6$
12. $d = 9.7$ m
13. $d = 33$ m
14. $r = 7.5$ cm
15. $r = 0.4$ km
16. $r = 16$ cm
17. $r = 18.4$ km

In Exercises 18–23, find the radius of a circle with the given circumference.

SAMPLE 2 $C = 12\pi$

$$C = 2\pi r$$
$$12\pi = 2\pi r \quad \leftarrow \text{Divide each side by } 2\pi.$$
$$6 = r$$

18. $C = 16\pi$
19. $C = 2\pi$
20. $C = 61\pi$ cm
21. $C = 25\pi$ m
22. $C = 46\pi$ cm
23. $C = 53\pi$ m

In Exercises 24–29, find an approximation, correct to the nearest tenth, of the circumference of a circle with the given radius. Use 3.14 for π.

B

24. $r = 8$ cm
25. $r = 15$ m
26. $r = 8.6$ cm
27. $r = 2.1$ km
28. $r = 37$ m
29. $r = 6.5$ m

In Exercises 30–35, find an approximation, correct to the nearest tenth, of the radius of a circle with the given circumference. Use 3.14 for π.

30. $C = 31.4$ m
31. $C = 314$ cm
32. $C = 1.57$ m
33. $C = 20$ cm
34. $C = 27.5$ cm
35. $C = 9.4$ km

36. Draw two circles. Then construct a circle whose circumference is equal to the sum of the circumferences of the given circles.

Strategy:
A. Let the radii of the given circles be s and t.
B. $2\pi s + 2\pi t = 2\pi(s + t)$
C. Construct a segment with length $s + t$.
D. Construct a circle with radius $s + t$.

In Exercises 37–39, use 3.14 for π. Write each answer correct to the nearest centimeter.

C **37.** The radii of two circles are 10 cm and 11 cm. By how many centimeters is the circumference of the larger circle greater than the circumference of the smaller circle?

38. Repeat Exercise 37, using circles with radii 100 cm and 101 cm.

39. Wheels revolving in opposite directions have a drive belt as shown. Find the total length of the belt.

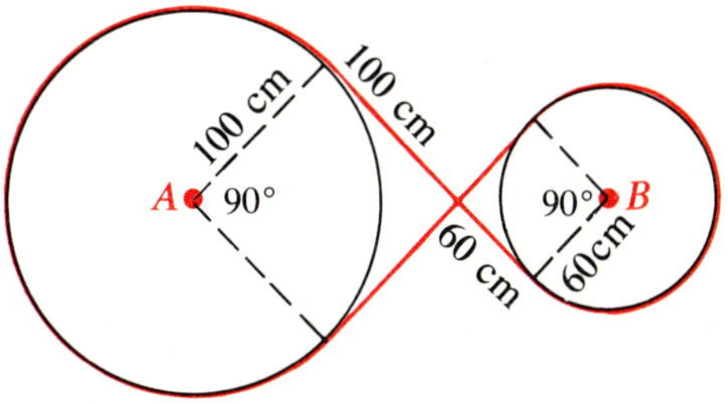

Suggested strategy:
A. $360° - 90° = 270°$
The belt covers $\frac{270}{360}$, or $\frac{3}{4}$ of $\odot A$.
B. The belt covers $\frac{3}{4}$ of $\odot B$.
C. Add the parts to find the total length of the belt.

40. Look back at Activity 2 on page 4 of Chapter 1. Now you can show that the distance between the earth and the stretched hoop is about 32 cm. Here is one way to do it.

a. Suppose the radius of the earth is r meters.
Explain why the stretched hoop has circumference $2\pi r + 2$.

b. Let R be the radius of the stretched hoop.
Explain why $2\pi R = 2\pi r + 2$.

c. Conclude that $2\pi(R - r) = 2$.
Therefore, $$R - r = \frac{2}{2\pi} \approx \frac{2}{2(3.14)} \approx 0.32.$$
The distance between the earth and the hoop is 0.32 m = __?__ cm.

d. Make a sketch showing the stretched hoop around the earth. Show R, r, and $R - r$ on your sketch.

8 Areas of Circles

The diagram represents an apple pie cut into eight equal pieces. Suppose you rearrange these eight pieces as shown below. Notice that the figure formed resembles a parallelogram.

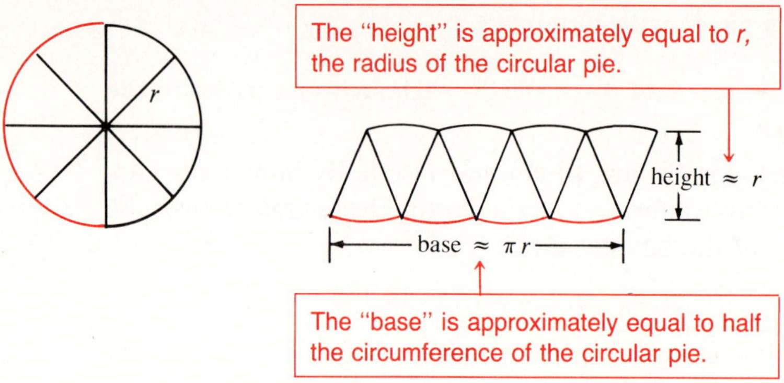

$$\begin{aligned}\text{Area of eight pieces of pie} &\approx \text{Area of parallelogram}\\ &= \text{base} \times \text{height}\\ &\approx \pi r \times r\\ &= \pi r^2\end{aligned}$$

Now suppose you cut the pie into sixteen equal pieces, instead of eight.

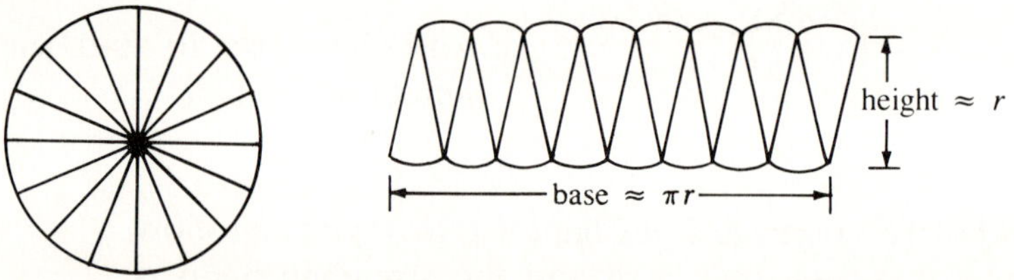

The figure formed looks even more like a parallelogram than the first one does. This means that the base is more nearly equal to πr. The area is more nearly equal to πr^2.

The discussion above suggests that πr^2 is approximately equal to the area of a circle with radius r. It can be proved, in an advanced geometry course, that πr^2 is the *exact* area.

The area of a circle is given by the formula:

$$A = \pi r^2$$

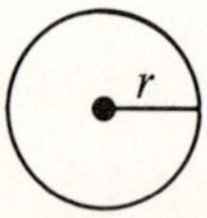

EXAMPLE

Find an approximation, correct to tenths, of the area of a circle with diameter 6 m.

Since the diameter is 6 m, the radius is 3 m.

$A = \pi r^2$

$A = \pi \times 3^2 = \pi \times 9 = 9\pi$ ← Exact area: 9π m^2

$A = 9\pi \approx 9 \times 3.14 = 28.26$ ← Approximate area: 28.3 m^2

Classroom Practice

Exercises 1–8 refer to a circle with radius *r*, diameter *d*, circumference *C*, and area *A*. Find the missing measures.

	1.	2.	3.	4.	5.	6.	7.	8.
r	5	9	?	?	?	?	?	?
d	?	?	8	16	?	?	?	?
C	?	?	?	?	28π	12π	?	?
A	?	?	?	?	?	?	4π	π

9. Suppose two circles have equal circumferences. Must their radii be equal? Must their areas be equal?

Written Exercises

Find the area of a circle with the given radius or diameter. Express your answers in terms of π.

A

1. $r = 5$ **2.** $r = 9$ **3.** $d = 6$

4. $d = 8$ cm **5.** $r = 1.2$ cm **6.** $r = 3.5$ m

7. $d = 7$ cm **8.** $d = 2.6$ m **9.** $r = 5\frac{5}{6}$

Find the radius of a circle with the given area.

SAMPLE

$A = 49\pi$ cm^2

$$A = \pi r^2$$
$$49\pi = \pi r^2$$
$$49 = r^2$$
$$7 = r$$

Answer: 7 cm

Find the radius of a circle with the given area.

10. $A = 36\pi$ **11.** $A = 25\pi$ **12.** $A = 9\pi\ m^2$

13. $A = 81\pi\ cm^2$ **14.** $A = 100\pi\ cm^2$ **15.** $A = 64\pi\ cm^2$

Find an approximation, to the nearest tenth, of the area of a circle with the given radius or diameter. Use 3.14 for π.

B **16.** $r = 6$ **17.** $r = 7$ **18.** $r = 3$ m

19. $r = 1.2$ cm **20.** $d = 1.4$ km **21.** $d = 17$ cm

Find the exact area of a circle with the given circumference.

SAMPLE $C = 8\pi$

1. Find r.

$$2\pi r = 8\pi$$

$$r = \frac{8\pi}{2\pi}$$

$$r = 4$$

2. Find the area.

$$A = \pi r^2$$

$$A = \pi(4^2) = 16\pi$$

22. $C = 14\pi$ **23.** $C = 48\pi$ **24.** $C = 16\pi$ **25.** $C = 11\pi$

26. The area of a circle is $4\pi\ m^2$. Find the circumference.

27. The area of a circle is $\frac{49}{4}\pi\ cm^2$. Find the circumference.

Find the area in terms of π of the region shaded in red.

C **28.**

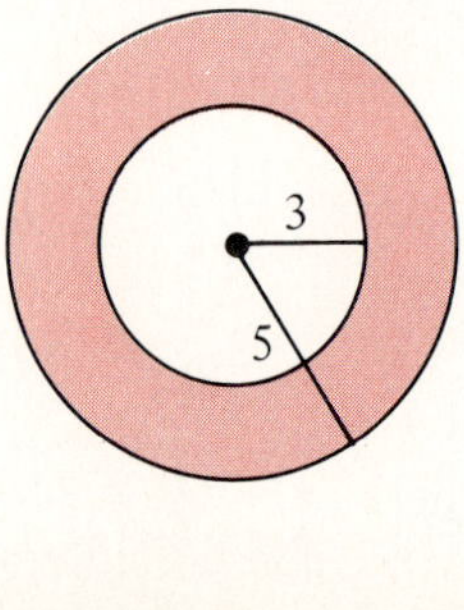

29.

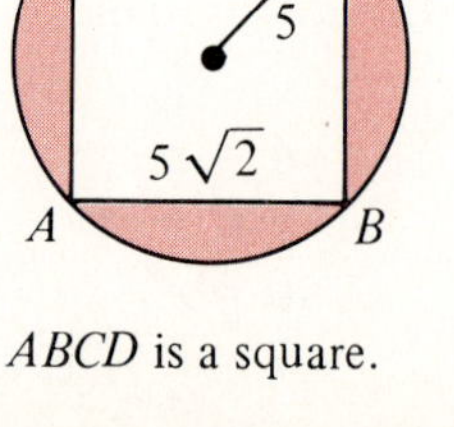

ABCD is a square.

30.

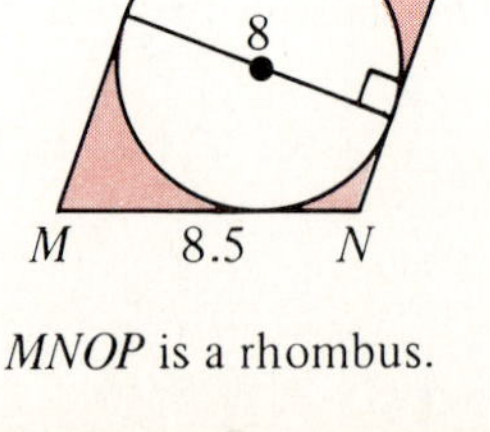

MNOP is a rhombus.

31.

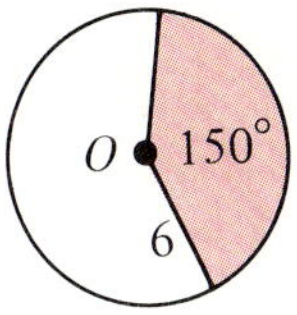

32.

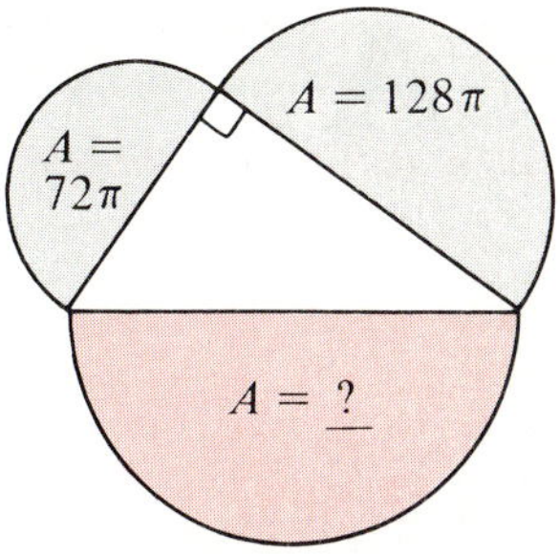

33. Draw two circles. Then construct a circle whose area is equal to the sum of the areas of the given circles.

Strategy:

A. Let the radii of the given circles be s and t.

B. $\pi s^2 + \pi t^2 = \pi(s^2 + t^2)$

C. Construct a right triangle with legs s and t. Let the hypotenuse be r. Then $r^2 = s^2 + t^2$.

D. Construct a circle with radius r.

SELF-TEST

Vocabulary

circle (p. 263)
center of a circle (p. 263)
circumference (p. 263)
radius (p. 263)
diameter (p. 263)

1. Complete: By definition, $\pi = \frac{?}{?}$. *(6-7)*

2. Write an approximation for π correct to the nearest hundredth.

Copy and complete the table about circles. Some answers may be given in terms of π.

	3.	4.	5.	6.
radius r	5	?	?	?
diameter d	?	16	?	?
circumference C	?	?	20π	?
area A	?	?	?	36π

(6-8)

APPLICATIONS

MIRRORS AND BILLIARDS

When a ray of light strikes a mirror, it is reflected at the same angle at which it arrives. For example, in the diagram, $\angle 1 = \angle 2$.

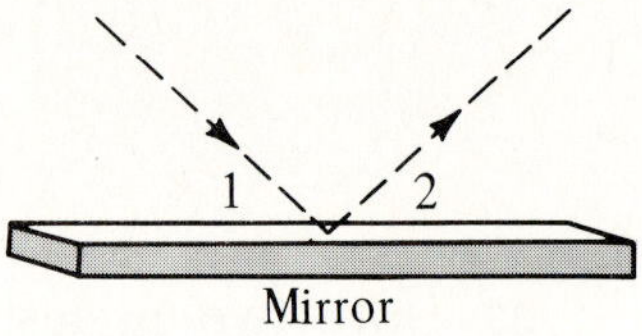

This principle of reflection is applied in the operation of a simple periscope. In the cross section of a periscope shown here, light is reflected twice so that light rays entering the periscope are parallel to light rays leaving the periscope.

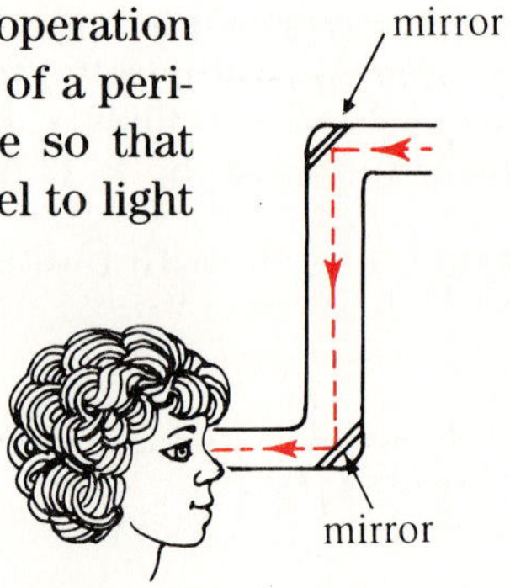

The same principle of reflection also applies to a ball bouncing off the sides of a pocket-billiard table or a miniature-golf course. The ball bounces back at the same angle at which it arrives. A good player uses this principle to sink a shot or to make a hole-in-one.

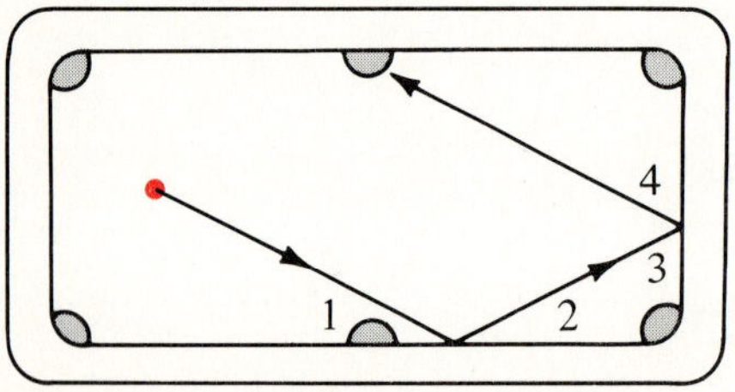

$\angle 1 = \angle 2$
$\angle 3 = \angle 4$

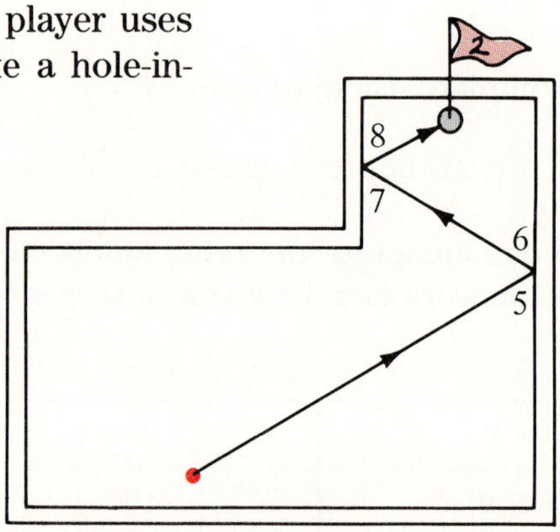

$\angle 5 = \angle 6$
$\angle 7 = \angle 8$

EXERCISES

Two mirrors form a 90° angle as shown.

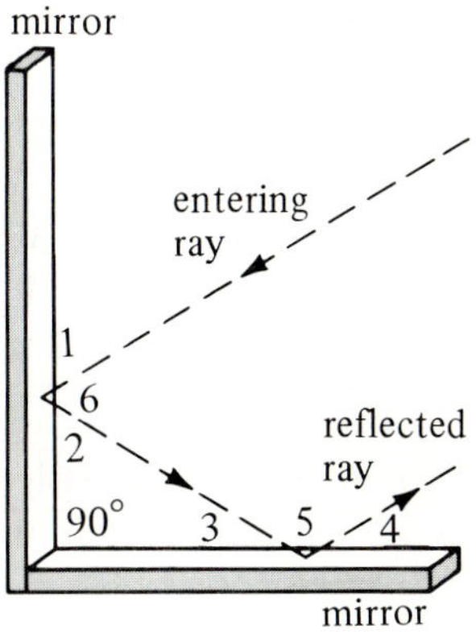

1. If $\angle 1 = 60°$ and $\angle 2$ is complementary to $\angle 3$, find the measure of each numbered angle.
2. Complete: $\angle 5 + \angle 6 =$ _?_°.
3. Explain why the entering ray is parallel to the reflected ray.

In the diagram at the right, suppose $\angle 1 = 50°$ and $\angle 2$ and $\angle 3$ are complementary angles.

4. Find the measure of each numbered angle.
5. Is the entering ray still parallel to the reflected ray?

A ball on a billiard table rolls along the path shown. $\angle 1 = 70°$.

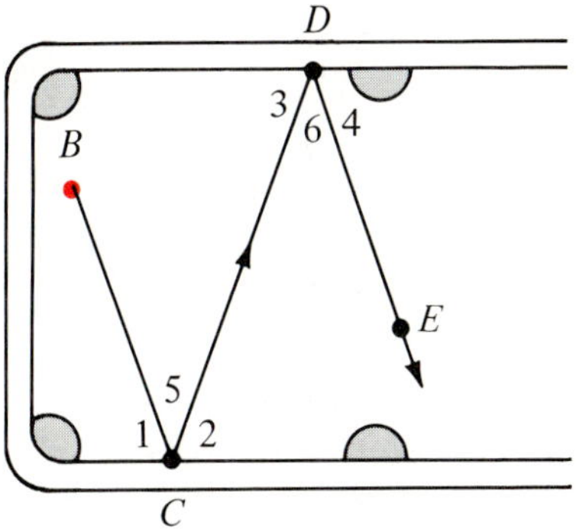

6. Find the measure of $\angle 2$.
7. Explain why $\angle 3 = 70°$.
8. Find the measures of $\angle 5$ and $\angle 6$.
9. Explain why $\overrightarrow{DE} \parallel \overrightarrow{BC}$.

10. A billiard ball rolls from point X to point Y as shown. If the ball is hit hard enough, will it go into a pocket?

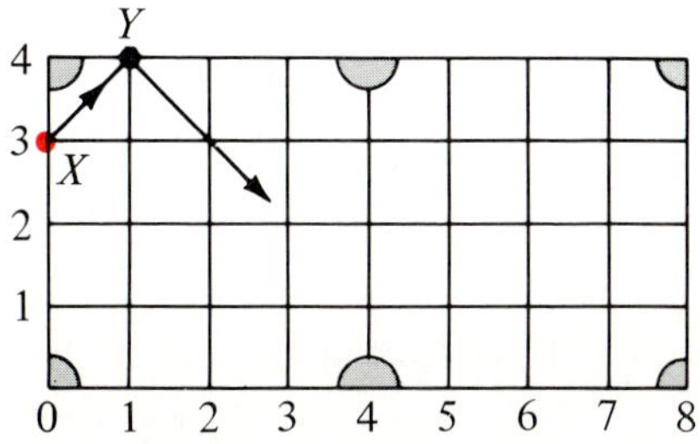

PROBLEM SOLVING STRATEGIES

WORKING BACKWARDS

If a problem has a specific goal, and you have to find how to get there, it is sometimes helpful to begin at the goal and then work backwards. You trace your way backwards step by step until you reach the beginning point.

EXERCISES

1. Two people play a game with the understanding that the loser is to double the money of the winner.

   They play the game twice, each person wins once, and they both end up with $6.00.

   How much did each one have at the beginning of the game?

   Let's work backwards. Begin at the end of the game. Here is a diagram showing A and B with $6.00 each at the end of two games.

	Beginning		*After Game 1*		*After Game 2*
A	?	←	?	←	$6.00
		Game 1		Game 2	
B	?	←	?	←	$6.00
	Winner has twice as much money as at the beginning.			Winner has twice as much money as after Game 1.	

   a. Assume that A lost the last game.

   How much did A have to pay B so that they both end up with $6.00?

   On a copy of the diagram, fill in the amounts that A and B must have had after Game 1.

   b. Now assume that B lost the first game.

   How much did B have to pay A in order that A and B end up with the amounts you found in part (a)?

   Enter the amounts each one must have had at the beginning of the game.

2. Three people play a game with the understanding that the loser will double the money of the other two.

At the end of three games each person has lost once, and they each end up with \$12.00.

How much did each person have at the beginning of the game?

3. A taxi cab driver charged Steve Yamamoto a total of \$5.65 for a trip from his home to the local airport. She told Steve that this amount included a flat fee of \$2.05 plus \$.15 for every $\frac{1}{6}$ of a mile.

How far is it from Steve's home to the airport?

a. Subtract the flat fee from the total charge to find the amount Steve paid for mileage.

b. Divide your answer to part (a) by \$.15 to find out how many sixths of a mile Steve was charged for.

c. Multiply your answer to part (b) by $\frac{1}{6}$ to find the distance from Steve's home to the airport.

4. The value of a car goes down each year so that it is only worth about $\frac{4}{5}$ of what it was the year before. Marta Ramirez owns a car worth \$9600. What was its value last year? You know that last year's value was multiplied by $\frac{4}{5}$ to get \$9600. Reverse the operation to find the value of Marta's car last year.

5. A state engineering department determines that each day they should be able to lay $\frac{3}{8}$ mile of a new highway. How many working days should the construction crew allow for 27 miles of highway?

6. *Given:* $AB = ED$
$AE = BC$
$\angle 1 = \angle 2$

Prove: $ABDE$ is a parallelogram.

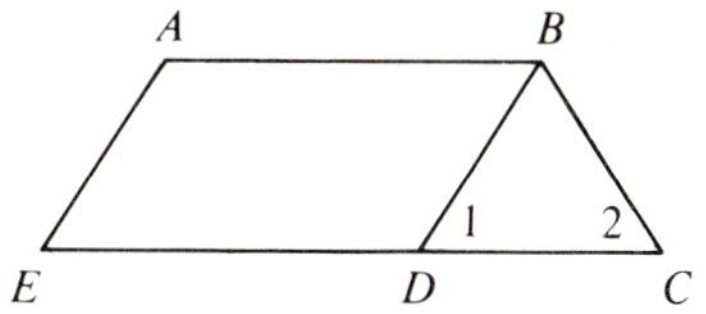

a. There are lots of ways to prove that $ABDE$ is a parallelogram. Since you know that $AB = ED$, what additional information would you need to know in order to say that $ABDE$ is a parallelogram? There are two choices: $AE = BD$ or $\overline{AB}$ is parallel to $\overline{DE}$. Which do you think seems the most promising?

b. Prove that $AE = BD$.

COMPUTER ACTIVITIES

RANDOM NUMBERS AND π (PI)

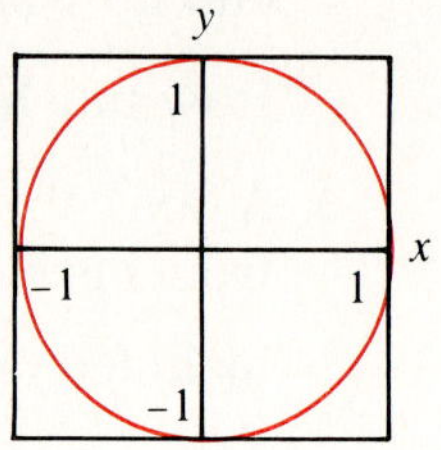

Can your computer throw darts? Not really, but it can pick numbers at random. This will simulate dart throwing. The target is a circle on a square board. The computer will use random numbers to show where a "dart" lands. Each dart has an equal chance of landing anywhere on the board. The computer will keep track of whether a dart lands inside or outside the circle.

The program below lets you decide how many darts to throw. RND means "random number." Different computers handle random numbers differently. This program is written to work on Apple computers. Modifications are given for TRS-80 and IBM computers. You may need to check on how your computer uses RND. The random numbers in this program are decimals between 0 and 1.

```
10  REM***SIMULATED DART THROWING
20  LET X = RND(-PEEK(78)-256*PEEK(79))
30  IN = 0
40  T = 0
50  PRINT "HOW MANY DARTS";
60  INPUT T
70  FOR C = 1 TO T
80  X = RND(1)
90  Y = RND(1)
100  REM***IGNORE DARTS LANDING ON
110  REM***OR OUTSIDE THE CIRCLE
120  IF X*X + Y*Y > = 1 THEN 150
130  REM***COUNT DARTS IN CIRCLE
140  IN = IN + 1
150  NEXT C
160  PRINT "I THREW  ";T;"  DARTS"
170  PRINT IN;"  DARTS WENT IN CIRCLE"
180  PRINT "RATIO OF DARTS IN CIRCLE"
190  PRINT "TO TOTAL IS  ";IN/T
200  END
```

On TRS-80 computers, substitute: 20 RANDOM

On IBM computers, substitute: 20 RANDOMIZE VAL(MID$(TIME$,7,2))

On TRS-80 computers, use RND (0) instead of RND(1).

EXERCISES

1. RUN the program. Time the computer. How long does it take your computer to throw
   a. 100 darts? b. 1000 darts? c. 10,000 darts?

2. a. What is the radius of the circle? b. What is the area of the circle in terms of π?
   c. How long is a side of the square? d. What is the area of the square?

3. RUN the program 6 times. Complete the chart.

Number inside circle	?	?	?	?	?	?
Number thrown	100	100	500	500	1000	1000
Ratio: $\frac{\text{darts inside}}{\text{total darts}}$	?	?	?	?	?	?

In Exercise 3, you probably got values between 0.77 and 0.82. The darts are thrown at random, so nobody can be sure what you will get.

The ratio $\frac{\text{darts inside circle}}{\text{total darts thrown}}$ tells what fractional part of the darts landed inside the circle. Suppose the darts land at random on the board. Then the ratio $\frac{\text{darts inside circle}}{\text{total darts thrown}}$ should be about the same as the ratio $\frac{\text{area of circle}}{\text{area of square}}$. Multiply the ratio by 4 (the area of the square). The answer is an approximation of the area of the circle.

4. For each of the RUNs in Exercise 3, copy the value of the ratio. Multiply this by 4 to get the approximate area of the circle. Round areas to the nearest hundredth.

Ratio: $\frac{\text{darts inside}}{\text{total darts}}$	?	?	?	?	?	?
Area of circle	?	?	?	?	?	?

5. In Exercise 2b, you found that the area of the circle is exactly π. Give the decimal value of π to the nearest hundredth.

6. Compare the areas you figured in Exercise 4 to the decimal approximation for π.

SKILLS REVIEW

ALGEBRAIC EXPRESSIONS AND EQUATIONS

Write an expression using the given variable or variables.

1. The quotient when 86 is divided by n
2. The third power of k
3. The sum of 13 and three times x
4. The product of 4, a, and b
5. 18 subtracted from half of c
6. m increased by 41
7. $(a + b)$ multiplied by 7
8. The ratio of 15 to w

Write an equation or an inequality for each statement. Let n be the variable.

9. When 27 is subtracted from some number, the difference is 15.
10. The product of some number and 7 is not equal to zero.
11. One fifth of some number, minus 13, is equal to 27.
12. When 7 is subtracted from half of some number, the difference is 18.
13. Three times some number is greater than 16.
14. The fifth power of some number, increased by 9, is 41.
15. The sum of some number and 43 is less than 75.
16. 16 is the quotient when 480 is divided by some number.

Solve.

SAMPLE

$$2x - 3 = 7$$
$$2x - 3 + 3 = 7 + 3$$ ← Add 3 to both sides.
$$2x = 10$$
$$x = 5$$ ← Divide both sides by 2 to get $x = 5$.

17. $b + 17 = 41$
18. $x - 9 = -12$
19. $14 = n - 36$
20. $8y = 72$
21. $12m = -48$
22. $-5z = 95$
23. $16 + x = 3$
24. $63 = 7g$
25. $\frac{t}{6} = -4$
26. $\frac{s}{12} = 30$
27. $\frac{a}{51} = 0$
28. $3x + 4 = 19$
29. $38 = 4k - 6$
30. $\frac{a}{6} + 2 = 7$
31. $-13 = \frac{k}{2} - 1$

CHAPTER REVIEW

CHAPTER SUMMARY

1. The following formulas for area can be used.

Rectangle
$A = bh$

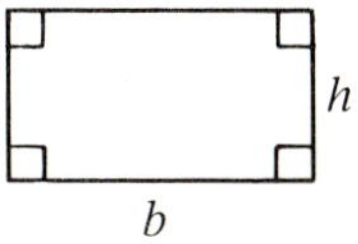

Square
$A = s^2$

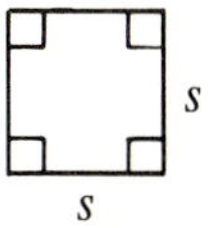

Parallelogram
$A = bh$

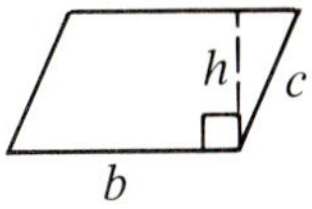

Unless a parallelogram is also a rectangle, $h \neq c$.

Triangle

$A = \frac{1}{2}bh$

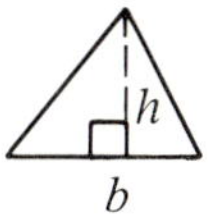

Trapezoid

$A = \frac{1}{2}h(b_1 + b_2)$

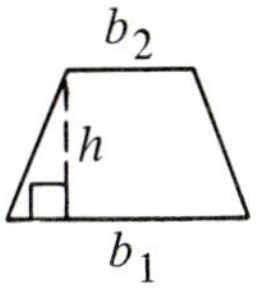

2. The Pythagorean Theorem states:
In any right triangle, the square of the hypotenuse is equal to the sum of the squares of the legs.
If $\triangle ABC$ is a right triangle, then $c^2 = a^2 + b^2$.

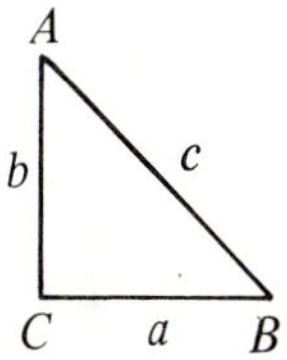

3. The converse of the Pythagorean Theorem is true.
If $c^2 = a^2 + b^2$, then $\triangle ABC$ is a right triangle.

4. For any circle: $C = 2\pi r$ and $A = \pi r^2$.

REVIEW EXERCISES

Copy and complete the table about rectangles. *(See pp. 238–242.)*

	1.	2.	3.	4.
base b	5	30	7	7.2
height h	2	20	?	4
perimeter p	?	?	?	?
area A	?	?	49	?

Find the area of the shaded region. *(See pp. 243–249.)*

5.

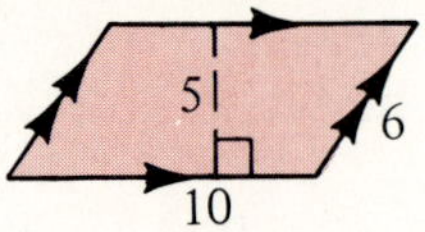

6.

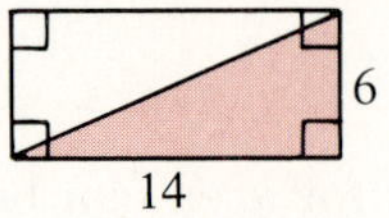

7.

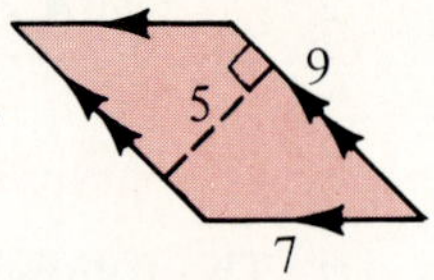

8.

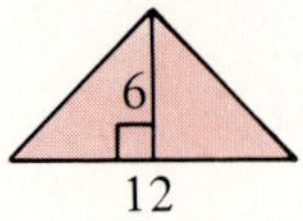

9.

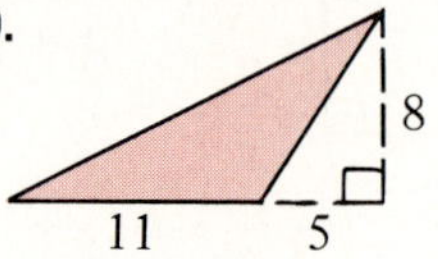

10.

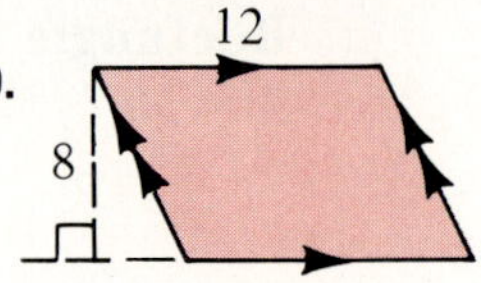

Copy and complete the table about trapezoids. *(See pp. 250–254.)*

	11.	**12.**	**13.**	**14.**
base length b_1	12	11	8.3	7
base length b_2	8	6	4.5	5
height h	5	4	5	?
area A	?	?	?	24

In each exercise, find the value of x. *(See pp. 255–258.)*

15.

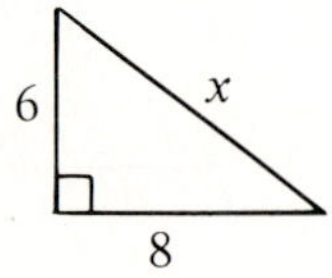

16.

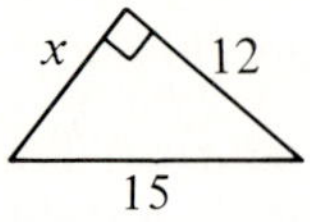

17.

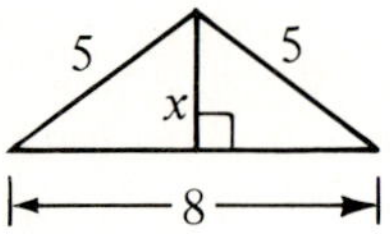

The lengths of the sides of a triangle are given. State whether the triangle is acute, right, or obtuse. *(See pp. 259–261.)*

18. 4, 5, 8 **19.** 11, 60, 61 **20.** 8, 8, 8

Complete the statements about circles. *(See pp. 263–271.)*

21. If $d = 12$, then $C =$ __?__. Answer in terms of π.

22. If $r = 5$, then $A =$ __?__. Answer in terms of π.

23. If $d = 8$, then $A =$ __?__. Use 3.14 for π and round your answer to tenths.

24. If $r = 12$, then $C =$ __?__. Use 3.14 for π and round your answer to tenths.

25. If $A = 81\pi$, then $r =$ __?__. **26.** If $C = 18\pi$, then $r =$ __?__.

CHAPTER TEST

1. Find the perimeter and area of a rectangle 10 cm long and 8.5 cm wide. *(6-1)*

2. Find the area of a square with perimeter 12 m.

3. Find the perimeter and area of the shaded parallelogram. *(6-2)*

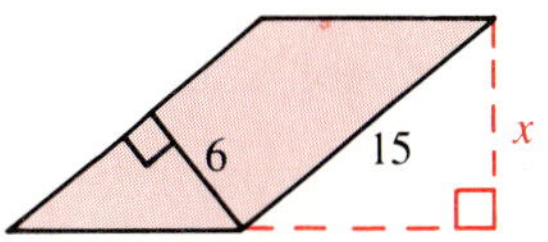

4. Use your answer from Exercise 3 to find the value of x.

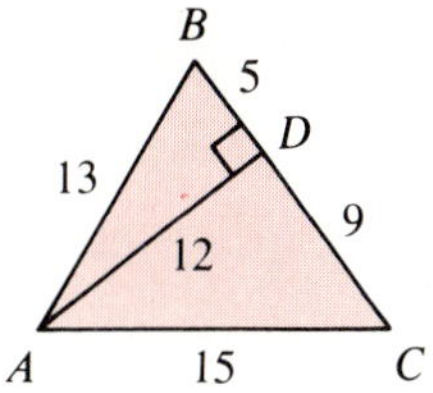

5. Find the area of $\triangle ABC$. *(6-3)*

6. Find the area of $\triangle ADC$.

7. A triangle has base $4x$ and height x. Find its area in terms of x.

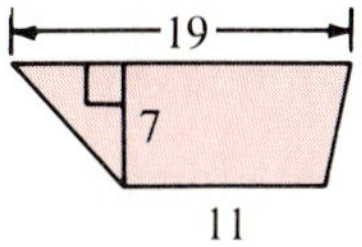

8. Find the area of the trapezoid shown. *(6-4)*

9. A trapezoid with bases 16 and 12 has area 126. Find its height.

Find the value of *x*. *(6-5)*

10.

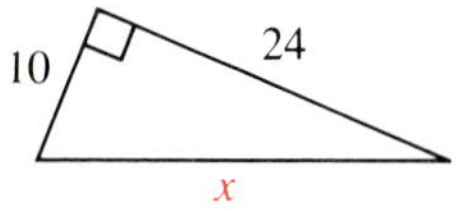

11.

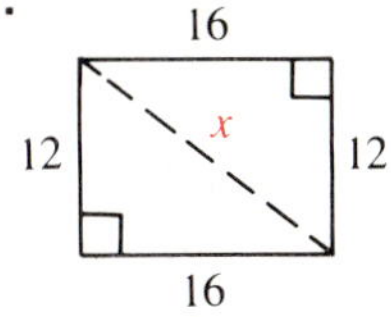

12.

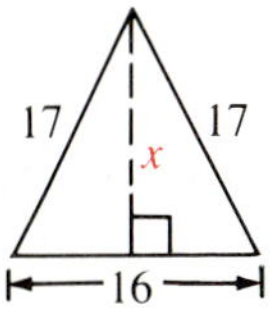

13. A triangle has sides 5, 6, and 9. Is it acute, right, or obtuse? *(6-6)*

14. A parallelogram has sides of lengths 6 and 8 and a diagonal of length 10. Draw a diagram and explain why the parallelogram must be a rectangle.

15. A circle has diameter 20. Find its circumference to the nearest tenth. (Use 3.14 for π.) *(6-7)*

16. Find the radius of a circle with circumference $\frac{4}{3}\pi$.

17. Find the area, in terms of π, of a circle with radius $1\frac{1}{2}$. *(6-8)*

18. Find the diameter of a circle with an area of 16π m^2.

UNIT C CUMULATIVE REVIEW

Complete using *always, sometimes,* or *never*.

1. An equilateral triangle is __?__ a regular polygon.
2. The interior angle sum of a pentagon is __?__ 540°.
3. A parallelogram is __?__ a square.
4. A rhombus is __?__ a parallelogram.
5. Opposite angles of a quadrilateral are __?__ equal.
6. Consecutive angles of a parallelogram are __?__ complementary.
7. The diagonals of a quadrilateral __?__ bisect each other.
8. The diagonals of a rectangle are __?__ equal.

Complete each statement.

9. In rhombus $JKLM$, $\overline{MK}$ __?__ $\overline{LJ}$.
10. In $\triangle ABC$ below, $MN =$ __?__.

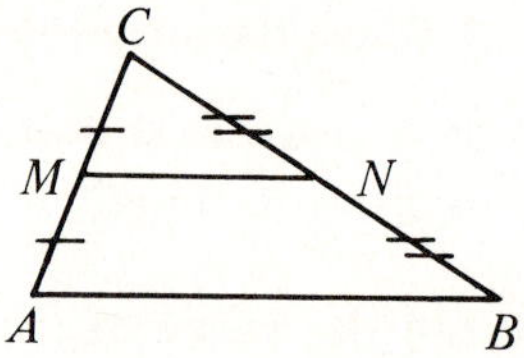

11. In parallelogram $QRST$, $QR =$ __?__ and $QT =$ __?__.
12. The median of a trapezoid is __?__ to the bases.
13. If M is the midpoint of hypotenuse $\overline{DE}$ in right $\triangle DEF$, then $MD =$ __?__ $=$ __?__.
14. A quadrilateral must be a parallelogram if one pair of opposite sides are both __?__ and __?__.

Find the area. Use 3.14 for π and round to the nearest tenth.

15.
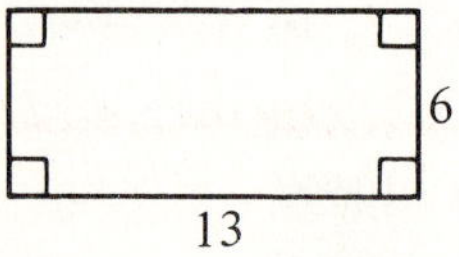

16.
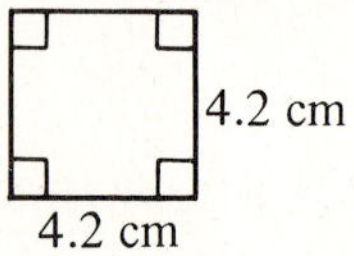

17.
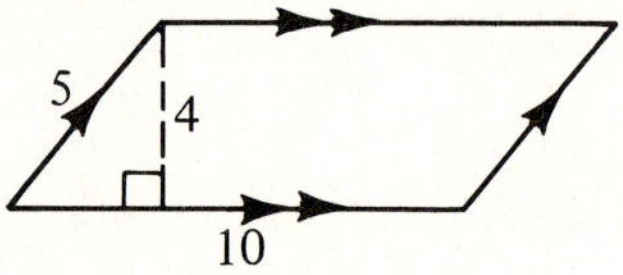

18.
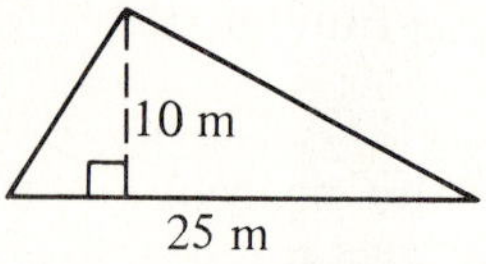

19.
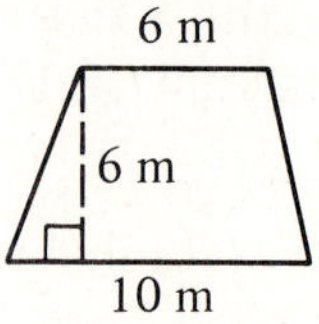

20.
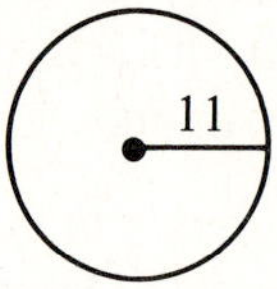

21. Is a triangle with sides 3 m, 4 m, and 6 m a right triangle?
22. Find the exact circumference of a circle with radius 6 m.

UNIT D

Here's what you'll learn in this chapter:

To find the ratio of two numbers.

To write proportions in several forms.

To solve problems using proportions.

To make maps and scale drawings.

The very large sand castle, shown above, was built at Pacific Beach in San Diego, California. Although all the measurements in a real castle would be much larger, they would be in proportion to those of the sand castle.

Chapter 7

Ratios and Proportions

1 Ratios

In this chapter and the next, we'll see how ratios and proportions are used by mapmakers, technical artists, and commercial photographers. Before doing so, however, we must review what is meant by *ratio.* In the next section, we'll review proportions.

The **ratio** of two numbers is the quotient of the two numbers.

A ratio is usually expressed in *simplest form.*

$\frac{4}{8} = \mathbf{\frac{1}{2}}$ ← simplest form

$\frac{2x^2}{2xy} = \frac{2 \cdot x \cdot x}{2 \cdot x \cdot y} = \boldsymbol{\frac{x}{y}}$ ← simplest form

Ratio	Written Forms
2 to 3	$\frac{2}{3}$ or 2:3
5 to 1	$\frac{5}{1}$ or 5:1
a to b, $b \neq 0$	$\frac{a}{b}$ or $a:b$

EXAMPLE 1 There are 60 streetcars and 72 subway cars in a city transportation system.

a. Find the ratio of streetcars to subway cars.
b. Find the ratio of subway cars to streetcars.

a. $\frac{60}{72} = \frac{5}{6}$

b. $\frac{72}{60} = \frac{6}{5}$

EXAMPLE 2 A rectangle is 3 m long and 80 cm wide. Find the ratio of its length to its width.

Method 1

Express the length in centimeters. 3 m = 300 cm

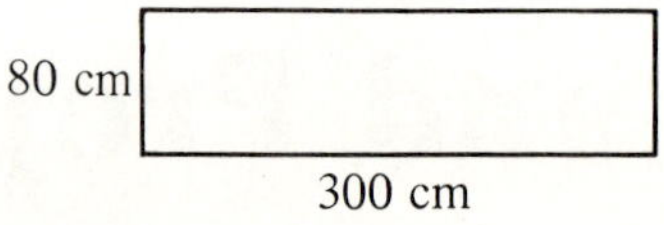

$\frac{\text{length}}{\text{width}} = \frac{300}{80} = \frac{15}{4}$

Method 2

Express the width in meters. 80 cm = 0.8 m

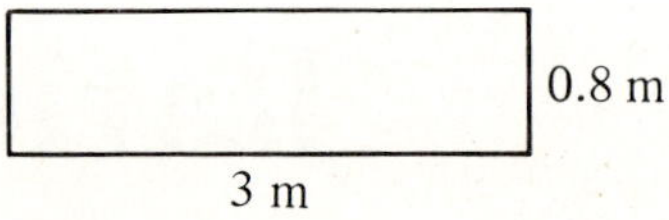

$\frac{\text{length}}{\text{width}} = \frac{3}{0.8} = \frac{30}{8} = \frac{15}{4}$

Notice that the ratio is 15:4, no matter which unit is used. Notice also that the ratio contains no units. The ratio is 15:4, and not 15:4 cm or 15:4 m.

In the preceding examples, ratios were used to compare two numbers. Ratios can also be used to compare three or more numbers. For example, in $\triangle ABC$ shown below, a, b, and c are in the ratio 3 to 4 to 5.

This means that

$$a:b = 3:4$$
$$a:c = 3:5$$
$$b:c = 4:5$$

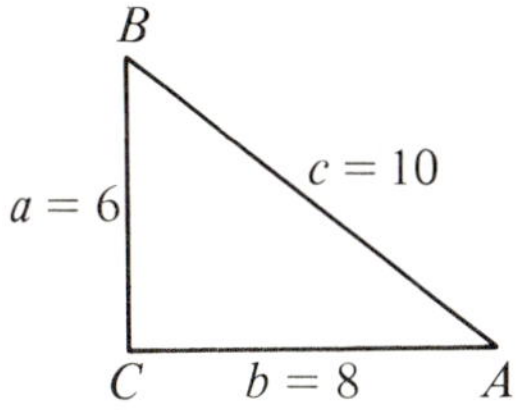

We combine these steps by writing

$$a:b:c = 3:4:5.$$

Classroom Practice

Express each ratio in simplest form.

1. $\frac{5}{15}$ 2. $\frac{10}{15}$ 3. $\frac{9}{12}$ 4. $\frac{500}{1000}$

5. 6:8 6. 20:25 7. 2:4:6 8. 30:40:50

Given $a = 3$ and $b = 5$, find each ratio.

9. $\frac{a}{b}$ 10. $\frac{b}{a}$ 11. $\frac{a}{a+b}$ 12. $\frac{b}{a+b}$

Use the figure to find each ratio in simplest form.

13. $\frac{AB}{BC}$ 14. $\frac{AB}{AC}$ 15. $\frac{AC}{BC}$

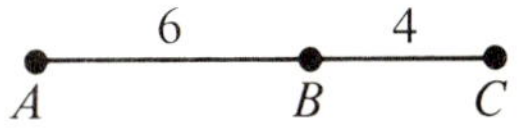

Written Exercises

Express each ratio in simplest form.

A 1. $\frac{6}{12}$ 2. $\frac{6}{9}$ 3. $\frac{18}{24}$ 4. $\frac{36}{72}$

5. 21:28 6. 25:45 7. 10:20:30 8. 15:25:35

Given $a = 2$ and $b = 3$, find each ratio.

9. $\frac{a}{b}$ 10. $\frac{b}{a}$ 11. $\frac{a}{a+b}$ 12. $\frac{b}{a+b}$

Given $c = 4$ and $d = 5$, find each ratio.

13. $\frac{c}{d}$ **14.** $\frac{d}{c}$ **15.** $\frac{d}{c + d}$ **16.** $\frac{d - c}{d + c}$

Use the figures below to find each ratio in simplest form.

17. $\frac{AD}{DB}$ **18.** $\frac{AD}{AB}$ **19.** $\frac{DB}{AB}$

20. $\frac{AE}{AC}$ **21.** $\frac{AC}{AE}$ **22.** $\frac{EC}{AC}$

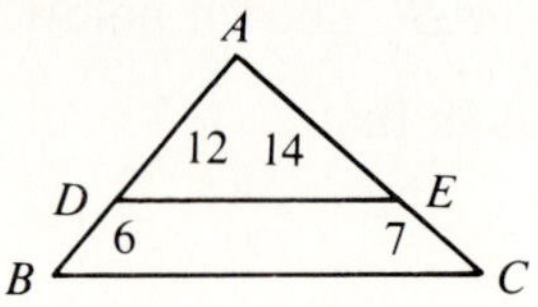

23. $\frac{PO}{OS}$ **24.** $\frac{PO}{PS}$ **25.** $\frac{OS}{PS}$

26. $\frac{OR}{QO}$ **27.** $\frac{OR}{QR}$ **28.** $\frac{QR}{PS}$

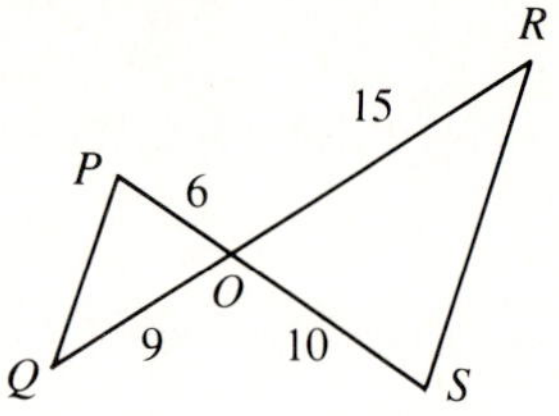

29. $\frac{\text{length of } A}{\text{length of } B}$ **30.** $\frac{\text{width of } A}{\text{width of } B}$

31. $\frac{\text{area of } A}{\text{area of } B}$ **32.** $\frac{\text{perimeter of } A}{\text{perimeter of } B}$

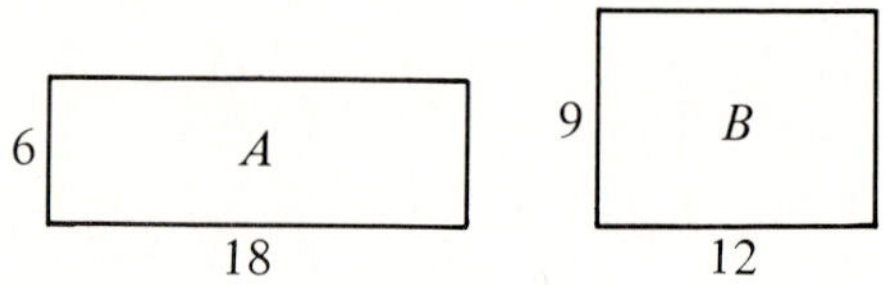

In Exercises 33–36, find, in simplest form:
a. the ratio of the area of the shaded part to the area of the unshaded part;
b. the ratio of the area of the shaded part to the total area.

B **33.**

34.

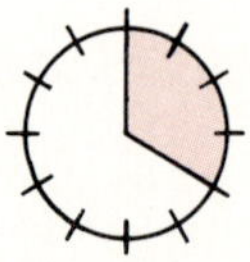

35.

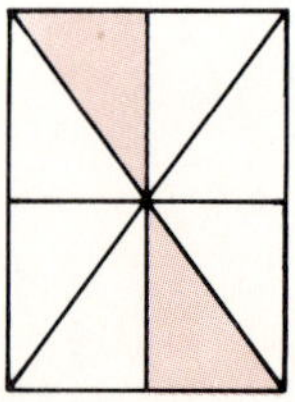

36.

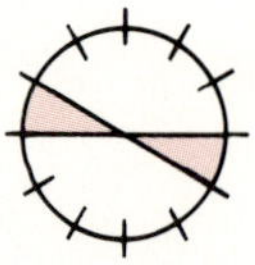

Express each ratio in simplest form.

37. $\frac{30 \text{ cm}}{1 \text{ m}}$ **38.** $\frac{4 \text{ mm}}{10 \text{ m}}$ **39.** $\frac{2 \text{ m}}{3 \text{ km}}$

C **40.** The measures of the angles of a triangle are in the ratio 3:4:5. Find the measures of the angles. (*Hint:* Let $3x$ represent the measure of the smallest angle.)

2 Proportions

A **proportion** is an equation which states that two ratios are equal. Here are some examples of proportions.

$\frac{4}{6} = \frac{2}{3}$ $\qquad$ $1:5 = 6:30$ $\qquad$ $\frac{a}{b} = \frac{c}{d}$ $(b \neq 0, d \neq 0)$

The first and last numbers in a proportion are called the *extremes.* The middle numbers are called the *means.*

$\frac{4}{6} = \frac{2}{3}$ $\qquad$ $1:5 = 6:30$ $\qquad$ $\frac{a}{b} = \frac{c}{d}$ $(b \neq 0, d \neq 0)$

means extremes $\qquad$ means extremes $\qquad$ **means extremes**

It is sometimes helpful to change a proportion from one form to another. The following properties, based on the postulates of equality, provide ways of changing proportions. The variables used in each property represent nonzero numbers.

PROPERTY 1 (The Cross-Multiplying Property)

In a proportion, the product of the means equals the product of the extremes.

$\frac{5}{10} = \frac{1}{2}$ $\qquad$ $\frac{a}{b} = \frac{c}{d}$

$10 \times 1 = 5 \times 2$ $\qquad$ $b \times c = a \times d$

PROPERTY 2 (The Switching Property)

In a proportion, the means (or the extremes) may be switched.

switch the means: $\frac{5}{10} = \frac{1}{2}$ → $\frac{5}{1} = \frac{10}{2}$

switch the extremes: $\frac{a}{b} = \frac{c}{d}$ → $\frac{d}{b} = \frac{c}{a}$

PROPERTY 3 (The Overturning Property)

In a proportion, the ratios may be turned upside-down.

$$\frac{5}{10} = \frac{1}{2} \qquad \frac{a}{b} = \frac{c}{d}$$

$$\frac{10}{5} = \frac{2}{1} \qquad \frac{b}{a} = \frac{d}{c}$$

PROPERTY 4 (The Numerator-Changing Property)

If $\frac{a}{b} = \frac{c}{d}$, *then* $\frac{a+b}{b} = \frac{c+d}{d}$.

Also, if $\frac{a+b}{b} = \frac{c+d}{d}$, *then* $\frac{a}{b} = \frac{c}{d}$.

In the following exercises, all variables represent nonzero numbers.

Classroom Practice

Tell whether each of the following is a proportion.

1. $3:6 = 1:2$ **2.** $\frac{2}{5} = \frac{4}{10}$ **3.** $\frac{5}{6} = \frac{1}{2} + \frac{1}{3}$ **4.** $\frac{1}{3} = \frac{3}{9}$

Use the Cross-Multiplying Property to complete each statement.

5. If $\frac{6}{8} = \frac{3}{4}$, then $8 \times 3 =$ __?__.

6. If $\frac{1}{7} = \frac{2}{14}$, then __?__ $= 1 \times 14$.

7. If $\frac{a}{b} = \frac{c}{d}$, then $bc =$ __?__.

8. If $\frac{x}{y} = \frac{p}{q}$, then $py =$ __?__.

Use the Switching Property to complete each statement.

9. If $\frac{4}{6} = \frac{2}{3}$, then $\frac{4}{2} =$ __?__ and $\frac{3}{6} =$ __?__.

10. If $\frac{1}{4} = \frac{2}{8}$, then $\frac{1}{2} =$ __?__ and $\frac{8}{4} =$ __?__.

Use the Overturning Property to complete each statement.

11. If $\frac{a}{b} = \frac{2}{3}$, then $\frac{b}{a} =$ __?__.

12. If $\frac{2}{9} = \frac{x}{6}$, then $\frac{9}{2} =$ __?__.

Use the Numerator-Changing Property to complete each statement.

13. If $\frac{7}{3} = \frac{x}{6}$, then $\frac{7+3}{3} = \underline{\ \ ?\ \ }$.

14. If $\frac{w}{x} = \frac{y}{z}$, then $\frac{w+x}{x} = \underline{\ \ ?\ \ }$.

Find the value of x.

15. $\frac{2}{5} = \frac{x}{20}$

$5x = \underline{\ \ ?\ \ }$

$x = \underline{\ \ ?\ \ }$

16. $\frac{5}{6} = \frac{x}{12}$

$6x = \underline{\ \ ?\ \ }$

$x = \underline{\ \ ?\ \ }$

17. $\frac{3}{8} = \frac{x}{10}$

$8x = \underline{\ \ ?\ \ }$

$x = \underline{\ \ ?\ \ }$

18. $\frac{6}{x} = \frac{9}{6}$

$9x = \underline{\ \ ?\ \ }$

$x = \underline{\ \ ?\ \ }$

Written Exercises

Complete each statement.

A

1. If $\frac{a}{2} = \frac{b}{3}$, then $2b = \underline{\ \ ?\ \ }$.

2. If $\frac{x}{y} = \frac{3}{4}$, then $3y = \underline{\ \ ?\ \ }$.

3. If $\frac{5}{r} = \frac{3}{5}$, then $\underline{\ \ ?\ \ } = 25$.

4. If $\frac{7}{c} = \frac{d}{5}$, then $\underline{\ \ ?\ \ } = 35$.

5. If $\frac{a}{3} = \frac{b}{4}$, then $\frac{a}{b} = \underline{\ \ ?\ \ }$.

6. If $\frac{p}{q} = \frac{5}{6}$, then $\frac{6}{q} = \underline{\ \ ?\ \ }$.

7. If $\frac{r}{s} = \frac{2}{3}$, then $\frac{r+s}{s} = \underline{\ \ ?\ \ }$.

8. If $\frac{y}{z} = \frac{7}{9}$, then $\underline{\ \ ?\ \ } = \frac{9}{7}$.

9. If $\frac{x}{12} = \frac{3}{4}$, then $\frac{x}{3} = \underline{\ \ ?\ \ }$.

10. If $\frac{1}{8} = \frac{9}{x}$, then $\underline{\ \ ?\ \ } = \frac{9+x}{x}$.

11. If $\frac{e}{f} = \frac{8}{5}$, then $5e = \underline{\ \ ?\ \ }$.

12. If $\frac{k}{j} = \frac{9}{4}$, then $4k = \underline{\ \ ?\ \ }$.

Suppose $\frac{a}{b} = \frac{3}{4}$ and $\frac{c}{d} = \frac{5}{7}$. Find each ratio in simplest form.

13. $\frac{b}{a}$

14. $\frac{a+b}{b}$

15. $\frac{a+b}{a}$

16. $\frac{2a}{2b}$

17. $\frac{d}{c}$

18. $\frac{c+d}{d}$

19. $\frac{c-d}{c}$

20. $\frac{3d}{3c}$

Find the value of x.

21. $\frac{x}{6} = \frac{1}{2}$

22. $\frac{4}{3} = \frac{16}{x}$

23. $\frac{1}{3} = \frac{10}{x}$

24. $\frac{9}{x} = \frac{9}{7}$

25. $3:8 = x:16$

26. $9:x = 3:1$

27. $x:10 = 4:5$

28. $3:5 = 24:x$

Find the value of x.

B 29. $\frac{10}{3x} = \frac{5}{6}$

30. $\frac{5}{2x} = \frac{25}{1}$

31. $\frac{1}{2} = \frac{x+1}{8}$

32. $\frac{x-5}{6} = \frac{4}{9}$

C 33. $\frac{3x-2}{4} = \frac{x}{3}$

34. $\frac{1}{2} = \frac{5x+4}{3}$

35. $\frac{x}{3} = \frac{x+2}{5}$

36. $\frac{x-5}{x} = \frac{3}{4}$

37. $\frac{x+2}{x-1} = \frac{x+3}{x-3}$

38. $\frac{(2x-3)^2}{(2x+1)(x-3)} = \frac{2}{1}$

CALCULATOR ACTIVITIES

Suppose the sides of a triangle have lengths a, b, and c. The area of the triangle is given by the formula

$$A = \sqrt{s(s-a)(s-b)(s-c)} \text{ where } s = \frac{a+b+c}{2}.$$

To use this formula:

1. First compute s. Then compute $s - a$, $s - b$, and $s - c$.
2. Substitute in the formula $\sqrt{s(s-a)(s-b)(s-c)}$.

EXAMPLE

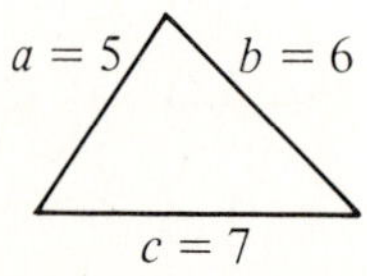

Find the area.

1. $s = \frac{5+6+7}{2}$; $s = 9$.
   $s - a = 4$, $s - b = 3$, and $s - c = 2$.
2. $\sqrt{s(s-a)(s-b)(s-c)} = \sqrt{9 \times 4 \times 3 \times 2} \approx 14.696938$

To the nearest tenth, the area is 14.7.

Find the area of the triangle specified.
Round your answers to the nearest tenth.

	1.	2.	3.	4.	5.	6.	7.	8.
a	6	7	6	8	11	21	16	13.1
b	8	5	6	6	15	17	19	12.0
c	10	10	6	6	17	10	12	14.8

3 Using Proportions

Proportions can be used to solve many kinds of problems. You probably remember, from studying algebra, how to solve Example 1.

EXAMPLE 1

Four cans of paint cost \$30.
How much will six cans of paint cost?

1. Summarize your data.

Number of cans	4	6
Cost	\$30	\$$x$

2. Write a proportion.

$$\frac{4}{30} = \frac{6}{x}$$

$$4x = 180$$
$$x = 45$$

Answer: \$45

EXAMPLE 2

John Kaylen's summer job is painting houses and barns. He has just used three cans of paint to paint the front of the barn shown. How much more paint will he need for the remaining sides?

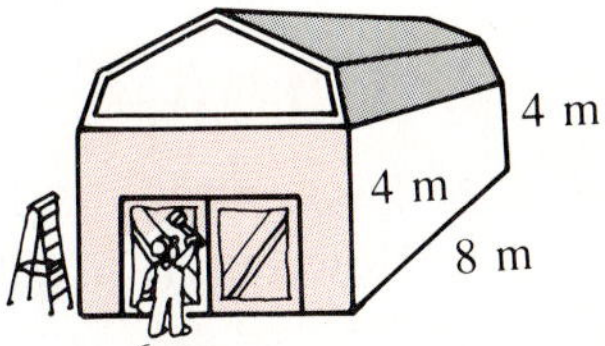

1. Estimate the area of the painted and unpainted sides. (Ignore the areas of doors.)

   Area of painted side $= 6 \times 4 = 24 \text{ m}^2$

   Area of unpainted sides $= \underbrace{8 \times 4}_{\text{side}} + \underbrace{6 \times 4}_{\text{back}} + \underbrace{8 \times 4}_{\text{side}} = 88 \text{ m}^2$

2. Summarize your data.

	Area	Number of cans
Painted	24 m^2	3
Unpainted	88 m^2	x

3. Write a proportion.

$$\frac{24}{88} = \frac{3}{x}$$

$$24x = 264$$
$$x = 11$$

Answer: He will need about 11 more cans of paint.

Classroom Practice

Make up a problem to go with each table below. Then write a proportion which can be solved for *x*.

1.

Number of cans	4	6
Cost	\$7	\$$x$

2.

Number of boxes	8	20
Cost	\$3	\$$x$

3.

Painted area	10 m^2	65 m^2
Cost	\$4	\$$x$

4.

Cost of gasoline	\$6	\$9
Distance traveled	280 km	x km

5. Six grapefruits cost \$1.92. How many can you buy for \$4.80?
6. Three tickets to a track meet cost \$15.90. How many can you buy for \$26.50?

Written Exercises

A

1. Two concert tickets cost \$34. How much would five tickets cost?
2. Three cans of cat food cost 93¢. How much would ten cans cost?
3. Three copies of *Sports Today* cost \$5.85. How much would five copies cost?
4. Four containers of yogurt cost \$2.52. How much would six containers cost?
5. Two books of stamps cost \$8.80. How many books can you buy with \$22.00?
6. Two packages of mints cost 69¢. How many packages can you buy for \$2.25?
7. It costs \$11 to paint the front of the barn shown. How much will it cost to paint the right side?
8. Two cans of paint are needed to paint the front of the barn shown. How much paint is needed to paint the other three sides?

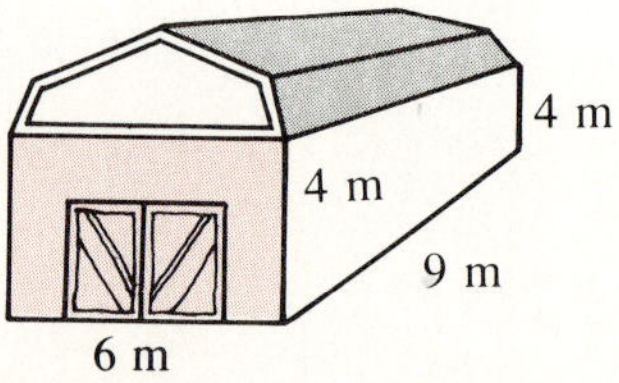

B

9. The Fongs would like to redo their kitchen floor. They find that linoleum will cost \$16 per square meter. Carpeting will cost \$20 per square meter. The new kitchen floor will cost \$192 if linoleum is used. How much will it cost if carpeting is used?

10. It has taken Alexis 30 minutes to mow a 6 m by 32 m strip of lawn. How much more time will she need to finish the job? (Assume that she mows at a constant rate.)

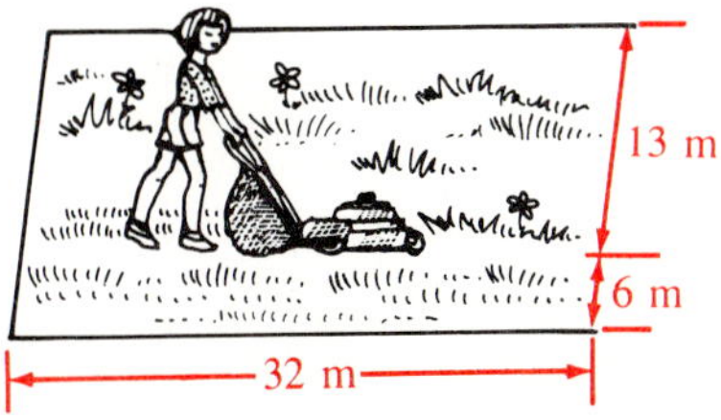

11. The Carmona Construction Company is laying a brick patio. Thirty rows of bricks have been installed. Twenty more rows are needed to finish the job. If 600 bricks have already been used, how many will be used for the entire job?

SELF-TEST

Vocabulary

ratio (p. 286)
proportion (p. 289)
means (p. 289)
extremes (p. 289)

Express each ratio in simplest form.

1. $\frac{8}{12}$ 2. $\frac{25}{15}$ 3. $25:40$ 4. $9:18:27$ *(7-1)*

Find the value of x.

5. $\frac{x}{6} = \frac{7}{3}$ 6. $\frac{10}{x} = \frac{8}{12}$ 7. $\frac{4}{6} = \frac{6}{x}$ 8. $x:6 = 3:9$ *(7-2)*

Solve.

9. Three loaves of bread cost \$2.67. How much will 5 loaves cost? *(7-3)*

10. A rug 4 m wide and 5 m long costs \$132. How much would a 3 m by 4 m rug of the same material cost?

PUZZLE ◆ PROBLEMS

Ron, Jo, and Bobbie each won a contest during class day at their high school. One student won the photography exhibit, another, the math contest, and the third, the track race. The photographer, who is an only child, does not know how to play tennis. Bobbie is a friend of the math whiz. Jo often plays tennis with Bobbie's sister. Which student won each event?

4 Maps and Scale Drawings

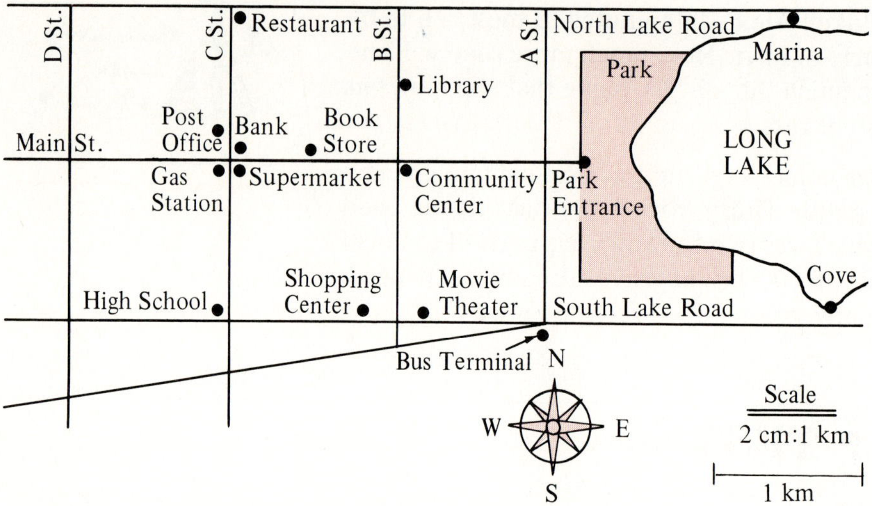

The map above shows part of the town of Lakeville. Notice that the **scale** of the map is

$$2 \text{ cm} : 1 \text{ km}.$$

This means that a distance of 2 cm *on the map* represents an *actual* distance of 1 km in Lakeville. For example, the distance between the high school and the post office is about 2.4 cm. To find the actual distance, we write a proportion:

$$\frac{2 \text{ cm}}{1 \text{ km}} = \frac{2.4 \text{ cm}}{x \text{ km}}$$

$$\frac{2}{1} = \frac{2.4}{x}$$

$$2x = 2.4$$

$$x = 1.2$$

The actual distance is about 1.2 km.

A *scale drawing* of a plot of land is very much like a map. So is the floor plan of a house. In the floor plan on the next page, the scale is listed as 1:200. This means that

$$\frac{\text{distance in drawing}}{\text{corresponding distance in house}} = \frac{1}{200}.$$

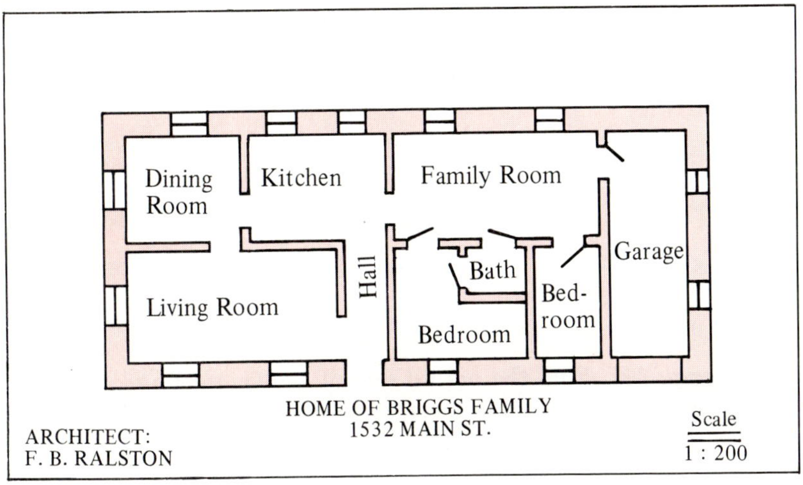

Classroom Practice

Exercises 1–4 refer to the floor plan above.

1. A scale of 1:200 means that

   1 cm in the floor plan = 200 cm in the house.

   Then 1 cm in the floor plan = __?__ m in the house.

Use a ruler to find each distance to the nearest centimeter. Then find the actual distance in the house.

2. The length, the width and the perimeter of the living room
3. The length and the width of the family room
4. The perimeter of the entire house

Refer to the map of Lakeville on page 296. Use a ruler to find the distances referred to in the first row of the table. Measure to the nearest centimeter. Then copy and complete the table.

		Post Office to Park Entrance	Gas Station to Marina	Community Center to High School
5.	Distance on the map in cm	?	?	?
6.	Actual distance when scale is 2 cm:1 km	?	?	?

Written Exercises

Use the map of Lakeville on page 296, with scale 1 cm:800 m. Find each actual distance in the town.

A

1. Gas station to community center
2. Bus terminal to movie theater
3. Shopping center to library
4. Park entrance to supermarket
5. Restaurant to shopping center
6. High school to bookstore
7. Marina to cove by boat
8. Marina to cove by car

Use the floor plan on page 297. Find the actual length, width, perimeter, and area of each room. Use the scale 1 cm:250 cm.

9. Dining room
10. The smaller bedroom
11. Kitchen
12. Garage

Use the map of Sequoia National Park to find the distance between the following places to the nearest kilometer.

B

13. Crystal Cave and Silver City
14. Mt. Whitney and Triple Divide Peak
15. Park Headquarters and Kern Canyon Ranger Station
16. Giant Forest and Garfield Grove of Big Trees

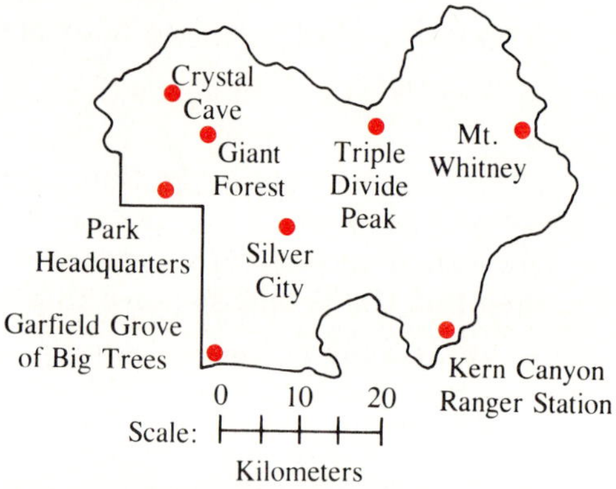

17. The diagram shows three cities in New Hampshire.
    a. Measure the map distances between the cities to the nearest millimeter.
    b. Write a proportion to find the actual distance between Portsmouth and Concord.
    c. Find the distance between Manchester and Concord.

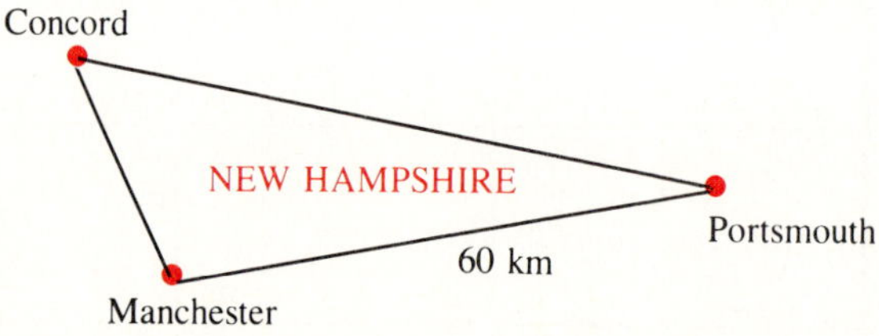

5 Making Maps and Scale Drawings

When making a map or a scale drawing, you must first decide on the scale you are going to use. For example, suppose you are making a scale drawing of your classroom.

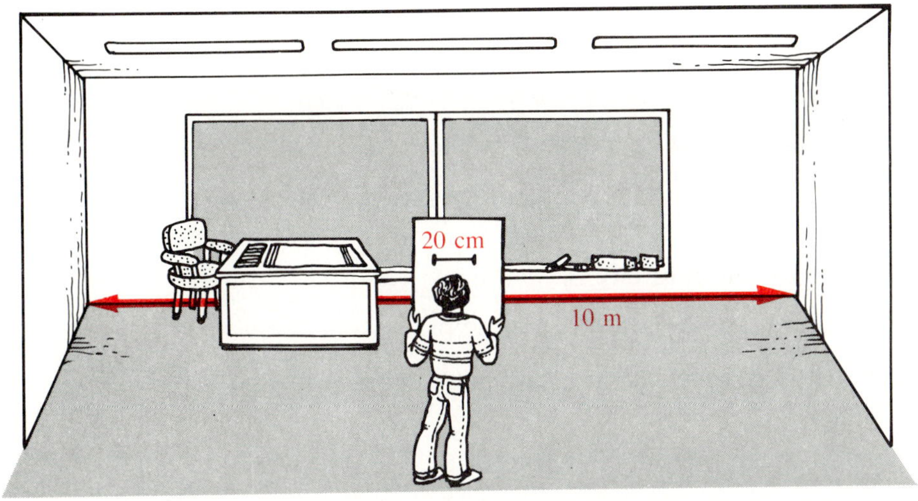

Suppose your classroom is 10 m long and you choose to represent this distance by a line segment 20 cm long. Then your scale is:

$$\frac{\text{length of classroom in drawing}}{\text{actual length of classroom}} = \frac{20 \text{ cm}}{10 \text{ m}} = \frac{20 \text{ cm}}{1000 \text{ cm}} = \frac{1}{50}$$

Measure the width and other distances in meters. Express these measurements in centimeters and multiply each by the *scale ratio*, $\frac{1}{50}$, to find the corresponding distance in your drawing.

Classroom Practice

Make a scale drawing of your classroom. Follow the instructions above.

Written Exercises

A

1. Make a scale drawing of an Olympic-sized swimming pool which is 50 m long and 21 m wide.
2. Draw the cover of this book, using the scale 1:2.
3. Repeat Exercise 2, using the scale 1:4.

4. In Frazier Park, the high school is 3 km east of the junior high school.

The shopping mall is 4 km south of the junior high school and is 5 km from the high school.

Draw a map showing the high school, junior high, and mall.

Be sure to include the scale on your map.

5. The recreation center pool measures 42 m by 15 m.

The pool is surrounded by a deck 3 m wide.

Make a scale drawing of the pool and deck with scale 1 cm:6 m.

6. Town A is 6 km south of city B.

City C is 3 km northwest (NW) of A.

Find the distance, to the nearest kilometer, from B to C.

7. Boothbay is 12 km west of New Harbor.

Monhegan Island is 20 km southeast (SE) of New Harbor.

Make a scale drawing and find the distance, to the nearest kilometer, from Boothbay to Monhegan Island.

B 8. Dallas, Texas, is about 300 km west of Shreveport, Louisiana.

San Antonio, Texas, is located about 620 km southwest (SW) of Shreveport.

To the nearest 50 km, how far is it from Dallas to San Antonio?

9. Campsite A is 6 km north of campsite B.

From A, a fire tower is sighted on a bearing of 75°. From B, the same fire tower is sighted on a bearing of 60°.

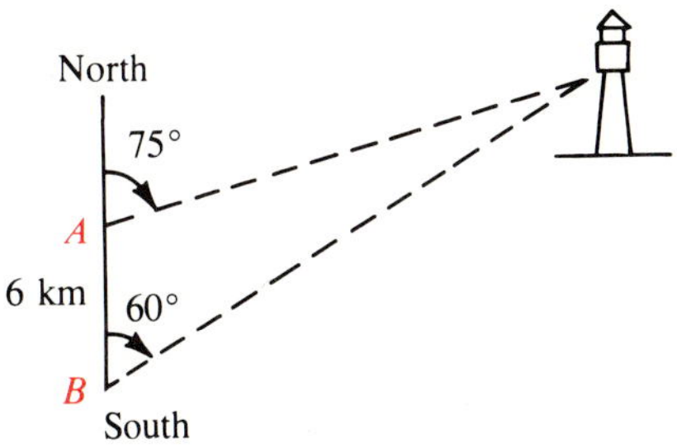

a. To the nearest kilometer, how far is it from A to the fire tower?
b. To the nearest kilometer, how far is it from B to the fire tower?

10. Campsite C is 2 km north of campsite D.

From C, a waterfall is sighted at a point W, directly east of C. From D, the same waterfall is sighted on a bearing of 70°.

a. Make a scale drawing showing C, D, and W.
b. Find the distances to the nearest kilometer from C and D to the waterfall.

C 11. Feingold's Furniture wants to put a new display in its front window. Design a furnished living room by making a scale drawing.

SELF-TEST

Vocabulary

scale (p. 296) scale drawing (p. 296)

Use the diagram at the right.

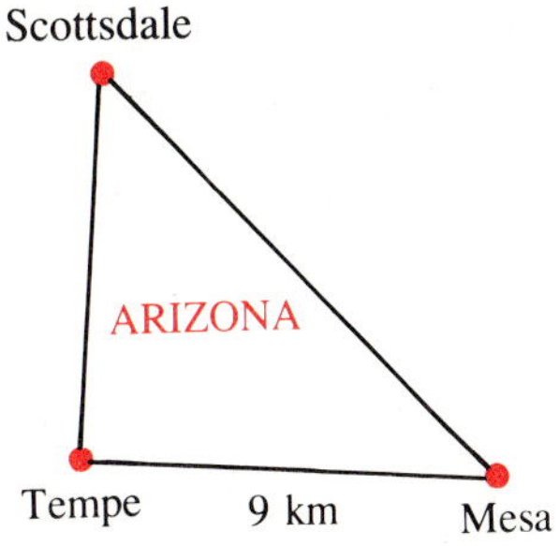

1. Measure the map distance between Mesa and Tempe to the nearest centimeter. Then give the scale of the map. *(7-4)*

2. Find the actual distance between Mesa and Scottsdale.

3. Find the actual distance between Scottsdale and Tempe.

4. Make a scale drawing of an ice hockey rink that is 61 m long and 30.5 m wide. *(7-5)*

APPLICATIONS

USING A COMPASS

A magnetic compass is a basic navigational tool for finding direction of travel with respect to magnetic north. A magnetic compass measures direction in degrees, like the protractors you have used to measure angles. Although many protractors only indicate measures from 0° to 180°, a compass includes measures from 0° to 360°, that is, a compass is a complete circle.

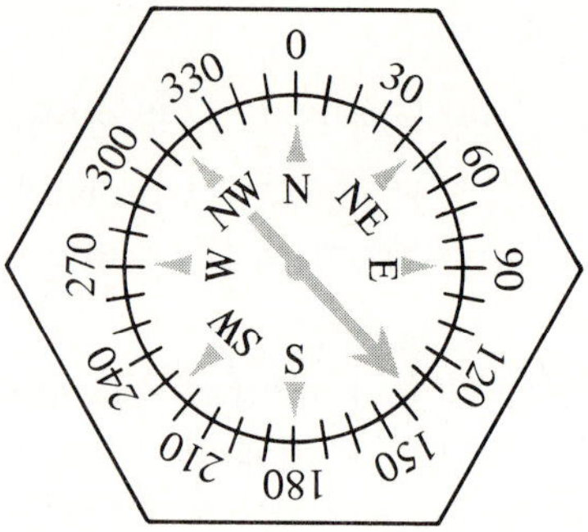

Look at the compass shown here. Notice that 0° indicates a northerly reading while 180° indicates a southerly direction. If you head toward 135°, your direction of travel is southeast. See if you can determine the reading for a southwesterly direction of travel.

Because magnetic north is not located at the north pole, navigators make corrections to accommodate for the difference between the actual direction of travel they wish to take and the magnetic compass bearing they will use. The difference between magnetic north and actual north in any particular area of the earth is called *magnetic variation.* As this variation is not the same everywhere, it is important to know the magnetic variation for your particular location before you use a compass to find your direction.

Because a magnetic compass is such a basic navigational tool, navigational charts are printed with one compass showing actual or true north, and another showing magnetic north. The example shown below is taken from a navigational chart. It shows a magnetic variation of 10° east. This means that the magnetic compass needle will point 10° east of true north. In this location, a magnetic compass reading due west or 270° indicates a true direction of 280°.

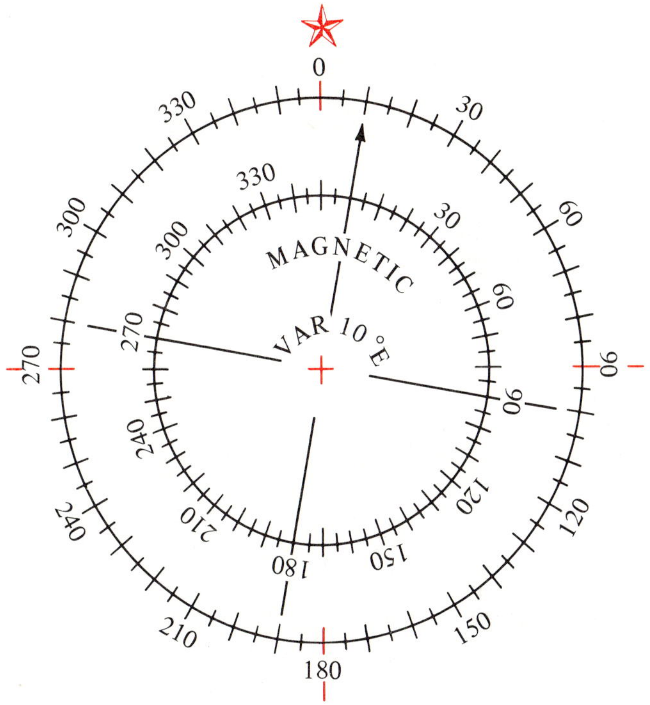

Suppose you want to head in the true direction of 45°. Find the compass heading by subtracting the magnetic variation, 10°, from the true direction. The magnetic compass heading that you should use is $45° - 10° = 35°$.

EXERCISES

If there is no magnetic variation, find the compass heading for:

1. An easterly course
2. A northwesterly course

Find the proper magnetic compass heading for the following:

3. A southwesterly course; magnetic variation 10°E
4. A westerly course; magnetic variation 7°W

SKILLS REVIEW

FRACTIONAL AND DECIMAL EQUATIONS

Solve.

SAMPLES

$$\frac{2x}{5} + 1 = 9$$
$$\frac{2x}{5} + 1 - 1 = 9 - 1$$
$$\frac{2x}{5} = \frac{8}{1}$$
$$2x = 40$$
$$x = 20$$

$$0.12x = 3$$
$$100 \times 0.12x = 100 \times 3$$
$$12x = 300$$
$$x = 25$$

$$\frac{n}{2} + \frac{n}{5} = 7$$
$$\frac{5n}{10} + \frac{2n}{10} = 7$$
$$\frac{5n + 2n}{10} = 7$$
$$\frac{7n}{10} = \frac{7}{1}$$
$$7n = 70$$
$$n = 10$$

Solve.

1. $\frac{a}{9} + 3 = 13$
2. $30 + \frac{d}{4} = 51$
3. $\frac{b}{2} + 17 = 0$
4. $0.6x = 42$
5. $0.25y = 3.5$
6. $78 = 1.3m$
7. $\frac{2n}{3} = \frac{4}{7}$
8. $\frac{3x}{5} + 2 = 11$
9. $\frac{5b}{2} - 12 = -7$
10. $\frac{k}{2} + \frac{k}{4} = 9$
11. $\frac{c}{3} - \frac{2c}{9} = \frac{1}{6}$
12. $\frac{3m}{10} - \frac{m}{2} = 4$
13. $\frac{1}{4} + \frac{2}{3} = \frac{a}{6}$
14. $\frac{2m}{3} + \frac{m}{2} = \frac{7}{2}$
15. $\frac{4k}{5} - \frac{2k}{3} = -4$

Write an equation. Then solve.

16. When half a number is subtracted from 4 times the number, the difference is 35. What is the number?
17. Regina is 105 cm tall. She is 5 cm more than $\frac{5}{6}$ as tall as her brother Ramon. How tall is Ramon?
18. One week Wally's Wheel City sold 27 ten-speed bicycles. This was $\frac{3}{5}$ of all the bicycles sold that week. How many bicycles did the shop sell in all?

CHAPTER REVIEW

CHAPTER SUMMARY

1. A ratio is a quotient of two numbers.

   The ratio of 2 to 3 is written $\frac{2}{3}$ or $2:3$.

2. To find the ratio of two measures, first express both measures in terms of the same unit. Note that there are *no* units in the final answer.

3. A proportion is an equation which states that two ratios are equal. The first and the last numbers in a proportion are called the extremes. The middle numbers are called the means.

4. Four helpful properties for changing the form of a proportion:
   a. The Cross-Multiplying Property
   b. The Switching Property
   c. The Overturning Property
   d. The Numerator-Changing Property

5. Proportions are used in problem solving.

6. Proportions are used in making and reading maps and scale drawings. The scale of a map is the ratio of the map distance to the actual distance.

REVIEW EXERCISES

Express each ratio in simplest form. *(See pp. 286–288.)*

1. $\frac{9}{15}$
2. $\frac{56}{16}$
3. $\frac{24x}{9x}$
4. $21:14$
5. $ab:ac$
6. $48:30:24$
7. 8 m:40 cm
8. 20 m:2 km
9. 20 min to 3 h

Use the figure to find each ratio in simplest form. *(See pp. 286–288.)*

10. $\frac{AB}{DC}$
11. $\frac{AE}{ED}$
12. $\frac{ED}{FC}$
13. $\frac{AD}{BC}$

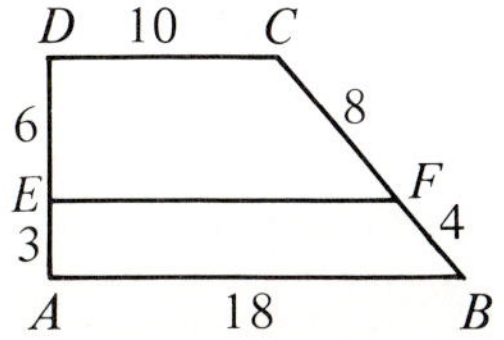

Complete each statement. All variables represent nonzero numbers. *(See pp. 289–292.)*

14. If $\frac{a}{4} = \frac{b}{5}$, then $\frac{a}{b} = $ __?__.

15. If $\frac{m}{6} = \frac{3}{2}$, then __?__ $= 18$.

16. If $\frac{p}{4} = \frac{2}{5}$, then $\frac{p+4}{4} = $ __?__.

17. If $\frac{x}{9} = \frac{4}{y}$, then $\frac{9}{x} = $ __?__.

Find the value of *x*. *(See pp. 289–292.)*

18. $\frac{x}{8} = \frac{3}{4}$

19. $\frac{5}{6} = \frac{x}{18}$

20. $x:8 = 5:2$

Solve. *(See pp. 293–295.)*

21. Two granola bars cost 90¢. What is the cost of five bars?

22. It costs $27.50 to paint a fence 3 m high and 25 m long. How much would it cost to paint another fence of the same type if it is 2 m high and 15 m long?

Use the diagram at the right. *(See pp. 296–298.)*

23. Measure the map distance between Bartlesville and Stillwater to the nearest centimeter. Then give the scale of the map.

24. Find the actual distance between Stillwater and Tulsa.

25. Find the actual distance between Bartlesville and Tulsa.

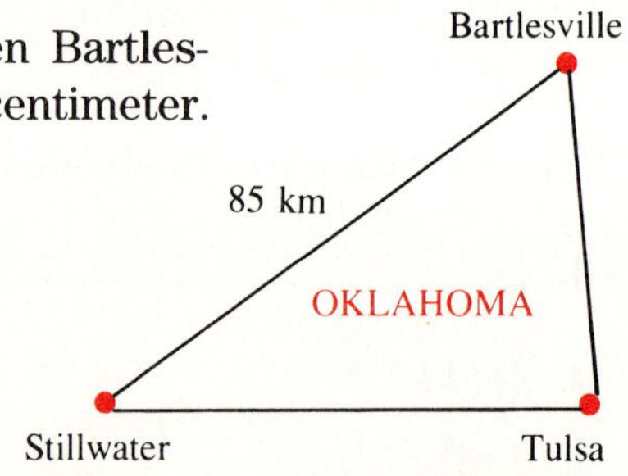

Draw a map. Be sure to include the scale. *(See pp. 299–301.)*

26. The Masconomet Regional High School is 1 km south of the county courthouse, and the local business district is 3 km west of the school. Draw a map showing the school, courthouse, and business district.

CHAPTER TEST

If $a = 6$ and $b = 4$, find each ratio in simplest form.

1. $\frac{a}{a+b}$

2. $a:b:(a-b)$ *(7-1)*

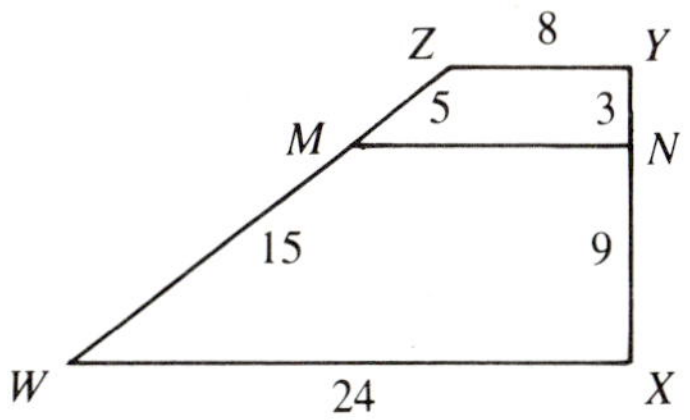

Find each ratio in simplest form.

3. $\frac{XY}{WZ}$

4. $\frac{YZ}{WX}$

5. $\frac{WM}{XY}$

6. If $\frac{e}{f} = \frac{j}{k}$, then $\frac{k}{f} = \underline{\ ?\ }$ and $fj = \underline{\ ?\ }$. *(7-2)*

7. If $\frac{r}{s} = \frac{2}{5}$, then $\frac{s}{r} = \underline{\ ?\ }$.

8. If $\frac{r}{s} = \frac{2}{5}$, then $\frac{r+s}{s} = \underline{\ ?\ }$.

9. If $\frac{x}{28} = \frac{3}{4}$, find the value of x.

10. One week Carlos earned \$85.50 for working 18 hours at Yamamoto's Furniture Store. How much did Carlos earn the following week when he worked 25 hours? *(7-3)*

11. Elaine found that a car trip of 136 km required 16 L of gas. How much gas would a trip of 61.2 km at the same speed require?

Use the diagram at the right.

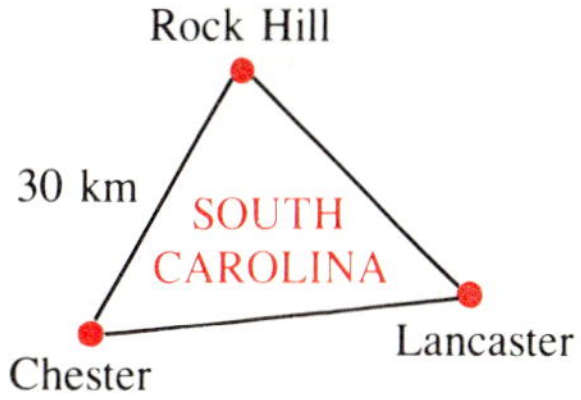

12. Measure the distance between Chester and Rock Hill to the nearest centimeter. The scale is 1 cm: $\underline{\ ?\ }$ km. *(7-4)*

13. Find the actual distance between Chester and Lancaster.

14. Find the map distance in centimeters that would represent an actual distance of 80 km.

Brenda and Larry attend a school that is 3 km northwest of their house. The post office is 1 km west of the school.

15. Draw a map showing the house, the school, and the post office. Include the scale that you use. *(7-5)*

16. Find the distance, to the nearest kilometer, from the post office to the house.

MIXED REVIEW

1. Express in "If . . . then" form and tell if the statement is true: "Every triangle with a 60° angle is equilateral."

2. Supply the reasons to complete the proof.

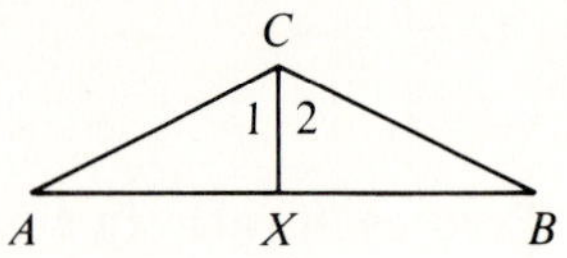

Given: $\overline{CX}$ is the perpendicular bisector of $\overline{AB}$.

Prove: $\overline{CX}$ is the angle bisector of $\angle ACB$.

STATEMENTS	REASONS
1. $\overline{CX}$ is the $\perp$ bisector of $\overline{AB}$.	1. ?
2. $\overline{CX} \perp \overline{AB}$; $AX = BX$	2. ?
3. $\angle CXA = 90° = \angle CXB$	3. ?
4. $CX = CX$	4. ?
5. $\triangle CXA \cong \triangle CXB$	5. ?
6. $\angle 1 = \angle 2$	6. ?
7. $\overline{CX}$ is the bisector of $\angle ACB$.	7. ?

3. An isosceles triangle has a base of 10 cm and a height of 12 cm. Find the length of each leg.

4. Find the diameter of a circle with circumference 17π cm.

5. $\angle 1$ and $\angle 2$ are same-side interior angles formed by two parallel lines. If $\angle 1 = 75°$, find the measure of $\angle 2$.

6. In quadrilateral $ABCD$, the diagonals intersect at point O. If $AO = BO = CO = DO$, what is the best name for $ABCD$: parallelogram, rectangle, rhombus, or square?

7. An isosceles triangle has a vertex angle of 28°. Find the measures of the exterior angles of the triangle.

8. Find the perimeter of $\triangle JKL$.

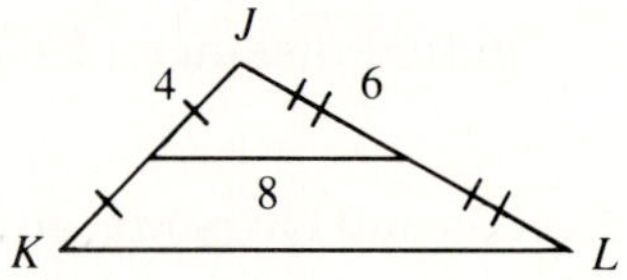

9. Find the value of x: $\frac{x-5}{6} = \frac{3}{4}$.

10. Find the area of a triangle with sides of lengths 15, 17, and 8.

11. In $\triangle GWC$, $\angle G = \angle W$. If $GW = 5t + 1$, $WC = 4t + 3$, and $CG = 6t - 3$, find the value of t.

Write SSS, SAS, ASA, AAS, or HL to indicate the method you would use to prove two triangles congruent. If no method applies, write *none*.

12. 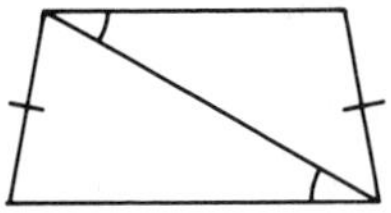

13. 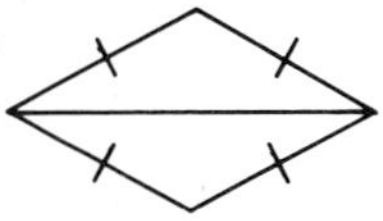

14. 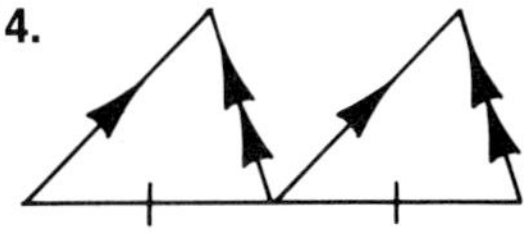

15. Construct a 60° angle. Construct the angle bisector.

16. Find the area, to the nearest tenth, of a circle with radius 2.4. Use 3.14 for π.

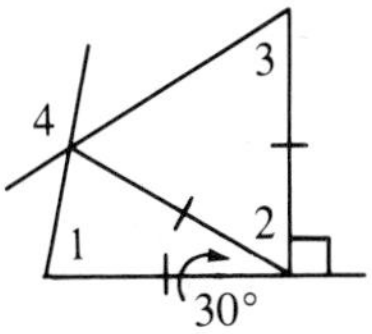

17. Find the measure of each numbered angle shown.

18. Draw a large $\triangle PQR$ with obtuse $\angle P$. Construct altitude $\overline{PX}$ of the triangle.

Find the area of the parallelogram, rectangle, and trapezoid.

19. 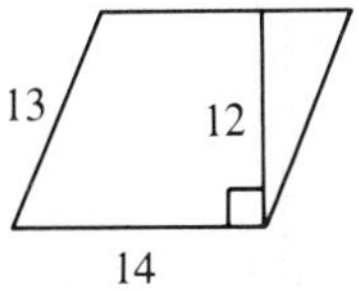

20. 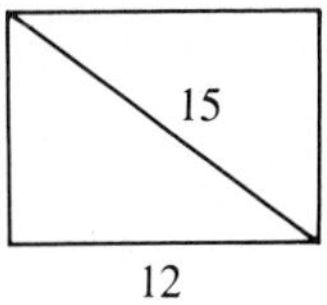

21. 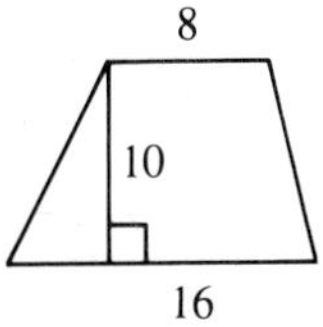

22. Write a two-column proof.
Given: $\angle P = \angle R$; $\angle Q = \angle S$.
Prove: $PS = QR$.
(*Hint:* Use the result of Exercise 15 on page 219.)

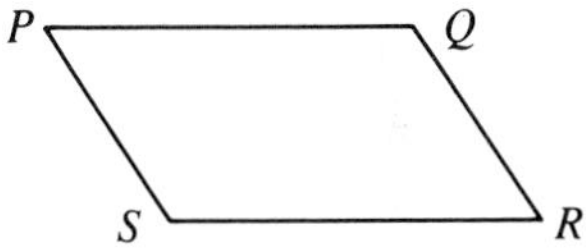

23. Eight oranges cost $1.92. How many can you buy for $1.20?

24. In a regular polygon with 15 sides, find the interior angle sum and the measure of *each* exterior angle.

25. A distance on a map is 5.6 cm. If the scale is 1 cm : 400 km, find the actual distance in kilometers.

26. Find the area of a square with perimeter $8x$.

27. Find the ratio 80¢ : $2 in simplest form.

Given: $\overline{SI} \perp \overline{KI}$; $\overline{SP} \perp \overline{KP}$; $SI = SP$

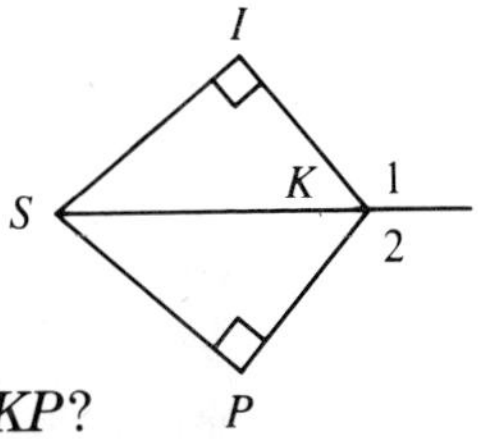

28. Write the statement that supports each statement.
a. $\angle SKI \cong \angle SKP$ **b.** $\angle 1 = \angle 2$

29. What method could be used to prove that $\triangle SKI \cong \triangle SKP$?

Here's what you'll learn in this chapter:

To tell whether two polygons are similar.

To find missing parts of similar triangles.

To use properties related to parallel lines.

To find the perimeters and areas of two similar polygons.

Nuts, bolts, pipes, and many other objects used in building remind us of geometric figures. The hexagonal nuts shown in the photograph are of different sizes, but they are similar because they have the same shape.

Chapter 8

Similar Polygons

1 Defining Similar Polygons

Two polygons with the same shape are called *similar* polygons. Quadrilaterals *ABCD* and *WXYZ* are similar polygons.

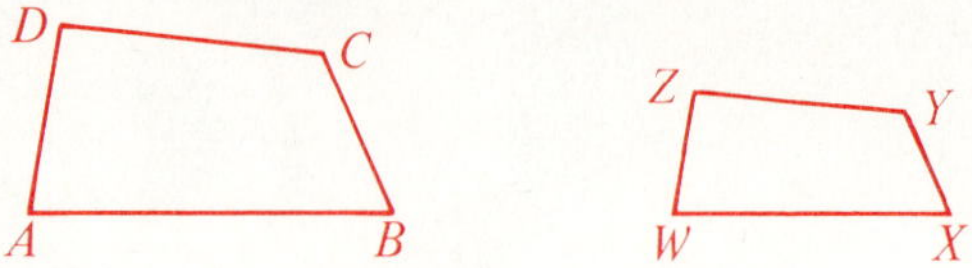

polygon *ABCD* ~ polygon *WXYZ*

This means "is similar to."

Recall that we name congruent polygons by listing their vertices in corresponding order. We name similar polygons the same way:

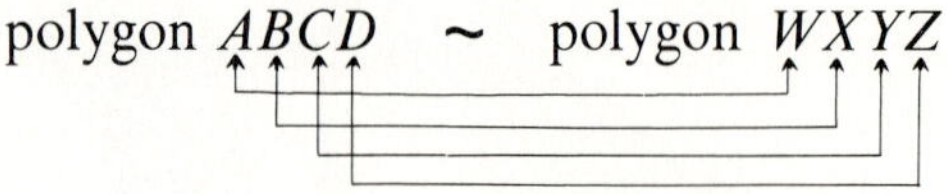

When two polygons are similar, the following are true:

1. Corresponding angles are equal.
2. Corresponding sides are in proportion (have equal ratios).

Since quadrilateral *ABCD* ~ quadrilateral *WXYZ*, we can conclude:

1. $\angle A = \angle W$, $\angle B = \angle X$, $\angle C = \angle Y$, $\angle D = \angle Z$
2. $\frac{AB}{WX} = \frac{BC}{XY} = \frac{CD}{YZ} = \frac{DA}{ZW}$

Suppose that you wish to show that two polygons are similar. To do this, you must show that *both* conditions of the definition are true.

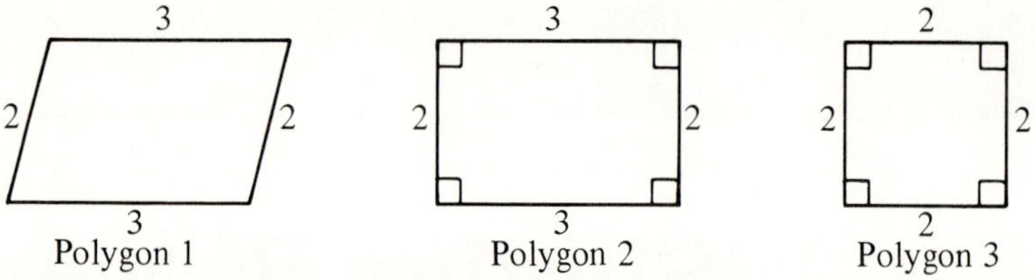

The corresponding sides of Polygons 1 and 2 are in proportion. However, corresponding angles are not equal. Therefore the polygons are not similar.

The corresponding angles of Polygons 2 and 3 are equal. However, their corresponding sides are not in proportion. Therefore Polygons 2 and 3 are not similar.

EXAMPLE $\triangle PQR \sim \triangle STU$.
Find the values of x and y.

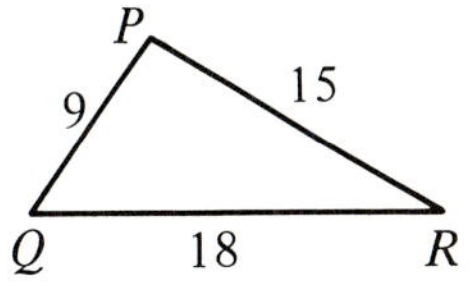

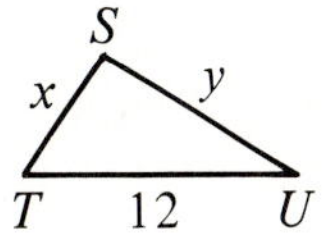

Since corresponding sides are in proportion:

$$\frac{PQ}{ST} = \frac{QR}{TU} = \frac{RP}{US}$$

$$\frac{9}{x} = \frac{18}{12} = \frac{15}{y}$$

$$\downarrow$$

$$\frac{9}{x} = \frac{3}{2} = \frac{15}{y}$$

$$\frac{9}{x} = \frac{3}{2} \qquad\qquad \frac{3}{2} = \frac{15}{y}$$

$$18 = 3x \qquad\qquad 3y = 30$$

$$6 = x \qquad\qquad y = 10$$

If two polygons are similar, the ratio of two corresponding sides is called the *scale factor.* The scale factor of $\triangle ABC$ to $\triangle DEF$ is $\frac{5}{10}$, or $\frac{1}{2}$. Likewise, the scale factor of $\triangle DEF$ to $\triangle ABC$ is $\frac{2}{1}$. Notice that the ratio of the perimeters of these triangles is equal to the scale factor:

$$\frac{\text{perimeter of } \triangle DEF}{\text{perimeter of } \triangle ABC} = \frac{10 + 8 + 6}{5 + 4 + 3} = \frac{24}{12} = \frac{2}{1}.$$

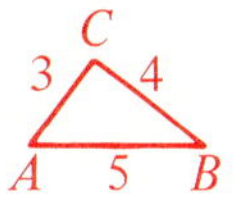

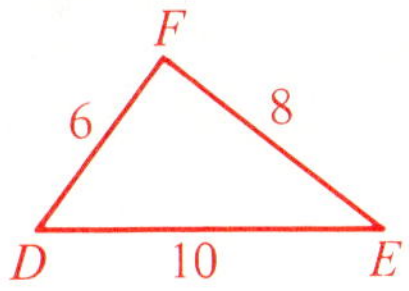

Classroom Practice

In Exercises 1–3, state why the two quadrilaterals are, or are not, similar.

1.

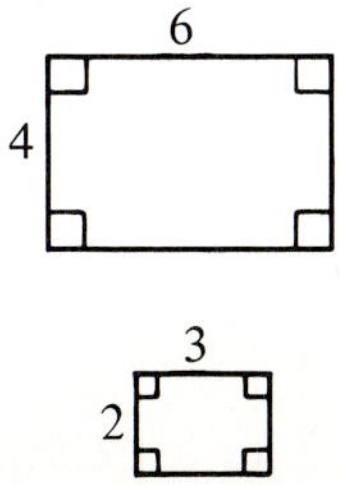

2.

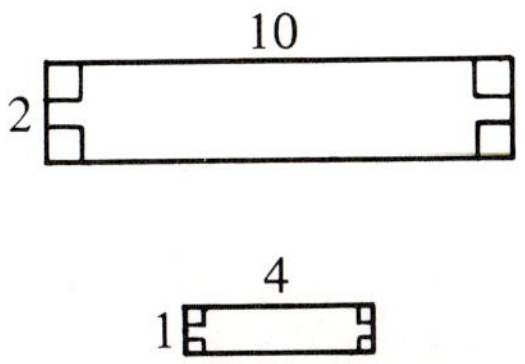

3.

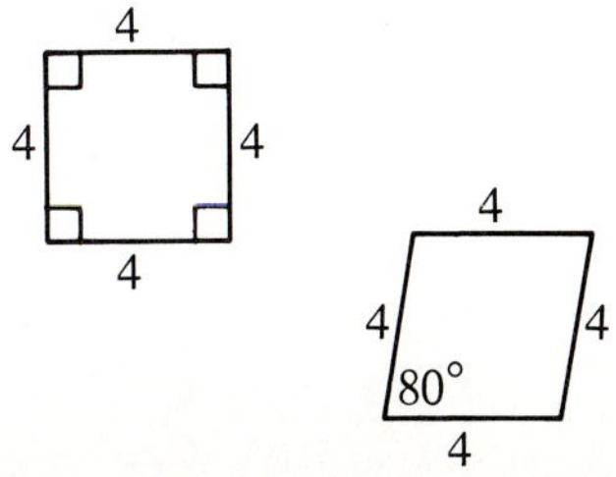

$\triangle ABC \sim \triangle DEF$. **Complete each statement.**

4. The scale factor of $\triangle ABC$ to $\triangle DEF$ is __?__.

5. $\frac{x}{9} = \frac{10}{?}$

6. $\frac{8}{y} = \frac{x}{?}$

7. $\angle C = \angle$ __?__

8. $\angle$ __?__ $= \angle D$

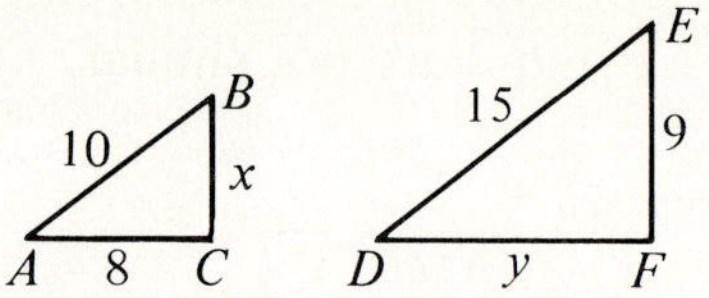

Tell whether each statement is true or false.

9. All squares are similar.
10. All rectangles are similar.
11. All equilateral triangles are similar.
12. All isosceles triangles are similar.
13. All rhombuses are similar.
14. Every polygon is similar to itself.

Written Exercises

Match the vertices of rectangles *ABCD* and *WXYZ* like this:

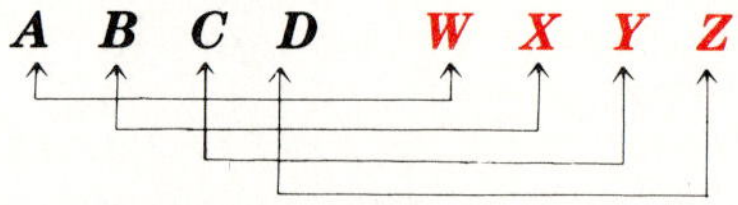

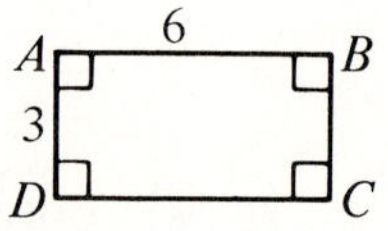

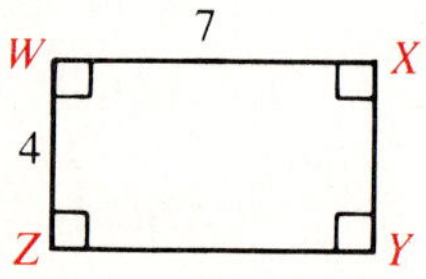

A

1. Are corresponding angles equal?
2. Are corresponding sides in proportion?
3. Are the rectangles similar?

Match the vertices of quadrilaterals *EFGH* and *PQRS* like this:

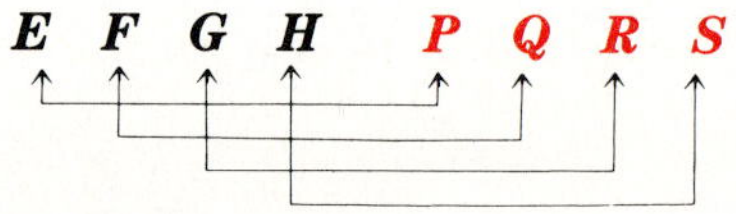

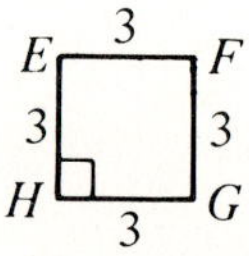

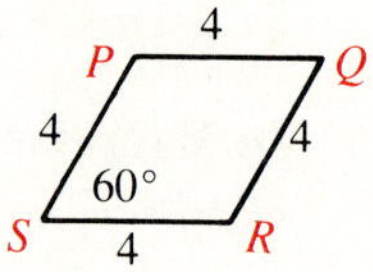

4. Are corresponding angles equal?
5. Are corresponding sides in proportion?
6. Are the quadrilaterals similar?

In Exercises 7–9, state why the two polygons are, or are not, similar.

7. 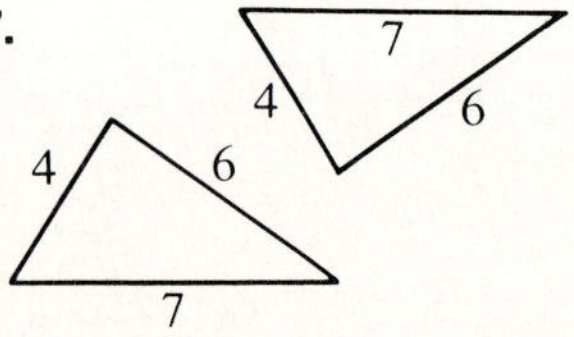

8. 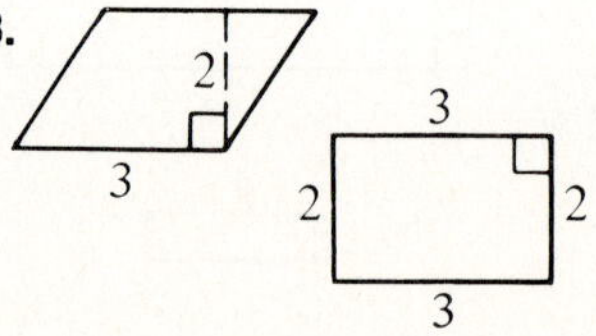

9. 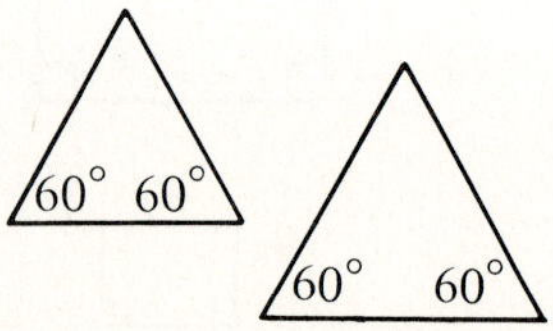

In Exercises 10–12, the two polygons are similar. Complete each statement.

10.

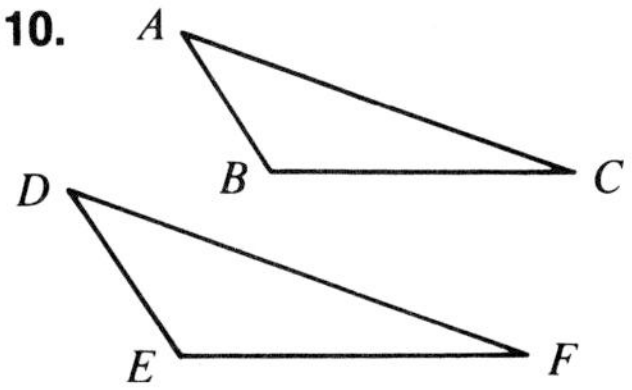

$\triangle ABC \sim$ __?__

$\frac{AB}{DE} = \frac{BC}{?} = \frac{CA}{?}$

11.

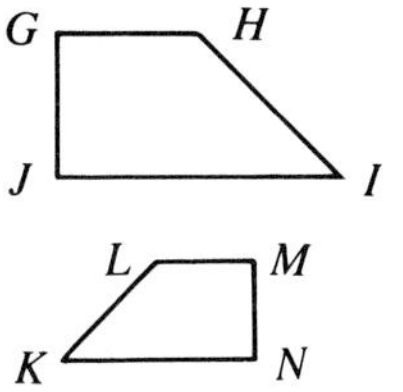

trapezoid $GHIJ \sim$ __?__

$\frac{GH}{ML} = \frac{HI}{?} = \frac{IJ}{?}$

12.

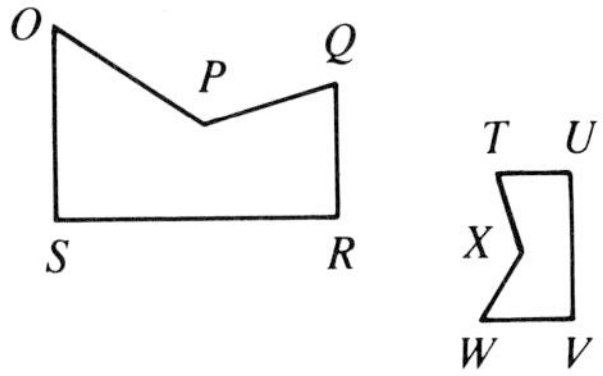

pentagon $OPQRS \sim$ __?__

$\angle Q = \angle$ __?__; $\frac{RS}{?} = \frac{SO}{?}$

13. Are rectangles 1 and 2 similar?

14. What is the scale factor of rectangle 1 to rectangle 2?

15. a. $\frac{\text{perimeter of rectangle 1}}{\text{perimeter of rectangle 2}} = \frac{2+4+2+4}{?+?+?+?} = \frac{12}{?}$

b. Is the ratio of the perimeters equal to the scale factor?

16. a. $\frac{\text{area of rectangle 1}}{\text{area of rectangle 2}} = \frac{8}{?}$

b. Is the ratio of the areas equal to the scale factor?

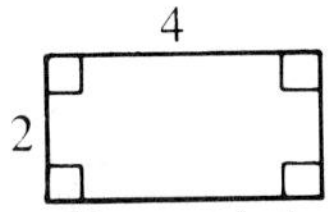

Rectangle 1

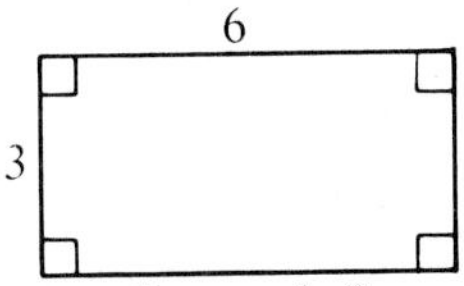

Rectangle 2

In Exercises 17–19, the two polygons shown are similar. Find the values of *x*, *y*, and *z*.

B **17.**

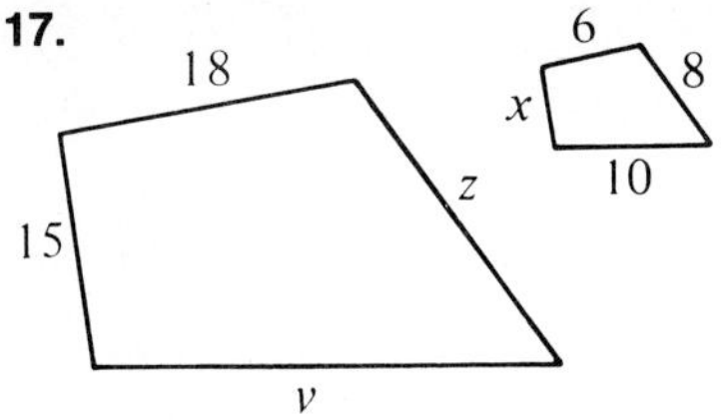

18.

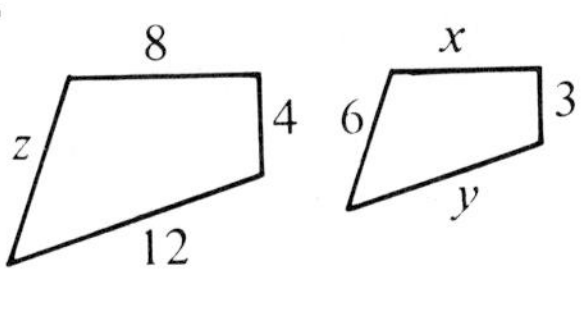

19.

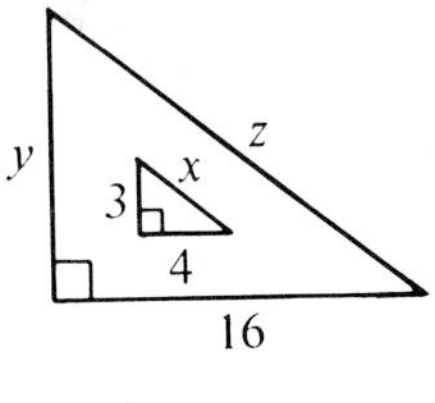

20. The length and width of a rectangle are 18 and 12. A similar rectangle has length 27. What is its width?

21. The sides of a triangle are 6, 8, and 9. The shortest side of a similar triangle is 15. How long are its other sides?

22. a. Plot points $A(-4, 8)$, $B(-4, 2)$, and $C(5, 2)$ on graph paper. Draw $\triangle ABC$.

b. Plot points $D(5, -3)$ and $E(5, -5)$ on the same graph.

c. Locate a point F so that $\triangle ABC \sim \triangle DEF$.

23. Repeat Exercise 22, using points $A(4, 10)$, $B(4, 4)$, $C(0, 4)$, $D(6, 1)$, and $E(6, 13)$.

C 24. The length and width of a rectangle are represented by x and y $(x \neq y)$. The length and width of another rectangle are represented by $x + 2$ and $y + 2$. Are the rectangles similar?

25. The length and width of a rectangle are represented by x and y $(x \neq y)$. The length and width of another rectangle are represented by $2x$ and $2y$. Are the rectangles similar?

26. When a card 8 cm long is cut in half, each piece has the same shape as the original card. Find the *exact* value of x. (The answer involves a square root.)

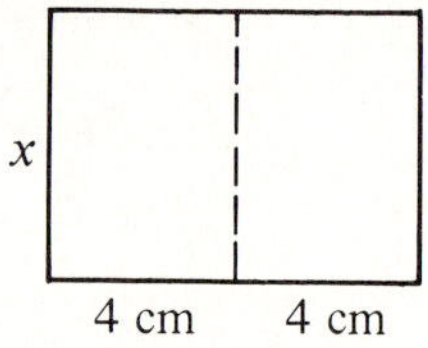

CAREER NOTES

ACTUARY

Did you know that in a recent year, a residential fire occurred in the United States about once every 49 seconds? Or that there were 2379 structural fires per day? This means that there were 868,335 structural fires per year, 643,592 (or about 74%) of which were residential fires!

Actuaries use information like that given above to establish the policy rates charged by insurance companies. First they collect all kinds of information about the circumstances in which certain types of losses occur. Then they organize this data into tables and charts.

Graphs often help actuaries to show the relationship between different factors. For instance, people who smoke often die younger than nonsmokers. On the other hand, those who exercise regularly often live longer than those who don't. To find out more about life insurance, read the Consumer Application on page 349.

Often actuaries use computers to analyze data on the frequency of actual losses. This helps them to estimate the probability of other losses that might occur in the future. Policy-rate scales are based on such probabilities.

2 The AA Postulate

You can prove that two polygons are similar by using the definition of *similar*. That is, by showing that

(1) corresponding angles are equal, and
(2) corresponding sides are in proportion.

There is, however, a more direct method for proving two *triangles* similar. The exploration below will help you discover this method.

EXPLORATIONS

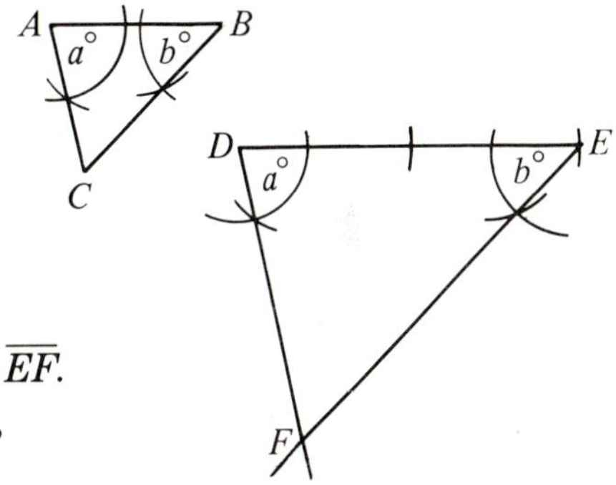

1. Draw any scalene $\triangle ABC$.
2. Construct $\overline{DE}$ with length $2(AB)$.
3. Copy $\angle A$ at D. Copy $\angle B$ at E. Let F be the point shown.
4. Measure $\overline{AC}$ and $\overline{DF}$. Then measure $\overline{BC}$ and $\overline{EF}$.
   Remember that $\frac{AB}{DE} = \frac{1}{2}$. Is $\frac{AC}{DF} \approx \frac{1}{2}$? Is $\frac{BC}{EF} \approx \frac{1}{2}$?
   Do you think that corresponding sides are in proportion?
5. Remember that $\angle A = \angle D$ and $\angle B = \angle E$. Is $\angle C = \angle F$? Do you think that corresponding angles are equal?
6. Does it seem that $\triangle ABC \sim \triangle DEF$?

If you repeat the exploration using different-looking triangles, you will obtain the same result.

POSTULATE 13 (The AA Postulate)

If two angles of one triangle are equal to two angles of another triangle, then the triangles are similar.

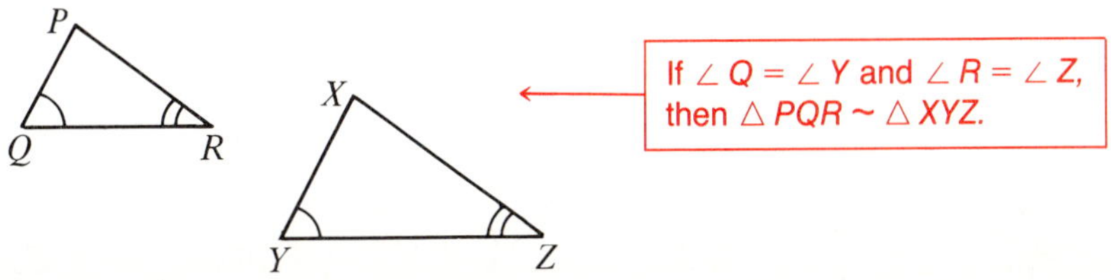

EXAMPLE

Given: $\angle B = \angle E$

Prove: $\triangle ABC \sim \triangle DEC$

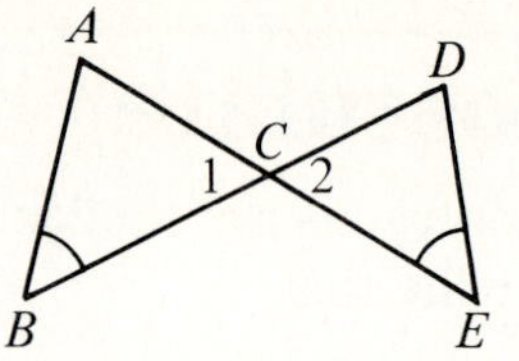

STATEMENTS	REASONS
1. $\angle B = \angle E$	1. Given
2. $\angle 1 = \angle 2$	2. Vertical angles are equal.
3. $\triangle ABC \sim \triangle DEC$	3. AA Postulate

Suppose, in the proof above, we were asked to prove $\frac{AB}{DE} = \frac{BC}{EC}$.
We would need just one additional step:

4. $\frac{AB}{DE} = \frac{BC}{EC}$	4. Corresponding sides of ~ $\triangle$s are in proportion.

The definition of similar triangles may be used in a proof in much the same way that the definition of congruent triangles was used.

The definition of ***congruent triangles*** *tells us:*

1. *Corresponding angles of congruent triangles are equal.*
2. *Corresponding sides of congruent triangles are equal.*

The definition of ***similar triangles*** *tells us:*

1. *Corresponding angles of similar triangles are equal.*
2. *Corresponding sides of similar triangles are in proportion.*

Classroom Practice

In each exercise, state whether the two triangles are similar.

1.

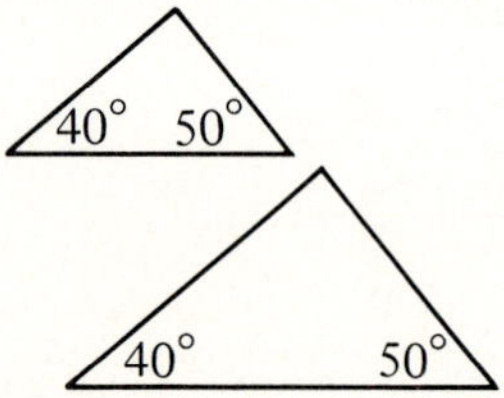

2.

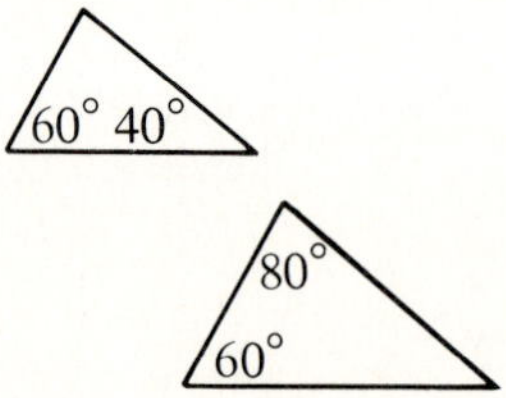

3.

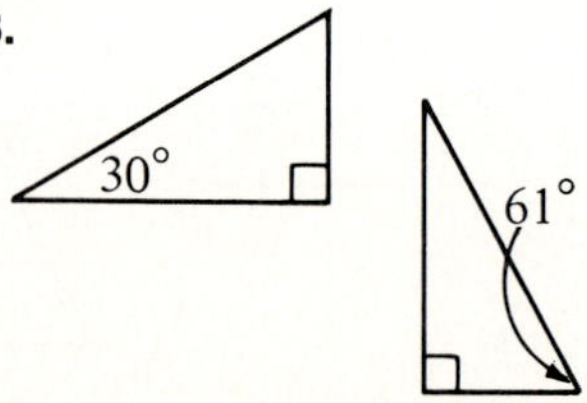

4. 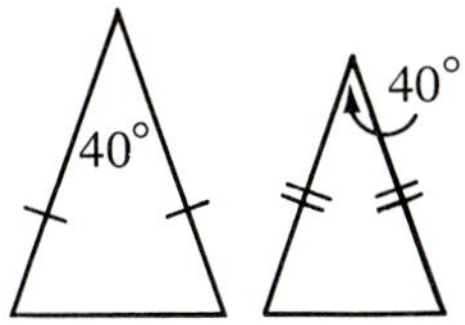

5. 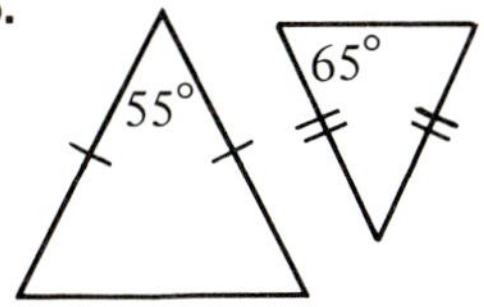

6. 

7. Exercise 4 suggests: If the vertex angles of two isosceles $\triangle$s $\underline{\quad ? \quad}$.

Complete each proportion.

8. $\triangle ABC \sim \triangle DEF$
$\frac{AB}{?} = \frac{BC}{?} = \frac{AC}{?}$

9. $\triangle RST \sim \triangle XYZ$
$\frac{XY}{?} = \frac{XZ}{?} = \frac{YZ}{?}$

10. $\triangle KLM \sim \triangle RST$
$\frac{?}{KL} = \frac{?}{LM} = \frac{?}{KM}$

11. Complete the proof.

Given: $\angle 1 = \angle 2$

Prove: $\frac{AB}{AD} = \frac{AC}{AE}$

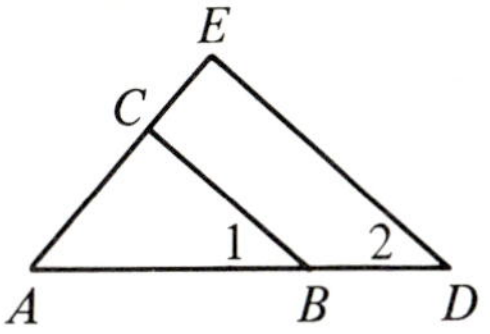

STATEMENTS	REASONS
1. $\angle 1 = \angle 2$	1. $\underline{\quad ? \quad}$
2. $\angle A = \angle A$	2. $\underline{\quad ? \quad}$
3. $\triangle ABC \sim \triangle ADE$	3. $\underline{\quad ? \quad}$
4. $\frac{AB}{AD} = \frac{AC}{AE}$	4. $\underline{\quad ? \quad}$

Use the figure from Exercise 11. Assume that $\triangle ABC \sim \triangle ADE$.

12. If $AB = 8$, $AD = 12$, and $BC = 6$, find DE.

13. If $AB = 15$, $AC = 9$, and $AE = 15$, find AD.

Written Exercises

In each exercise, state whether the two triangles are similar.

A 1. 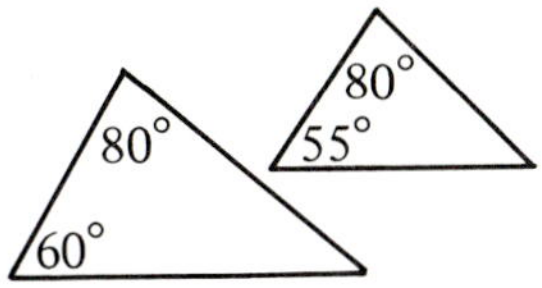

2. 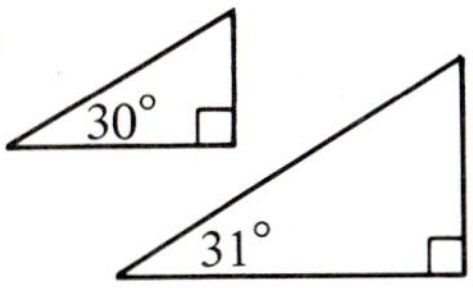

3. 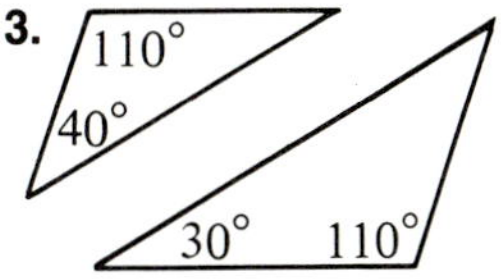

4. 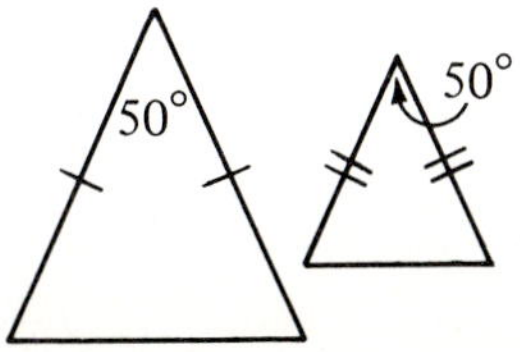

5. 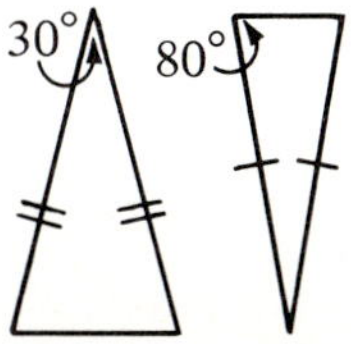

6. 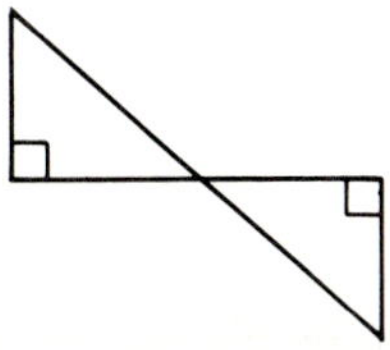

State whether the two triangles are similar.

7. 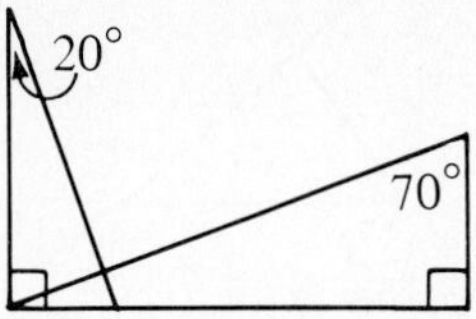

8. 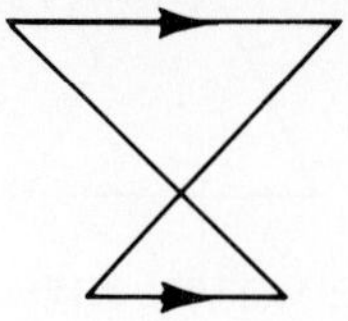

9. 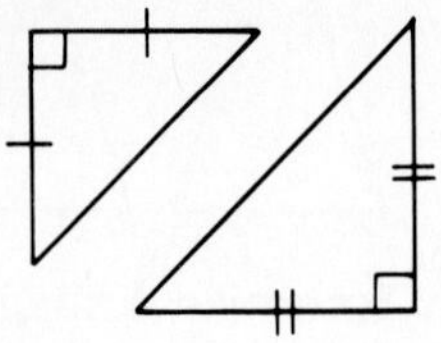

Complete each statement.

10. 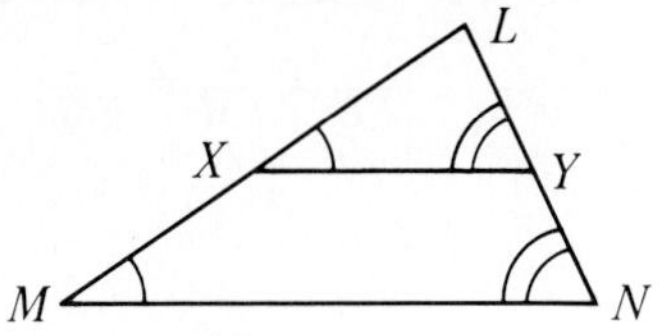

a. $\triangle LMN \sim \triangle$ ___?___

b. $\frac{LX}{LM} = \frac{LY}{?} = \frac{XY}{?}$

11. 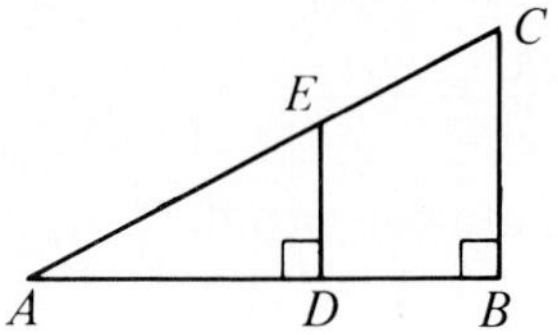

a. $\triangle ABC \sim \triangle$ ___?___

b. $\frac{AB}{AD} = \frac{BC}{?} = \frac{CA}{?}$

12. 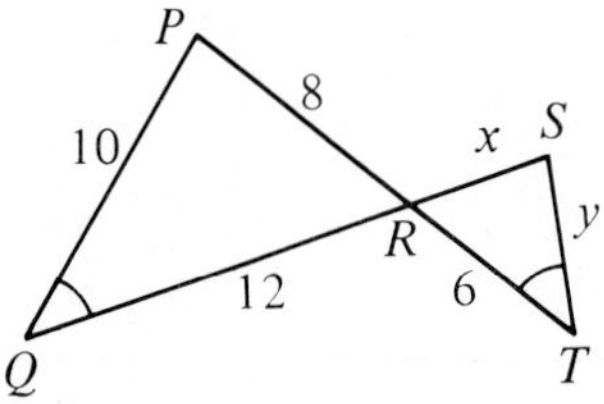

a. $\triangle PQR \sim \triangle$ ___?___

b. $\frac{12}{6} = \frac{?}{x}$ and $\frac{12}{6} = \frac{?}{y}$

c. $x =$ ___?___ and $y =$ ___?___

13. 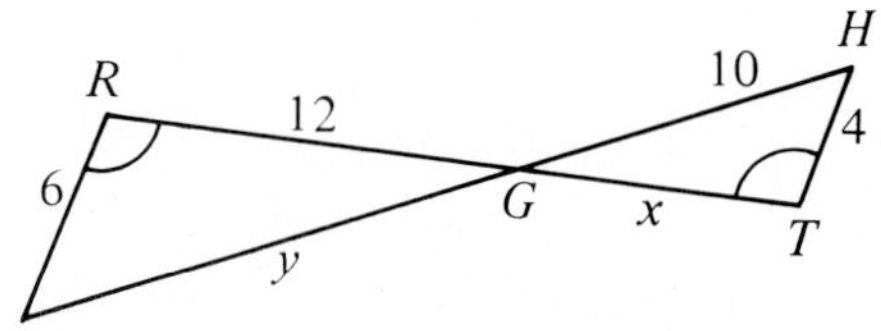

a. $\triangle RIG \sim \triangle$ ___?___

b. $\frac{4}{6} = \frac{x}{?}$ and $\frac{4}{6} = \frac{?}{y}$

c. $x =$ ___?___ and $y =$ ___?___

14. 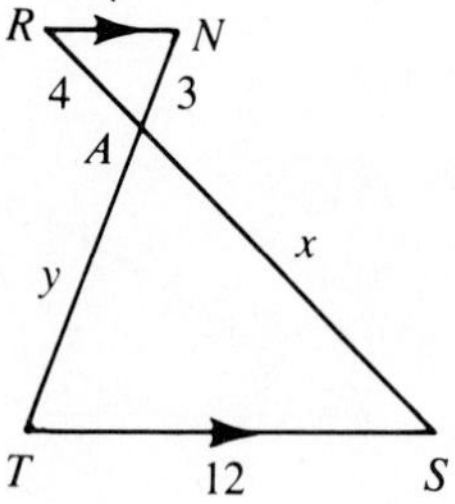

a. $\triangle SAT \sim \triangle$ ___?___

b. $x =$ ___?___

c. $y =$ ___?___

15. 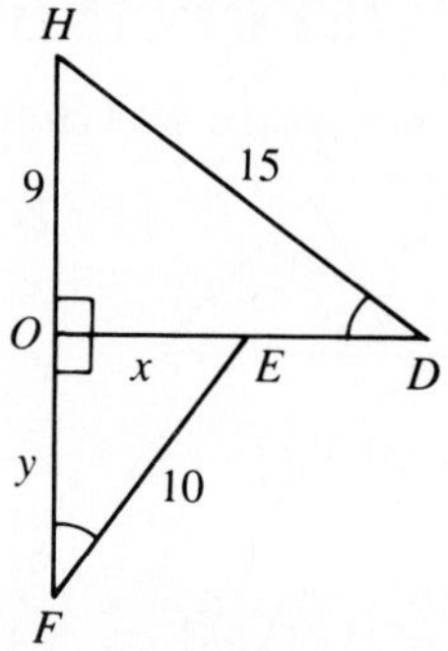

a. $\triangle FOE \sim \triangle$ ___?___

b. $x =$ ___?___

c. $y =$ ___?___

Supply the reasons to complete each proof.

B **16.**

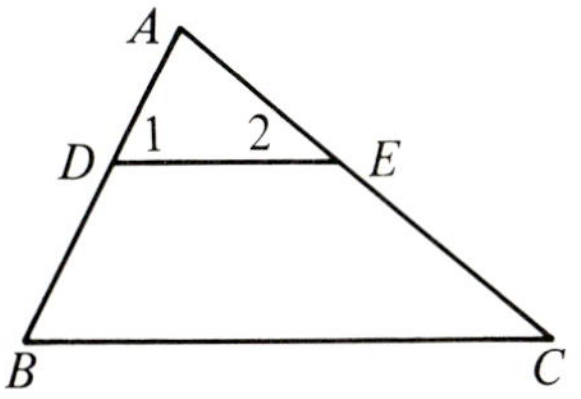

Given: $\overline{DE} \parallel \overline{BC}$

Prove: $\frac{AD}{AB} = \frac{DE}{BC}$

1. $\overline{DE} \parallel \overline{BC}$
2. $\angle B = \angle 1$
3. $\angle C = \angle 2$
4. $\triangle ADE \sim \triangle ABC$
5. $\frac{AD}{AB} = \frac{DE}{BC}$

17.

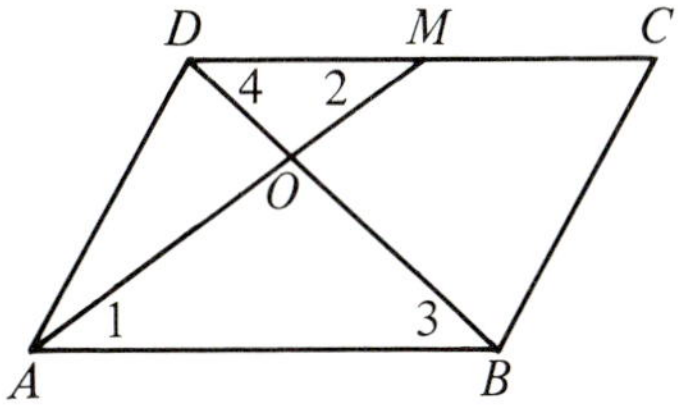

Given: $ABCD$ is a $\square$.

Prove: $\frac{AO}{MO} = \frac{AB}{MD}$

1. $ABCD$ is a $\square$.
2. $\overline{DM} \parallel \overline{AB}$
3. $\angle 1 = \angle 2$; $\angle 3 = \angle 4$
4. $\triangle AOB \sim \triangle MOD$
5. $\frac{AO}{MO} = \frac{AB}{MD}$

In the figure shown, *ABCD* is a parallelogram.

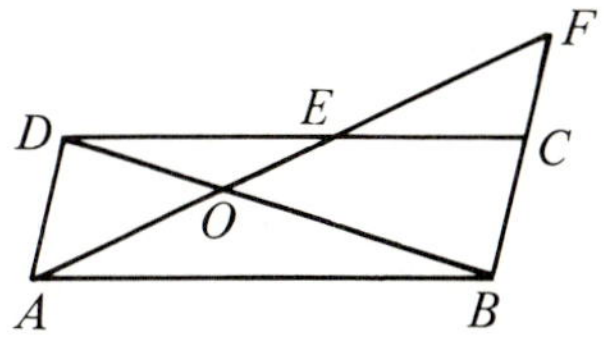

18. $\triangle ABO \sim \triangle$ __?__ **19.** $\triangle DOA \sim \triangle$ __?__

20. $\triangle AFB \sim \triangle$ __?__; also $\triangle AFB \sim \triangle$ __?__

21. $\triangle AED \sim \triangle$ __?__; also $\triangle AED \sim \triangle$ __?__

22. *Given:* $\angle X = \angle Y$

Prove: $\frac{XS}{YS} = \frac{ST}{SR}$

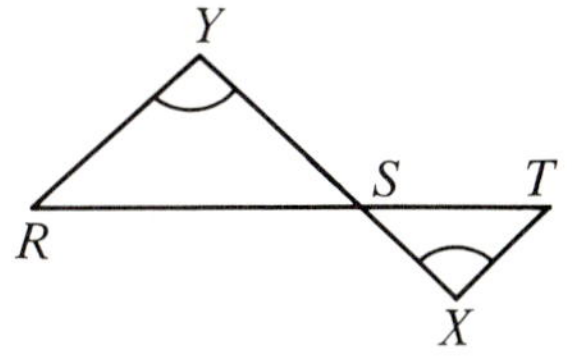

C **23.** *Given:* $\overline{EA} \parallel \overline{PL}$

Prove: $\frac{SE}{SA} = \frac{SP}{SL}$

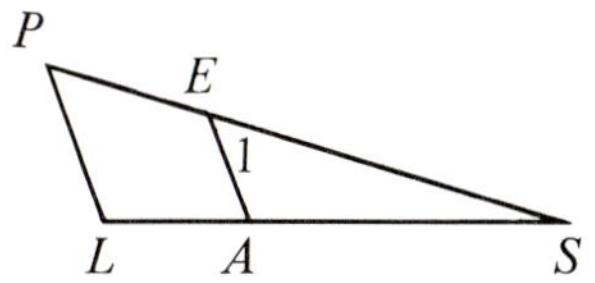

24. Prove the following: If two similar triangles have scale factor *k*, then their corresponding altitudes have scale factor *k*.

Given: $\triangle ABC \sim \triangle DEF$; $\frac{AC}{DF} = k$

$\overline{CX}$ and $\overline{FY}$ are corresponding altitudes.

Prove: $\frac{CX}{FY} = k$

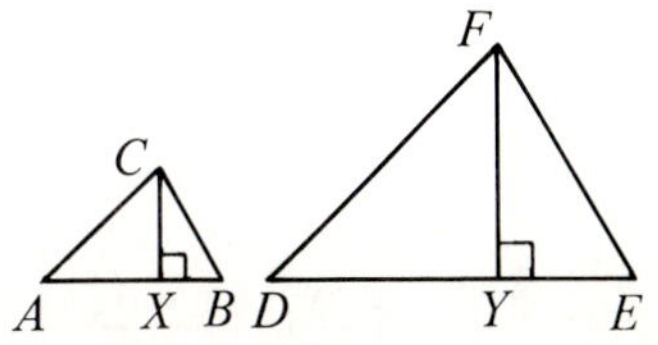

SELF-TEST

The two polygons given are similar. Complete each statement.

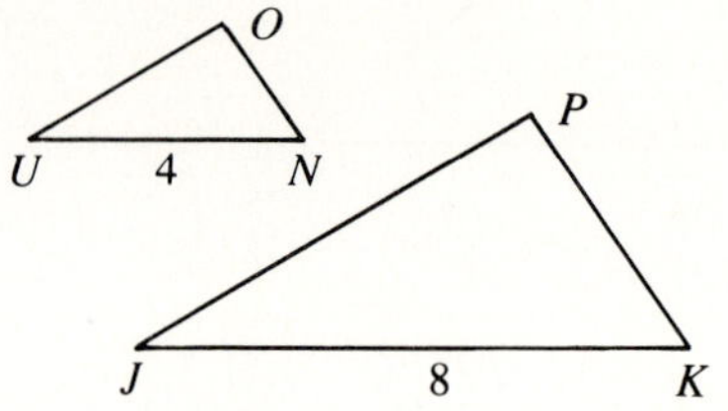

1. $\triangle UNO \sim \triangle$ __?__ *(8-1)*

2. $\frac{UN}{JK} = \frac{NO}{?} = \frac{OU}{?}$

3. The scale factor is __?__ : __?__.

4. $\angle U = \angle$ __?__

State whether the two polygons are, or are not, similar. *(8-2)*

5.

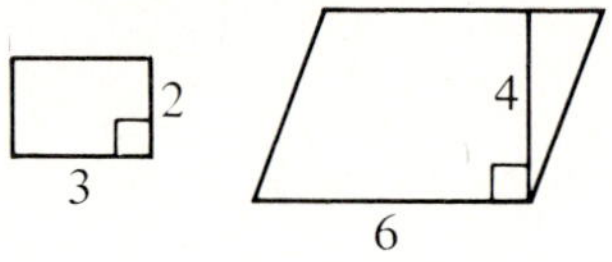

6.

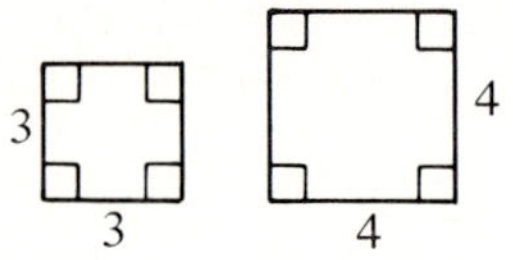

7.

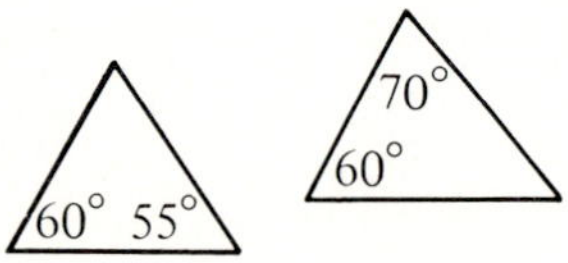

8.

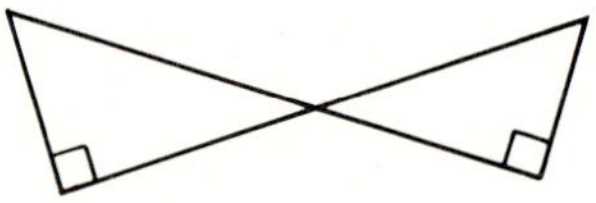

CALCULATOR ACTIVITIES

Can you see that any two equilateral triangles are similar polygons?

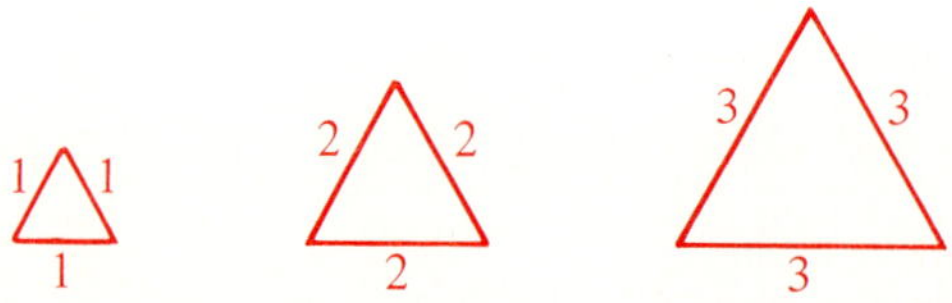

The area of an equilateral triangle is given by the formula $A = \frac{s^2\sqrt{3}}{4}$, where s is the length of a side.

Use a calculator to compute the perimeter and area of each triangle described. ($\sqrt{3} \approx 1.732$) Round each area to the nearest tenth.

	s	1	2	3	4	5	6	7
1.	perimeter	3	?	?	?	?	?	?
2.	area	0.4	?	?	?	?	?	?

3 A Special Case of Similar Triangles

In the figure at the right, $\overline{DE} \parallel \overline{BC}$. Do you see that $\triangle ADE \sim \triangle ABC$? (For a proof, see Exercise 16 on page 321.) Since corresponding sides of these triangles are in proportion, we can find the values of x and y.

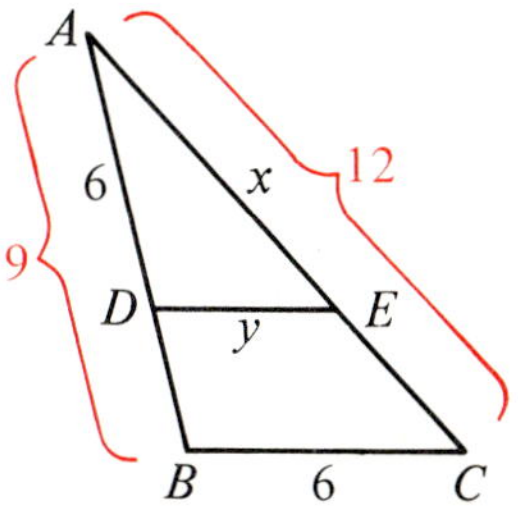

△ ADE ~ △ ABC

$\frac{AD}{AB} = \frac{AE}{AC}$	$\frac{AD}{AB} = \frac{DE}{BC}$
$\frac{6}{9} = \frac{x}{12}$	$\frac{6}{9} = \frac{y}{6}$
$72 = 9x$	$36 = 9y$
$8 = x$	$4 = y$

Classroom Practice

Complete each statement. Then find the values of x and y.

1.

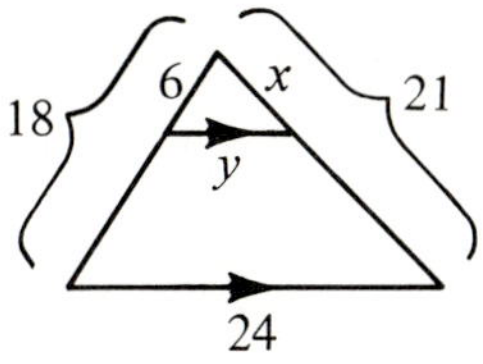

$$\frac{6}{18} = \frac{x}{?} = \frac{y}{?}$$

2.

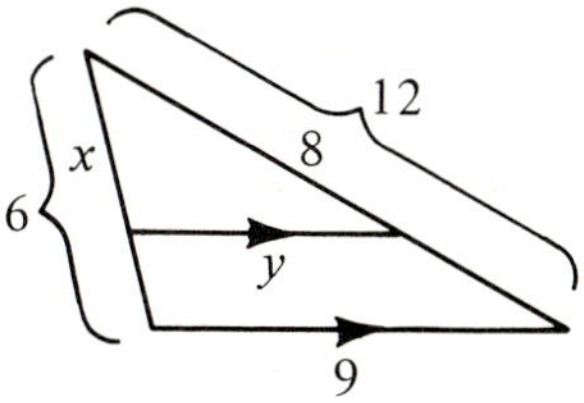

$$\frac{x}{6} = \frac{8}{?} = \frac{y}{?}$$

3.

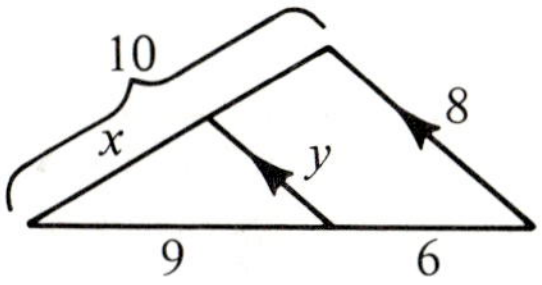

$$\frac{10}{x} = \frac{15}{?} = \frac{?}{y}$$

Written Exercises

Complete each statement. Then find the values of x and y.

A **1.**

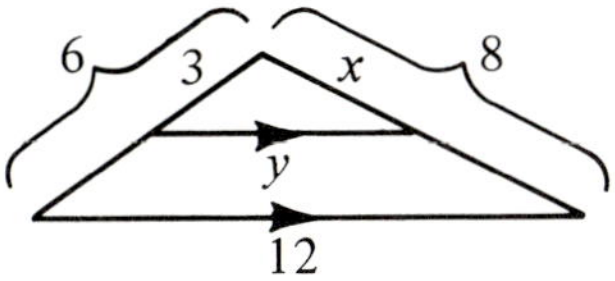

$$\frac{3}{6} = \frac{x}{?} = \frac{y}{?}$$

2.

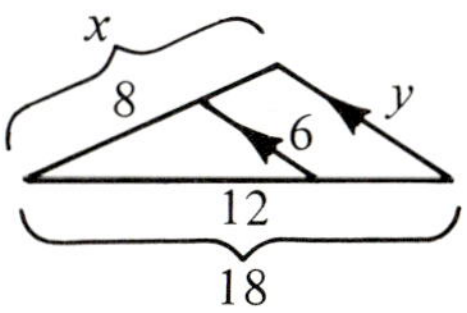

$$\frac{18}{12} = \frac{x}{?} = \frac{y}{?}$$

3.

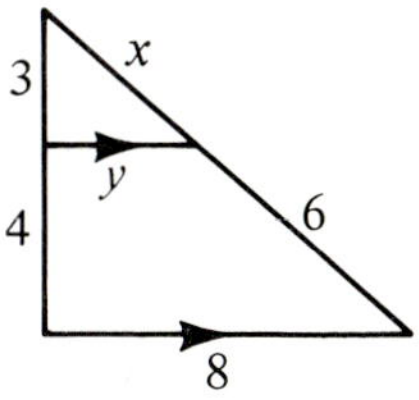

$$\frac{x}{x+6} = \frac{3}{?} = \frac{y}{?}$$

Find the values of x and y.

4.

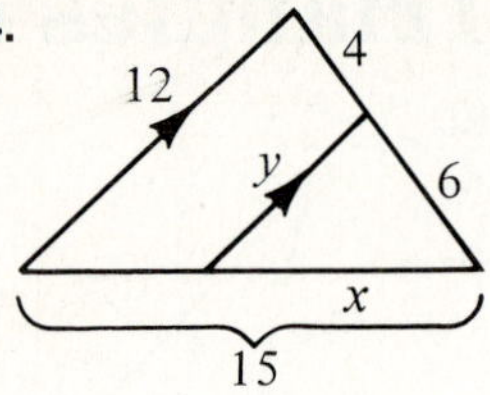

5.

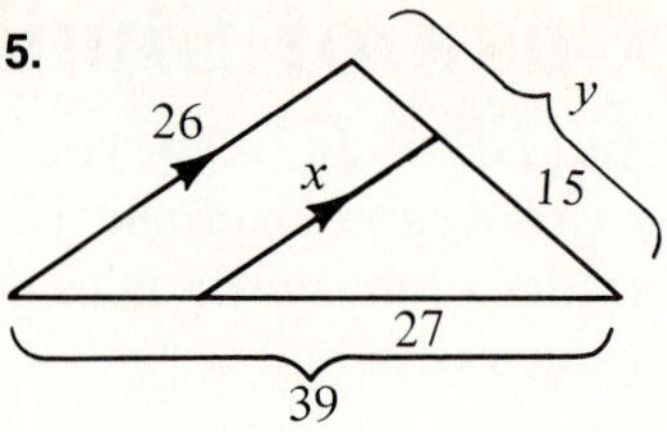

6.

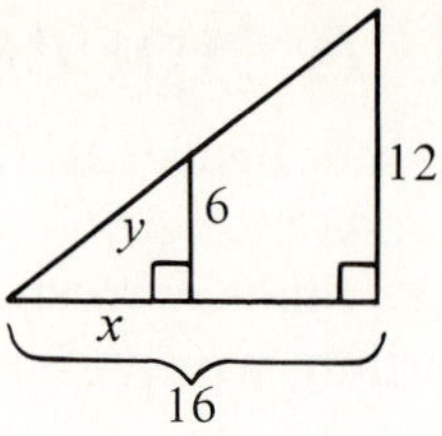

7.

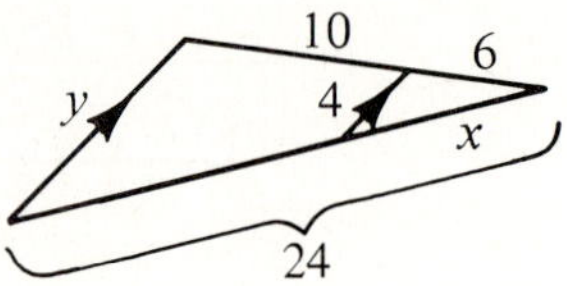

8.

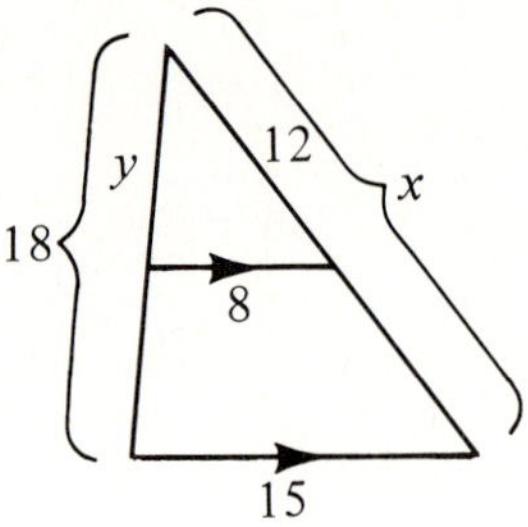

9.

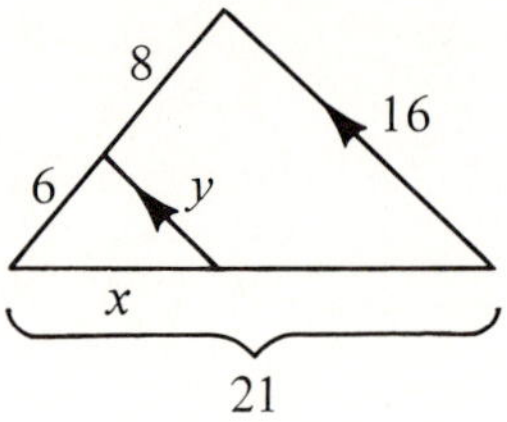

Find the value of x.

B **10.**

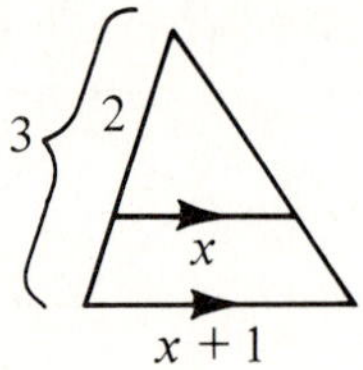

11.

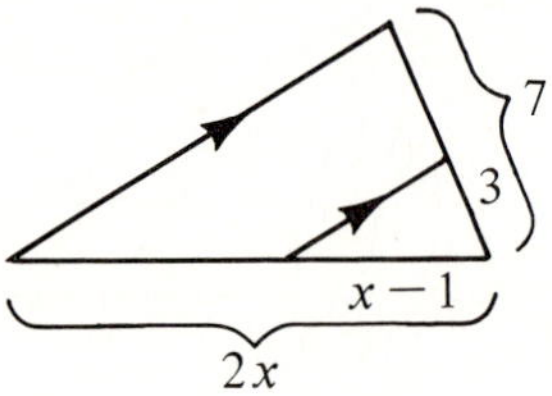

12.

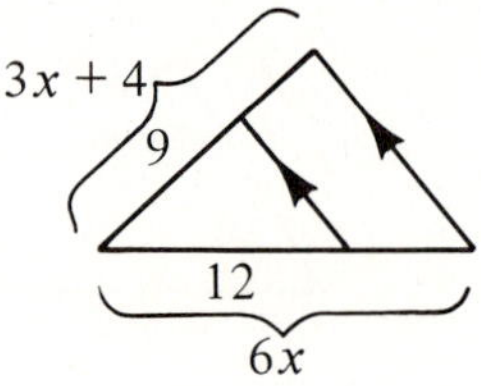

PUZZLE ◆ PROBLEMS

Try to draw each figure below without lifting your pencil from the paper and without retracing any line. It is impossible to do this for one of the figures shown. Which one?

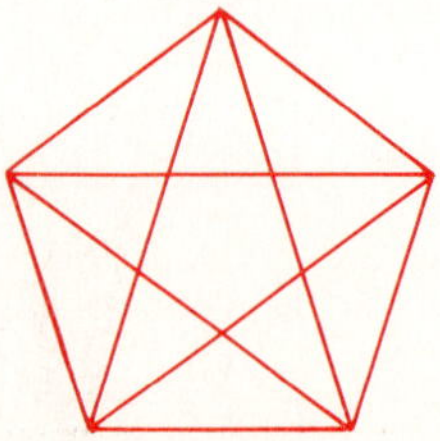

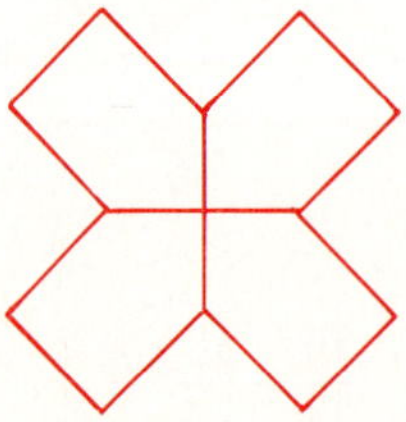

4 The Triangle Proportionality Theorem

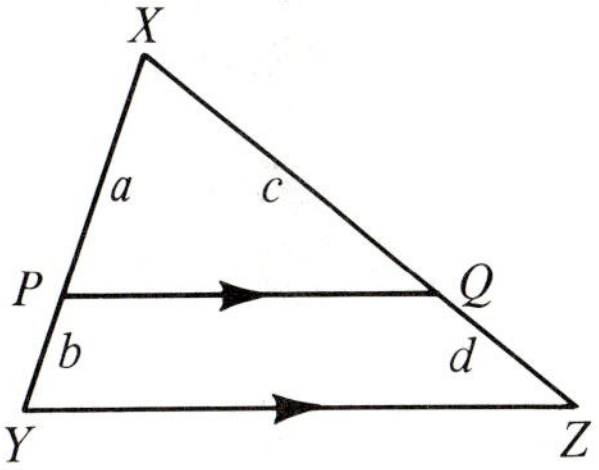

In the figure at the right, $\overline{PQ} \parallel \overline{YZ}$. We have seen that this means that $\triangle XYZ \sim \triangle XPQ$. Since corresponding sides of these similar triangles are in proportion,

$$\frac{XY}{XP} = \frac{XZ}{XQ}$$

or $$\frac{a+b}{a} = \frac{c+d}{c}$$

$$\frac{b}{a} = \frac{d}{c}$$

$$\frac{a}{b} = \frac{c}{d}$$

Use properties of proportions.

The discussion above proves the following useful theorem.

THEOREM 1 (The Triangle Proportionality Theorem)

If a line intersects a triangle and is parallel to one side, then it divides the other two sides proportionally.

If 

, then $\frac{a}{b} = \frac{c}{d}$.

EXAMPLE

Find the values of x and y.

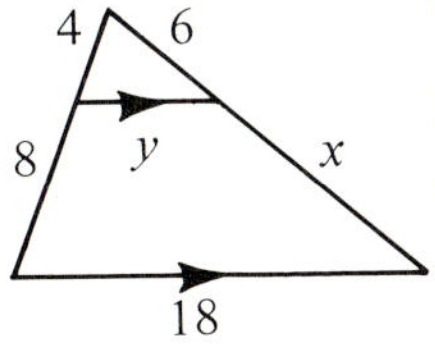

Caution! Notice that $\frac{4}{8} = \frac{6}{x}$, but $\frac{4}{8} \neq \frac{y}{18}$. (Why?)

Use the Triangle Proportionality Theorem to find x and use corresponding sides of similar triangles to find y.

$$\frac{4}{8} = \frac{6}{x} \qquad \frac{4}{4+8} = \frac{y}{18}$$

$$4x = 48 \qquad 72 = 12y$$

$$x = 12 \qquad 6 = y$$

Using properties of proportions, it can be shown that the following proportions are true when $\overline{PQ} \parallel \overline{YZ}$:

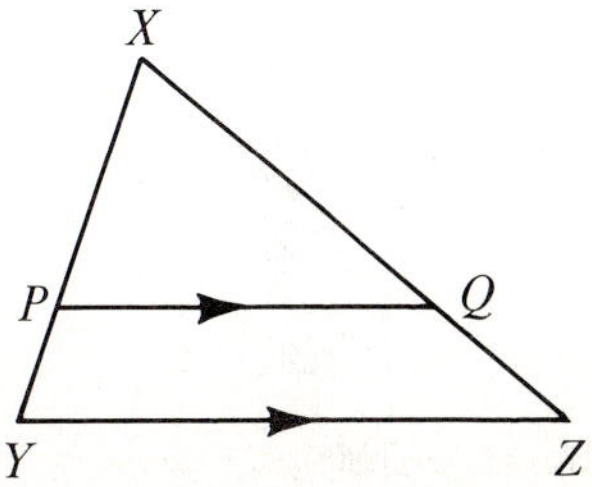

$$\frac{XP}{XY} = \frac{XQ}{XZ} \qquad \frac{XP}{PY} = \frac{XQ}{QZ} \qquad \frac{PY}{XY} = \frac{QZ}{XZ}$$

We shall agree that Theorem 1 may be used to justify each of these proportions.

Theorem 1 has a corollary. For a proof, see Exercise 19.

COROLLARY

If three parallel lines intersect two transversals, they divide the transversals proportionally.

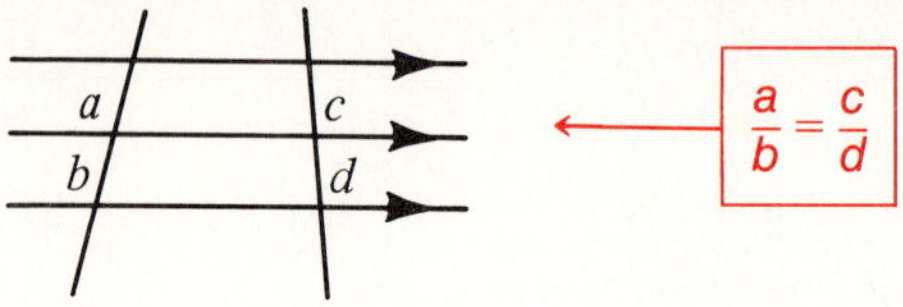

Classroom Practice

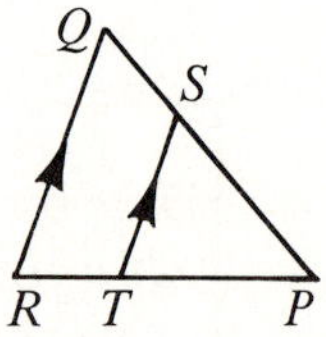

Complete each statement.

1. $\frac{PS}{SQ} = \frac{PT}{?}$
2. $\frac{PQ}{PS} = \frac{PR}{?}$
3. $\frac{ST}{QR} = \frac{PS}{?}$
4. $\frac{SQ}{PQ} = \frac{TR}{?}$

Use the figure below to find the value of each of the following.

5. $\frac{BD}{BA}$
6. $\frac{BE}{EC}$
7. $\frac{BE}{BC}$
8. $\frac{EC}{BC}$
9. $\frac{DE}{AC}$
10. $\frac{BA}{DA}$

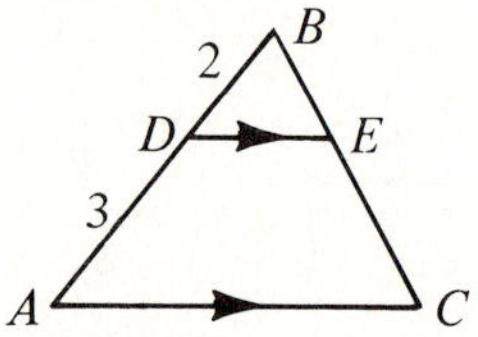

Complete each statement. Then find the values of x and y.

11. $\frac{x}{4} = \frac{9}{?}$ and $\frac{y}{16} = \frac{9}{?}$

12. $\frac{x}{3} = \frac{9}{?}$ and $\frac{y}{20} = \frac{9}{?}$

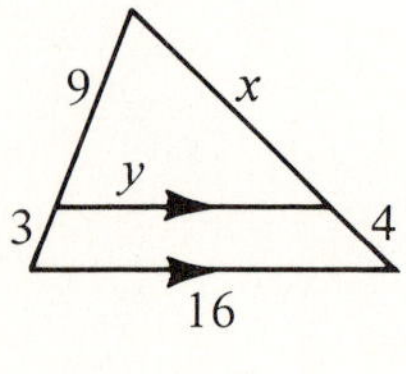

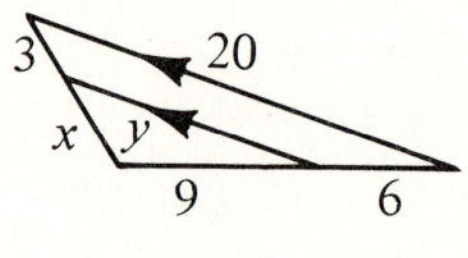

13. If $\frac{AB}{BC} = \frac{5}{7}$, then $\frac{DE}{EF} = \underline{\quad ? \quad}$.

14. If $\frac{AB}{BC} = \frac{5}{7}$, then $\frac{DF}{DE} = \underline{\quad ? \quad}$.

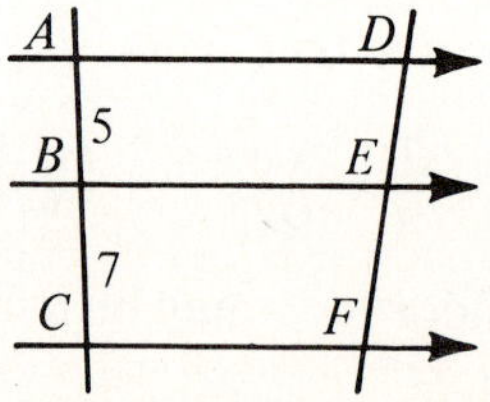

Written Exercises

For each figure below, complete the statement: $\frac{a}{b} = \frac{?}{?}$.

A

1.

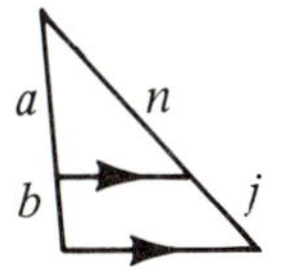

2.

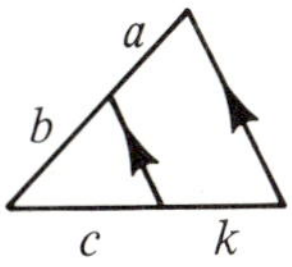

3.

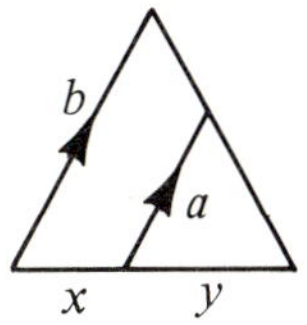

4.

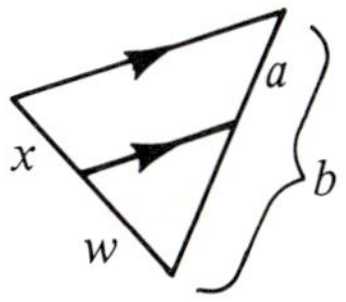

Complete each statement.

5. $\frac{AX}{XB} = \frac{AY}{?}$

6. $\frac{AX}{AB} = \frac{AY}{?}$

7. $\frac{XY}{BC} = \frac{AY}{?}$

8. $\frac{XB}{AB} = \frac{YC}{?}$

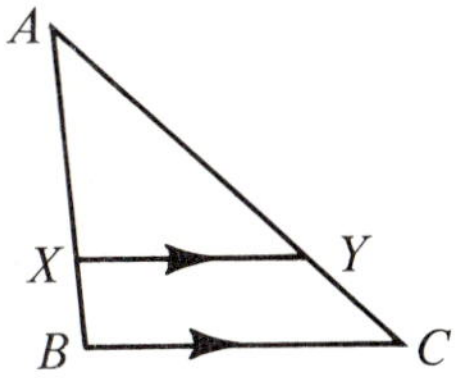

Find the values of *x* and *y*.

9.

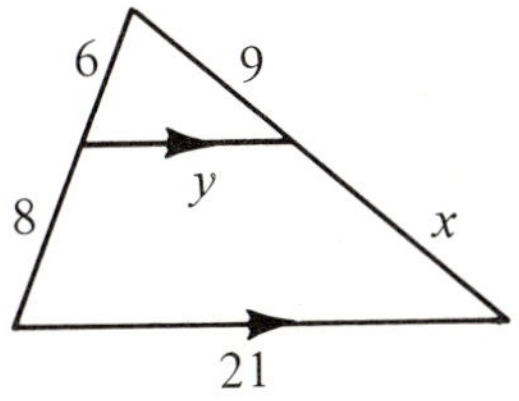

10.

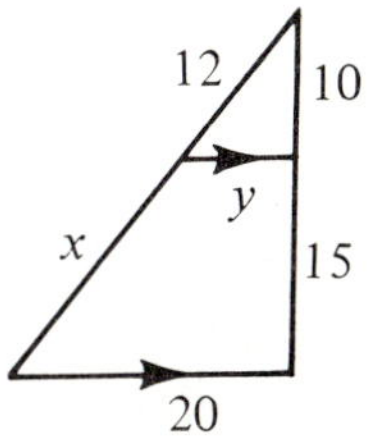

11.

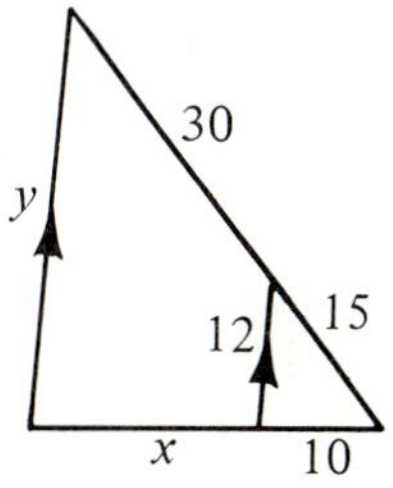

12.

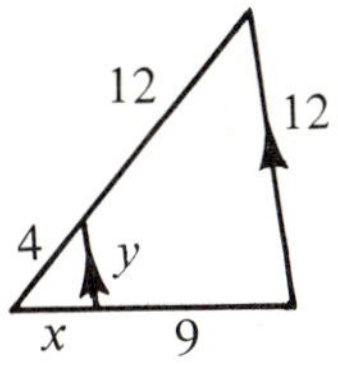

13.

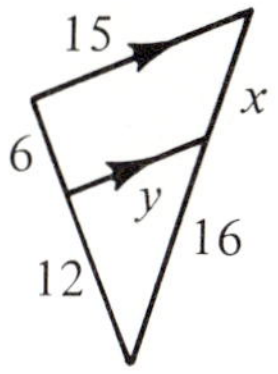

14.

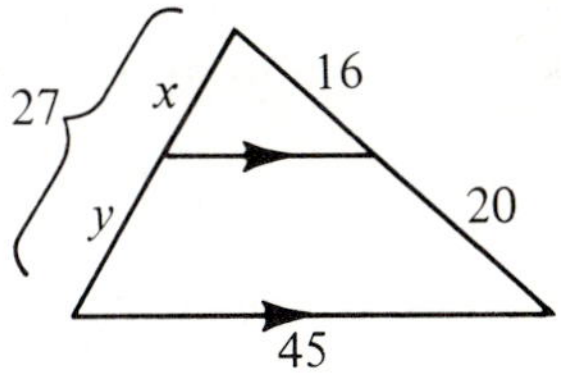

B

15.

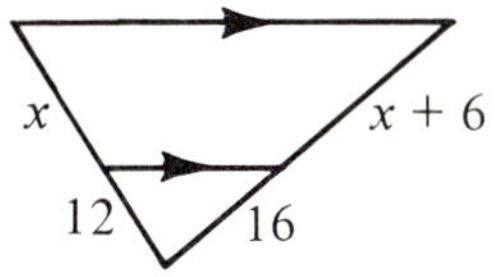

16.

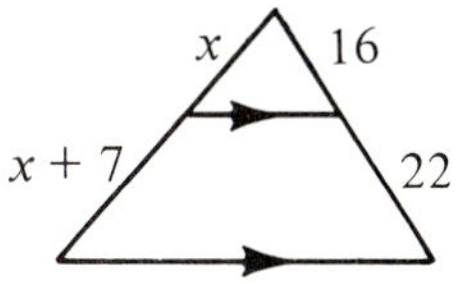

17.

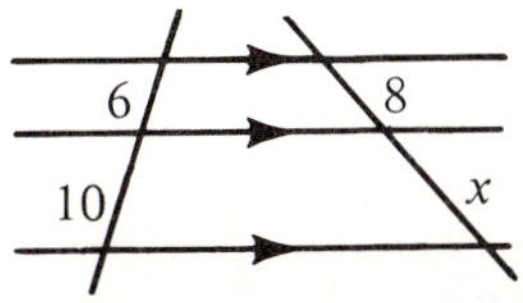

18.

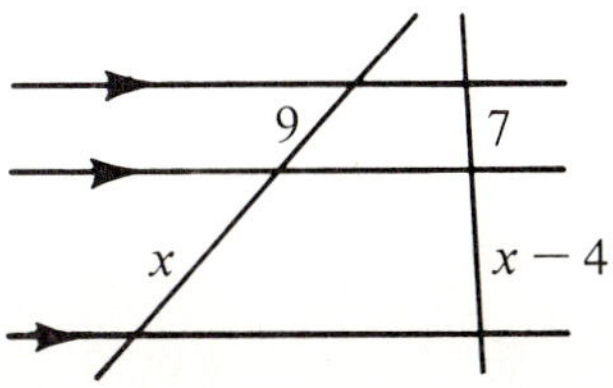

19. Supply the reasons for the following proof of the Corollary to Theorem 1.

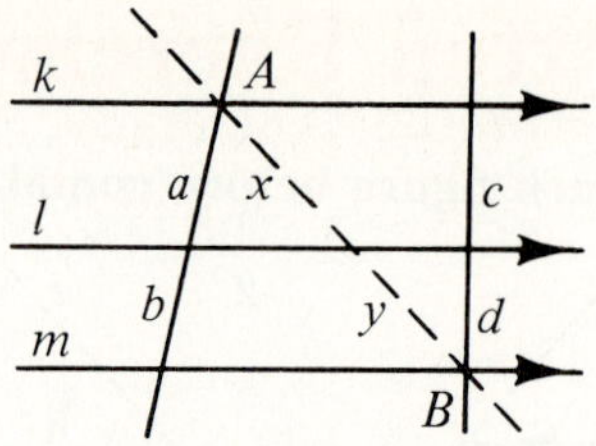

Given: $k \parallel l \parallel m$

Prove: $\frac{a}{b} = \frac{c}{d}$

STATEMENTS	REASONS
1. $k \parallel l \parallel m$	1. __?__
2. Draw $\overleftrightarrow{AB}$.	2. Through any two points there is __?__.
3. $\frac{a}{b} = \frac{x}{y}$; $\frac{x}{y} = \frac{c}{d}$	3. __?__
4. $\frac{a}{b} = \frac{c}{d}$	4. __?__

20. As shown, lots A and B are bounded on either side by Hill Street and River Road. Use the diagram to find x and y, the distance each lot has along River Road.

21. Which lot has the greater area? How much greater?

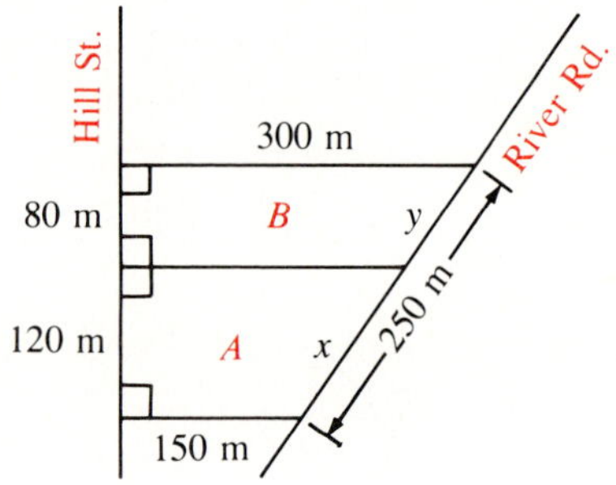

In Exercises 22 and 23, use the statement: The bisector of an angle of a triangle divides the opposite side in the same ratio as the other two sides of the triangle.

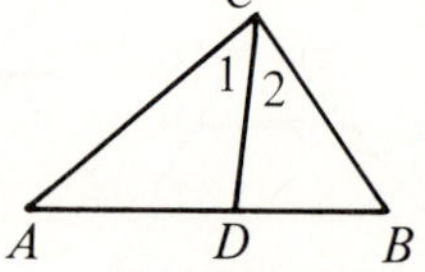

If $\angle 1 = \angle 2$, then $\frac{DA}{DB} = \frac{CA}{CB}$.

Complete each statement. Then find the value of x.

C **22.**

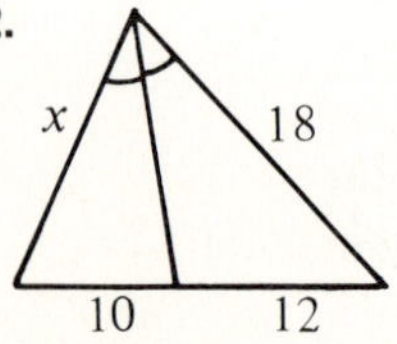

$\frac{10}{12} = \frac{x}{?}$

23.

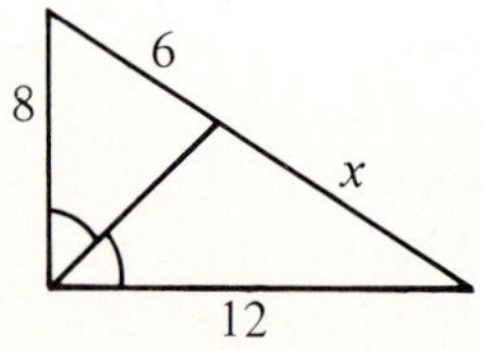

$\frac{x}{6} = \frac{?}{?}$

5 Perimeters and Areas

Have you ever seen a billboard like this one?

The increased view is represented by pictures of two similar windows. The dimensions of the larger window are *twice* the corresponding dimensions of the smaller one. Yet the larger window seems huge in comparison with the smaller one. This is because its area is *four times as great.* Do you see why?

Area of a rectangle = base × height
Area of smaller rectangle = xy
Area of larger rectangle = $(2x)(2y)$
$= 4xy$
= 4(area of smaller rectangle)

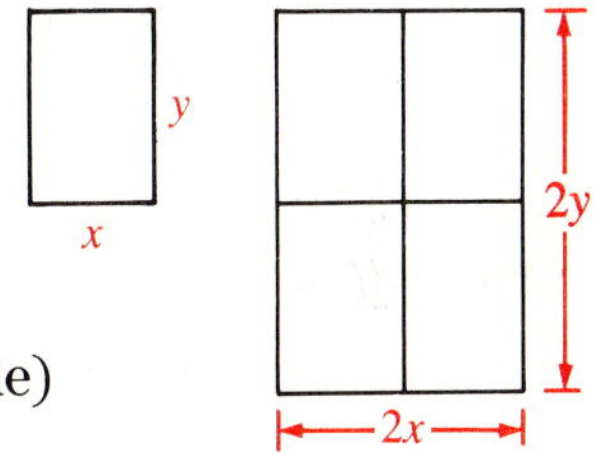

Suppose that the dimensions of the larger rectangle were three times those of the smaller rectangle. Then the larger area would be *nine* times the smaller one:

Area of smaller rectangle = xy
Area of larger rectangle = $(3x)(3y)$
$= 9xy$
= 9(area of smaller rectangle)

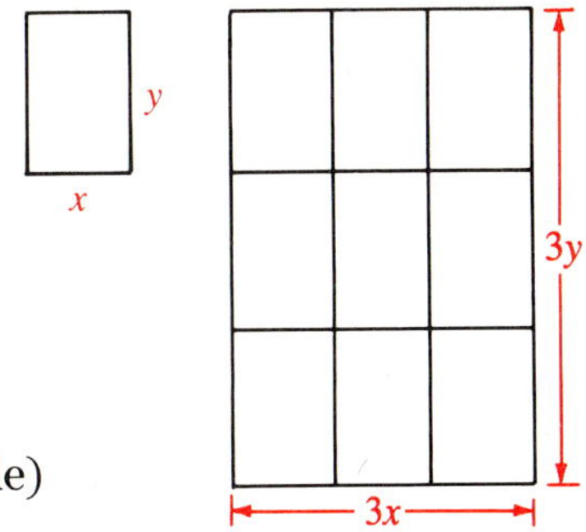

Our work in Section 1 suggested that the perimeters of similar polygons are related in a special way. The discussion above suggests that the areas of similar polygons are also related.

THEOREM 2
If two similar polygons have a scale factor $a:b$, then *(1) the ratio of their perimeters is $a:b$;* *(2) the ratio of their areas is $a^2:b^2$.*

Theorem 2 is true, not only for polygons, but for figures such as circles, which are not polygons.

EXAMPLE

The similar figures below have the scale factor 3:2. Find
a. the ratio of their perimeters;
b. the ratio of their areas.

Using Theorem 2:
a. the ratio of their perimeters is 3:2;
b. the ratio of their areas is $3^2:2^2$, or 9:4.

Classroom Practice

Each exercise of the table refers to two similar figures. Copy and complete the table.

	1.	2.	3.	4.	5.	6.	7.	8.
Scale factor	1:4	1:5	2:5	3:7	?	?	?	?
Ratio of perimeters	?	?	?	?	2:3	3:8	?	?
Ratio of areas	?	?	?	?	?	?	1:4	36:25

The purpose of Exercises 9 and 10 is to prove Theorem 2 for similar triangles. The triangles shown have the scale factor $a:b$.

9. $\dfrac{\text{perimeter of triangle 1}}{\text{perimeter of triangle 2}} = \dfrac{at + ak + an}{bt + bk + bn}$

Show that this ratio is equal to $a:b$.

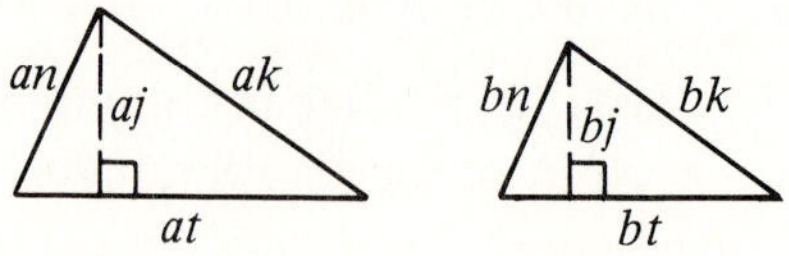

Triangle 1 Triangle 2

10. Recall from Exercise 24, page 321, that corresponding altitudes of similar triangles have the same ratio as corresponding sides. Write an expression for the ratio of the areas, and show that this ratio is equal to $a^2:b^2$.

Written Exercises

Each exercise refers to two similar figures with the given scale factor. Find the ratio of the perimeters and the ratio of the areas.

A

1. 1:2 **2.** 1:3 **3.** 2:3 **4.** 5:4

Each exercise refers to two similar figures. Copy and complete the table.

	5.	6.	7.	8.	9.	10.
Scale factor	?	?	?	?	?	?
Ratio of perimeters	3:5	7:4	?	?	?	?
Ratio of areas	?	?	4:9	81:25	9:49	100:1

By measuring each pair of similar figures, find: a. their scale factor; b. the ratio of their perimeters; c. the ratio of their areas.

11.

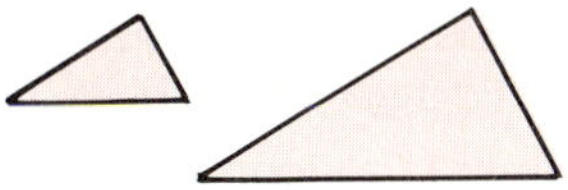

12.

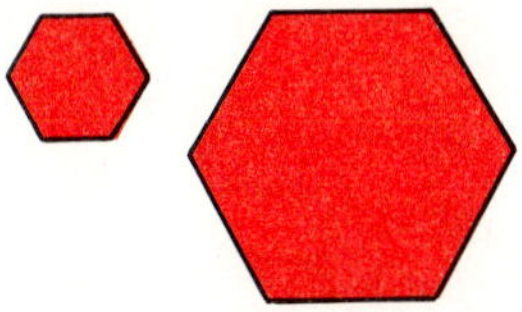

13.

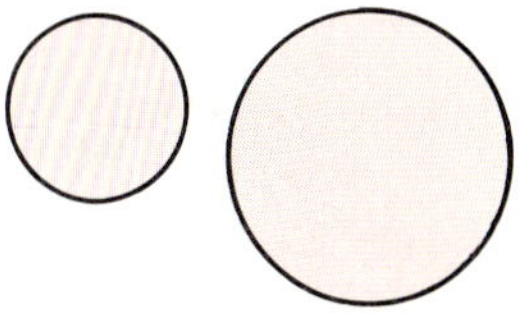

B

14. Refer to the floor plan of a house on page 297. Find the ratio of the floor area in the actual house to the floor area in the drawing.

15. The widths of two similar rectangles are 15 and 25. Find the ratio of their perimeters and the ratio of their areas.

16. The bases of two similar triangles are 18 and 27. Find the ratio of their perimeters and the ratio of their areas.

17. *Given:* $\triangle ABC \sim \triangle DEF$ with scale factor 3:4.
a. If $\triangle ABC$ has perimeter 15, then $\triangle DEF$ has perimeter __?__.
b. If $\triangle ABC$ has area 18, then $\triangle DEF$ has area __?__.

18. *Given:* $\triangle PQR \sim \triangle XYZ$ with scale factor 3:5.
a. If $\triangle PQR$ has perimeter 30, then $\triangle XYZ$ has perimeter __?__.
b. If $\triangle PQR$ has area 36, then $\triangle XYZ$ has area __?__.

19. *Given:* $\triangle LMN$; O is the midpoint of $\overline{LN}$; P is the midpoint of $\overline{MN}$.

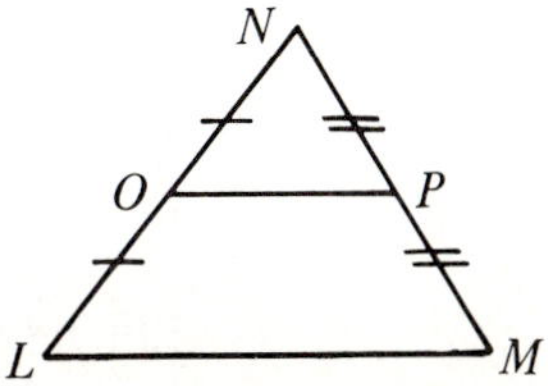

a. $\dfrac{\text{perimeter of } \triangle OPN}{\text{perimeter of } \triangle LMN} =$ __?__ **b.** $\dfrac{\text{area of } \triangle OPN}{\text{area of } \triangle LMN} =$ __?__

20. The ceilings of two rooms are similar rectangles with scale factor 2 : 5. It costs \$9 to paint the ceiling of the smaller room. How much does it cost to paint the ceiling of the larger room?

C 21. The perimeters of two similar hexagons are in the ratio 3 : 4. The sum of the areas of the hexagons is 70 cm^2. Find the area of each hexagon.

22. The areas of two similar pentagons are 45 cm^2 and 80 cm^2. The sum of the perimeters of the pentagons is 35 cm. Find the perimeter of each pentagon.

CALCULATOR ACTIVITIES

A landscape architect designs a park pathway with round stepping stones. The stones come in four different sizes.

Recall the formulas for the circumference and the area of a circle:

$$C = 2\pi r$$
$$C = \pi d$$
$$A = \pi r^2$$

Use a calculator to find the missing measures. ($\pi \approx 3.14$) Round your answers to one decimal place.

		Radius	Diameter	Circumference	Area
1.	Small stones	6.2 cm	?	?	?
2.	Medium stones	?	23 cm	?	?
3.	Large stones	?	?	100 cm	?
4.	Extra large stones	?	?	?	1200 cm^2

Experiments

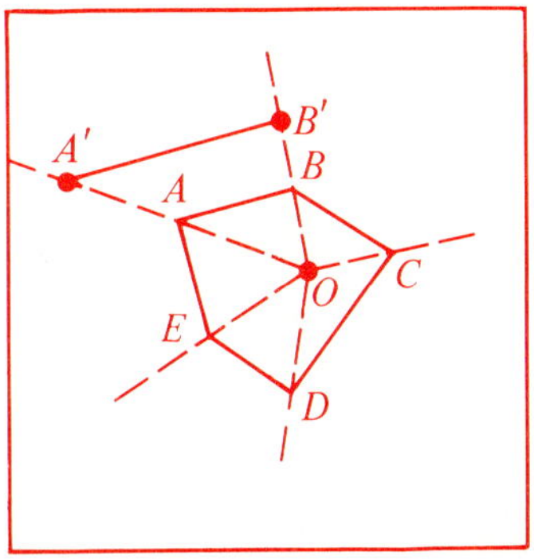

1. Draw a small pentagon *ABCDE* near the center of a large sheet of paper.
2. Choose a point O inside the pentagon. Draw dotted rays from O through each vertex.
3. As shown, locate point A' on $\overrightarrow{OA}$ so that
$$OA' = 2(OA).$$
Locate point B' on $\overrightarrow{OB}$ so that $OB' = 2(OB)$. Locate points C', D', and E' in the same way.
4. Draw pentagon $A'B'C'D'E'$.
5. Using a protractor, compare the measures of corresponding angles of the pentagons.
6. Using a ruler, compare the lengths of corresponding sides of the pentagons.
7. What can you conclude about pentagons *ABCDE* and $A'B'C'D'E'$? Can you prove your conclusion?

SELF-TEST

Find the values of x and y.

1. 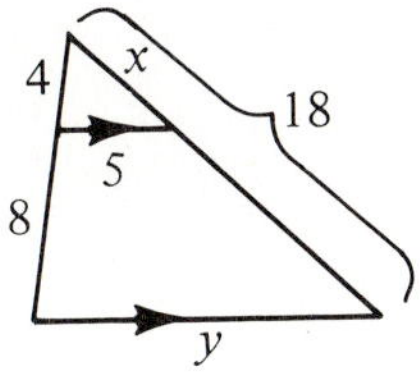

(8-4)

2. 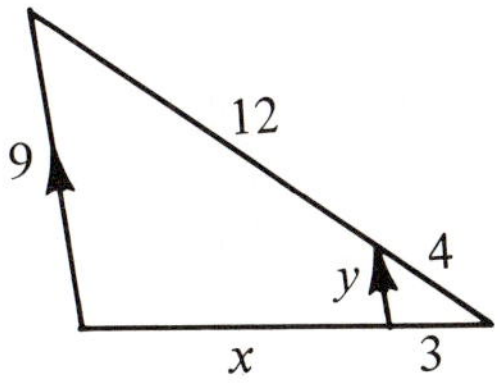

(8-5)

The table refers to two similar figures. Copy and complete the table.

	3.	**4.**	**5.**
Scale factor	1:3	?	?
Ratio of perimeters	?	3:7	?
Ratio of areas	?	?	25:36

APPLICATIONS

INDIRECT PROOF

The method we have been using to prove statements is called *direct proof.* Sometimes a direct proof is very difficult or impossible to use, so we use an indirect method of reasoning instead. We call this method of reasoning *indirect proof.* We often use it in everyday situations.

Suppose you are watching the basketball playoffs on television. The Pivots lead the Chargers by one point in the final seconds of the game. One of the Chargers takes a shot and misses. A teammate tips up the rebound for a basket just as the final buzzer goes off. No fouls are called, but suddenly the television goes blank. Did the final basket count?

Later you learn that the Pivots won the game. You conclude that the final basket did not count. The next paragraph outlines the reasoning used to arrive at this conclusion.

First, assume that the final basket counts. That basket puts the Chargers ahead by one point, and they win the game. This result contradicts the fact that the Pivots won the game. Our assumption must be incorrect. We conclude that the final basket did not count.

To write an indirect proof:

1. Begin by assuming that what you wish to prove is not true.
2. Reason logically until you reach a contradiction of a known fact.
3. Conclude that your assumption is *false* and that what you wish to prove is *true.*

EXAMPLE 1 *Given:* Lines r and s with transversal t, $\angle 1 \neq \angle 2$.

Prove: r is not parallel to s.

Proof: Assume that r is parallel to s. Then, since $\angle 1$ and $\angle 2$ are alternate interior angles, $\angle 1 = \angle 2$. This contradicts the given information, $\angle 1 \neq \angle 2$. Our assumption, r is parallel to s, must be false. We conclude that r is not parallel to s.

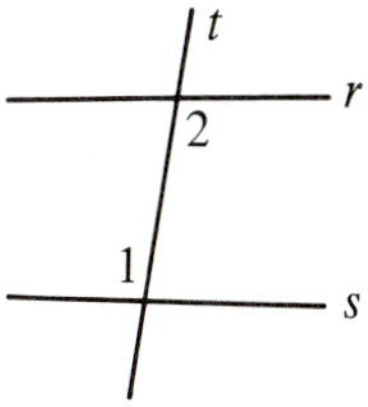

EXAMPLE 2 *Given:* line l and point P.

Prove: There is only one line through P perpendicular to l.

Proof: We know from Construction 3, page 26, that we can draw a line through P perpendicular to l. Assume that there is another line n through P perpendicular to l.

If two lines are perpendicular to a third line, they are parallel to each other, so line m is parallel to line n. But m and n intersect at P. This contradicts the fact that m is parallel to n. We conclude that there is only one line through P perpendicular to l.

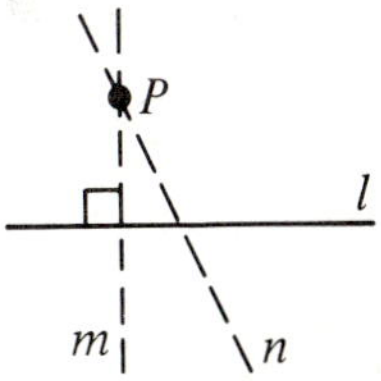

Notice that an indirect proof depends on our ability to reason until we have found a contradiction. Indirect proof is sometimes called proof by contradiction.

EXERCISES

Write an indirect proof for the following.

1. *Given:* Lines a and b with transversal t; $\angle 1 \neq \angle 2$.

   Prove: a is not parallel to b.

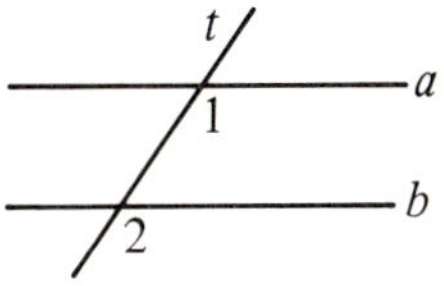

2. *Given:* Lines l and m; l is not perpendicular to m.

   Prove: $\angle 1 \neq \angle 2$

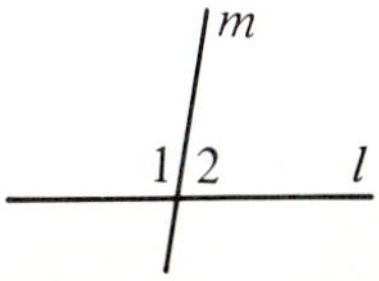

PROBLEM SOLVING STRATEGIES

AUXILIARY LINES

In solving a geometry problem, you often first draw a clear diagram. Then you mark in the given information on your diagram. Next you focus your attention on what you have to prove. Finally you try to figure out a way to get from the known to the unknown. To do this, it is often necessary to draw in auxiliary, or helping lines.

You have already seen auxiliary lines used in your study of geometry. For example, the proof of Theorem 1 on page 92 depended on drawing a line through C, parallel to $\overline{AB}$. The proof of Theorem 3 on page 167 depended on drawing the bisector of $\angle C$.

Auxiliary lines were also used often in Chapter 5. They helped us to find the interior and exterior angle sums of a polygon on pages 194–195. The key steps used to prove Theorems 10, 11, and 12 on pages 214–215 also use auxiliary lines.

EXERCISES

1. In the figure at the right, $AB = 5$, $BC = 6$, and $CD = 7$. The angles at B and C are both 90°. Find the distance AD.

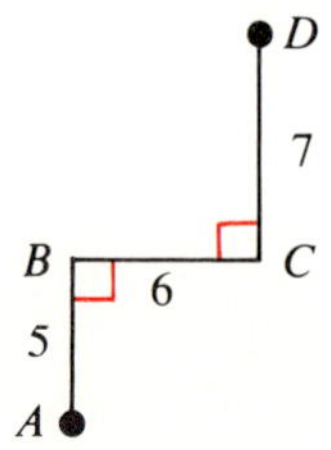

One way to solve the problem is to see $\overline{AD}$ as the hypotenuse of a right triangle.
Draw a line through A, parallel to $\overline{BC}$.
Extend $\overline{DC}$ until it meets the first line you drew.
Use the right triangle ADX to find the length of $\overline{AD}$.

2. Prove Theorem 3 on page 167 by drawing the median from C to $\overline{AB}$ (instead of the bisector of $\angle C$).

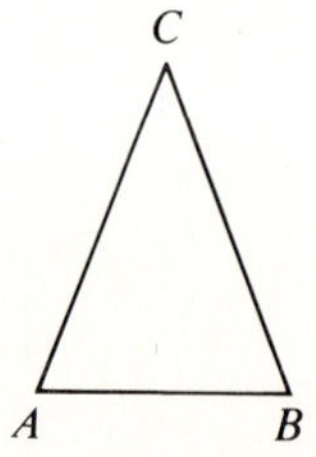

Given: $AC = BC$

Prove: $\angle A = \angle B$

3. Prove Theorem 1 on page 92 by drawing the auxiliary lines shown in the diagram below.

Given: $\triangle ABC$

Prove: $\angle 1 + \angle 2 + \angle 3 = 180°$

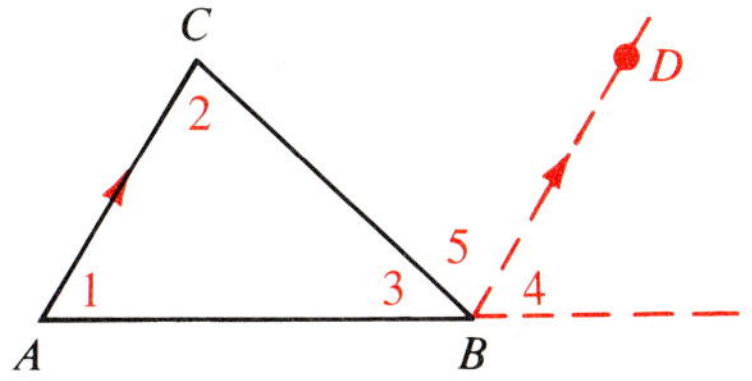

4. The figure below shows $\triangle ABC$ with medians $\overline{BE}$ and $\overline{CD}$ intersecting at T.

If $AB = 6$ and $AC = 8$, find the ratio $DT:TC$.

Let's draw an auxiliary line.

Since we need the ratio $DT:TC$, draw a line through D, parallel to TE, and intersecting $\overline{AE}$ at X.

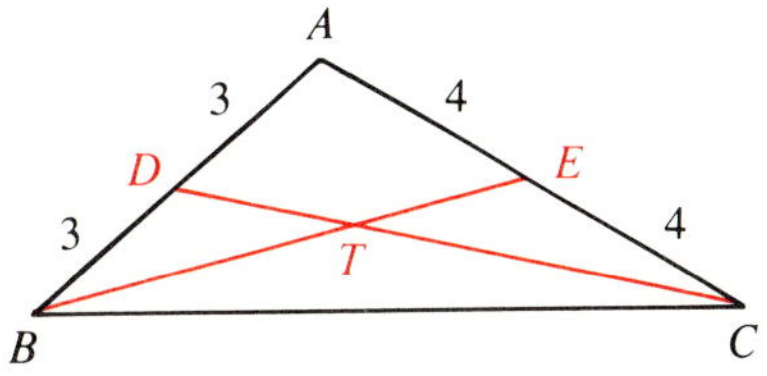

We now have $\overline{TE} \parallel \overline{DX}$ and intersecting two sides of $\triangle CDX$.

a. Find the length of $\overline{AX}$ and of $\overline{XE}$.

b. Find the ratio $CE:EX$.

c. Find the ratio $CT:TD$.

Look ahead to Chapter 9 where you will work with circles. Name the auxiliary lines shown in the diagrams that go with each of the exercises below.

5. Exercise 30 on page 363

6. Exercise 31 on page 367

7. Exercise 32 on page 367

8. Exercise 33 on page 367

9. Name the auxiliary line used in the example on page 437 of Chapter 11.

SKILLS REVIEW

THE DISTRIBUTIVE PROPERTY

Simplify. Use the distributive property.

SAMPLE

$$3(2 + 4x) - 8x + 1$$
$$6 + 12x - 8x + 1$$
$$4x + 7$$

1. $5(x + 8) + 3x - 7$
2. $2(3 - 6k^2) + 9k^2$
3. $-(a + 9 - 6a) - 4a$
4. $8b^3 + 4(6 - 2b^3)$
5. $5 - 3(n + 1) - 6n$
6. $x^2 - (12 - x^2) + 16$
7. $0.2(5 - 3m) + m$
8. $4(9 - 1.75x) + 7x$
9. $10(1.5a + 6.13) - 7a$

Solve.

SAMPLE

$$5(m + 4) = 35$$
$$5m + 20 = 35$$
$$5m = 15$$
$$m = 3$$

10. $2(x - 10) = 6$
11. $5(w + 3) = 45$
12. $3(2n + 1) = 21$
13. $11(a + 7) = 0$
14. $7x = 4(x - 9)$
15. $6(3b - 4) = 12$
16. $39 = 13(x + 4)$
17. $3(4 + 2m) = 3m$
18. $6(2k + 1) - 9k = 0$
19. $9y = 4(2y - 12)$
20. $7(9 + 2c) = 5c$
21. $13 + 10d = 9(d + 5)$
22. $5(32 - x) = 3x$
23. $a + 7 = 2(5a - 1)$
24. $7x - 3(5x + 1) = 13$

Write an equation. Then solve.

25. At a garage sale, Tanya bought 5 records that cost $(6x - 5)$ cents each. She spent $7.25 in all. Find the value of x. Then find the cost of each record.

26. A rectangular window has length and width as shown. The perimeter is 5.1 m. How many meters wide is the window? How many meters long?

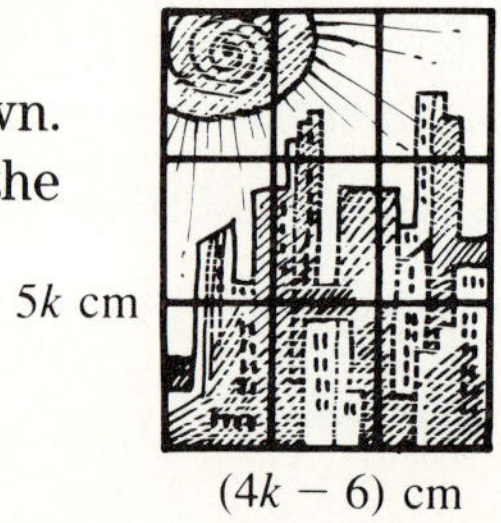

27. The perimeter of a square is 128. The length of a side is $3n + 5$. Find the value of n.

CHAPTER REVIEW

CHAPTER SUMMARY

1. If two figures have the same shape, they are similar. Two things are true of similar polygons:
   (a) Corresponding angles are equal.
   (b) Corresponding sides are in proportion (have equal ratios).

2. To show that two polygons are similar, you must show that both (a) and (b) above are true. However, to show that two *triangles* are similar, you only need to show that two angles of one triangle are equal to two angles of the other triangle.

3. The Triangle Proportionality Theorem states:

   If 

, then $\frac{a}{b} = \frac{c}{d}$.

4. If three parallel lines intersect two transversals, they divide the transversals proportionally.

5. The scale factor of the similar figures shown is 3:2. This is the ratio of a pair of corresponding sides. The ratio of the perimeters is also 3:2. The ratio of the areas is $3^2:2^2$, or 9:4.

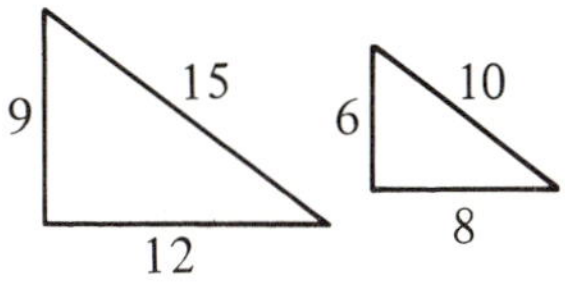

REVIEW EXERCISES

In Exercises 1–6, the two polygons given are similar. Complete each statement. ***(See pp. 312–316.)***

1. $\triangle LAF \sim \triangle$ __?__

2. $\frac{LA}{JI} = \frac{LF}{?} = \frac{AF}{?}$

3. $\angle A = \angle$ __?__

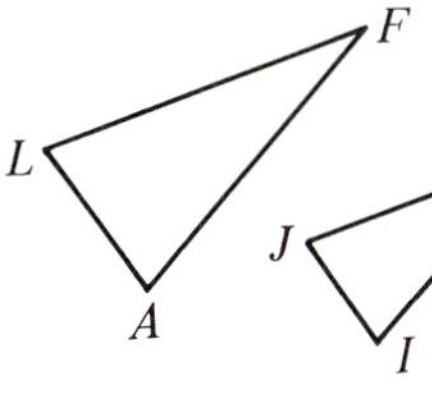

4. Trapezoid *SKON* ~ trapezoid __?__

5. The scale factor is __?__ : __?__.

6. $\frac{14}{21} = \frac{x}{?}$

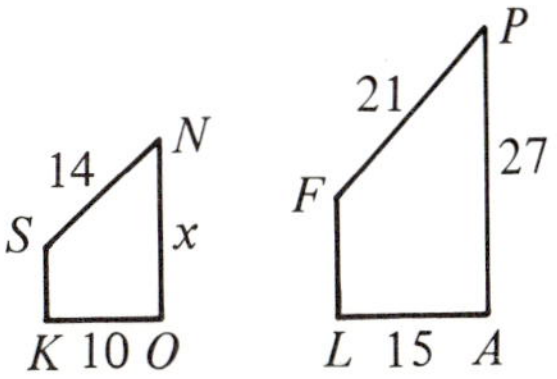

In the diagram at the right, $\overline{UV} \parallel \overline{RS}$. Complete each statement. *(See pp. 317–322.)*

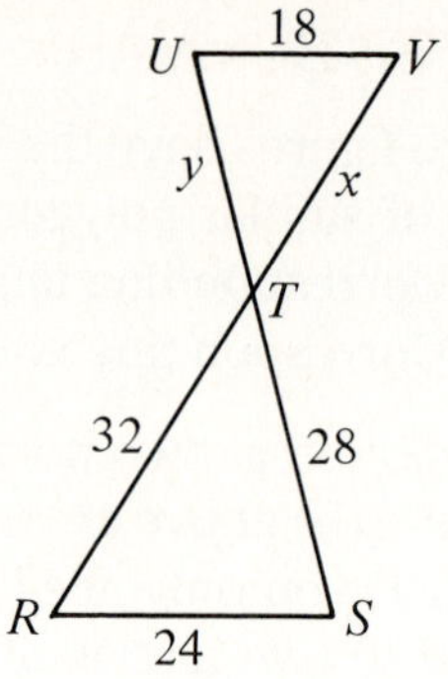

7. $\angle R = \angle$ __?__ and $\angle S = \angle$ __?__

8. $\triangle RST \sim \triangle$ __?__

9. $\frac{18}{24} = \frac{x}{?}$

10. $x =$ __?__

11. $y =$ __?__

For each figure, complete the statement $\frac{a}{b} = \frac{?}{?}$. *(See pp. 323–324.)*

12.

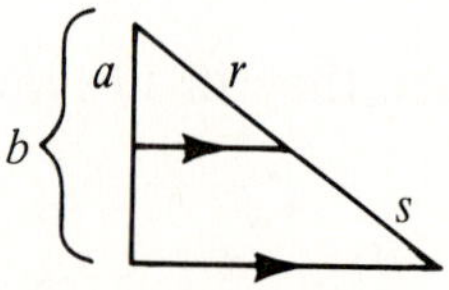

13.

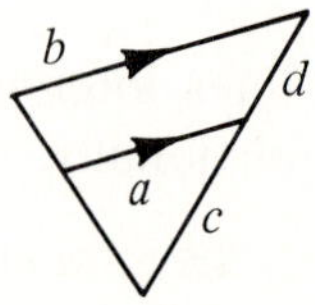

Find the value of x. *(See pp. 325–328.)*

14.

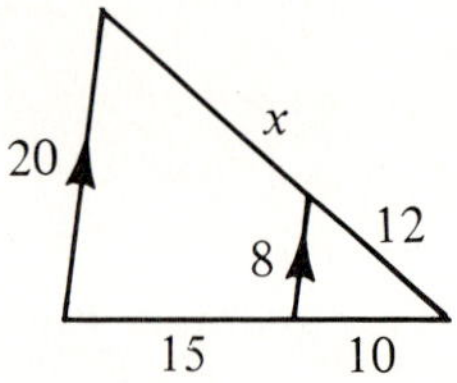

15.

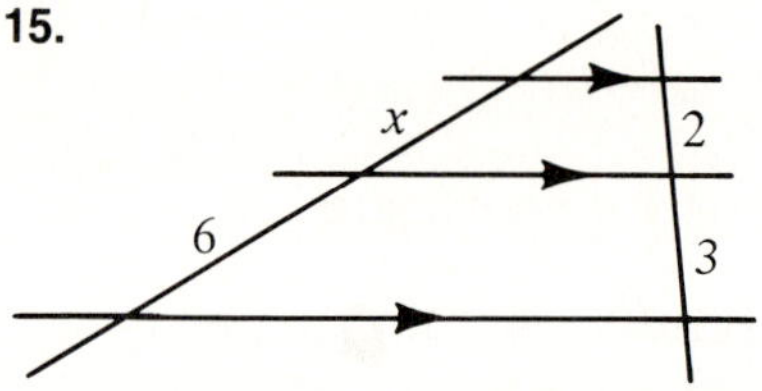

For the pair of similar rectangles shown, complete each statement. *(See pp. 329–332.)*

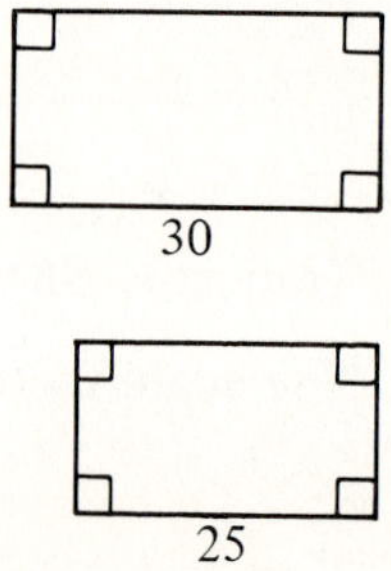

16. The scale factor is __?__ : __?__.

17. The ratio of their perimeters is __?__ : __?__.

18. The ratio of their areas is __?__ : __?__.

CHAPTER TEST

The two triangles are similar.

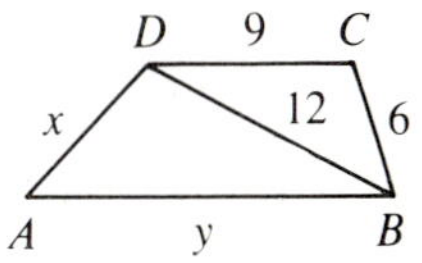

1. $\triangle BCD \sim \triangle$ __?__ 2. $\angle CBD = \angle$ __?__ *(8-1)*

3. Find the value of x. 4. Find the value of y.

5. Complete the proof. *(8-2)*

Given: $\angle 1 = \angle 2$

Prove: $\frac{RU}{RS} = \frac{UV}{ST}$

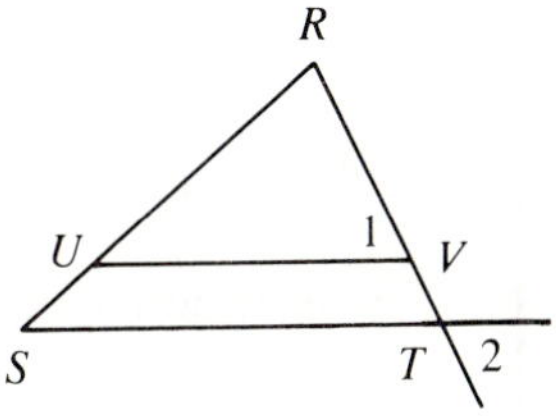

STATEMENTS	REASONS
1. $\angle 1 = \angle 2$	1. __?__
2. $\angle 2 = \angle VTS$	2. __?__
3. $\angle 1 = \angle VTS$	3. __?__
4. $\angle R = \angle R$	4. __?__
5. $\triangle RUV \sim \triangle RST$	5. __?__
6. $\frac{RU}{RS} = \frac{UV}{ST}$	6. __?__

In Questions 6–8, refer to the diagram.

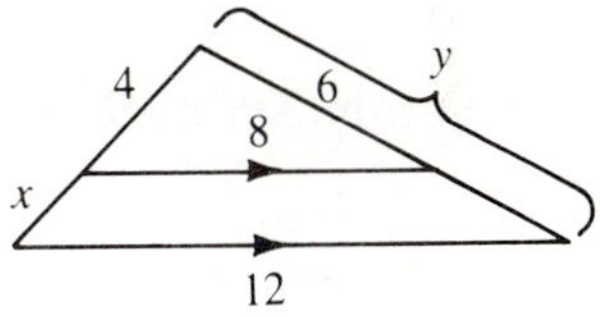

6. $\frac{8}{12} = \frac{4}{?} = \frac{6}{?}$ 7. $x =$ __?__ 8. $y =$ __?__ *(8-3)*

For each figure, complete the statement $\frac{a}{b} = \frac{?}{?}$. *(8-4)*

9.

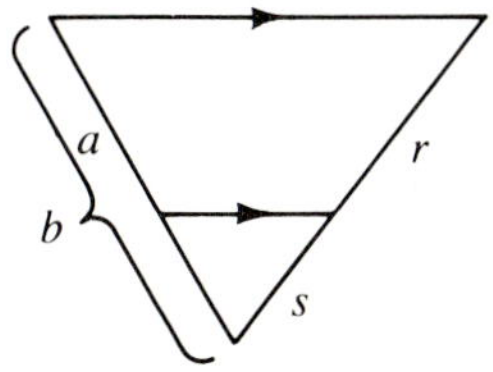

10.

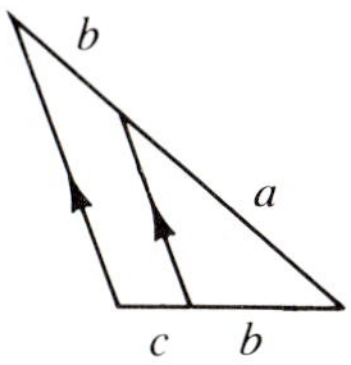

11.

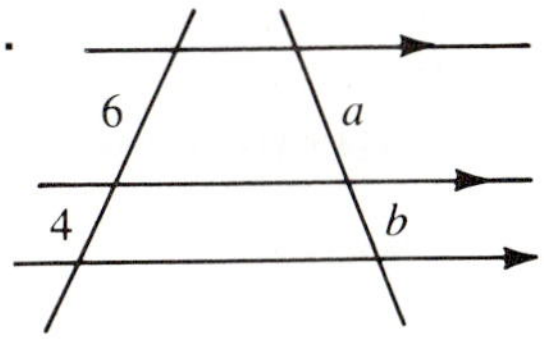

The bases of two similar parallelograms are 9 cm and 12 cm.

12. Find their scale factor. 13. Find the ratio of their areas. *(8-5)*

14. If the perimeter of the smaller parallelogram is 30 cm, what is the perimeter of the larger parallelogram?

UNIT D CUMULATIVE REVIEW

Express each ratio in simplest form.

1. $\frac{11}{77}$
2. $\frac{45}{100}$
3. $\frac{20 \text{ cm}}{4 \text{ m}}$
4. $8:16:24$

Find the value of x.

5. $\frac{x}{2} = \frac{6}{3}$
6. $\frac{1}{7} = \frac{2}{x}$
7. $\frac{2}{x} = \frac{4}{22}$
8. $10:15 = x:3$
9. Three cassette tapes cost \$24. How much will five tapes cost?
10. At a shopping center, Hal's Hardware is 40 m south of Sound City. The supermarket is 60 m east of Sound City. Draw a map showing the hardware store, music store, and supermarket. Be sure to include the scale.

State whether the two polygons are, or are not, similar.

11.

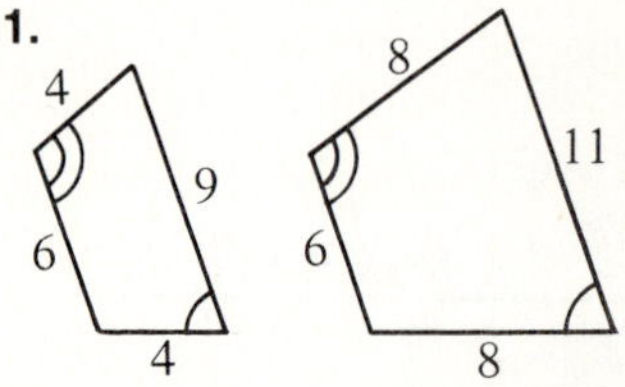

12.

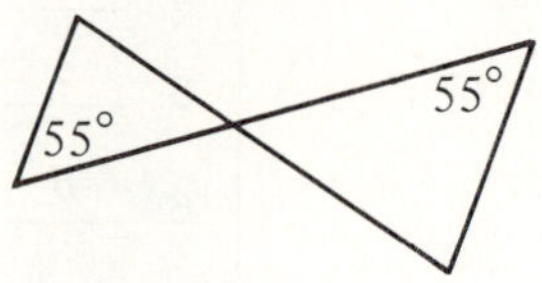

13.

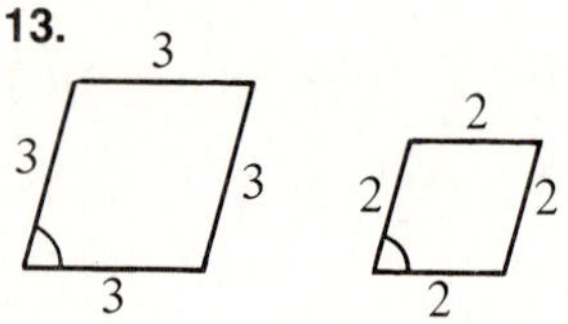

In the diagram, $\overline{OS} \parallel \overline{HE}$. Complete each statement.

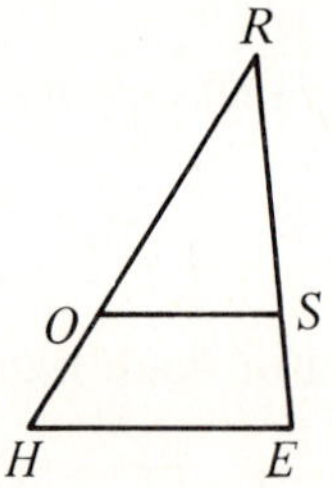

14. $\triangle ROS \sim \triangle$ __?__
15. $\angle RHE = \angle$ __?__
16. $\frac{?}{RH} = \frac{RS}{RE}$
17. $\frac{RO}{OH} = \frac{RS}{?}$
18. If $RS = 8$, $SE = 4$, and $OS = 6$, then $HE =$ __?__.

The quadrilaterals shown are similar. Complete each statement.

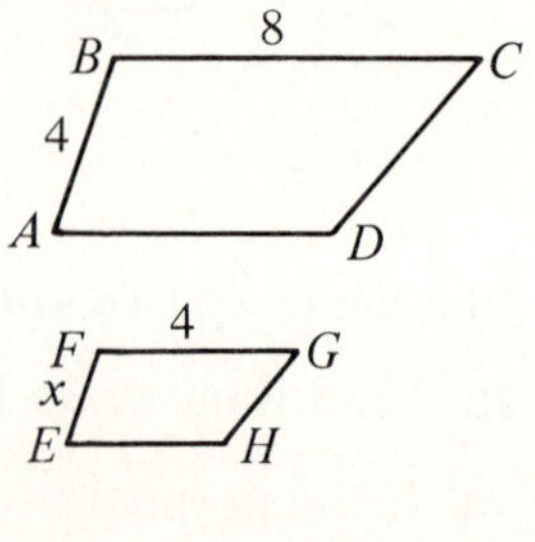

19. $\angle E = \angle$ __?__
20. The scale factor is __?__:__?__.
21. $x =$ __?__
22. The ratio of the perimeters is __?__:__?__.
23. The ratio of the areas is __?__:__?__.

UNIT

E

Here's what you'll learn in this chapter:

To define a circle and the points, lines, and segments related to circles.

To apply theorems relating tangents and radii.

To construct a tangent to a circle at a given point on the circle.

To classify and measure arcs.

To use theorems involving the chords of a circle.

To draw a circle inscribed in a polygon and a circle circumscribed about a polygon.

To use theorems relating angle measure and arc measure.

Notice the circular pattern in which the reflecting mirrors shown above are arranged. These movable mirrors focus sunlight on a boiler, in order to make steam for generating electricity.

Chapter 9

Circles

1 Basic Terms

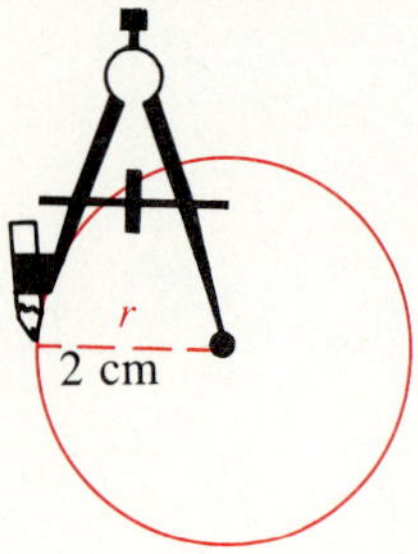

Set your compass for a radius of 2 cm.
Now draw a complete curve.
You know, of course, that the curve is a *circle.*
The pin point is the *center* of the circle.

This experiment should remind you that a **circle** is a figure, in a plane, whose points are the same distance from a particular point in the plane. The table below shows some important definitions relating to circles.

radius	1. a segment that joins the center and a point on the circle 2. the length of such a segment
chord	a segment that joins two points on the circle
diameter	1. a chord that passes through the center 2. the length of such a chord
secant	a line that contains a chord
tangent	a line, in the plane of the circle, that intersects the circle in exactly one point
point of tangency	the point in which a tangent intersects the circle

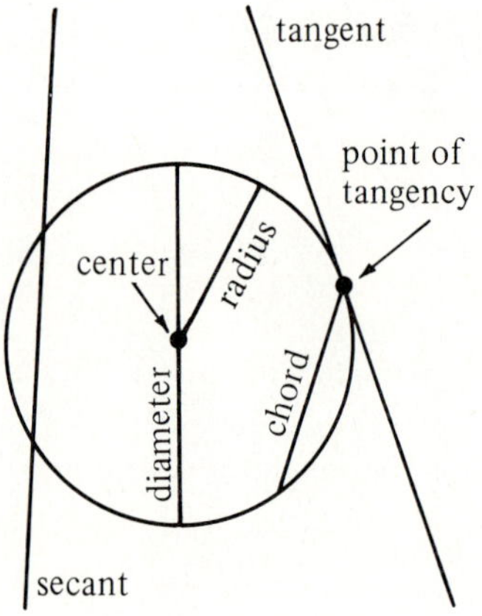

From the definition of a circle, you can see that:

(1) All radii of a circle are equal.

(2) All diameters of a circle are equal.

Many of the terms used for circles are also used for *spheres.* A **sphere** is a figure in space whose points are the same distance from a particular point. That point is called the *center* of the sphere.

$\overline{PA}$, $\overline{PB}$, and $\overline{PC}$ are radii of the sphere.
$\overline{BC}$ is a diameter.
$\overleftrightarrow{AD}$ is a tangent.
$\overleftrightarrow{BC}$ is a secant.

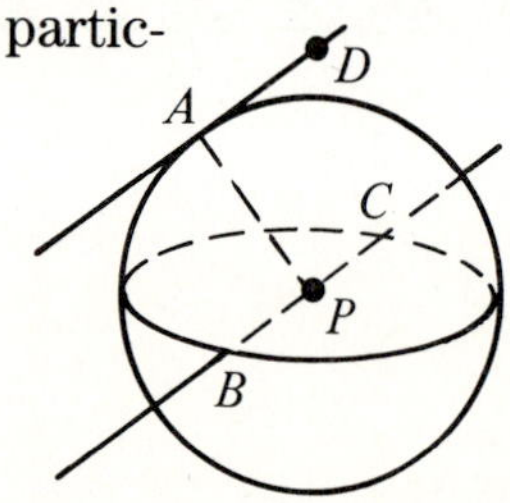

Classroom Practice

Exercises 1–10 refer to $\odot O$ (circle with center O) in the diagram.

1. Name all the radii shown.
2. Name a secant shown.
3. Is $\overleftrightarrow{CD}$ a chord?
4. Is $\overline{CD}$ a chord?
5. Explain why $\overline{AF}$ is not a chord.
6. Name a tangent to $\odot O$.
7. Is $\overline{BD}$ a chord? a diameter?
8. Name a point of tangency.
9. How many diameters that contain point C can be drawn? that contain point O?

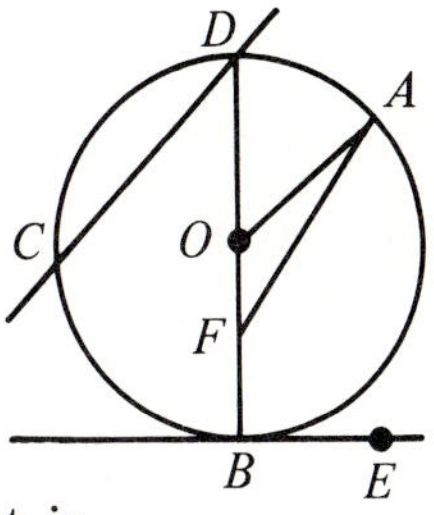

10. How many secants containing both A and C can be drawn?
11. State a definition of a radius of a sphere.
12. State a definition of a tangent to a sphere.
13. Set a basketball on top of a desk and you see a sphere that is tangent to a plane. Hold a piece of cardboard to illustrate another plane tangent to the sphere. In how many positions can you hold the cardboard to show a tangent plane?
14. Refer to Exercise 13. Hold the piece of cardboard so that it is both tangent to the ball and parallel to the desk top. Consider the point where the cardboard touches the ball and the point where the desk top touches the ball. What can you say about the segment that joins these two points?

Written Exercises

Exercises 1–10 refer to $\odot A$. Name each of the following.

A

1. A circle
2. Four radii
3. Two diameters
4. Two tangents
5. A chord that is not a diameter
6. Two secants, each containing a diameter
7. A secant that does not contain a diameter
8. Two points of tangency
9. How many tangents containing both point A and point D can be drawn?
10. How many chords containing both point R and point C can be drawn?

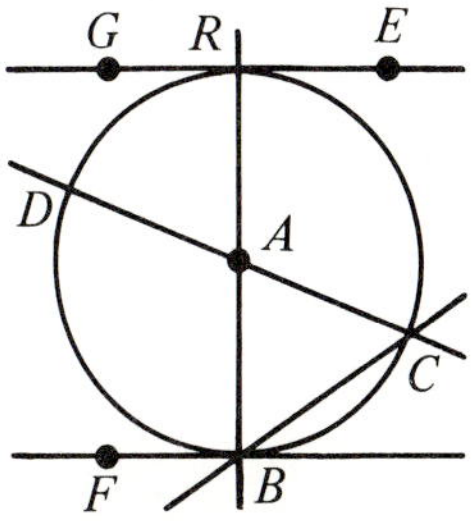

The length of a radius is given. Find the length of a diameter.

11. 19 **12.** 5.3 **13.** $7\frac{1}{2}$ **14.** $3k$

The length of a diameter is given. Find the length of a radius.

15. 26 **16.** 2.4 **17.** $8\frac{1}{2}$ **18.** $5k$

Refer to the diagram of the sphere with center *O*.
Name each of the following.

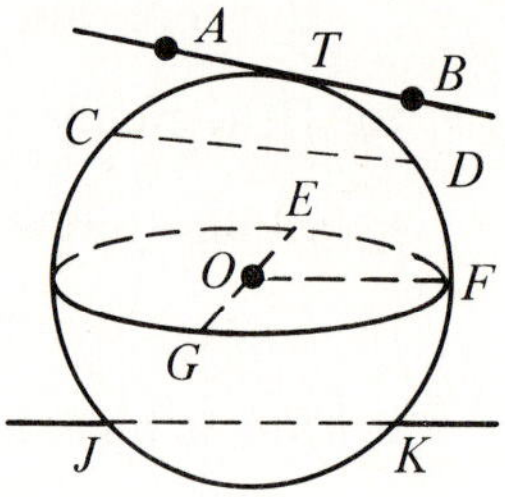

B **19.** Eight points on the sphere

20. Three radii of the sphere

21. A diameter of the sphere

22. Three chords of the sphere

23. A secant of the sphere

24. A tangent to the sphere

25. Write a definition of a diameter of a sphere.

26. Write a definition of a chord of a sphere.

For each exercise draw a circle with radius 8.
Draw two points *R* and *S* that satisfy the conditions stated.
If the conditions cannot be satisfied, write *not possible*.

C **27.** $\overline{RS}$ is a chord and $RS = 10$.

28. R and S both lie inside the circle and $RS = 17$.

29. R and S both lie outside the circle and $RS = 7$.

30. R lies inside the circle.
S lies outside the circle.
$\overleftrightarrow{RS}$ intersects the circle in exactly one point.

31. $\overline{OA}$ and $\overline{OB}$ are radii.
$\angle A = 5x - 3$; $\angle B = 3x + 17$.
Find the value of x.

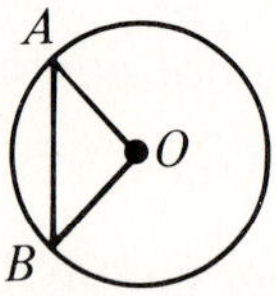

32. Point P is the center of the circle.
Find the measure of $\angle E$.

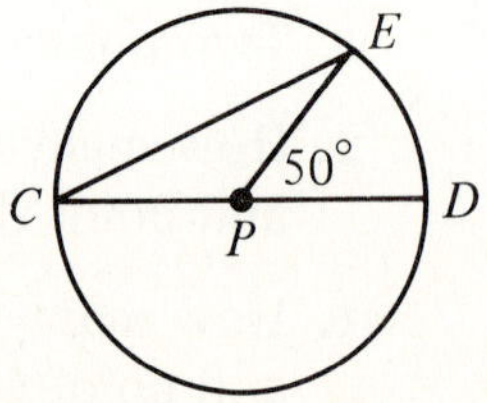

Experiments

1. Copy the diagram shown.
2. Construct a line, through O, perpendicular to chord $\overline{AB}$. Let C be the point of intersection with $\overline{AB}$.
3. What appears to be true about AC and BC?
4. Prove that your conclusion is true.

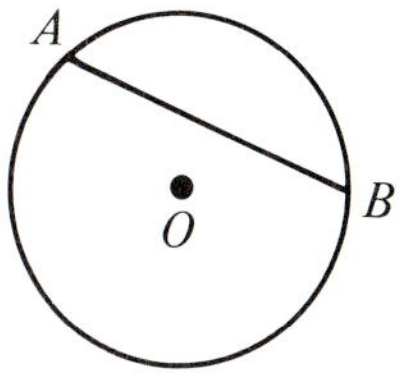

CONSUMER APPLICATIONS

LIFE INSURANCE

A life insurance policy is an agreement that you and your insurance company make with each other. You agree to pay the insurance company a certain fee (or *premium*) each year. In return, the company agrees to pay your family a certain amount when you die.

In order to make money, life insurance companies must take in more money by selling premiums than they pay out. Statisticians who compute insurance risks and premiums are called actuaries. To find out more about what they do, read the Career Notes on page 316.

Many different life insurance policies are offered by many different companies. There are two general types of life insurance with which you should be familiar.

Permanent life insurance is the kind most people know. This type of insurance is part financial protection in case of death and part savings plan. The face amount of the policy is paid out by the insurance company either when the insured person dies or when the insured person reaches an age (often 65 or 70) specified in the policy.

Term life insurance does not provide a savings feature. The principal is paid out by the insurance company only if the insured person dies within the term of years specified in the policy.

Which type of insurance is better? Term insurance costs less, but there is no guarantee of ever collecting any money from the insurance company. Permanent life insurance costs more, but provides a savings plan in addition to financial protection against sudden death. Before you buy insurance, you should consult an insurance broker or a banker.

2 Tangents

EXPLORATIONS

1. Draw circles and tangents like those shown.

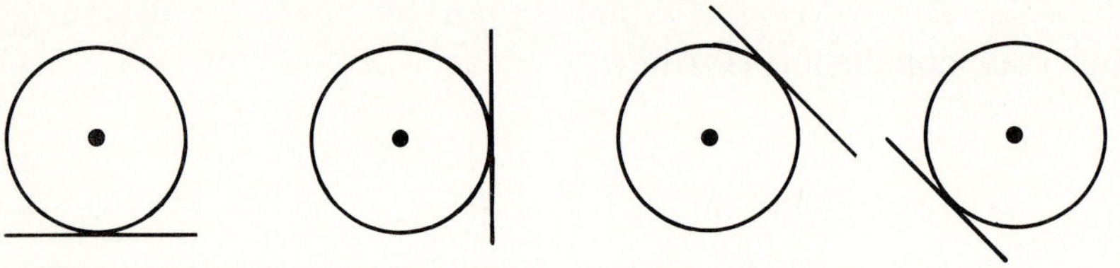

In each case draw the radius to the point of tangency. What do you observe?

2. Draw circles and radii like those shown.

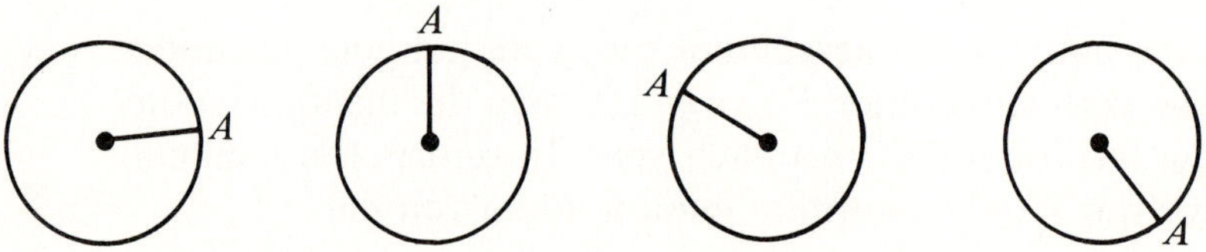

In each case draw a line that passes through A and is perpendicular to the radius. What do you observe?

These explorations suggest two theorems that we state but do not prove.

THEOREM 1
A radius drawn to a point of tangency is perpendicular to the tangent.
THEOREM 2
If a line lies in the plane of a circle and is perpendicular to a radius at its outer endpoint, the line is tangent to the circle.

If line t is tangent to $\odot O$, then $\overline{OA} \perp t$.

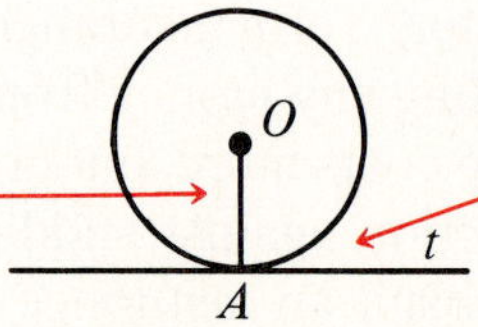

If line $t \perp \overline{OA}$ at point A, then t is tangent to $\odot O$.

A tangent to a circle has been defined as a line. However, we sometimes say that a *segment* is tangent to a circle. For example, since $\overleftrightarrow{PX}$ is tangent to the circle at X, $\overline{PX}$ may also be called a tangent to the circle.

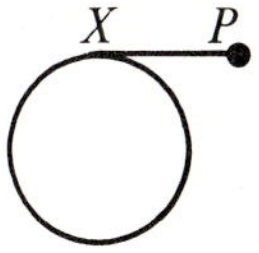

A line that is tangent to each of two coplanar circles is called a **common tangent.**

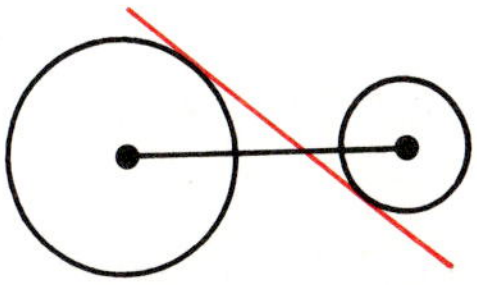

A common *internal* tangent intersects the segment joining the centers.

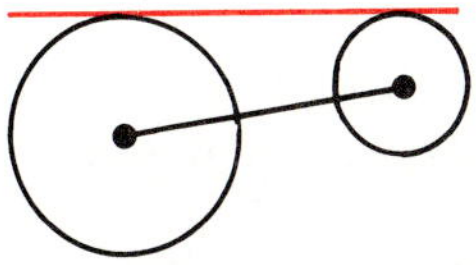

A common *external* tangent does not intersect the segment joining the centers.

When two coplanar circles are tangent to a line at one point, we say that the *circles are tangent to each other.*

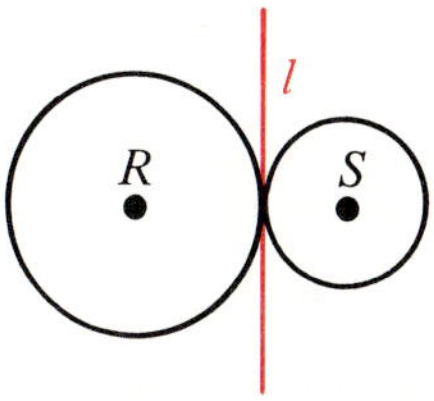

Circles *R* and *S* are *externally* tangent. Line *l* is a common internal tangent.

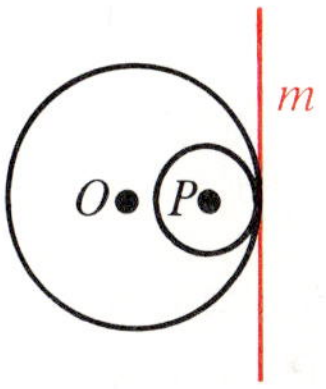

Circles *O* and *P* are *internally* tangent. Line *m* is a common external tangent.

CONSTRUCTION 10

Given: *Point A on* $\odot O$
Construct: *A tangent to* $\odot O$ *at point A*

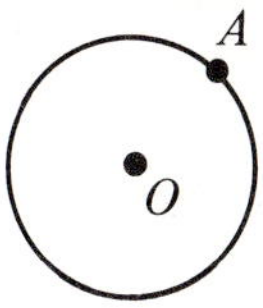

1. You are given $\odot O$ with point A.
2. Draw $\overrightarrow{OA}$.
   Using Construction 2, page 26, construct a line perpendicular to $\overrightarrow{OA}$ at point A. Call the line l.
3. Line l is tangent to $\odot O$.

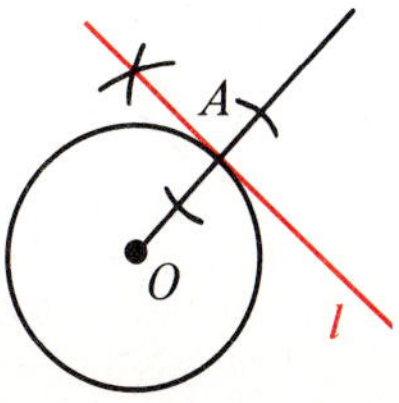

Classroom Practice

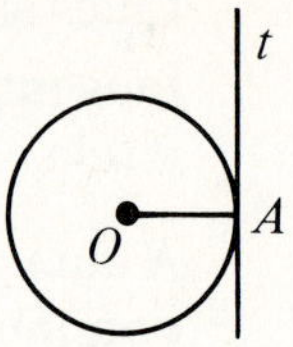

1. If line t is tangent to $\odot O$ at point A, what must be true about $\overline{OA}$ and line t? State the theorem that supports your conclusion.

2. If line t and $\odot O$ lie in the same plane, and if line t is perpendicular to $\overline{OA}$, what must be true about line t and $\odot O$? State the theorem that supports your conclusion.

In each diagram, which lines are common external tangents and which are common internal tangents?

3.
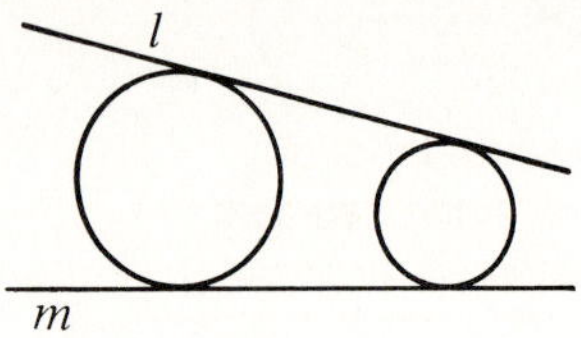

4.
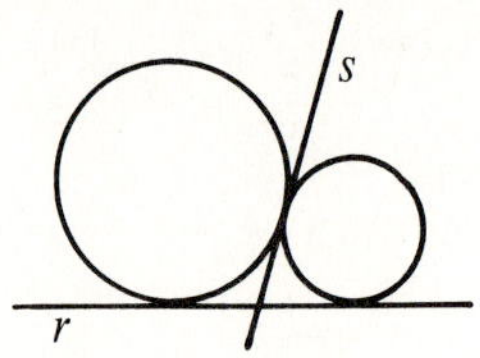

5.
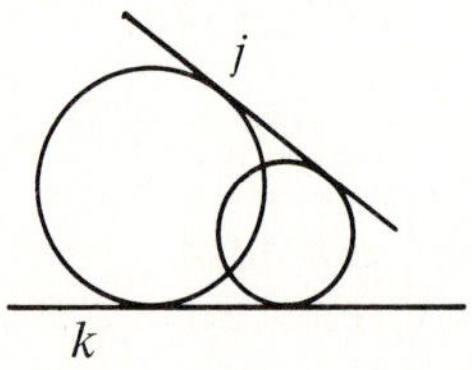

6.
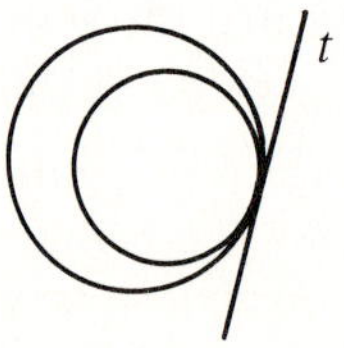

7. In the figures for Exercises 3–6, state whether the two circles are externally tangent, internally tangent, or are not tangent.

Draw a circle O on the chalkboard. Choose a point P outside the circle. Draw tangents, $\overline{PX}$ and $\overline{PY}$, to the circle.

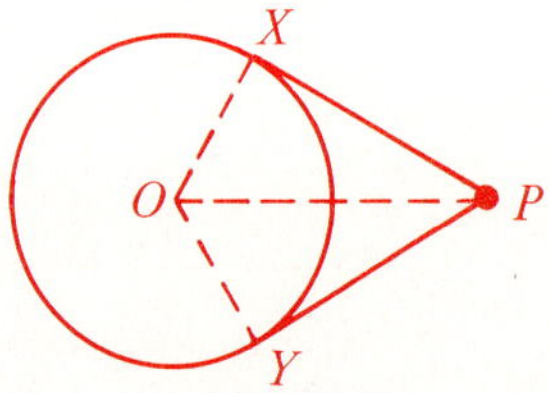

Exercises 8–14

8. What appears to be true about PX and PY?

9. Complete the listing of what is given.
   Given: $\odot O$; $\overline{PX}$ and $\overline{PY}$ are . . .
   Prove: $PX = PY$

By drawing $\overline{OX}$, $\overline{OY}$, and $\overline{OP}$, we form two triangles that appear to be congruent.

10. State a theorem that supports the statements: $\overline{OX} \perp \overline{PX}$; $\overline{OY} \perp \overline{PY}$.

11. State a reason that supports the statement: $OX = OY$.

12. Explain why right $\triangle PXO \cong$ right $\triangle PYO$.

13. State a reason that supports the statement: $PX = PY$.

14. Complete this statement of the theorem that has just been proved: If two tangent segments are drawn to . . .

Written Exercises

In each exercise, state the number of common internal tangents and the number of common external tangents that can be drawn to the two circles.

A

1. 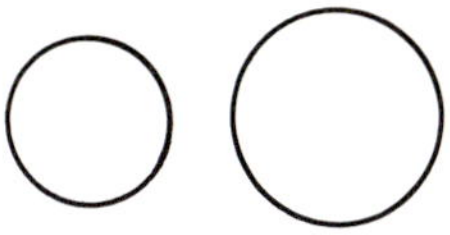

2. 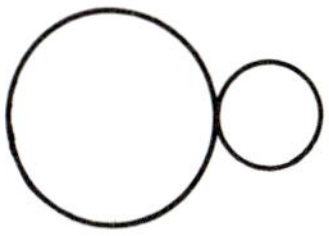

3. 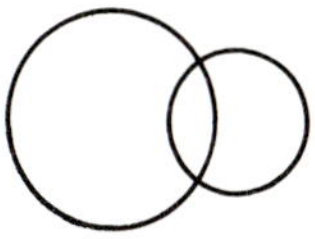

4. 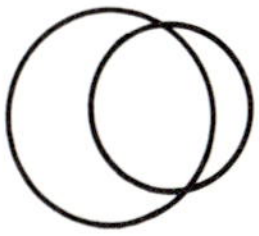

5. 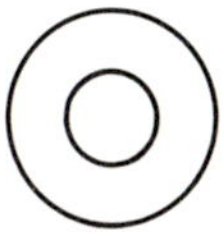

6. 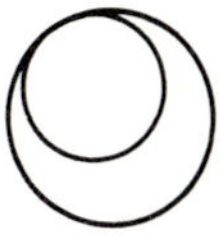

7. Which of the six diagrams above shows two circles that are externally tangent? two circles that are internally tangent?

8. Draw a figure showing two circles that are tangent to a third circle but are not tangent to each other.

In Exercises 9–12, line t lies in the plane of the three circles and intersects each of the circles in exactly one point, X. Classify each statement as true or false.

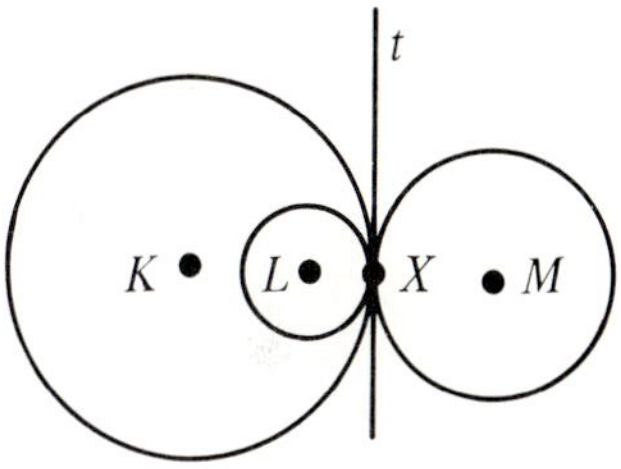

9. Line t is a common tangent of $\odot L$ and $\odot M$.
10. Line t is a common external tangent of $\odot L$ and $\odot M$.
11. Line t is a common external tangent of $\odot K$ and $\odot L$.
12. $\odot M$ and $\odot K$ are internally tangent circles.

In Exercises 13 and 14, draw a figure roughly like the one shown, but larger.

13. Construct a line tangent to $\odot P$ at point X.

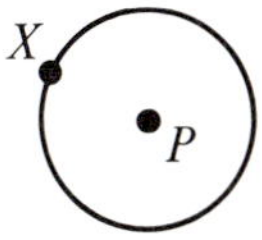

14. Construct lines tangent to $\odot O$ at points D and E.

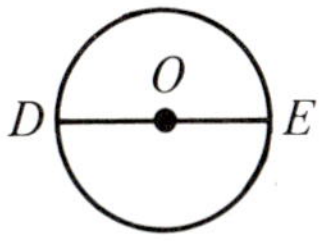

If possible, draw a diagram to illustrate the conditions.

B

15. Two circles with five common tangents
16. Two tangent circles with three common tangents

If possible, draw a diagram to illustrate the conditions.

17. Two externally tangent circles with two common internal tangents

18. Two internally tangent circles with one common external tangent

Find the required length.

19.

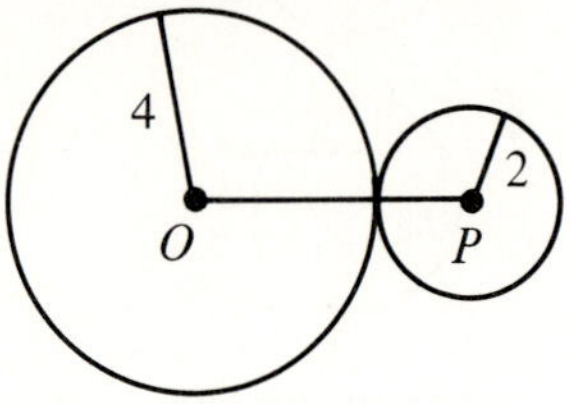

$OP = \underline{\ ?\ }$

20.

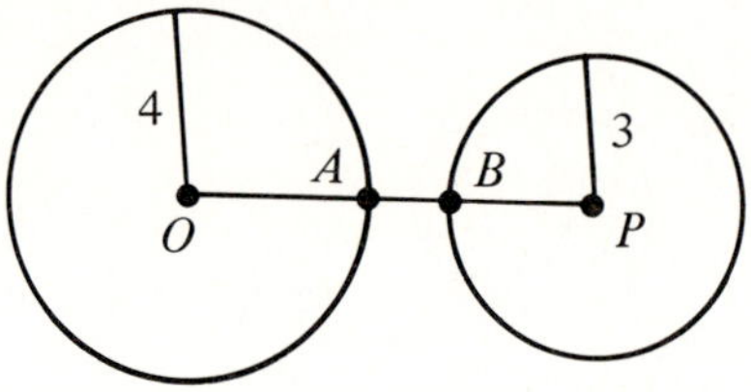

$OP = 9 \quad AB = \underline{\ ?\ }$

21.

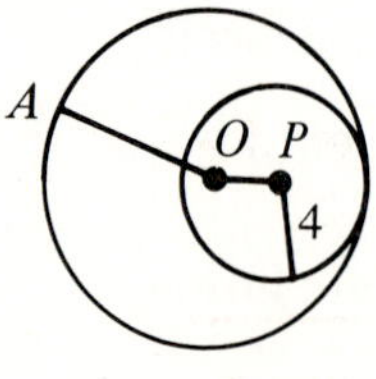

$OA = 7 \quad OP = \underline{\ ?\ }$

22.

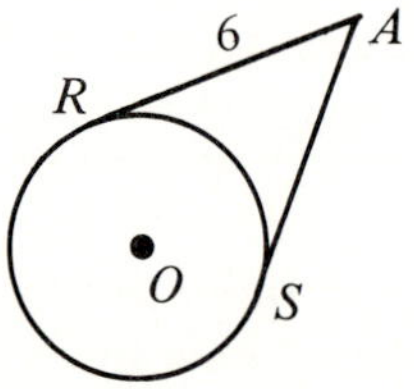

$AS = \underline{\ ?\ }$
(*Hint:* See Exercise 14, page 353.)

C **23.**

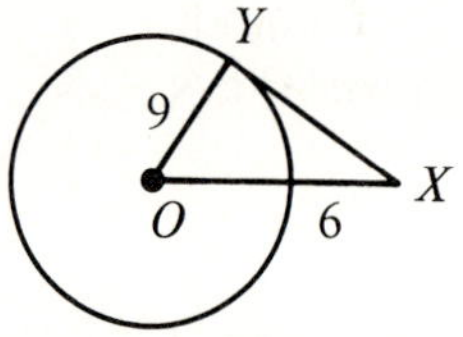

$XY = \underline{\ ?\ }$

24.

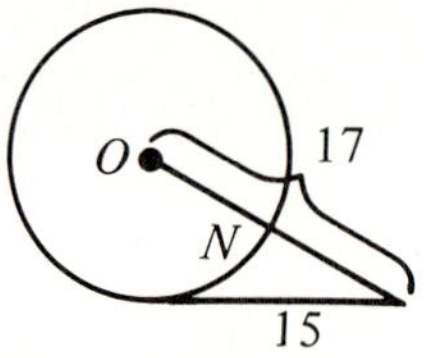

$ON = \underline{\ ?\ }$

25. Refer to the diagram at the bottom of page 351. Explain why line l must be tangent to $\odot O$.

PUZZLE ♦ PROBLEMS

Using ordinary household waxed paper, cut out a circle with as large a radius as convenient. Mark the center. Fold the paper so that some point of the circle falls on the center O. Make a sharp crease, then unfold. Make about twenty different creases in the same way. Hold the paper at arm's length. The creases suggest a circle!

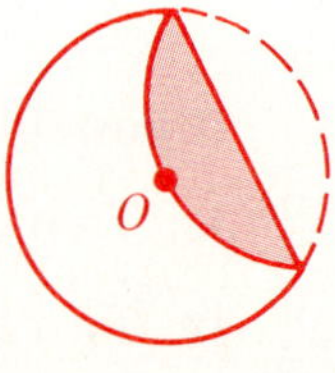

3 Arcs and Central Angles

An *arc* is part of a circle.

Classifying Arcs:

Semicircle

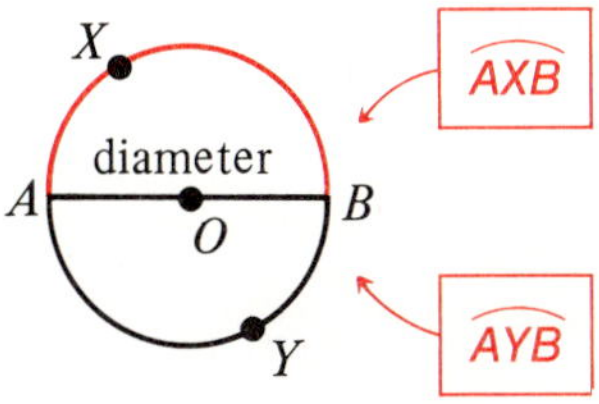

Exactly half a circle

Minor Arc

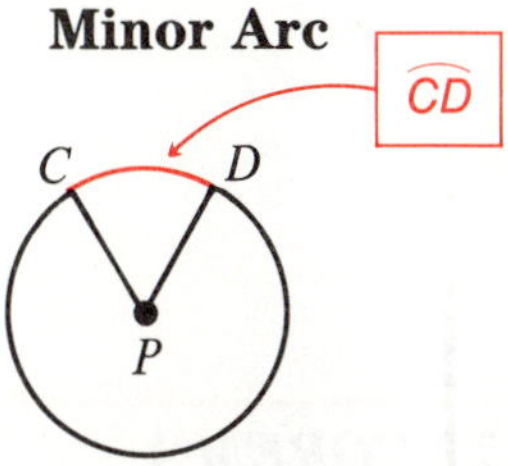

Less than a semicircle

Major Arc

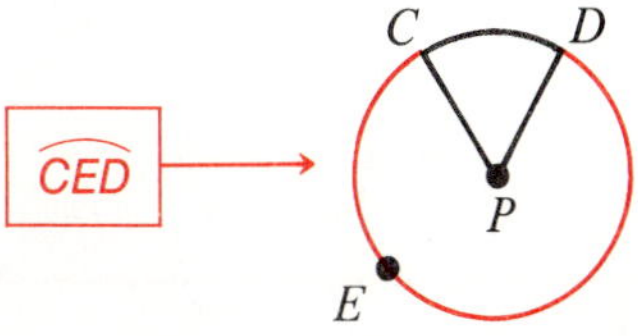

Greater than a semicircle

Notice that you need three letters to name a semicircle or a major arc. $\overset{\frown}{AXB}$ is read "arc *AXB*."

A **central angle** of a circle is an angle whose vertex is the center of the circle.

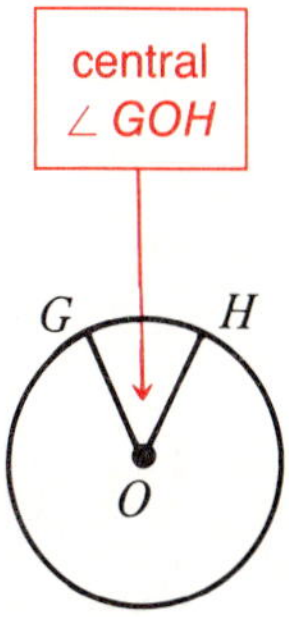

> *The degree measure of a minor arc is defined to be the measure of its central angle.*

By definition, $\overset{\frown}{GH} = \angle GOH$. You may read this as "arc *GH* equals angle *GOH*." Just remember that this means that the *measure* of the angle equals the *measure* of the arc.

Finding the Measure of an Arc:

Semicircle

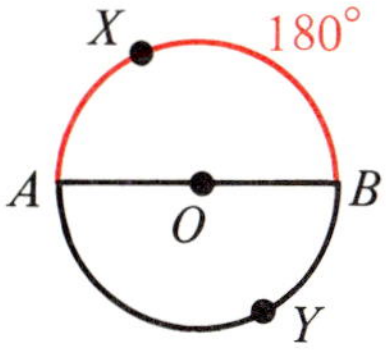

$\overset{\frown}{AXB} = \overset{\frown}{AYB} = 180°$

Minor Arc

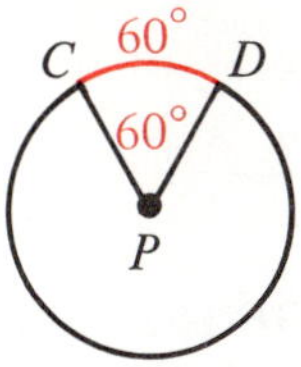

$\overset{\frown}{CD} = \angle CPD = 60°$

Major Arc

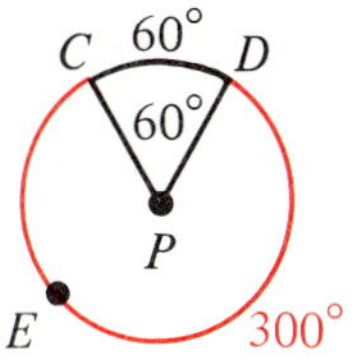

$\overset{\frown}{CED} = 360° - \overset{\frown}{CD} = 300°$

If two arcs of a circle have equal measures, they are called **equal arcs.**

Thus, in the diagram, $\overset{\frown}{AB} = \overset{\frown}{CD}$ because each arc has measure 70°.

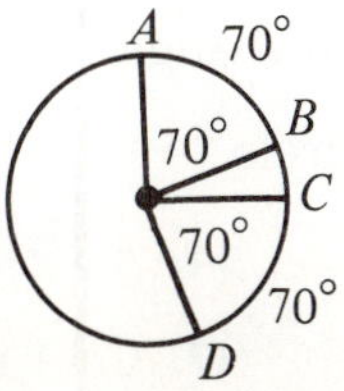

THEOREM 3

In a circle, equal central angles have equal minor arcs.

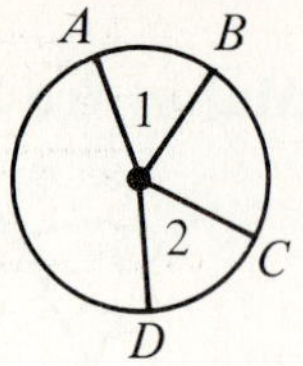

If $\angle 1 = \angle 2$, then $\overset{\frown}{AB} = \overset{\frown}{CD}$.

Strategy for proof:

$\overset{\frown}{AB} = \angle 1$ (Why?)

$\angle 1 = \angle 2$ (Why?)

$\angle 2 = \overset{\frown}{CD}$ (Why?)

$\overset{\frown}{AB} = \overset{\frown}{CD}$

THEOREM 4

In a circle, equal minor arcs have equal central angles.

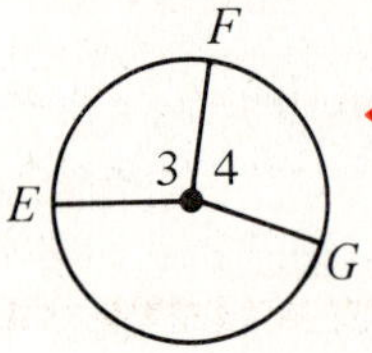

If $\overset{\frown}{EF} = \overset{\frown}{FG}$, then $\angle 3 = \angle 4$.

The proof is left as Written Exercise 26 on page 358.

Classroom Practice

Exercises 1–6 refer to the diagram shown.

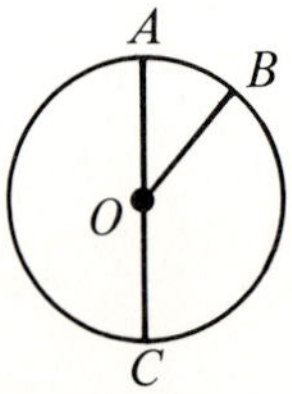

1. Name a minor arc.

2. Name a major arc.

3. Name a semicircle.

4. Name two central angles.

5. If $\angle AOB = 40°$, then $\overset{\frown}{AB} = \underline{\ \ ?\ \ }°$.

6. If $\angle AOB = 40°$, then $\overset{\frown}{ACB} = \underline{\ \ ?\ \ }°$.

$\overline{AD}$ and $\overline{EC}$ are diameters of $\odot O$. $\overline{OB}$ is a radius. Classify each statement as true or false.

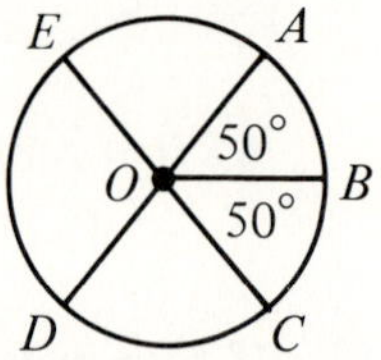

7. $\angle AOC = 100°$

8. $\angle AOE = 50°$

9. $\angle AOE$ is a central angle.

10. $\overset{\frown}{ABC}$ is a minor arc.

11. $\overset{\frown}{AED}$ is a semicircle.

12. $\overset{\frown}{BCD}$ is a major arc.

13. $\overset{\frown}{AB} = 50°$

14. $\overset{\frown}{ABD} = 180°$

15. $\overset{\frown}{AB} = \overset{\frown}{BC}$

16. $\overset{\frown}{BD} = 100°$

17. $\overset{\frown}{CDA}$ is a major arc.

18. $\overset{\frown}{DC} = \overset{\frown}{EA}$

Written Exercises

In Exercises 1–9, radii, diameters, and chords of ⊙ *O* are shown. State the measure of ∠ 1.

A

1.

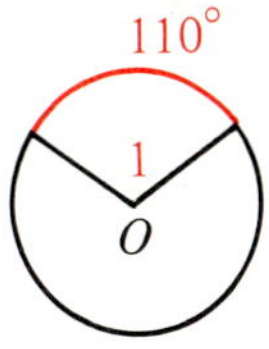

2.

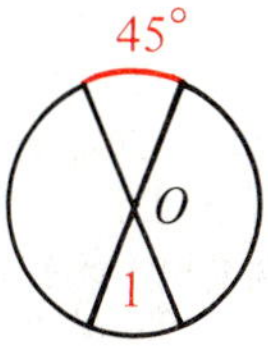

3.

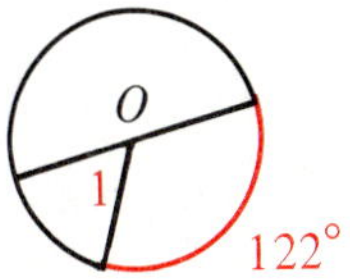

4.

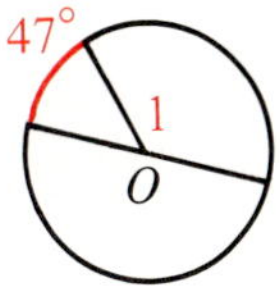

5.

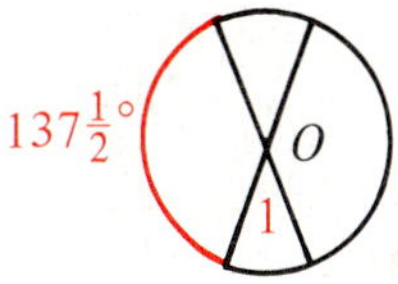

6.

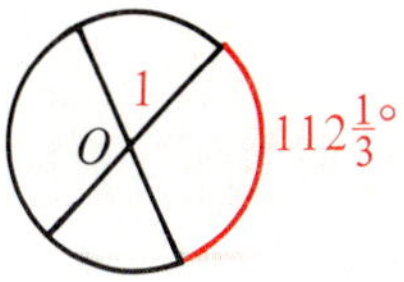

7.

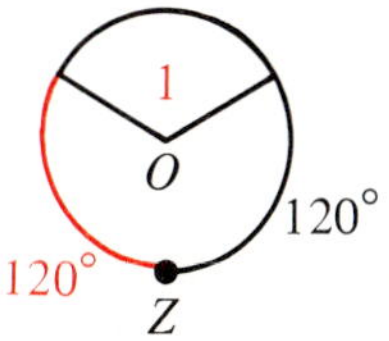

8.

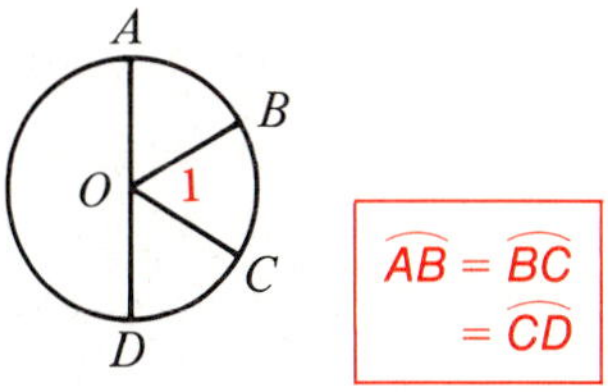

9.

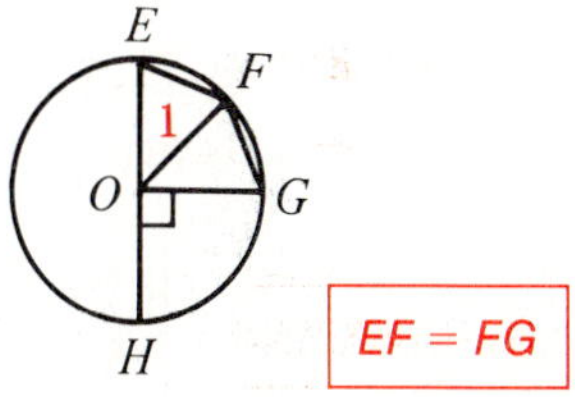

In Exercises 10–12, find the measure of the indicated arc.

10. $\overset{\frown}{AB}$ = ___?___ °

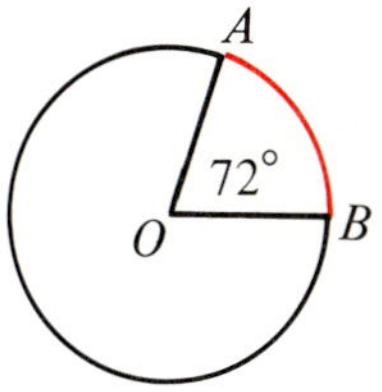

11. $\overset{\frown}{JL}$ = ___?___ °

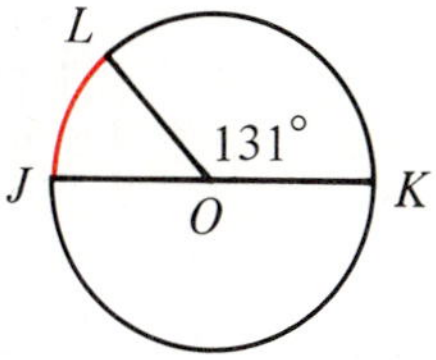

12. $\overset{\frown}{RSQ}$ = ___?___ °

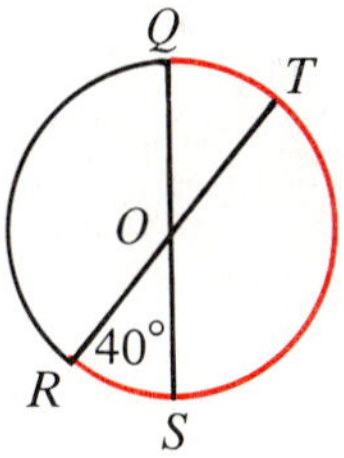

In the figure, diameter $\overline{XY}$ bisects acute ∠ *POQ*.

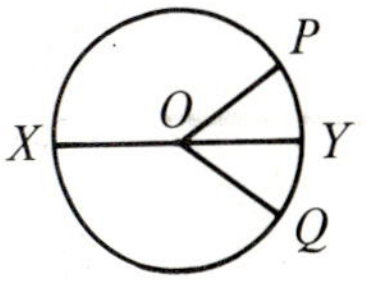

13. Name five minor arcs.

14. Name five major arcs.

15. Name two semicircles.

16. Name three *pairs* of equal arcs.

In the figure, $\overline{BC}$ is a diameter of ⊙*O*.

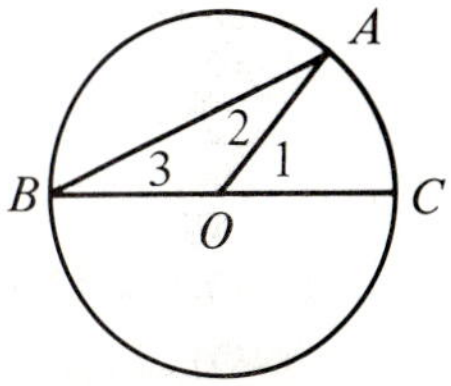

B

17. If ∠ 1 = 40°, then $\overset{\frown}{AB}$ = ___?___°.

18. If ∠ 1 = 36°, then ∠ 3 = ___?___°. (*Hint: AO = BO*)

19. If ∠ 1 = 42°, then ∠ 2 = ___?___°.

20. If ∠ 2 = 25°, then ∠ 1 = ___?___°.

21. If ∠ 3 = 28°, then ∠ 1 = ___?___°.

22. If ∠ 2 = 31°, then $\overset{\frown}{ACB}$ = ___?___°.

In Exercises 23–25, $\overline{AB}$ is a diameter of $\odot O$.

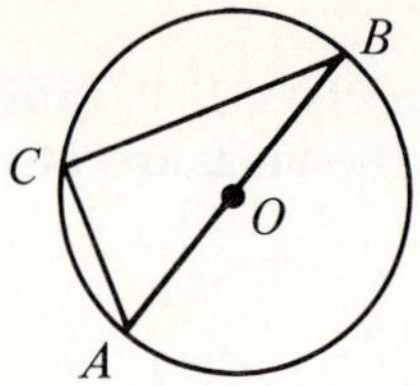

C 23. If $\overset{\frown}{AC} = 50°$, then $\angle ACB =$ __?__ °. (*Hint:* Draw $\overline{OC}$. Find $\angle ACO$ and $\angle BCO$.)

24. If $\overset{\frown}{BC} = 120°$, then $\angle ACB =$ __?__ °.

25. If $\overset{\frown}{AC} = j°$, then $\angle ACB =$ __?__ °.

26. Write a proof of Theorem 4.

SELF-TEST

Vocabulary

circle (p. 346)
chord (p. 346)
diameter (p. 346)
secant (p. 346)
tangent (p. 346)
point of tangency (p. 346)
sphere (p. 346)
common tangent (p. 351)
internal tangent (p. 351)
external tangent (p. 351)
semicircle (p. 355)
minor arc (p. 355)
major arc (p. 355)
central angle (p. 355)

Name each of the following.

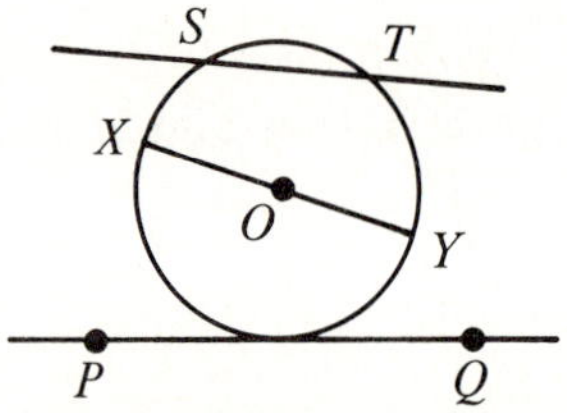

1. A radius
2. A diameter
3. A secant (9-1)
4. A tangent
5. A chord that is not a diameter

6. Construct a circle O. Use X to label a point on the circle. Construct a tangent to circle O at X. (9-2)

7. How many common tangents can be drawn to the two circles?

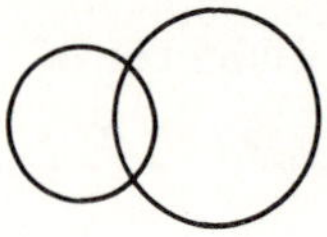

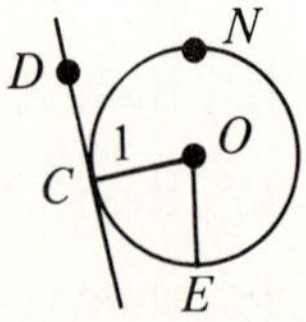

8. If $\angle 1 = 90°$, then $\overleftrightarrow{CD}$ is __?__ to $\odot O$. (9-3)

9. If $\overset{\frown}{CE} = 80°$, then $\angle COE =$ __?__ °.

10. If $\overset{\frown}{CE} = 80°$, then $\overset{\frown}{CNE} =$ __?__ °.

4 Chords

Chord $\overline{JK}$ separates $\odot O$ into two arcs: minor arc JK and major arc JZK. We call $\overset{\frown}{JK}$ *the* arc of chord $\overline{JK}$.

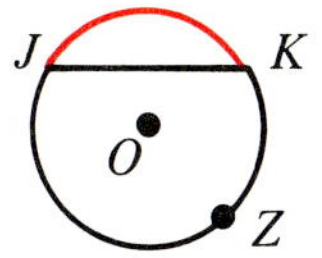

If $\overset{\frown}{CM} = \overset{\frown}{MD}$, then M is the **midpoint** of $\overset{\frown}{CD}$. A line, ray, or segment that contains M **bisects** $\overset{\frown}{CD}$. In the diagram, line j bisects $\overset{\frown}{CD}$.

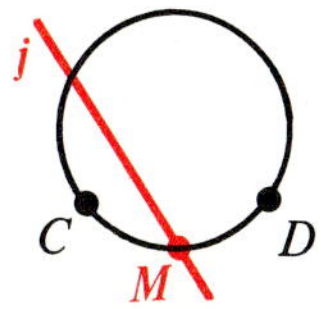

The *distance from center O to chord* $\overline{AB}$ is the length of the shortest segment that can be drawn from O to a point on $\overline{AB}$. It can be shown that the shortest segment is always the *perpendicular* segment from the point to the line.

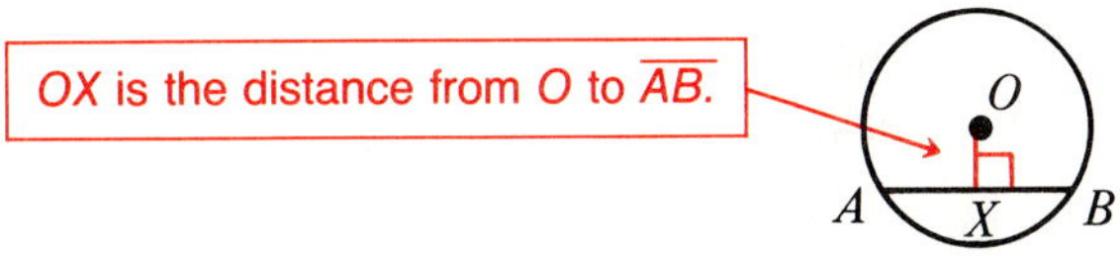

EXPLORATIONS

1. Draw a circle with center O. Then draw equal chords $\overline{WX}$ and $\overline{YZ}$ as shown. Draw four additional segments and prove that $\triangle WOX \cong \triangle YOZ$. Then use corresponding angles of the congruent triangles to prove that $\overset{\frown}{WX} = \overset{\frown}{YZ}$.

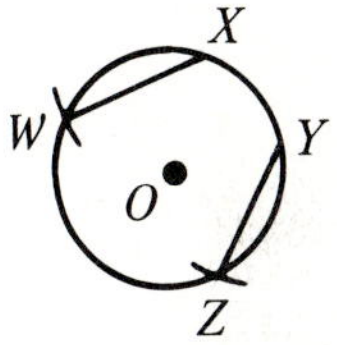

2. Draw a circle with center P. On $\odot P$ mark off an arc, $\overset{\frown}{EF}$. Choose a point G on the circle, and then a point H so that it looks as if $\overset{\frown}{GH} = \overset{\frown}{EF}$. Assume that $\overset{\frown}{GH}$ really does equal $\overset{\frown}{EF}$. Draw $\overline{EF}$ and $\overline{GH}$. Try to prove that $EF = GH$.

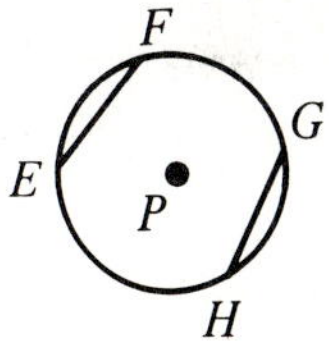

3. Draw a circle with center Q. Then draw equal chords $\overline{RS}$ and $\overline{TV}$. Draw a segment that will help you think of the distance from Q to $\overline{RS}$. Also draw a segment that will remind you of the distance from Q to $\overline{TV}$. What appears to be true about the segments that you have drawn?

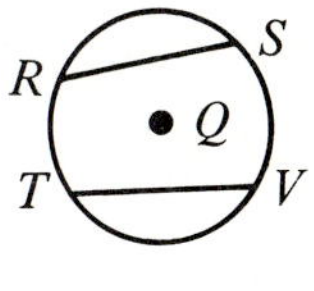

4. Draw a large circle. Then draw a chord and the diameter that is perpendicular to the chord. Do you see any equal segments? any equal arcs?

These explorations lead to the following theorems.

THEOREM 5

In a circle, equal chords have equal arcs and equal arcs have equal chords.

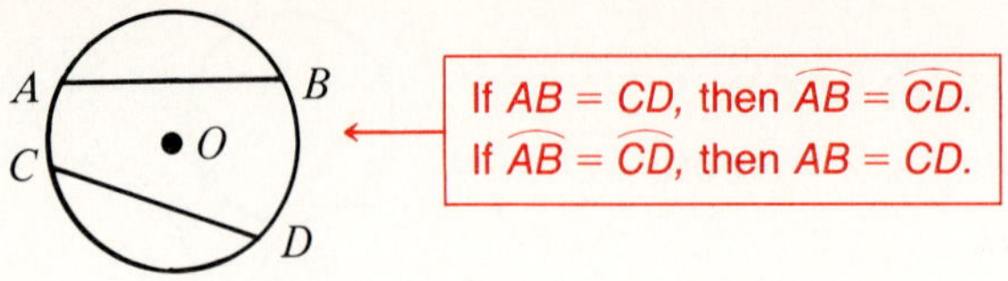

Do you see that Theorem 5 is a "double" theorem? It states both a theorem and its converse. Theorem 6 is also a "double" theorem.

THEOREM 6

In a circle, equal chords are equidistant from the center. Chords that are equidistant from the center are equal.

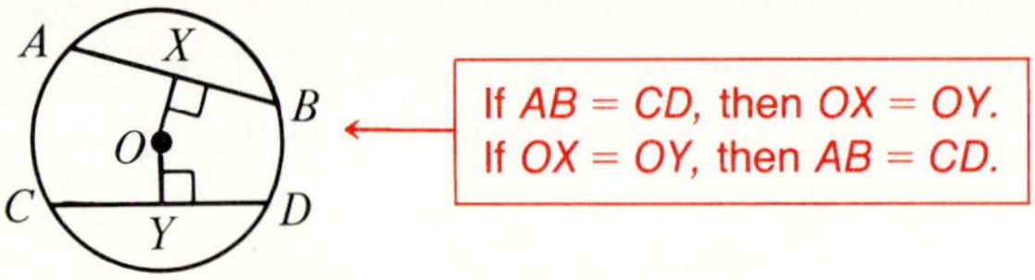

We omit the proof.

THEOREM 7

A diameter that is perpendicular to a chord bisects the chord and its arc.

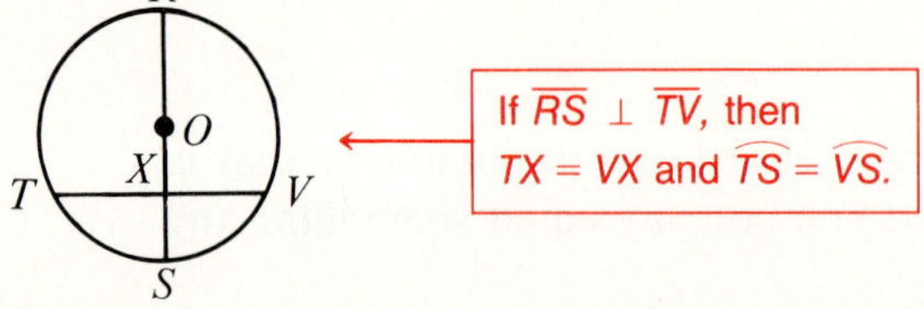

The strategy for a proof is left as Exercise 30 on page 363.

We have used the terms *inscribed* and *circumscribed* with respect to triangles and circles. As the diagrams show, the terms are applied to other polygons, too.

A polygon is *inscribed in* a circle when each vertex lies on the circle. A polygon is *circumscribed about* a circle when each side is tangent to the circle.

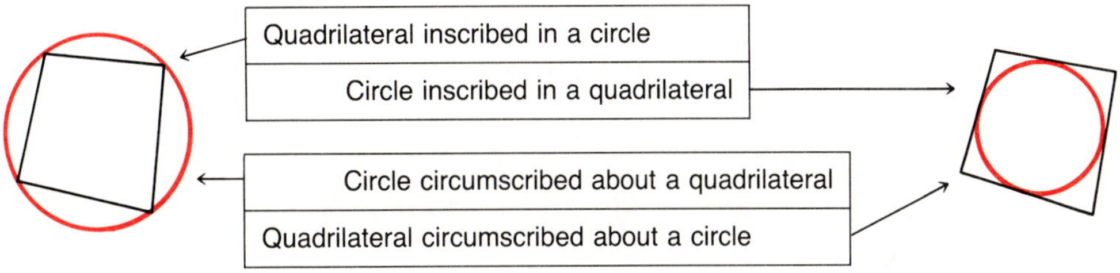

In Chapter 4, you learned how to circumscribe a circle about a triangle (Construction 7). You also learned how to inscribe a circle in a triangle (Construction 8).

Classroom Practice

$\overline{AB}$ is a diameter of $\odot O$. $\overline{AB} \perp \overline{XY}$.

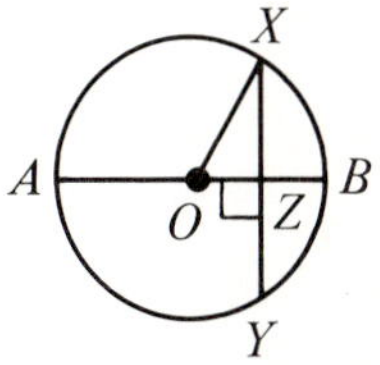

1. $XZ =$ __?__ 2. $\widehat{XB} = \widehat{\text{?}}$

3. The midpoint of $\widehat{XY}$ is __?__.

4. $\widehat{AX} = 180° - \widehat{\text{?}}$ and $\widehat{AY} = 180° - \widehat{\text{?}}$

5. From Exercises 3 and 4, we conclude: $\widehat{\text{?}} = \widehat{\text{?}}$.

Exercises 6–10 refer to the figure.

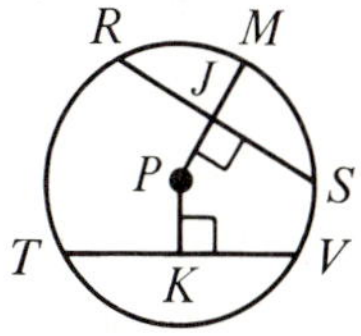

6. $\overline{PM}$ is a __?__ of $\widehat{RS}$.

7. If $PJ = PK$, then $RS =$ __?__.

8. Which is longer: $\overline{PS}$ (not drawn) or $\overline{JS}$?

9. Compare the length of $\overline{TK}$ with the length of $\overline{TV}$.

10. If $RS = TV$, then $\widehat{RS} = \widehat{\text{?}}$.

Tell whether each statement is true or false.

11. If a polygon is inscribed in a circle, the sides of the polygon are chords of the circle.

12. If a rectangle is circumscribed about a circle, then the rectangle must be a square.

Tell whether each statement is true or false.

13. A triangle inscribed in a circle must be isosceles.

14. It is possible to draw an inscribed quadrilateral in such a way that one of its sides is a diameter of the circle.

Written Exercises

State whether the polygon is inscribed in the circle, is circumscribed about the circle, or neither.

A

1. 

2. 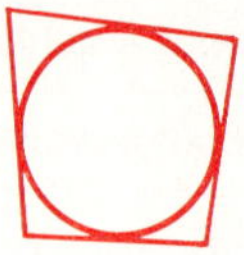

3. 

4. 

5. 

6. 

Exercises 7–12 refer to $\odot O$ with $\overline{OR} \perp \overline{AB}$.

7. If $RB = 5$, $AB =$ __?__.

8. If $AB = 14$, $AR =$ __?__.

9. If $RB = 4$ and $OR = 3$, $OB =$ __?__.

10. If $OB = 10$ and $RB = 8$, $OR =$ __?__.

11. If $OB = 10$ and $AR = 6$, $OR =$ __?__.

12. If $OB = 17$ and $AB = 30$, $OR =$ __?__.

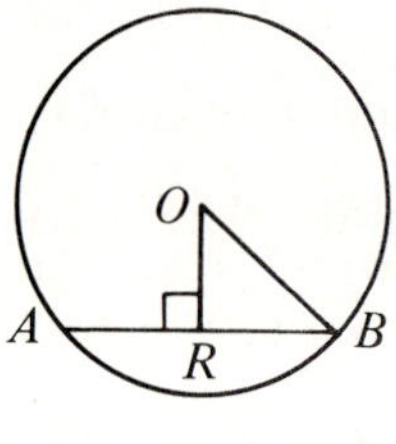

In each exercise, a circle O is shown. Can you conclude that $\widehat{AB} = \widehat{BC}$? Write *yes* or *no*.

13. 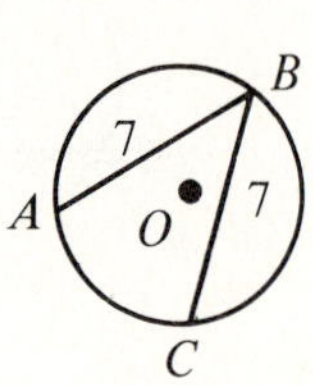

14. 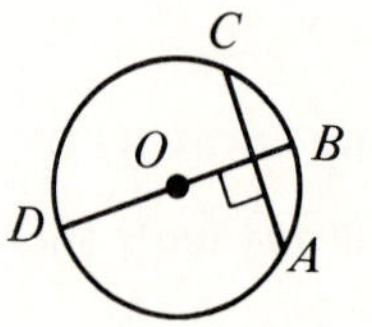

15. 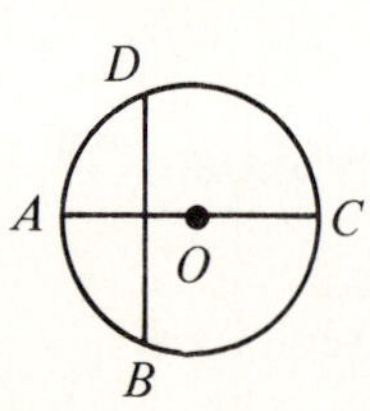

16. 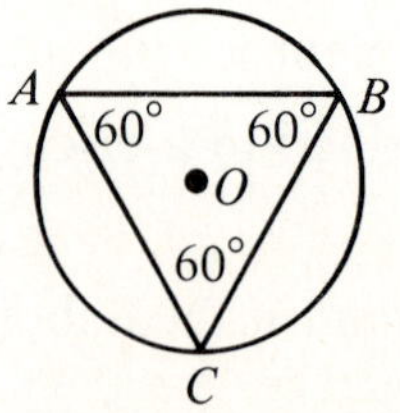

B 17. Here is a way to construct an inscribed regular hexagon:
   a. Construct a circle.
   b. Without changing your compass, mark off points around the circle. You should end up at the starting point.
   c. Draw the hexagon.

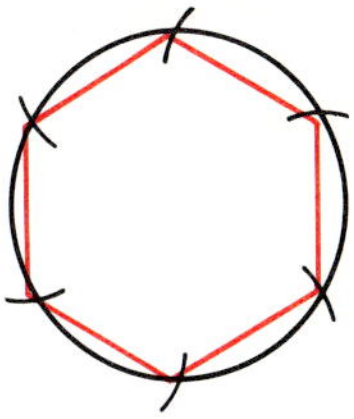

18. Construct a circle. Then inscribe an equilateral triangle in the circle. (*Hint:* See Exercise 17.)

19. Construct a circle. Inscribe a square in the circle. (*Hint:* Draw any diameter. Use Construction 2, page 26, to get a diameter that is perpendicular to the first one. Connect endpoints.)

20. Construct a circle. Inscribe a regular octagon in the circle. (*Hint:* Construct perpendicular diameters as in Exercise 19. Then construct two additional diameters that bisect central angles.)

In each exercise, a circle *O* is shown. Find *ST*.

21. 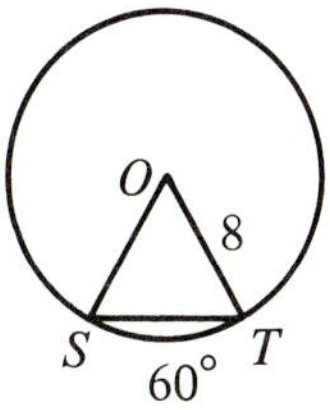

22. 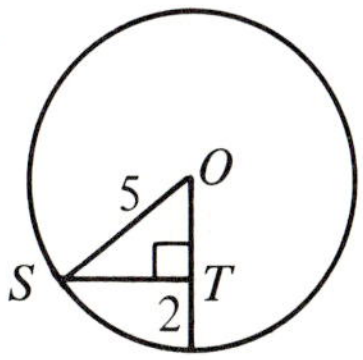

23. 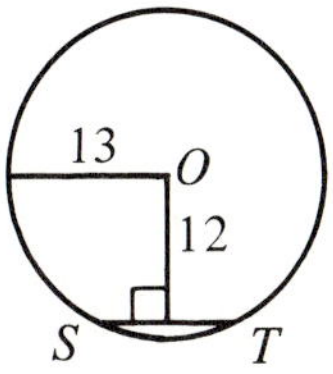

Draw the figure described. If the figure cannot be drawn, write *not possible*. (You need not construct.)

24. A rectangle inscribed in a circle

25. A trapezoid circumscribed about a circle

26. A parallelogram, not a rectangle, inscribed in a circle

C 27. Explain why the construction described in Exercise 17 works.

28. Construct a circle. Then construct a square that is circumscribed about the circle.

29. One student insisted: A diameter that bisects a chord must be perpendicular to the chord. Draw a figure that would show that the student is mistaken.

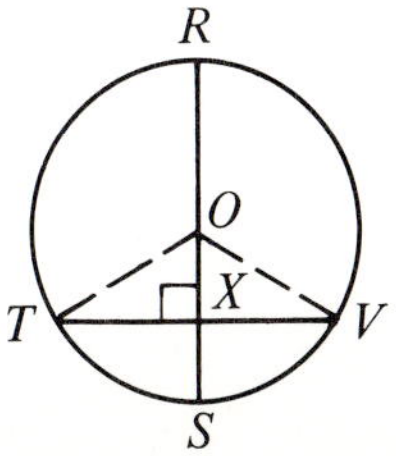

30. *Given:* $\odot O$ with $\overline{RS} \perp \overline{TV}$.
State a strategy you could use to show that
$$TX = VX \text{ and } \widehat{TS} = \widehat{VS}.$$
You need not write out the proof.

5 Inscribed Angles

An angle is *inscribed in* a circle and is an **inscribed angle** if its vertex lies on the circle and its sides are chords of the circle. Each diagram below shows an $\angle C$ inscribed in $\odot O$.

Inscribed Angles

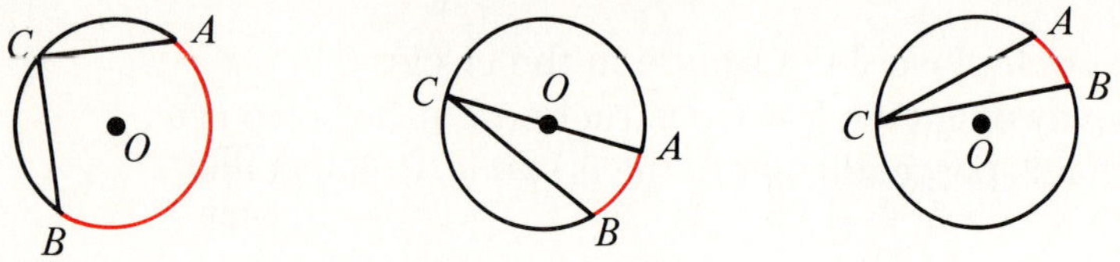

In each case, $\angle C$ **intercepts** an arc, $\widehat{AB}$.

EXPLORATIONS

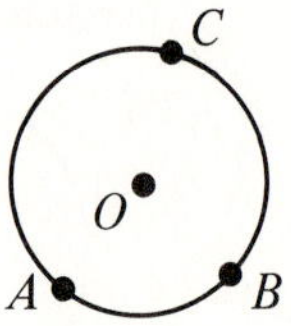

1. Draw a circle O with a radius at least 5 cm long.
2. Choose three points A, B, and C on the circle so that $\widehat{ACB}$ is a major arc.
3. Draw $\overline{CA}$ and $\overline{CB}$. Use a protractor to measure $\angle ACB$.
4. Draw $\overline{OA}$ and $\overline{OB}$. Use a protractor to measure $\angle AOB$.

How does the measure of $\angle ACB$ compare with that of $\angle AOB$?
Recall that $\angle AOB = \widehat{AB}$.
How does the measure of $\angle ACB$ compare with that of $\widehat{AB}$?

Even if your angle measures are not exact, your results should suggest the following theorem.

THEOREM 8
An inscribed angle is equal to half its intercepted arc.

$\angle 1 = \frac{1}{2}\widehat{AB}$ →

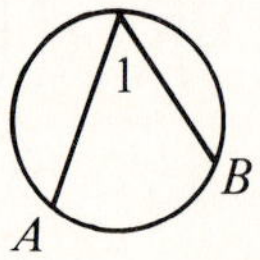

See Written Exercises 31–33.

EXAMPLE Find the measures of $\angle 1$, $\angle 2$, and $\angle 3$.

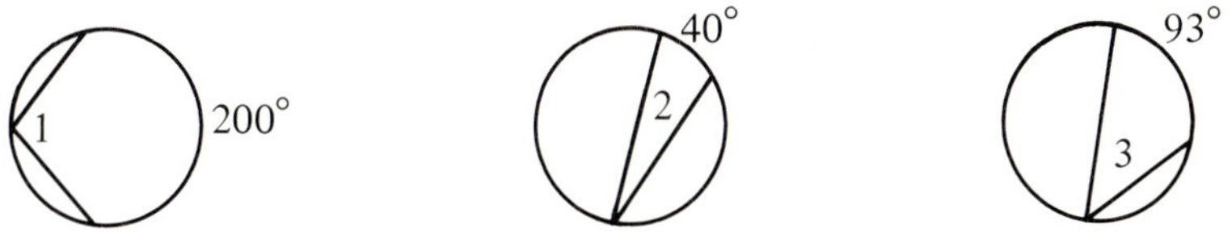

$\angle 1 = \frac{1}{2}(200°) = 100°$ $\angle 2 = \frac{1}{2}(40°) = 20°$ $\angle 3 = \frac{1}{2}(93°) = 46.5°$

In the first three figures below, $\angle A$ is an inscribed angle formed by chords $\overline{AX}$ and $\overline{AY}$. As point Y moves along $\overset{\frown}{XA}$ toward point A, $\overline{AY}$ becomes more and more like a tangent to the circle.

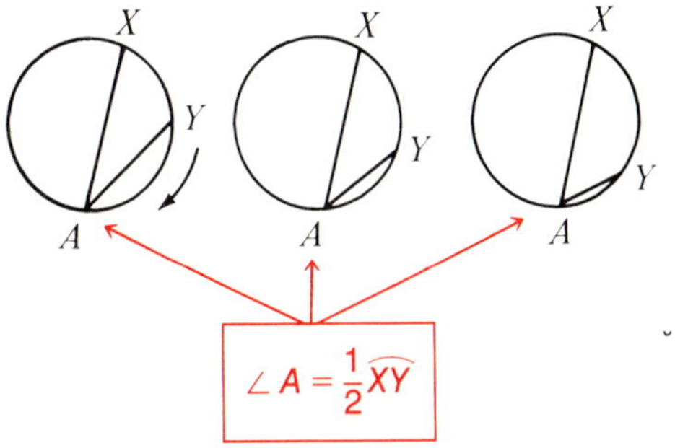

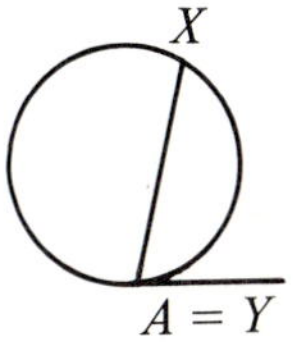

What seems to be true about the measure of $\angle A$?

THEOREM 9

An angle formed by a chord and a tangent is equal to half its intercepted arc.

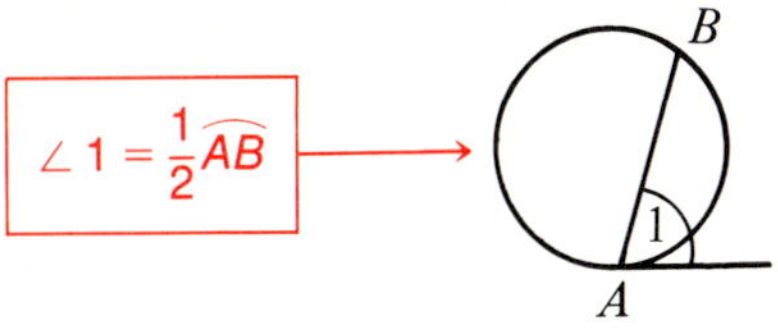

Classroom Practice

State whether or not $\angle 1$ is an inscribed angle.

1.

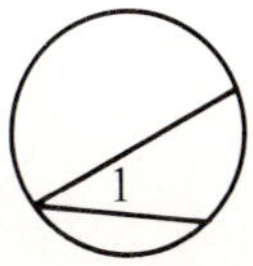

2.

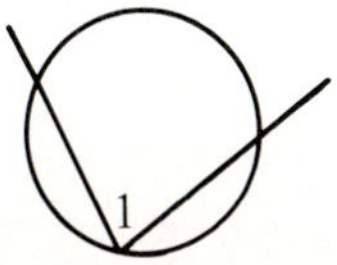

3.

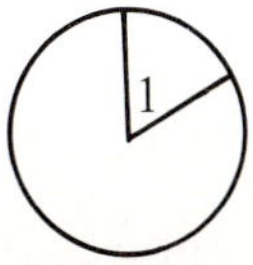

State whether or not ∠ 1 is an inscribed angle.

4. 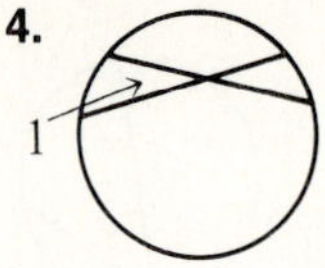

5. 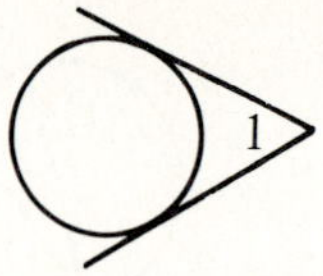

6. 

In the diagram, $\overrightarrow{CD}$ is a tangent. Complete each statement.

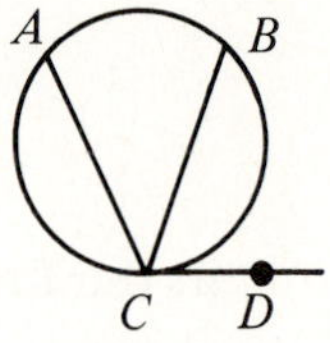

7. The inscribed angle pictured is ∠ ___?___.

8. ∠ ACB intercepts $\overset{\frown}{\_\_?\_\_}$.   9. ∠ BCD intercepts $\overset{\frown}{\_\_?\_\_}$.

10. ∠ $ACB = \frac{1}{2}\overset{\frown}{\_\_?\_\_}$ 11. ∠ $BCD = \frac{1}{2}\overset{\frown}{\_\_?\_\_}$ 12. ∠ $ACD = \frac{1}{2}\overset{\frown}{\_\_?\_\_}$

Written Exercises

In Exercises 1–8, use the diagram.

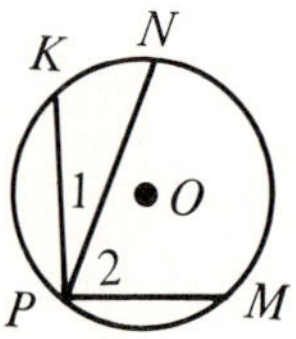

A 1. If $\overset{\frown}{KN} = 50°$, then ∠ 1 = ___?___°.

2. If $\overset{\frown}{NM} = 110°$, then ∠ 2 = ___?___°.

3. If $\overset{\frown}{KN} = 70°$ and $\overset{\frown}{NM} = 115°$, then $\overset{\frown}{KNM}$ = ___?___°.

4. If $\overset{\frown}{KPM} = 170°$ and $\overset{\frown}{KP} = 100°$, then $\overset{\frown}{PM}$ = ___?___°.

5. If ∠ 1 = 28°, then $\overset{\frown}{KN}$ = ___?___°.

6. If ∠ 2 = 76°, then $\overset{\frown}{NM}$ = ___?___°.

7. If ∠ 1 = 24° and ∠ 2 = 68°, then $\overset{\frown}{KNM}$ = ___?___°.

8. If ∠ 1 = 28° and ∠ 2 = 69°, then $\overset{\frown}{KPM}$ = ___?___°.

In the diagram, $\overleftrightarrow{DE}$ is tangent to ⊙O at point A.

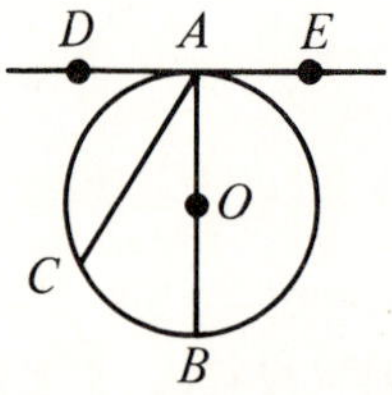

9. ∠ BAE = ___?___°   10. ∠ DAB = ___?___°

11. If ∠ CAB = 30°, then ∠ DAC = ___?___°.

12. If ∠ DAC = 55°, then ∠ CAB = ___?___°.

13. If ∠ CAB = 32°, then $\overset{\frown}{BC}$ = ___?___°.

14. If $\overset{\frown}{AC} = 100°$, then ∠ CAB = ___?___°.

Exercises 15–20 refer to the diagram.

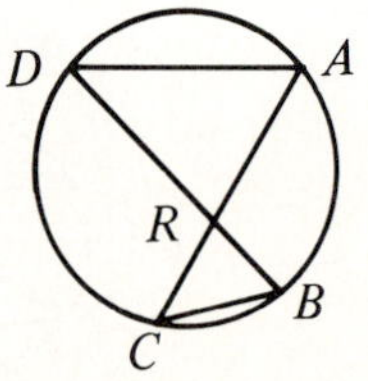

B 15. If $\overset{\frown}{AB} = 96°$, then ∠ D = ___?___° and ∠ C = ___?___°.

16. If $\overset{\frown}{DC} = 118°$, then ∠ A = ___?___° and ∠ B = ___?___°.

17. If $\angle D = 44°$, then $\widehat{AB} =$ __?__° and $\angle C =$ __?__°.

18. If $\angle D = 47°$, then $\widehat{AB} =$ __?__° and $\angle C =$ __?__°.

19. Is $\angle D = \angle C$? Is $\angle A = \angle B$?

20. Is $\triangle ADR \sim \triangle BCR$?

In each diagram, find the measure of $\angle 1$.

21.

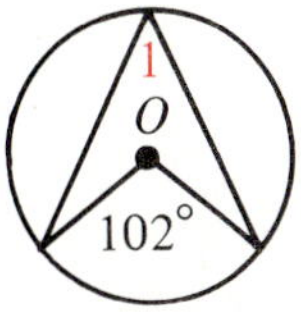

22.

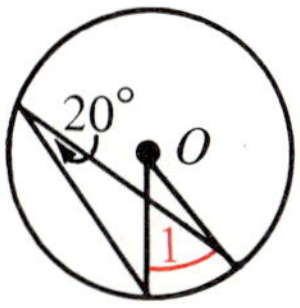

23.

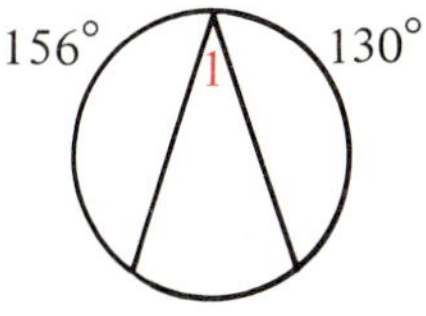

24.

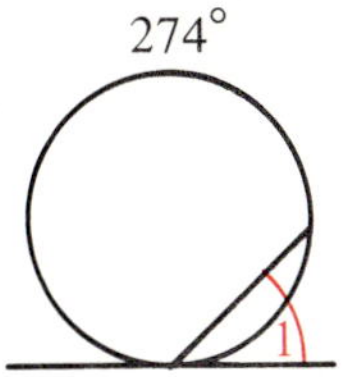

25.

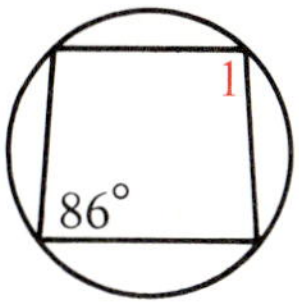

26.

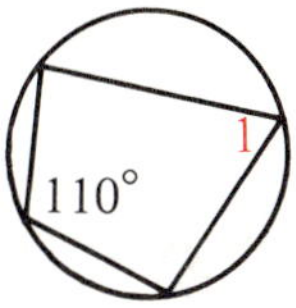

$\triangle CAB$ is isosceles with $AC = AB$.

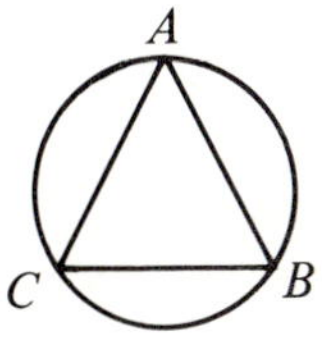

27. If $\angle A = 55°$, then $\widehat{AB} =$ __?__°.

28. If $\widehat{CB} = k°$, then $\angle B =$ __?__°. (Answer in terms of k.)

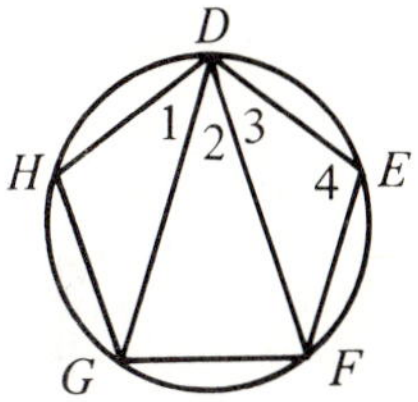

29. *DEFGH* is a regular pentagon.
   a. $\widehat{DE} =$ __?__° and $\widehat{EF} =$ __?__°
   b. Find the measure of each numbered angle.

C 30. Explain why the opposite angles of a quadrilateral inscribed in a circle must be supplementary.

In Exercises 31–33, write a strategy for proving the three cases of Theorem 8. In each case, $\overline{CA}$ and $\overline{CB}$ are chords of $\odot O$. Show that $\angle ACB = \frac{1}{2}\widehat{AB}$. You need not write out the proof.

31. *Case 1*

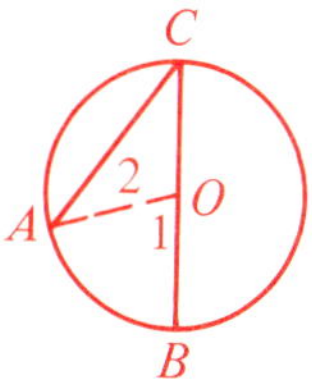

Hint:
$\angle ACB + \angle 2 = \angle 1$

32. *Case 2*

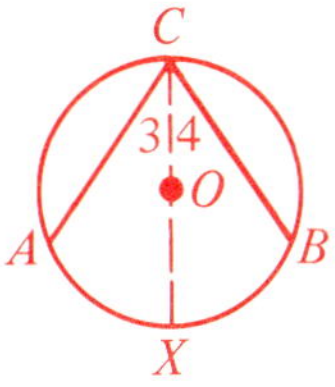

Hint: Use Case 1.
$\widehat{AB} = \widehat{AX} + \widehat{XB}$

33. *Case 3*

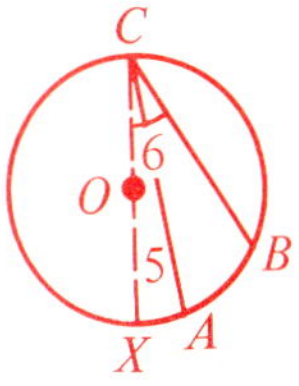

Hint: Use Case 1.
$\widehat{AB} = \widehat{XB} - \widehat{XA}$

SELF-TEST

Vocabulary

midpoint of an arc (p. 359)
bisector of an arc (p. 359)
inscribed angle (p. 364)
intercepted arc (p. 364)

(9-4)

1. If $AB = 12$, then $AX =$ __?__.
2. If $\widehat{BY} = 32°$, then $\widehat{AY} =$ __?__ °.
3. If $AB = 12$ and $OW = OX$, then $UV =$ __?__.

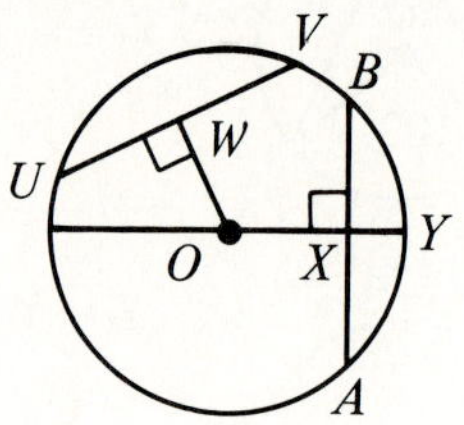

(9-5)

4. If $\widehat{RS} = 80°$, then $\angle 1 =$ __?__ °.
5. If $\angle 2 = 38°$, then $\widehat{ST} =$ __?__ °.
6. If $\angle 1 = 42°$ and $\angle 2 = 36°$, then $\widehat{RST} =$ __?__ ° and $\widehat{RKT} =$ __?__ °.
7. If $\widehat{TK} = 60°$, then $\angle 3 =$ __?__ °.

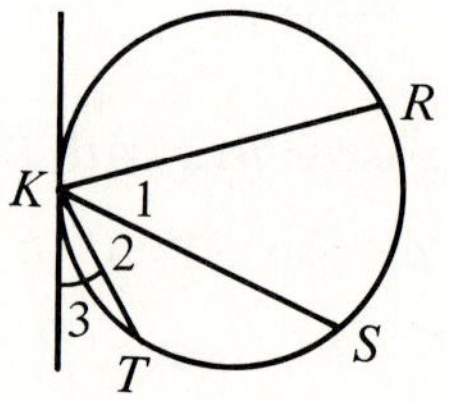

8. Draw a rhombus inscribed in a circle.

PUZZLE ◆ PROBLEMS

The project on page 354 involved folding waxed paper to form a circle. You can fold paper to form other curves. For this project, use a rectangular sheet of waxed paper.

Select a side $\overline{XY}$ and a point P.

Fold the paper so that some point of $\overline{XY}$ falls on point P.

Make a sharp crease, then unfold.

Make about twenty different creases in the same way.

Hold the paper at arm's length and the creases will suggest a *parabola*.

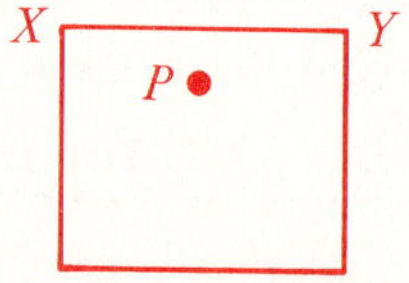

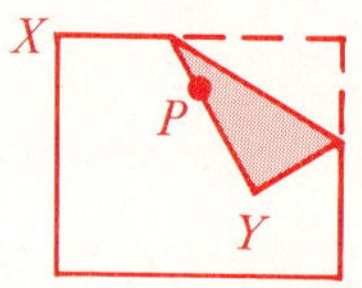

6 Other Angles

EXPLORATIONS

A. We wish to find the measure of $\angle 1$ without using a protractor.

1. Draw $\overline{AD}$. You can easily find the measures of $\angle ADC$ and $\angle DAB$.
2. Notice that $\angle 1$ is an exterior angle of $\triangle AXD$.

   $\angle 1 = \frac{1}{2}(\underline{\ \ ?\ \ }°) + \frac{1}{2}(\underline{\ \ ?\ \ }°)$

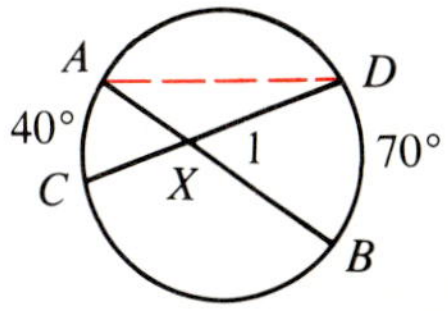

B. We wish to find the measure of $\angle P$ without using a protractor.

1. Draw $\overline{ST}$. You can easily find the measures of $\angle RST$ and $\angle STV$.
2. Notice that $\angle RST$ is an exterior angle of $\triangle TSP$.

   $\angle P = \frac{1}{2}(\underline{\ \ ?\ \ }°) - \frac{1}{2}(\underline{\ \ ?\ \ }°)$

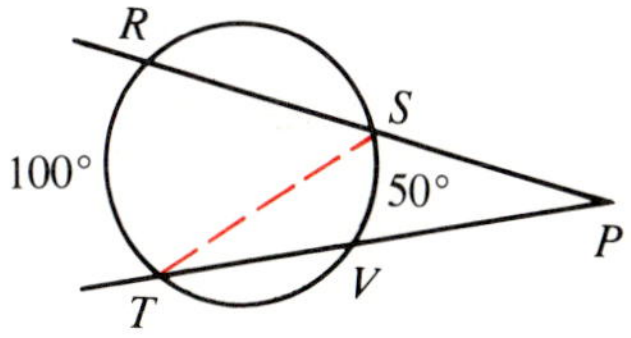

These explorations suggest two theorems—and their proofs.

THEOREM 10

An angle formed by two chords is equal to half the sum of the intercepted arcs.

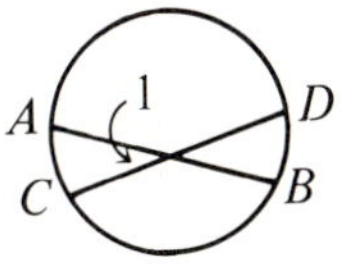

$\angle 1 = \frac{1}{2}(\overset{\frown}{AC} + \overset{\frown}{BD})$

THEOREM 11

An angle formed by two secants is equal to half the difference of the intercepted arcs.

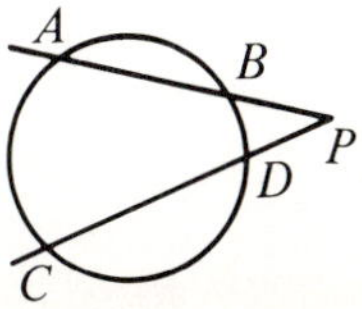

$\angle P = \frac{1}{2}(\overset{\frown}{AC} - \overset{\frown}{BD})$

EXAMPLE In each diagram, find the measure of $\angle 1$.

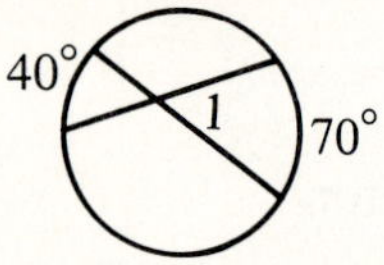

$$\angle 1 = \frac{1}{2}(40° + 70°) = \frac{1}{2}(110°) = 55°$$

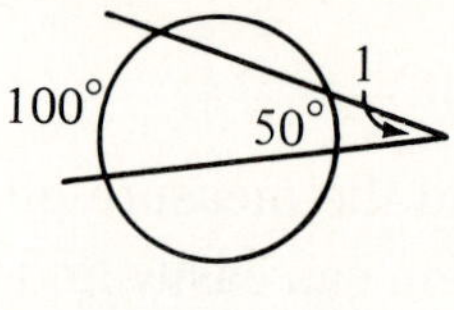

$$\angle 1 = \frac{1}{2}(100° - 50°) = \frac{1}{2}(50°) = 25°$$

Did you get these answers when you tried the explorations on the previous page?

SUMMARY OF ANGLE AND ARC RELATIONSHIPS

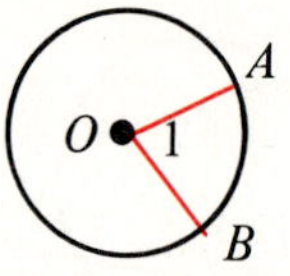

$\angle 1$ is a central angle.

$\angle 1 = \overset{\frown}{AB}$

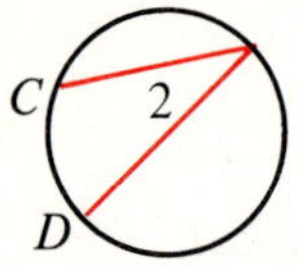

$\angle 2$ is an inscribed angle.

$\angle 2 = \frac{1}{2}\overset{\frown}{CD}$

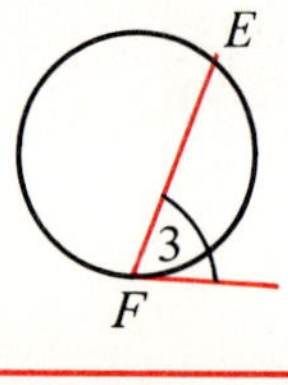

$\angle 3$ is formed by a chord and a tangent.

$\angle 3 = \frac{1}{2}\overset{\frown}{EF}$

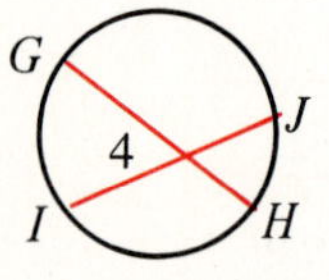

$\angle 4$ is formed by two chords.

$\angle 4 = \frac{1}{2}(\overset{\frown}{GI} + \overset{\frown}{JH})$

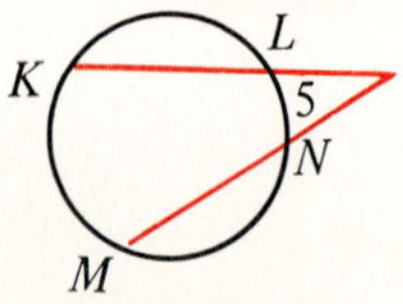

$\angle 5$ is formed by two secants.

$\angle 5 = \frac{1}{2}(\overset{\frown}{KM} - \overset{\frown}{LN})$

Classroom Practice

Find the measure of each numbered angle.

1.

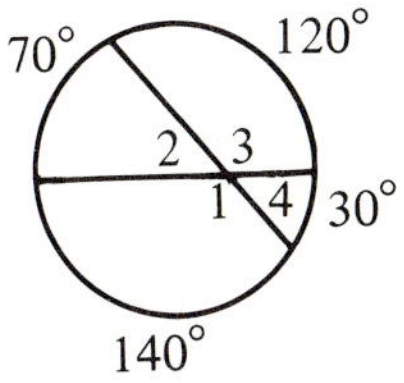

2.

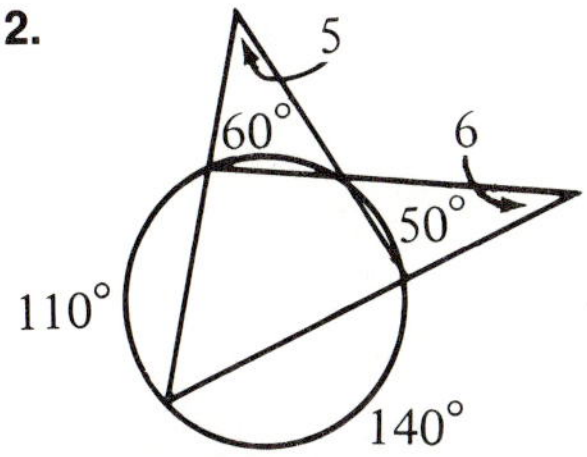

3.

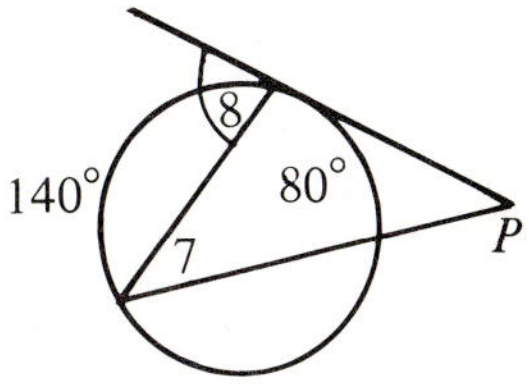

4. Using the diagram for Exercise 3, find ∠ *P*. (*Hint:* ∠ 8 is an exterior angle of the triangle.)

Written Exercises

Find the measure of ∠ 1.

A

1.

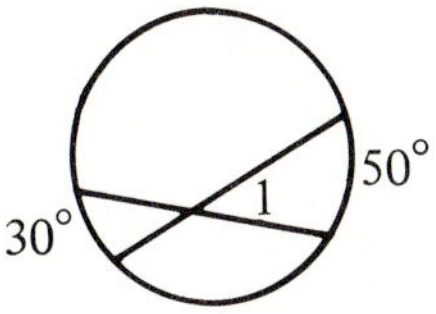

2.

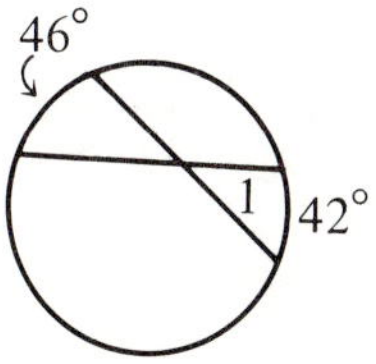

3.

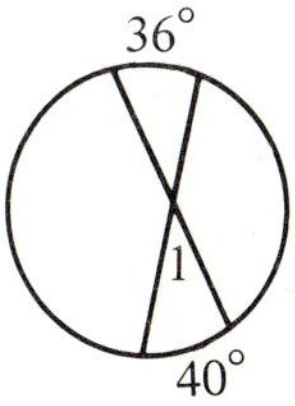

4.

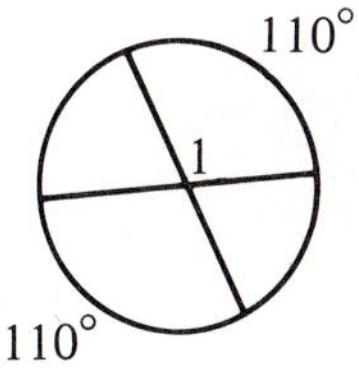

5.

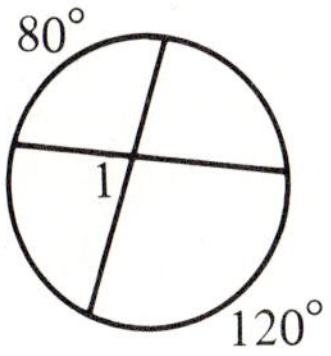

6.

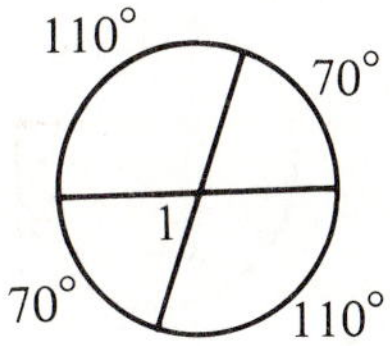

7.

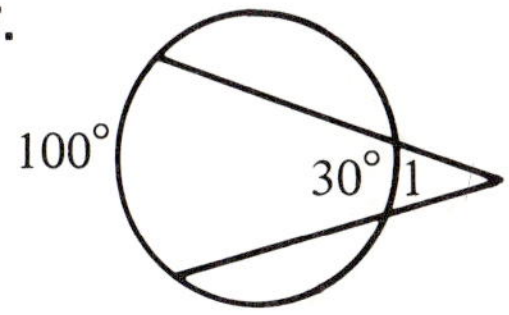

8.

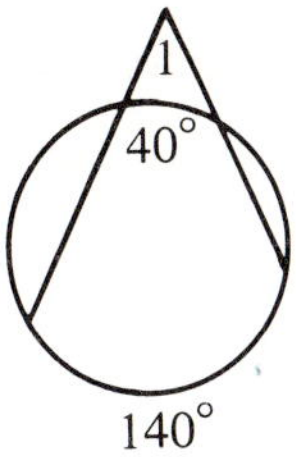

9.

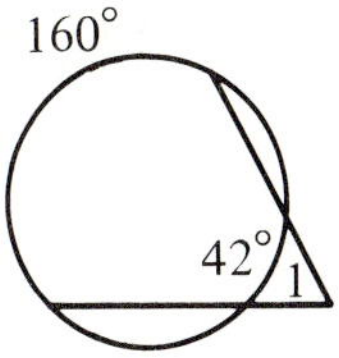

10.

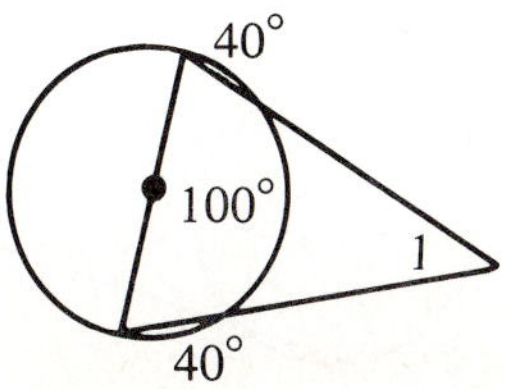

11.

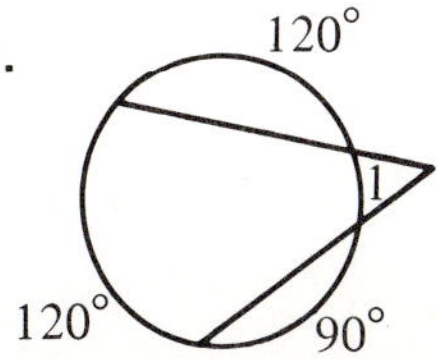

12.

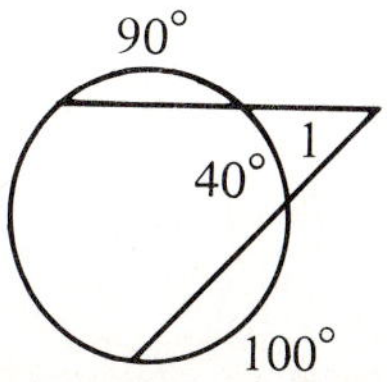

Find the value of x.

SAMPLE

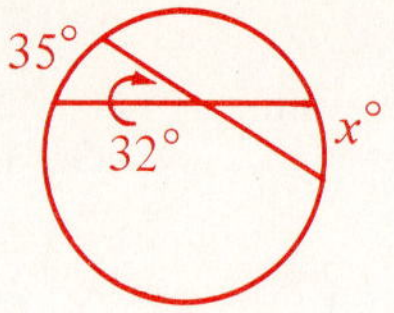

$$32 = \frac{1}{2}(35 + x)$$

$$64 = 35 + x$$

$$29 = x$$

B **13.**

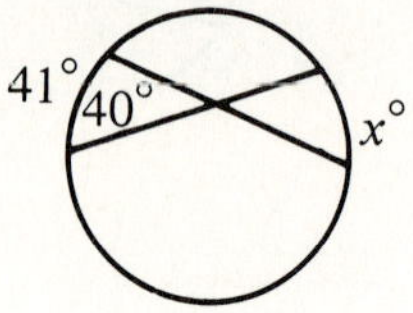

14.

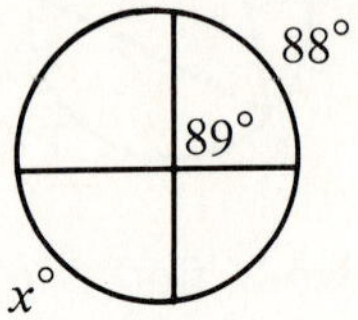

15.

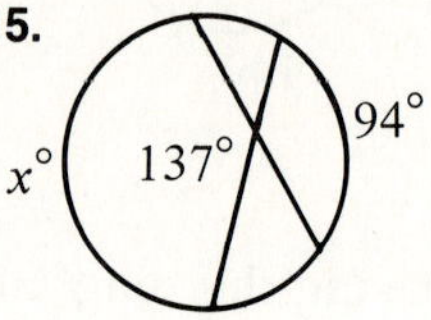

16.

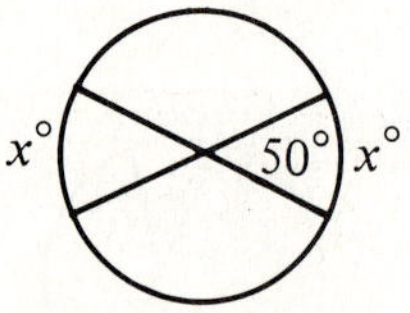

17.

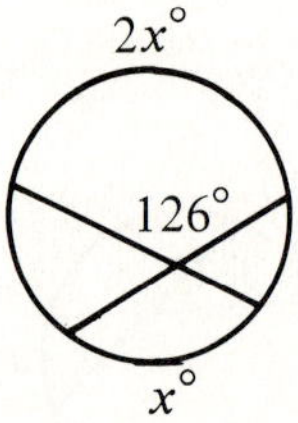

18.

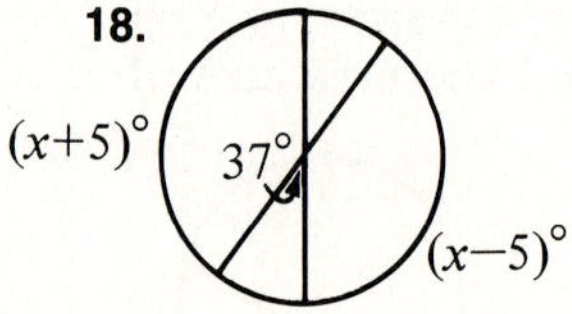

19.

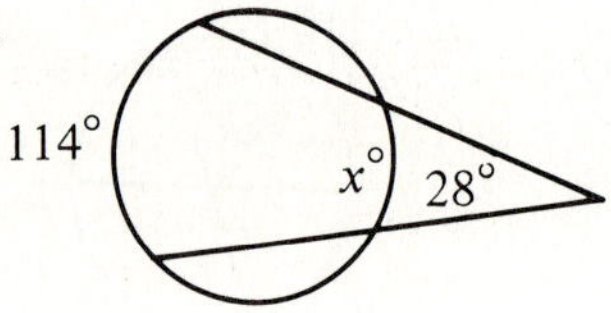

20.

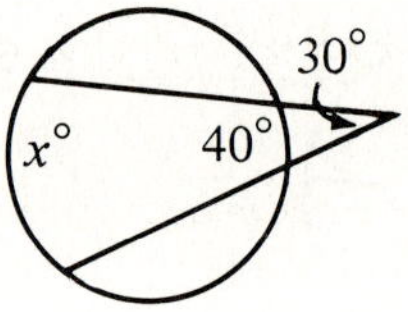

C **21.**

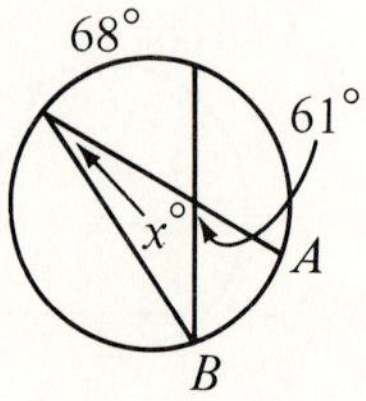

22.

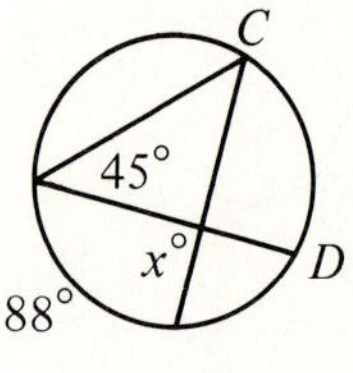

23.

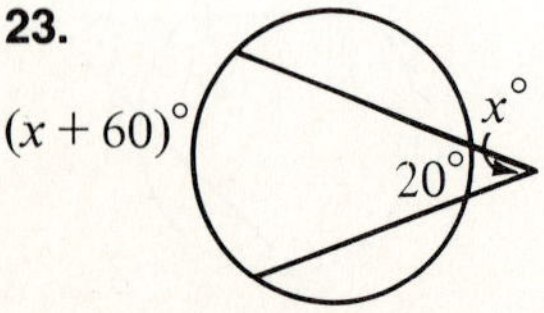

24.

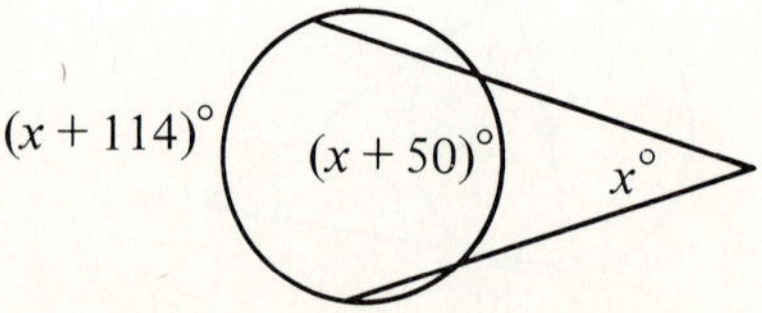

7 Segments of Chords

Do you believe that the equation

$$AX \cdot XB = CX \cdot XD$$

can possibly be correct? Some people do not believe it until they see a proof.

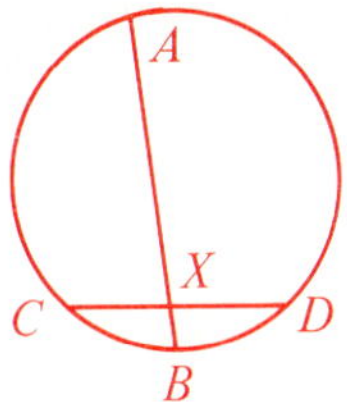

THEOREM 12

If two chords intersect inside a circle, the product of the lengths of the segments of one chord equals the product of the lengths of the segments of the other.

Given: Chords $\overline{AB}$ and $\overline{CD}$ intersect at point X inside the circle.

Prove: $AX \cdot XB = CX \cdot XD$

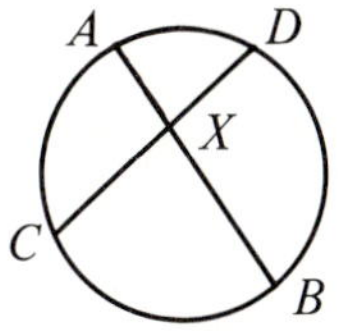

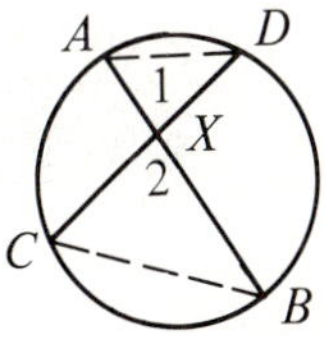

STATEMENTS	REASONS
1. Draw $\overline{AD}$ and $\overline{BC}$.	1. Through any two points there is exactly one line.
2. $\angle A = \frac{1}{2}\widehat{BD}$; $\angle C = \frac{1}{2}\widehat{BD}$	2. An inscribed angle is equal to __?__.
3. $\angle A = \angle C$	3. Substitution Postulate
4. $\angle 1 = \angle 2$	4. __?__
5. $\triangle AXD \sim \triangle CXB$	5. AA Postulate
6. $\frac{AX}{CX} = \frac{XD}{XB}$	6. Corresponding sides of ~ $\triangle$s are in proportion.
7. $AX \cdot XB = CX \cdot XD$	7. A property of proportions

Sometimes it is simpler to apply Theorem 12 by thinking of a figure like the one shown.

$p \times q = r \times s$

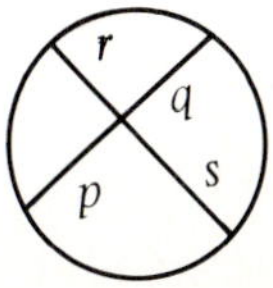

Classroom Practice

In each exercise, find the length x.

1. 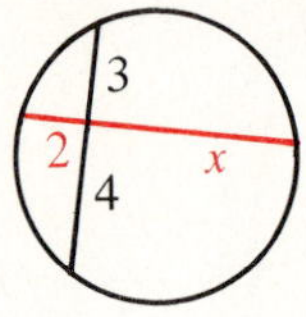

$2x = \underline{\ ?\ }$
$x = \underline{\ ?\ }$

2. 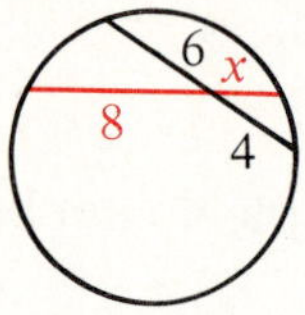

$8x = \underline{\ ?\ }$
$x = \underline{\ ?\ }$

3. 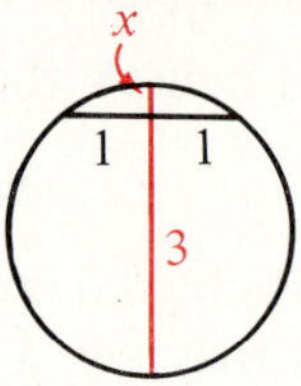

$3x = \underline{\ ?\ }$
$x = \underline{\ ?\ }$

Exercises 4–7 lead to a formula about the lengths associated with secants to a circle from an outside point.

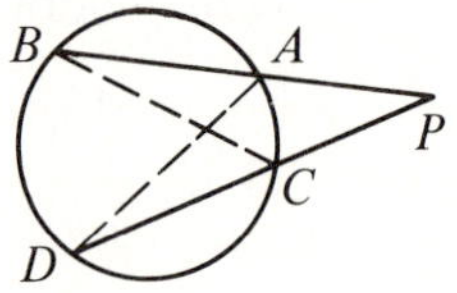

4. Name two angles of $\triangle PBC$ that are equal to two angles of $\triangle PDA$.

5. $\triangle PBC \sim \triangle \underline{\ ?\ }$.

6. Write a proportion that involves PB, PA, PD, and PC.

7. $PB \cdot \underline{\ ?\ } = PD \cdot \underline{\ ?\ }$.

8. Look at the figures below. As point B moves along $\overset{\frown}{BA}$ toward point A, $\overrightarrow{PB}$ becomes more and more like a tangent to the circle.

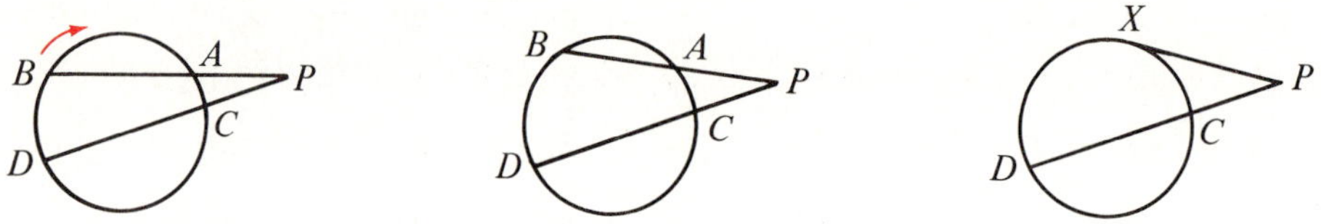

In the third diagram, A and B have become the same point, X. Look at the equation in Exercise 7. Suggest an equation involving PX, PC, and PD.

Written Exercises

In each exercise, find the length x.

SAMPLE

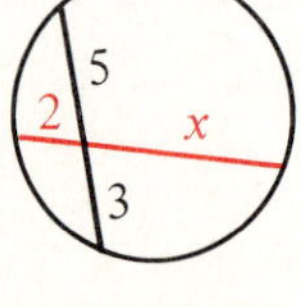

$2x = 3 \cdot 5$
$2x = 15$
$x = \frac{15}{2} = 7\frac{1}{2}$

A

1.

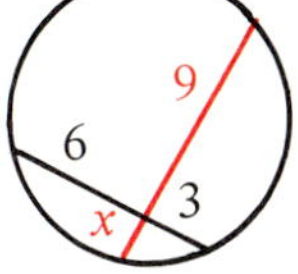

2.

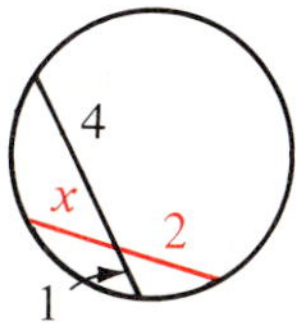

3.

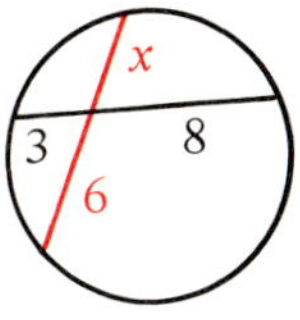

4.

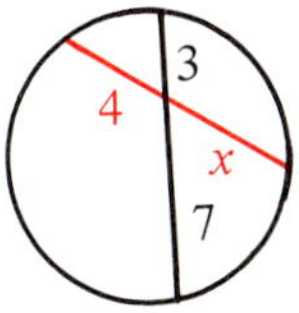

5.

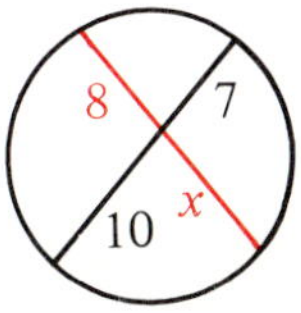

6.

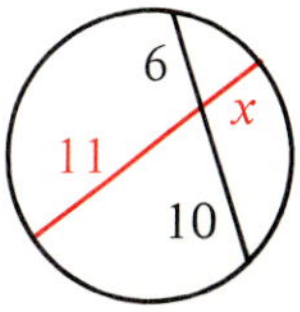

7.

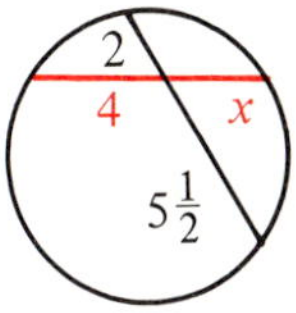

8.

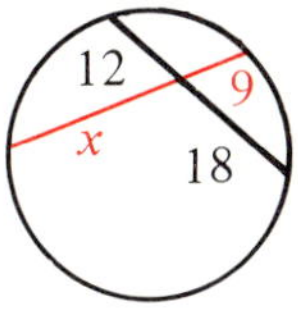

9.

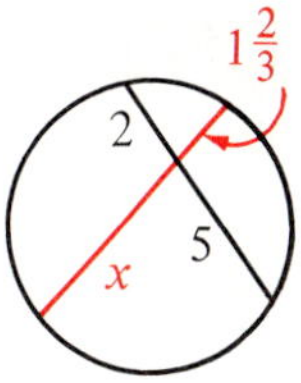

B

10.

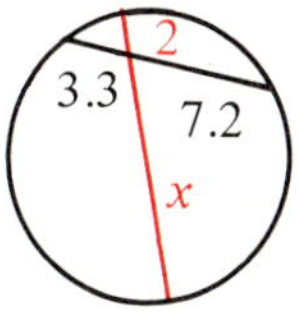

11.

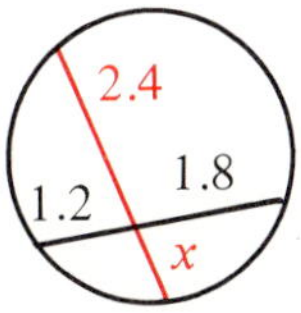

12.

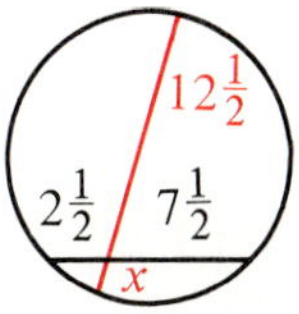

Exercises 13–16 lead to the same equation that was developed in Exercise 8, page 374. In the diagram, $\overline{PX}$ is a tangent segment and $\overleftrightarrow{PD}$ is a secant.

13. $\angle PXC = \frac{1}{2}\ \underline{\ \widehat{?}\ }$; $\angle PDX = \frac{1}{2}\ \underline{\ \widehat{?}\ }$

$\angle PXC = \angle PDX$; $\angle P = \angle P$

14. $\triangle PXC \sim \triangle$ __?__

15. $\frac{PX}{?} = \frac{PC}{?}$

16. $PX \times$ __?__ $= PC \times$ __?__

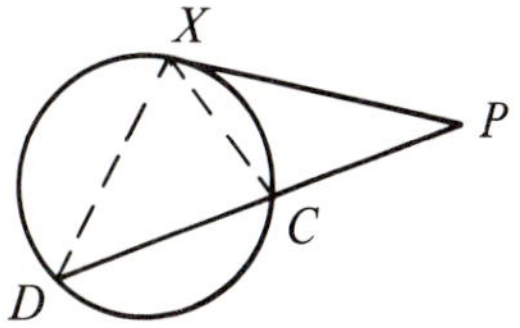

Secants of circles are shown. Find the value of *x*.

C

17.

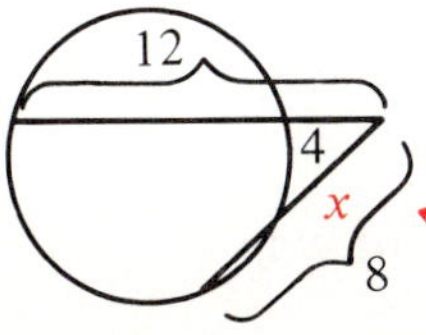

Hint: $4 \times 12 = 8x$

18.

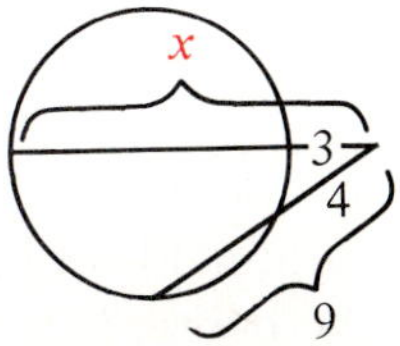

Secants and tangents of circles are shown. Find the value of x.

19.

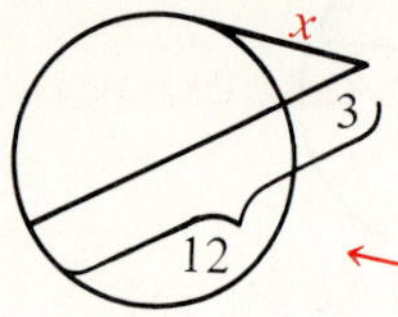

20.

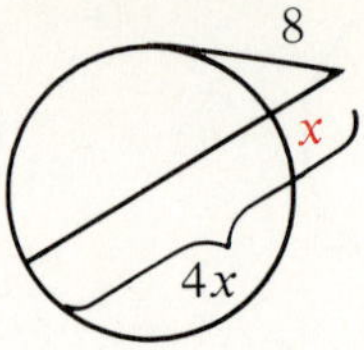

Hint: See Exercise 16.

SELF-TEST

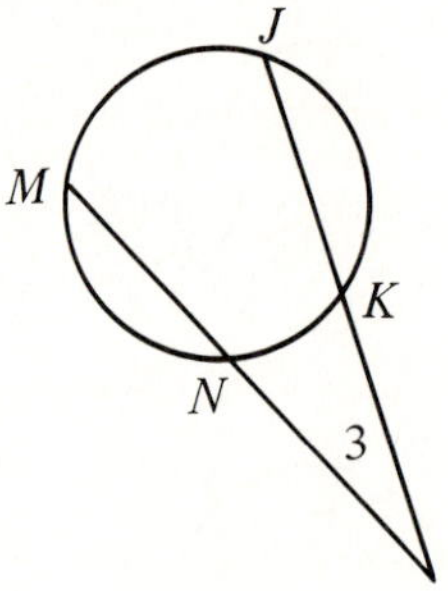

(9-6)

1. If $\overparen{JM} = 100°$ and $\overparen{KN} = 50°$, then $\angle 3 =$ __?__°.
2. If $\overparen{JK} = 108°$, $\overparen{JM} = 96°$, and $\overparen{MN} = 94°$, then $\angle 3 =$ __?__°.
3. If $\angle 3 = 40°$ and $\overparen{KN} = 30°$, then $\overparen{JM} =$ __?__°.

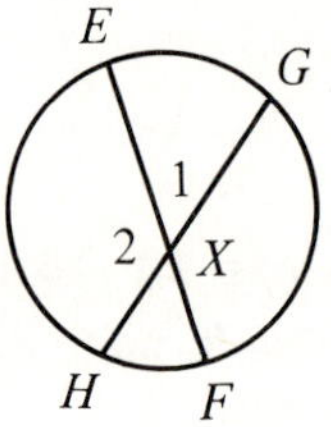

(9-7)

4. If $\overparen{GE} = 60°$ and $\overparen{HF} = 40°$, then $\angle 1 =$ __?__°.
5. If $\overparen{GE} = 67°$ and $\overparen{HF} = 41°$, then $\angle 2 =$ __?__°.
6. If $\angle 2 = 130°$ and $\overparen{EH} = 140°$, then $\overparen{GF} =$ __?__°.
7. If $EX = 15$, $XF = 8$, and $GX = 12$, then $XH =$ __?__.
8. If $GX = 9$, $XH = 8$, and $EX = 12$, then $FX =$ __?__.

PUZZLE ◆ PROBLEMS

Suppose you have a circle, but have lost track of its center. How can you find the center?

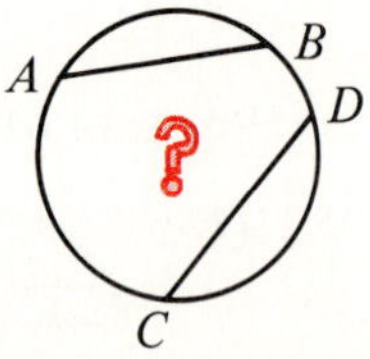

Draw two chords, $\overline{AB}$ and $\overline{CD}$. Construct the perpendicular bisector of each chord. The two perpendicular bisectors intersect at the center of the circle.

Try the construction when $\overline{AB}$ and $\overline{CD}$ are parallel. Why doesn't the construction work?

Return to the original problem. Can you discover a different way to construct the center?

APPLICATIONS

USING CIRCLES TO SET UP SCHEDULES

Five teams enter a tournament. Each team is to play each other team once. Begin arranging a schedule by dividing a circle into five equal arcs. Call the points, and the teams, *A*, *B*, *C*, *D*, and *E*.

FIRST ROUND

Draw the diameter containing *A*.
Connect other points as shown.
Games: A-bye *B-E* *C-D*
Bye means that the team does not play.

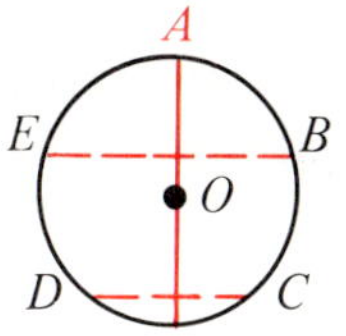

SECOND ROUND

Draw the diameter containing *B*.
Games: B-bye *C-A* *D-E*

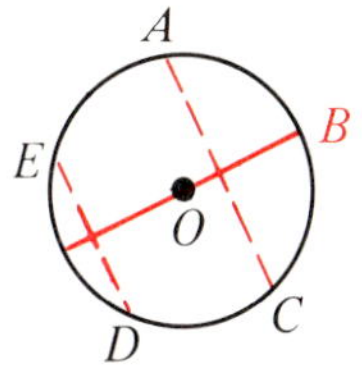

THIRD, FOURTH, AND FIFTH ROUNDS

Draw diameters containing *C*, *D*, and *E*. In the five rounds there will be a total of ten games, not counting byes.

For a tournament of six teams, use five equal arcs as above. But when you draw the diameter containing *A*, let that mean that *A* will play the sixth team, *O*. There won't be any byes. In the five rounds there will be a total of fifteen games.

For a tournament of seven teams—or of eight teams—use seven equal arcs. You should get 21 games and 28 games, respectively.

COMPUTER ACTIVITIES

TEN-SPEED BICYCLES

A ten-speed bike doesn't have ten gears! It has two sprockets (gears) in front and five gears in back. Each front sprocket can be connected to each back gear with the chain. The pedals turn the front sprocket. When the pedals turn once, the front sprocket also turns once.

The back gears have fewer teeth than the front sprockets. Therefore, they turn faster. For example, a bike might have 40 front sprocket teeth and 20 back gear teeth. Every time the pedals turned once, the gear would turn twice. The back wheel is connected to a rear gear. So for one pedal turn, you would have two turns of the wheels.

Different bikes have different gear setups. Here are the numbers of teeth on one typical bike.

FRONT: 52,40 REAR: 28,24,20,17,14

"First gear" is easiest to pedal. The chain is on the smaller front sprocket (40 teeth) and the biggest gear (28 teeth). This is the gear for climbing hills. "Tenth gear" is the hardest to pedal. The chain is on the larger front sprocket (52 teeth) and the smallest gear (14 teeth). This gear is used for fast cruising on the level or downhill.

The gear ratio indicates the degree of effort needed to use each gear. The program below will figure the gear ratio for each of the ten gear combinations. This is found by dividing the number of front sprocket teeth by the number of back gear teeth. The lower the gear ratio, the easier it is to pedal.

```
10  REM***FINDING GEAR RATIOS
20  DATA 28,24,20,17,14
30  DATA 28,24,20,17,14
40  FOR G = 1 TO 10
50  READ B
60  REM***FIGURE GEAR RATIOS
70  REM***FOR 40 FRONT TEETH
80  IF G < = 5 THEN R = 40/B
90  REM***FOR 52 FRONT TEETH
100  IF G > 5 THEN R = 52/B
110  PRINT "GEAR ";G;" HAS RATIO ";R
120  NEXT G
200  END
```

Notice the DATA and READ statements in lines 20, 30, and 50. This is a way to give the computer information it needs when the values are known ahead of time. Then the computer doesn't have to stop to get the values, as it does for INPUT statements.

EXERCISES

1. RUN the program. Complete the chart. Round gear ratios to the nearest hundredth.

Gear		1	2	3	4	5	6	7	8	9	10
Teeth	Front	40	40	40	40	40	52	52	52	52	52
	Rear	28	24	20	17	14	28	24	20	17	14
Gear ratio		?	?	?	?	?	?	?	?	?	?

2. List the gears in order. Go from easiest (1st) to hardest (10th).
3. Suppose your bike had 36 and 50 teeth on the two front sprockets. What would you have to change in the program? (Ignore REM statements.)
4. Your tire is 27 inches in diameter.
   a. What is the diameter in feet?
   b. What is the circumference? Use $\pi \approx 3.14$. (Round to the nearest tenth of a foot.) This is how far the bike goes when the wheel turns once.
5. How many times does the wheel turn in a mile (5280 feet)? Divide 5280 by the answer to Exercise 4(b). (Round to the nearest whole number.)

Make the following changes in the program. Put the answer to Exercise 5 in the blank. The revised program finds the number of times you must turn the pedals to go a mile in each gear.

```
105 T = __?__ / R
115 PRINT "GEAR: ";G;"TURNS/MILE: ";INT(T)
```

6. RUN the program. Complete the chart.

Gear	1	2	3	4	5	6	7	8	9	10
Pedal turns per mile	?	?	?	?	?	?	?	?	?	?

7. How many times do you turn the pedals to go
   a. five miles in 1st gear?
   b. five miles in 10th gear?
   c. ten miles in 5th gear?
   d. ten miles in 8th gear?

SKILLS REVIEW

QUOTIENTS OF SQUARE ROOTS

Do you remember this rule? $\sqrt{\frac{a}{b}} = \frac{\sqrt{a}}{\sqrt{b}}$ and $\frac{\sqrt{a}}{\sqrt{b}} = \sqrt{\frac{a}{b}}$

Find the square root. Express the result in simplest form. Assume that any variable represents a number greater than zero.

SAMPLES

$$\sqrt{\frac{100}{49}} = \frac{\sqrt{100}}{\sqrt{49}} = \frac{10}{7} \text{ or } 1\frac{3}{7}$$

$$\sqrt{\frac{5}{64}} = \frac{\sqrt{5}}{\sqrt{64}} = \frac{\sqrt{5}}{8}$$

$$\sqrt{\frac{8n^2}{9}} = \frac{\sqrt{4 \cdot 2 \cdot n^2}}{\sqrt{9}} = \frac{2n\sqrt{2}}{3}$$

1. $\sqrt{\frac{1}{81}}$ **2.** $\sqrt{\frac{4}{49}}$ **3.** $\sqrt{\frac{121}{36}}$ **4.** $\sqrt{\frac{9}{25}}$ **5.** $\sqrt{\frac{169}{16}}$

6. $\sqrt{\frac{2}{49}}$ **7.** $\sqrt{\frac{3}{100}}$ **8.** $\sqrt{\frac{7}{36}}$ **9.** $\sqrt{\frac{5}{16}}$ **10.** $\sqrt{\frac{11}{64}}$

11. $\sqrt{\frac{36x^2}{25}}$ **12.** $\sqrt{\frac{4b^2}{81}}$ **13.** $\sqrt{\frac{2c^2}{100}}$ **14.** $\sqrt{\frac{7a^2}{16}}$ **15.** $\sqrt{\frac{18n^2}{49}}$

16. $\sqrt{\frac{20k^2}{81}}$ **17.** $\sqrt{\frac{75x^2}{36}}$ **18.** $\sqrt{\frac{28a^2}{25}}$ **19.** $\sqrt{\frac{45d^2}{144}}$ **20.** $\sqrt{\frac{32b^2}{49}}$

Express in simplest radical form.

SAMPLE

$$\frac{\sqrt{2}}{\sqrt{5}} = \frac{\sqrt{2} \cdot \sqrt{5}}{\sqrt{5} \cdot \sqrt{5}} = \frac{\sqrt{10}}{5} \text{ or } \frac{1}{5}\sqrt{10}$$

Multiply the numerator and denominator by a number that will make the denominator a whole number.

21. $\frac{\sqrt{3}}{\sqrt{7}}$ **22.** $\frac{\sqrt{6}}{\sqrt{3}}$ **23.** $\frac{\sqrt{5}}{\sqrt{8}}$ **24.** $\frac{\sqrt{11}}{\sqrt{2}}$ **25.** $\frac{\sqrt{7}}{\sqrt{5}}$

26. $\sqrt{\frac{1}{8}}$ **27.** $\sqrt{\frac{4}{11}}$ **28.** $\sqrt{\frac{16}{7}}$ **29.** $\sqrt{\frac{3}{2}}$ **30.** $\sqrt{\frac{2}{5}}$

31. $\frac{5\sqrt{6}}{\sqrt{3}}$ **32.** $\frac{3\sqrt{8}}{\sqrt{2}}$ **33.** $\frac{6\sqrt{5}}{\sqrt{3}}$ **34.** $\frac{10\sqrt{2}}{\sqrt{5}}$ **35.** $\frac{4\sqrt{7}}{\sqrt{2}}$

CHAPTER REVIEW

CHAPTER SUMMARY

1. Many of the terms used with circles are discussed on page 346.

2. If line t is tangent to $\odot O$ at point A, then $\overline{OA} \perp t$. If line $t \perp \overline{OA}$ at point A, then t is tangent to $\odot O$. Construction 10 shows how to construct a tangent to a circle at a given point.

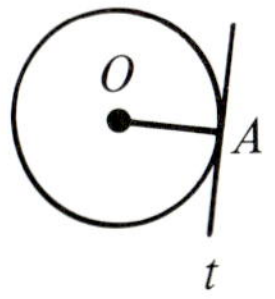

3. Given two circles, a common internal tangent intersects the segment joining the centers, and a common external tangent does not. Two circles may be externally tangent or internally tangent.

4. Arcs are classified on page 355.

5. In a circle, equal central angles have equal arcs, and equal minor arcs have equal central angles.

6. Given the figure at the right. When any one of these three statements is true, both of the others are also true.

$QR = ST$
$\overset{\frown}{QR} = \overset{\frown}{ST}$
$OJ = OK$

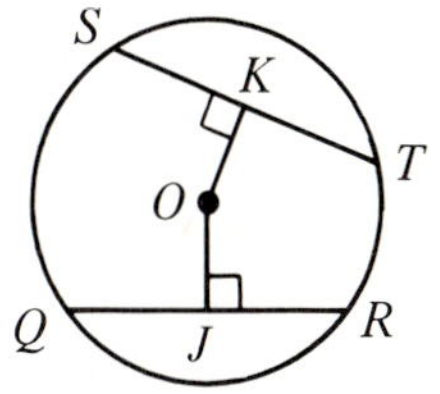

7. A diameter that is perpendicular to a chord bisects the chord and its arc.

8. A polygon is inscribed in a circle when each vertex lies on the circle. A polygon is circumscribed about a circle when each side is tangent to the circle.

9. A summary of angle and arc relationships is given on page 370.

10. If two chords intersect inside a circle, then the product of the lengths of the segments of one chord equals the product of the lengths of the segments of the other.

REVIEW EXERCISES

Refer to the figure and use letters to name the following.
(See pp. 346–349)

1. A radius
2. A diameter
3. A secant
4. A tangent

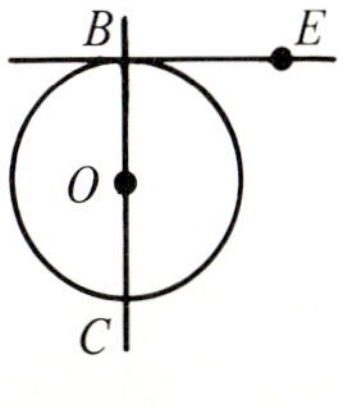

Refer to the diagram to complete each statement. *(See pp. 350–354.)*

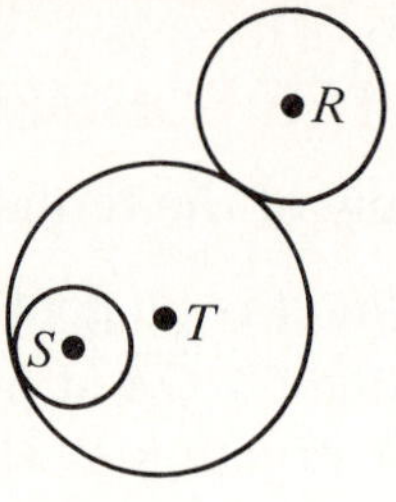

5. The number of common external tangents that can be drawn to $\odot R$ and $\odot S$ is __?__.

6. The number of common internal tangents that can be drawn to $\odot R$ and $\odot S$ is __?__.

7. $\odot T$ and $\odot$ __?__ are externally tangent.

8. $\odot T$ and $\odot$ __?__ are internally tangent.

Write numerical values to complete the statements.
(See pp. 355–358.)

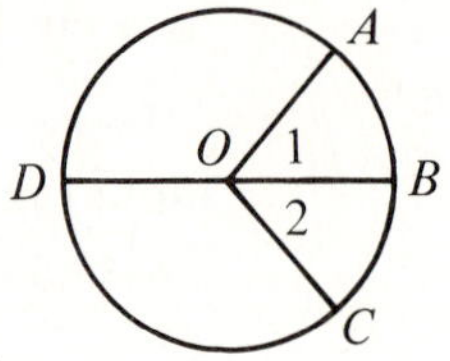

9. $\overset{\frown}{DAB} =$ __?__ °

10. If $\angle 1 = 50°$, $\overset{\frown}{AB} =$ __?__°.

11. If $\overset{\frown}{BC} = 48°$, $\angle 2 =$ __?__°.

12. If $\overset{\frown}{AD} = 126°$, $\angle 1 =$ __?__°.

13. If $\overset{\frown}{AB} = \overset{\frown}{BC}$ and $\angle 1 = 47°$, then $\angle 2 =$ __?__°.

Write numerical values to complete the statements.
(See pp. 359–363.)

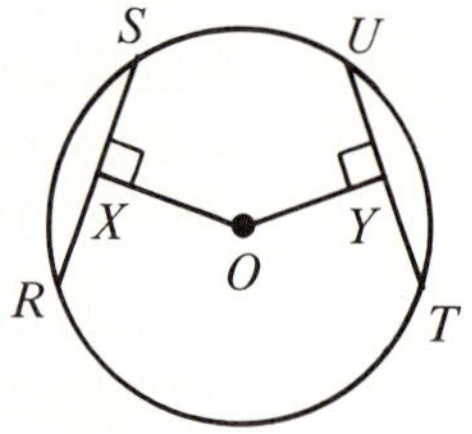

14. If $TU = 7$ and $\overset{\frown}{RS} = \overset{\frown}{TU}$, then $RS =$ __?__.

15. If $\overset{\frown}{RS} = 70°$ and $TU = RS$, then $\overset{\frown}{TU} =$ __?__°.

16. If $RS = TU$ and $OX = 5$, then $OY =$ __?__.

17. If $OX = OY$ and $TU = 4$, then $RS =$ __?__.

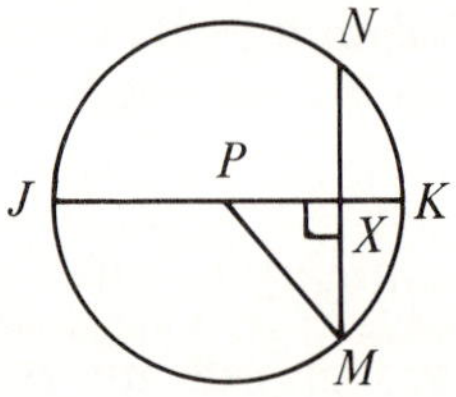

18. If $\overset{\frown}{MN} = 100°$, $\overset{\frown}{MK} =$ __?__°.

19. If $\overset{\frown}{MK} = 48°$, $\overset{\frown}{JM} =$ __?__°.

20. If $MN = 14$, $MX =$ __?__.

21. If $MN = 16$ and $PX = 6$, $MP =$ __?__.

Find the value of *x*. *(See pp. 364–376.)*

22.

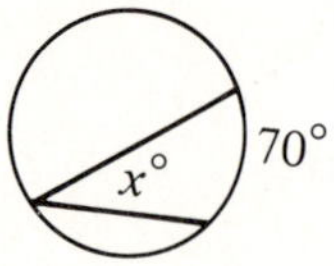

23.

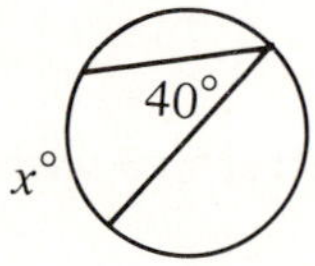

24.

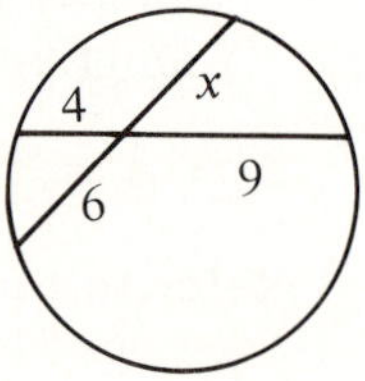

25.

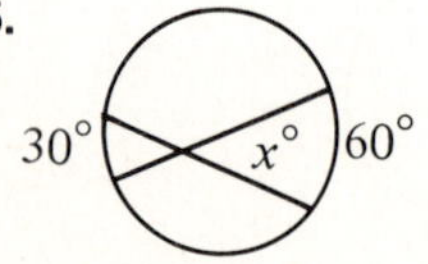

26.

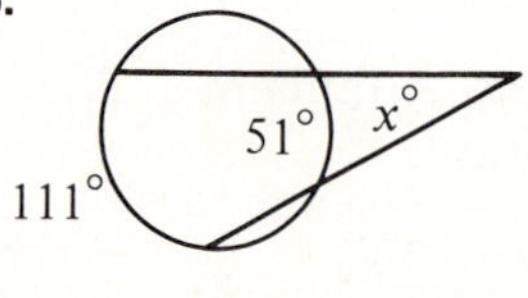

27.

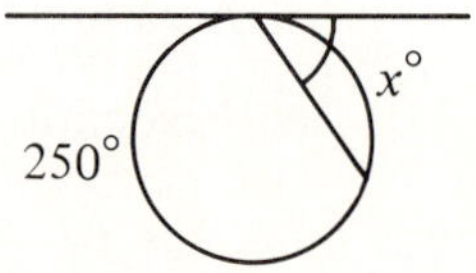

CHAPTER TEST

1. Sketch a circle with center P and a diameter $\overline{AB}$. *(9-1)*

2. Sketch a tangent to $\odot P$ at point A.

3. Sketch two circles so that the circles have exactly one common tangent. Are the circles internally or externally tangent? *(9-2)*

4. Construct a tangent t to a $\odot O$ at any point X on the circle.

In $\odot O$, $\angle 1 = \angle 2 = \angle 3$. *(9-3)*

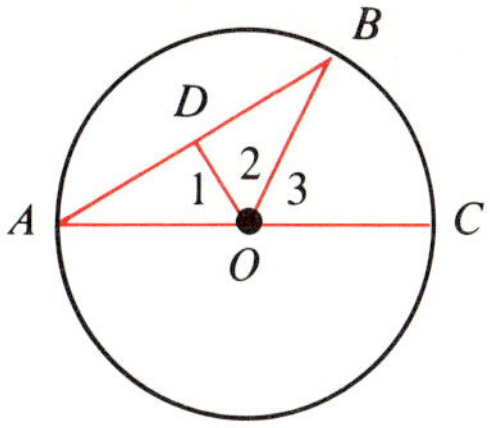

5. Name a minor arc and its measure.

6. Name a major arc and its measure.

7. If you draw $\overrightarrow{OD}$, which intersects $\odot O$ at point X, then __?__ = __?__ = __?__.

Find the value of x. *(9-4)*

8.
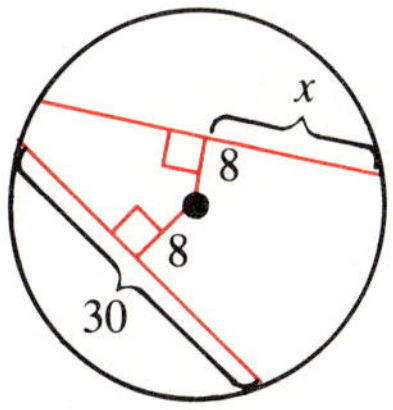

9.
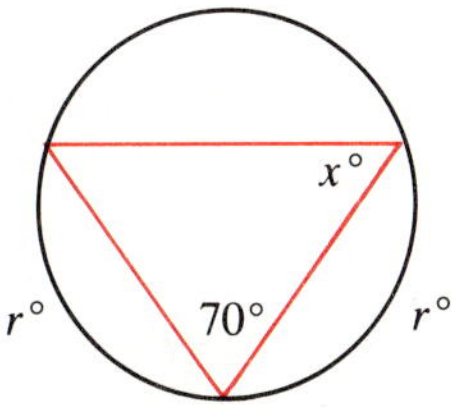

10.
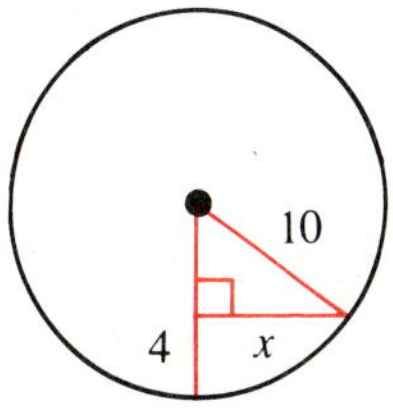

11. $\angle STR =$ __?__° *(9-5)*

12. $\angle STQ =$ __?__°

13. Name an angle equal to $\angle RTP$.

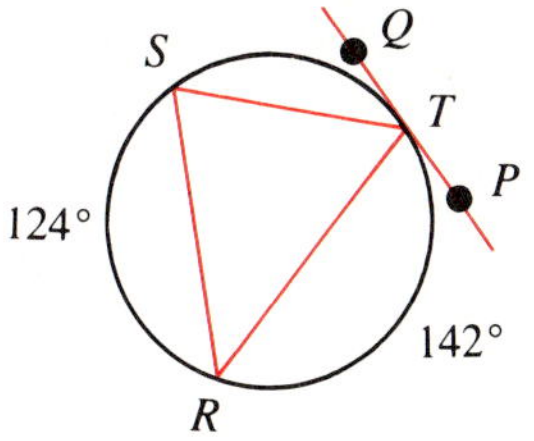

Find the value of x. *(9-6)*

14.
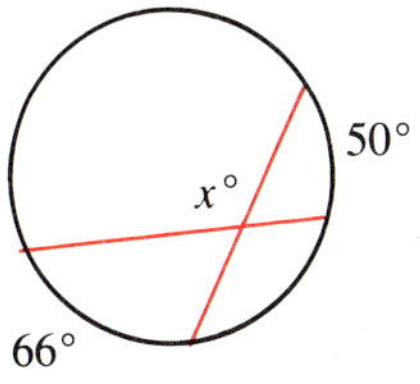

15.
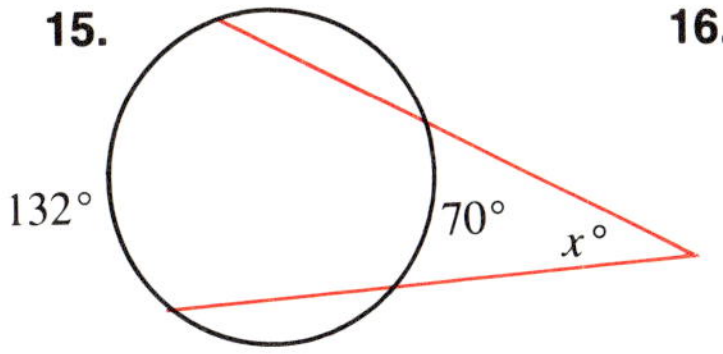

16.
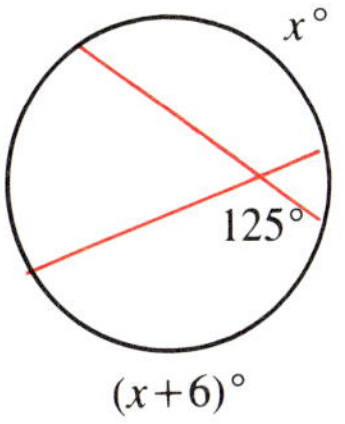

$\overset{\frown}{AB}$ and $\overset{\frown}{CD}$ are arcs of $\odot O$. $\overline{AB}$ and $\overline{CD}$ intersect at point J.

17. If $AJ = 9$, $BJ = 8$, and $CJ = 6$, then $CD =$ __?__ *(9-7)*

18. If $AJ = x$, $BJ = x + 5$, and $CJ = DJ = x + 2$, find the value of x.

MIXED REVIEW

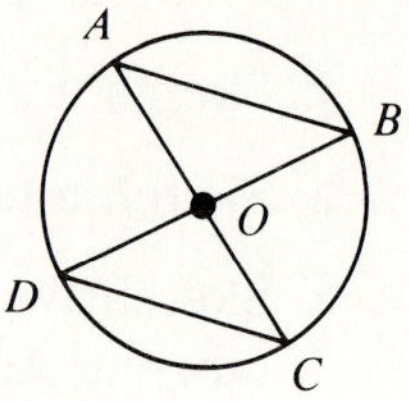

1. An angle inscribed in a circle intercepts a semicircle. Is the angle acute, right, or obtuse?

2. Supply the reason to support each statement about $\odot O$.
   a. $\angle AOB = \angle COD$ **b.** $\overset{\frown}{AB} = \overset{\frown}{CD}$ **c.** $AB = CD$

3. The measure of each interior angle of a regular polygon is 144°. How many sides does the polygon have?

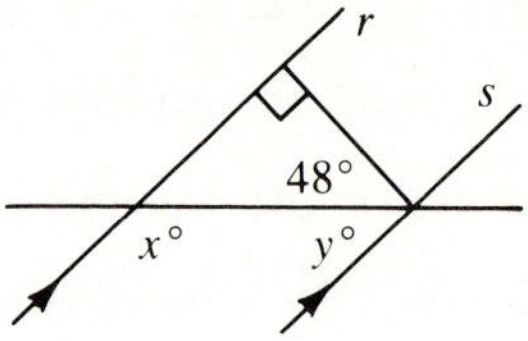

4. If $r \parallel s$, find the value of x and y.

5. *Given:* $WXYZ$ is a ▱.
   $\overline{WY}$ and $\overline{XZ}$ intersect at P.

   Prove: $\triangle WPX \cong \triangle YPZ$

6. Draw a large acute triangle and construct a circle passing through the vertices of the triangle.

7. A triangle has sides of lengths 7, 12, and 14. Is it acute, right, or obtuse?

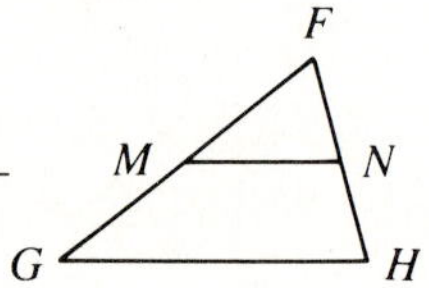

In the figure, $FM = MG$ and $FN = NH$.

8. Why must $\overline{MN}$ and $\overline{GH}$ be parallel?
9. $\triangle MNF \sim \triangle$ ___?___
10. If $\triangle MNF$ has area 12, find the area of $\triangle FGH$.
11. A circle has radius 8 cm. Find its circumference and area.
12. A rhombus has diagonals of 6 and 8. Find its perimeter.

Find the value of x.

13.

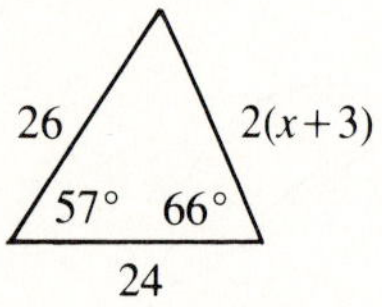

14.

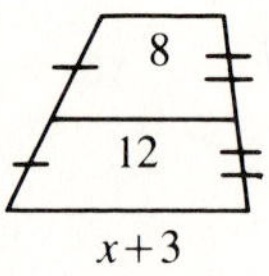

15.

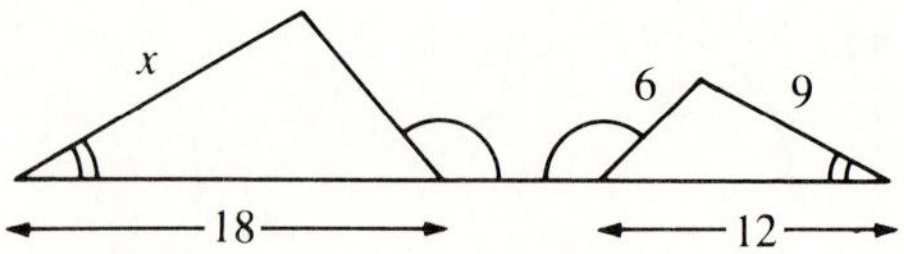

16. If $\frac{j}{k} = \frac{3}{5}$, then $\frac{k}{j} =$ ___?___, $\frac{k+j}{j} =$ ___?___, $\frac{5}{k} =$ ___?___, and $3k =$ ___?___.

17. Construct a circle. Then construct a circumscribed triangle by constructing tangents to the circle at three appropriately chosen points on the circle.

18. The three __?__ of any triangle always meet inside the triangle.

Find the area of the triangle, parallelogram, and trapezoid shown.

19.

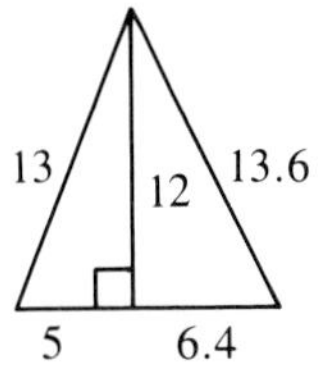

20.

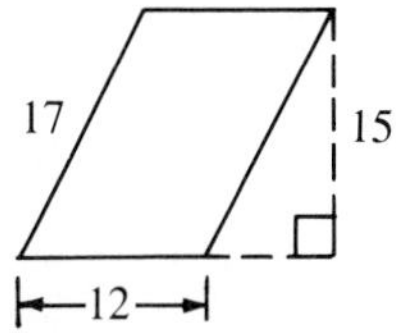

21.

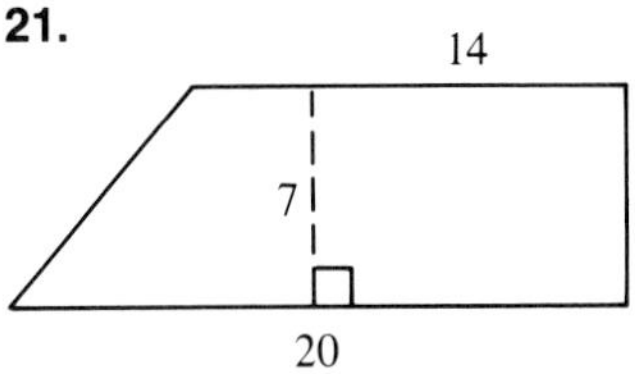

22. Find the measure of each numbered angle shown.

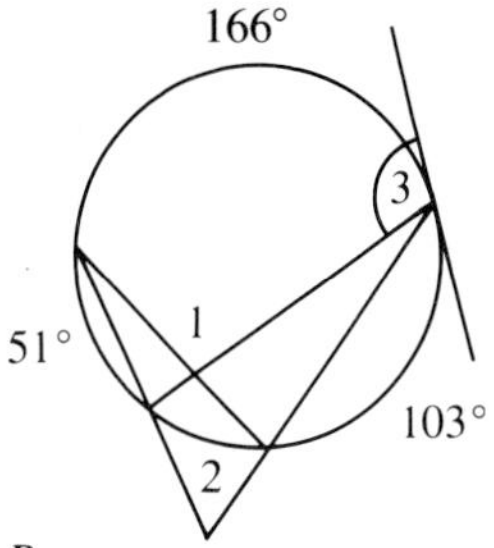

23. A bank account of \$600 earned \$3.50 for one month's interest. How much interest did an account of \$420 earn?

24. Consider the statement "If a polygon is regular, then it is equilateral." Write the converse and tell if the converse is true.

25. *Given:* $\overline{RT}$ bisects $\angle QRS$.
$\overline{TR}$ bisects $\angle QTS$.

Prove: $QT = ST$

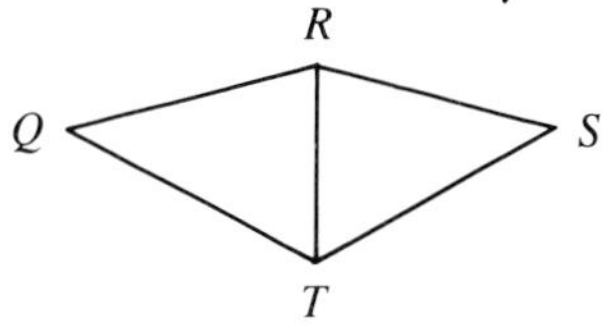

26. In $\triangle RST$, $RS = ST$. If $RS = 3x - 1$, $RT = 4x - 6$, and $ST = 2x + 4$, find the numerical lengths of the sides.

27. How many tangent lines contain a given point inside a circle?

28. Could the information be used to prove that quadrilateral $GRIN$ is a parallelogram?

a. $RI = GN$ and $\overline{GR} \parallel \overline{NI}$

b. $\angle G = \angle I$ and $\angle R = \angle N$.

c. $GR = NI$ and $RI = GN$

d. Consecutive angles are supplementary

Find the value of x.

29.

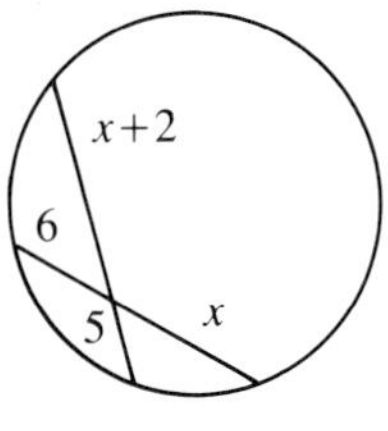

30.

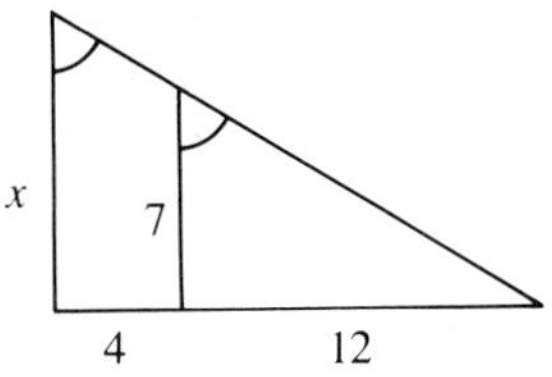

31.

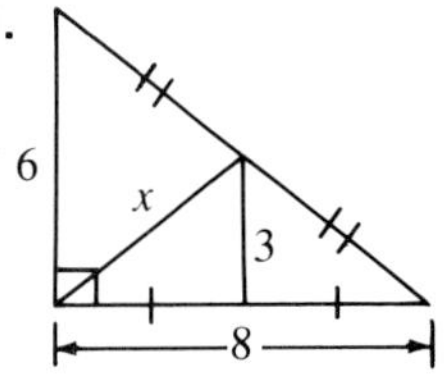

32. Draw a large scalene, obtuse triangle. Sketch and label the median and the altitude to the longest side.

Here's what you'll learn in this chapter:

To state the positions of points and lines with respect to each other.

To find the lateral area, total area, and volume of a right prism.

To find the lateral area, total area, and volume of a right circular cylinder.

To find the lateral area, total area, and volume of a regular pyramid.

To find the lateral area, total area, and volume of a right circular cone.

To find the area and volume of a sphere.

Both surface areas and volumes are important in the design and construction of buildings. Architects and contractors must make many accurate measurements. Surface area is also important to the window washers shown above.

Chapter 10

Areas and Volumes of Solids

1 Lines and Planes in Space

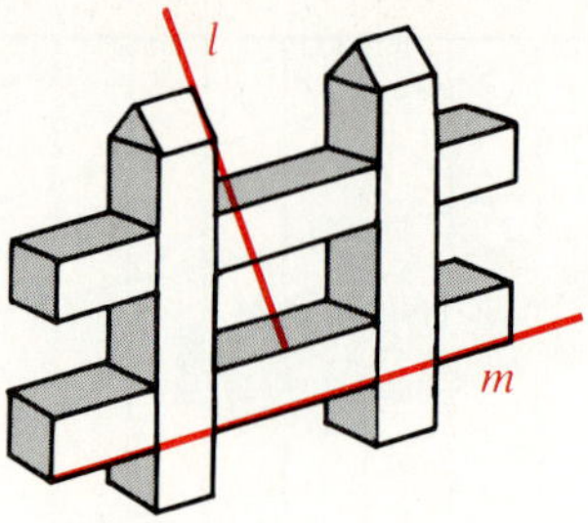

Look at lines l and m in the diagram. The lines do not meet, but they are not parallel. They are *skew* lines. **Skew lines** are two lines that do not lie in any one plane.

Possible positions of two lines:

Intersecting lines are coplanar.

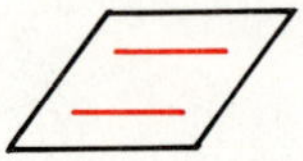

Parallel lines are coplanar.

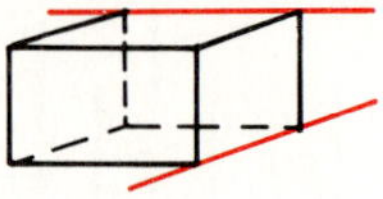

Skew lines are not coplanar.

Look at line t and plane P at the right. The line and the plane are *parallel.* A line and a plane are **parallel** when they do not have any point in common. Any segment of the line is also said to be parallel to the plane.

Possible positions of a line and a plane:

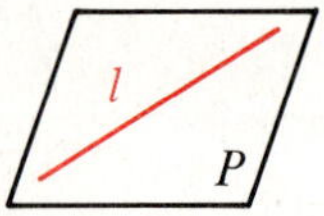

l lies in P.
P contains l.

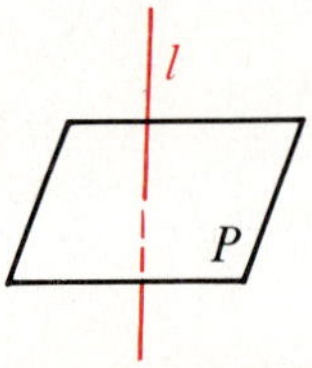

l intersects P.
P intersects l.

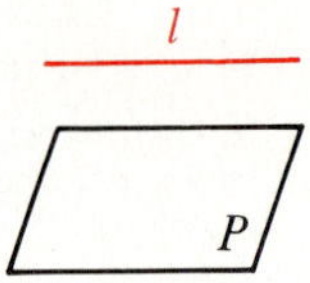

l and P are parallel.
$l \parallel P$

A line and a plane can intersect in a special way. Notice in the diagram that an axle is perpendicular to each spoke of a wagon wheel. The diagram suggests a line *perpendicular* to a plane.

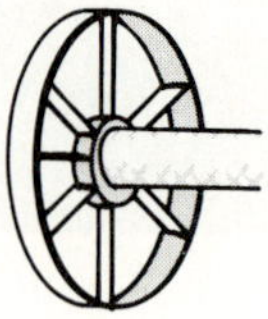

Line l is **perpendicular** to plane P when l is perpendicular to *every* line in P that passes through the point of intersection.

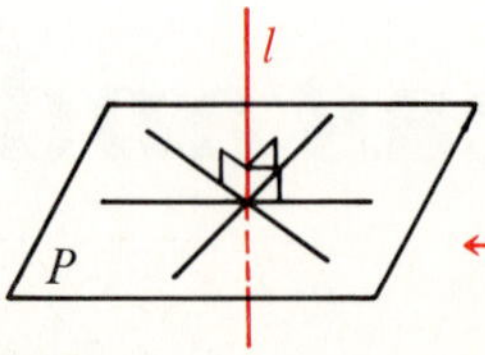

$l \perp P$

Suppose that $l \perp P$ and that plane M contains line l. Then plane M is said to be **perpendicular** to plane P. Perpendicular planes are one special kind of intersecting planes.

If $l \perp P$, then $M \perp P$.

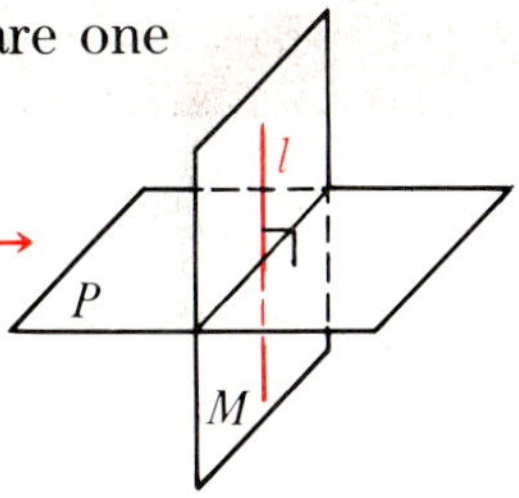

Possible positions of two planes:

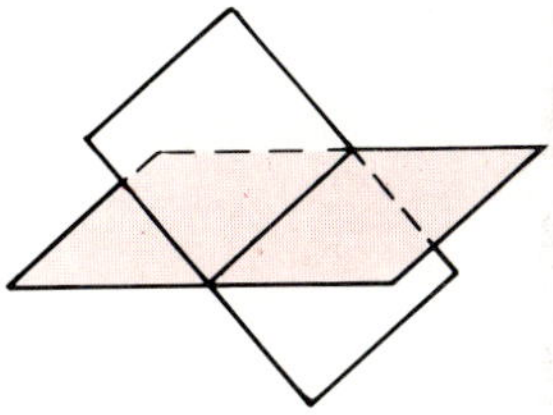

Intersecting

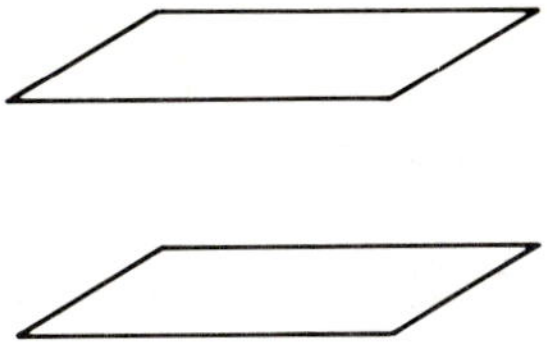

Parallel

Classroom Practice

Use your classroom and its furnishings to find examples that suggest each figure named.

1. Two lines that intersect
2. Two lines that are parallel
3. Two lines that are skew
4. Two planes that intersect
5. Two planes that are parallel
6. Two planes that are perpendicular
7. A line and a plane such that the line
   a. lies in the plane
   b. intersects the plane and is perpendicular to the plane
   c. intersects the plane but is not perpendicular to the plane
   d. is parallel to the plane
8. a. Four points that all lie in one plane
   b. Four points that do not all lie in any one plane
9. Two parallel planes that are intersected by a third plane
10. Two lines that are both perpendicular to a third line and
    a. are parallel to each other b. are perpendicular to each other
11. Alonzo said: "If two planes are perpendicular, every line in one of the planes must be perpendicular to the other plane." Explain why Alonzo was wrong.

12. Lee wanted to set a flagpole upright, and said to a friend: "Stand over there and tell me when it's straight up." Could the friend tell for sure that the pole was upright?

Written Exercises

Use a straightedge to draw each figure.

A

1. A line in a plane
2. A line intersecting a plane
3. A line parallel to a plane
4. A line perpendicular to a plane

A rectangular solid is shown. Name the figures described. In some cases, more than one answer is possible.

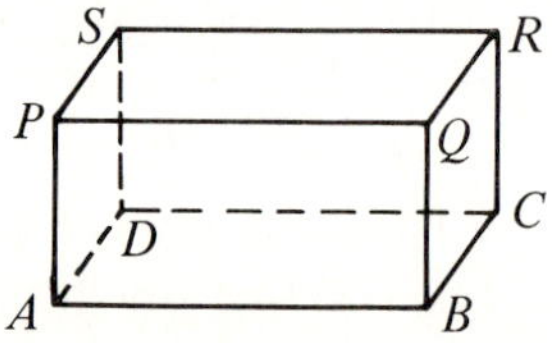

Exercises 5–14

5. Three lines parallel to $\overleftrightarrow{AB}$
6. Four lines perpendicular to $\overleftrightarrow{BQ}$
7. Four lines skew to $\overleftrightarrow{AD}$
8. A plane parallel to plane *ADSP*
9. Four planes, each one perpendicular to plane *ADSP*
10. Two lines that are perpendicular to $\overleftrightarrow{PS}$ and are parallel to each other
11. Two lines that are perpendicular to $\overleftrightarrow{PS}$ and are skew to each other
12. Two lines that are perpendicular to $\overleftrightarrow{PS}$ and are perpendicular to each other
13. $\overleftrightarrow{BD}$ and $\overleftrightarrow{QS}$ (not drawn) are __?__ lines.
14. $\overleftrightarrow{PA}$ and $\overleftrightarrow{AC}$ are __?__ lines.

Quadrilateral *JKLM* is a square. The lengths of $\overline{VJ}$, $\overline{VK}$, $\overline{VL}$, and $\overline{VM}$ are equal. Tell whether each statement is true or false.

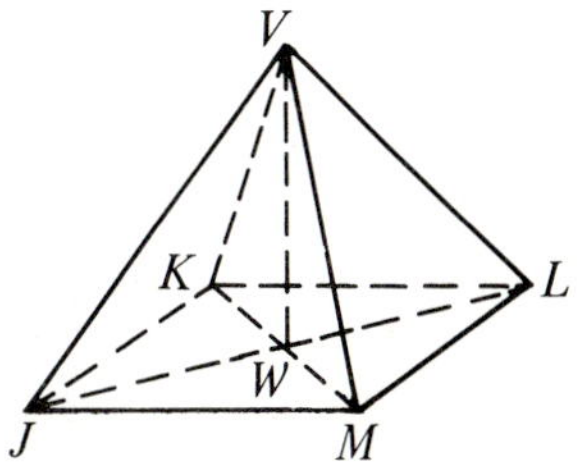

15. $JW = WL$

16. $JW = KW$

17. $\angle JKL > 90°$

18. $\angle VJM < \angle VMJ$

19. $\overleftrightarrow{JL}$ and $\overleftrightarrow{VK}$ are skew.

20. $\triangle MVL \cong \triangle WVL$

21. $\angle JWM > 90°$

22. $\triangle VKJ$ is a right $\triangle$.

23. $VM = VW$

24. $\overline{VW} \perp$ plane $JKLM$

Draw a figure to illustrate each statement. Some statements have more than one solution.

B **25.** Two lines that are perpendicular to the same line l do not have to be parallel to each other.

26. Two lines that are skew to the same line l do not have to be skew to each other.

27. Line j lies in a plane P and line k lies in a plane Q, but lines j and k do not have to be skew lines.

28. Two planes both perpendicular to plane X do not have to be parallel to each other.

29. A line that is perpendicular to a line in a plane does not have to be perpendicular to the plane.

Line *b* and plane *Q* have the positions described. Tell how many planes can be drawn that contain line *b* and are, at the same time, perpendicular to plane *Q*.

30. b lies in Q.

31. $b \parallel Q$

32. $b \perp Q$

33. b intersects Q, but b is not $\perp$ to Q.

C **34.** In the figure:
$j \parallel X$; V contains j;
V intersects X in k.
Explain why j and k must be parallel lines.

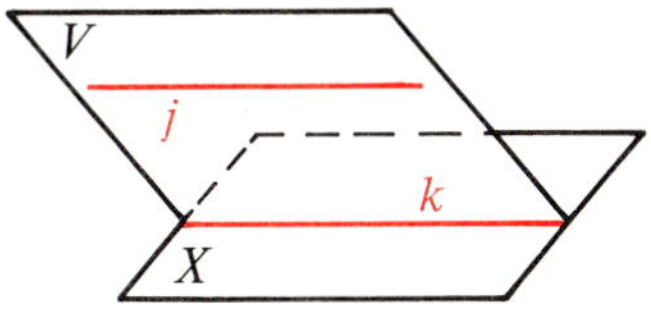

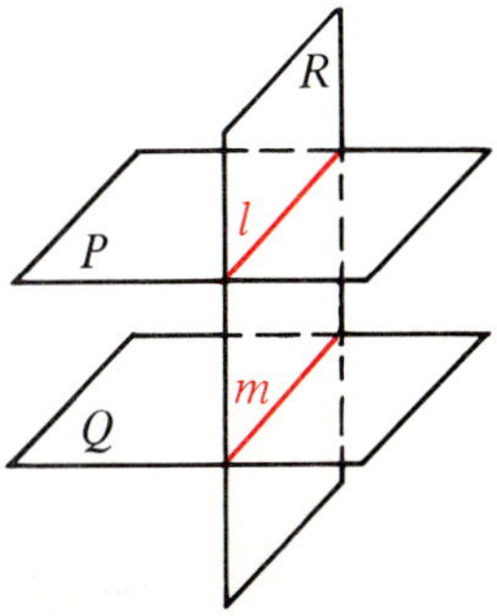

35. In the figure:
$P \parallel Q$; R intersects P in l, and Q in m.
Explain why l and m must be parallel lines.

Experiments

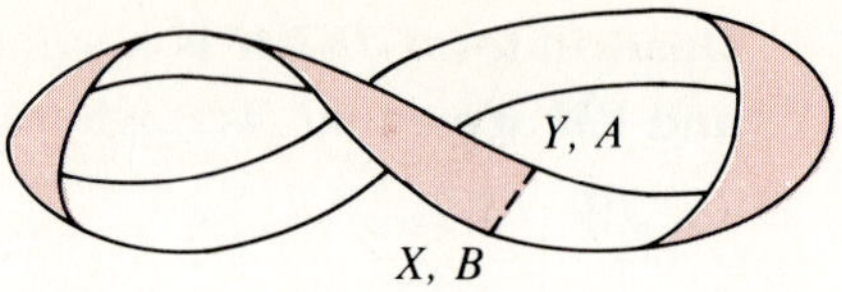

To make an interesting surface like the one pictured at the right, take a long strip of paper as shown below. Give the strip a half-twist.

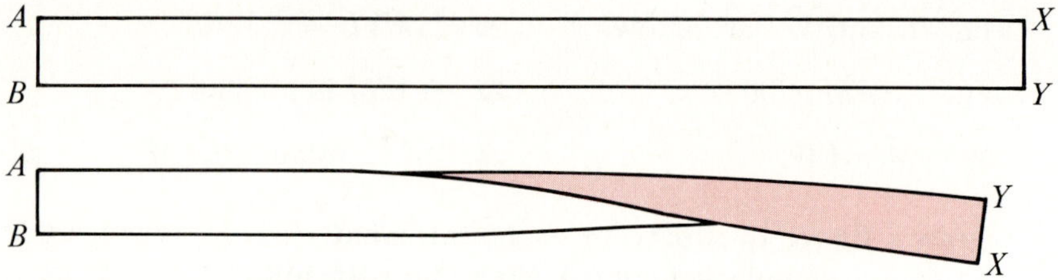

Tape the ends together, joining *X* and *B*, *Y* and *A*. The resulting figure is called a *Möbius band.*

1. Draw a pencil line down the middle. Do you return to your starting point? How many sides does the band have?
2. Cut along your pencil line. What happens to the band?
3. Take another long strip of paper. Give the paper a full twist and tape the ends together. Then cut the band down the middle, as before. What happens to the band?

CONSUMER APPLICATIONS

CALLING LONG DISTANCE

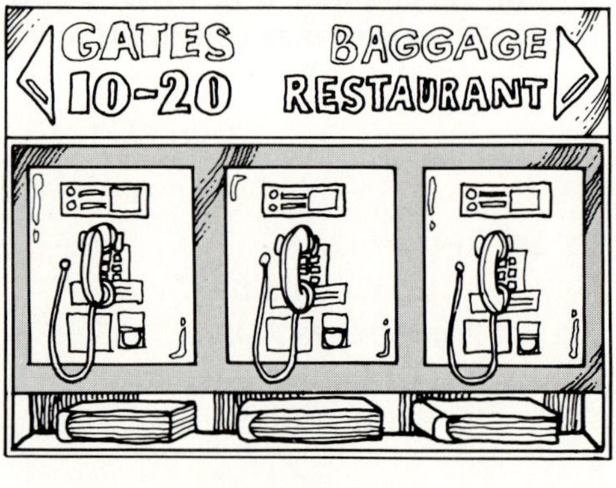

It's fun to make long distance calls to friends or family members, but a long distance call costs more than one across town. You can save on telephone bills by knowing how long distance rates vary at different times of the day. The chart below shows a sample rate structure for a long distance direct-dial call.

BUSINESS HOURS (8 A.M.–5 P.M.)		EVENINGS (5 P.M.–11 P.M.)		NIGHTS (11 P.M.–8 A.M.)	
First min	Each add. min	First min	Each add. min	First min	Each add. min
\$0.56	\$0.23	\$0.36	\$0.14	\$0.22	\$0.09

2 Right Prisms

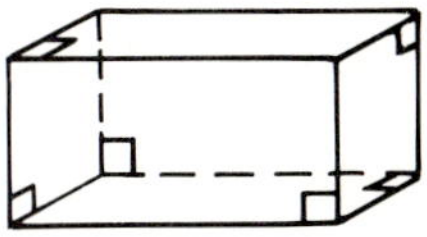

The figure shows a *rectangular solid.* You see many rectangular solids every day, for example, bricks and cereal boxes.

Some people have trouble drawing rectangular solids. One way to proceed is this:

Draw a parallelogram.
Draw a second parallelogram directly below the first one.

Draw four vertical segments.

When you look at a box, you cannot see some of the edges. Because of this fact, some people draw dashed lines to show hidden segments. Other people don't show the hidden edges at all.

A rectangular solid is the most common kind of **right prism.** Two other right prisms are pictured below. The shaded *faces,* called *bases,* are congruent and parallel. The other faces of each prism are called *lateral faces.* Other parts of a prism are named as shown.

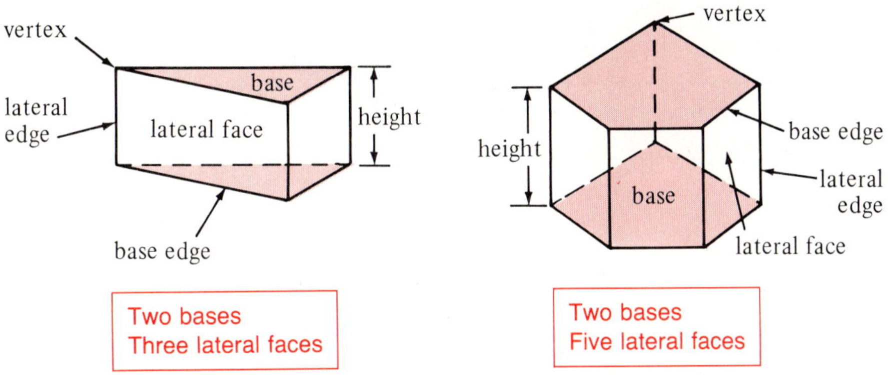

In a right prism, each lateral edge is perpendicular to the bases. Any segment, such as a lateral edge, that is perpendicular to the bases and has an endpoint on each base is called an *altitude* of the prism. The length of an altitude is the *height, h,* of the prism.

The **lateral area** (L.A.) of a prism is the sum of the areas of the lateral faces. Can you see that the lateral area of the prism shown is the sum of the areas of six rectangles?

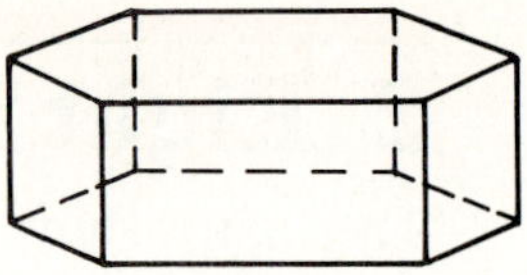

The **total area** (T.A.) of a prism is the sum of the areas of all the faces. Using B to represent the area of a base,

Total Area = Lateral Area + 2 × Area of a Base
T.A. = L.A. + $2B$

EXAMPLE 1

The base of a right prism is a 5 by 3 rectangle. The height is 2. Find:

a. the lateral area; **b.** the total area.

a. The area of the closest face is $5 \times 2 = 10$.
The area of an end face is $3 \times 2 = 6$.
L.A. $= 10 + 6 + 10 + 6 = 32$

b. T.A. = L.A. + $2B$
T.A. $= 32 + 2(5 \times 3) = 32 + 30 = 62$

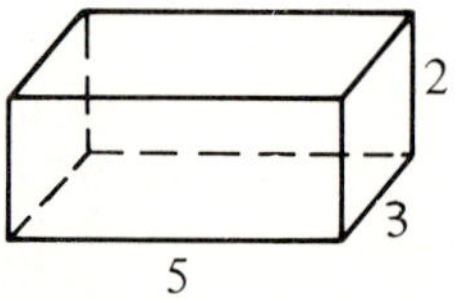

A prism has *volume* as well as area. When each edge of a rectangular solid is one unit long, the volume of the solid is one cubic unit. Can you see that the rectangular solid shown below can be filled with blocks that are 1 cm cubes?

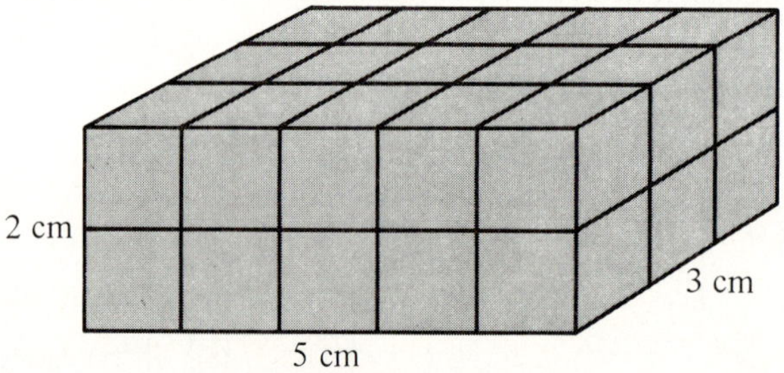

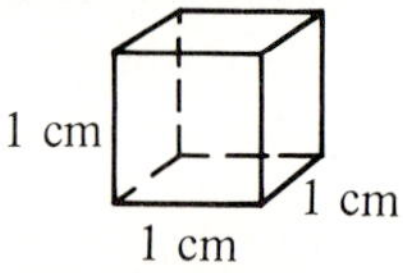

1 cubic centimeter
1 cm^3

Fifteen blocks form a bottom layer, fifteen more a top layer. The box can be filled with 30 blocks, and the volume, V, is 30 cm^3. Notice that the volume is the product of the base area, B, and the height, h:

$$V = 15 \times 2$$

In any right prism: *Volume = Base × height*
$V = Bh$

EXAMPLE 2 The base of a rectangular solid is a 6 cm square. The height of the solid is 10 cm. Find the volume.

$B = s^2 = 6^2 = 36$
$V = Bh$
$V = 36 \times 10 = 360$
Answer: 360 cm^3

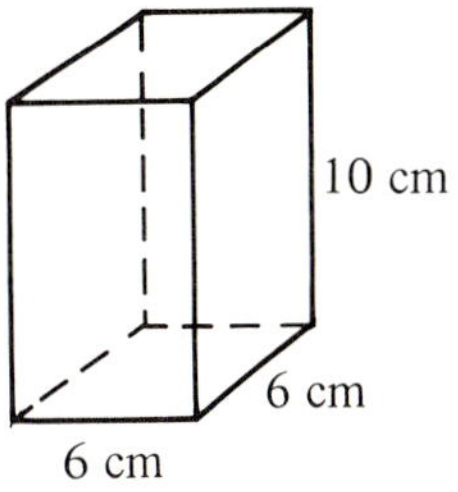

Classroom Practice

For each right prism, state the number of:

a. lateral edges **b. lateral faces** **c. bases**
d. base edges **e. faces** **f. vertices**

1.

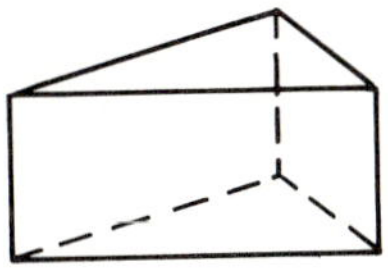

2.

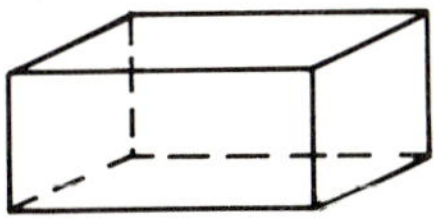

3.

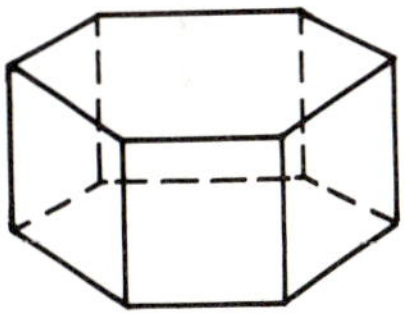

Tell why the solid pictured is not a right prism.

4.

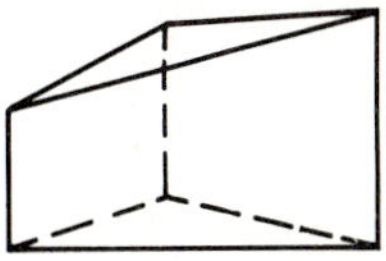

5.

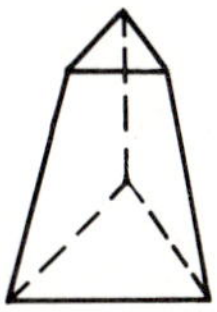

6.

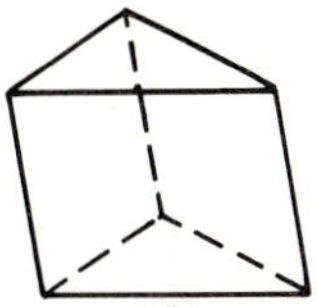

7. One student said that the diagram at the left shows a right prism, but the one at the right does not. Was the student correct? Explain.

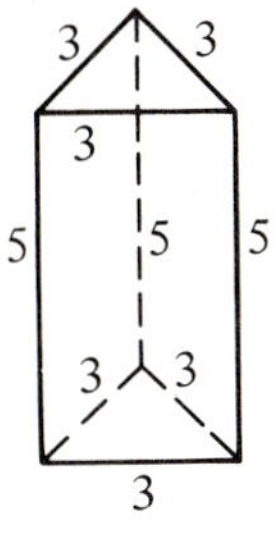

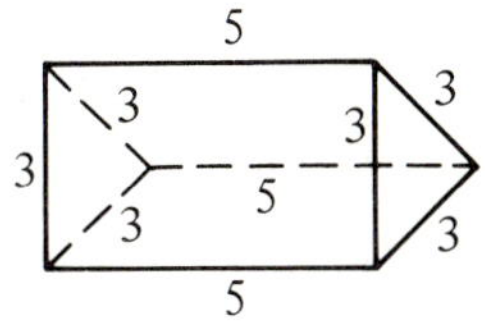

8. The bases of the right prism shown are isosceles trapezoids. Find the lateral area, the total area, and the volume of the prism.

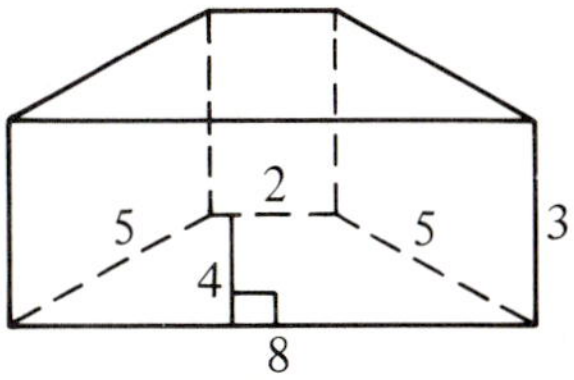

Tell whether each statement is true or false.

9. Every right prism has two bases.
10. Every right prism has a rectangular base.
11. Each base of a right prism must have at least four edges.

Tell whether each statement is true or false.

12. Every lateral face of a right prism must be a rectangle.

13. The length of an altitude of a prism is called its height.

14. In any right prism, T.A. = 2(L.A.) + B.

Written Exercises

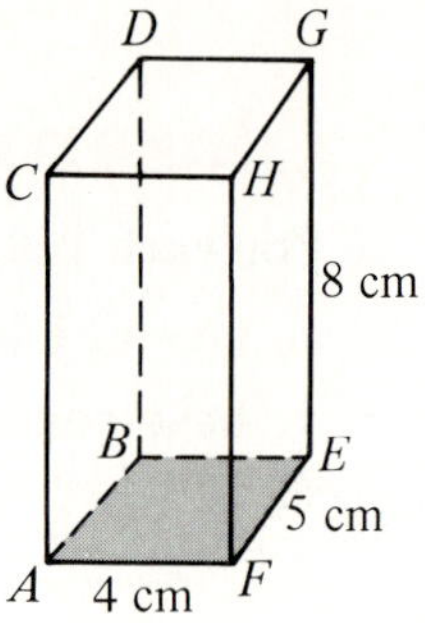

Exercises 1–10 refer to the rectangular solid.

A 1. BE = _?_ 2. DG = _?_

3. DB = _?_ 4. AB = _?_

5. The perimeter of quadrilateral $AFHC$ = _?_.

6. The perimeter of a base = _?_.

7. The area of a base = _?_.

8. The lateral area of the prism = _?_.

9. The total area of the prism = _?_. 10. The volume of the prism = _?_.

Exercises 11–14 deal with right prisms. The number of sides in the base is not specified. Complete the table.

	B	h	V
11.	30 cm^2	4 cm	?
12.	?	5 cm	100 cm^3
13.	?	2 cm	100 cm^3
14.	64 cm^2	?	256 cm^3

Exercises 15–18 deal with a rectangular solid having dimensions *l*, *w*, and *h*. Draw a figure if you need one. Complete the table.

	l	w	h	L.A.	T.A.	V
B 15.	6	2	3	?	?	?
16.	8	3	4	?	?	?
17.	10	4	?	?	?	120
18.	?	6	6	?	?	540

Exercises 19–22 deal with *cubes*. A cube is a rectangular solid with all edges equal. Complete the table.

	Edge	B	L.A.	T.A.	V
19.	2	?	?	?	?
20.	?	16	?	?	?
21.	5	?	?	?	?
22.	?	?	?	?	1000

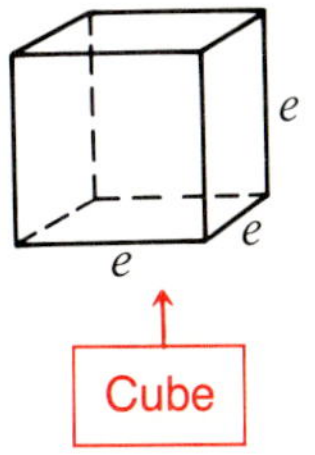

23. Examine your answers for Exercises 19 and 20 as a pair. Then look at those for Exercises 21 and 22 as a pair. These exercises suggest that when each edge of a cube is doubled:
a. the area is multiplied by __?__;
b. the volume is multiplied by __?__.

C **24.** One cube has edges 10 cm long. Another cube has edges 30 cm long. Find the ratio of the:
a. total area of the smaller cube to that of the larger cube;
b. volume of the smaller cube to that of the larger cube.

25. A rough-cut piece of lumber has a 2-inch by 4-inch end. Before being sold, the end is planed down to $1\frac{1}{2}$ inches by $3\frac{1}{2}$ inches. What percent of the wood is lost?

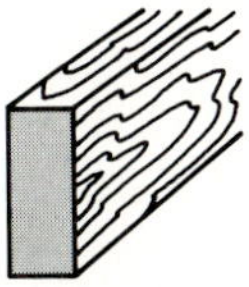

26. a. How many square centimeters of aluminum foil are needed to cover the block of cheese pictured? (Ignore overlap.)
b. Suppose the block is cut into cubes with 2 cm edges. How many square centimeters of foil are needed to wrap all the individual cubes?

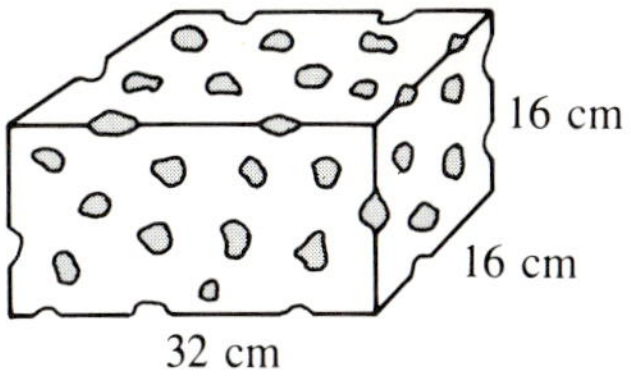

27. How many liters of water are needed to fill the swimming pool shown to within 0.2 m of the top edge? (*Hints:* Think of $PQRST$ as the base of a right prism. 1 m^3 contains 1000 L.)

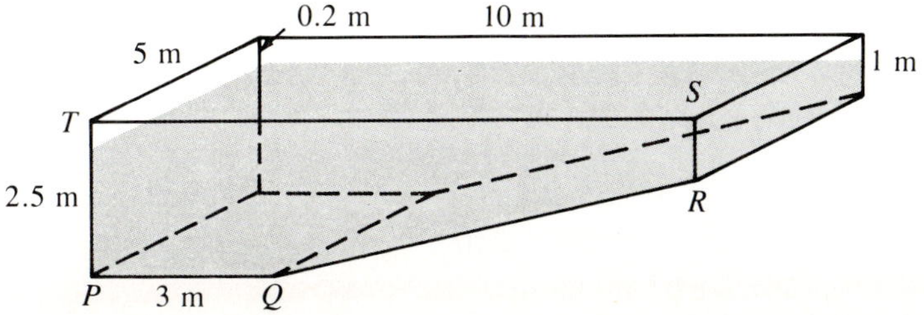

3 Right Circular Cylinders

A **right circular cylinder** is like a right prism except that its bases are congruent circles instead of congruent polygons. The *radius* of a base is also called the radius of the cylinder.

As in right prisms, an *altitude* of a right circular cylinder is a segment that is perpendicular to the bases and has an endpoint in each base. The length of an altitude is the *height, h,* of the cylinder. We'll use the word *cylinder* to mean a right circular cylinder.

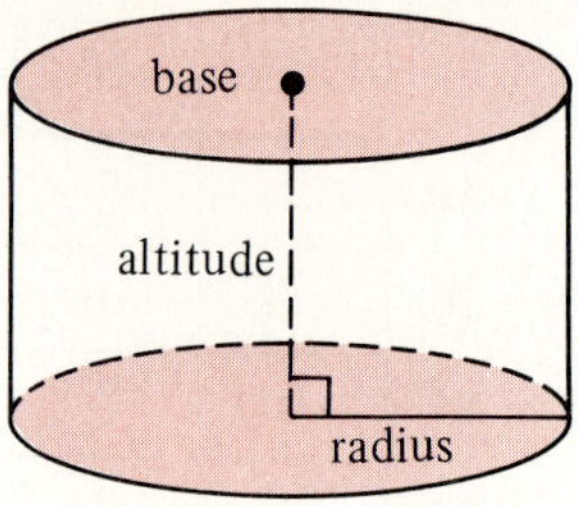

Right Circular Cylinder

To find the lateral area of a cylinder, imagine that the diagram shows a tin can with its ends removed. If you cut along $\overline{AB}$, you can unroll the metal and lay it out flat.

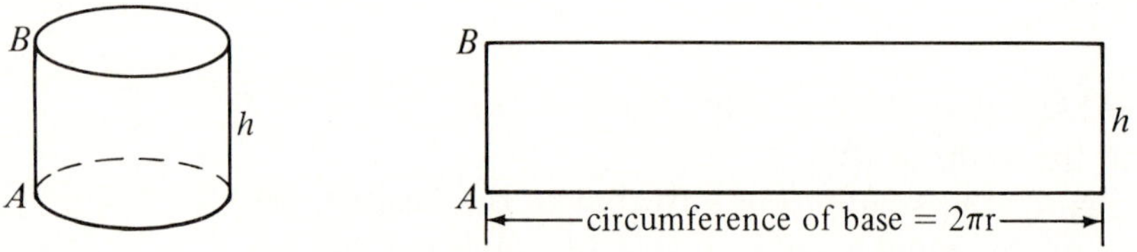

The lateral surface of the cylinder becomes a rectangular region.

$$\begin{aligned}\text{L.A. of cylinder} &= \text{area of rectangle}\\ &= 2\pi r \times h\\ &= 2\pi rh\end{aligned}$$

The formula for the total area of a cylinder is much like the formula for the total area of a right prism:

$$\text{T.A.} = \text{L.A.} + 2B$$

$$\text{T.A.} = 2\pi rh + 2 \times \pi r^2$$

You can find the volume of a cylinder in a similar way:

$$V = Bh$$

$$V = \pi r^2 \times h$$

In any (right circular) cylinder:

L.A. $= 2\pi rh$

T.A. $=$ L.A. $+ 2B$
$= 2\pi rh + 2\pi r^2$

$V = Bh$
$= \pi r^2 h$

EXAMPLE A cylinder has a radius of 3 cm and a height of 4 cm. Find:
a. the lateral area **b.** the total area **c.** the volume

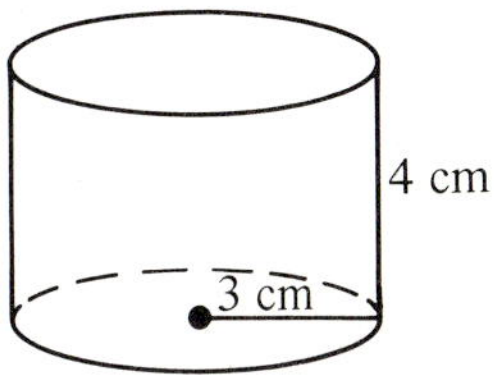

a. L.A. $= 2\pi rh = 2\pi \times 3 \times 4 = 24\pi$ cm^2
b. T.A. $=$ L.A. $+ 2B$
$= 24\pi + 2\pi r^2$
$= 24\pi + (2\pi \times 3^2)$
$= 24\pi + 18\pi = 42\pi$ cm^2
c. $V = \pi r^2 h = \pi \times 3^2 \times 4 = 36\pi$ cm^3

Classroom Practice

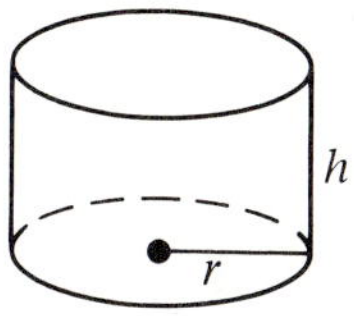

1. When you use the formula T.A. = L.A. + $2B$, what should you substitute for L.A.? (Use the figure shown.)
2. When you use the formula $V = Bh$, what formula do you use to find B? (Use the figure shown.)
3. The lateral area of a new piece of chalk is much greater than the area of a base. Name an object, shaped like a cylinder, in which:
   a. the area of a base is much greater than the lateral area;
   b. the area of a base is roughly equal to the lateral area.

4. Why is it incorrect to call the solid shown a cylinder?

5. Why is it incorrect to call the solid shown a right cylinder?
6. For the cylinder shown, find:
   a. the lateral area;
   b. the total area;
   c. the volume.

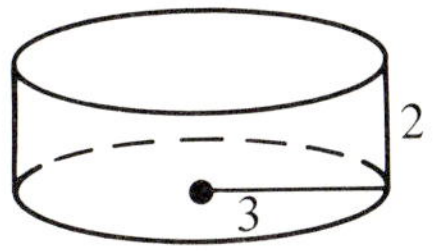

Written Exercises

Find the indicated values for the cylinder shown.

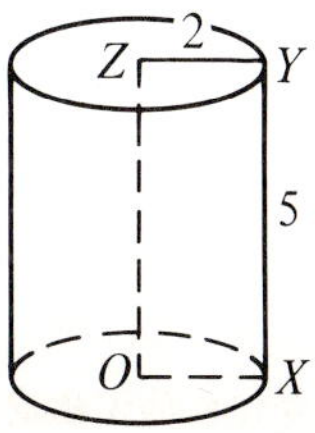

A

1. OX
2. OZ
3. Height of cylinder
4. Circumference of base
5. Area of base
6. Lateral area
7. Total area
8. Volume

Complete the table, which refers to cylinders.
***C* represents the circumference of a base.**

	r	h	C	B	L.A.	T.A.	V
9.	5	2	?	?	?	?	?
10.	8	3	?	?	?	?	?
11.	1	6	?	?	?	?	?
12.	6	1	?	?	?	?	?
B **13.**	?	2	6π	?	?	?	?
14.	?	4	?	36π	?	?	?
15.	3	?	?	?	?	?	36π
16.	?	5	?	?	?	?	20π

Use 3.14 for π. Write your answers correct to the nearest integer.

17. In a cylinder: $d = 6$; $h = 8$. $V =$ _?_

18. In a cylinder: $r = 10$; $h = 3$. T.A. = _?_

19. One can of nuts is twice as tall as another, but only half as wide. Which can holds more nuts?

20. If the can on the left costs \$1.09 and the one on the right costs \$2.19, which is the better buy?

C **21.** A chemical company has developed an additive that cuts down on friction, permitting water to flow through a hose twice as fast. Using this additive, a fire department can replace 7 cm hose with 5 cm hose, handle a hose considerably lighter, yet deliver the same amount of water.

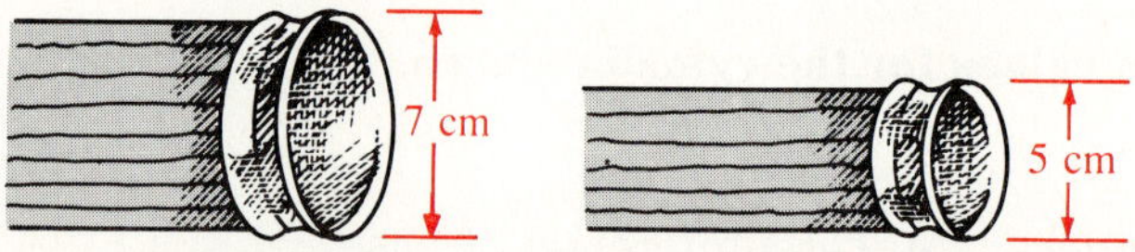

Show that the amount of water in a 7 cm hose is about twice the amount in a 5 cm hose. (Use the same length for each hose.)

SELF-TEST

Vocabulary

skew lines (p. 388)
parallel planes (p. 388)
perpendicular planes (p. 388)
rectangular solid (p. 393)
right prism (p. 393)
lateral area (p. 394)
right circular cylinder (p. 398)

Refer to the rectangular solid shown.

1. Name two planes that are parallel to $\overline{CR}$. *(10-1)*
2. Name two planes that are perpendicular to $\overline{DE}$.
3. Name any line shown that is skew to $\overleftrightarrow{UT}$.

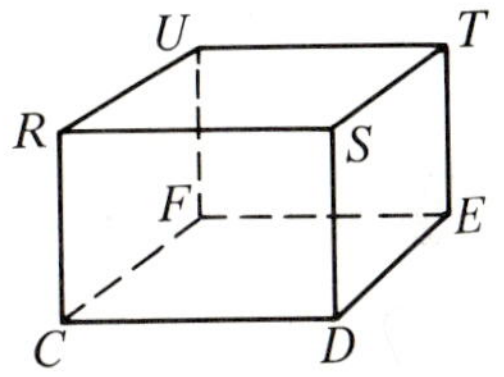

In the rectangular solid shown, let $CD = 8$, $DE = 5$, and $DS = 4$.

4. Total area = __?__ *(10-2)*
5. Volume = __?__

In a certain cylinder, $r = 8$ and $h = 4$. Find the indicated values in terms of π.

6. L.A. = __?__ *(10-3)*
7. T.A. = __?__
8. V = __?__

PUZZLE ◆ PROBLEMS

The faces of a cube are marked with these six symbols:

Three views of the cube are shown below.
Which symbols are on opposite sides of the cube?

4 Regular Pyramids

The ancient Egyptians built *pyramids* that still stand. Because the bases are squares, the pyramids are called *square* pyramids. We say that a pyramid is *regular* if its base is a regular polygon and the top lies directly over the center of the base.

The table and diagram below explain some of the terms used for regular pyramids.

vertices	points A, B, C, D, and V
the vertex	point V
the base	square $ABCD$
lateral edges	$\overline{VA}$, $\overline{VB}$, $\overline{VC}$, and $\overline{VD}$
base edges	$\overline{AB}$, $\overline{BC}$, $\overline{CD}$, and $\overline{DA}$
faces	$\triangle VDC$ is one of four congruent lateral faces. Base $ABCD$ is also a face.
center of base	point T
altitude	$\overline{VT}$ (Note: $\overline{VT} \perp$ plane $ABCD$)
height	VT, also h, the length of $\overline{VT}$
slant height	$\overline{VM}$, also l, the length of $\overline{VM}$ (Note: $\overline{VM} \perp \overline{CD}$; M is the midpoint of $\overline{CD}$.)

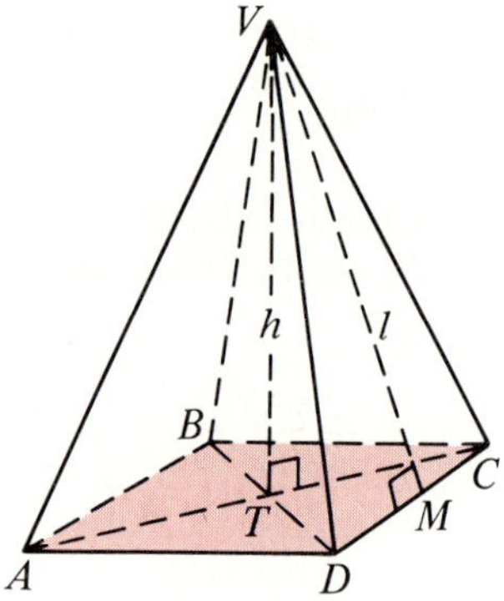

Regular Square Pyramid

The lateral faces of a pyramid are always triangular. To find the lateral area of a pyramid, use the formula for the area of a triangle. To find the total area, you simply add the lateral area and the area of the base. To find the volume, use the formula that we state without proof.

In any regular pyramid: $\text{T.A.} = \text{L.A.} + B \qquad V = \frac{1}{3}Bh$

EXAMPLE The base of a regular square pyramid has edges of length 6. The height of the pyramid equals 4 and the slant height equals 5. Find: **a.** the lateral area; **b.** the total area; **c.** the volume.

Draw a diagram.

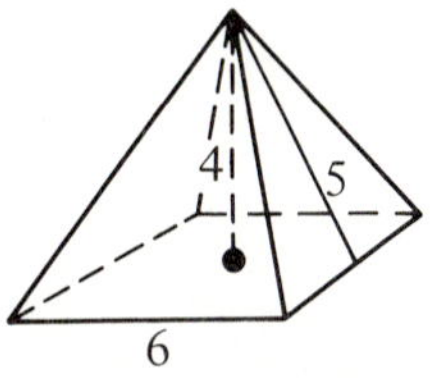

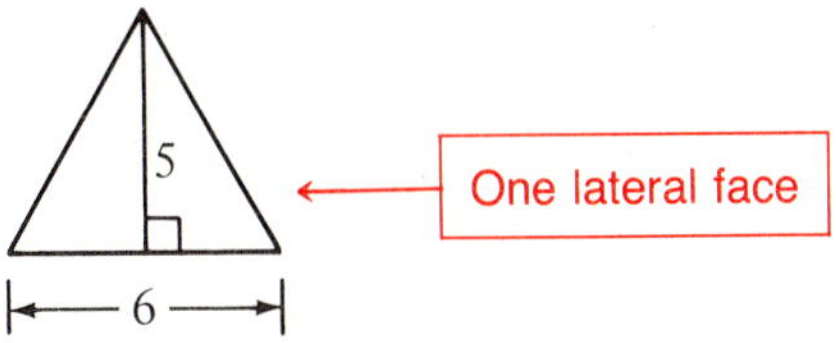

a. The area of one lateral face $= \frac{1}{2} \times 6 \times 5 = 15$

L.A. $= 4 \times 15 = 60$

b. $B = 6^2 = 36$

T.A. $=$ L.A. $+ B = 60 + 36 = 96$

c. $V = \frac{1}{3}Bh = \frac{1}{3} \times 36 \times 4 = 48$

Classroom Practice

Use the diagram of a regular triangular pyramid to name:

1. A lateral edge **2.** The vertex

3. The altitude **4.** The base

5. A slant height **6.** A lateral face

7. Name all the edges that equal $\overline{VY}$.

8. Which is longer: $\overline{VW}$ or $\overline{VM}$?

9. Which is longer: $\overline{VM}$ or $\overline{VX}$?

10. Name the hypotenuse of right $\triangle YVM$.

11. Does $\triangle VYZ$ have to be isosceles? equilateral?

12. $\angle XYZ =$ _?_ ° $\quad \angle XWY =$ _?_ ° $\quad \angle WXY =$ _?_ °

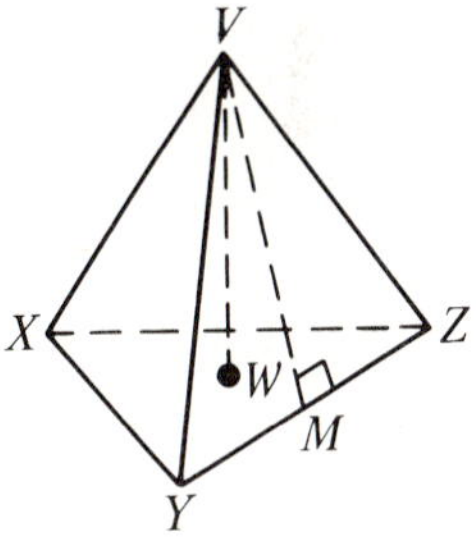

Exercises 1–12

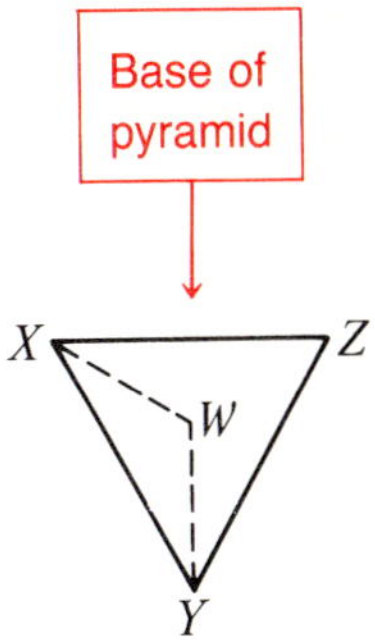

Exercise 12

The diagram shows a regular square pyramid.

13. Find the lateral area.

14. Find the total area.

15. Find the volume.

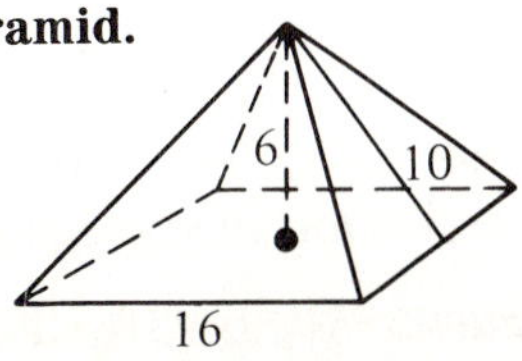

Written Exercises

Refer to the regular square pyramid shown.
Tell whether each statement is true or false.

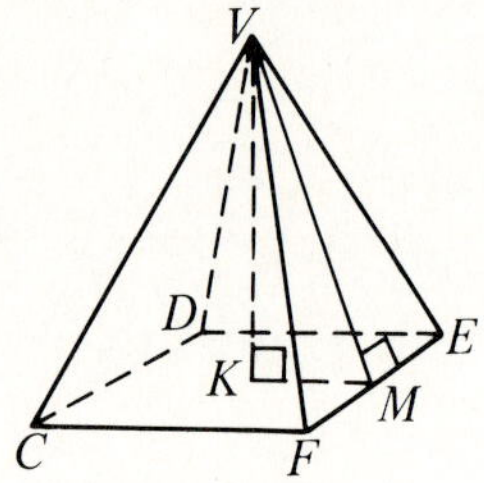

A **1.** The pyramid has five faces.

2. All eight edges of the pyramid must be equal.

3. *VM* is the height of the pyramid.

4. $KM = \frac{1}{2} CF$ **5.** $KM = ME$

6. If $\overline{KC}$ were drawn, $\angle VKC$ would be a right angle.

7. L.A. = 4 × area of quadrilateral *CDEF* **8.** T.A. = L.A. + 2*B*

9. *VM* must be greater than *VK*. **10.** *VM* must be greater than *KM*.

Draw a regular pyramid whose base is the polygon named.

11. Triangle **12.** Square **13.** Pentagon **14.** Hexagon

Exercises 15–20 refer to a regular pyramid whose base is an 8 by 8 square. The height equals 3 and the slant height equals 5.

15. Draw a figure. **16.** Perimeter of the base = __?__

17. Area of the base = __?__ **18.** Lateral area of the pyramid = __?__

19. Total area of the pyramid = __?__ **20.** Volume of the pyramid = __?__

In Exercises 21 and 22, a regular pyramid is described. Find the lateral area, the total area, and the volume.

B **21.** The base is a 12 cm by 12 cm square. The height equals 8 cm. The slant height equals 10 cm.

22. The base is a 16 cm by 16 cm square. The height equals 15 cm. The slant height equals 17 cm.

23. The base of a regular pyramid is a hexagon with sides 4 m long. The slant height equals 10 m. Find the lateral area.

24. In a certain regular pyramid: $B = 52$; $h = 6$; $l = 7$. Decide which you can find, the lateral area or the volume. Then find it.

C **25.** A cube has a 6 cm edge. A pyramid is formed by joining the vertices of one face to *X*, the center of the opposite face. Draw a figure. Find the volume of the pyramid.

26. Repeat Exercise 25, using *Y*, the center of the cube, instead of *X*.

Experiments

The purpose of this experiment is to find a relation between the number of vertices, the number of faces, and the number of edges that a solid has. Divide the class into ten committees. Each committee can work with one of the solids below.

1. prism with triangular base
2. pyramid with triangular base
3. prism with square base
4. pyramid with square base
5. prism with hexagonal base
6. pyramid with hexagonal base

7. 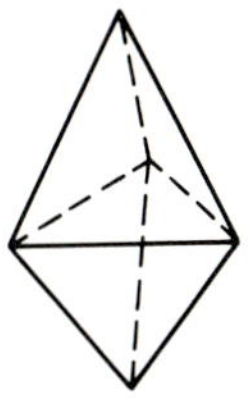

8. 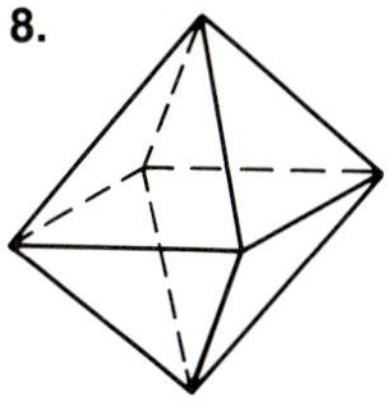

9. 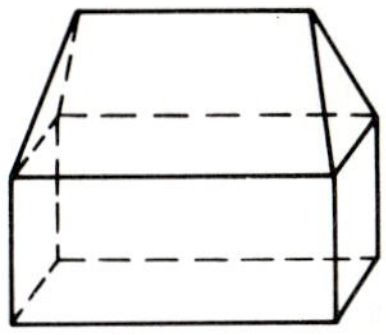

10. 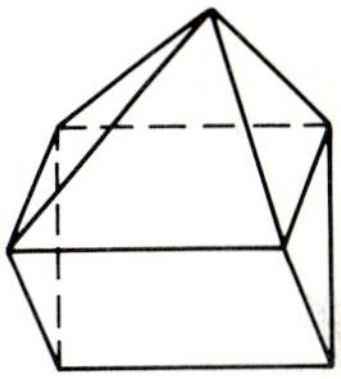

For each figure, count:

V, the number of vertices;
F, the number of faces;
E, the number of edges.

Copy and complete the table at the chalkboard.

	1.	2.	3.	4.	5.	6.	7.	8.	9.	10.
V	?	?	?	?	?	?	?	?	?	?
F	?	?	?	?	?	?	?	?	?	?
E	?	?	?	?	?	?	?	?	?	?

Study the columns.
Find a formula that tells how V, F, and E are related.

PUZZLE ◆ PROBLEMS

Sam stared at a cactus in the Painted Desert until he lost his memory. The cactus is 0.5 m high and its shadow is 1 m long. Sam's shadow is 3.5 m long. Can you tell him how tall he is?

5 Right Circular Cones

A **right circular cone** is very much like a regular pyramid. Notice that the base of a cone is circular. As the diagram shows, we use many of the same terms to describe a cone as we use for other solids.

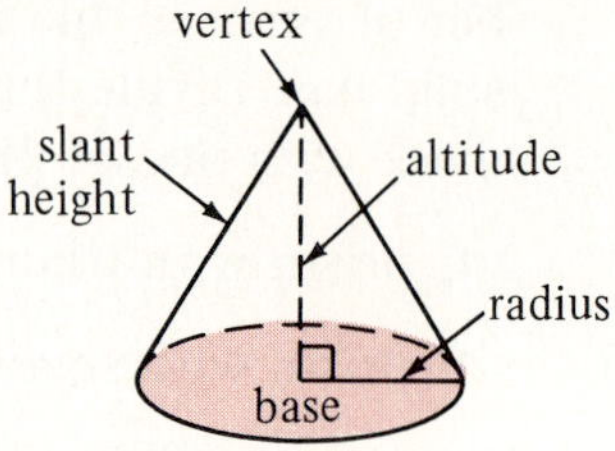

Right Circular Cone

The formula for the volume of a pyramid suggests the formula for the volume of a cone:

$$V = \frac{1}{3}Bh;\ V = \frac{1}{3} \times \pi r^2 \times h$$

The formula, stated below, for the lateral area of a cone will be verified in the exercises.

In any right circular cone: $\text{L.A.} = \pi r l$

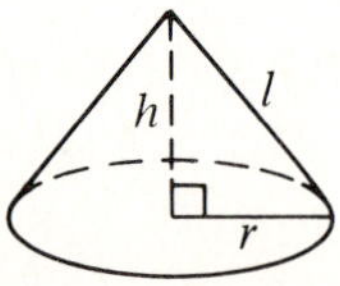

$$\text{T.A.} = \text{L.A.} + B \qquad V = \frac{1}{3}Bh$$
$$= \pi r l + \pi r^2 \qquad = \frac{1}{3}\pi r^2 h$$

In the remainder of the course, the word *cone* will mean right circular cone.

EXAMPLE For the cone pictured, find:
a. the lateral area **b.** the total area **c.** the volume

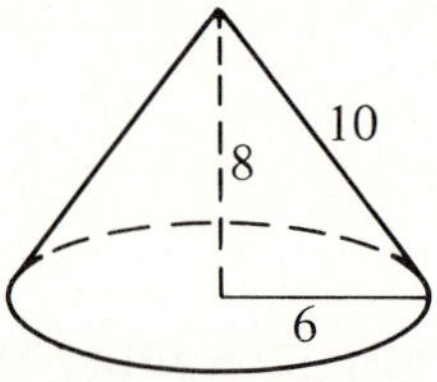

a. $\text{L.A.} = \pi r l$
$= \pi \times 6 \times 10 = 60\pi$

b. $\text{T.A.} = \text{L.A.} + B$
$= 60\pi + \pi r^2$
$= 60\pi + (\pi \times 6^2)$
$= 60\pi + 36\pi = 96\pi$

c. $V = \frac{1}{3}Bh$
$= \frac{1}{3} \times \pi r^2 \times h$
$= \frac{1}{3} \times 36\pi \times 8 = 96\pi$

Classroom Practice

1. In the example on page 406, the number of cubic units in the volume is equal to the number of square units in the total area. Do you think that this is true for all cones? Find the lateral area, the total area, and the volume for a cone with $r = 8$, $h = 6$, and $l = 10$.

Draw a circle whose radius is actually 6 cm.
Draw two radii that form a 120° angle.

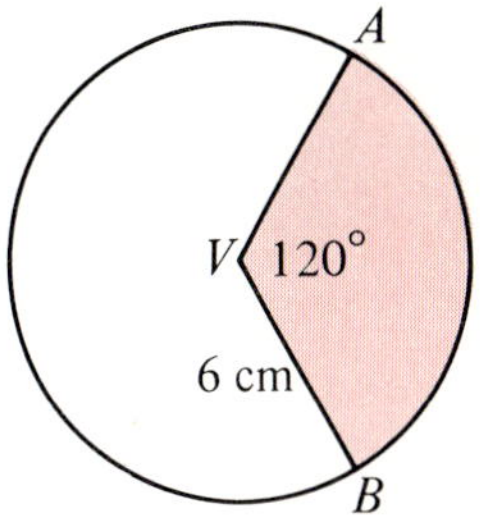

2. The area of the shaded region is what fractional part of the area of the circle?
3. Find the area of the shaded region.
4. The length of $\overset{\frown}{AB}$ is what fractional part of the circumference of the circle?
5. Find the length of $\overset{\frown}{AB}$.

Cut out the shaded region. Curve the paper to form a cone, letting $\overline{VA}$ and $\overline{VB}$ come together.

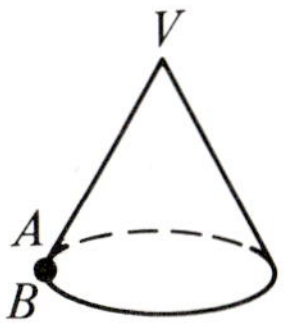

6. The slant height of the cone equals __?__.
7. The lateral area of the cone equals __?__. (Answer this the easy way. See Exercise 3.)
8. The circumference of the base of the cone equals __?__. (Again, the answer is easy. See Exercise 5.)
9. Use your answer for Exercise 8 to find the radius of the base of the cone.
10. Show that your answers for Exercises 6, 7, and 9 satisfy the formula L.A. $= \pi rl$.

Written Exercises

Find the lateral area of each cone.

A 1.
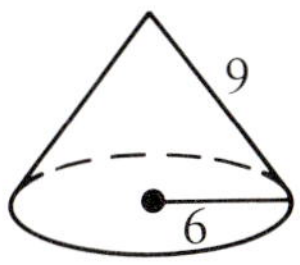

2.
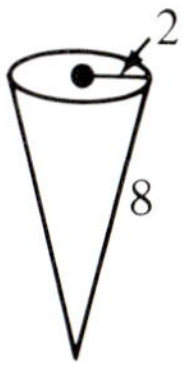

3.
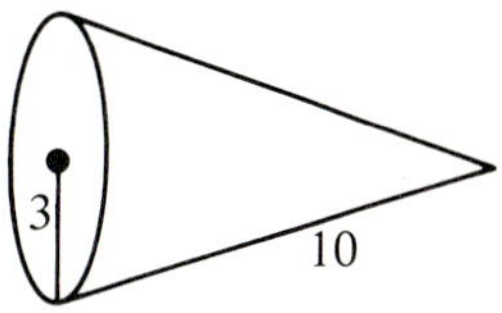

4–6. Find the total areas of the cones shown in Exercises 1–3.

Find the volume of each cone.

7. 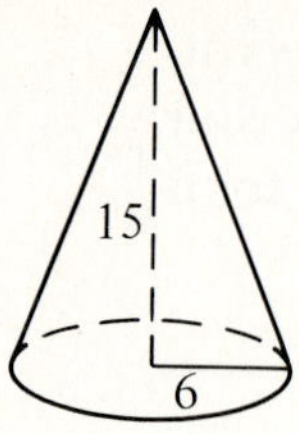

8. 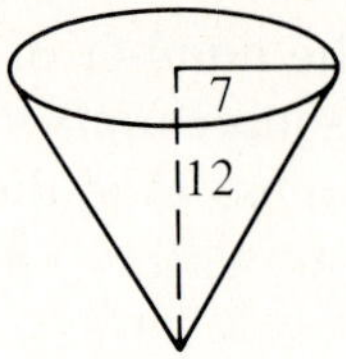

9. 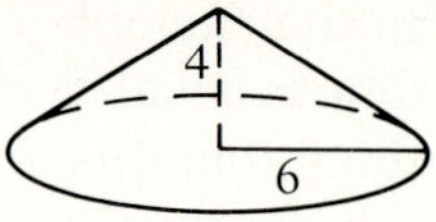

Copy and complete the table, which deals with cones. (*C* represents the circumference of the base.)

	r	*C*	*B*	*h*	*l*	L.A.	T.A.	*V*
10.	3	?	?	4	5	?	?	?
11.	12	?	?	9	15	?	?	?
12.	8	?	?	15	17	?	?	?
B **13.**	?	12π	?	8	10	?	?	?
14.	?	?	16π	3	5	?	?	?
15.	5	?	?	?	?	?	?	100π

16. A piece of sheet metal is cut and formed into the lateral surface of the cone pictured. There are 0.24 g in 1 cm^2 of the metal. Find the total number of grams of metal. (Use 3.14 for π.)

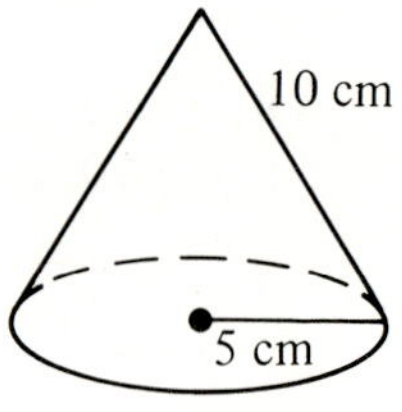

C **17.** The solid hardwood cone pictured has a 16 cm radius and a 35 cm altitude. There are 0.62 g of wood in 1 cm^3 of the hardwood used. Find the total number of grams of wood, to the nearest 10 g. (Use 3.14 for π.)

Refer to the diagram below of a cone and a regular square pyramid.

18. Find the ratio of the lateral areas of the solids.

19. Find the ratio of the total areas of the solids.

20. Find the ratio of the volumes of the solids.

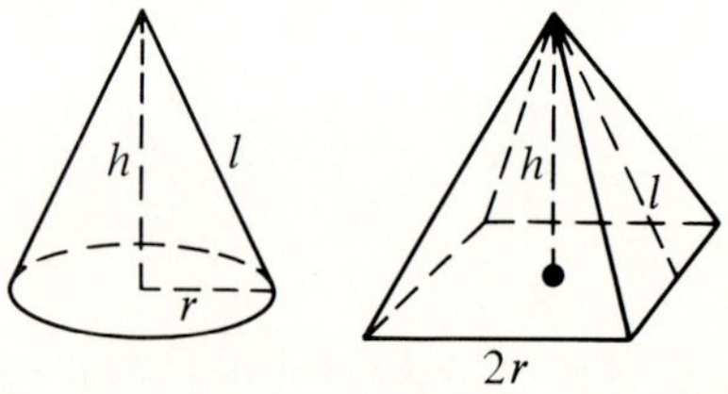

Experiments

For this experiment you need a small piece of cardboard with a bright, solid-colored surface. Close your left eye. Gaze straight ahead. Hold the card in your right hand with your right arm stretched out to the side. The color of the card cannot be accurately judged. Slowly move your unbent arm forward. At a certain stage you begin to see the color clearly. Why?

The back of the retina of your eye is lined with *rods* and *cones.* The rods—long, thin cylinders—register only in black and white. An object that you view from the side is seen primarily by the rods. The cones, which are able to distinguish color, are concentrated near the center of your eye.

CALCULATOR ACTIVITIES

Solid metal cones are to be made of various metals. How many cones of the size specified can be made from 10 kg (10,000 g) of the metal named? Round your answer to the nearest 10 cones.

EXAMPLE

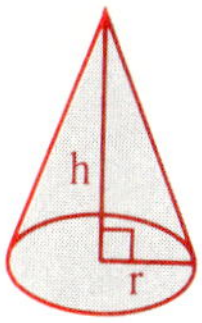

Bronze cones: $r = 1.5$ cm; $h = 4$ cm.
There are 8.8 g in 1 cm^3 of bronze.

$$\text{Number of cones} = \begin{pmatrix}\text{total number of}\\ \text{grams of metal}\end{pmatrix} \div \begin{pmatrix}\text{total number of}\\ \text{grams in one cone}\end{pmatrix}$$

$$= 10{,}000 \div \left(\frac{1}{3}\pi r^2 h \times 8.8\right)$$

$$= 10{,}000 \div \left(\frac{1}{3}\pi \times 1.5^2 \times 4 \times 8.8\right) \approx 120$$

Copy and complete the table.

		r (in cm)	h (in cm)	Grams in 1 cm^3	No. of cones
1.	Aluminum	2	3	2.7	?
2.	Brass	2.6	4	8.6	?
3.	Iron	5.2	7.6	7.9	?
4.	Magnesium	3.2	4.1	1.7	?
5.	Nickel	3	8	8.9	?

6 Spheres

Look at the circle and the sphere. Each has radius r.

Recall that the area of the circle equals πr^2. It happens that the surface area of the sphere is four times the area of the circle, or $4\pi r^2$.

The formula for the volume of a sphere is stated without proof.

> *In any sphere:* $\quad A = 4\pi r^2 \qquad V = \frac{4}{3}\pi r^3$

EXAMPLE

The diameter of a sphere is 8.

a. Find the area and the volume in terms of π.

b. Find approximations for the area and the volume. Round your answers to the nearest integer.

Since $d = 8$, $r = 4$.

a.

$$A = 4\pi r^2 \qquad V = \frac{4}{3}\pi r^3$$

$$= 4\pi \times 4^2 \qquad = \frac{4}{3}\pi \times 4^3$$

$$= 64\pi \qquad = \frac{256\pi}{3}$$

The area is 64π, and the volume is $\frac{256\pi}{3}$.

b.

$$A = 64\pi \qquad V = \frac{356\pi}{3}$$

$$A \approx 64 \times 3.14 = 200.96 \qquad V \approx \frac{256 \times 3.14}{3} = 267.95$$

$$A \approx 201 \qquad V \approx 268$$

To the nearest integer, the area is 201, and the volume is 268.

Classroom Practice

Find the area and the volume of a sphere with the given radius. Express your answers in terms of π.

1. $r = 2$ cm **2.** $r = 5$ cm **3.** $r = 0.1$ cm **4.** $r = \frac{1}{4}$

5. If the area of a sphere is 36π, what is its radius?

6. "Half" of a sphere is called a *hemisphere.* Find the area and the volume of the hemisphere shown.

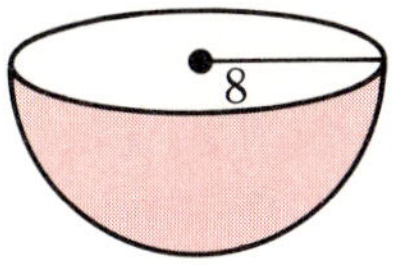

Written Exercises

Find the area and the volume of a sphere with the given radius. Express your answers in terms of π.

A **1.** $r = 6$ **2.** $r = 12$ **3.** $r = 8$ **4.** $r = \frac{1}{2}$

Find the area and the volume of a sphere with the given diameter. Use 3.14 for π. Round your answers to the nearest integer.

5. $d = 18$ **6.** $d = 2$ m **7.** $d = 12$ cm **8.** $d = 0.6$

Copy and complete the table about spheres. Write your answers in terms of π.

B

	9.	**10.**	**11.**	**12.**	**13.**	**14.**	**15.**	**16.**
r	$\frac{1}{8}$	?	?	?	$\frac{3}{2}$	?	?	?
d	?	$\frac{3}{4}$	?	?	?	?	?	$2\sqrt{3}$
A	?	?	100π	?	?	144π	?	?
V	?	?	?	288π	?	?	$\frac{1372\pi}{3}$	?

A sphere fits inside a cylinder as shown.

C **17.** Show that the area of the sphere equals the lateral area of the cylinder.

18. Find the ratio of the volume of the sphere to the volume of the cylinder.

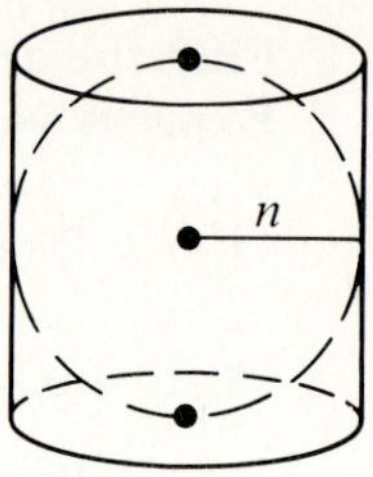

SELF-TEST

Vocabulary

regular pyramid (p. 402):

vertices	faces
the vertex	center of base
the base	altitude
lateral edges	height
base edges	slant height

right circular cone (p. 406)
sphere (p. 410)
hemisphere (p. 411)

Exercises 1–4 refer to a regular pyramid whose base is an 18 by 18 square. The height equals 12 and the slant height equals 15.

1. Draw a figure. *(10-4)*

2. L.A. = _?_

3. T. A. = _?_

4. V = _?_

Refer to the cone shown. Answer in terms of π.

5. Slant height = _?_ *(10-5)*

6. L.A. = _?_

7. T.A. = _?_

8. V = _?_

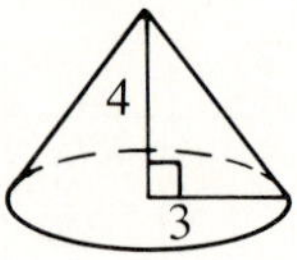

The radius of a certain sphere is 4. Use 3.14 for π. Round to the nearest integer.

9. A = _?_ *(10-6)*

10. V = _?_

APPLICATIONS

SURFACE AREA AND VOLUME

Geometry tells us that, in general, a small object has more surface area *per unit of volume* than a large object has. Let us compare a baseball with radius 3.5 cm and a basketball with radius 12.4 cm.

Baseball

$$\frac{\text{Surface area}}{\text{Volume}} \approx \frac{153.9}{179.5} \approx 0.86$$

Basketball

$$\frac{\text{Surface area}}{\text{Volume}} \approx \frac{1931.2}{7982.4} \approx 0.24$$

Now let us see how this principle affects a relatively small animal, a mouse, and a larger animal, a human being. Having more skin area per unit of volume, a mouse loses body heat faster than a person does. Just to keep warm, a mouse must eat and burn more food relative to its size. A mouse needs to eat about one half of its weight in food each day. How much food would you have to consume if you ate at that rate?

The relationship between surface area and volume also helps to explain the difference between a mouse's lung and a human's lung. In order to do its job well, the lung must have enough surface area in its air sacs. A small animal like a mouse needs only a simple lung to provide enough air-sac surface. A larger animal like a human being needs a more complex lung to provide enough surface area for breathing. A giant mouse with a simple lung could not breathe enough oxygen to keep it alive.

EXERCISE

Compute and compare the ratio of total surface area to volume for a small cube with edges 2 cm long and a larger cube with edges 60 cm long.

7 Similar Solids (Optional)

Two solids are called *similar solids* if they have the same shape. The triangular prisms shown are similar.

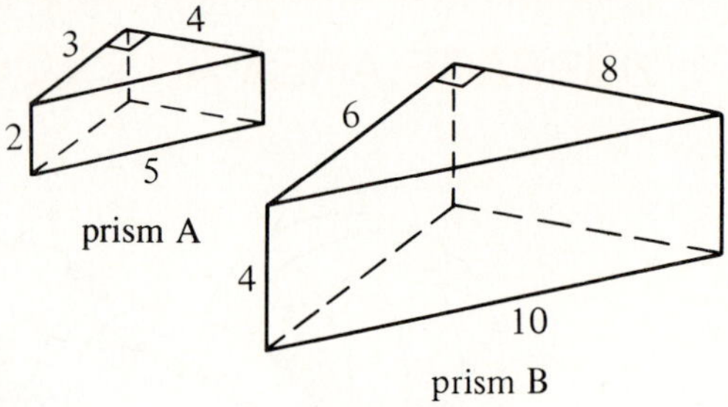

> Notice that in similar solids:
>
> (1) corresponding angles are equal;
>
> (2) corresponding segments are in proportion.

EXPLORATIONS

Recall, from Chapter 8, that the perimeters and areas of similar figures are related. Let's see whether the surface areas and volumes of similar solids are related. Use the similar triangular prisms shown above. Copy and complete the table.

	prism A	prism B	ratio in simplest form
hypotenuse of base	5	10	1:2
height	?	?	?
base area	?	?	?
lateral area	?	?	?
total area	?	?	?
volume	?	?	?

You should have found that corresponding lengths are in the ratio 1:2. Now notice: the base areas, lateral areas, and total areas are all in the ratio 1:4, or $1^2:2^2$. The volumes are in the ratio 1:8, or $1^3:2^3$. This exploration suggests the following theorem.

THEOREM 1

Suppose two similar solids have scale factor $a:b$. Then:

(1) the ratio of corresponding segments is $a:b$;

(2) the ratio of the areas is $a^2:b^2$;

(3) the ratio of the volumes is $a^3:b^3$.

EXAMPLE

Suppose the two right cylinders shown are similar. Find:

a. the scale factor

b. the height of the larger cylinder

c. the ratio of the base areas

d. the ratio of the lateral areas

e. the ratio of the volumes

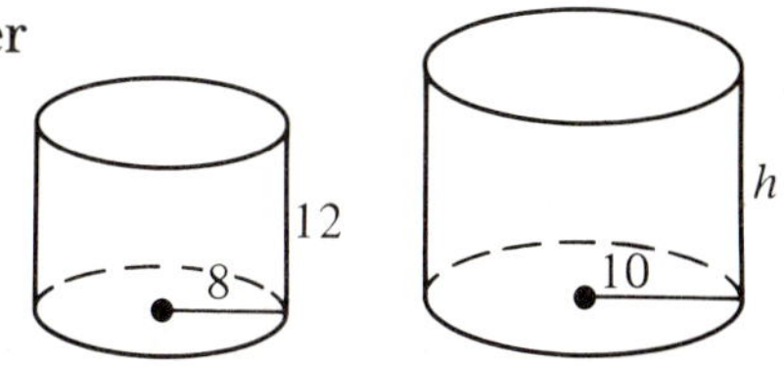

a. The scale factor is 8:10, or 4:5.

b. Write a proportion and solve for h. Then $h = 15$.

$$\frac{4}{5} = \frac{12}{h}$$
$$4h = 60$$
$$h = 15$$

c. ratio of base areas $= 4^2:5^2 = 16:25$

d. ratio of lateral areas $= 16:25$

e. ratio of volumes $= 4^3:5^3 = 64:125$

Classroom Practice

Refer to the similar pyramids shown.

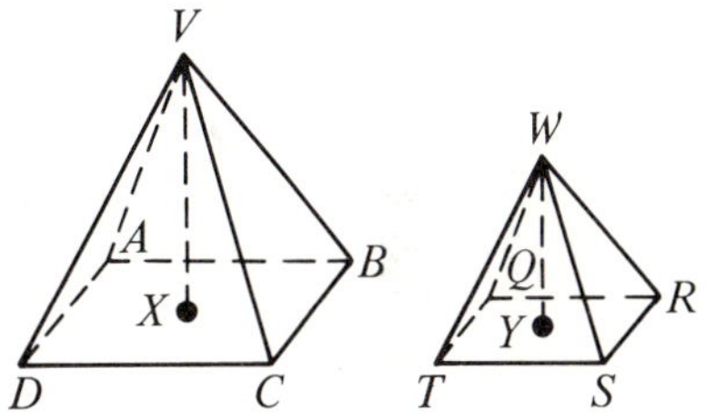

1. $\angle DVC$ corresponds to $\angle$ __?__.
2. $\overline{BC}$ and __?__ are corresponding segments.
3. $\frac{VC}{WS} = \frac{DA}{?}$
4. $\frac{AB}{QR} = \frac{?}{WY}$
5. Suppose the pyramids have scale factor 10:7. Find the ratio of:
   a. the base areas **b.** the total areas **c.** the volumes

Written Exercises

Copy and complete the table about similar solids.

A

	1.	2.	3.	4.	5.	6.	7.	8.
ratio of corresponding segments	3:4	4:3	1:6	7:9	?	?	?	?
ratio of areas	?	?	?	?	4:25	?	?	100:1
ratio of volumes	?	?	?	?	?	27:8	1:64	?

Exercises 9–11 refer to two similar prisms with scale factor 3:5.

B 9. The shortest edge of the smaller prism is 6 cm long. How long is the shortest edge of the larger prism?

10. The base area of the larger prism is 50 cm^2. Find the base area of the smaller prism.

11. The volume of the smaller prism is 216 cm^3. Find the volume of the larger prism.

12. The total areas of two similar cones have the ratio 9:4. Find the ratio of their volumes.

13. Any two spheres are similar solids. The volumes of two spheres are 288π and 7776π, respectively. Find the ratios of their areas.

A town has two water towers that are similar solids. Two corresponding lengths are labeled.

C 14. Twenty liters of paint were needed to paint the smaller tower. How much paint is needed for the larger tower?

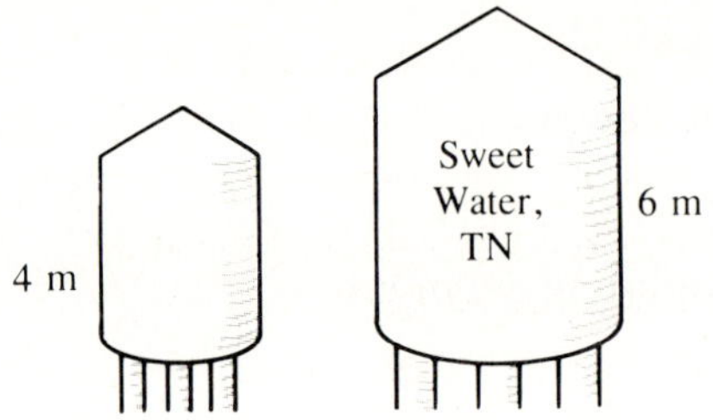

15. The capacity of the larger tank is 180,000 liters. What is the capacity of the smaller tank?

16. For two spheres with scale factor $a:b$, prove:
   a. the ratio of their areas is $a^2:b^2$;
   b. the ratio of their volumes is $a^3:b^3$.

17. A leg bone of a certain animal has a cross-section area of 15 cm^2. The bone supports about 50 kg. Imagine a similar, but larger, animal with scale factor 10:1.
   a. In the smaller animal, the bone supports _?_ kg/cm^2.
   b. In the larger animal, the corresponding bone would have a cross-section area of _?_, and the bone would have to support _?_ kg. Thus it would have to support _?_ kg/cm^2.
   c. Comparing (a) and (b), comment on the statement "Bigger is better."

READING GEOMETRY

THREE-DIMENSIONAL FIGURES

Most of your work in this book has involved plane figures. Since plane figures have only two dimensions, they can easily be pictured by diagrams on a flat piece of paper. Figures in space have three dimensions, so it isn't as easy to represent them in this way. For example, you already know that a plane can only be suggested by a drawing. You have to imagine it as extending endlessly.

When studying figures in space, models can help you see relationships more clearly. The walls, floor, and ceiling of your classroom can be used as models of parallel planes and intersecting planes. Large pieces of cardboard can be held to represent planes in different positions, and pencils or pieces of stiff wire can be used for lines. Boxes and cans make good models of rectangular solids, cubes, and cylinders. Balls are good models of spheres. Some sets of building blocks contain cones and pyramids as well as more common shapes. Test your powers of observation by looking for models of solids around you.

Making models of solid figures can be fun. To do this, you need patterns. A pattern for a square pyramid is shown at the right. Make a large copy of it on stiff paper or cardboard, cut it out, fold on the dashed lines, and tape the edges together. Your school library may have a book of patterns for other models, or try making your own.

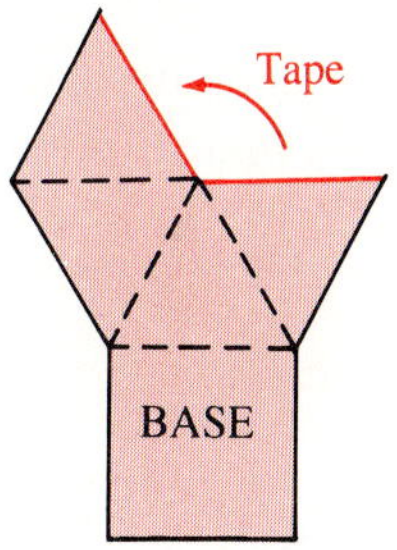

Page 393 explains an easy way to draw a rectangular solid. How could you use this method to draw a cylinder? Try copying some of the diagrams in Chapter 10.

EXERCISES

1. Use a milk carton to make a pattern for a rectangular solid. Look carefully at the carton in order to cut it so that you can spread it out flat in one piece.

2. Look at the diagram of the triangular pyramid on page 403. Try making a model of this figure. (What kind of triangle forms the base of this pyramid?)

3. Try the Mechanical Drawing Application on page 431. Go to it now and do Exercises 1–3.

PROBLEM SOLVING STRATEGIES

COUNTEREXAMPLES

One way you can disprove a statement is to find an example for which the statement is not true. We call this a *counterexample.*

If someone says to you that all triangles are acute, you could draw an obtuse triangle. If someone says that the diagonals of a parallelogram bisect the angles of the parallelogram, you could draw a picture similar to the one below.

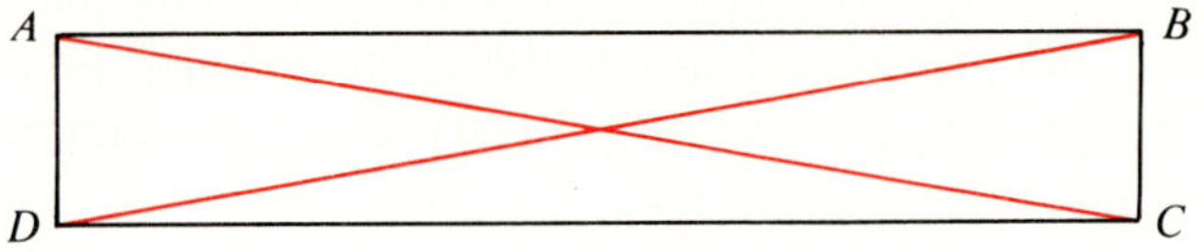

ABCD is a parallelogram.
$\angle ADB$ is not congruent to $\angle BDC$.

To disprove a statement, you need to find an example that shows the conclusion false when the hypothesis is true.

EXERCISES

Draw a picture that shows that each of the following statements is FALSE.

1. If a triangle is isosceles, then it can't be a right triangle.
2. The diagonals of a parallelogram must always be perpendicular to each other.
3. If a quadrilateral is equilateral, then it is equiangular.
4. If a quadrilateral is equiangular, then it is equilateral.
5. If the diagonals of a trapezoid are perpendicular, the trapezoid is a rhombus.
6. All isosceles triangles are acute.
7. If a figure is a polygon, then it is a triangle.
8. If $\triangle XYZ$ is obtuse, then $\angle XYZ$ is obtuse.

9. Two circles which do not intersect have two common external tangents.

10. If two chords are perpendicular, then one of them must pass through the center of the circle.

Give a counterexample to disprove each statement.

SAMPLE

All even integers are divisible by 4.

The even integers are made up of 2, 4, 6, 8, 10, . . . and −2, −4, −6, −8, −10, ...

Many of these numbers are not divisible by 4. You need only come up with one of them to have a counterexample. For instance:

10 is an even integer.
10 is not divisible by 4.

Therefore, not all even integers are divisible by 4.

11. The altitudes of a triangle meet inside the triangle.

12. $(a + b)^2$ is always equal to $a^2 + b^2$.

13. If the ratio of the sides of two similar polygons is 1:2, then the ratio of their areas is 1:2.

14. The square of any integer is odd.

15. If $x > 5$, then $x = 6$.

16. If $AM = MB$, then M is the midpoint of $\overline{AB}$.

17. If $x^2 = y^2$, then $x = y$.

18. If two angles are congruent, they must be vertical angles.

19. If two angles are complementary, their measures are each 45°.

20. Equal fractions have the same denominator.

21. If $m^2 > n^2$, then $m > n$.

22. If $(x - 4)(x + 6) = 0$, then x must equal 4.

23. If $\angle 1 = \angle 2$, then $\angle 1$ and $\angle 2$ are vertical angles.

SKILLS REVIEW

WORKING WITH POLYNOMIALS

Multiply. The samples will help you remember some patterns.

SAMPLES

1. $2x^2(x - 3y) = (2x^2)(x) - (2x^2)(3y) = 2x^3 - 6x^2y$

2. $(a - 4b)(2a + 9b) = 2a^2 + 9ab - 8ab - 36b^2 = 2a^2 + ab - 36b^2$

3. $(c + 7d)(c + 7d) = c^2 + 7cd + 7cd + 49d^2 = c^2 + 14cd + 49d^2$

4. $(12 + 2k)(12 - 2k) = 144 - 24k + 24k - 4k^2 = 144 - 4k^2$

1. $a^3(3a^2 - ab - b^2)$
2. $(a + 5)(a - 3)$
3. $(r + 4)(r + 4)$
4. $(m + 8)(m - 8)$
5. $(10 - 3x)(10 - 3x)$
6. $c^2d^3(cd^2 - 7cd)$
7. $(b - 13)(b - 13)$
8. $(m + 2n)(m - n)$
9. $(x - 6)(x - 9)$
10. $(6w + 2x)(6w - 2x)$
11. $-6k(5h + 4hk - k^4)$
12. $(2d + 7c)(2d + 7c)$
13. $(a + 12b)(a - 12b)$
14. $(3y + 4z)(5y + 7z)$
15. $(8k - h)(3k + 2h)$
16. $(2b + 6)(b - 5)$
17. $(3m + 2n)(3m - 2n)$
18. $-ab^4(2a^2b^2 - 5b^3)$

Factor.

SAMPLES

1. $4d^6 + 14cd^8 = 2d^6(2 + 7cd^2)$
2. $n^2 - 11n - 26 = (n - 13)(n + 2)$
3. $x^2 - 6x + 9 = (x - 3)(x - 3)$, or $(x - 3)^2$
4. $100c^2 - 9d^2 = (10c + 3d)(10c - 3d)$

19. $15a^2x - 10ax^2$
20. $x^2 + 6x + 8$
21. $y^2 - 20y + 100$
22. $4m^2 - 9$
23. $a^2 + 2a - 35$
24. $14b^7 - 7b^6 + 28b^5$
25. $k^2 + 14k + 49$
26. $m^2 + 17m + 16$
27. $c^2 - 13c + 30$
28. $25x^2 - 121y^2$
29. $27a^3b^3 + 6ab^2c$
30. $b^2 - 8b + 12$
31. $144 + 24n + n^2$
32. $1 - 36x^2y^2$
33. $h^2 + 6h - 16$
34. $r^2 - 16r + 64$
35. $n^2 - 12n + 27$
36. $a^2 - a - 72$
37. $k^2 - 9k - 22$
38. $c^2 + 18c + 81$
39. $r^2 + 11r + 28$
40. $144 - 25y^2$
41. $d^2 - 20d + 19$
42. $m^2 - 2m - 63$

CHAPTER REVIEW

CHAPTER SUMMARY

1. Two lines can intersect, be parallel, or be skew. A line can lie in a plane, intersect the plane, or be parallel to the plane. Two planes intersect, or else they are parallel.

   If line $j \perp$ plane M, and if plane N contains line j, then $N \perp M$.

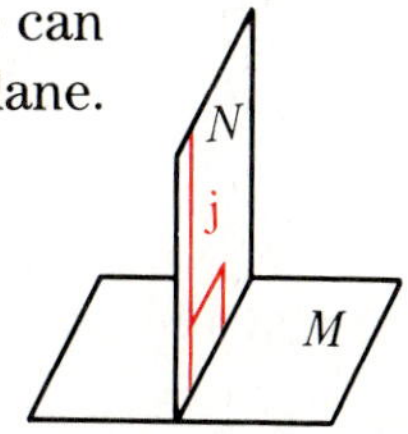

2. In any **right prism**:
   L.A. = sum of the areas of the lateral faces
   T.A. = L.A. + $2B$
   $V = Bh$

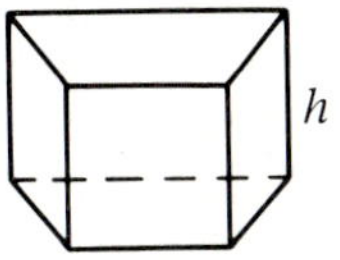

3. In any **right circular cylinder:**
   L.A. = $2\pi rh$
   T.A. = L.A. + $2B = 2\pi rh + 2\pi r^2$
   $V = Bh = \pi r^2 h$

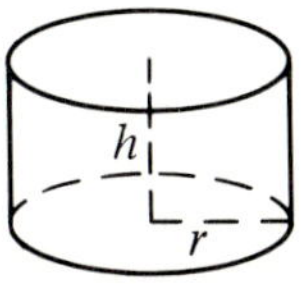

4. In any **regular pyramid:**
   L.A. = sum of the areas of the lateral faces
   T.A. = L.A. + B

   $V = \frac{1}{3}Bh$

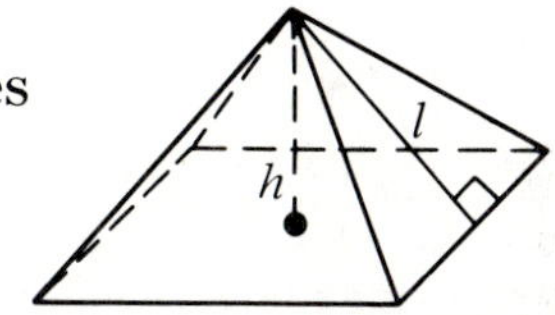

5. In any **right circular cone:**
   L.A. = πrl
   T.A. = L.A. + $B = \pi rl + \pi r^2$

   $V = \frac{1}{3}Bh = \frac{1}{3}\pi r^2 h$

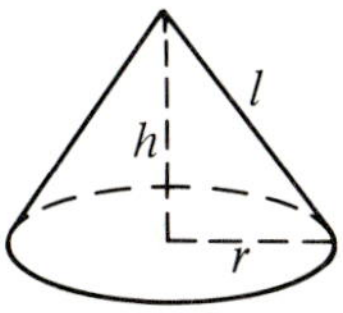

6. In any **sphere:**
   $A = 4\pi r^2$

   $V = \frac{4}{3}\pi r^3$

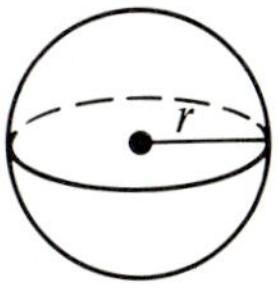

7. If two similar solids have scale factor $a:b$, then:
   a. the ratio of corresponding segments is $a:b$;
   b. the ratio of the areas is $a^2:b^2$;
   c. the ratio of the volumes is $a^3:b^3$.

REVIEW EXERCISES

Tell whether each statement is true or false. ***(See pp. 388–391.)***

1. If two planes do not intersect, they must be parallel.
2. If two lines do not intersect, they must be parallel.
3. If a line doesn't lie in a plane, the line must be parallel to the plane.
4. If line $\overleftrightarrow{AB}$ is perpendicular to plane R, then $\overleftrightarrow{AB}$ is perpendicular to every line in R.

Refer to the right prism shown. ***(See pp. 393–397.)***

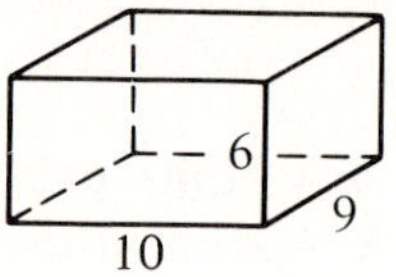

5. $B =$ ___?___
6. L.A. = ___?___
7. T.A. = ___?___
8. $V =$ ___?___

Refer to the cylinder shown. Answer in terms of π. ***(See pp. 398–400.)***

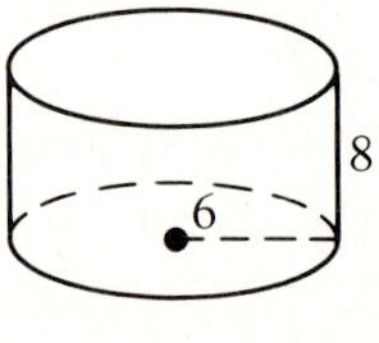

9. $C =$ ___?___
10. L.A. = ___?___
11. T.A. = ___?___
12. $V =$ ___?___

Refer to the regular square pyramid shown. ***(See pp. 402–404.)***

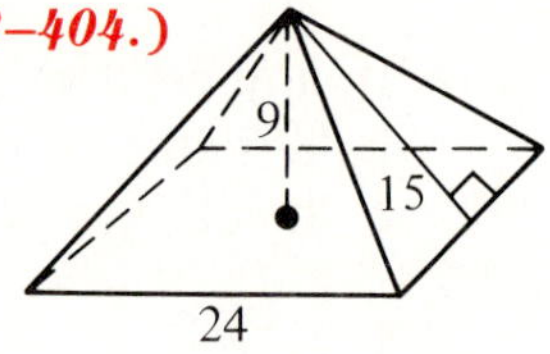

13. $B =$ ___?___
14. L.A. = ___?___
15. T.A. = ___?___
16. $V =$ ___?___

Refer to the right circular cone shown. Answer in terms of π. ***(See pp. 406–408.)***

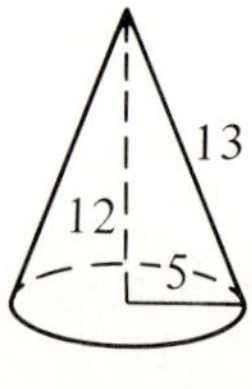

17. $C =$ ___?___
18. L.A. = ___?___
19. T.A. = ___?___
20. $V =$ ___?___

Answer the questions about spheres in terms of π. ***(See pp. 410–412.)***

21. When $r = 5$, $A =$ ___?___.
22. When $d = 8$, $V =$ ___?___.

(Optional) Copy and complete the table which deals with similar solids. ***(See pp. 414–416.)***

	23.	24.	25.
ratio of corresponding segments	2:5	?	?
ratio of areas	?	49:81	?
ratio of volumes	?	?	64:27

Chapter Test

Complete each statement with *always*, *sometimes*, or *never*.

1. Two planes that do not intersect are _?_ parallel. *(10-1)*
2. Two lines that do not intersect are _?_ parallel.
3. If line j is perpendicular to plane P, then line j is _?_ perpendicular to every line in P.

A rectangular solid has length 10, width 8.5, and height 6.

4. L.A. = _?_ 5. T.A. = _?_ 6. $V =$ _?_ *(10-2)*
7. If a cube has sides of length x, find its total area and volume in terms of x.

A cylinder has a radius of 7 and a height of 5. Find the indicated values in terms of π.

8. The circumference of a base 9. The area of a base *(10-3)*
10. The lateral area and the total area 11. The volume

Refer to the regular square pyramid shown.

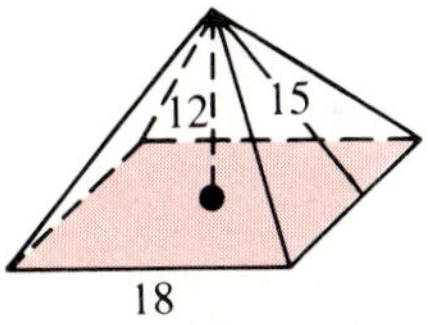

12. $B =$ _?_ 13. L.A. = _?_ *(10-4)*
14. T.A. = _?_ 15. $V =$ _?_

Refer to the right circular cone shown. Give answers in terms of π.

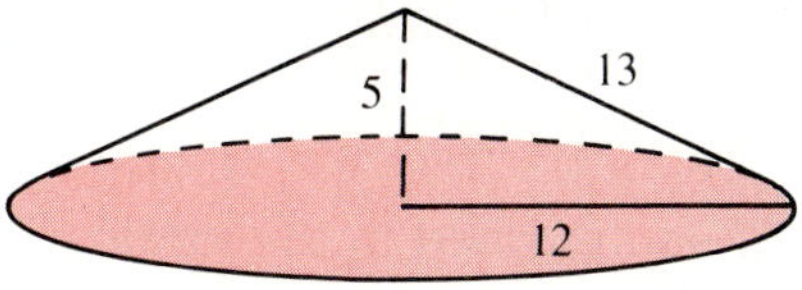

16. $B =$ _?_ 17. L.A. = _?_ *(10-5)*
18. T.A. = _?_ 19. $V =$ _?_

A sphere has a diameter of 9 cm.

20. Find the area in terms of π. *(10-6)*
21. Find the volume to the nearest integer. Use 3.14 for π.
22. Find the exact volume of a sphere with area 16π.

(Optional) Two similar cones have heights of 6 cm and 8 cm.

23. Find the ratio of their total areas. *(10-7)*
24. If the volume of the smaller cone is 96π cm^3, find the volume of the larger cone.

UNIT E CUMULATIVE REVIEW

Complete each statement.

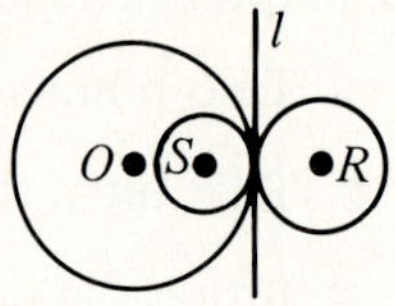

1. Line l is a common internal tangent of $\odot S$ and $\underline{\ ?\ }$.
2. Line l is a common external tangent of $\underline{\ ?\ }$ and $\odot S$.

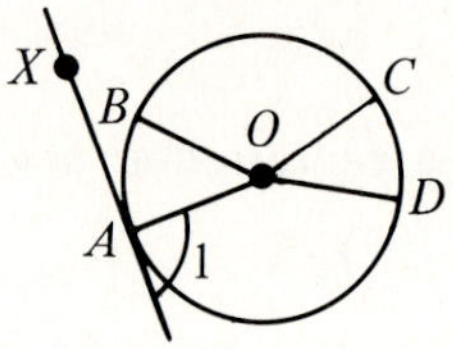

3. If $\overleftrightarrow{AX}$ is a tangent of $\odot O$, then $\angle 1 = \underline{\ ?\ }°$.
4. If $\angle AOD = 150°$, then $\overset{\frown}{AD} = \underline{\ ?\ }°$.
5. If $\angle BOA = \angle COD$ and $\overset{\frown}{AB} = 47°$, then $\overset{\frown}{CD} = \underline{\ ?\ }°$.

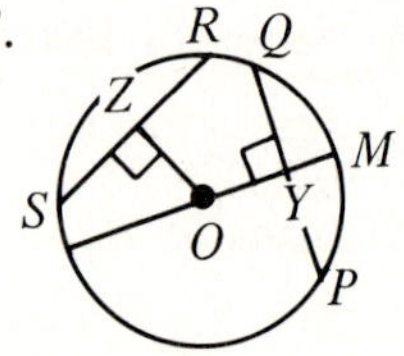

6. If $PQ = RS$ and $\overset{\frown}{RS} = 100°$, then $\overset{\frown}{PQ} = \underline{\ ?\ }°$.
7. If $OY = OZ$, then $PQ = \underline{\ ?\ }$.
8. If $RS = 10$, then $RZ = \underline{\ ?\ }$.
9. If $\overset{\frown}{PQ} = 98°$, then $\overset{\frown}{PM} = \underline{\ ?\ }°$.

Tell whether each statement is true or false.

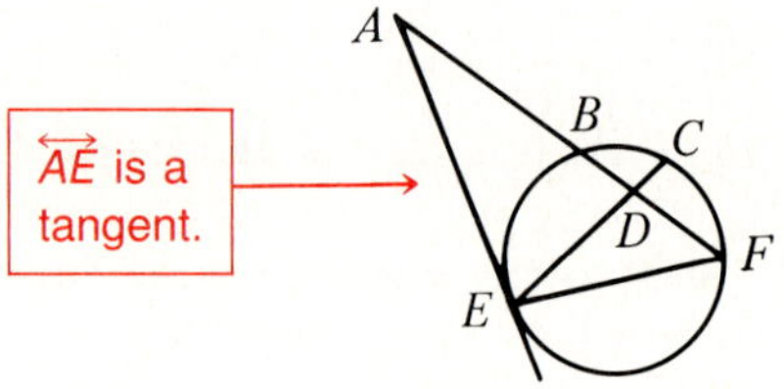

Exercises 10–15

10. If $\overset{\frown}{CF} = 60°$, then $\angle CEF = 30°$.
11. If $\angle CEA = 70°$, then $\overset{\frown}{EBC} = 140°$.
12. $\angle BDE = \frac{1}{2}(\overset{\frown}{BE} + \overset{\frown}{EF})$
13. $\angle EAF = \frac{1}{2}(\overset{\frown}{EF} - \overset{\frown}{BE})$
14. $BD \cdot DF = ED \cdot DC$
15. $\triangle DEF$ is inscribed in the circle.
16. Skew lines sometimes intersect.
17. Two intersecting lines must be coplanar.
18. Find the volume of a right prism with base area 25 and height 7.
19. Find the lateral area of a cylinder with radius 4 and height 7.
20. A regular square pyramid has base edges that are 4 m long. The height is 6 m. Find the volume of the pyramid.
21. Find the total area of a cone with radius 3 and slant height 5.
22. Find the area and volume of a sphere with radius 3 cm.

UNIT F

Here's what you'll learn in this chapter:

To solve problems involving some common right triangle lengths.

To apply theorems about 45°-45°-90° triangles and 30°-60°-90° triangles.

To find the length of the diagonal of a rectangular solid.

To use right triangles to solve problems involving pyramids and cones.

To use the tangent, sine, and cosine ratios of an acute angle of a right triangle.

You can see right angles almost everywhere you look in the photograph above. Most of the triangles shown are right triangles.

Chapter 11

Right Triangles

1 Reviewing Right Triangles

In this section, we shall review right triangles. The Pythagorean Theorem and its converse are the most important statements to remember when working with right triangles. Certain right triangles appear often in geometry and are listed at the right.

Some Common Right Triangle Lengths	
3-4-5	5-12-13
6-8-10	
9-12-15	8-15-17

The first entry in the table, 3-4-5, represents the triangle shown. Remember: The triangle must be a right triangle because

$$3^2 + 4^2 = 5^2.$$

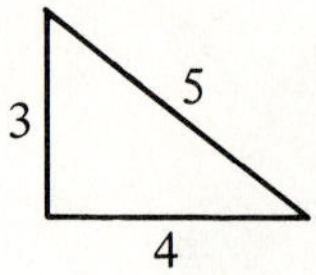

EXAMPLE

Find the exact values of x and y in the diagram.

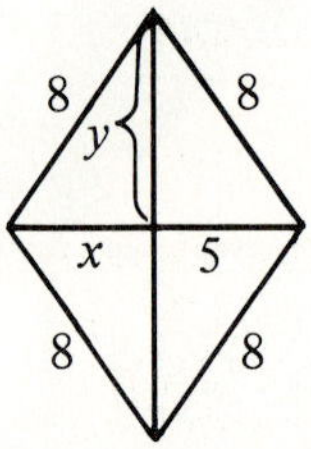

1. Since the four sides are equal, the quadrilateral is a rhombus.
2. The diagonals of a rhombus bisect each other. Therefore $x = 5$.
3. The diagonals of a rhombus are perpendicular. Therefore:

$$5^2 + y^2 = 8^2$$
$$25 + y^2 = 64$$
$$y^2 = 39$$
$$y = \sqrt{39}$$

Take the square root of both sides.

Classroom Practice

Show that each equation is true.

1. $6^2 + 8^2 = 10^2$
2. $5^2 + 12^2 = 13^2$
3. $8^2 + 15^2 = 17^2$

Find the exact value of *x*.

4. 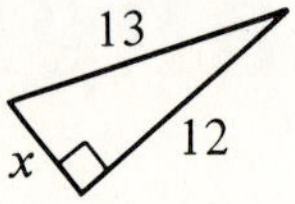

5. 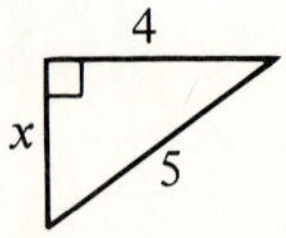

6. 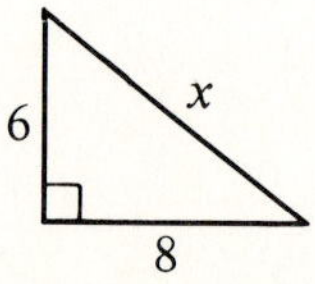

7.

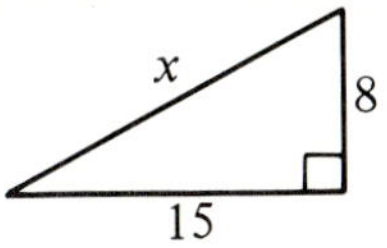

8.

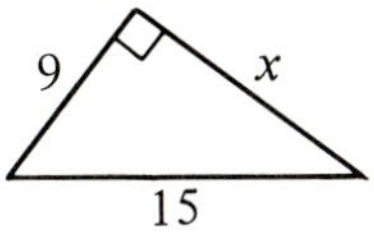

9.

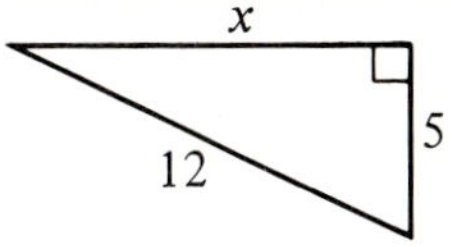

Find the exact values of x and y.

10.

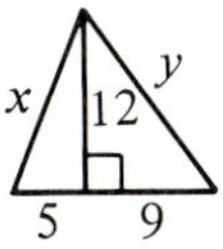

11.

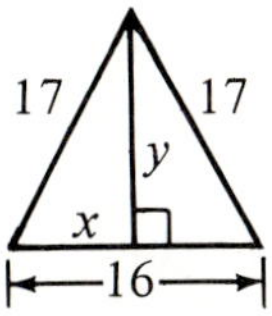

12.

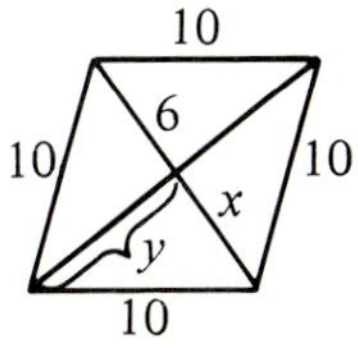

Written Exercises

Find the exact value of x.

A 1.

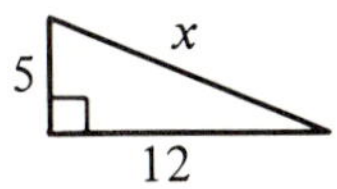

2.

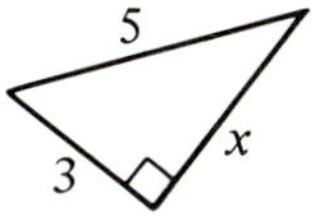

3.

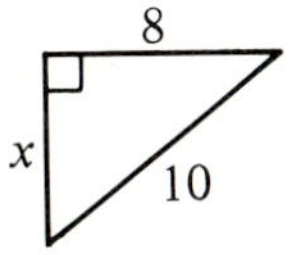

4.

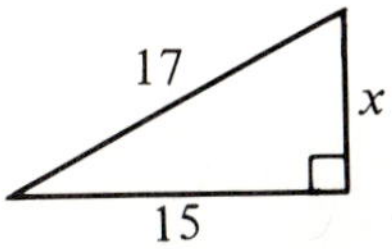

5.

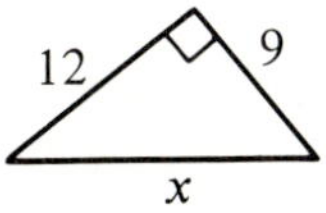

6.

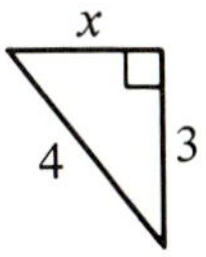

Find the exact values of x and y.

7.

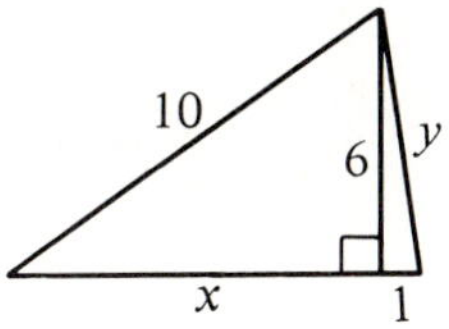

8.

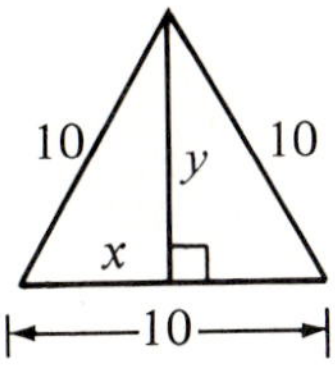

9.

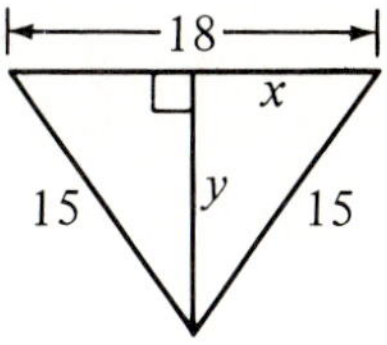

10.

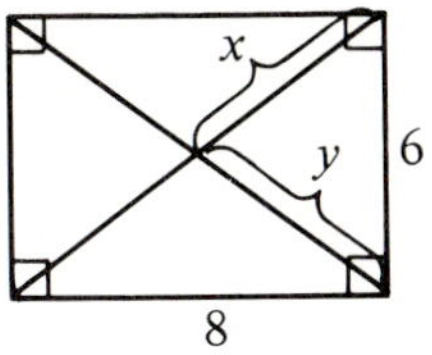

11.

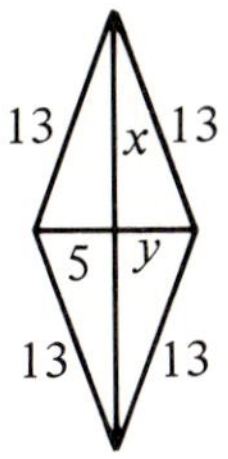

12.

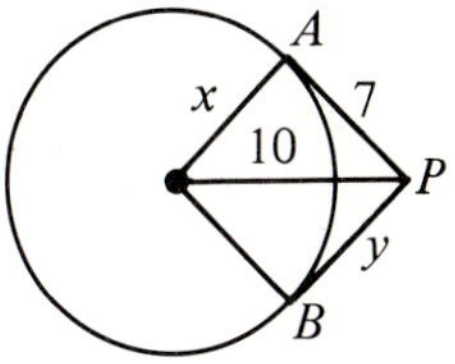

$\overline{PA}$ and $\overline{PB}$ are tangents.

13.

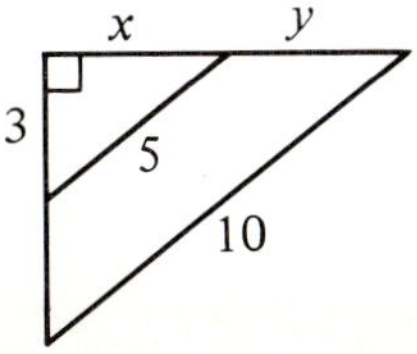

14.

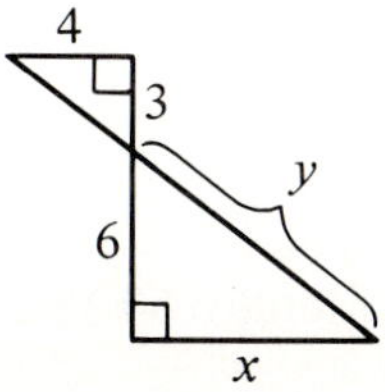

15.

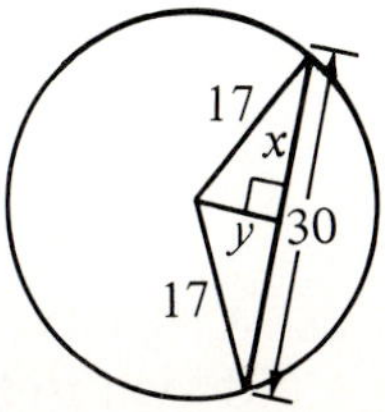

Find the area. (*Hint:* First find *x*.)

16.

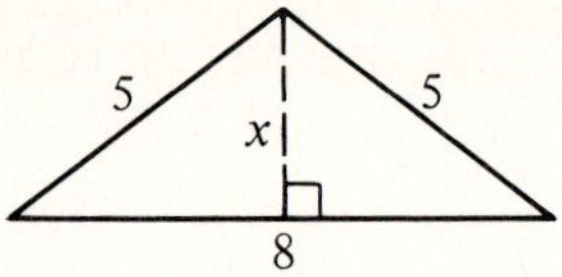

17.

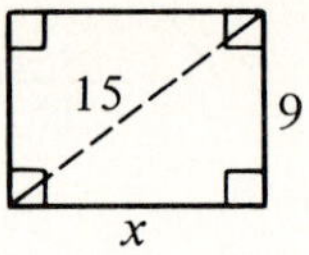

18.

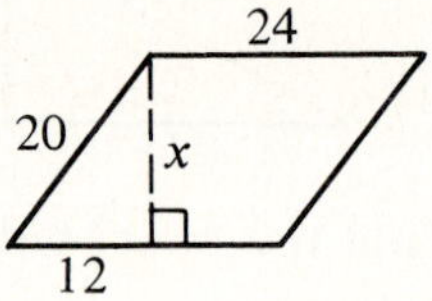

B **19.** Is $\angle C$ a right angle? Explain.

20. Why must $\overline{AB}$ be a diameter of the circle? (*Hint:* What is the measure of $\widehat{ADB}$?)

21. What is the area of the circle?

22. What is the circumference of the circle?

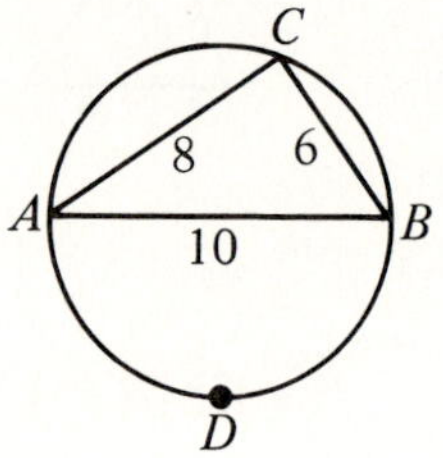

23. Find the values of p and q.

24. Is $p^2 + q^2 = r^2$?

25. Is $\triangle JKL$ a right triangle?

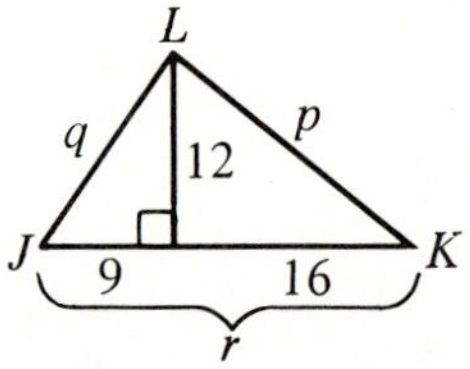

26. Find the values of x and y.

27. Is $x^2 + y^2 = z^2$?

28. Is $\triangle STU$ a right triangle?

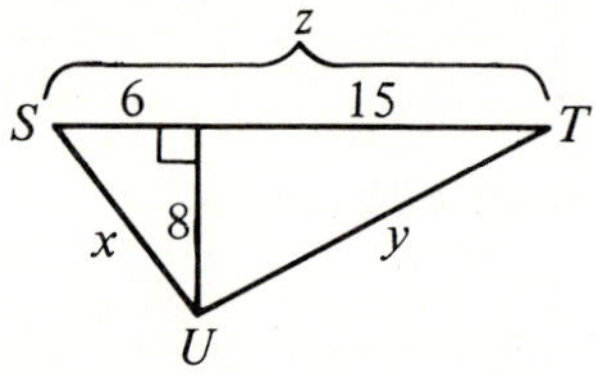

***ABCD* is a parallelogram. *AB* = 15, *AC* = 24, and *BD* = 18.**

29. $AO =$ __?__ **30.** $BO =$ __?__

31. Is $\triangle AOB$ a right triangle? Explain.

32. What special kind of parallelogram is *ABCD*? Explain.

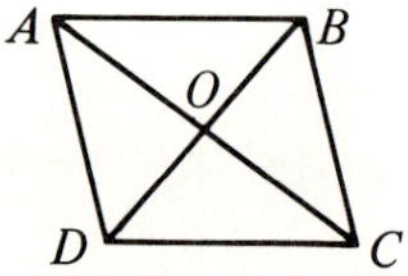

Find the exact area of each shaded region.

C **33.**

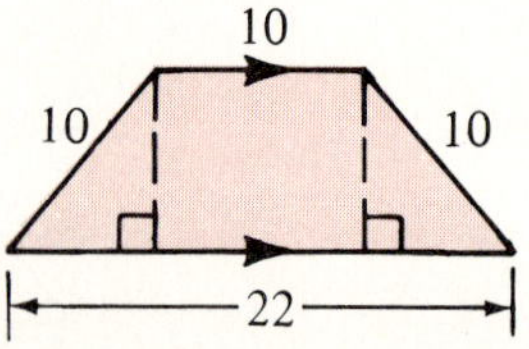

34.

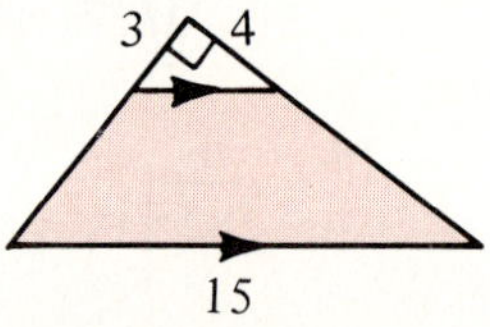

35.

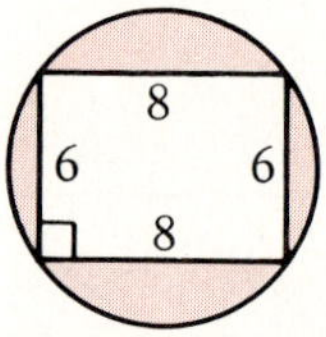

36. A right triangle has legs of length 30 and 40. How long is the median drawn to the hypotenuse?

APPLICATIONS

MECHANICAL DRAWING

Technical artists prepare mechanical drawings for architects and machinists. Many mechanical drawings show a building, a machine, or a tool, as seen from different views. For example, a calculator might be drawn like this.

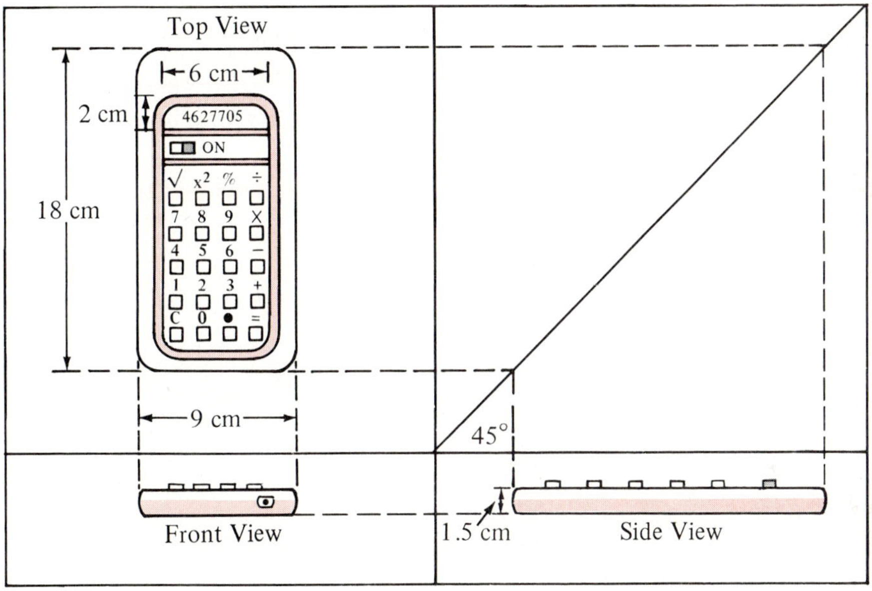

Each length in a mechanical drawing is drawn to scale. Suppose the actual calculator is 9 cm wide. If this width is represented by a segment 3 cm long, then the scale is 3:9, or 1:3.

EXERCISES

Sketch each object as seen from the top, front, and side.

1. 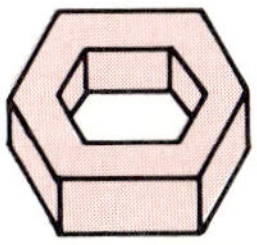

2. 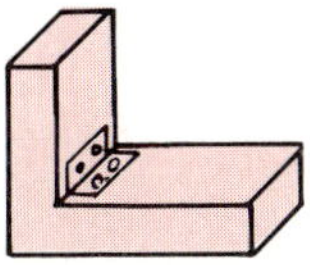

3. 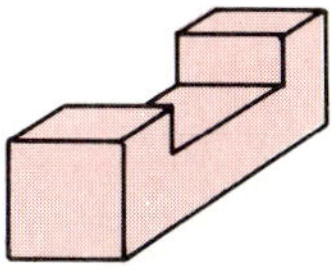

4. Make a mechanical drawing of a milk container, showing the top, the front, and the side.

2 Special Right Triangles

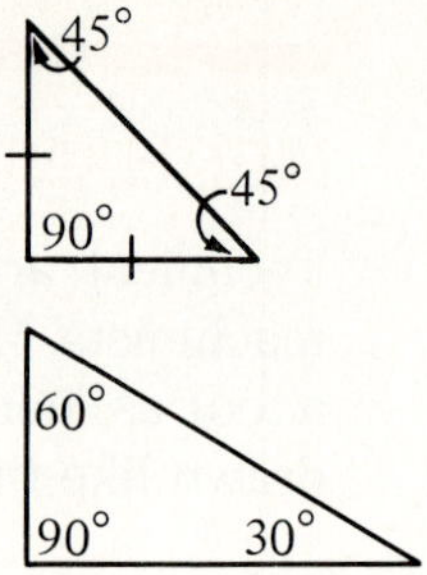

The two triangles shown are special right triangles. The first one is an isosceles right triangle. It is often called a 45°-45°-90° triangle. The other special triangle is a 30°-60°-90° triangle.

The Pythagorean Theorem can be used to prove two theorems about these special triangles.

THEOREM 1

In a 45°-45°-90° triangle, the hypotenuse is $\sqrt{2}$ times as long as a leg.

Given: A 45°-45°-90° triangle

Prove: $h = l\sqrt{2}$

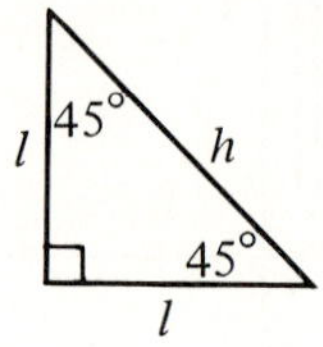

Proof:

$$h^2 = l^2 + l^2$$
$$h^2 = 2l^2$$
$$h = \sqrt{2l^2}$$
$$h = l\sqrt{2}$$

$\sqrt{2} \approx 1.4$

EXAMPLE 1 For each diagram below, find the value of t.

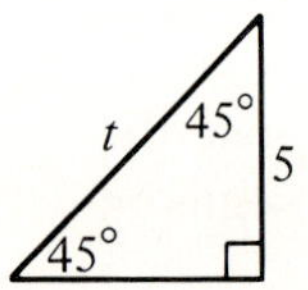

hypotenuse = leg × $\sqrt{2}$
$t = 5\sqrt{2}$

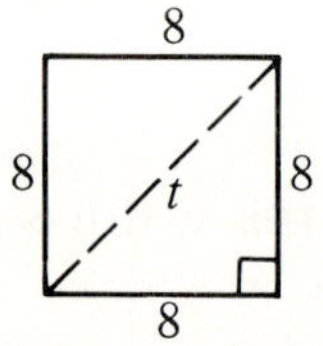

hypotenuse = leg × $\sqrt{2}$
$t = 8\sqrt{2}$

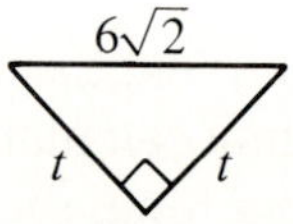

hypotenuse = leg × $\sqrt{2}$
$6\sqrt{2} = t\sqrt{2}$
$6 = t$

THEOREM 2

In a 30°-60°-90° triangle,
(1) the hypotenuse is twice as long as the shorter leg;
(2) the longer leg is $\sqrt{3}$ times as long as the shorter leg.

Given: A 30°-60°-90° triangle

Prove: **a.** $h = 2s$ **b.** $l = s\sqrt{3}$

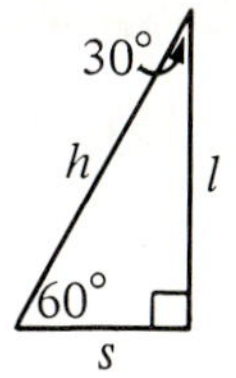

Proof: Notice that a 30°-60°-90° triangle is "half" of an equilateral triangle.

a. Since $\triangle ABC$ is equilateral, $h = 2s$.

b. To find l, use the Pythagorean Theorem.

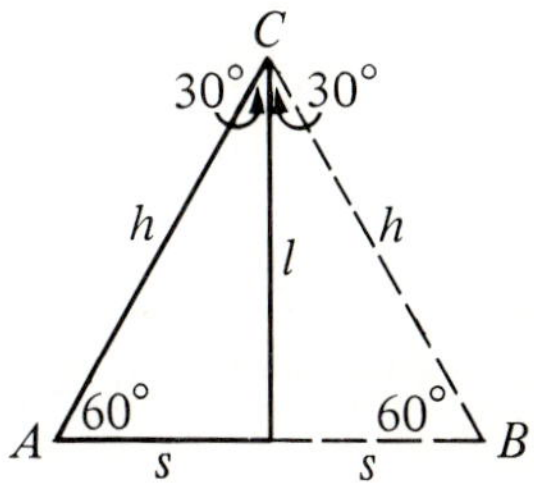

$$l^2 + s^2 = h^2$$
$$l^2 + s^2 = (2s)^2$$
$$l^2 + s^2 = 4s^2$$
$$l^2 = 3s^2$$
$$l = \sqrt{3s^2}$$
$$l = s\sqrt{3}$$

$\sqrt{3} \approx 1.7$

EXAMPLE 2 In each triangle below, the length of one side is given. Find the lengths of the other two sides.

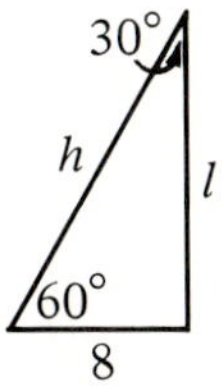

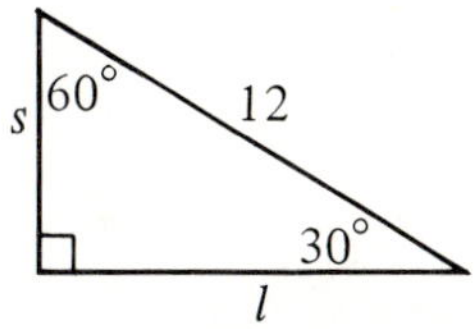

hypotenuse = 2 × shorter leg
$h = 2 \times 8$
$h = 16$

longer leg = shorter leg × $\sqrt{3}$
$l = 8\sqrt{3}$

hypotenuse = 2 × shorter leg
$12 = 2s$
$6 = s$

longer leg = shorter leg × $\sqrt{3}$
$l = 6\sqrt{3}$

Classroom Practice

$\triangle ABC$ is an isosceles right triangle.

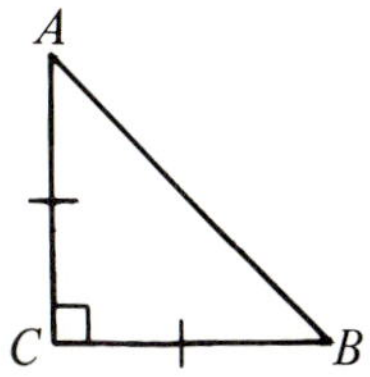

1. $\angle A =$ __?__ °
2. $\angle B =$ __?__ °
3. If $AC = 3$, then $BC =$ __?__ and $AB =$ __?__.
4. If $AC = 4$, then $BC =$ __?__ and $AB =$ __?__.
5. If $BC = 9$, then $AB =$ __?__.
6. If $AB = 7\sqrt{2}$, then $BC =$ __?__.
7. If $AB = 2\sqrt{2}$, then $AC =$ __?__.

△*XYZ*, shown below, is a 30°-60°-90° triangle.

8. Which leg is the shorter leg?

9. In a 30°-60°-90° triangle, the shorter leg is *always* opposite the __?__° angle.

10. Which leg is the longer leg?

11. In a 30°-60°-90° triangle, the longer leg is *always* opposite the __?__° angle.

12. If $YZ = 5$, then $XY =$ __?__ and $XZ =$ __?__.

13. If $YZ = 6$, then $XY =$ __?__ and $XZ =$ __?__.

14. If $XY = 8$, then $YZ =$ __?__ and $XZ =$ __?__.

15. If $XY = 6$, then $YZ =$ __?__ and $XZ =$ __?__.

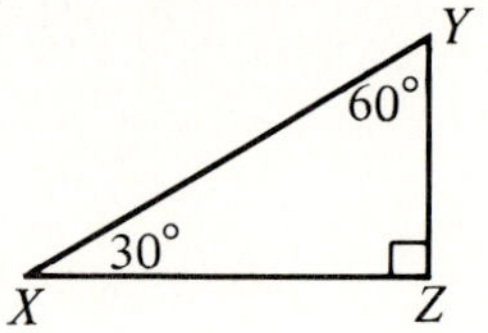

Written Exercises

In Exercises 1–9, find the exact values of *x* and *y*.

A

1.

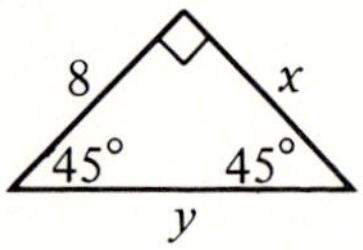

2.

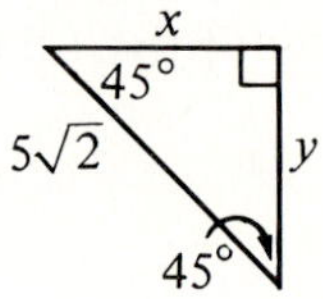

3.

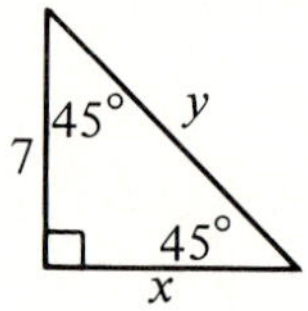

4.

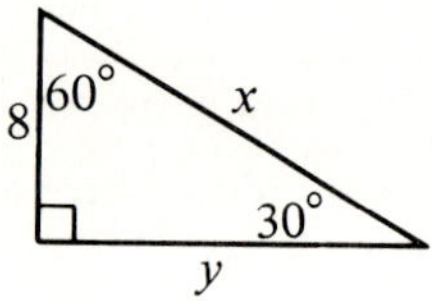

5.

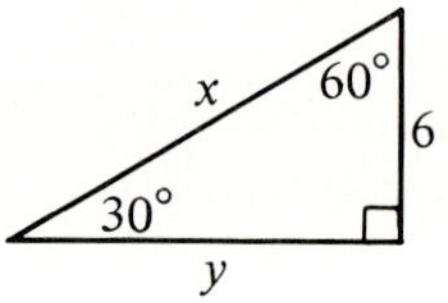

6.

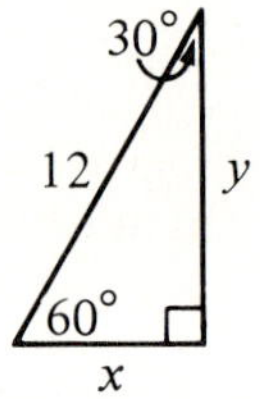

7.

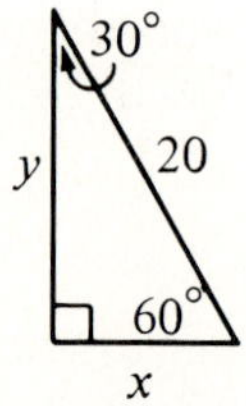

8.

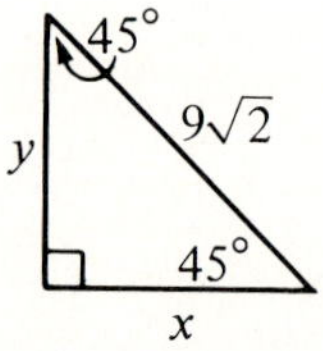

9.

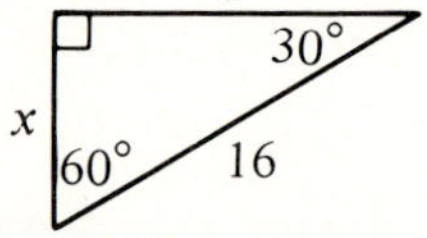

In the diagram shown, △*ABC* is equilateral. $\overline{AD}$ is an altitude.

10. $\angle ABD =$ __?__°

11. $\angle ADB =$ __?__°

12. $\angle BAD =$ __?__°

13. $\angle CAD =$ __?__°

Suppose *AB* = 6. Find:

14. *BD*

15. *AD*

16. Area of △*ABC*

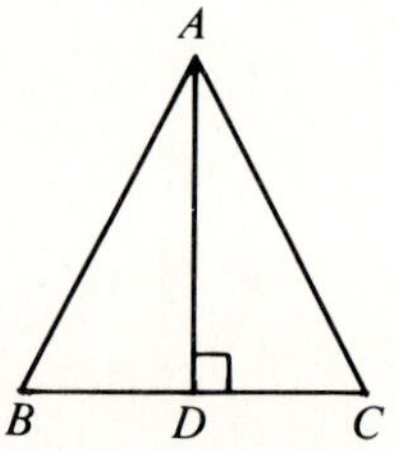

Exercises 10–16

In the diagram shown, *ABCD* is a square. Suppose $AB = 6$. Find:

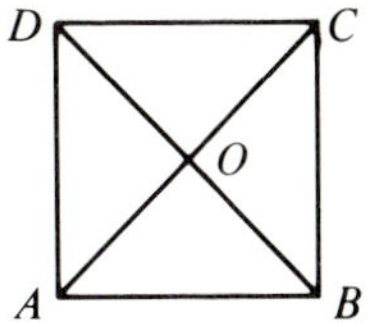

Exercises 17–25

17. AC **18.** AO **19.** BO

Suppose $AO = 4$. Find:

20. BO **21.** Area of $\triangle AOB$ **22.** Area of $ABCD$

Suppose $DO = 5$. Find:

23. DC **24.** Area of $\triangle DOC$ **25.** Area of $ABCD$

In Exercises 26–28, find the exact value of *t*.

SAMPLE

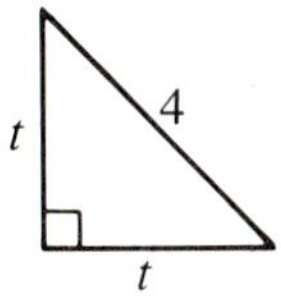

Method 1

Hypotenuse = leg × $\sqrt{2}$ (Thm. 1)

$$4 = t\sqrt{2}$$

$$\frac{4}{\sqrt{2}} = t$$

Simplify.

$$t = \frac{4}{\sqrt{2}} = \frac{4 \cdot \sqrt{2}}{\sqrt{2} \cdot \sqrt{2}} = \frac{4\sqrt{2}}{2} = 2\sqrt{2}$$

Method 2

$$t^2 + t^2 = 4^2 \text{ (Pythagorean Thm.)}$$

$$2t^2 = 16$$

$$t^2 = 8$$

$$t = \sqrt{8}$$

Simplify.

$$t = \sqrt{8} = \sqrt{4 \times 2} = 2\sqrt{2}$$

B **26.**

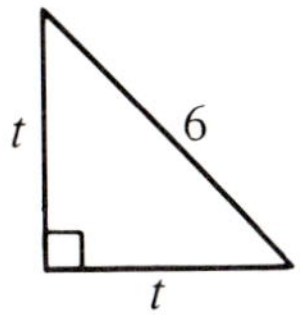

27.

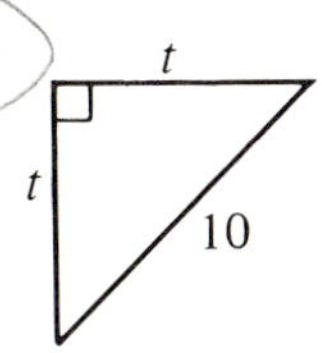

28.

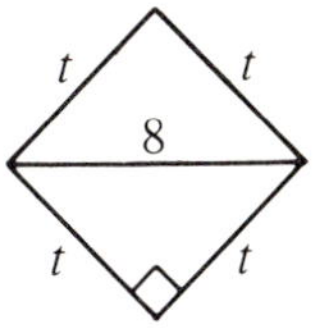

29. Each edge of the cube shown has length 6. Find the length of each side of the red triangle.

30. What is the measure of each angle of the red triangle?

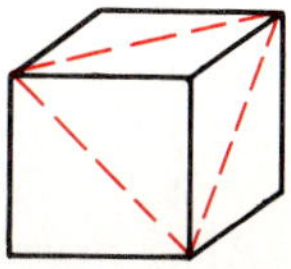

$\triangle ABC$ is a 30°-60°-90° triangle. $\overline{CD}$ is an altitude.
Copy and complete the table.

	AB	BC	AC	CD	AD	DB
Sample	4	2	$2\sqrt{3}$	$\sqrt{3}$	3	1
31.	12	?	?	?	?	?
32.	?	10	?	?	?	?
33.	?	?	$4\sqrt{3}$	?	?	?
34.	?	?	?	?	?	3
C **35.**	?	?	?	$7\sqrt{3}$	?	?
36.	?	?	?	?	9	?

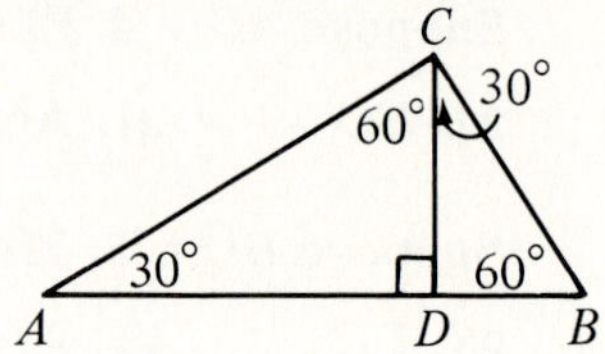

CALCULATOR ACTIVITIES

A jewelry designer plans to make earrings that are hollow spheres. Find the total number of grams in each earring described. Round your answers to the nearest tenth of a gram.

EXAMPLE The outer diameter is 1 cm, and the inner diameter is 0.9 cm. The alloy to be used has 8.6 g/cm³.

Total number of grams =

$$\left[\frac{4}{3}\pi(0.5)^3 - \frac{4}{3}\pi(0.45)^3\right]8.6 \approx 1.2$$

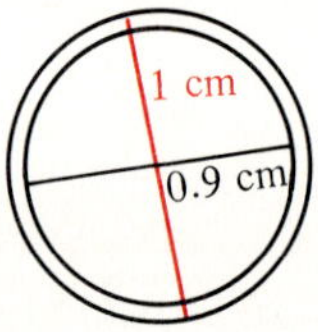

Copy and complete the table.

		Outer diam. (in cm)	Inner diam. (in cm)	Grams in 1 cm³	Total no. of grams
1.	Bronze	0.8	0.6	8.8	?
2.	German Silver	0.8	0.66	8.3	?
3.	Silver	0.8	0.65	10.5	?
4.	Gold	0.7	0.54	19.3	?

3 Using Special Right Triangles

In this section, we will apply the theorems about 45°-45°-90° triangles and 30°-60°-90° triangles to many figures you have studied in this course. Here is an example.

EXAMPLE

Find the area of ▱*ABCD* if $\angle A = 60°$, $AD = 8$, and $AB = 10$.

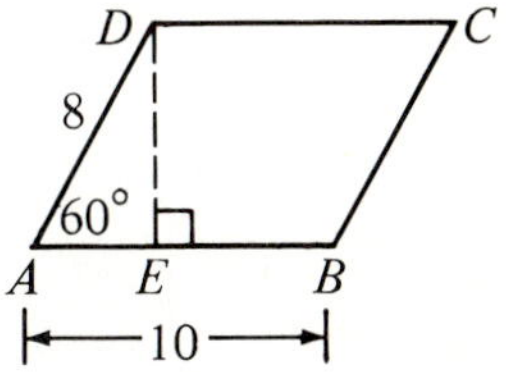

1. Draw altitude $\overline{DE}$.
2. Notice that $\triangle ADE$ is a 30°-60°-90° triangle.
3. $AE = \frac{1}{2} \times 8 = 4$ $\qquad DE = 4\sqrt{3}$
4. Area = base × height = $10 \times 4\sqrt{3} = 40\sqrt{3}$

Classroom Practice

Exercises 1–6 refer to the diagram.

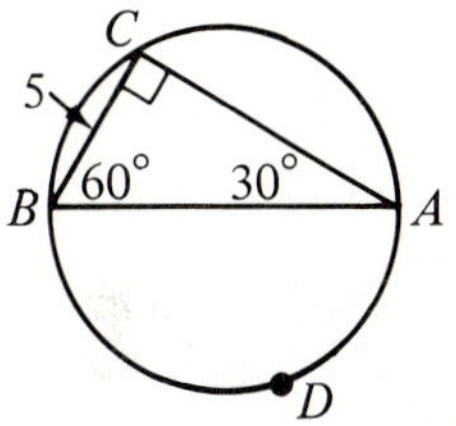

1. $\triangle ABC$ is a __?__°-__?__°-__?__° triangle.
2. Since $BC = 5$, $AB =$ __?__.
3. Since $\angle C = 90°$, $\overset{\frown}{BDA} =$ __?__°.
4. $\overline{AB}$ is a __?__ of the circle.
5. Find the radius of the circle.
6. Find the area of the circle.

***WXYZ* is a rectangle. *XZ* = 12 and ∠ *ZXW* = 30°. Find:**

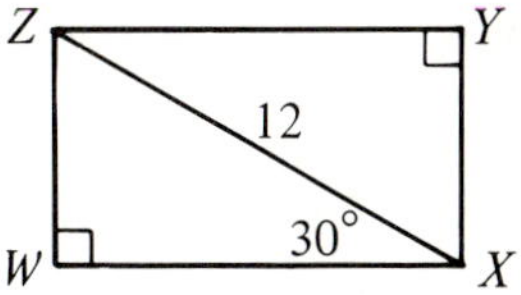

7. *ZW*   8. *WX*
9. The area of rectangle *WXYZ*

Written Exercises

Find the area of each triangle.

A

1.

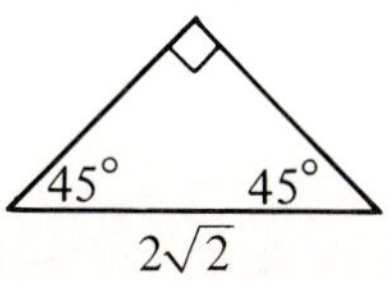

2.

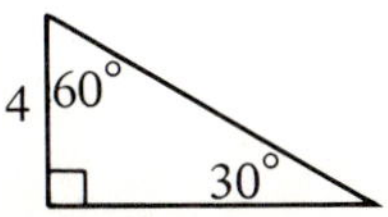

3.

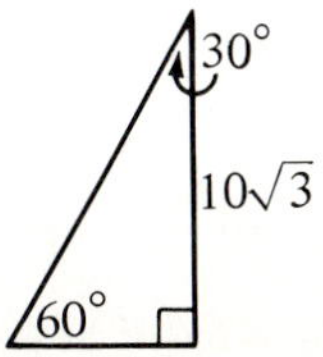

Find the area of each parallelogram. Refer to the example, page 437.

4.

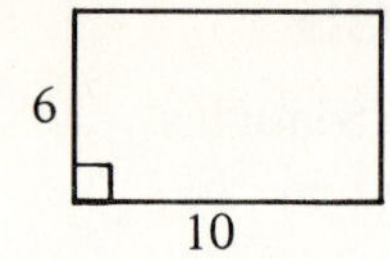

5.

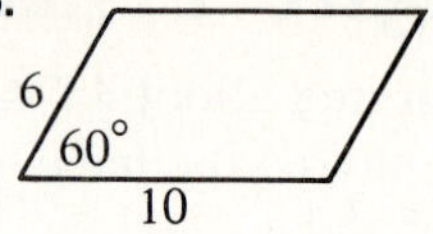

6.

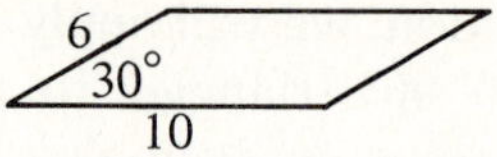

For each triangle shown, find the value of x. Then find the area of the triangle.

7.

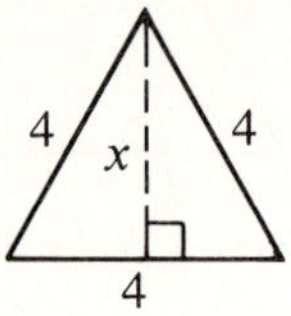

8.

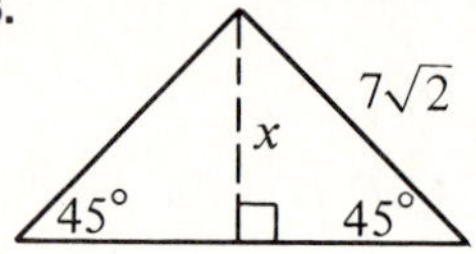

9.

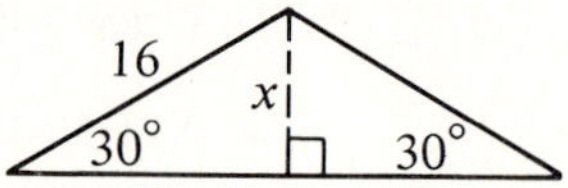

$\overline{TA}$ and $\overline{TB}$ are tangents to $\odot O$.
$OA = 4$ and $\angle ATB = 60°$.
Find each of the following.

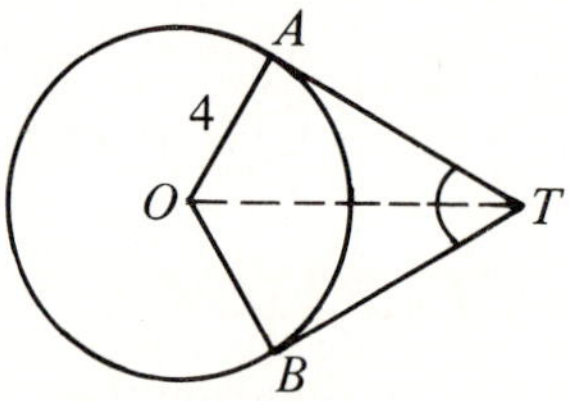

10. $\angle OAT$ **11.** $\angle OTA$

12. AT **13.** BT

Study the diagram. Then find each of the following.

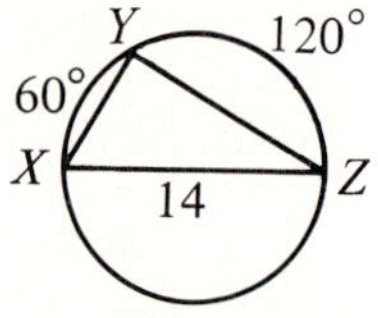

14. $\angle Z$ **15.** $\angle X$ **16.** $\angle Y$

17. XY **18.** YZ **19.** The area of $\triangle XYZ$

***PQRS* is a rhombus. $\angle SPQ = 120°$ and $SP = 8$.**
Find each of the following.

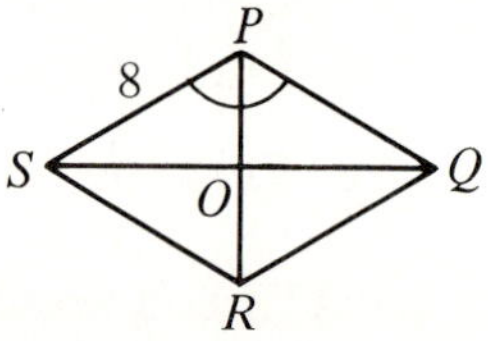

B **20.** $\angle SPO$ **21.** $\angle POS$ **22.** $\angle PSO$

23. PO **24.** SO **25.** The area of $PQRS$

26. Is $\triangle EFG \sim \triangle HIJ$?

27. What is the scale factor of $\triangle EFG$ to $\triangle HIJ$?

28. What is the ratio of their perimeters?

29. What is the ratio of their areas?

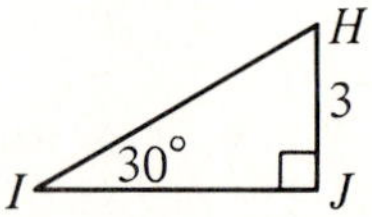

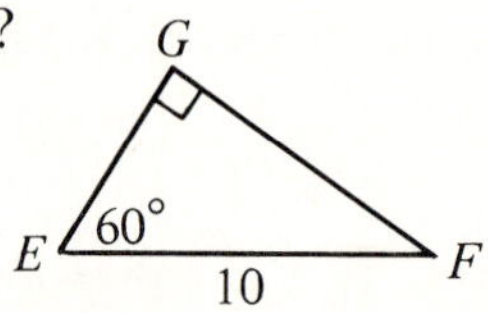

***QRST* is a square inscribed in $\odot O$. The radius of $\odot O$ is 4.**

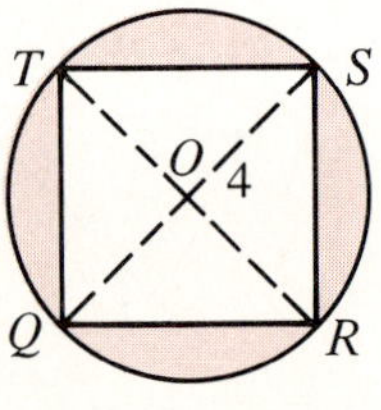

30. Find the area of the circle.

31. Find the area of the square.

32. Find the area of the shaded region.

C 33. Find the area of trapezoid $ABCD$ if $\angle A = \angle B = 45°$, $AB = 16$, and $DC = 6$.

34. Find the area of a regular hexagon with sides 2 cm long.

35. The wrench shown just fits the nut. Find the value of x.

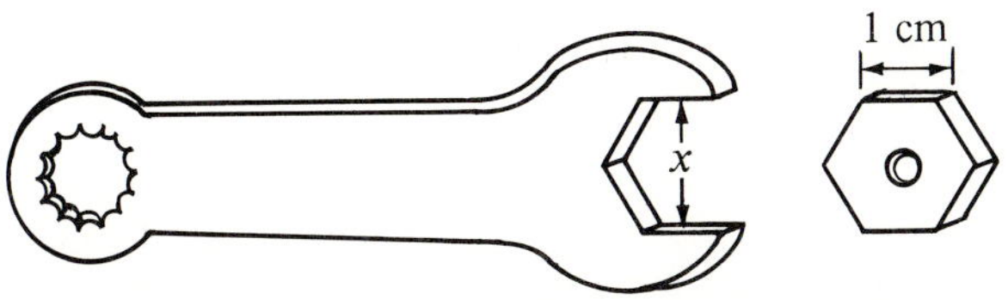

$\triangle ABC$ is an equilateral triangle with sides 6 cm long. OD is the radius of the inscribed circle. OA is the radius of the circumscribed circle.

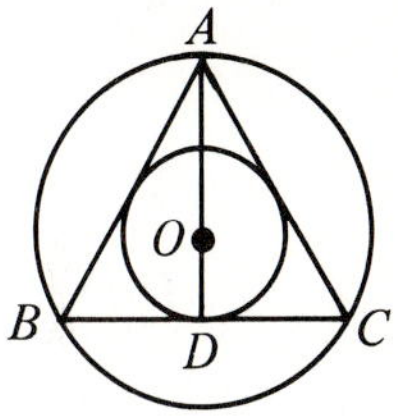

36. $AD = \underline{\ ?\ }$

37. $OD = \underline{\ ?\ }$

38. $OA = \underline{\ ?\ }$

Hint: See the Experiments on page 162.

SELF-TEST

Find the exact values of x and y.

1. (11-1)

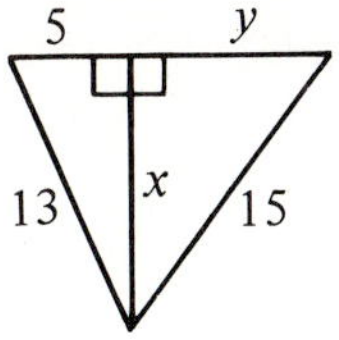

2. (11-2)

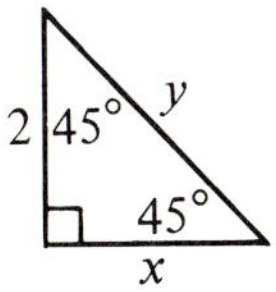

3.

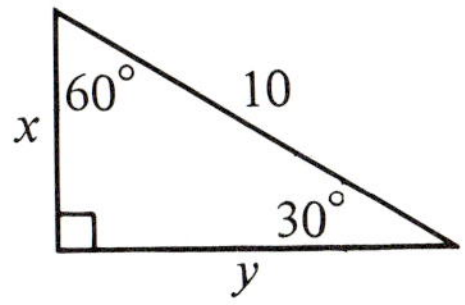

Find the area of each polygon.

4. (11-3)

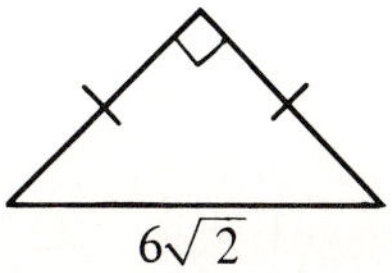

5.

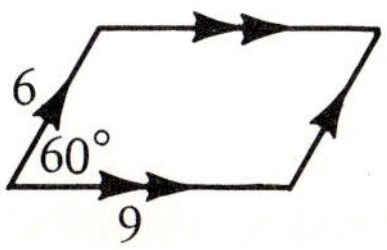

6.

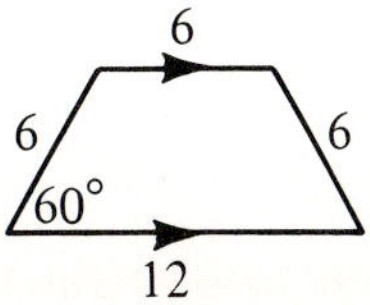

4 Diagonals of Rectangular Solids

The rectangular solid shown is 3 cm long, 1 cm wide, and 2 cm high. $\overline{HB}$ is called a *diagonal* of the solid. A rectangular solid has four diagonals, which are all equal in length. The other diagonals, not drawn in the figure, are $\overline{EC}$, $\overline{GA}$, and $\overline{FD}$.

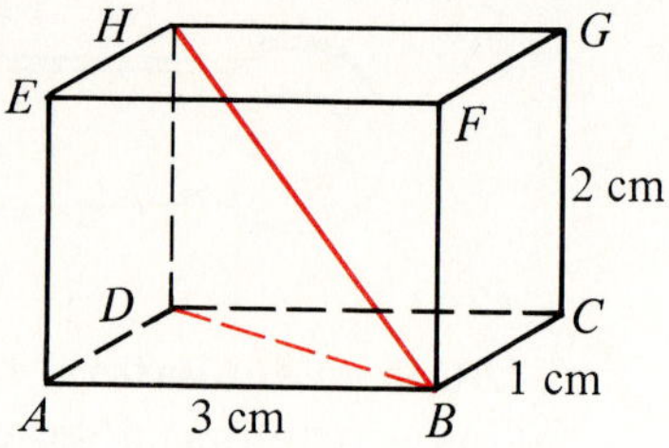

We can use the Pythagorean Theorem to find the length of $\overline{HB}$.

1. First study rectangle *ABCD*.

$$(DB)^2 = 1^2 + 3^2$$
$$(DB)^2 = 1 + 9 = 10$$
$$DB = \sqrt{10}$$

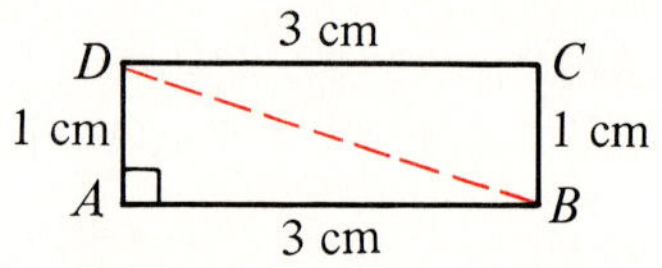

2. Now study right $\triangle HDB$. Notice

$$(HB)^2 = (HD)^2 + (DB)^2$$
$$(HB)^2 = 2^2 + 10$$
$$(HB)^2 = 14$$

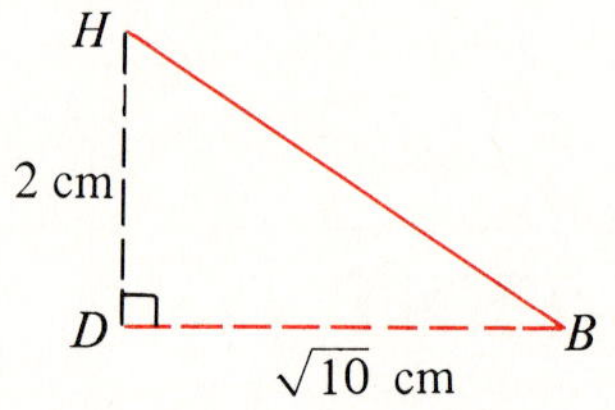

3. $\overline{HB}$ is exactly $\sqrt{14}$ cm long. If we use the square root table on page 516, we find that $\overline{HB}$ is approximately 3.742 cm long.

The method illustrated above can be used to find the length of a diagonal of any rectangular solid.

1. $x^2 = a^2 + b^2$
2. $d^2 = x^2 + c^2$
   $d^2 = (a^2 + b^2) + c^2$ ← Substitute using Step 1.
3. $d = \sqrt{a^2 + b^2 + c^2}$

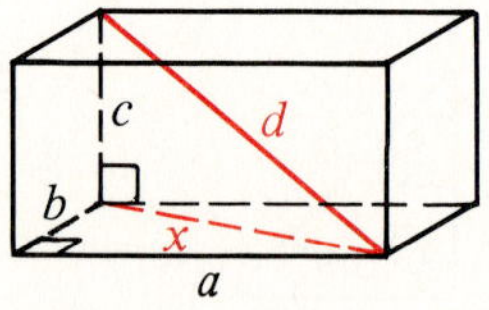

If a rectangular solid has dimensions a, b, and c, then a diagonal has length

$$d = \sqrt{a^2 + b^2 + c^2}.$$

EXAMPLE A rectangular solid is shown. Find the length of a diagonal of the solid.

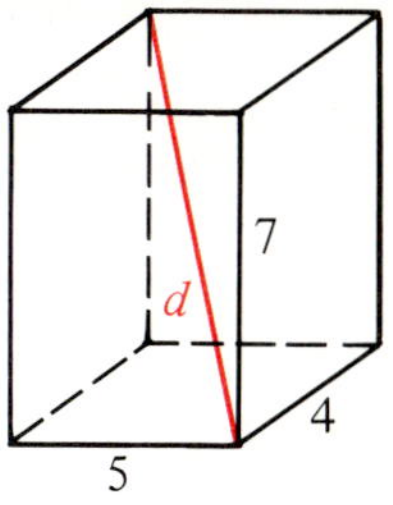

$$\begin{aligned} d &= \sqrt{5^2 + 4^2 + 7^2} \\ &= \sqrt{25 + 16 + 49} \\ &= \sqrt{90} \\ &= \sqrt{9 \cdot 10} \\ &= 3\sqrt{10} \end{aligned}$$

Classroom Practice

The rectangular solid shown has dimensions 4, 2, and 3.

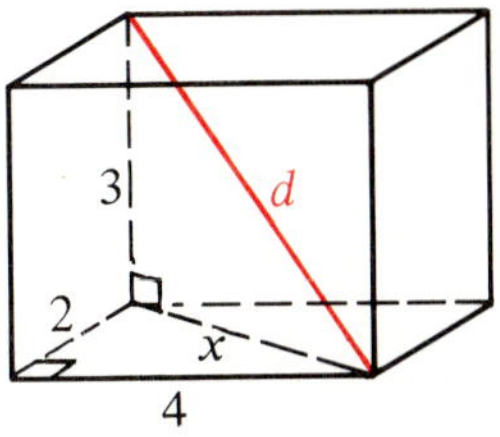

1. $x^2 = 2^2 + 4^2$
   $x^2 = \underline{\ \ ?\ \ }$
2. $d^2 = 3^2 + x^2$
   $d^2 = \underline{\ \ ?\ \ } + \underline{\ \ ?\ \ }$
3. $d = \sqrt{\underline{\ \ ?\ \ }}$
4. Using the formula for a diagonal of a rectangular solid:

$$d = \sqrt{\underline{\ \ ?\ \ }^2 + \underline{\ \ ?\ \ }^2 + \underline{\ \ ?\ \ }^2}$$
$$d = \sqrt{\underline{\ \ ?\ \ } + \underline{\ \ ?\ \ } + \underline{\ \ ?\ \ }}$$
$$d = \sqrt{\underline{\ \ ?\ \ }}$$

The rectangular solid shown has dimensions 6, 2, and 3.

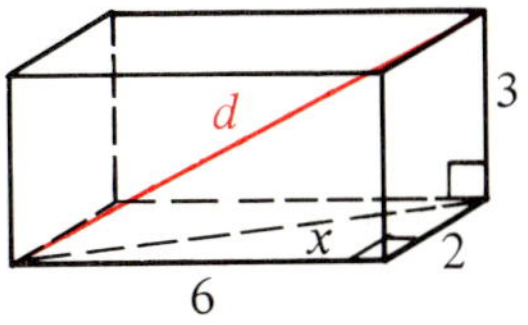

5. $x^2 = \underline{\ \ ?\ \ }^2 + \underline{\ \ ?\ \ }^2$
   $x^2 = \underline{\ \ ?\ \ }$
6. $d^2 = x^2 + \underline{\ \ ?\ \ }^2$
   $d^2 = \underline{\ \ ?\ \ } + \underline{\ \ ?\ \ }$
7. $d = \sqrt{\underline{\ \ ?\ \ }}$
8. Use the formula for a diagonal of a rectangular solid to show that your answer for Exercise 7 is correct.

Exercises 9–11 refer to the rectangular solid shown.

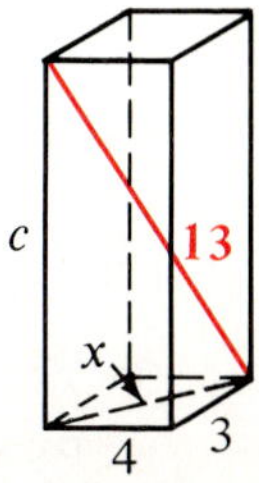

9. Find the value of x.
10. Find the height c.
11. Find the volume of the solid.

Written Exercises

Copy and complete the table about the rectangular solid shown.

	a	b	c	x	d
A **1.**	2	4	5	?	?
2.	3	4	5	?	?
3.	2	1	3	?	?
4.	3	2	6	?	?
5.	5	2	6	?	?
6.	6	2	5	?	?
B **7.**	9	12	?	?	17
8.	3	?	?	5	13

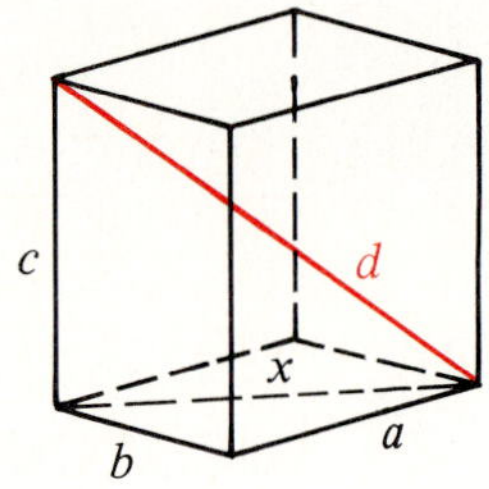

The figure shows a cube with edges 1 cm long.

9. Show that $x = \sqrt{2}$.

10. Show that $d = \sqrt{3}$.

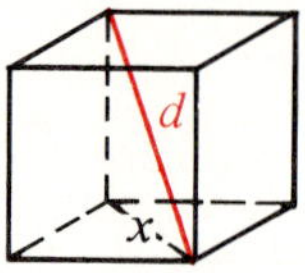

11. A rectangular solid is 21 cm long, 20 cm wide, and 8 cm high. Find the length of a diagonal to the nearest tenth.

12. A rectangular solid is 20 cm long, 15 cm wide, and 10 cm high. Find the length of a diagonal to the nearest tenth.

C **13.** Will a very thin metal rod, 70 cm long, fit inside a box that is 51 cm long, 43 cm wide, and 21 cm high? Explain.

PUZZLE ◆ PROBLEMS

Make a 3-4-5 triangle with 12 straws of equal length.

a. By moving only 3 straws, make a polygon with area 4.

b. By moving only 2 straws, make a polygon with area 5.

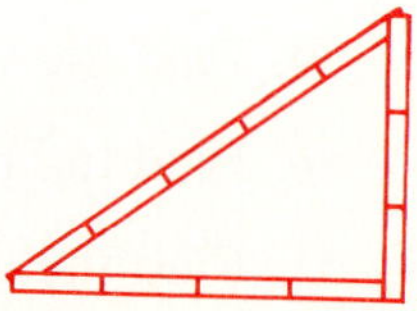

5 Right Triangles in Pyramids and Cones

In the previous section, we used right triangles to solve problems involving rectangular solids. In this section, we'll use right triangles to work with pyramids and cones.

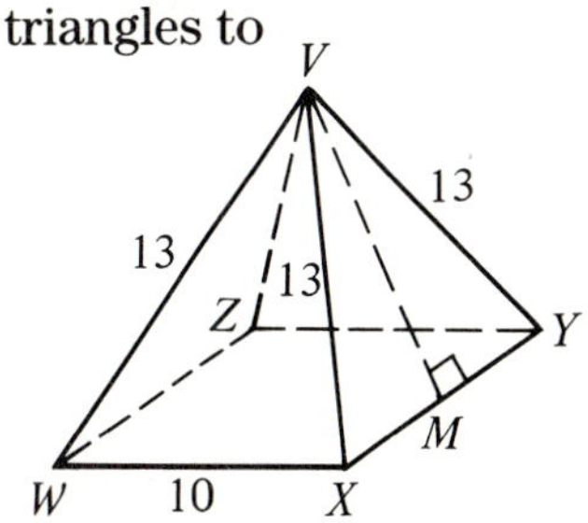

EXAMPLE 1 For the regular pyramid shown, find:

a. the lateral area;
b. the total area;
c. the volume.

a. Notice that $\triangle VXM$ is a 5-12-13 right triangle with $VM = 12$.

The area of one lateral face $= \frac{1}{2} \times 10 \times 12 = 60$

L.A. $= 4 \times 60 = 240$

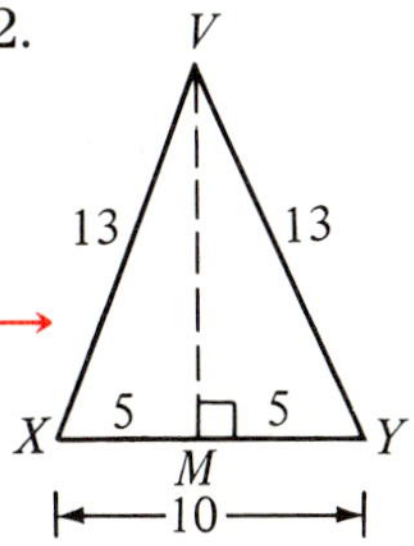

b. T.A. = L.A. + B

$$= 240 + 10^2$$
$$= 340$$

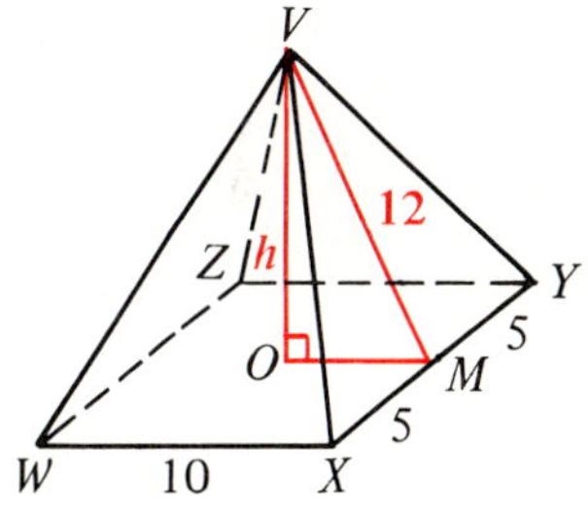

c. To find the volume, we need to find the height, h.
To do this, we use right $\triangle VOM$.
We know from part (a) that $VM = 12$.
We also know that $OM = 5$. (Why?)

Now we can use the Pythagorean Theorem to find h.

$$h^2 + 5^2 = 12^2$$
$$h^2 + 25 = 144$$
$$h^2 = 119$$
$$h = \sqrt{119}$$

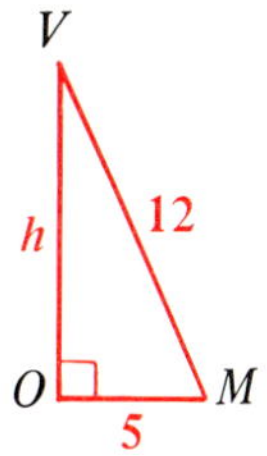

Recall the formula for the volume of a pyramid:

$$V = \frac{1}{3}Bh$$
$$= \frac{1}{3} \times 10^2 \times \sqrt{119}$$
$$= \frac{100}{3}\sqrt{119}$$

EXAMPLE 2

For the cone shown, find:

a. the lateral area;
b. the total area;
c. the volume.

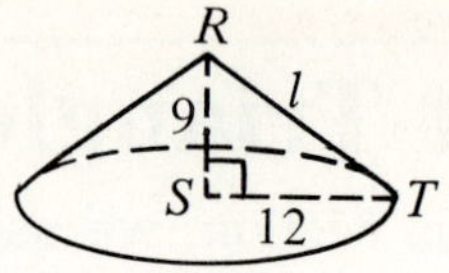

a. To find the lateral area, we must find l, the slant height of the cone. Notice that $\triangle RST$ is a 9-12-15 triangle with $l = 15$.

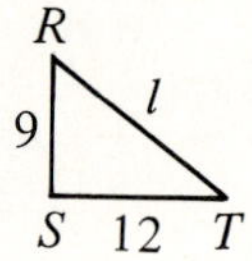

$$\begin{aligned} \text{L.A.} &= \pi r l \\ &= \pi \times 12 \times 15 \\ &= 180\pi \end{aligned}$$

b.
$$\begin{aligned} \text{T.A.} &= \text{L.A.} + B \\ &= 180\pi + \pi r^2 \\ &= 180\pi + (\pi \times 12^2) \\ &= 324\pi \end{aligned}$$

c. $V = \frac{1}{3}Bh = \frac{1}{3}\pi r^2 h = \frac{1}{3}\pi \times 12^2 \times 9 = 432\pi$

Classroom Practice

Exercises 1–7 refer to the regular square pyramid shown.

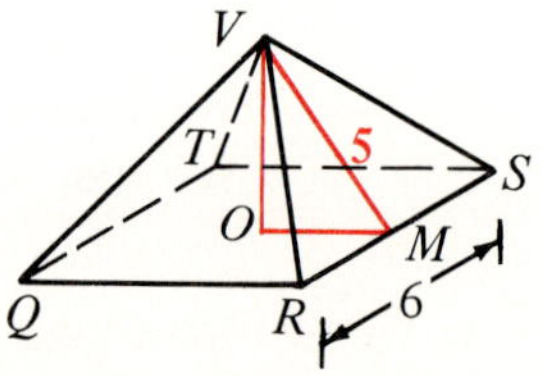

1. The area of lateral face VRS = ___?___.
2. The lateral area of the pyramid = ___?___.
3. The area of the base = ___?___.
4. The total area of the pyramid = ___?___.
5. OM = ___?___
6. The height, VO, of the pyramid = ___?___.
7. The volume of the pyramid = ___?___.

Exercises 8–11 refer to the cone pictured below.

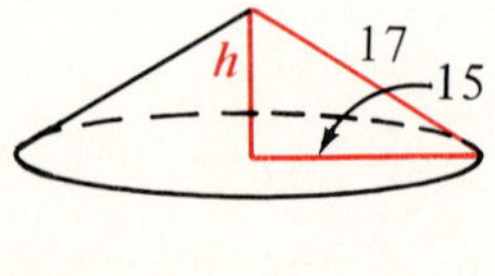

8. The lateral area of the cone = ___?___.
9. The total area of the cone = ___?___.
10. The height, h, of the cone = ___?___.
11. The volume of the cone = ___?___.

Written Exercises

Copy and complete the table about regular square pyramids. Draw an accurate diagram for each exercise if you wish.

	s	h	l	L.A.	T.A.	V
A **1.**	8	3	?	?	?	?
2.	10	12	?	?	?	?
3.	12	8	?	?	?	?
4.	24	?	13	?	?	?
5.	12	?	10	?	?	?
6.	?	6	12	?	?	?

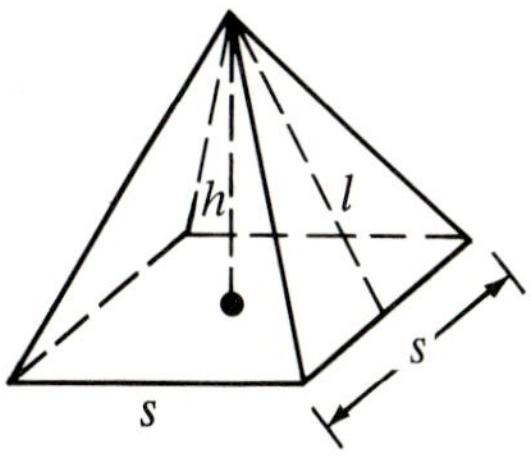

Copy and complete the table about cones. Draw an accurate diagram for each exercise if you wish.

	r	h	l	L.A.	T.A.	V
7.	8	6	?	?	?	?
8.	5	12	?	?	?	?
9.	15	8	?	?	?	?
10.	9	12	?	?	?	?
11.	?	8	17	?	?	?
12.	1	?	2	?	?	?

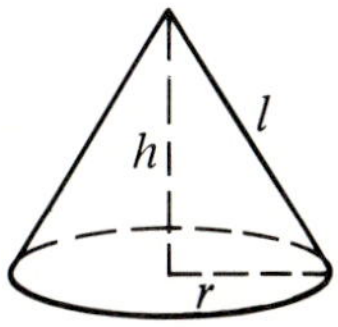

The four faces of the pyramid shown are equilateral triangles.

B **13.** Find the slant height l.

14. Find the lateral area of the pyramid.

15. Find the total area of the pyramid.
Remember: the four faces are congruent.

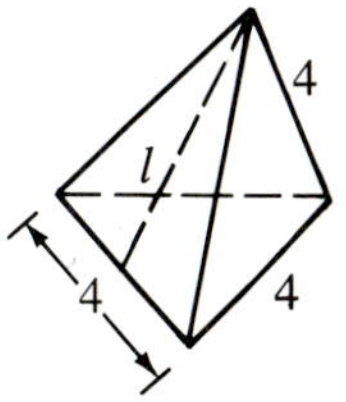

A regular square pyramid has base edges that are 16 cm long and lateral edges that are 17 cm long. Find each of the following.

16. The slant height of the pyramid

17. The lateral area of the pyramid

18. The total area of the pyramid

19. The volume of the pyramid

20. A regular square pyramid has base edges that are 18 cm long and lateral edges that are 15 cm long. Find the lateral area, the total area, and the volume of the pyramid.

A cone with radius 6 is inscribed in a sphere with radius 10.

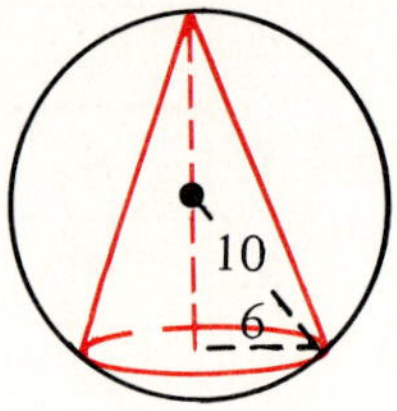

C 21. Find the height of the cone.

22. Find the volume of the cone.

23. Find the volume of the sphere.

SELF-TEST

1. A rectangular solid is 8 cm long, 8 cm wide, and 4 cm high. Find the length of a diagonal. *(11-4)*

2. A cone has height 3 and slant height 5. What is its radius?

3. Find the volume of the cone described in Exercise 2.

4. A regular square pyramid has height 12 and base edges 10. Find its lateral area. *(11-5)*

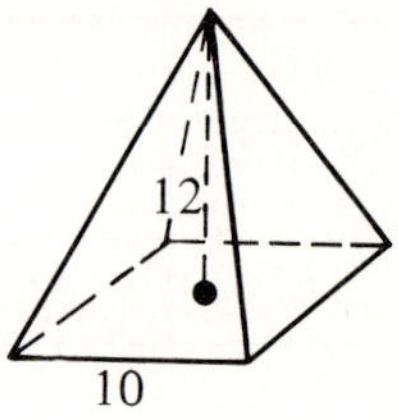

PUZZLE ◆ PROBLEMS

An ant sitting at the corner of an open box smells some sugar at the center of the bottom. The dotted path shows the shortest route to the sugar. How far does the ant have to crawl?

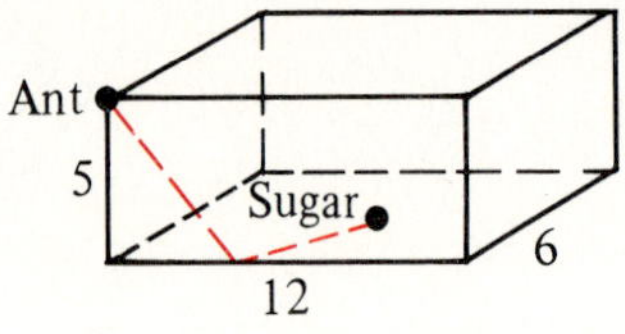

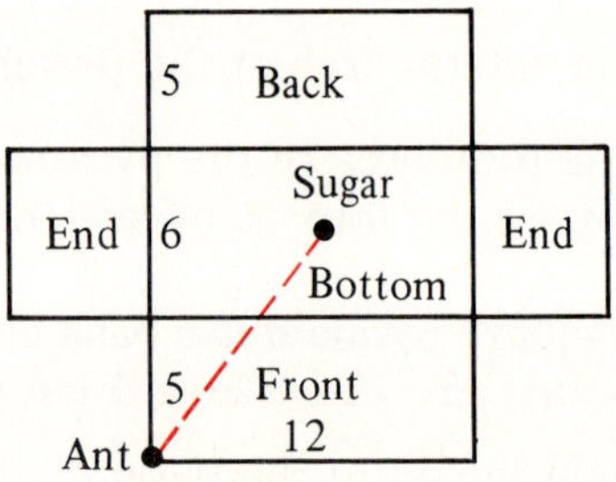

Hint: Imagine unfolding the box as shown.

6 The Tangent Ratio

The word *trigonometry* comes from Greek words which mean "triangle measurement." In this course, the study of trigonometry will be limited to *right* triangles.

In each right triangle below, an angle is marked in color. The diagram indicates the leg *adjacent to* (next to) this angle and the leg *opposite* this angle.

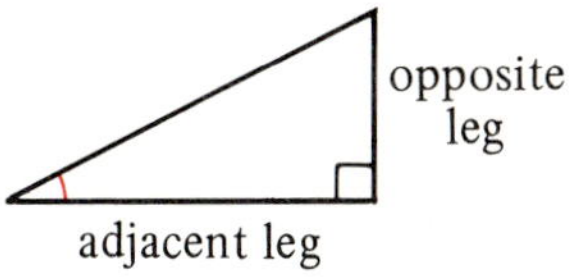

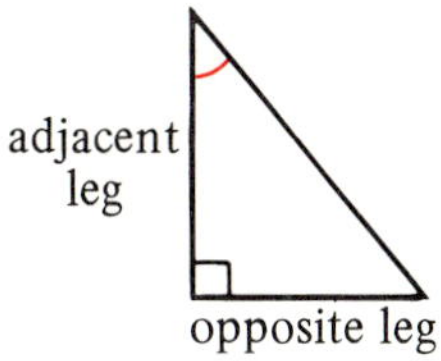

We are now ready to define the first important trigonometric ratio.

$$\textbf{tangent of } \angle A = \frac{\text{leg opposite } \angle A}{\text{leg adjacent to } \angle A}$$

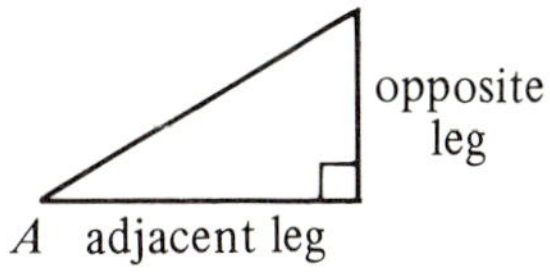

We can abbreviate this by writing:

$$\tan A = \frac{\text{opposite}}{\text{adjacent}}$$

EXAMPLE 1 In the right triangle shown, find tan A and tan B.

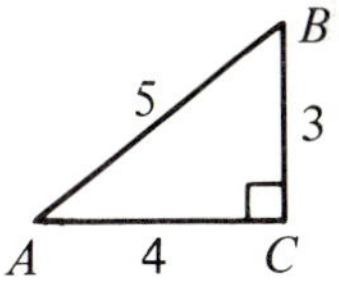

$$\tan A = \frac{\text{leg opposite } \angle A}{\text{leg adjacent to } \angle A} = \frac{3}{4}$$

$$\tan B = \frac{\text{leg opposite } \angle B}{\text{leg adjacent to } \angle B} = \frac{4}{3}$$

In the right triangles shown, $\angle R = \angle X$.
Do you see that $\triangle RST \sim \triangle XYZ$? (Why?)
Therefore,

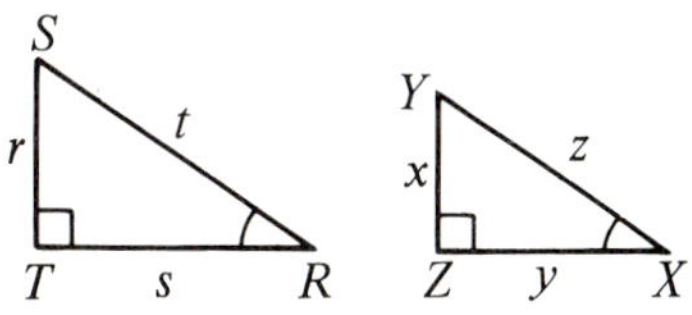

$$\frac{r}{x} = \frac{s}{y}$$ ← Corr. sides of ~ △ are proportional.

$$\frac{r}{s} = \frac{x}{y}$$ ← Use the Switching Property, page 289.

$$\tan R = \tan X$$

The discussion above shows that if $\angle R = \angle X$, then $\tan R = \tan X$. This means that the value of the tangent ratio does not depend on the size of a right triangle, but only on the size of an angle.

On page 517, there is a table that lists values of the tangent ratio for selected acute angles. Most of the values in the table are not exact. They are rounded to four decimal places.

Part of the table is shown at the right. Notice that:

Angle	Tangent
5°	0.0875
10°	0.1763
15°	0.2679
20°	0.3640

$$\tan 15° = 0.2679.$$

For convenience, we'll use = rather than ≈.

EXAMPLE 2 Find the value of y to the nearest tenth.

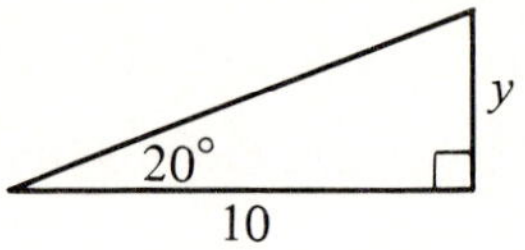

$$\tan 20° = \frac{y}{10}$$

From the table → $0.3640 = \frac{y}{10}$

$$10 \times 0.3640 = y$$
$$3.64 = y$$

To the nearest tenth, $y = 3.6$.

Classroom Practice

Use the table on page 517 to complete the following.

1. tan 15° = __?__ **2.** tan 30° = __?__ **3.** tan 65° = __?__

4. tan __?__ = 0.3640 **5.** tan __?__ = 1.7321 **6.** tan __?__ = 0.0875

Use the definition of tangent to find the value of tan *A* for each right triangle shown. Express your answers in simplest form.

7.

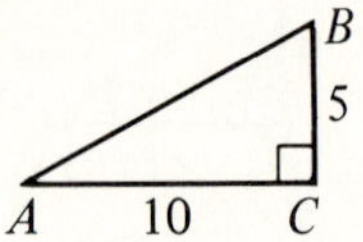

8.

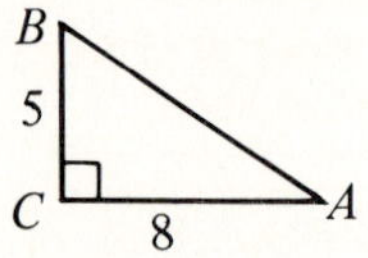

9.

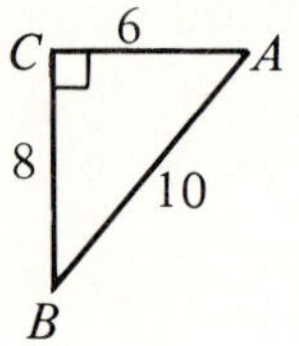

10–12. Find tan B for each right triangle above. Express your answers in simplest form.

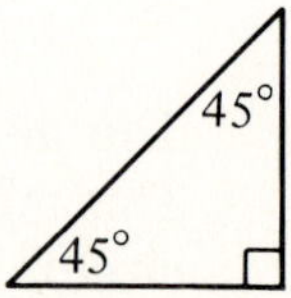

13. What do you know about the legs of a 45°-45°-90° triangle?

14. Without using the table, state the value of tan 45°.

15. The diagram shows a 30°-60°-90° triangle. The length of the longer leg is exactly $\sqrt{3}$. This is approximately 1.732.
   a. Using the diagram, find tan 60°.
   b. Using the table, find tan 60°.

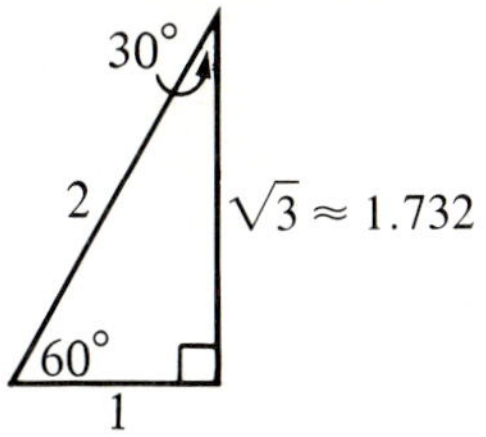

There are two ways to find the value of *t*.

16. a. From the figure, $\tan 55° = \frac{?}{?}$.
   b. Use the table: tan 55° = __?__.
   c. From parts (a) and (b) we have $1.4281 = \frac{4}{t}$.
   Then $1.4281t = 4$, and $t = \frac{4}{1.4281}$.
   Correct to tenths, $t =$ __?__.

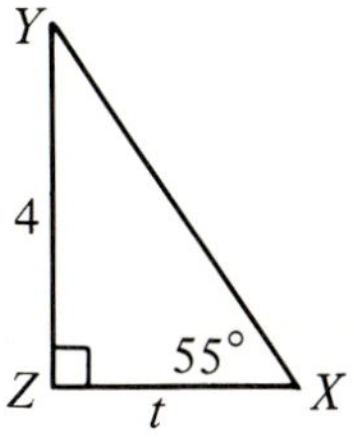

17. Here is an easier way to find the value of *t*.
   a. Find the measure of $\angle Y$.
   b. From the figure, $\tan Y = \frac{?}{?}$.
   c. Use the table to find the value of tan *Y*.
   d. From parts (b) and (c) we have $0.7002 = \frac{t}{4}$.
   Correct to tenths, $t =$ __?__.
   e. Compare your answer for part (d) with your answer for Exercise 16, part (c).

Written Exercises

Use the table on page 517 to complete the following.

A 1. tan 5° = __?__
2. tan 25° = __?__
3. tan 70° = __?__
4. tan __?__ = 3.7321
5. tan __?__ = 0.5774
6. tan __?__ = 0.1763

Find tan *A* and tan *B* for each right triangle shown. Express your answers in simplest form.

7. 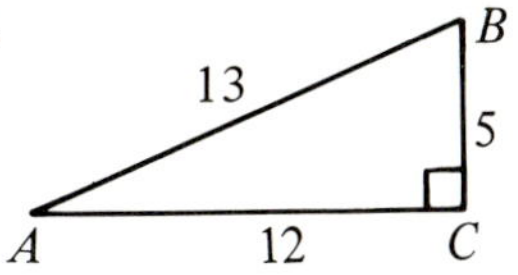

8. 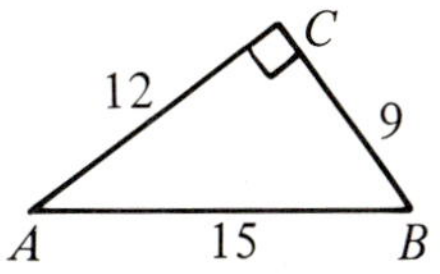

9. 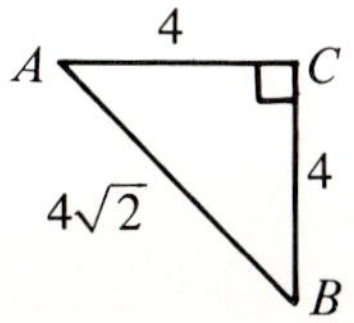

10. 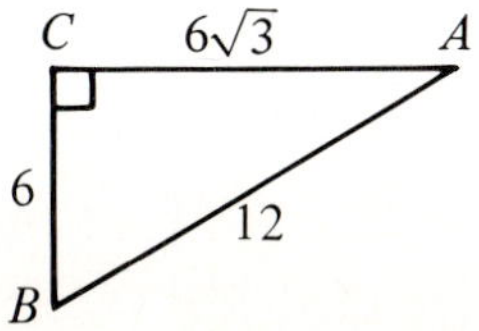

Find the value of y to the nearest tenth. Use the table on page 517.

11.

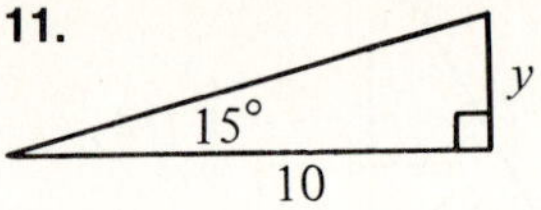

12.

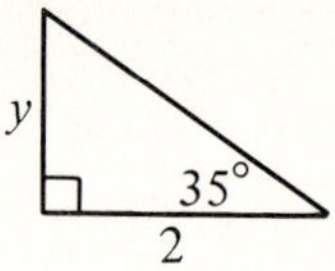

13.

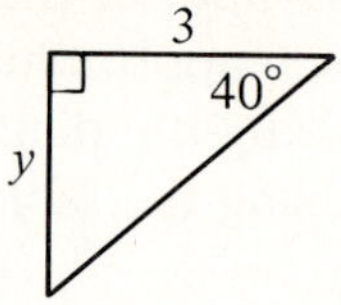

B **14.**

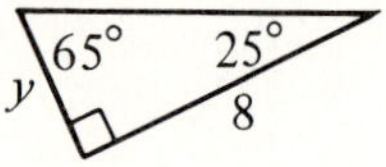

15.

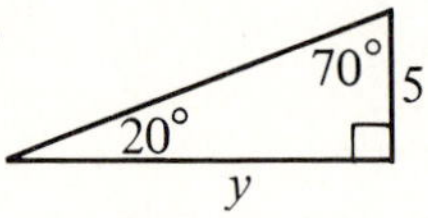

16.

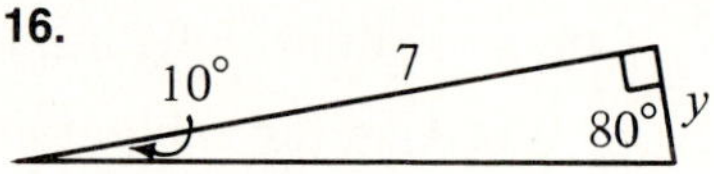

17. The diagram shows a loading ramp. Find x, the distance the floor of the truck is from the ground, to the nearest tenth of a meter.

18. The shadow of a lighthouse is 30 m long. The angle between the ground and the line to the sun is 35°. Find x, the height of the lighthouse to the nearest meter.

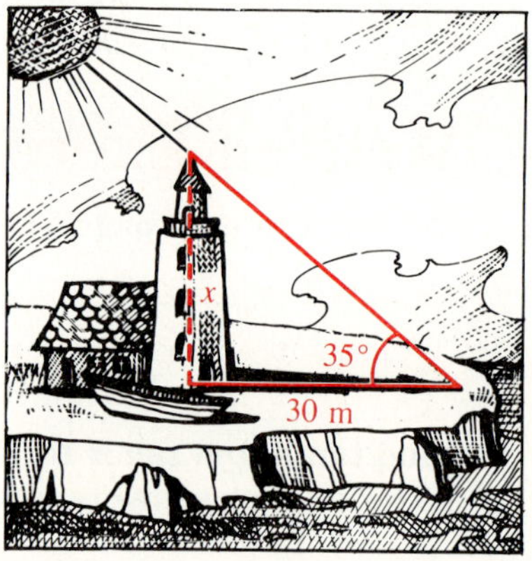

For each right triangle shown, do three things:
a. find tan A as a fraction;
b. find tan A as a decimal;
c. find the measure of $\angle A$ to the nearest degree by using the table on page 517.

C **19.**

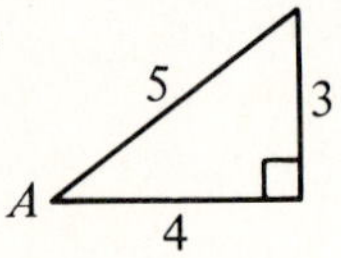

20.

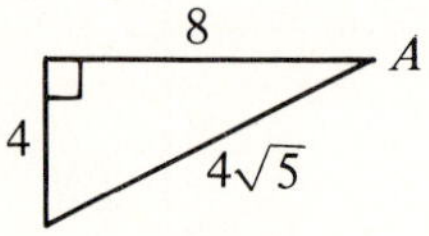

21.

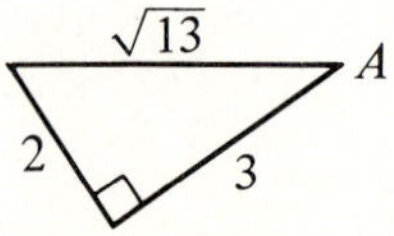

Exercises 22-24 refer to the rectangular solid shown.

22. $x = \underline{\ \ ?\ \ }$

23. $\tan \angle ABC = \underline{\ \ ?\ \ }$

24. To the nearest degree, $\angle ABC = \underline{\ \ ?\ \ }°$.

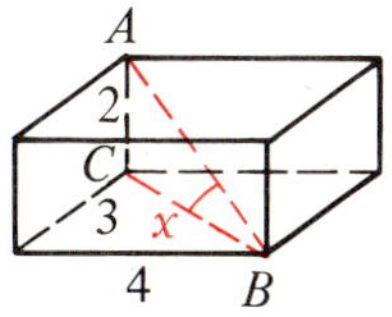

CAREER NOTES

AIRCRAFT ASSEMBLY TECHNICIAN

Suppose someone gave you a model airplane kit with forty parts inside. How long do you think it would take you to put it together? Four hours? Eight hours? Two days? Can you imagine how long it would take you to assemble an actual airplane with its thousands or even millions of parts?

Teams of aircraft assembly technicians divide the job of airplane construction into a large variety of highly specialized tasks. Some assembly technicians join complete sections of the aircraft. For example, they may use bolts, rivets, drills, and solder to join the landing gear to the fuselage. Other technicians work on a smaller scale, assembling engine or auxiliary component parts. Still others specialize in electronic equipment including tiny circuits and modules.

All aircraft assembly technicians must interpret engineering specifications on blueprints for both mechanical and electronic assemblies. Skilled assemblers need high school or vocational training with courses in mathematics, electronics, and mechanical drawing. Two to four years of plant experience are necessary to fully master aircraft assembly skills.

7 The Sine and Cosine Ratios

The tangent ratio involves the two legs of a right triangle. Two other important trigonometric ratios involve the hypotenuse and a leg of a right triangle. These ratios are called the *sine* and the *cosine.*

TRIGONOMETRIC RATIOS

tangent *of* $\angle A = \dfrac{\textit{leg opposite } \angle A}{\textit{leg adjacent to } \angle A}$

sine *of* $\angle A = \dfrac{\textit{leg opposite } \angle A}{\textit{hypotenuse}}$

cosine *of* $\angle A = \dfrac{\textit{leg adjacent to } \angle A}{\textit{hypotenuse}}$

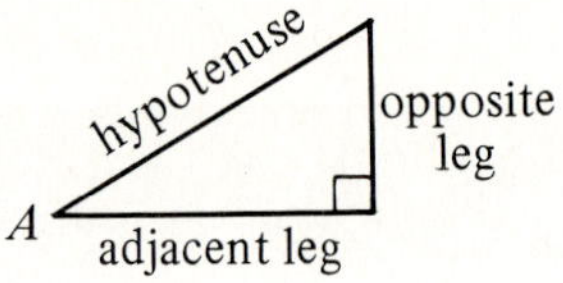

We write:

$\tan A = \dfrac{\text{opposite}}{\text{adjacent}}$ $\sin A = \dfrac{\text{opposite}}{\text{hypotenuse}}$ $\cos A = \dfrac{\text{adjacent}}{\text{hypotenuse}}$

In the previous section, we used the table on page 517 in our work with tangents. Notice that the table lists values of the sine ratio and the cosine ratio, in addition to those of the tangent ratio. Remember that most of the values are approximate. We shall use the trigonometric table in the examples which follow.

EXAMPLE 1 Find the value of y to the nearest tenth.

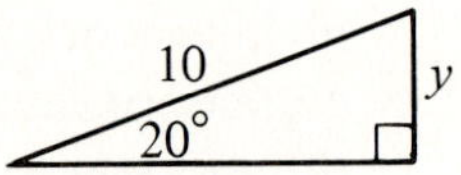

$$\sin 20° = \frac{\text{opposite}}{\text{hypotenuse}}$$

$$\sin 20° = \frac{y}{10}$$

From the table → $$0.3420 = \frac{y}{10}$$

$$10 \times 0.3420 = y$$

$$3.42 = y$$

To the nearest tenth, $y = 3.4$.

EXAMPLE 2

A guy wire is stretched from the top of a tower to a point 10 m from the base of the tower. The wire makes a 65° angle with the ground. Find z, the length of the wire, to the nearest meter.

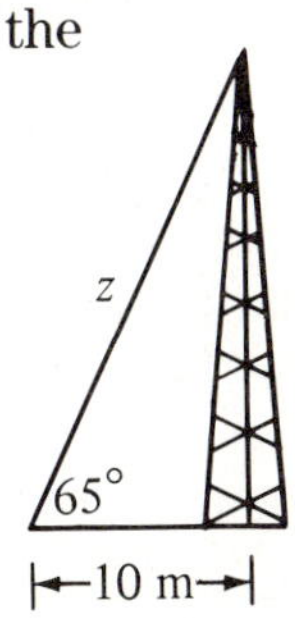

$$\cos 65^\circ = \frac{\text{adjacent}}{\text{hypotenuse}}$$

$$0.4226 = \frac{10}{z}$$

$$0.4226z = 10$$

$$z = \frac{10}{0.4226}$$

To the nearest meter, $z = 24$ m.

EXAMPLE 3

Find the measure of $\angle A$ to the nearest five degrees.

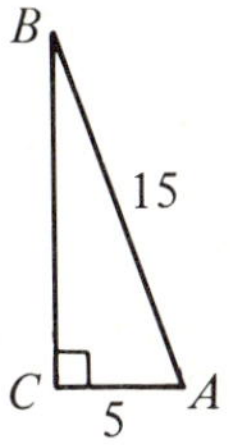

The given sides of the triangle are the leg adjacent to $\angle A$ and the hypotenuse. Therefore, we use the cosine ratio.

1. Find $\cos A$ as a fraction: $\cos A = \frac{\text{adjacent}}{\text{hypotenuse}}$

   $\cos A = \frac{5}{15} = \frac{1}{3}$

2. Express $\cos A$ as a decimal:

   $\cos A = 0.3333 \ldots$ ← $3\overline{)1.000}$ = 0.3333 . . .

3. Look in the table under *Cosine* to find the entry closest to 0.3333.

To the nearest five degrees, $\angle A = 70°$.

Classroom Practice

Match each name with the correct expression.

1. sine **a.** $\frac{\text{opposite}}{\text{adjacent}}$

2. cosine **b.** $\frac{\text{opposite}}{\text{hypotenuse}}$

3. tangent **c.** $\frac{\text{adjacent}}{\text{hypotenuse}}$

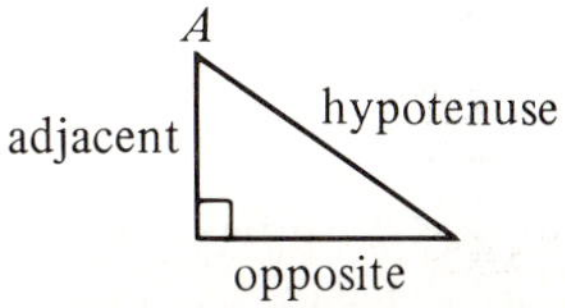

Refer to △*PQR*. Find each ratio.

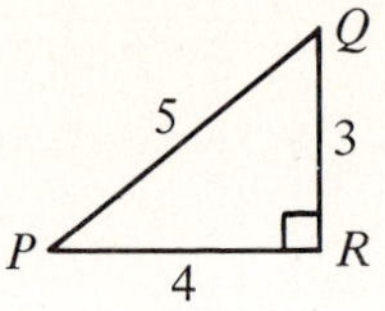

4. sin P 5. cos P 6. tan P

7. sin Q 8. cos Q 9. tan Q

Use the table on page 517 to find the following.

10. sin 25° 11. cos 40° 12. tan 75° 13. sin 50°

Use the table on page 517 to find the measure of each angle to the nearest degree.

14. $\sin A = 0.2588$ 15. $\cos P = 0.6428$ 16. $\tan R = 0.3750$ 17. $\sin S = 0.3500$

In each exercise, state an equation you could use to find the value of *x*.

18.

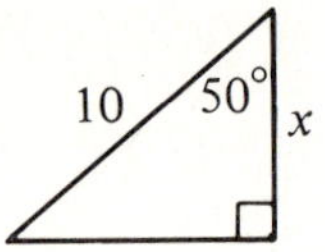

19.

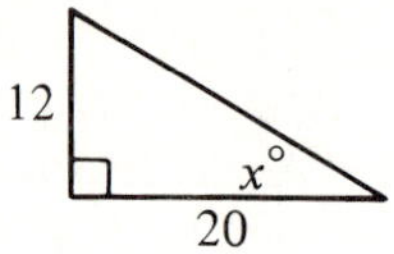

20.

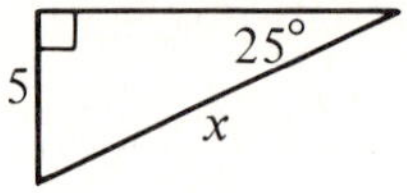

Written Exercises

Use the table on page 517 to find the following.

A 1. sin 5° 2. cos 15° 3. tan 45° 4. cos 80°

Use the table on page 517 to find the measure of ∠*A* to the nearest degree.

5. $\sin A = 0.9659$ 6. $\cos A = 0.5736$ 7. $\tan A = 2.1400$ 8. $\cos A = 0.4900$

Refer to △*ABC*. Find each ratio.

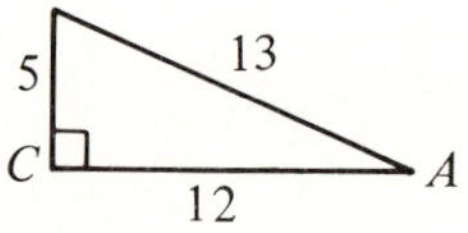

9. sin A 10. cos A 11. tan A

12. sin B 13. cos B 14. tan B

15. In △*ABC* above, $\tan B = \frac{12}{5} = 2.4000$. Use the trigonometric table to find the measure of ∠*B* to the nearest degree.

Refer to △*XYZ*. Find each ratio.

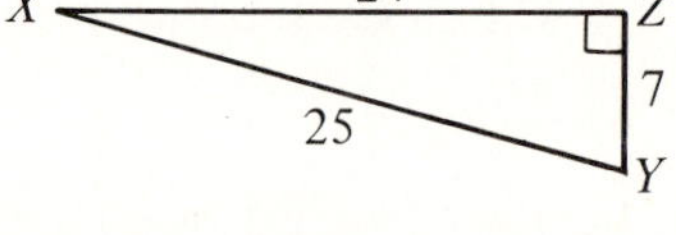

16. sin X 17. cos X 18. tan X

19. sin Y 20. cos Y 21. tan Y

22. In △*XYZ* above, $\sin X = \frac{7}{25} = 0.2800$. Use the trigonometric table to find the measure of ∠*X* to the nearest degree.

23. Find sin A as a fraction.
24. Find sin A as a decimal.
25. Use the table on page 517 to find the measure of $\angle A$ to the nearest degree.

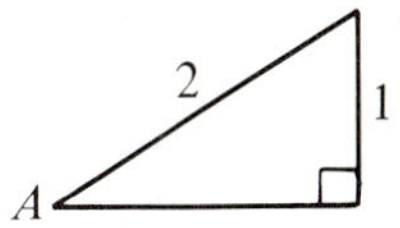

26. Find cos B as a fraction.
27. Find cos B as a decimal.
28. Use the table on page 517 to find the measure of $\angle B$ to the nearest degree.

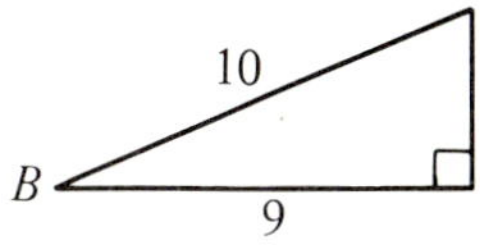

Find the value of y to the nearest tenth. Use the table on page 517.

29.

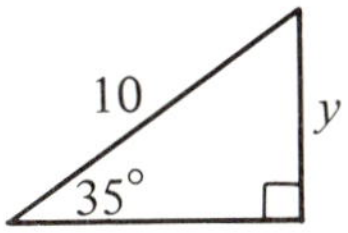

30.

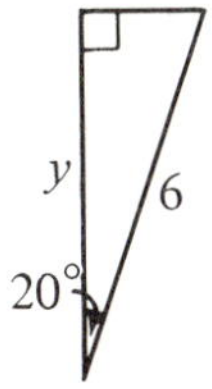

31.

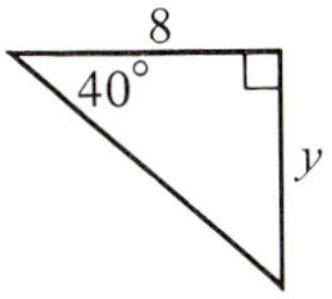

B 32.

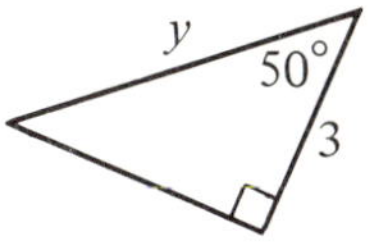

33.

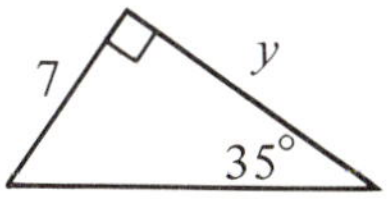

34.

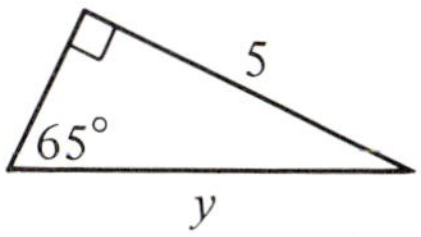

35. A tree trimmer has a ladder 4 m long. It is leaning against a palm tree at a 70° angle with the ground. How far up the tree does the ladder reach?

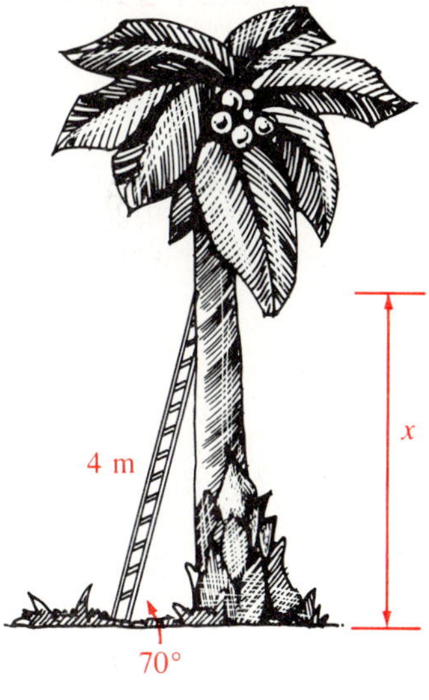

36. A golfer stands 80 m from the Number 5 hole. The golfer's aim is poor, and the ball goes 15° off course and lands at the left of the hole.
   a. How far is the ball from the hole?
   b. How far is the ball from the golfer?

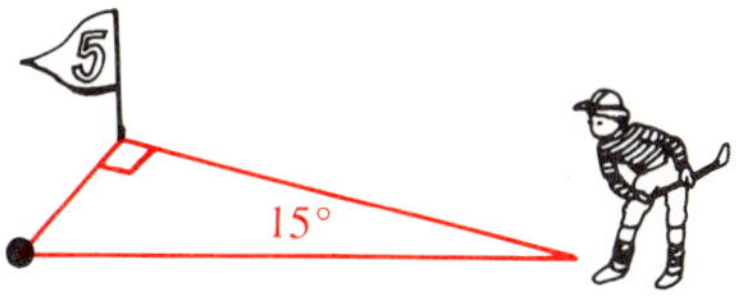

Find the measure of $\angle A$ to the nearest degree.

C 37.

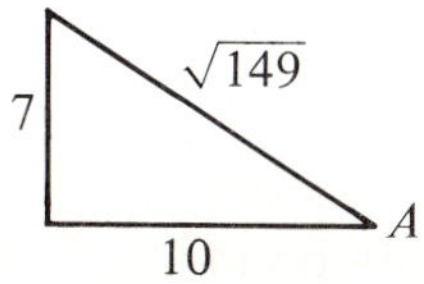

38.

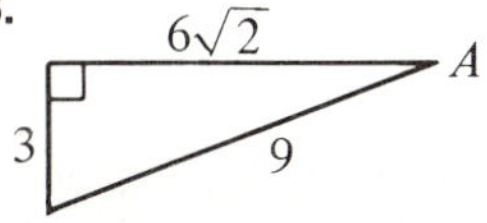

39.

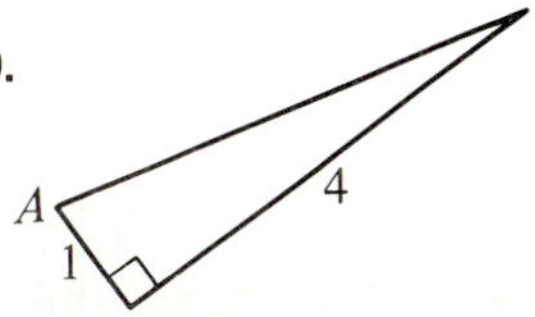

SELF-TEST

Vocabulary

tangent (p. 447) sine (p. 452) cosine (p. 452)

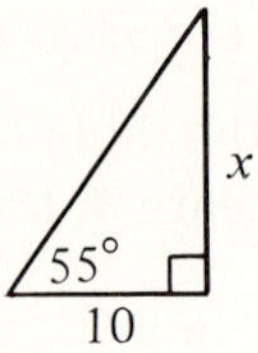

1. Find the value of x to the nearest tenth. Use the table on page 517. *(11-6)*

Complete each statement.

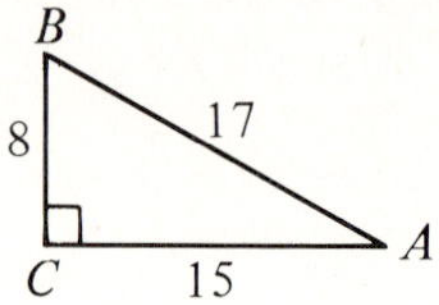

2. $\tan A =$ __?__
3. $\sin A =$ __?__ *(11-7)*
4. $\cos B =$ __?__
5. $\sin B =$ __?__

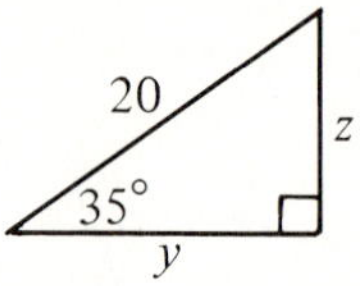

6. Find the values of y and z to the nearest tenth. Use the table on page 517.

PUZZLE ◆ PROBLEMS

1. ENCOUNTER is a game that is played on graph paper. The players choose any two starting points, A and B, on the grid.

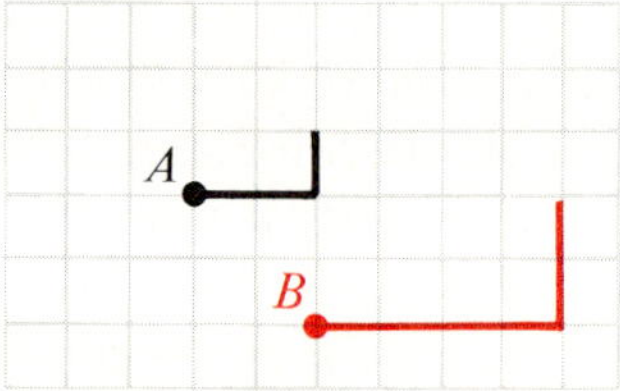

2. Player A draws a horizontal segment from point A to any other point of the grid. (The segment should not be very long.)
3. Player B starts at point B and draws a segment in the same direction as player A's segment, but twice as long.

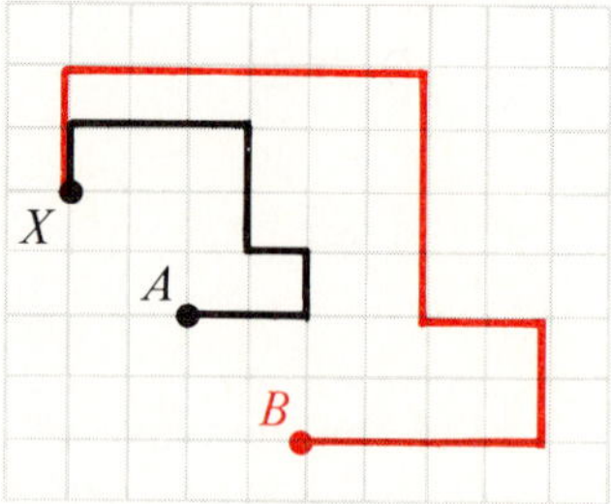

4. Player A joins a vertical segment to the end of the first segment drawn.
5. Player B draws a segment in the same direction as player A's second segment, but twice as long.
6. Taking turns, the players draw horizontal segments. Then they both draw vertical segments. Each time, player B draws a segment in the same direction as player A's, but twice as long.

The object of the game is for player A to move, in as few moves as possible, to a point X that will force B to move to the same point.

APPLICATIONS

ANGLES OF ELEVATION AND DEPRESSION

Sometimes you can not make measurements using a meter stick, yard stick, or tape measure. Airplanes and ships often have sophisticated electronic devices or computers that figure out altitudes, ocean depth, and other distances. Trigonometry can also be used to find distances that are hard to measure physically.

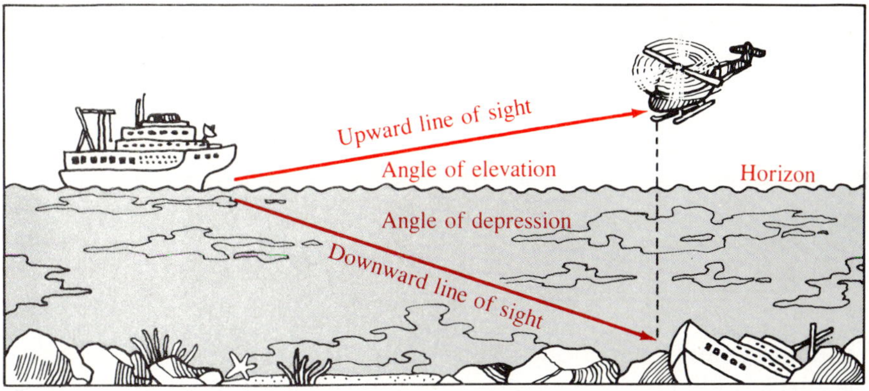

The ship in the diagram is taking scientists and historians on a search for sunken treasure. Their helicopter has spotted what appears to be a shipwreck on the ocean floor. While it hovers above the wreck, the ship's crew determines the relative positions of the ship, helicopter, and wreck. They look out over the ocean into the horizon. Then they adjust their line of sight upward and downward.

If the crew looks *up*, they see the helicopter. Their line of sight forms an **angle of elevation** with the horizon. Looking *downward*, their line of sight forms an **angle of depression** with the horizon.

Now consider the perspectives looking *from* the helicopter and wreck *to* the ship. Notice the angles that are formed by these lines of sight and the vertical line from H to W. These are also called angles of elevation and depression. The helicopter pilot looks *down* to the ship. The angle indicated at H is called an *angle of depression* of S. A scuba diver exploring the wreck would have to look *up* to the ship. The angle indicated at W is called an *angle of elevation* of S.

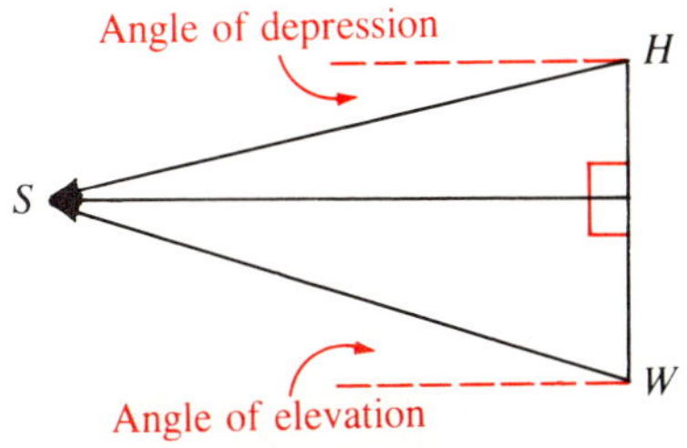

Notice that the line of sight is always the hypotenuse of a right triangle. The vertical and horizontal distances are the legs of the triangle. Therefore, given an angle of elevation or depression and one of these distances, we can use one of the trigonometric ratios to find a missing distance.

EXAMPLE An Olympic swimming pool is 50 m long. From the opposite end of the pool, the angle of elevation of the highboard diving platform is about 11°. How high is the platform?

Step 1 Draw a diagram to illustrate the problem.
Include all the given information in relation to a right triangle.

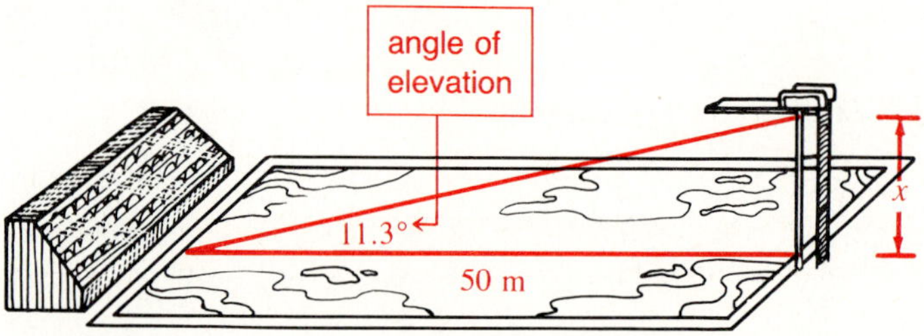

Step 2 Write an equation using one of the trigonometric ratios.
The platform ladder is opposite the angle of elevation.
The length of the pool is adjacent to the angle of elevation.

$$\tan 11° = \frac{\text{opposite}}{\text{adjacent}} = \frac{x}{50}$$

Step 3 Solve the equation.

$$\tan 11° = \frac{x}{50}$$
$$0.1944 = \frac{x}{50}$$
$$50(0.1944) = x$$
$$9.72 = x$$

To the nearest whole number, the height of the diving platform is 10 m.

EXERCISES

1. $\overleftrightarrow{PR}$ is horizontal
   a. For a person at Q, what is the angle of elevation of S?
   b. From S, what is the angle of depression of Q?
   c. Someone at P measures the angle of elevation of S. Is the measure greater than or less than 15°?

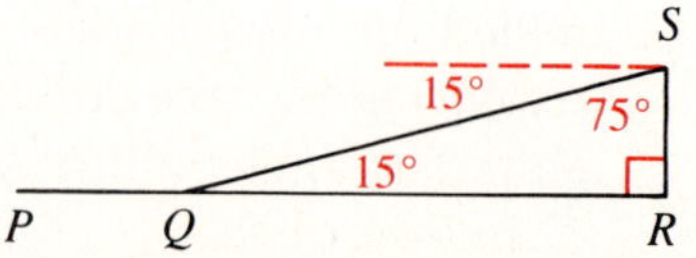

2. The lines shown are horizontal and vertical except for $\overleftrightarrow{AD}$ and $\overleftrightarrow{CD}$. Give the number of the angle and its special name when:
   a. An observer at D sights A.
   b. An observer at D sights C.
   c. An observer at A sights D.
   d. An observer at C sights D.

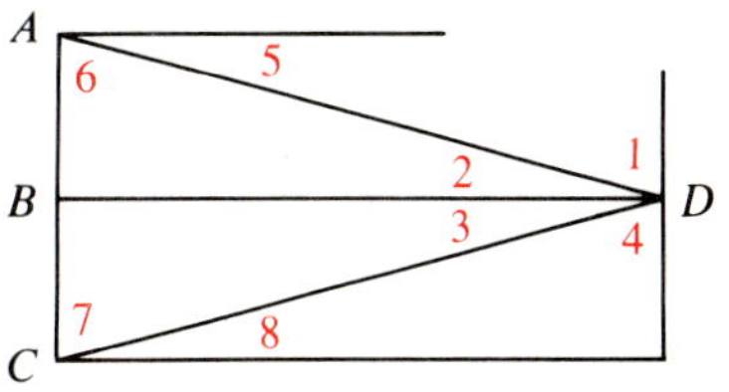

Draw a diagram to illustrate each situation below.

3. A kite flies at an angle of elevation of 50°. All 120 m of string have been let out. You want to find the kite's height.

4. The height of a lighthouse is 16 m. From the top of the lighthouse, the angle of depression of a nearby boat is 40°. You want to find the distance from the base of the lighthouse to the boat.

5. A sloping theater aisle is 80 m long. From the lowest point next to the stage, the angle of elevation of the highest point is 10°. You want to find the *horizontal* distance between the stage and the rear of the theater.

6–8. Write an equation to find the missing distances in Exercises 3–5. (*Hint:* Each exercise requires a different trigonometric ratio.)

9–11. Solve the equations you wrote for Exercises 6–8. Round your answers to the nearest whole number.

PUZZLE ◆ PROBLEMS

Begin with a right triangle.
Draw squares, as shown, and cut them out.

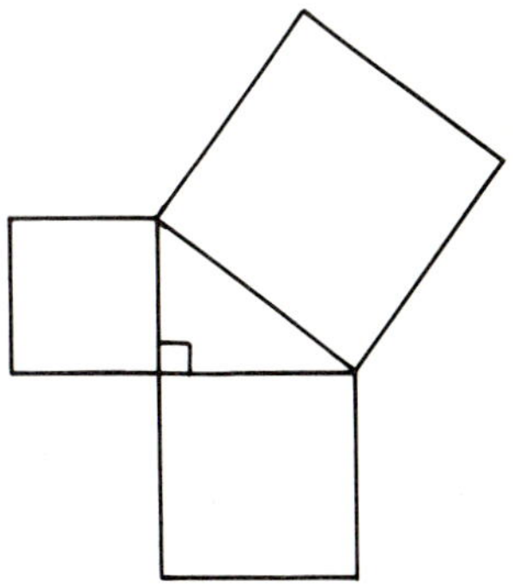

Place the two smaller squares next to one another.
Locate point X so that $VX = AB$.
Cut along $\overline{AX}$ and $\overline{WX}$.
Try to rearrange the five pieces formed to cover the largest square.

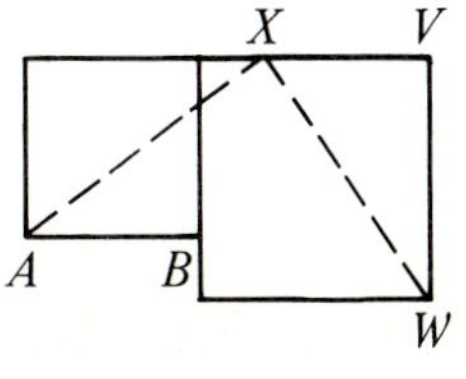

SKILLS REVIEW

ALGEBRAIC FRACTIONS: MULTIPLICATION AND DIVISION

Simplify.

SAMPLE 1 $\frac{-5x^2}{20xy} = \frac{-1 \cdot \mathbf{5} \cdot \boldsymbol{x} \cdot x}{4 \cdot \mathbf{5} \cdot \boldsymbol{x} \cdot y} = \frac{-x}{4y} \text{ or } -\frac{x}{4y}$

SAMPLE 2 $\frac{7x - 21}{x^2 - 9} = \frac{7\boldsymbol{(x - 3)}}{(x + 3)\boldsymbol{(x - 3)}} = \frac{7}{x + 3}$

1. $\frac{-3a^2b^2}{12ab^2}$
2. $\frac{2h^4k^5}{10h^3k}$
3. $\frac{11b^2c^4}{-33bc^2}$
4. $\frac{-4r^3s}{-16rt}$
5. $\frac{2m}{am + bm}$
6. $\frac{6x^2}{3xy - 9x^2}$
7. $\frac{2a + 2b}{a^2 - b^2}$
8. $\frac{x + 4}{x^2 + 6x + 8}$
9. $\frac{a^2 - 64}{a + 8}$
10. $\frac{6r - 6s}{r^2 - s^2}$
11. $\frac{d^2 - 2d - 15}{13d + 39}$
12. $\frac{3a - 21}{a^2 - 14a + 49}$

Multiply.

SAMPLE

$$\frac{a^2 - 2ab + b^2}{10} \cdot \frac{5}{a - b} = \frac{5(a^2 - 2ab + b^2)}{10(a - b)} = \frac{\mathbf{5}\boldsymbol{(a - b)}(a - b)}{2 \cdot \mathbf{5}\boldsymbol{(a - b)}} = \frac{a - b}{2}$$

13. $\frac{x}{y} \cdot \frac{5y}{2x}$
14. $\frac{8a^2}{b} \cdot \frac{b^2}{2a}$
15. $\frac{a + b}{3} \cdot \frac{9}{ab + b^2}$
16. $\frac{7x}{9y} \cdot \frac{18}{35x^2}$
17. $\frac{4}{x + 2} \cdot \frac{x^2 - 4}{12}$
18. $\frac{7x - 21}{3} \cdot \frac{15}{x - 3}$
19. $\frac{x - 5}{y} \cdot \frac{y^3}{x^2 - 25}$
20. $\frac{m - 3}{m + 3} \cdot \frac{3 + m}{3 - m}$
21. $\frac{x^2 + 4x + 4}{28} \cdot \frac{7}{x + 2}$

Divide.

SAMPLE $\frac{a^3}{3} \div \frac{a^4}{3a^2 + 3} = \frac{a^3}{3} \cdot \frac{3a^2 + 3}{a^4} = \frac{\boldsymbol{a \cdot a \cdot a \cdot} \mathbf{3}(a^2 + 1)}{\mathbf{3} \boldsymbol{\cdot a \cdot a \cdot a} \cdot a} = \frac{a^2 + 1}{a}$

22. $\frac{b}{3} \div \frac{1}{9}$
23. $\frac{m^2}{n^2} \div 2m$
24. $3r^2s^2 \div \frac{12s^3}{5t^2}$
25. $\frac{a + b}{12} \div \frac{a + b}{6b}$
26. $\frac{2\pi r - 2\pi s}{9} \div \frac{2\pi}{3}$
27. $\frac{4c - 12}{7} \div (c - 3)$
28. $\frac{n^2 - 64}{5} \div (n + 8)$
29. $\frac{d^2 + d^3}{2c + 4} \div \frac{d^2}{6}$
30. $\frac{3y}{y^2 - 49} \div \frac{1}{y + 7}$

CHAPTER REVIEW

CHAPTER SUMMARY

1. Right triangle lengths you should know are included in the table on page 428.
2. In a 45°-45°-90° triangle, the hypotenuse is $\sqrt{2}$ times as long as a leg.
3. In a 30°-60°-90° triangle, the hypotenuse is twice as long as the shorter leg, and the longer leg is $\sqrt{3}$ times as long as the shorter leg.
4. If a rectangular solid has dimensions a, b, and c, then a diagonal has length $d = \sqrt{a^2 + b^2 + c^2}$.
5. You can often use the Pythagorean Theorem to find the slant height or the height of a regular pyramid or cone.
6. Three trigonometric ratios are defined for any acute angle of a right triangle as follows:

$$\tan A = \frac{\text{opposite}}{\text{adjacent}}$$

$$\sin A = \frac{\text{opposite}}{\text{hypotenuse}}$$

$$\cos A = \frac{\text{adjacent}}{\text{hypotenuse}}$$

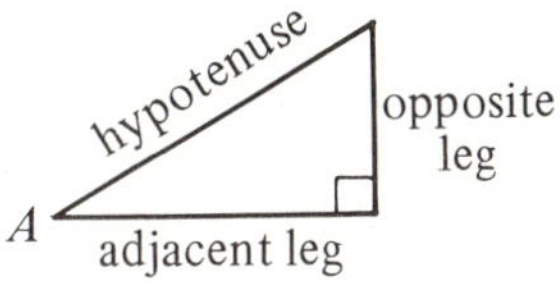

REVIEW EXERCISES

Find the exact values of x and y. ***(See pp. 428–436.)***

1.

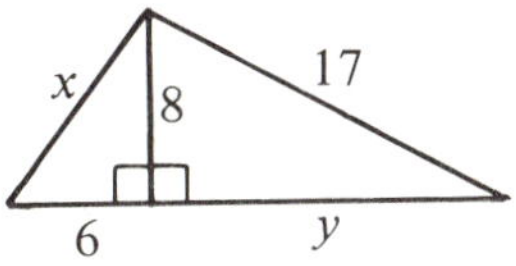

2.

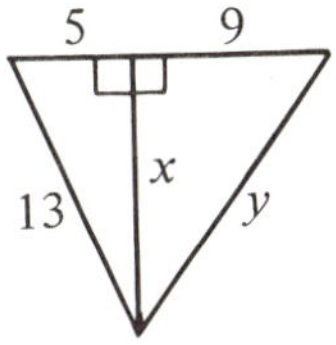

3.

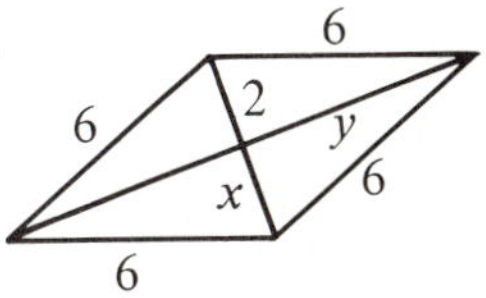

4.

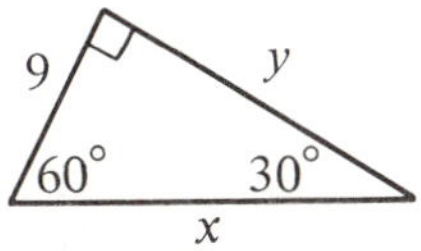

5.

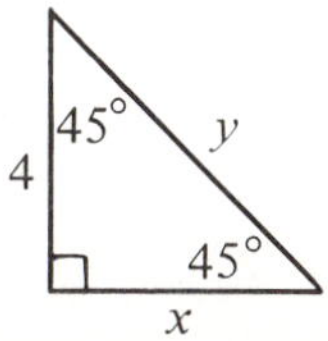

6.

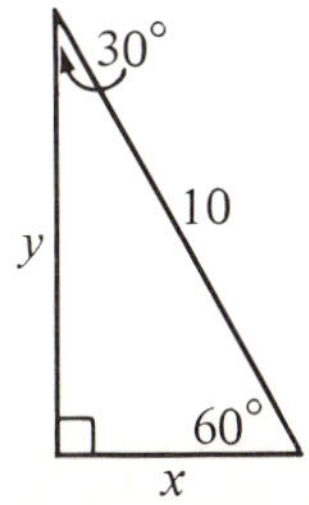

Find the exact area of each shaded figure. ***(See pp. 437–439.)***

7.

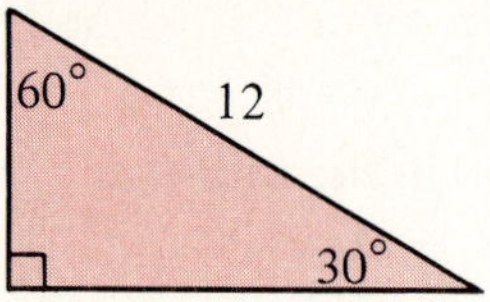

8.

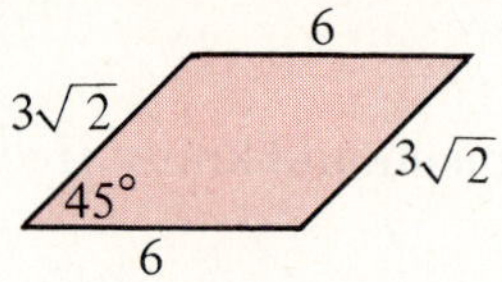

9.

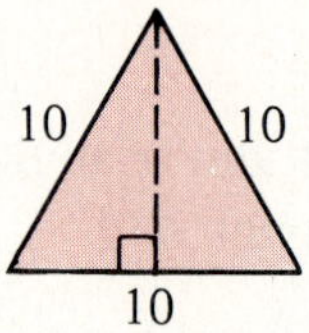

Refer to the rectangular solid shown. ***(See pp. 440–442.)***

10. Find the value of x.

11. Find the length of the diagonal d.

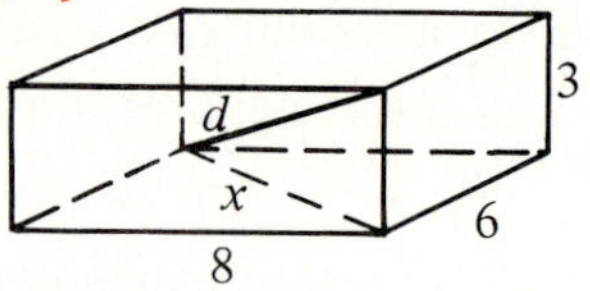

Refer to the regular square pyramid and cone shown at the right. ***(See pp. 443–446.)***

12. Find the lateral area of the pyramid.

13. Find the total area of the pyramid.

14. Find the height of the pyramid.

15. Find the volume of the pyramid.

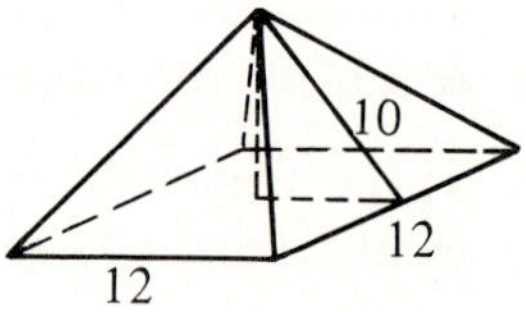

16. Find the slant height of the cone.

17. Find the lateral area of the cone.

18. Find the total area of the cone.

19. Find the volume of the cone.

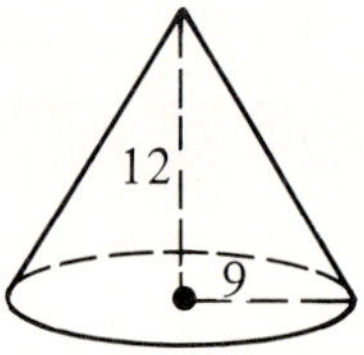

Use the diagram to find the following. ***(See pp. 447–455.)***

20. tan A **21.** cos A

22. sin B **23.** tan B

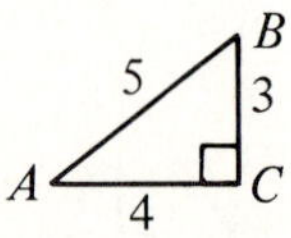

Find the value of x to the nearest tenth. Use the table on page 517. ***(See pages 447–455.)***

24.

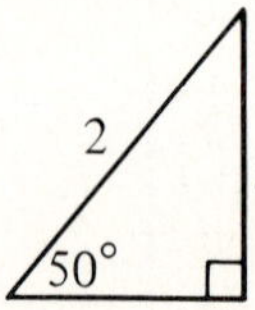

25.

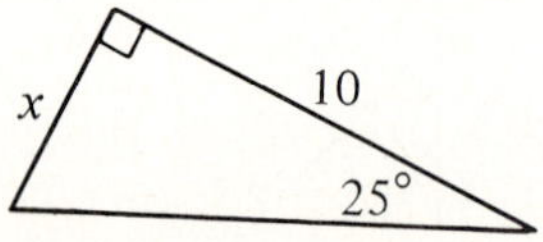

26.

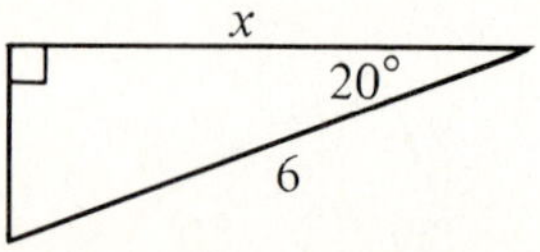

CHAPTER TEST

Find the exact values of *x* and *y*. *(11-1)*

1.
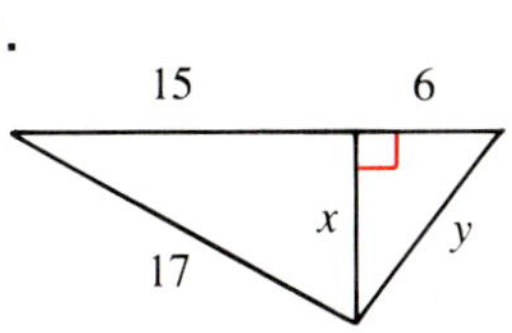

2.
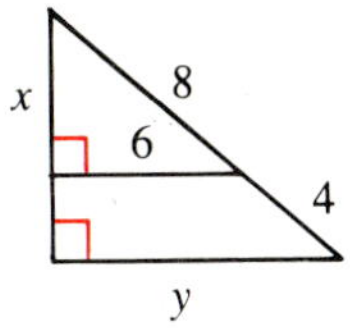

3.
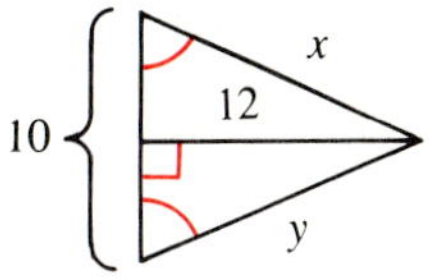

(11-2)

4.
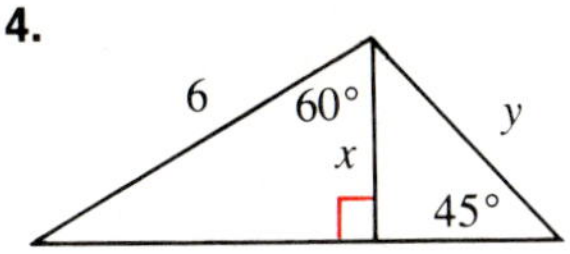

5.
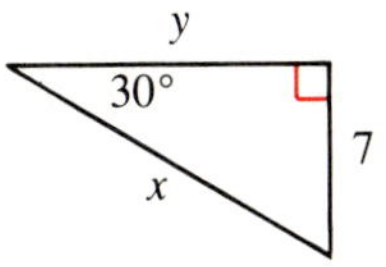

6.
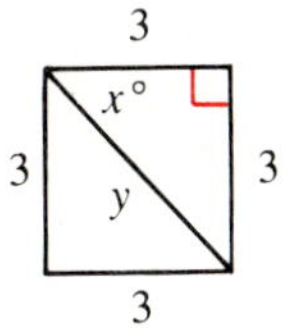

Find the area of each figure described. *(11-3)*

7. The parallelogram shown

8. An isosceles right triangle with a hypotenuse of $9\sqrt{2}$

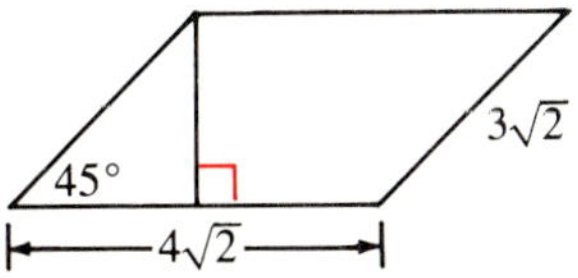

A rectangular solid has length 4, width 3, and height 12. *(11-4)*

9. Sketch the figure.

10. Find the length of a diagonal of a base.

11. Find the length of a diagonal of the rectangular solid.

A regular square pyramid has height 16 and base edges 24. *(11-5)*

12. l = ___?___ 13. L.A. = ___?___ 14. T.A. = ___?___ 15. V = ___?___

A cone has height 21 and slant height 29.

16. r = ___?___ 17. L.A. = ___?___ 18. T.A. = ___?___ 19. V = ___?___

20. In $\triangle ABC$, $AC = 3$, $BC = 4$, and $AB = 5$. Find $\tan A$ and $\tan B$. *(11-6)*

21. Write an equation that can be used to find the value of x.

22. If $\tan 50° = 1.1918$, find the value of x to the nearest tenth.

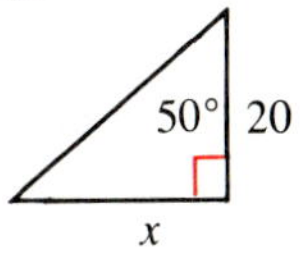

Find the value of *x* to the nearest tenth. Use the following as needed: $\sin 15° = 0.2588$, $\cos 15° = 0.9659$, and $\tan 15° = 0.2679$. *(11-7)*

23.
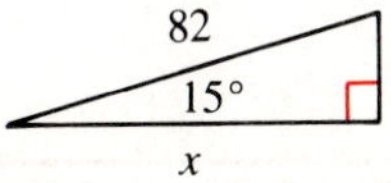

24.
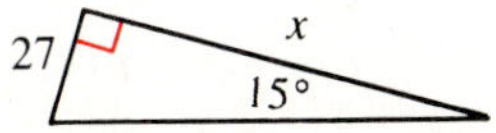

25.
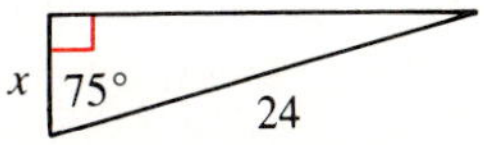

MIXED REVIEW

Tell whether each statement is true or false.

1. If a triangle has two equal angles, then it must be isosceles.
2. If the diagonals of a quadrilateral are perpendicular, then the quadrilateral is a rhombus.
3. The converse of a true "If . . . then" statement is not always true.
4. If two lines do not intersect, then they must be parallel.
5. The ratio of the areas of two similar polygons is equal to the ratio of two corresponding sides.
6. An exterior angle of a regular polygon is always obtuse.
7. Complete the following proof.

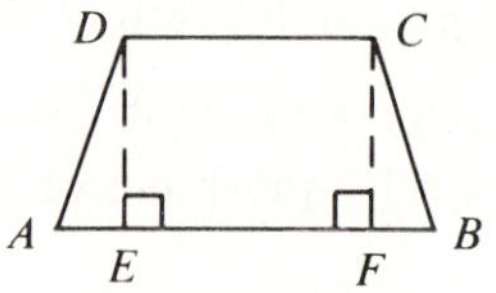

Given: $ABCD$ is an isosceles trapezoid
$\overline{DE} \perp \overline{AB}$; $\overline{CF} \perp \overline{AB}$

Prove: $\angle ADE = \angle BCF$

STATEMENTS	REASONS
1. $ABCD$ is an isos. trapezoid	1. ___?___
2. $AD = BC$	2. ___?___
3. $\angle A = \angle B$	3. ___?___
4. $\overline{DE} \perp \overline{AB}$; $\overline{CF} \perp \overline{AB}$	4. ___?___
5. $\angle DEA = 90° = \angle CFB$	5. ___?___
6. $\triangle AED \cong \triangle$ ___?___	6. ___?___
7. ___?___	7. ___?___

8. In $\triangle MLS$, $\angle M = 90°$, $LM = 15$, and $SM = 8$. Find the values of $\sin S$, $\cos S$, and $\tan S$.
9. Explain why the median of a triangle divides it into two triangles with equal areas. (*Hint:* The triangles have the same height.)
10. Draw a line l and a point P not on l. Construct a line through P and parallel to l. Label each pair of alternate interior angles.

11. $\angle 1$ is a central angle of $\odot O$ and $\angle 2$ is an inscribed angle of $\odot O$. If both $\angle 1$ and $\angle 2$ intercept $\widehat{AB}$ and $\widehat{AB} = 50°$, $\angle 1 =$ __?__ ° and $\angle 2 =$ __?__ °.

12. The bases of a right prism are 3-4-5 right triangles. The height of the prism is 10. Find the lateral area, the total area, and the volume of the prism.

Find the values of *x* and *y*.

13.

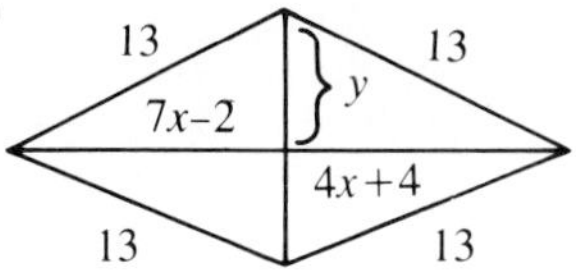

14.

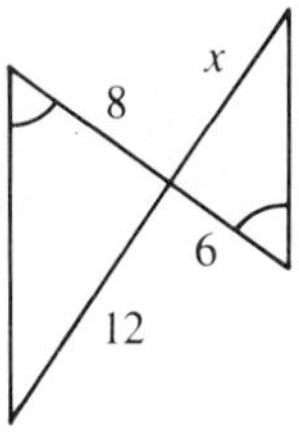

15.

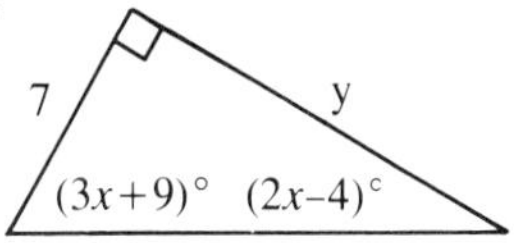

16. Which *one* of the following is *not* true?

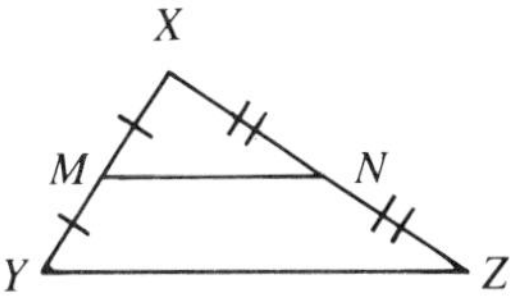

a. $\angle XMN = \angle Y$ b. $\frac{XM}{MY} = \frac{MN}{YZ}$

c. $YZ = 2(MN)$

17. A cone has slant height 29 and radius 21. Find its volume.

18. Find the area of a trapezoid with bases of 11 and 20, and height 8.

19. Draw a large obtuse triangle. Construct and label the altitude to the shortest side.

20. Suppose R and S lie on the perpendicular bisector of $\overline{FG}$. What postulate or theorem supports the statement?

a. $RF = RG$ and $SF = SG$

b. $\triangle RFS \cong \triangle RGS$ (*Hint:* $RS = RS$)

Find the value of *t*.

21.

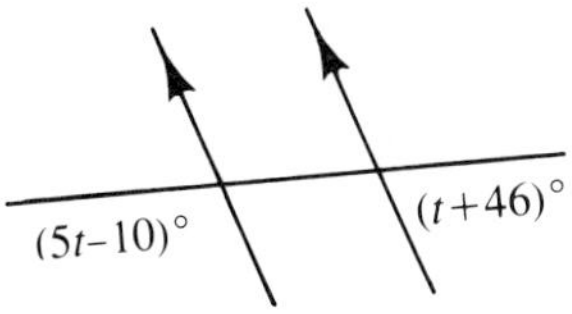

22.

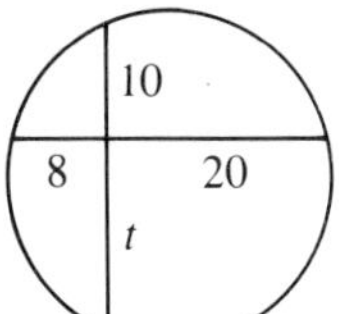

23.

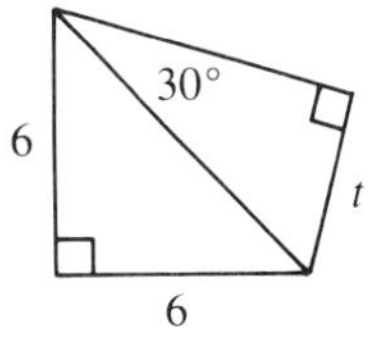

24. Find the sum of the measures of the angles of a decagon.

25. Find the diameter of a sphere with surface area 400π.

Here's what you'll learn in this chapter:

- To specify points by their coordinates in the coordinate plane.
- To find the distance between two points in the coordinate plane.
- To apply the midpoint formula.
- To find the slope of a line.
- To find the slopes of parallel and perpendicular lines.
- To graph the line specified by a given equation.
- To prove theorems using coordinate geometry.

The intersecting lines drawn to help you locate points on the coordinate plane are called a grid. Such a grid is suggested by the construction of the power switching yard in Vancouver, Washington, shown above.

Chapter 12

Coordinate Geometry

1 Points and Coordinates

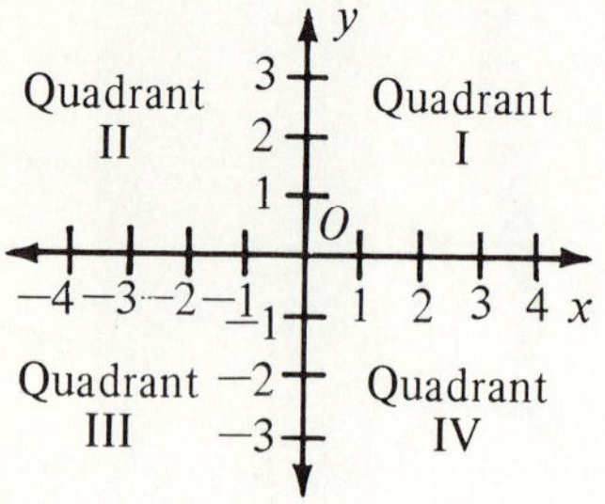

The *coordinate plane* has an x-axis and a y-axis, as shown in the diagram. These **coordinate axes** (**ax**-eez) are number lines that intersect in a point called the origin, point O. The axes separate the plane into four quadrants.

Every point in the coordinate plane can be named by two numbers, called the **coordinates** of the point. In the diagram, the ***x*-coordinate** of point A is 2. The ***y*-coordinate** is 4. We can refer to point A as the point (2, 4). Notice that the origin is the point (0, 0).

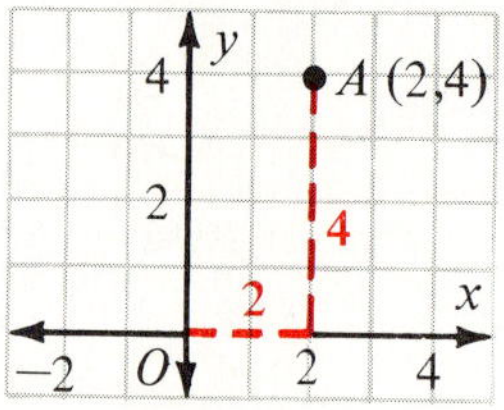

Point B, with coordinates -2 and 3, has been *plotted* (drawn) in this diagram. We can name point B as the point $(-2, 3)$. Can you name the coordinates of points C and D? (Remember that the x-coordinate is always named first.)

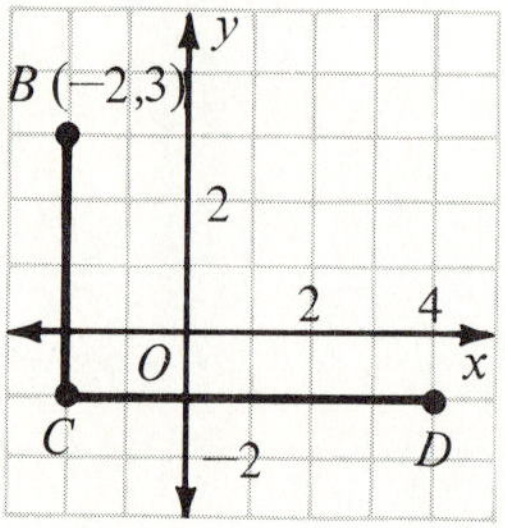

From the diagram you can conclude:

$BC = 4$ $\qquad CD = 6$ $\qquad \angle BCD = 90°$

Classroom Practice

Exercises 1–5 refer to points that are labeled in Figure 1.

1. Point B lies on an axis, not in a quadrant. Name two other points that do not lie in quadrants.
2. Name the points that lie in:
   a. quadrant I. b. quadrant II.
   c. quadrant III. d. quadrant IV.
3. Name the points that have:
   a. x-coordinate 0.
   b. a positive x-coordinate.
   c. a negative x-coordinate.

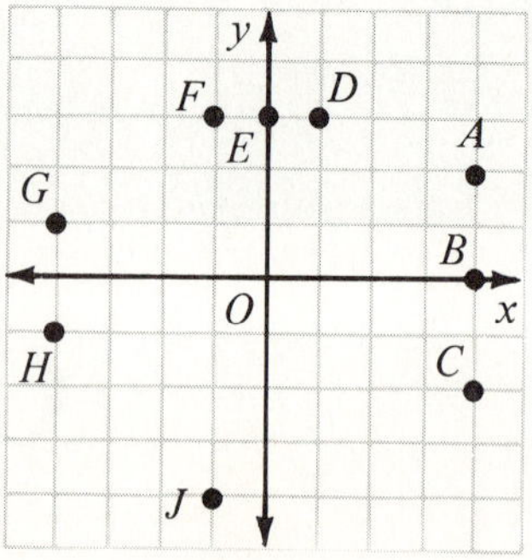

Figure 1

4. Name the points in Figure 1 that have:
   a. y-coordinate 0.
   b. a positive y-coordinate.
   c. a negative y-coordinate.

5. State the coordinates of each point in Figure 1.

6. Think of all the points that lie on the x-axis. What can you say about the y-coordinate of each of those points?

7. Think of all the points that have an x-coordinate equal to 0. What can you say about each of those points?

8. From the diagram, what can you conclude about ST? about $\angle RTS$?

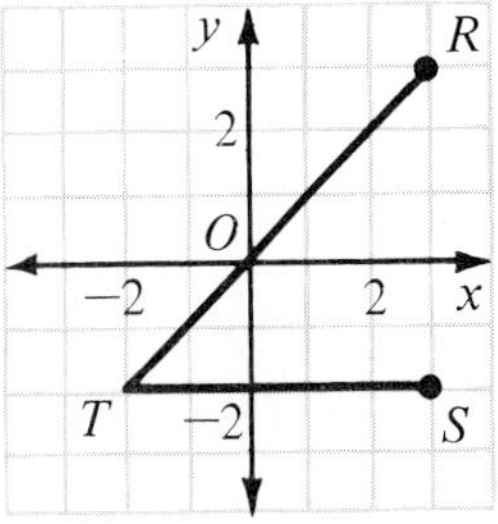

Exercise 8

The legs of the right triangle are parallel to the coordinate axes. State the coordinates of point C.

9.

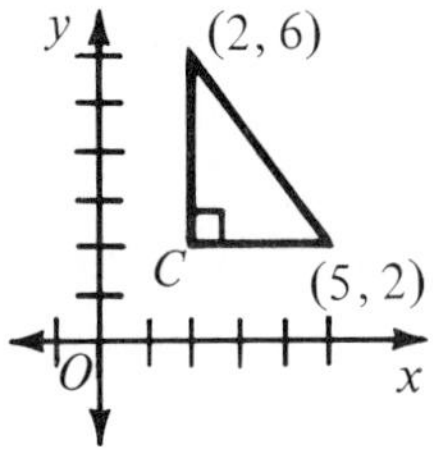

10.

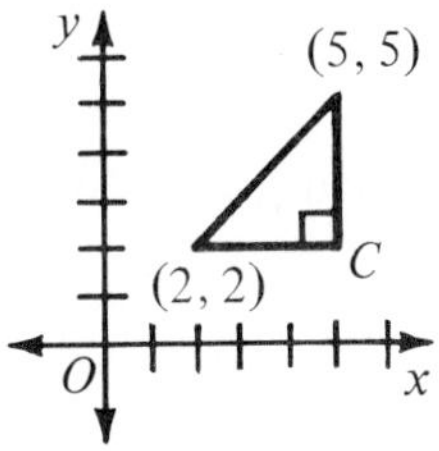

11.

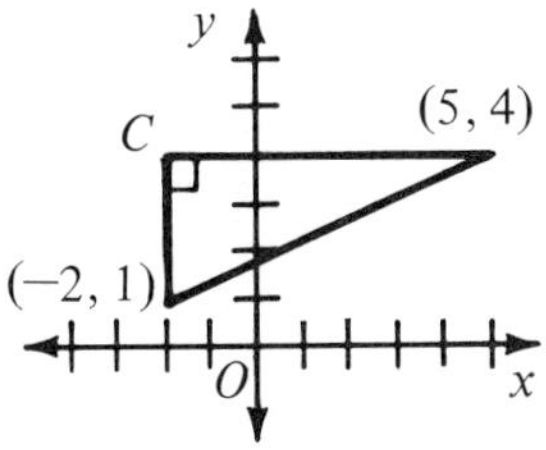

12.

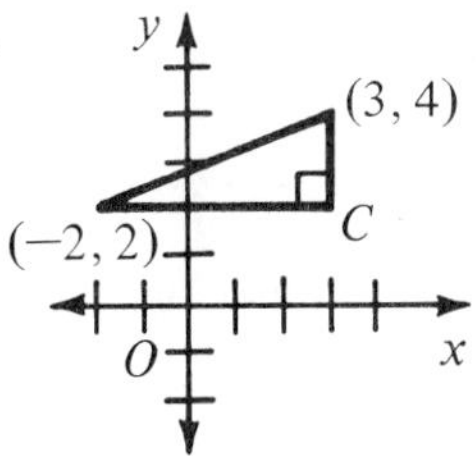

13.

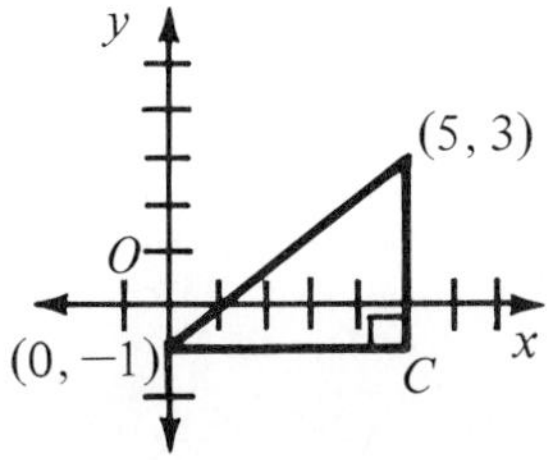

14.

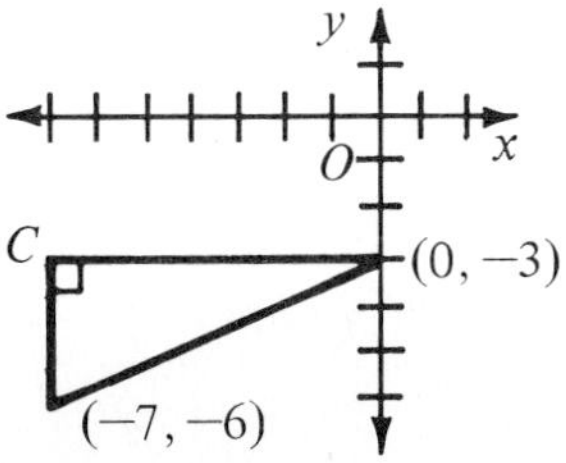

Written Exercises

A **1.** Point (3, −1) lies in quadrant __?__, and point (−3, 1) lies in quadrant __?__.

2. Are (−5, −4) and (−4, −5) the same point? Do the points lie in the same quadrant?

3. If the x-coordinate and the y-coordinate of a point are both negative numbers, the point must lie in quadrant __?__.

4. Point (−4, 0) lies on the __?__-axis.

5. If the x-coordinate of a point is 0, the point must lie on the __?__-axis.

6. Use the diagram to find the coordinates of each of the labeled points.

SAMPLE $Z(3, -2)$

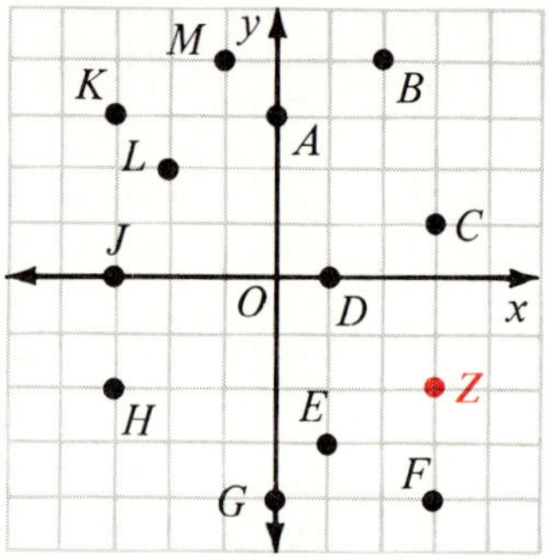

Exercise 6

For each exercise, draw a pair of coordinate axes.
a. Plot points R and S. Draw $\overleftrightarrow{RS}$. **b. State which point (A, B, or C) lies on $\overleftrightarrow{RS}$.**

	R	S	A	B	C
7.	(−2, −3)	(−2, 1)	(2, 2)	(4, −2)	(−2, −2)
8.	(0, 0)	(−3, 2)	(3, 2)	(−6, 4)	(1, 1)
9.	(1, 1)	(2, 2)	(5, 0)	(6, 7)	(−3, −3)
10.	(5, 5)	(−5, 0)	(−7, −1)	(−2, 1)	(7, 7)

In Exercises 11–16, a base of the rectangle is parallel to the x-axis. State the coordinates of points J and K.

11.
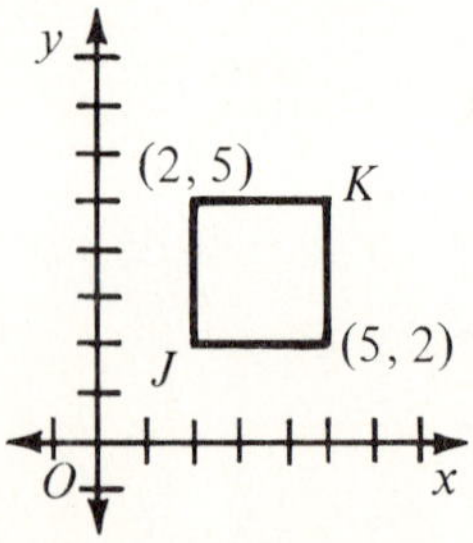

12.
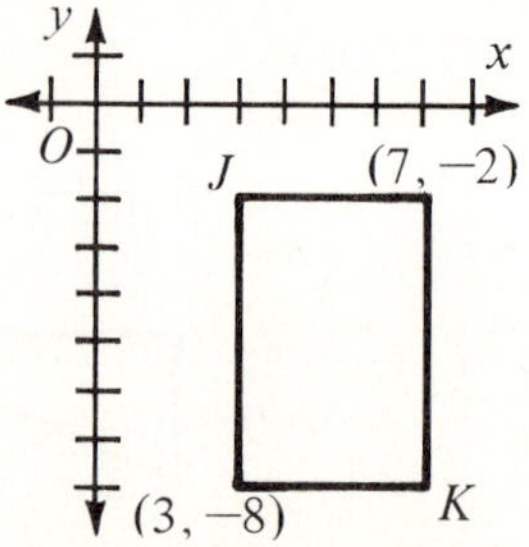

13.
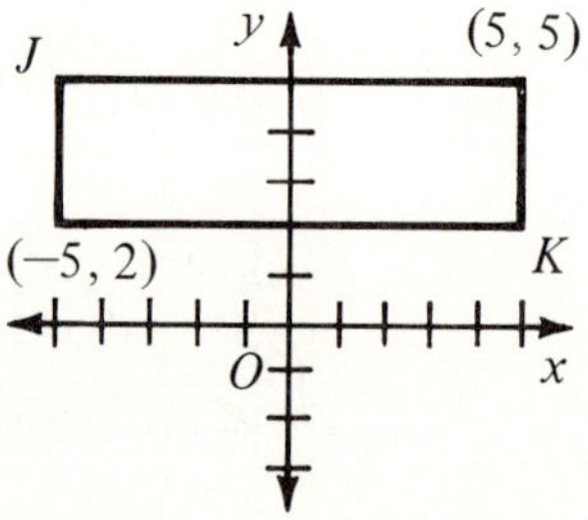

14.

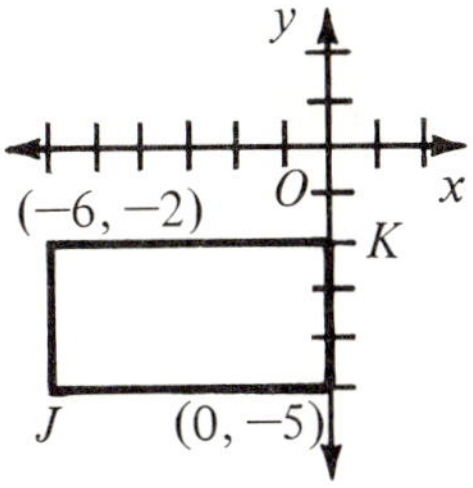

15.

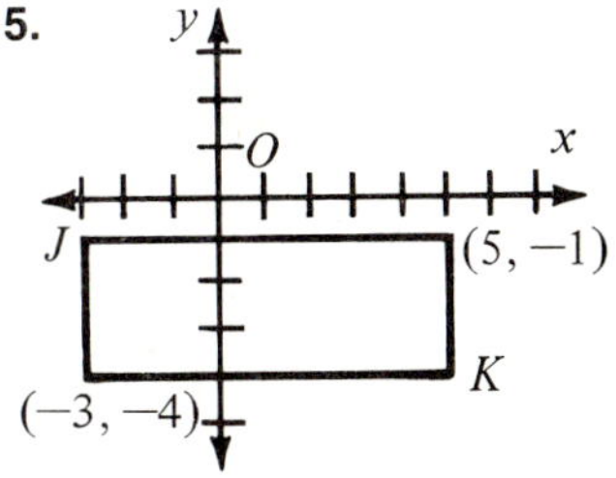

16.

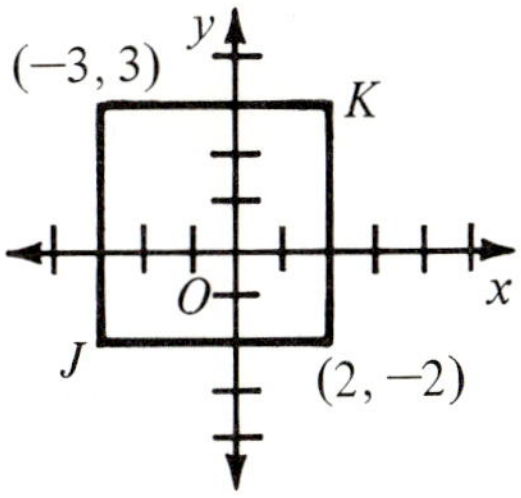

In Exercises 17–22, find the distance between the points named.

17. (4, 0) and (9, 0)
18. (0, 7) and (0, 2)
19. (5, 2) and (2, 2)
20. (4, 1) and (4, 6)
21. (1, 3) and (−5, 3)
22. (−2, 4) and (−2, −3)

Plot the four points and draw quadrilateral *ABCD*. State whether the quadrilateral is a square, rectangle, parallelogram, or trapezoid. (Use the name that gives the best description of the figure.)

	A	*B*	*C*	*D*
B **23.**	(−3, −2)	(6, −2)	(5, 1)	(−1, 1)
24.	(2, 1)	(6, 3)	(6, 0)	(2, −2)
25.	(−1, −1)	(1, 1)	(5, −3)	(3, −5)
26.	(3, 2)	(5, −1)	(8, 1)	(6, 4)

Plot points *R* and *S* and draw $\overleftrightarrow{RS}$. Point *T* is to lie on $\overleftrightarrow{RS}$, but only one coordinate of *T* is given. Find the other coordinate of *T*.

	R	*S*	*T*
27.	(0, −2)	(−3, 0)	(−6, ?)
28.	(9, 9)	(6, 4)	(0, ?)
29.	(6, −5)	(10, 5)	(?, 0)
30.	(−6, −1)	(9, −7)	(?, −3)

C **31.** Points (0, 0), (0, 5), and (3, 4) are three of the vertices of a parallelogram. Find three possible positions for the fourth vertex.

32. Repeat Exercise 31 but use points (−1, 2), (5, 0), and (7, 4).

2 Distance between Two Points

You can count squares to check the lengths of $\overline{RT}$ and $\overline{ST}$ marked in the diagram. You cannot find the length of $\overline{RS}$ by counting squares. However, you can think this way:

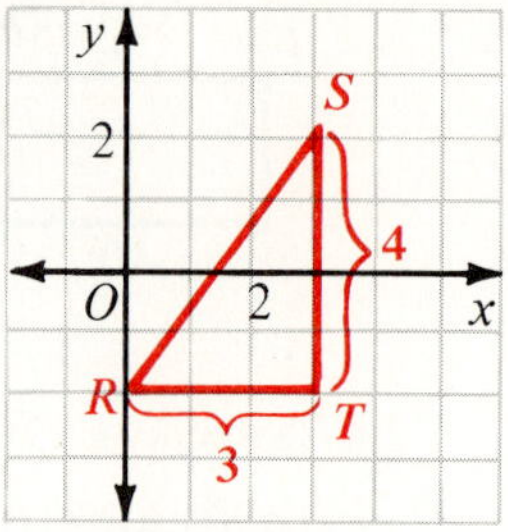

$\triangle RST$ is a right triangle.
The lengths of the legs are 3 and 4.
Thus, the length RS of the hypotenuse must be 5.

EXAMPLE Find the distance between $A(-2, 2)$ and $B(5, -1)$.

Plot points A and B. Then draw horizontal and vertical segments to form right $\triangle ACB$. Find AC and BC by counting.

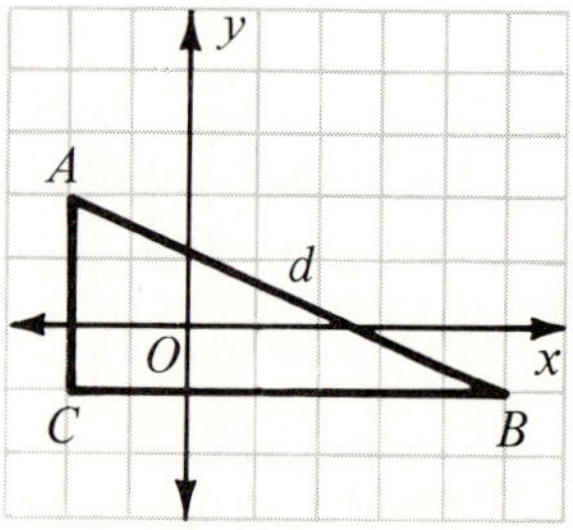

Use the Pythagorean Theorem.

$AC = 3 \qquad BC = 7$
$d^2 = 3^2 + 7^2$
$d^2 = 9 + 49 = 58$
$d = \sqrt{58}$

By using the square root table on page 516, you can approximate d to the nearest tenth.

$$d = \sqrt{58} \approx 7.6$$

Classroom Practice

State the lengths of the two legs of the right triangle.

1. 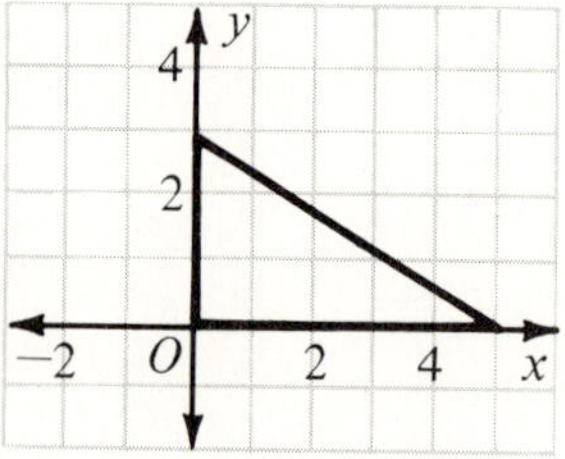

2. 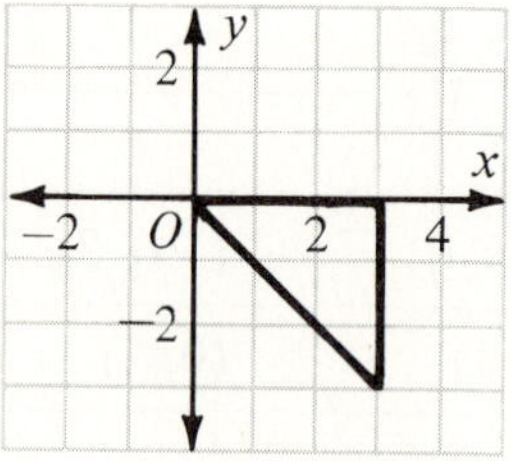

3. 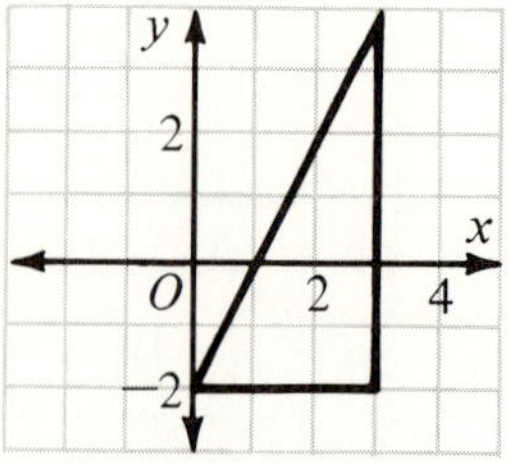

4. 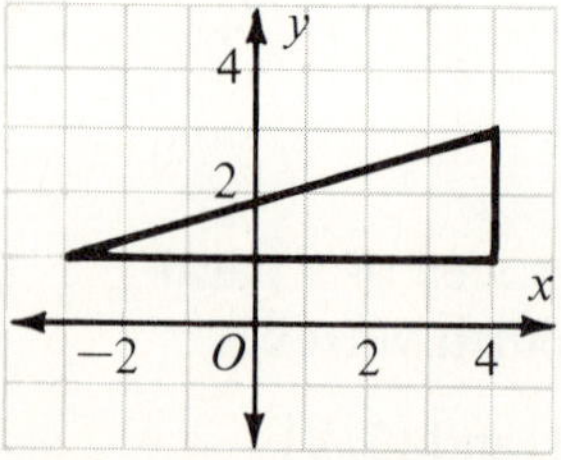

5. 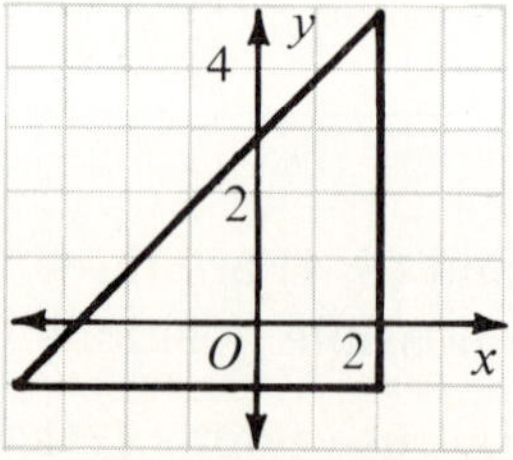

6. 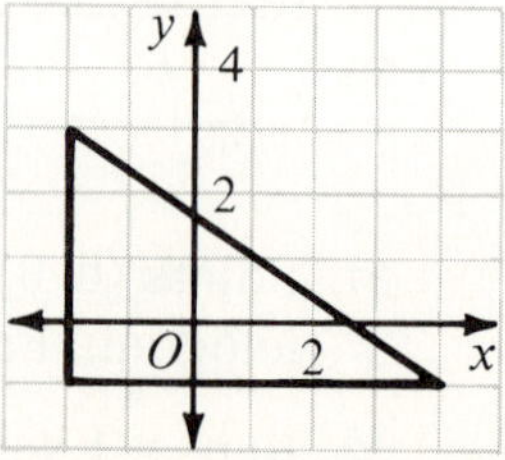

7. Leg $\overline{RT}$ of the right triangle is parallel to the x-axis. Answer in terms of a, b, and c.

   a. The x-coordinate of point T is __?__.

   b. The y-coordinate of point T is __?__.

   c. The length of $\overline{RT}$ is __?__.

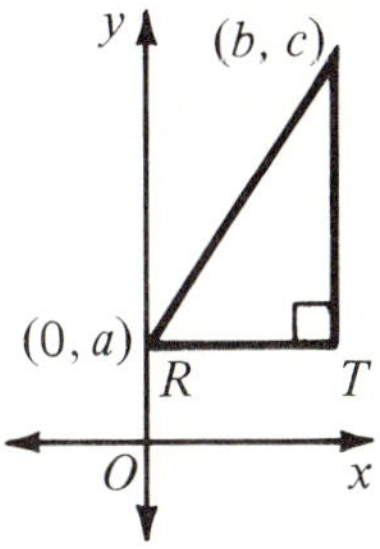

8. Leg $\overline{VW}$ of the right triangle is parallel to the x-axis. Answer in terms of j, k, p, and q.

   a. The x-coordinate of point W is __?__.

   b. The length of $\overline{VW}$ is __?__.

   c. The y-coordinate of point W is __?__.

   d. The length of $\overline{UW}$ is __?__.

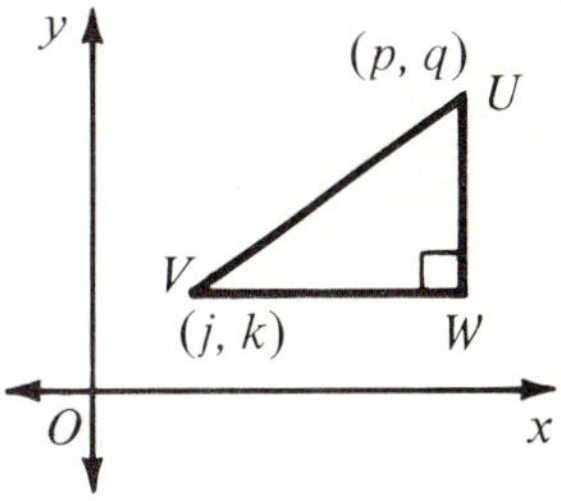

Written Exercises

In Exercises 1–12, draw axes and plot points A and B. Then draw a right triangle you can use to find AB. Then find AB. Leave your answer in radical form if it isn't a whole number.

SAMPLE $A(-3, 1)$ and $B(5, 4)$

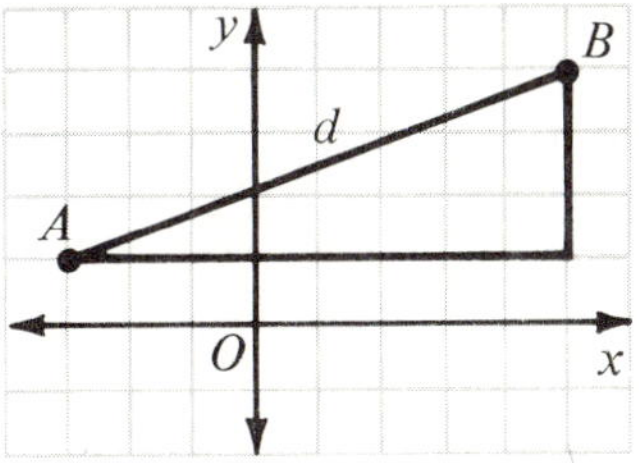

$d^2 = 8^2 + 3^2$
$d^2 = 64 + 9$
$d^2 = 73$
$d = \sqrt{73}$

A

1. $A(0, 4)$ and $B(5, 2)$
2. $A(2, 2)$ and $B(6, 7)$
3. $A(-2, -1)$ and $B(3, 5)$
4. $A(-5, 3)$ and $B(4, -4)$
5. $A(2, 1)$ and $B(4, 2)$
6. $A(3, 5)$ and $B(5, 8)$
7. $A(-2, 1)$ and $B(-5, 4)$
8. $A(-6, 7)$ and $B(-4, -2)$
9. $A(2, 4)$ and $B(-2, 1)$
10. $A(-3, 3)$ and $B(1, 6)$
11. $A(4, -2)$ and $B(-2, 3)$
12. $A(-2, -3)$ and $B(1, 5)$

Find the distance between points A and B correct to tenths. Use the square root table on page 516.

B **13.** $A(-2, -2)$ and $B(3, 3)$ **14.** $A(2, -2)$ and $B(-4, 4)$

15. $A(-4, -3)$ and $B(-6, -7)$ **16.** $A(0, 6)$ and $B(-4, -1)$

Find the perimeter and the area of the triangle shown.

17.

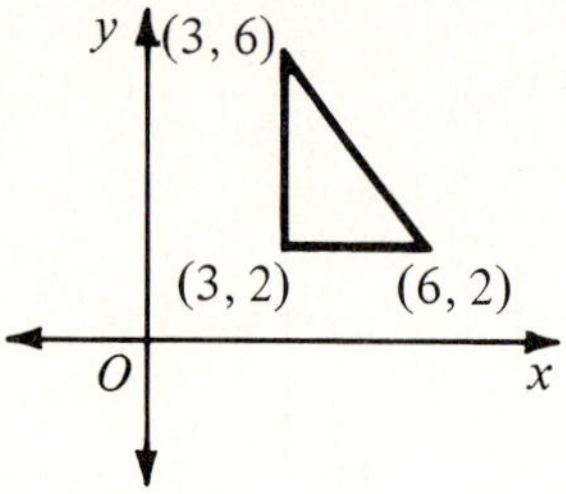

18.

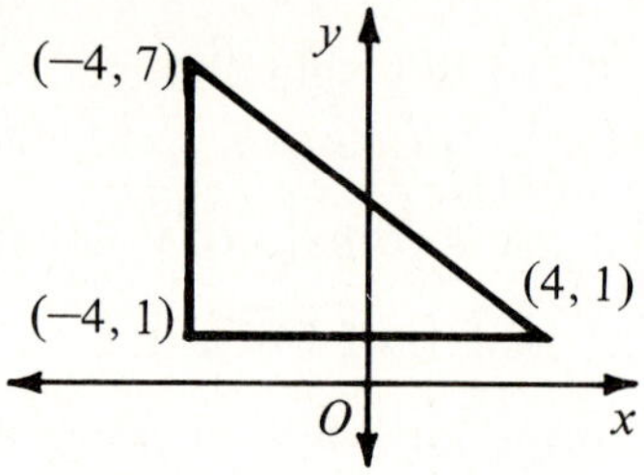

19.

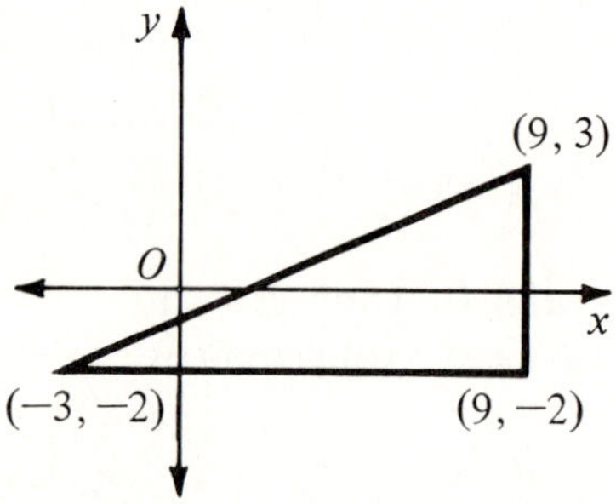

20.

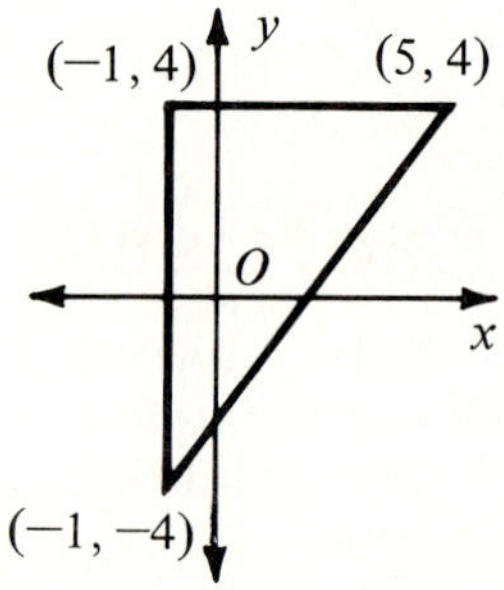

Find the perimeter and area of the square with the given vertices.

C **21.** Vertices: (0, 0); (4, 3); (1, 7); (−3, 4)

22. Vertices: (2, −1); (4, 1); (6, −1); (4, −3)

Find the perimeter and the area of the quadrilateral shown.

23.

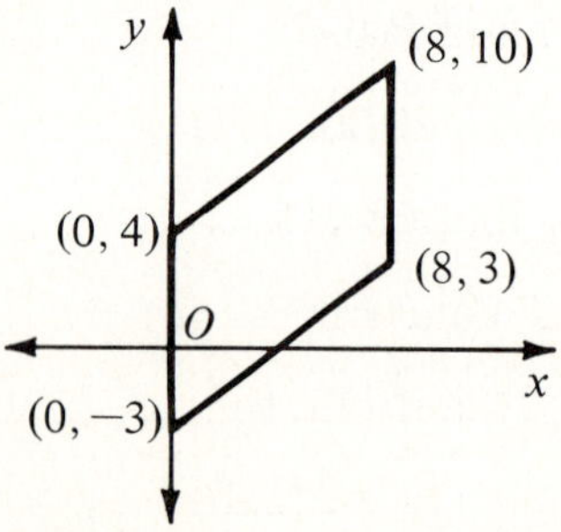

24.

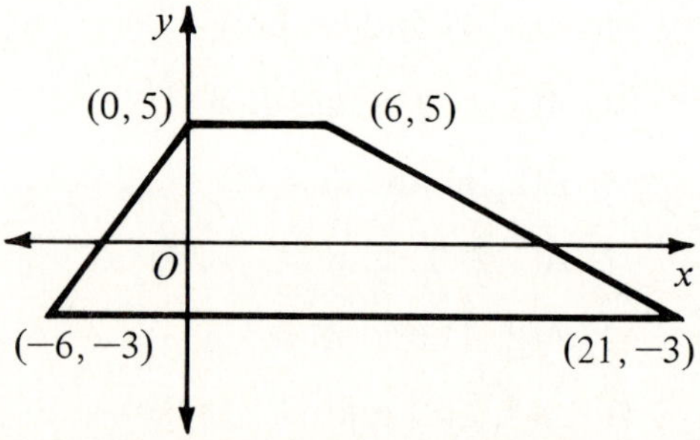

CALCULATOR ACTIVITIES

The rise, run, and length of a ramp are dependent upon each other. When one changes, the others do also.

The tangent ratio can be used to describe the slope of the ramp.

$$\text{Slope} = \frac{\text{rise}}{\text{run}}$$

$$= \frac{\text{leg opposite } \angle A}{\text{leg adjacent to } \angle A}$$

$$= \text{tangent of } \angle A$$

Similarly, $\frac{\text{rise}}{\text{ramp length}} = \text{sine of } \angle A$

and $\frac{\text{run}}{\text{ramp length}} = \text{cosine of } \angle A$.

Assume that $\angle A$ is known in each case. State which trigonometric ratio you would use to find the unknown length.

1. Known: rise — Unknown: **a.** run **b.** ramp length
2. Known: run — Unknown: **a.** rise **b.** ramp length
3. Known: ramp length — Unknown: **a.** rise **b.** run

Use a calculator to find the trigonometric ratios for each measure of $\angle A$ to four decimal places. Then calculate the missing lengths, rounding to one decimal place.

	$\angle A$	sin $\angle A$	cos $\angle A$	tan $\angle A$	Rise	Run	Ramp
4.	7°	?	?	?	2 m	?	?
5.	12°	?	?	?	?	10 m	?
6.	20°	?	?	?	?	?	5 m

3 Midpoint of a Segment

You should remember that the **average** of two numbers is equal to half the sum of the numbers.

EXAMPLE 1 The average of 4 and 10 is $\frac{4+10}{2} = \frac{14}{2} = 7$.

EXAMPLE 2 The average of 4 and -10 is $\frac{4+(-10)}{2} = \frac{-6}{2} = -3$.

EXPLORATIONS

Part A

1. Points R and S lie on a number line. Find the coordinate of the midpoint of $\overline{RS}$.

a. 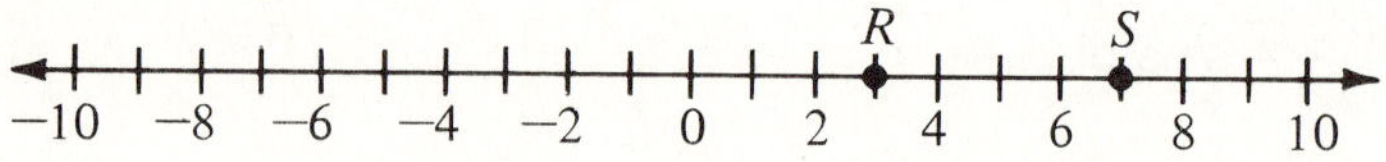

b. 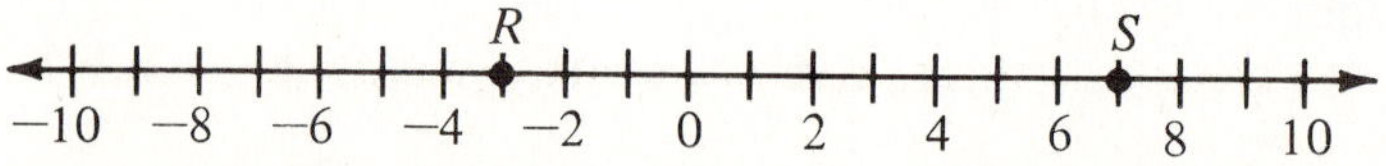

c. 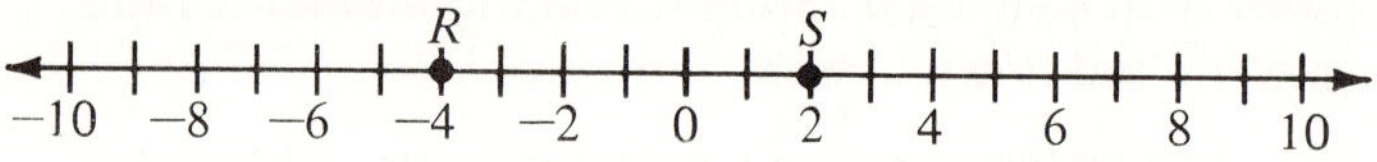

d. 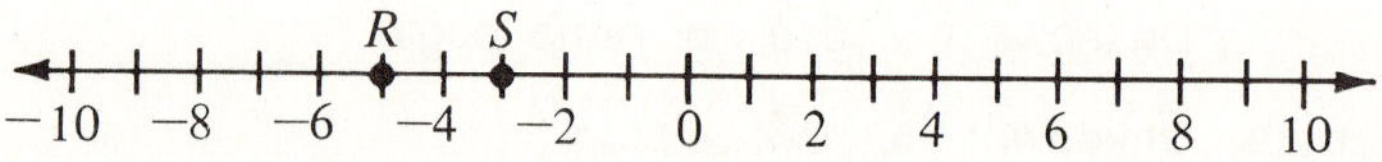

2. Suppose, in Exercise 1, that each number line were vertical rather than horizontal. Would the answers be any different? If necessary, draw vertical number lines to check your answer.

3. Find the average of each pair of numbers.
   a. 3 and 7 b. -3 and 7 c. -4 and 2 d. -5 and -3

4. Compare your answers to Exercises 1 and 3. Complete the following statement about points R and S on a number line:
   The coordinate of the midpoint of $\overline{RS}$ is equal to the __?__ of the coordinates of R and S.

Part B

Given: $\overline{AB}$ with midpoint M; the horizontal and vertical segments shown.

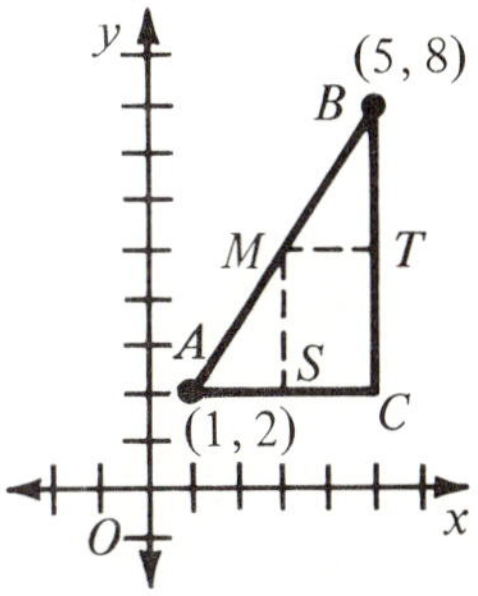

1. The x-coordinate of point C is __?__.
2. Because S is the midpoint of $\overline{AC}$, the x-coordinate of S is __?__.
3. The x-coordinate of M is __?__.
4. You know that the x-coordinate of A is 1, and the x-coordinate of B is 5. Then what is a simple way of finding the x-coordinate of midpoint M?
5. The y-coordinate of A is 2 and of B is 8. The y-coordinate of M is __?__. (*Hint:* Use point T.)

The explorations above suggest this strategy:

> *To find the coordinates of the midpoint of a segment when you know the coordinates of the endpoints:*
> *Find the average of the x-coordinates.*
> *Find the average of the y-coordinates.*

A convenient way to name points is suggested by the diagram at the right.

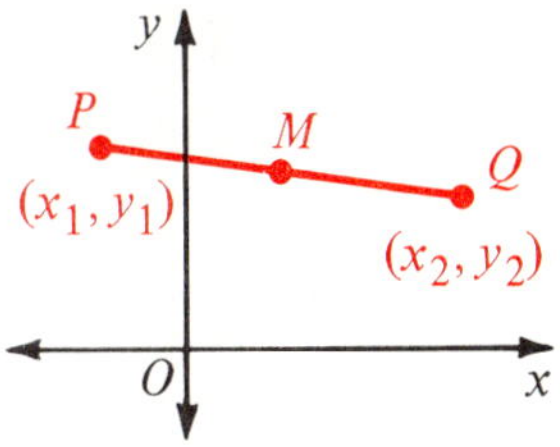

Point P has coordinates (x_1, y_1).

This is read "x-sub-1, y-sub-1."

> ***THE MIDPOINT FORMULA***
>
> *Let $P(x_1, y_1)$ and $Q(x_2, y_2)$ be any two points. Then the midpoint of $\overline{PQ}$ is the point* $\left(\frac{x_1 + x_2}{2}, \frac{y_1 + y_2}{2}\right)$.

EXAMPLE 1 Find the coordinates of the midpoint of the segment that joins points (1, 2) and (5, 8).

The coordinates are $\left(\frac{1+5}{2}, \frac{2+8}{2}\right)$, or (3, 5).

EXAMPLE 2 Find the coordinates of the midpoint of the segment that joins points $(-2, 4)$ and $(-6, 7)$.

The coordinates are $\left(\frac{-2+(-6)}{2}, \frac{4+7}{2}\right)$, or $\left(\frac{-8}{2}, \frac{11}{2}\right)$, or $\left(-4, 5\frac{1}{2}\right)$.

Classroom Practice

1. State the x-coordinate of the midpoint of each segment.

a.

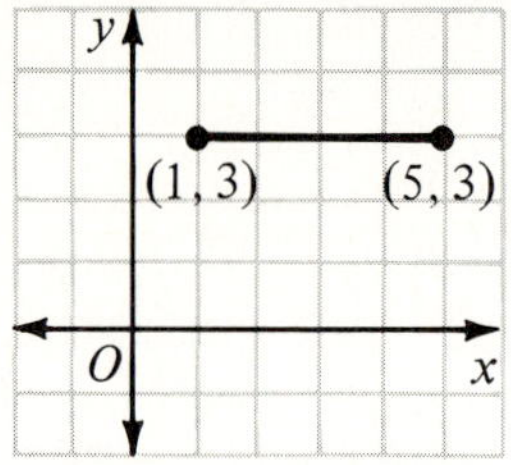

b.

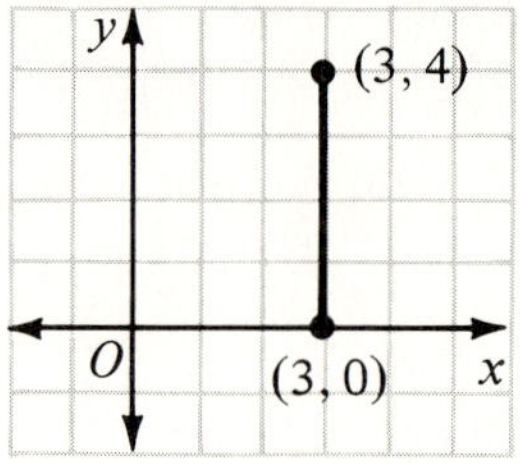

c.

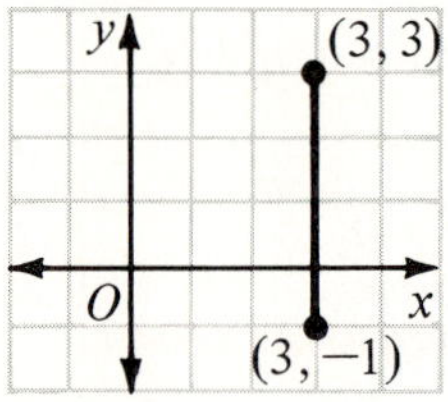

d.

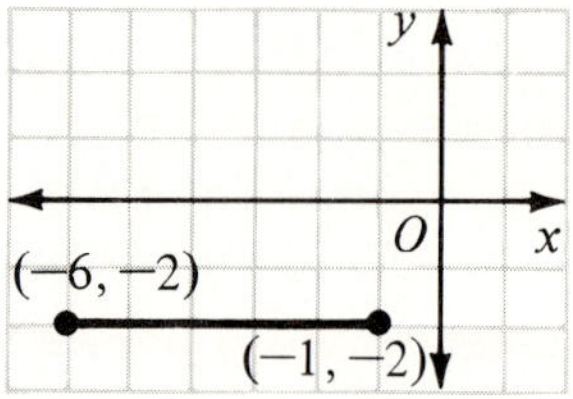

2. State the y-coordinate of the midpoint of each segment shown in Exercise 1.

3. State the coordinates of the midpoint of each segment.

a.

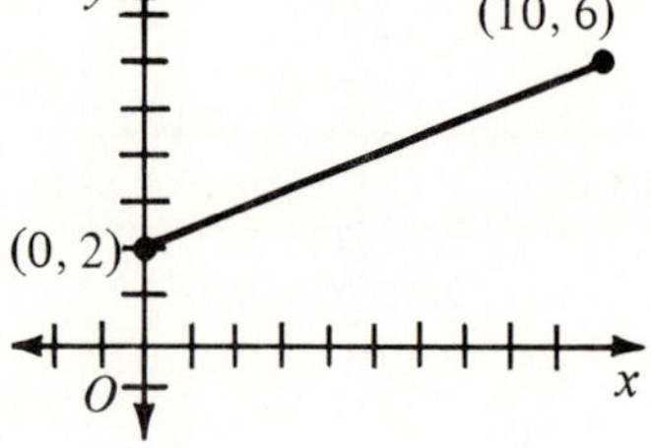

b.

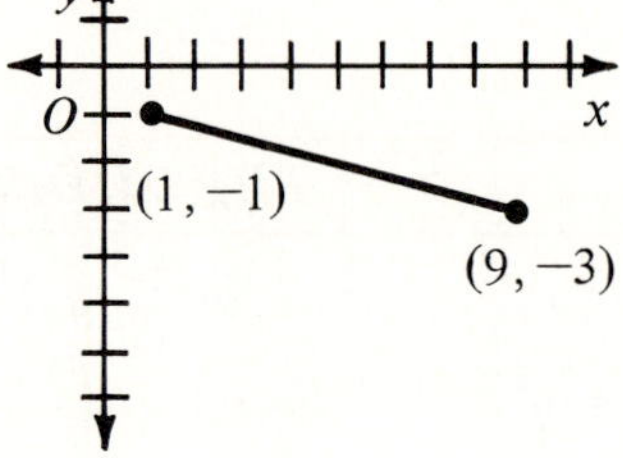

c.

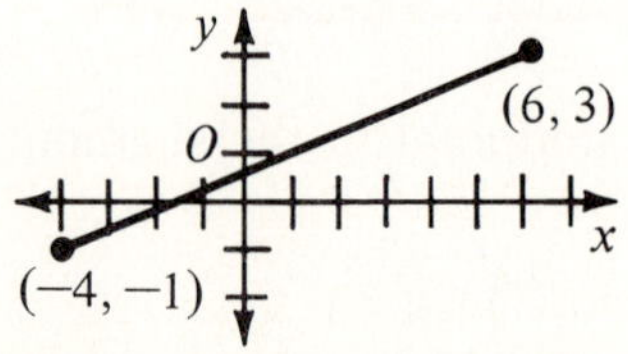

d.

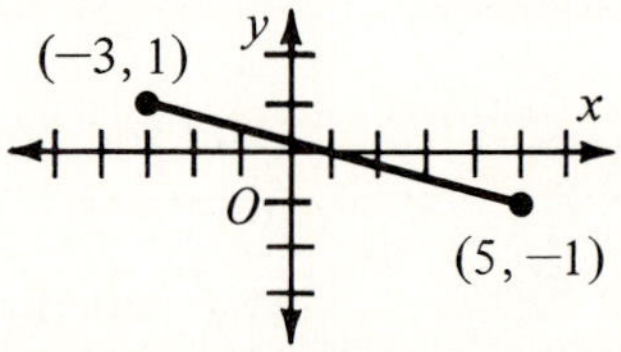

Written Exercises

Plot points *C* and *D*. Draw $\overline{CD}$. Mark the midpoint of $\overline{CD}$ and label it *M*. State the coordinates of *M*.

A

1. $C(2, 0)$ and $D(8, 6)$

2. $C(3, 1)$ and $D(-7, 1)$

3. $C(-5, 2)$ and $D(3, 6)$

4. $C(0, 1)$ and $D(6, 6)$

State the coordinates of the midpoint of the segment.

5.

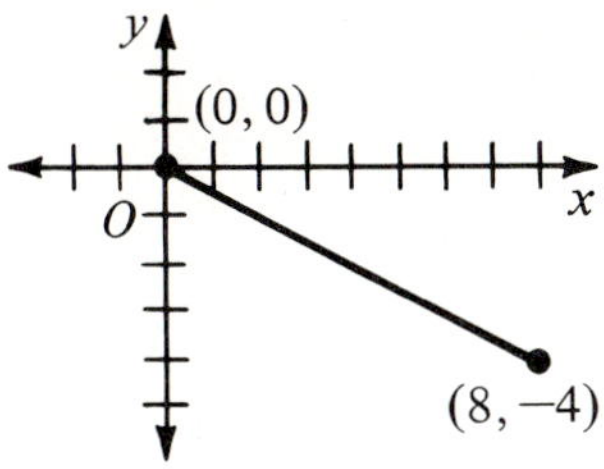

6.

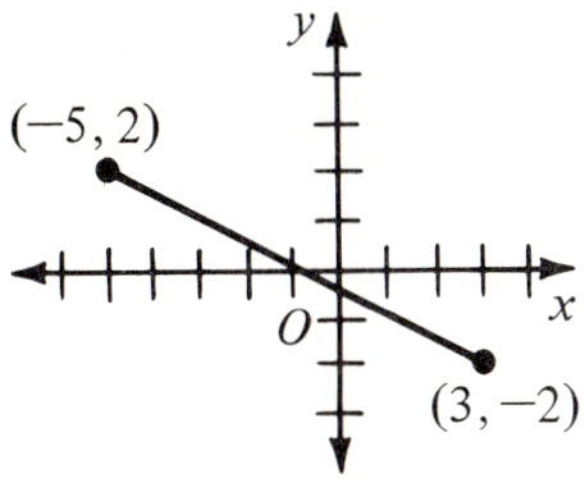

7.

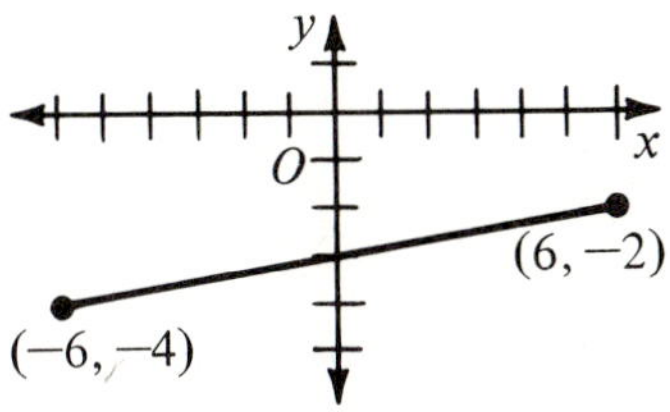

8.

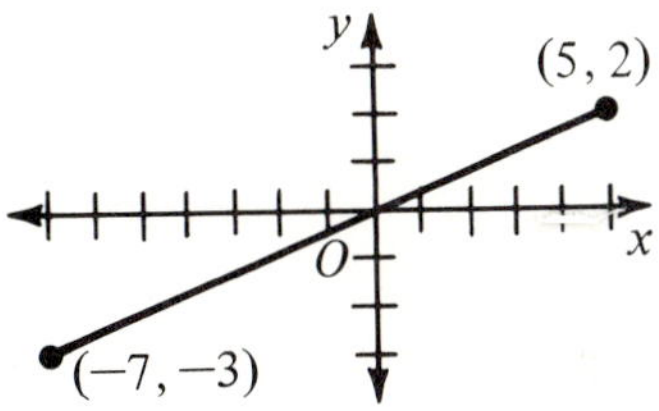

Plot points *E* and *M* and draw $\overleftrightarrow{EM}$. Mark point *F* in such a position that *M* is the midpoint of $\overline{EF}$. From your drawing state the coordinates of *F*.

9. $E(0, 3)$ and $M(4, 3)$

10. $E(5, -1)$ and $M(5, 2)$

11. $E(2, 2)$ and $M(5, 4)$

12. $E(-2, 1)$ and $M(1, 0)$

Use the midpoint formula to find the coordinates of the midpoint of the segment that joins the points named.

13. $(2, 5)$ and $(8, 1)$

14. $(0, -6)$ and $(4, -2)$

15. $(-3, -5)$ and $(7, -5)$

16. $(11, 5)$ and $(-1, 7)$

17. $(2, 1)$ and $(8, 8)$

18. $(0, 5)$ and $(-6, -8)$

B

19. $(1, 2.5)$ and $(5, 3.2)$

20. $(-1, 1.6)$ and $(2, 6.4)$

21. $(0, -3)$ and $\left(1\frac{1}{2}, -7\right)$

22. $\left(-\frac{1}{3}, 0\right)$ and $\left(4\frac{1}{3}, 5\right)$

M is the midpoint of $\overline{AB}$. The coordinates of A and M are given. Use the midpoint formula to find the coordinates of B.

SAMPLE

$A(4, -3)$ and $M(6, 1)$

Let (t, u) be the coordinates of B.

$$\frac{4+t}{2} = 6 \qquad \frac{-3+u}{2} = 1$$
$$4 + t = 12 \qquad -3 + u = 2$$
$$t = 8 \qquad u = 5$$

The coordinates of point B are (8, 5).

23. $A(0, 3)$ and $M(5, 4)$ **24.** $A(-1, 1)$ and $M(2, 5)$

25. $A(1, 5)$ and $M(2, 2)$ **26.** $A(4, 1)$ and $M(2, 0)$

SELF-TEST

Vocabulary

coordinate axes (p. 468)
coordinate(s) (p. 468)
x-coordinate (p. 468)
y-coordinate (p. 468)
average (p. 476)
midpoint of a segment (p. 477)

The sides of rectangle *RSTU* are parallel to the coordinate axes.

1. Point T has coordinates (_?_, _?_). *(12-1)*

2. $RU =$ _?_ *(12-2)*

3. $SU =$ _?_

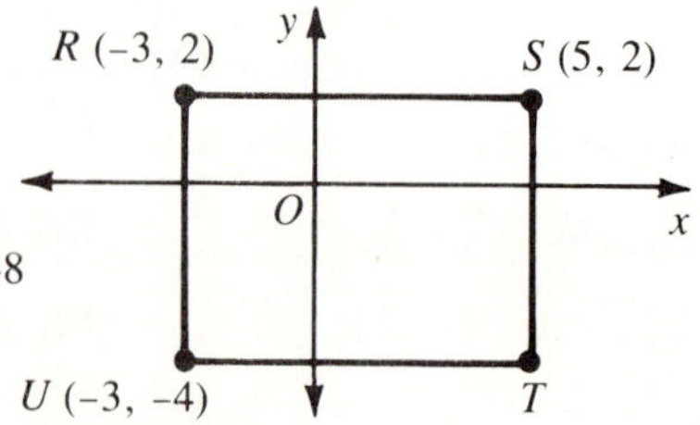

Exercises 1–3, 7–8

In Exercises 4–6, use points $A(-9, -5)$ and $B(3, 0)$.

4. The distance between the origin and point B is _?_.

5. The distance between A and B is _?_.

6. The midpoint of $\overline{AB}$ has coordinates (_?_, _?_). *(12-3)*

In Exercises 7–8, refer to the diagram above.

7. The midpoint of $\overline{RS}$ has coordinates (_?_, _?_).

8. The midpoint of $\overline{SU}$ has coordinates (_?_, _?_).

CALCULATOR ACTIVITIES

If you know two sides and the included angle of a triangle, you can find its area using this formula:

Area $= \frac{1}{2}$ (product of two sides) $\times$ (sine of included angle)

EXAMPLE

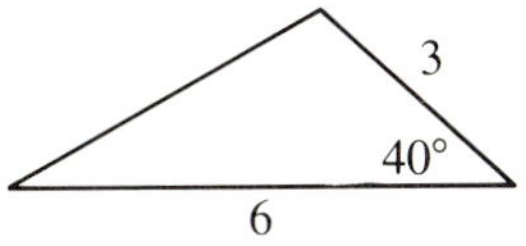

Area $= \frac{1}{2}(6 \times 3) \sin 40° = 5.787$

This formula can be used to find the area of any regular polygon. For example, the area of a regular octagon may be found by circumscribing a circle about it. Then the center of the circle is joined to the vertices of the polygon. The octagon is made up of 8 isosceles triangles.

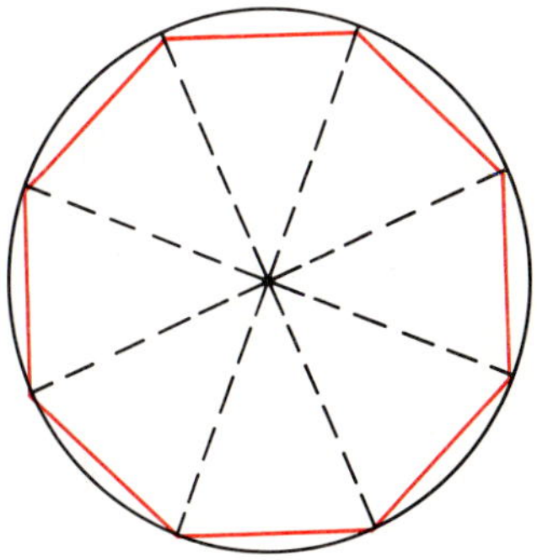

Each central angle measures $\frac{360°}{8} = 45°$.

Suppose the radius of the circle is 10.
Then,

$$\begin{aligned} \text{Area of octagon} &= 8 \times \text{Area of one triangle} \\ &= 8 \times \left[\frac{1}{2}(10 \times 10) \sin 45°\right] \\ &= 35.35 \end{aligned}$$

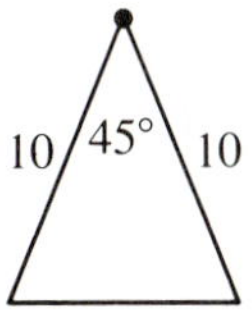

Use a calculator to find the area of each polygon. Round your answers to two decimal places.

1. Two sides of a triangle are 6 and 10 and the included angle is 70°.
2. Two sides of a triangle are 17 and 23 and the included angle is 41°.
3. A regular hexagon is inscribed in a circle with radius 10.
4. A regular polygon with 9 sides is inscribed in a circle with radius 1.
5. A regular polygon with 100 sides is inscribed in a circle with radius 4.

APPLICATIONS

LOCUS

Sometimes we need to identify the points that meet one or more requirements. For example, we might want to specify the service contour of a radio station, that is, the locations at which the radio signal can be received.

The mathematical name for specifying points is *locus*, the Latin word for "place." A **locus** is the group made up of all the points that satisfy some condition.

EXPLORATIONS

1. Draw any point P near the middle of a sheet of paper.
2. Locate any point Q that is 5 cm from P.
3. Locate several other points, each 5 cm from P.
4. Look at all the points that are 5 cm from P.
   What kind of figure do the points seem to form?
   Complete: The locus of points that are 5 cm from P is a __?__ with __?__ = 5 cm.

EXAMPLE 1 Specify the divider of State Highway 35 as a locus of points.

The divider line is the locus of points equidistant from the parallel edges of the road.

Example 1 shows that the locus of points equidistant from two parallel lines, and contained in the plane of the lines, is the line coplanar with the given lines that is halfway between them. The locus described in the Explorations, above, also consists of coplanar points.

Example 2 illustrates a locus of points that are not coplanar.

EXAMPLE 2 Find the locus of points *in space* that are 5 cm from a given point P.

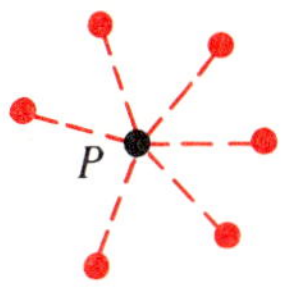

The locus is a sphere with center P and radius 5 cm.

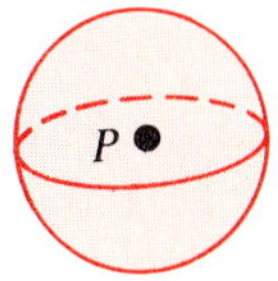

When you specify a locus you should include a clear description and a diagram of the locus.

Some loci (plural of *locus*) are determined by more than one condition, as illustrated in Example 3.

EXAMPLE 3 Mayberry Mall is shaped like a trapezoid, as shown. The architect decides to install a fountain in the mall. The fountain is to be equidistant from Store A and Store B. The fountain should also be the same distance from the specialty shops along the north and south sides of the mall. Find the location of the fountain.

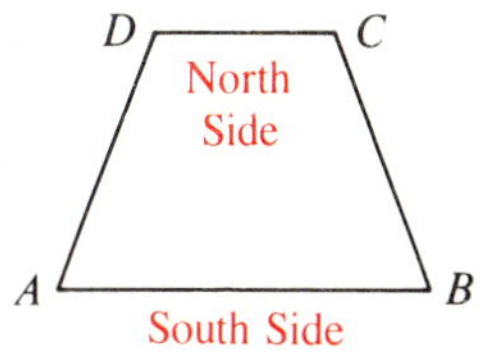

Step 1 The locus of points equidistant from A and B and in the plane of the mall is the perpendicular bisector of $\overline{AB}$.

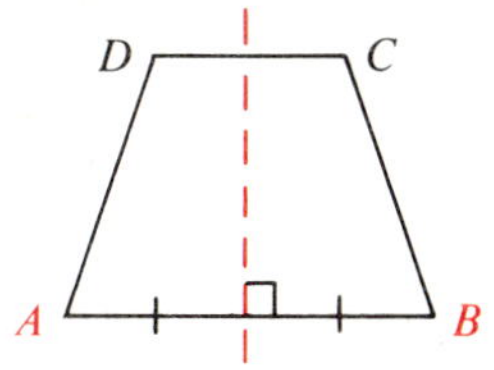

Step 2 The locus of points equidistant from the north ($\overline{CD}$) and south ($\overline{AB}$) sides of the mall is __?__. (*Hint:* $\overline{AB} \parallel \overline{CD}$)

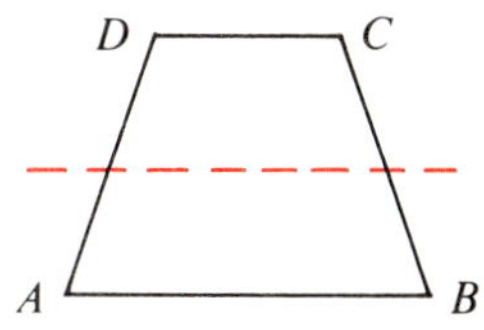

Step 3 The required locus must satisfy both conditions. The required locus is the intersection of the loci in Steps 1 and 2. The fountain should be located at point F.

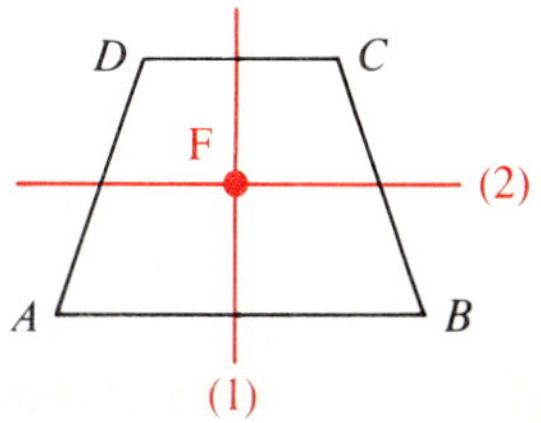

EXERCISES

1. a. Draw a line t on your paper.
   Locate many points that are 3 cm from t.
   b. Draw and label the locus of points that are in the plane of the paper and are 3 cm from t.
   c. Describe the locus.

2. Refer to Exercise 1. Draw and describe the locus of points in space that are 3 cm from t.

3. Sketch and describe the locus of points in space equidistant from two parallel planes R and S. (*Hint:* It may be helpful to use two pieces of paper to represent two parallel planes.)

4. a. Draw a large acute angle, $\angle ANG$.
   b. Find a point that is the same distance from $\overrightarrow{NA}$ as from $\overrightarrow{NG}$.
   c. Find many points that satisfy this condition.
   d. Describe the locus of points that are equidistant from the sides of an angle.

5. Refer to Step 1 of Example 3. State the theorem that allows you to conclude that the perpendicular bisector of $\overline{AB}$ contains the points that are equidistant from A and B.

6. Specify the plane of the net in a tennis court as a locus of points. (*Hint:* See Example 1.)

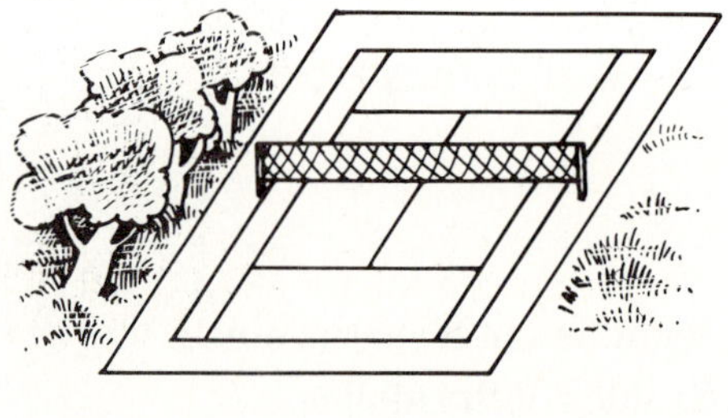

7. A chandelier is to be installed in a rectangular room.
   The light must be positioned at an equal distance from each of the four corners of the ceiling.
   Draw a diagram and write a description of the required position.

8. What is the locus of points in space that are equidistant from points J and K? Include a diagram and a description.

9. What is the locus in space of points that are 1 m from plane P?

10. Draw two points, R and S. Find the locus of points that are at a distance of RS from point R and lie in the plane of the paper.

11. A 7 m hose is attached to a 15 m by 12 m building at point X. Show the region that can be reached by the hose.

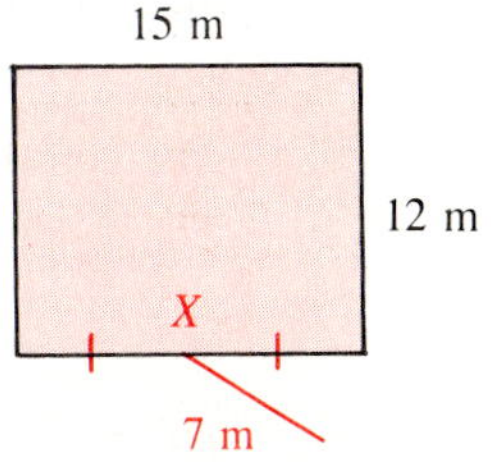

12. Repeat Exercise 11 if the hose is attached at point Y, shown at the right.

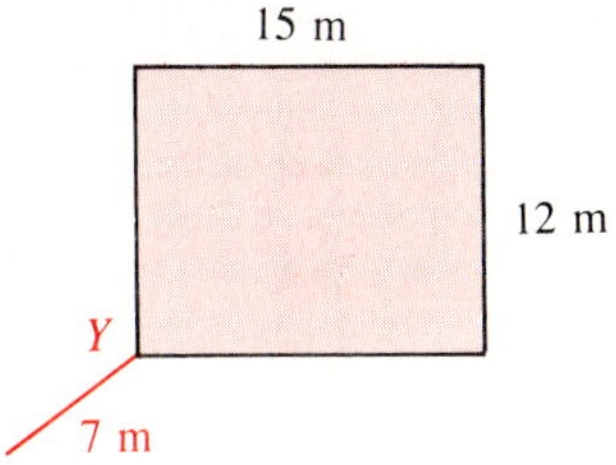

13. A regional recreation center is planned. The mayors of La Palma, Eastbury, and Azalia want the center to be an equal distance from the three towns. Find this location.

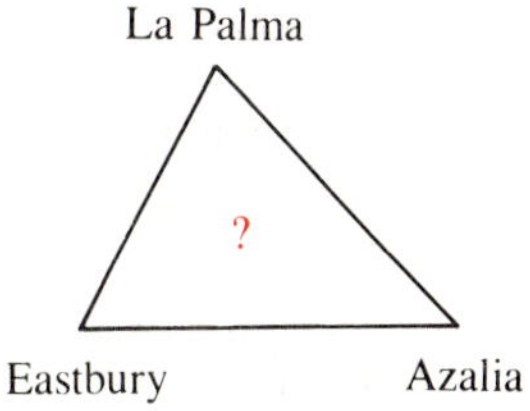

14. The Fongs have a triangular garden. They want to plant a rosebush in the garden. The rosebush should be an equal distance from the edges of the garden. Sketch the garden and determine the location of the rosebush. (*Hint:* The center of an inscribed circle is equidistant from the sides of a triangle.)

Cora and Octavio live 5 km from each other. How many possible meeting points do they have if they meet at a point that is:

15. equidistant from their houses?

16. 4 km from Cora's house and 3 km from Octavio's house?

17. 2 km from Cora's house and 1 km from Octavio's house?

18. 2.4 km from Cora's house and 2.6 km from Octavio's house?

19. Draw two points, X and Y. Construct (not draw) the locus of points in the plane of the paper that are equidistant from X and Y and at a distance of $2(XY)$ from point Y.

4 Slope of a Line

A car can make the climb at the left, but not the climb at the right.

Slope $= \frac{19}{100}$

Slope $= \frac{63}{100}$

The *slope* of a road is the ratio of *rise* to *run*.

Slope $= \dfrac{\textbf{rise}}{\textbf{run}}$

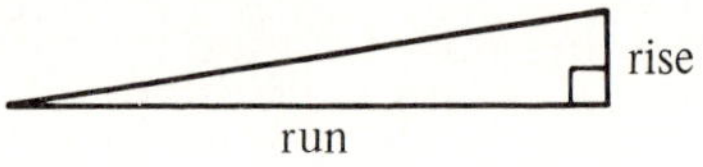

Lines have slopes, too.

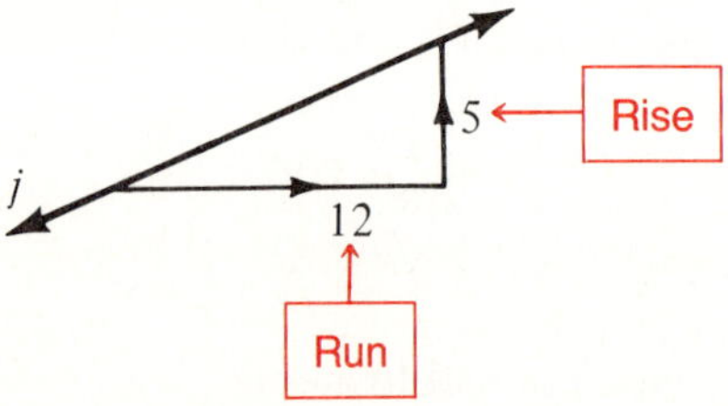

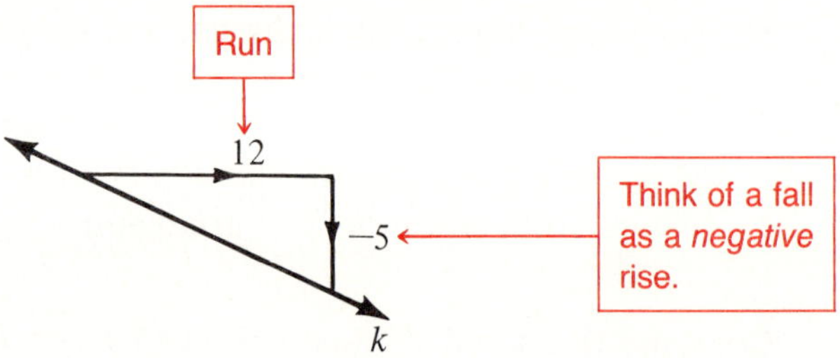

Slope of $j = \frac{\text{rise}}{\text{run}} = \frac{5}{12}$

Slope of $k = \frac{\text{rise}}{\text{run}} = \frac{-5}{12} = -\frac{5}{12}$

Count squares in this figure and you find that the rise is 4, the run is 3, and the slope is $\frac{4}{3}$. Instead of counting squares you can subtract coordinates, as shown in the figure.

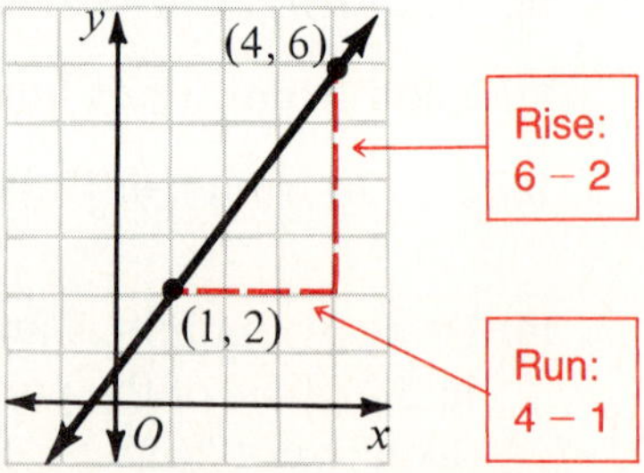

$$\text{Slope} = \frac{\text{rise}}{\text{run}} = \frac{6-2}{4-1} = \frac{4}{3}$$

The slope of the line that joins points (x_1, y_1) and (x_2, y_2) is defined by

$$\textit{Slope} = \frac{y_2 - y_1}{x_2 - x_1}$$

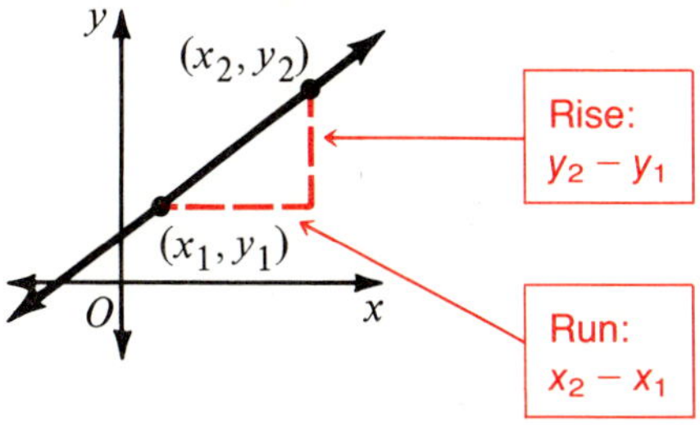

Here are some more facts that will help you understand the meaning of "slope."

1. When a line *rises from left to right,* the slope is a *positive* number.

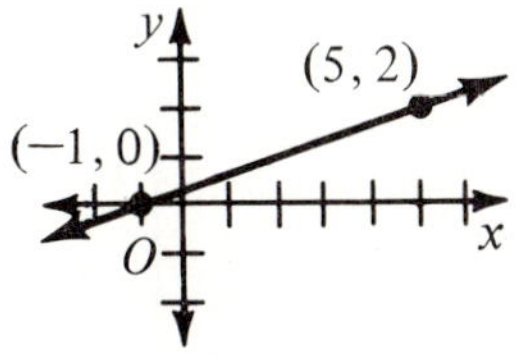

$$\frac{2-0}{5-(-1)} = \frac{2}{6} = \frac{1}{3}$$

2. When a line *falls from left to right,* the slope is a *negative* number.

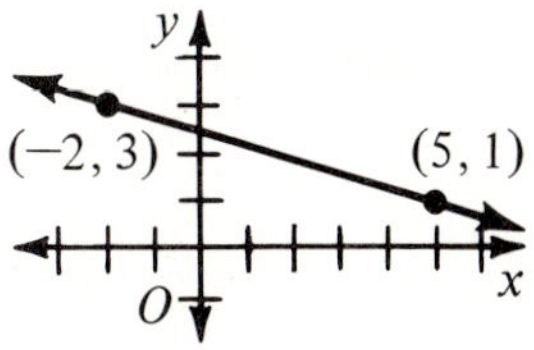

$$\frac{1-3}{5-(-2)} = \frac{-2}{7} = -\frac{2}{7}$$

3. A *horizontal* line does not rise or fall. The slope of a horizontal line is zero.

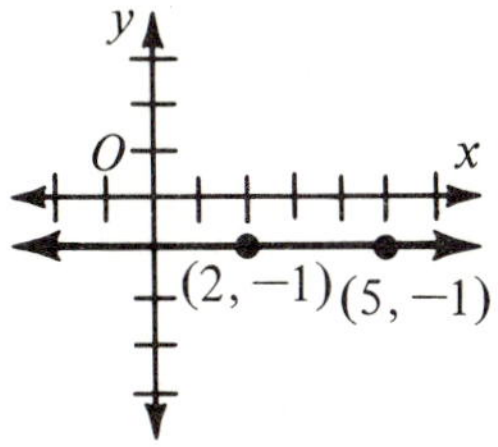

$$\frac{-1-(-1)}{5-2} = \frac{0}{3} = 0$$

4. The word "slope" does not apply to a *vertical* line. We say that the slope of a vertical line is *not defined.*

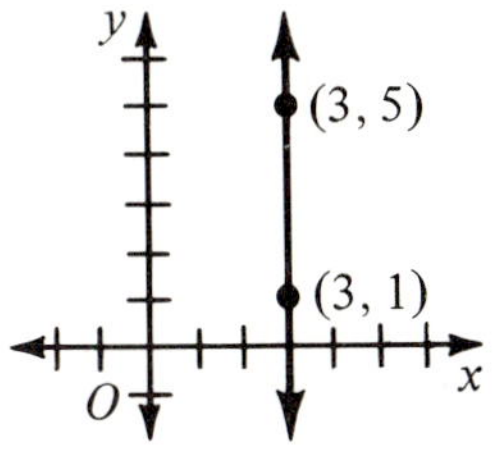

Note that the expression $\frac{5-1}{3-3}$ does not represent a number.

EXPLORATIONS

Part A

1. Think of going from A to B.

   Run = ___?___ Rise = ___?___

   Slope = $\frac{\text{rise}}{\text{run}}$ = ___?___

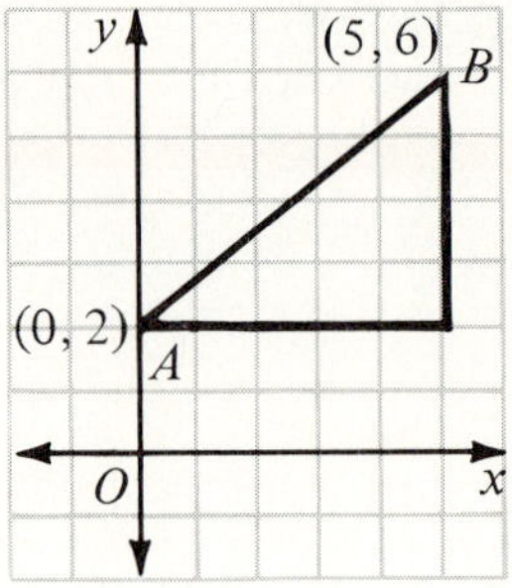

2. Think of going from B to A.

   Run = -5 Rise = -4

   Slope = $\frac{\text{rise}}{\text{run}}$ = ___?___

Part A suggests that when you compute the slope by using the ratio $\frac{\text{rise}}{\text{run}}$, you may start at *either* point. Test this with other points.

Part B

Copy and complete the table.

	(x_1, y_1)	(x_2, y_2)	$y_2 - y_1$	$x_2 - x_1$	Slope
1.	(0, 2)	(5, 6)	?	?	?
2.	(5, 6)	(0, 2)	?	?	?

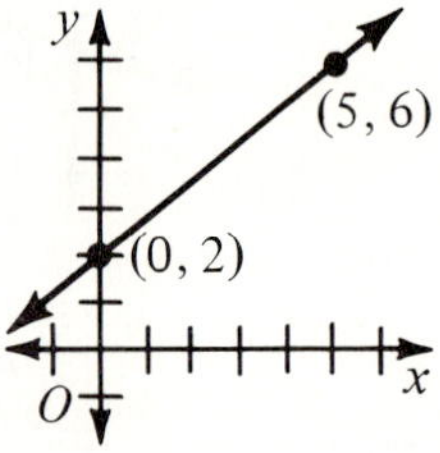

Part B suggests that when you compute the slope by using $\frac{y_2 - y_1}{x_2 - x_1}$, it doesn't matter which point you choose as (x_2, y_2).

Part C

Use the points named to compute the slope of the line.

Points	**Slope**
1. R and S	___?___
2. R and T	___?___
3. R and U	___?___
4. R and V	___?___
5. S and U	___?___
6. S and V	___?___

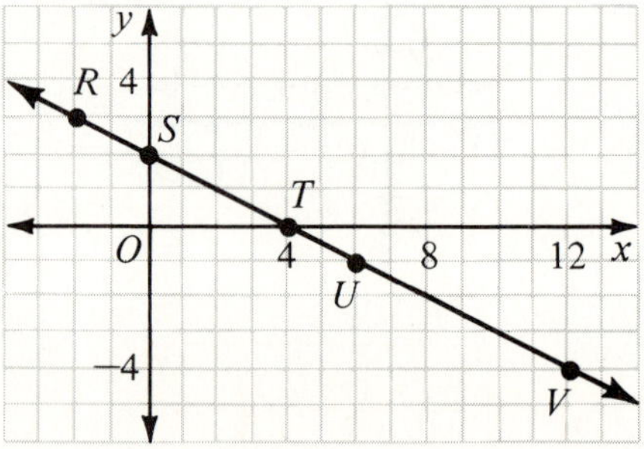

Part C suggests that when you compute the slope of a line, you may use *any* two points on the line.

Classroom Practice

For each diagram, think of going from point *R* to point *S*.

a. Tell whether the rise is positive, negative, or zero.

b. Tell whether the run is positive, negative, or zero.

c. Tell whether the slope is positive, negative, zero, or not defined.

1.

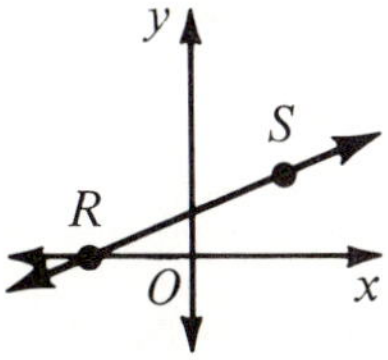

2.

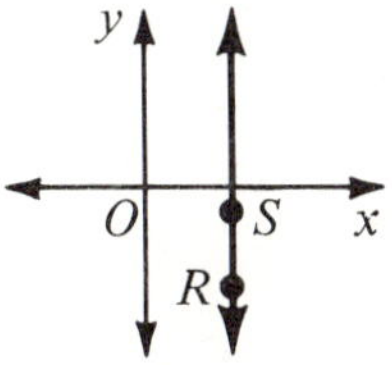

3.

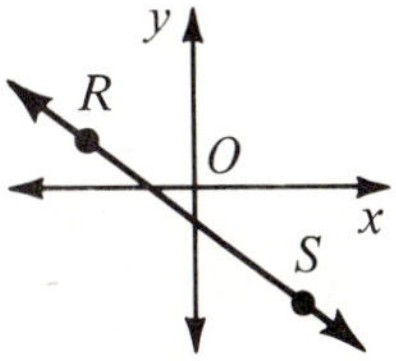

4.

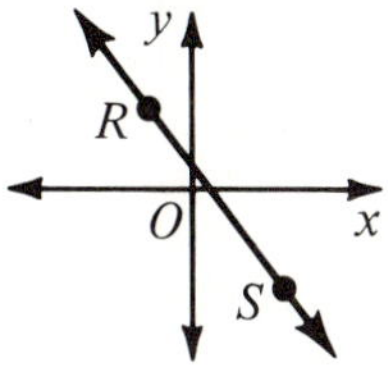

5.

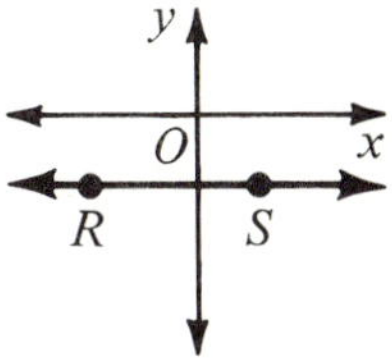

6.

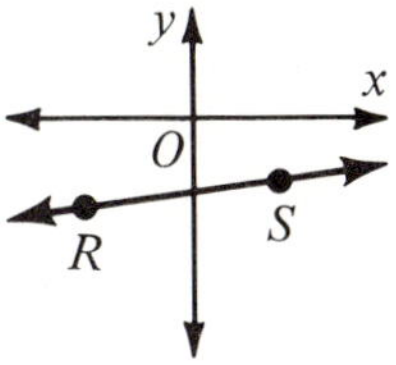

Written Exercises

A

1. Name the line that fits the description.

a. The slope is positive.
b. The slope is zero.
c. The slope is negative.
d. Slope is not defined for the line.

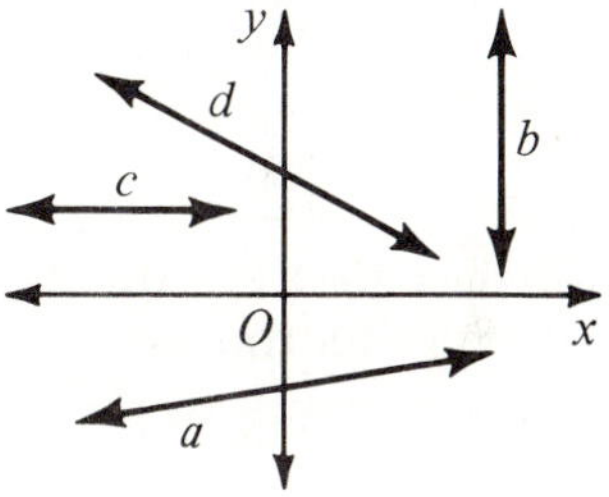

State the slope of the line shown. If the slope is not defined for the line, write *not defined*.

2.

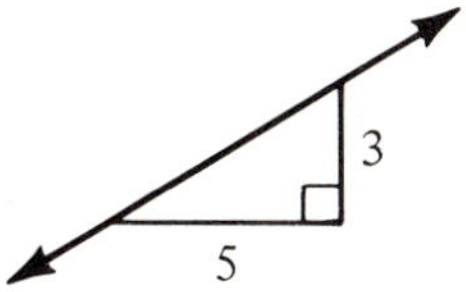

3.

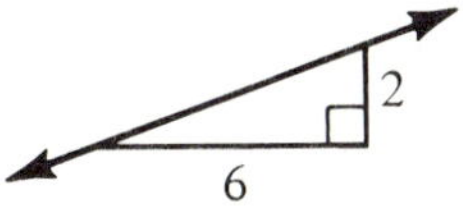

4.

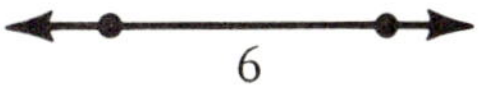

5.

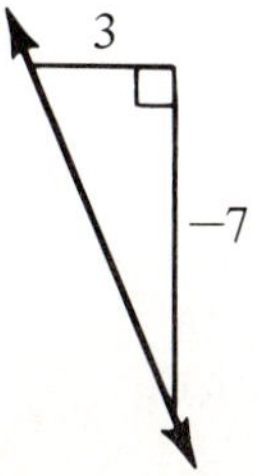

6.

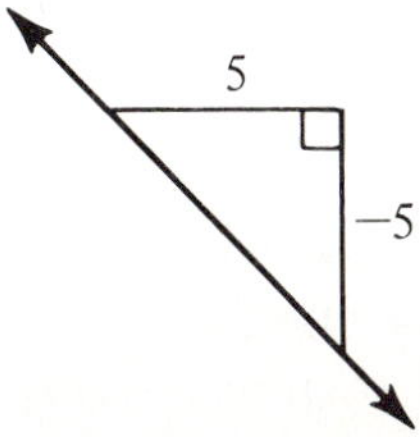

7.

Take P as point (x_1, y_1) and Q as point (x_2, y_2). Copy and complete the table. When $\overleftrightarrow{PQ}$ does not have a slope, write *not defined*.

	P	Q	$y_2 - y_1$	$x_2 - x_1$	Slope
8.	(0, 0)	(6, 5)	?	?	?
9.	(0, 0)	(−3, −7)	?	?	?
10.	(2, 1)	(6, −2)	?	?	?
11.	(6, −2)	(2, 1)	?	?	?
12.	(−3, −1)	(−5, 4)	?	?	?
13.	(−5, 4)	(−3, −1)	?	?	?

The coordinates of the vertices of quadrilateral *ABCD* are given. State which sides of the quadrilateral have positive slopes.

B

14. $A(-5, 2)$; $B(4, 6)$; $C(2, -3)$; $D(-1, -2)$

15. $A(3, 4)$; $B(6, -5)$; $C(-7, -1)$; $D(-2, 6)$

16. $A(8, -10)$; $B(-6, -2)$; $C(-4, 6)$; $D(10, -2)$

17. $A(-5, 0)$; $B(1, 1)$; $C(1, 6)$; $D(-4, 5)$

In Exercises 18–21, a line is described. State three points (any three points) that the line passes through. You may use graph paper, but write your answers in the form: Points (__?__, __?__), (__?__, __?__), and (__?__, __?__).

18. A line through the origin with slope equal to 1

19. A line through point (0, 4) with slope equal to 1

20. A line containing points (2, 0) and (0, 3)

21. A line through point (0, −4) with slope equal to 0

C

22. Rosalie says: "No matter how great a number you choose, I can find a line whose slope is equal to that number." Isabella chooses the number *one billion*. Rosalie could respond by saying: "Take the line that contains points (__?__, __?__) and (__?__, __?__)."

23. You are given points (a, b) and (c, d) on a line. Show that you get the same value for the slope whether you choose (a, b) to be (x_1, y_1) or (x_2, y_2). *Suggested strategy:* Write two expressions for the slope. Then use the algebraic property: $d - b = -(b - d)$.

5 Parallel and Perpendicular Lines

The purpose of this section is to show how you can use the slopes of two lines to decide if the lines are parallel or perpendicular. Since vertical and horizontal lines are special, they will be considered in the exercises, rather than the lesson itself.

Suppose lines p and q are parallel. Draw two vertical segments and label lengths as shown.

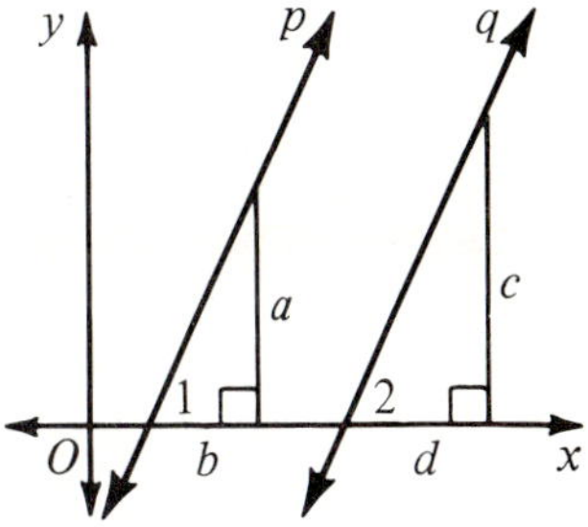

1. $\angle 1 = \angle 2$ (Why?)
2. The triangles are similar. (Why?)
3. $\frac{a}{c} = \frac{b}{d}$. (Why?)
4. $\frac{a}{b} = \frac{c}{d}$. (Why?)
5. Slope of line p = slope of line q.

We have just shown that if two lines are parallel, then their slopes are equal. The converse is also true. (See Exercise 21.)

If two lines are parallel, then they have equal slopes.

If two lines have equal slopes, then they are parallel.

EXPLORATIONS

Draw a line through the origin and point (2, 3). If your line is carefully drawn, you will see that it also contains points (−2, −3), (4, 6), and (6, 9).

1. Use a compass to construct, or use a protractor to draw, lines perpendicular to the given line at points (−2, −3), (0, 0), (2, 3), (4, 6), and (6, 9).
2. On each of the perpendiculars choose a convenient point and label it with its coordinates.
3. Find the slope of each perpendicular.
4. If you worked carefully, you found that each perpendicular has slope $-\frac{2}{3}$. Notice that the given line has slope $\frac{3}{2}$. Describe a relation between the numbers $\frac{3}{2}$ and $-\frac{2}{3}$.

Suppose a line p has slope $\frac{4}{5}$. From the Explorations on page 491, what would you guess to be the slope of a line perpendicular to p? The Explorations suggest that there is a simple relationship between the slopes of two perpendicular lines.

If two lines are perpendicular, then the product of their slopes is −1.
If the product of the slopes of two lines is −1, then the lines are perpendicular.

Classroom Practice

1. Line l has slope $\frac{2}{7}$.
   a. The slope of any line parallel to l equals __?__.
   b. The slope of any line perpendicular to l equals __?__.

2. Line j has slope -5.
   a. The slope of any line parallel to j equals __?__.
   b. The slope of any line perpendicular to j equals __?__.

3. Suppose one line has slope $\frac{3}{11}$ and another line has slope $\frac{4}{11}$. Can the lines be parallel? Explain.

4. Suppose one line has slope $\frac{2}{5}$ and another line has slope $\frac{5}{2}$. Are the lines perpendicular? Explain.

5. *Given:* The slopes of two lines are 1 and -1. Are the lines parallel? Are they perpendicular? Explain.

6. *Given:* Points $A(0, 1)$, $B(3, 3)$, $C(6, 5)$, and $D(12, 9)$.
   a. The slope of $\overleftrightarrow{AB}$ equals __?__.
   b. The slope of $\overleftrightarrow{CD}$ equals __?__.
   c. Do you think that $\overleftrightarrow{AB}$ and $\overleftrightarrow{CD}$ are parallel lines?
   d. Plot points A, B, C, and D. Do you need to change your answer to part (c)? Explain.

Written Exercises

A 1. A given line has slope $-\frac{5}{7}$.

a. Any line parallel to the given line has slope __?__.

b. Any line perpendicular to the given line has slope __?__.

2. A given line has slope 2.

a. Any line parallel to the given line has slope __?__.

b. Any line perpendicular to the given line has slope __?__.

3. A given line is vertical.

a. The slope of the given line is __?__ (0/not defined).

b. The slope of a line perpendicular to the given line is __?__ (0/not defined).

4. A given line rises to the right.

a. The slope of the line is __?__ (positive/negative).

b. The slope of a line perpendicular to the given line is __?__ (positive/negative).

The slopes of two lines are given. Are the lines parallel, perpendicular, or neither?

5. $\frac{4}{8}$ and $\frac{1}{2}$

6. $\frac{7}{2}$ and $\frac{2}{7}$

7. $-\frac{4}{3}$ and $\frac{3}{4}$

8. 3 and −3

Copy and complete the table.

	A	B	C	D	Slope of $\overleftrightarrow{AB}$	Slope of $\overleftrightarrow{CD}$	Is $\overleftrightarrow{AB} \parallel \overleftrightarrow{CD}$?	Is $\overleftrightarrow{AB} \perp \overleftrightarrow{CD}$?
Sample	(−1, 1)	(3, 5)	(0, −2)	(3, 2)	1	$\frac{4}{3}$	No	No
9.	(2, 2)	(5, 2)	(−4, 4)	(0, 4)	?	?	?	?
10.	(0, 0)	(3, 4)	(−1, −2)	(2, 2)	?	?	?	?
11.	(3, −1)	(2, 1)	(5, 0)	(3, 4)	?	?	?	?
12.	(0, 3)	(2, 0)	(0, 0)	(6, −4)	?	?	?	?

$\angle F$ **is a right angle. Find the slope of** $\overleftrightarrow{EF}$ **and the slope of** $\overleftrightarrow{FG}$**.**

B **13.**

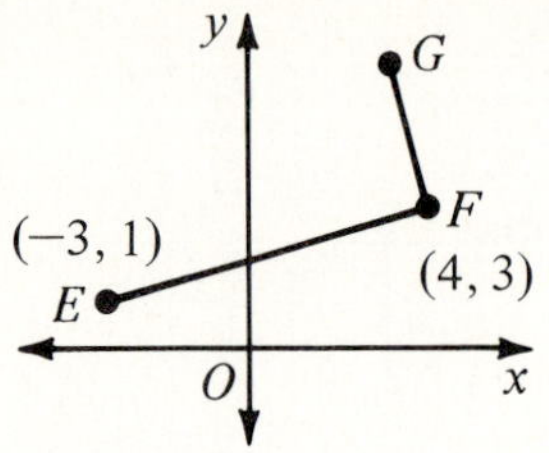

14.

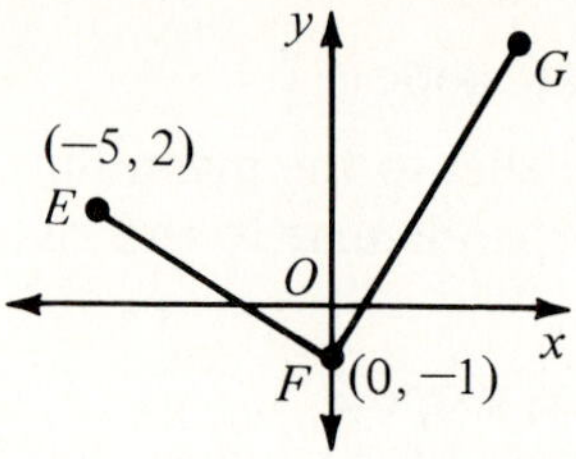

For Exercises 15–18, a strategy is:
Draw a diagram. Then find and use slopes as necessary.
Determine if the diagonals of quadrilateral *RSTV* are perpendicular.

15. $R(0\ ,0)$; $S(4, 3)$; $T(1, 7)$; $V(-3, 4)$

16. $R(-2, 6)$; $S(5, 5)$; $T(7, -3)$; $V(-3, -4)$

What special kind of figure is quadrilateral *ABCD*?

17. $A(-1, 0)$; $B(0, 3)$; $C(6, 3)$; $D(5, 0)$

18. $A(0, 0)$; $B(4, 2)$; $C(2, 3)$; $D(0, 2)$

C **19.** You are given the figure shown. M and N are the midpoints of $\overline{RV}$ and $\overline{ST}$.

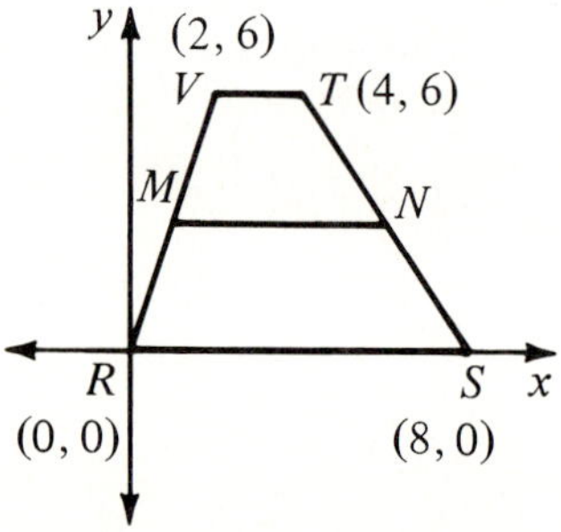

a. M has coordinates (__?__, __?__).
b. N has coordinates (__?__, __?__).
c. Slope of $\overline{MN}$ = __?__; slope of $\overline{RS}$ = __?__; slope of $\overline{VT}$ = __?__.
d. $\overline{MN} \parallel \overline{RS} \parallel \overline{VT}$. Why?
e. RS = __?__; VT = __?__; MN = __?__
Does $MN = \frac{1}{2}(RS + VT)$?

20. *Given:* $\triangle ABC$; M and N are the midpoints of $\overline{AC}$ and $\overline{BC}$.

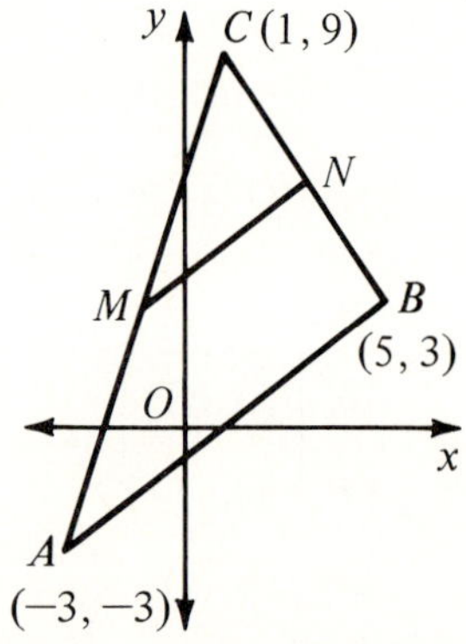

a. M has coordinates (__?__, __?__).
b. N has coordinates (__?__, __?__).
c. Slope of $\overline{MN}$ = __?__;
slope of $\overline{AB}$ = __?__.
d. Is $\overline{MN} \parallel \overline{AB}$?
e. MN = __?__ and AB = __?__
f. Does $MN = \frac{1}{2}AB$?

21. Our goal is to show that:
If two lines have equal slopes, then the lines are parallel.

Given: Lines p and q
Slope of p = slope of q
Prove: $p \parallel q$

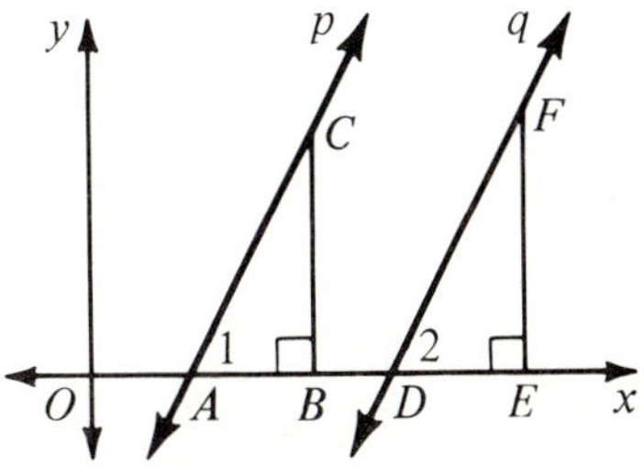

Suggested strategy:
Take points B and E so that $AB = DE$.
Draw vertical segments $\overline{BC}$ and $\overline{EF}$.
Show that $\frac{BC}{AB} = \frac{EF}{DE}$ and conclude that $BC = EF$.
Prove that $\triangle ABC \cong \triangle DEF$.
Since $\angle 1 = \angle 2$, you can conclude that $p \parallel q$.

22. The statements on pages 491 and 492 do not apply to the special case of vertical lines. Complete the sentence that follows to show that more complicated statements could cover this special case: When two lines are perpendicular, either the product of their slopes is -1, or . . . .

SELF-TEST

Vocabulary

slope of a line (p. 486) rise (p. 486) run (p. 486)

1. The slope of a $\underset{\text{horizontal/vertical}}{\underline{\quad ? \quad}}$ line equals 0. *(12-4)*

2. The slope of a $\underset{\text{horizontal/vertical}}{\underline{\quad ? \quad}}$ line is not defined.

3. You are given point $A(2, 0)$ and point $B(5, 4)$.
Find the slope of $\overleftrightarrow{AB}$.

4. You are given point $B(5, 4)$ and point $C(8, -2)$.
Find the slope of $\overleftrightarrow{BC}$.

5. If the slope of line j equals $\frac{3}{7}$, then the slope of any line parallel to j equals _?_. *(12-5)*

6. If the slope of line k equals $-\frac{5}{2}$, then the slope of any line perpendicular to k equals _?_.

6 Equations and Lines

Let's see how the line shown in the figure is related to the equation $2x + 3y = 12$. Do the coordinates of the four labeled points make the equation a true statement when they are substituted for x and y?

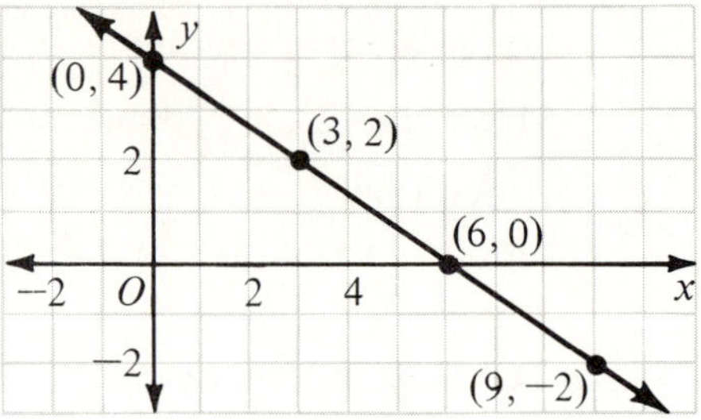

Point (0, 4)

$$2x + 3y = 12$$
$$2(0) + 3(4) \stackrel{?}{=} 12$$
$$0 + 12 \stackrel{?}{=} 12 \checkmark$$

Yes

Point (3, 2)

$$2x + 3y = 12$$
$$2(3) + 3(2) \stackrel{?}{=} 12$$
$$6 + 6 \stackrel{?}{=} 12 \checkmark$$

Yes

Point (6, 0)

$$2x + 3y = 12$$
$$2(6) + 3(0) \stackrel{?}{=} 12$$
$$12 + 0 \stackrel{?}{=} 12 \checkmark$$

Yes

Point (9, −2)

$$2x + 3y = 12$$
$$2(9) + 3(-2) \stackrel{?}{=} 12$$
$$18 - 6 \stackrel{?}{=} 12 \checkmark$$

Yes

Suppose you kept on testing the coordinates of different points. You would find that the coordinates of any point on the line make the equation a true statement. The coordinates of any other point do not. The line is called the **graph** of the equation $2x + 3y = 12$.

In general, it is true that:

If a point lies on the graph of an equation, then its coordinates make the equation a true statement.

If the coordinates of a point make an equation a true statement, then the point lies on the graph of the equation.

The facts you discovered about the equation $2x + 3y = 12$ are true for every equation in the form $ax + by = c$.

If an equation can be written in the form $ax + by = c$, then the graph of the equation is a line. (Assume that a and b are not both equal to zero.)

The examples that follow show how to graph an equation that can be written in the form $ax + by = c$.

EXAMPLE 1 Draw the graph of $2x + y = 6$.

Pick any three numbers to use for x or for y. Find the coordinates of three points.

Let $x = 0$:	Let $y = 0$:	Let $x = 1$:
$2x + y = 6$	$2x + y = 6$	$2x + y = 6$
$2(0) + y = 6$	$2x + 0 = 6$	$2(1) + y = 6$
$0 + y = 6$	$2x = 6$	$2 + y = 6$
$y = 6$	$x = 3$	$y = 4$

Now plot the points:
(0, 6); (3, 0); (1, 4).
Then draw the line.

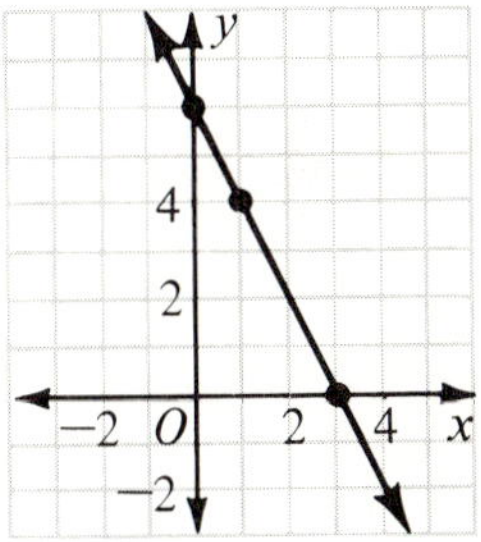

EXAMPLE 2 Draw the graph of $x - 3y = 3$.

Let $x = 0$:	Let $y = 0$:	Let $x = 7$:
$x - 3y = 3$	$x - 3y = 3$	$x - 3y = 3$
$0 - 3y = 3$	$x - 3(0) = 3$	$7 - 3y = 3$
$-3y = 3$	$x - 0 = 3$	$-3y = -4$
$y = -1$	$x = 3$	$y = 1\frac{1}{3}$

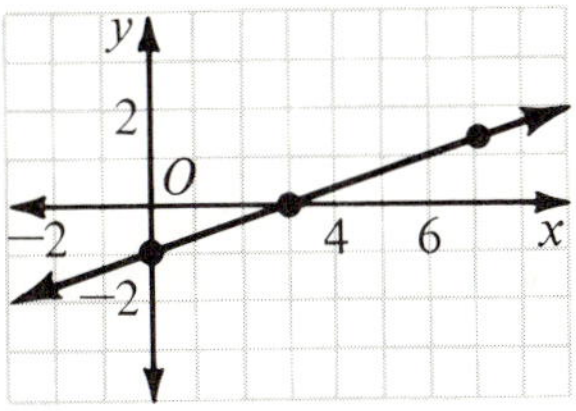

Notice in Example 2, that when you let $x = 7$, you get a fractional value for y. You have two choices:

(1) estimate the position of $\left(7, 1\frac{1}{3}\right)$; or

(2) forget that point and try other values for x and y until you get a pair of coordinates that doesn't involve fractions.

EXAMPLE 3

Draw the graph of $x = 2$.

The equation can be rewritten in the form $x + 0y = 2$.

If you try to let x be any number other than 2, you cannot find a value for y.

On the other hand, no matter what value you assign to y, the value of x is always 2.

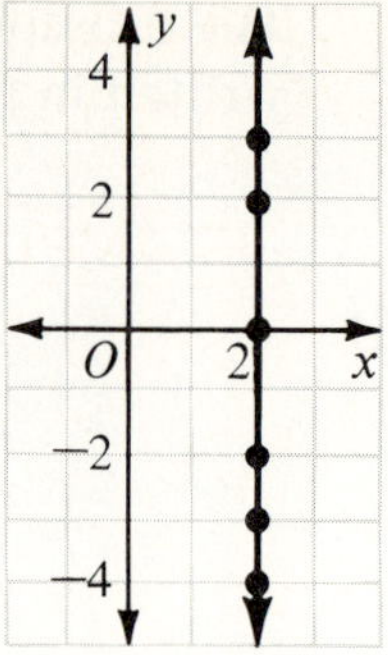

The graph of $x = 2$ is a vertical line.

Classroom Practice

1. The graph of the equation $y = x - 2$ is shown. Does the point lie on the line?

   a. (2, 0)
   b. (0, 2)
   c. (3, 1)
   d. (1, 3)
   e. (4, 4)
   f. (1, −1)
   g. (−1, −3)
   h. $\left(2\frac{1}{2}, \frac{1}{2}\right)$

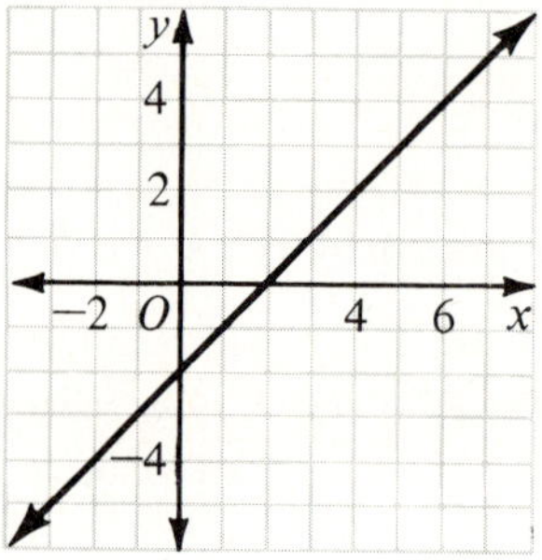

2. Do the coordinates of the point make the equation $2x - y = 4$ a true statement?

   a. (2, 2) b. (4, 0) c. (2, 0) d. (1, 1)
   e. (10, 16) f. (0, −4) g. (−3, −10) h. (−2, 0)

3. Name the coordinates of five points (any five) that lie on the graph of the equation $y = 2x$.

4. The equation $y = 5$ is given. Note that the equation can be written in the form $0x + y = 5$.

   a. When $x = 0$, $y =$ __?__.
   b. When $x = -13$, $y =$ __?__.
   c. No matter what value you assign to x, $y =$ __?__.
   d. The graph of $y = 5$ is a ____?____ (horizontal/vertical) line.

Written Exercises

A **1.** State the coordinates of four points (any four) that lie on the line at the left, below.

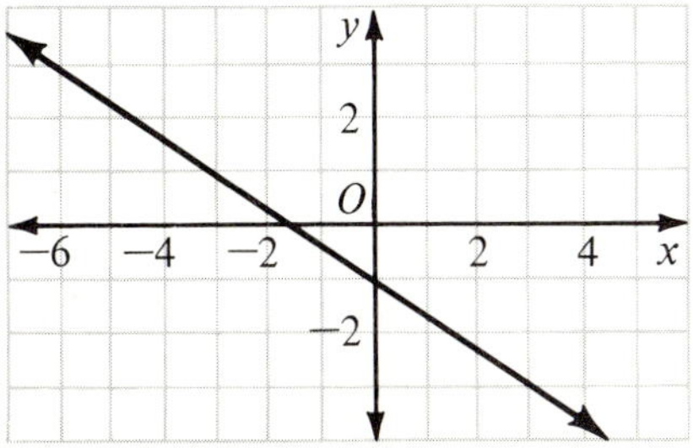

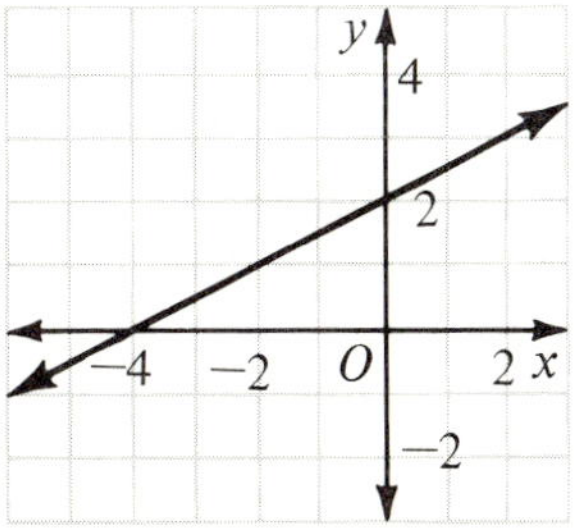

2. One of the points named does not lie on the line at the right, above. Which point?

$(-4, 0)$ $(1, 3)$ $\left(1, 2\frac{1}{2}\right)$ $(0, 2)$

Copy and complete the table of values.

3. $x + y = 6$

x	y
0	?
?	0
2	?

4. $x - y = 6$

x	y
0	?
?	0
4	?

5. $y = 3x$

x	y
0	?
?	0
3	?

6. $y = x + 4$

x	y
0	?
?	0
3	?

7. $2x - 3y = 12$

x	y
0	?
?	0
?	4

8. $3x - y = -6$

x	y
0	?
?	0
?	3

9–14. Draw the graphs of the equations in Exercises 3–8. If your points do not lie on a line, check the work you did earlier.

15. Draw the graph of $x = 5$. (*Hint:* See Example 3, page 498.)

16. Draw the graph of $y = -4$.

Make a table of values for plotting three points. Plot the points and draw the graph of the equation.

B **17.** $3x + 2y = 9$

18. $x - 4y = 6$

19. $4x + 3y = 6$

20. $2x - 3y = -12$

For each exercise, graph the two equations on the same pair of axes. Inspect your drawing and state the coordinates of the point where the graphs intersect.

21. $x + y = 6$
$x - y = 2$

22. $y = x + 2$
$y = -x$

23. $2x + y = 6$
$3x - y = 4$

24. $2x + y = -1$
$x + 2y = 4$

Graph the three equations on the same pair of axes. Find the area of the triangle formed by the three lines.

C **25.** $x = 1$
$y = -1$
$y = x + 2$

26. $x = 2$
$y = 2$
$x + y = 6$

27. $y = -1$
$y = 3x + 2$
$3x + y = 2$

CAREER NOTES

CARPENTER

Have you ever put together a picture frame or bookcase? Do you enjoy helping build props for your school play? If so, you have been using carpentry skills.

Carpenters build, remodel, and repair structures made of wood. Rough carpenters work on a building's framework. They put up scaffolding and make the forms into which concrete is poured. Finish carpenters do more detailed work. They put in paneling, cabinets, window sashes, doors, and molding.

Being able to make careful, accurate measurements is an important part of a carpenter's job. Carpenters often apply their knowledge of arithmetic, geometry, and informal algebra. For instance, you have learned in geometry that the diagonals of a rectangle are equal. The converse of this theorem is also true. Knowing this, carpenters can measure the diagonals of a house's foundation. If they are the same length, the foundation is a rectangle. This then tells the carpenters that the corners of the foundation are all right angles.

A four-year apprenticeship program with classes in drafting, blueprint reading, mathematics, and wood shop is good preparation for a career in carpentry.

7 Coordinate Geometry Proofs

Recall Theorem 14 in Chapter 5:

> The median of a trapezoid has two properties.
> (1) It is parallel to the bases.
> (2) Its length equals half the sum of the base lengths.

This theorem was stated without proof. We are now able to prove the theorem by using coordinates.

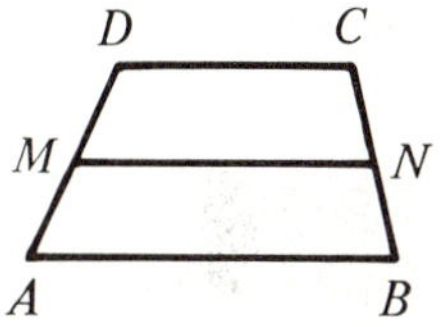

Given: Trapezoid $ABCD$ with median $\overline{MN}$

Prove: 1. $\overline{MN} \parallel \overline{AB} \parallel \overline{DC}$

2. $MN = \frac{1}{2}(AB + DC)$

Proof:

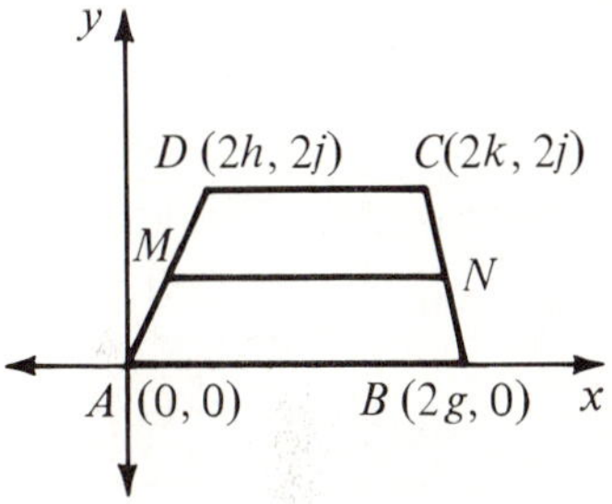

Draw coordinate axes as shown.
Then A is point (0, 0), and the y-coordinate of B is 0.
Assign coordinates to B, C, and D as shown.
(Because $\overline{DC}$ is horizontal, D and C must have equal y-coordinates.)

The coordinates of M are $\left(\frac{2h+0}{2}, \frac{2j+0}{2}\right)$, or (h, j).

The coordinates of N are $\left(\frac{2g+2k}{2}, \frac{0+2j}{2}\right)$, or $(g + k, j)$.

1. Slope of $\overline{MN} = \frac{j-j}{(g+k)-h} = \frac{0}{g+k-h} = 0$.
   Slope of $\overline{AB}$ = slope of $\overline{DC} = 0$.
   Therefore, $\overline{MN} \parallel \overline{AB} \parallel \overline{DC}$.

2. $MN = g + k - h$
   $AB = 2g - 0 = 2g$
   $DC = 2k - 2h$

   $AB + DC = 2g + 2k - 2h$, and $\frac{1}{2}(AB + DC) = g + k - h$.

   Because both MN and $\frac{1}{2}(AB + DC)$ equal $g + k - h$,

   $MN = \frac{1}{2}(AB + DC)$.

Classroom Practice

The purpose of these exercises is to use coordinates to prove Theorem 15 from Chapter 5: The segment joining the midpoints of two sides of a triangle is parallel to the third side and half as long.

Given: $\triangle ABC$;
midpoints D and E as shown

Prove: 1. $\overline{DE} \parallel \overline{AB}$
2. $DE = \frac{1}{2}AB$

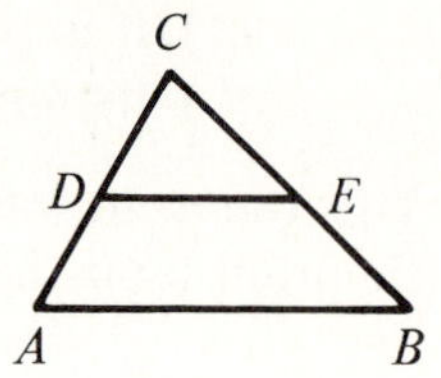

Proof:
Draw axes and assign coordinates as shown.

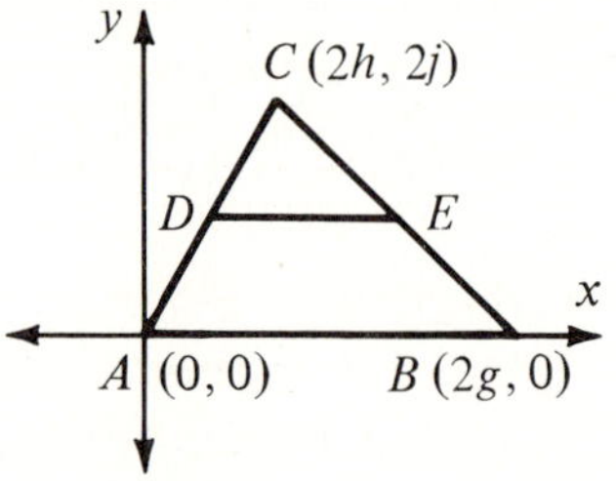

1. a. The coordinates of D are (? , ?).
b. The coordinates of E are (? , ?).
c. Slope of $\overline{AB} = \frac{0}{2g} =$? .
d. Slope of $\overline{DE} = \frac{j - j}{(g + h) - h} =$? .
e. $\overline{DE} \parallel \overline{AB}$ because ? .

2. a. $DE = (g + h) - h =$? .
b. Because $AB = 2g$, $\frac{1}{2}AB =$? .
c. Because both DE and $\frac{1}{2}AB$ equal ? , $DE = \frac{1}{2}AB$.

Written Exercises

A **1.** Axes have been drawn for a particular rectangle. Complete a proof that the diagonals are equal.

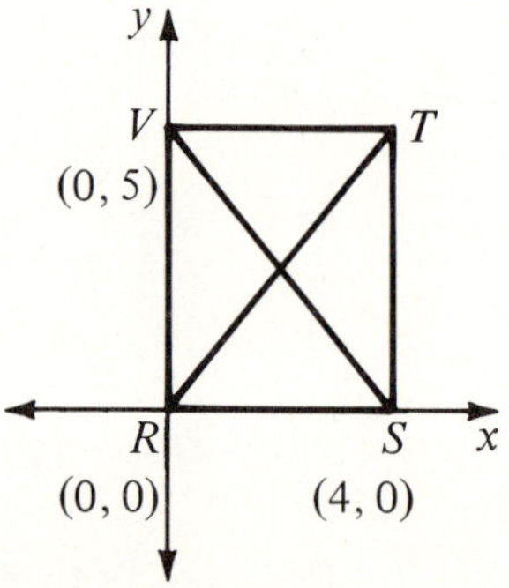

a. Because $\overline{VT}$ is horizontal, the y-coordinate of T is ? .
b. Because $\overline{ST}$ is vertical, the x-coordinate of T is ? .
c. In right $\triangle VRS$, $(VS)^2 = 5^2 + \underline{\ ?\ }^2$, and $VS = \sqrt{\underline{\ ?\ }}$.
d. In right $\triangle TSR$, $(RT)^2 = \underline{\ ?\ }^2 + \underline{\ ?\ }^2$, and $RT = \sqrt{\underline{\ ?\ }}$.
e. Because both VS and RT equal ? , $VS = RT$.

2. Axes have been drawn for a particular square. Complete a proof that the diagonals are perpendicular.

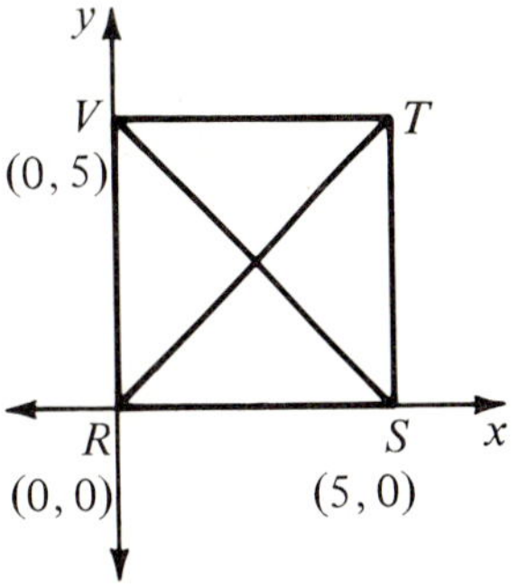

a. The coordinates of T are (__?__, __?__).

b. Slope of $\overline{RT} = \frac{5-0}{5-0} =$ __?__

c. Slope of $\overline{VS} = \frac{0-5}{5-0} =$ __?__

d. (Slope of $\overline{RT}$) · (Slope of $\overline{VS}$) = __?__ · __?__ = __?__, so $\overline{RT} \perp \overline{VS}$.

3. Axes have been drawn for a particular right triangle. Complete a proof that the midpoint M of the hypotenuse is equidistant from the three vertices.

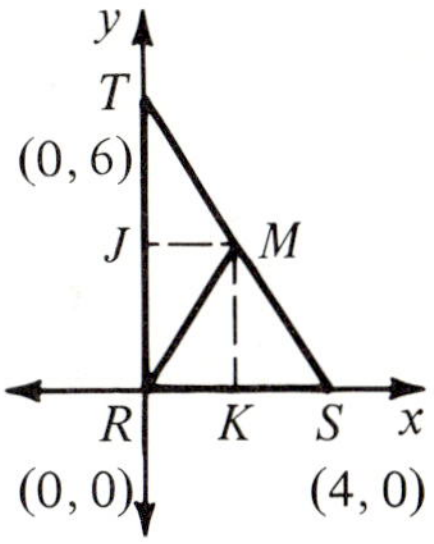

a. The coordinates of M are (__?__, __?__).

b. Draw horizontal and vertical segments from M as shown.

c. The coordinates of J are (__?__, __?__), and the coordinates of K are (__?__, __?__).

d. In right $\triangle TJM$, $(MT)^2 = 3^2 +$ __?__^{2}, and $MT = \sqrt{\text{?}}$.

e. In right $\triangle MKS$, $(MS)^2 =$ __?__^{2} + __?__^{2}, and $MS = \sqrt{\text{?}}$.

f. In right $\triangle MKR$, $(MR)^2 =$ __?__^{2} + __?__^{2}, and $MR = \sqrt{\text{?}}$.

g. Because MT, MS, and MR all equal $\sqrt{\text{?}}$, $MT = MS = MR$.

In Exercises 4–6 copy the diagram shown. See Exercises 1–3 for ideas about proofs.

B

4. *Prove:* The diagonals of any rectangle are equal.

5. *Prove:* The diagonals of any square are perpendicular.

6. *Prove:* The midpoint of the hypotenuse of any right triangle is equidistant from the three vertices.

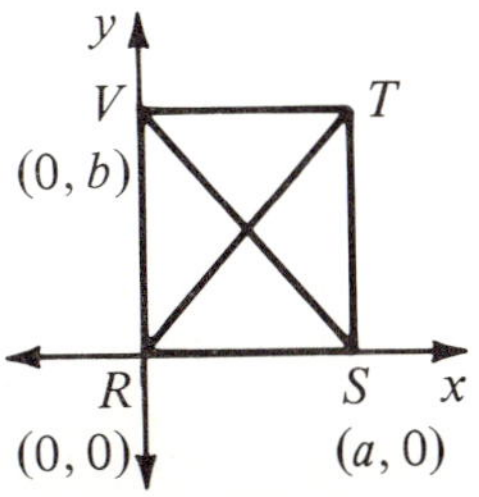

Exercise 4

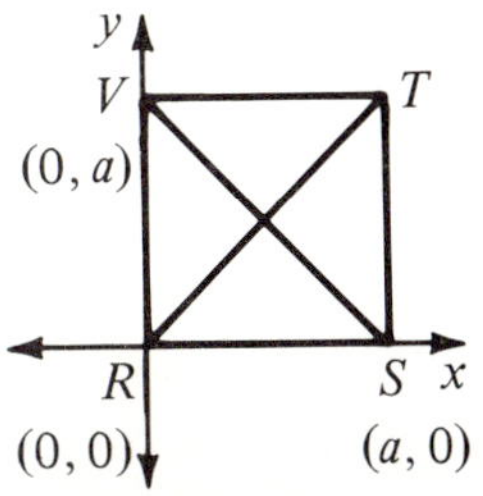

Exercise 5

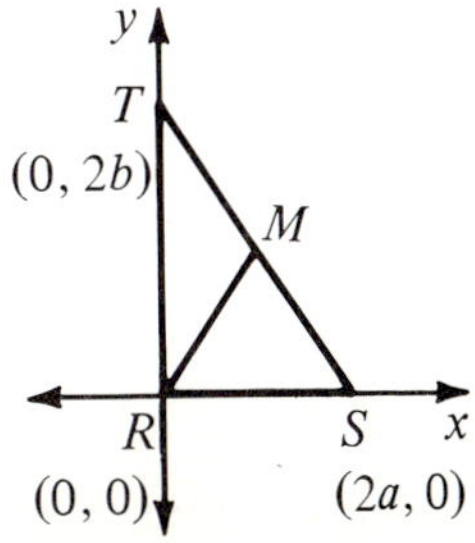

Exercise 6

In Exercises 7 and 8, copy the diagrams shown.

C 7. *Prove:* The medians drawn to the legs of an isosceles triangle are equal. (*Hint:* Notice the x-coordinate of T. What must the x-coordinate of S be?)

8. *Prove:* The segments joining, in order, the midpoints of the sides of any quadrilateral form a parallelogram.

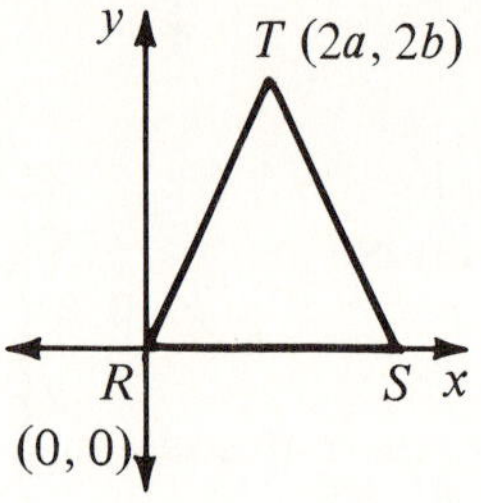

Exercise 7

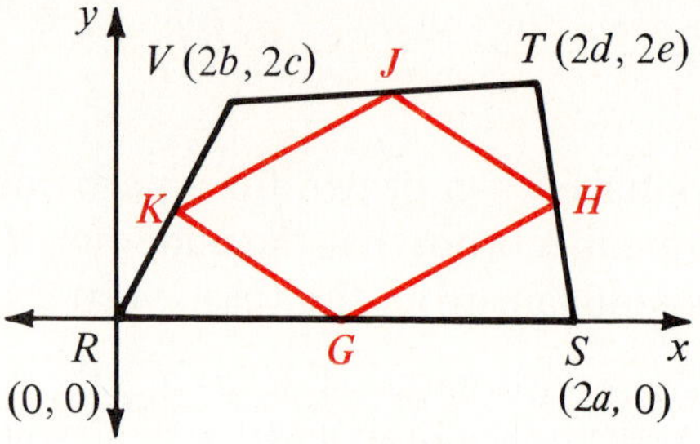

Exercise 8

SELF-TEST

Vocabulary

graph of an equation (p. 496)

1. Do the coordinates of the point (5, −3) make the equation $2x - 3y = 1$ a true statement? *(12-6)*

Copy and complete each table of values.

2. $x + y = 8$

x	y
0	?
?	0
3	?

3. $4x - 3y = 12$

x	y
0	?
?	0
5	?

4. Given the equation $2x + y = 10$, make a table of values, plot the points, and draw the graph of the equation.

5. You are given points $A(0, 0)$, $B(5, 0)$, $C(6, 4)$, and $D(1, 4)$. Prove that quadrilateral $ABCD$ is a parallelogram. *(12-7)*

APPLICATIONS

SOME SHAPES IN NATURE

Many of the shapes you have studied in this course occur in nature. Here are a few examples.

1. 

Did you guess that this is a much-enlarged picture of a snowflake? A snowflake is sometimes described as *hexagonal.*

2. 

A crystal of Iceland spar has the form of a *parallelepiped* (a solid with six faces, each a parallelogram). The crystal has an interesting property. Set a piece of Iceland spar over one segment and you see two segments.

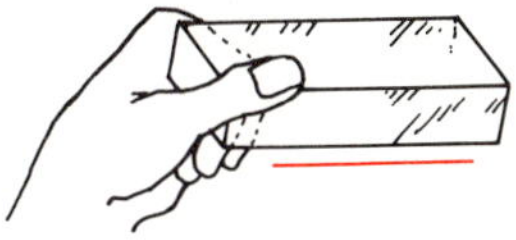

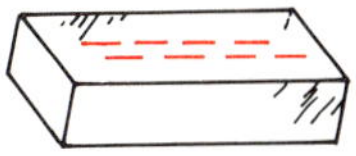

3. 

Quartz crystals suggest plane surfaces bounded by various kinds of polygons.

PROBLEM SOLVING STRATEGIES

PATTERNS

What is the sum of the interior angles of a polygon? On page 194 you looked at the sum of the interior angles of simple polygons and observed a pattern: The sum of the interior angles of a polygon with n sides is given by the formula

$$S = (n - 2) \times 180°.$$

EXERCISES

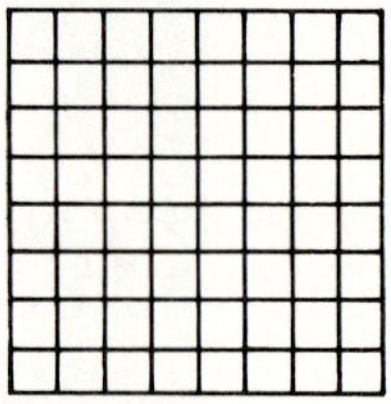

1. How many squares are there in this picture?

   Fill in the following chart. Use the pattern formed by the answers to predict the number of squares in question.

Number of Rows/Columns		Number of Squares
1		1
2	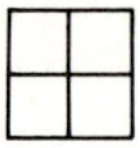	$5 = 4 + 1$
3	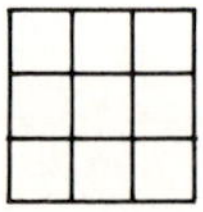	$14 = 9 + 4 + 1$
4	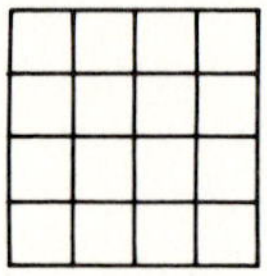	? = ? + ? + ? + ?

2. A square 12 units on each side has 100 congruent circles inscribed in it. What is the sum of the areas of the circles?

   To answer this question, let's look at some data. Remember that the area of a circle is πr^2.

Number of Circles		Radius of Each Circle	Sum of Areas
1	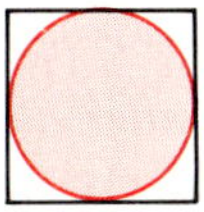 12 12	$\frac{1}{2}(12) = 6$	$1 \times 36\pi$
4	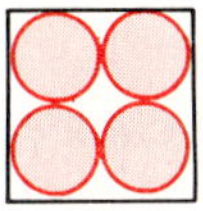 12 12	$\frac{1}{2}\left(\frac{12}{2}\right) = 3$	$4 \times 9\pi$
9	 12 12	$\frac{1}{2}\left(\frac{12}{3}\right) = 2$	$\underline{\ \ ?\ \ } \times \underline{\ \ ?\ \ }\pi$

3. If ten points are arranged around a circle and all the possible lines connecting the points in pairs are drawn, how many such lines will there be? Extend the table for five points and then predict the answer to the question.

Number of Points		Number of Lines
2	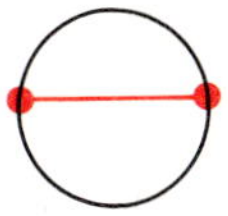	1
3	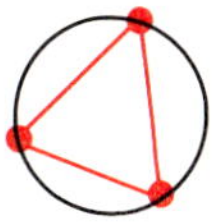	$3 = 2 + 1$
4	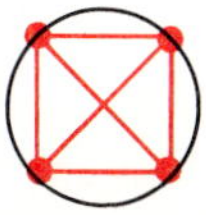	$6 = 3 + 2 + 1$

COMPUTER ACTIVITIES

THE BEST BOX

You are designing the package for a new product. The package is a cardboard box. You decide that the box will have a square base. It has to hold 1000 cm^3 of the new product. There are many possible choices for the dimensions of the box. For instance, it could be 5 cm by 5 cm by 40 cm. Or it could be 8 cm by 8 cm by about 15.6 cm. Either box would have a square base and hold 1000 cm^3 of the product.

However, you also have to think about the cost of the cardboard. The less cardboard you use, the less it will cost to make the box. You want the lowest cost box that will satisfy your requirements.

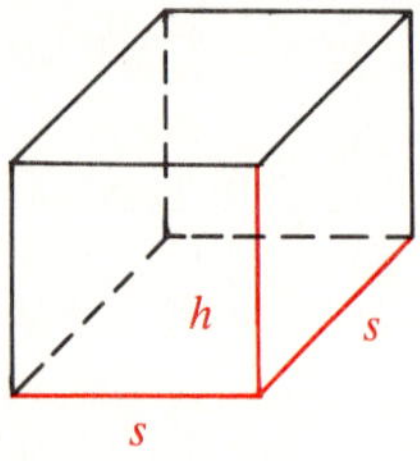

The volume of the box is the amount of product it can hold. The surface area of the box is the area of the cardboard. Your problem is to design the box with: (1) a square base, (2) a volume of 1000 cm^3, and (3) the smallest possible area.

In the program below, you choose the length of each side of the square base of the box. The computer figures the height and total area. *Note:* The complicated expression in line 60 is just the computer's way of rounding a decimal to thousandths.

```
10  REM***BOX VOLUME AND TOTAL AREA
20  PRINT "LENGTH OF A SIDE OF"
30  PRINT "THE SQUARE BASE = ";
40  INPUT S
50  REM***FIGURE HEIGHT IF VOLUME = 1000
60  H = INT(1000*(1000/S/S + .0005)) / 1000
70  REM***FIGURE THE TOTAL AREA
80  TA = 2*S*(S + 2*H)
90  REM***GIVE RESULTS
100  PRINT "HEIGHT: ";H
110  PRINT "TOTAL AREA: ";TA
120  PRINT
130  PRINT "DO YOU WANT TO TRY AGAIN";
140  INPUT Q$
150  IF Q$ = "YES" THEN 10
200  END
```

EXERCISES

1. RUN the program. Complete the chart.

Side of base	4	5	6	7	8	9	10	11	12	13	14
Height	?	?	?	?	?	?	?	?	?	?	?
Total area	?	?	?	?	?	?	?	?	?	?	?

2. a. What seems to be the smallest possible total area?
   b. What are the dimensions of the box with this area? Give length, width, and height.

3. a. RUN the program. Use a base side 0.5 cm more than in Exercise 2b.
   b. RUN the program. Use a base side of 0.5 cm less than in Exercise 2b.
   c. Do either of these boxes have a smaller total area than the box in Exercise 2?

Next you work on a box that has to hold 2000 cm^3. It will also have a square base. Change the program as shown.

```
50  REM***FIGURE HEIGHT IF VOLUME IS 2000
60  H = INT(1000*(2000/S/S + .0005)) / 1000
```

4. RUN the changed program. Experiment until you think you have found the box with the smallest possible total area.
   a. What is this area?
   b. What are the dimensions of the box? Give length, width, and height. Round to the nearest tenth of a centimeter.

5. a. What is special about the dimensions of the boxes in Exercises 2 and 4?
   b. What is the special name for this kind of box?

6. A box has to hold 8000 cm^3. You are to use the least cardboard possible. What should the dimensions be? Try to answer without using the computer. (*Hint:* If you decide to use the computer, you will need to change lines 50 and 60 again.)

SKILLS REVIEW

ALGEBRAIC FRACTIONS: ADDITION AND SUBTRACTION

Simplify.

SAMPLE 1

$$\frac{6ab}{cd} - \frac{3ab+2}{cd} = \frac{6ab-(3ab+2)}{cd}$$

The denominators are the same.

Be careful with signs!

$$= \frac{6ab-3ab-2}{cd}$$

$$= \frac{3ab-2}{cd}$$

1. $\frac{12x}{7} - \frac{8x}{7}$
2. $\frac{2ab}{3z} + \frac{4ab}{3z}$
3. $\frac{4xy}{5} - \frac{9xy}{5}$
4. $\frac{11d^2}{2c} - \frac{9d^2}{2c}$
5. $\frac{6b^3}{a} + \frac{2b}{a}$
6. $\frac{-8k}{7} + \frac{3k}{7}$
7. $\frac{11}{x+y} - \frac{10}{x+y}$
8. $\frac{4x+2}{k} - \frac{2+4x}{k}$
9. $\frac{4m-n}{5n} + \frac{n+m}{5n}$
10. $\frac{5h+2k}{h+k} + \frac{h+4k}{h+k}$
11. $\frac{b^2}{b+1} - \frac{1}{b+1}$
12. $\frac{a^2-6a}{a-3} + \frac{9}{a-3}$

SAMPLE 2

$$\frac{9}{8c} + \frac{5}{2d} = \frac{9d}{8cd} + \frac{20c}{8cd} = \frac{9d+20c}{8cd}$$

The denominators are different.

First rename the fractions.

The new denominator can be $8cd$.

13. $\frac{a}{5} + \frac{a}{2}$
14. $\frac{x}{3} - \frac{x}{8}$
15. $\frac{2a}{3} + \frac{a}{6}$
16. $\frac{n}{10} + \frac{4n}{5}$
17. $\frac{5}{2a} + \frac{1}{a}$
18. $\frac{3}{h} - \frac{1}{3h}$
19. $\frac{7}{ab} - \frac{5a}{b}$
20. $\frac{2}{d} + \frac{3}{d^2}$
21. $\frac{3m}{4} - \frac{2m}{3}$
22. $\frac{2a}{5} + \frac{a}{4}$
23. $\frac{7}{x} + \frac{1}{2}$
24. $\frac{1}{y} - \frac{1}{z}$
25. $\frac{4}{3n} - \frac{2}{n}$
26. $\frac{a^2c}{b} + \frac{a^2b}{c}$
27. $\frac{5x}{6} + \frac{x}{2}$
28. $\frac{3m}{5} - \frac{m}{10}$

CHAPTER REVIEW

CHAPTER SUMMARY

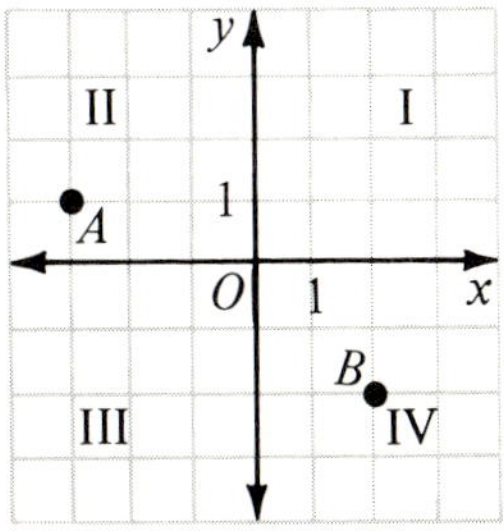

1. In the coordinate plane, the x-axis and the y-axis intersect at the origin, point (0, 0).

   Point $A(-3, 1)$ lies in quadrant II.

   Point $B(2, -2)$ lies in quadrant IV.

2. The Pythagorean Theorem can be used to find the distance between two points in the coordinate plane.

3. Given $P(x_1, y_1)$ and $Q(x_2, y_2)$.
   The midpoint M of $\overline{PQ}$ has coordinates $\left(\frac{x_1 + x_2}{2}, \frac{y_1 + y_2}{2}\right)$.

4. The line that joins points (x_1, y_1) and (x_2, y_2) has slope $\frac{y_2 - y_1}{x_2 - x_1}$.

   The slope of a vertical line is not defined.

5. The following statements apply to nonvertical lines:
   If two lines are parallel, then their slopes are equal. If two lines have equal slopes, then they are parallel.
   If two lines are perpendicular, then the product of their slopes is -1. If the product of the slopes of two lines is -1, then the lines are perpendicular.

6. If an equation can be written in the form $ax + by = c$ (a and b not both 0), then its graph is a straight line. We find the graph by first preparing a table of values.

7. Coordinate geometry can be used to prove theorems.

REVIEW EXERCISES

Refer to the diagram at the right. ***(See pp. 468–471.)***

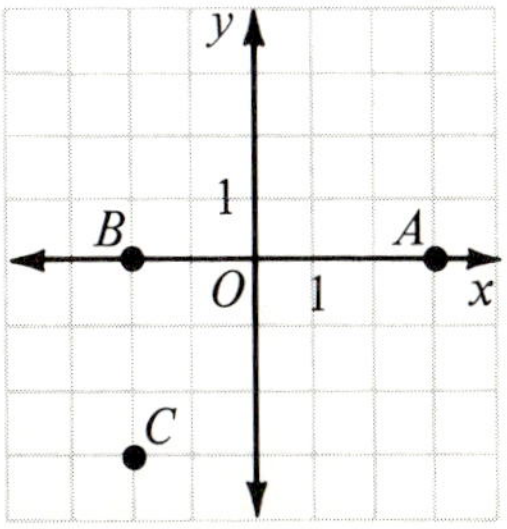

1. The coordinates of point A are __?__.

2. The coordinates of point B are __?__.

3. Point B lies on the __?__-axis.

4. Suppose D is located so that quadrilateral $ABCD$ is a rectangle. Point D has coordinates (__?__, __?__).

5. Point C lies in quadrant __?__.

Find the distance between the points named. Use graph paper if you wish. *(See pp. 472–474.)*

6. (0, 0) and (6, 8)
7. (3, 0) and (3, 7)
8. (0, 2) and (4, 5)
9. (−1, −1) and (7, −7)

State the coordinates of the midpoint of the segment that joins the points named. *(See pp. 476–480.)*

10. (2, 0) and (6, 0)
11. (3, −8) and (3, 0)
12. (3, 4) and (5, −4)
13. (3, 4) and (−5, 1)

State the slope of the line. If the slope is not defined for a particular line, write *not defined*. *(See pp. 486–490.)*

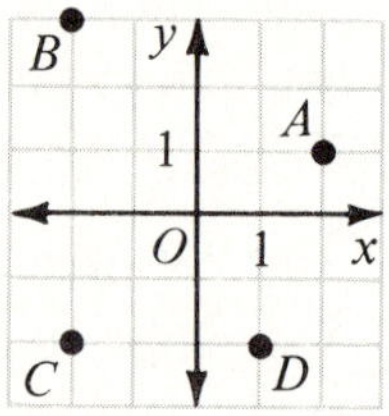

14. $\overleftrightarrow{OA}$
15. $\overleftrightarrow{OB}$
16. $\overleftrightarrow{BA}$
17. $\overleftrightarrow{CD}$
18. $\overleftrightarrow{CA}$
19. $\overleftrightarrow{CB}$

A certain line t has slope $\frac{2}{3}$. Tell whether each statement is true or false. *(See pp. 491–495.)*

20. Line t rises to the right.
21. Every line that is perpendicular to t has slope $\frac{3}{2}$.
22. The line that contains points (−1, 0) and (3, 6) is parallel to t.
23. Every line that is parallel to t has slope $\frac{2}{3}$.

Line j has the equation $2x - y = 10$. *(See pp. 496–500.)*

24. Does j pass through the origin?
25. Does j pass through the point (5, 1)?
26. On line j, when $y = 4$, $x =$ __?__.

Points *A*, *B*, *C*, and *D* have the coordinates shown. *(See pp. 501–504.)*

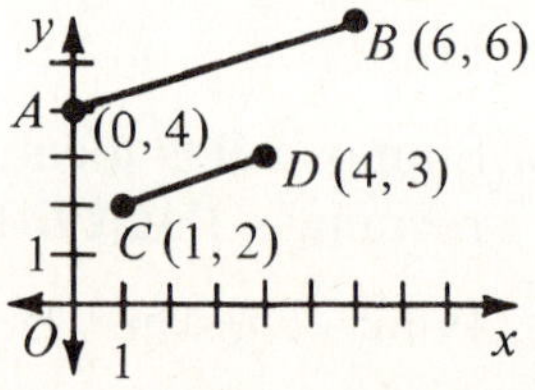

27. *Prove:* $\overleftrightarrow{AB} \parallel \overleftrightarrow{CD}$
28. *Prove:* $CD = \frac{1}{2}AB$

CHAPTER TEST

Consider points $A(-3, 0)$, $B(-2, -1)$, and $C(-2, 3)$.

1. Plot points A, B, and C.
2. Which point lies in quadrant III? *(12-1)*
3. Find the distance between points B and C.

Find the distance between the points named.

4. $A(0, 0)$ and $B(-9, -12)$
5. $A(2, 5)$ and $B(9, 4)$ *(12-2)*
6. Find the perimeter and area of the triangle with vertices $A(-3, 0)$, $B(3, 0)$, and $C(0, 4)$.

Find the coordinates of the midpoint of $\overline{AB}$.

7. $A(-7, -1)$ and $B(-3, 5)$
8. $A(0, 6)$ and $B(3, -10)$ *(12-3)*
9. M is the midpoint of $\overline{XY}$, X is the point $(2, 3)$, and M is point $(5, 1)$. Find the coordinates of point Y.

You are given points $R(2, -5)$, $S(-3, -5)$ and $T(2, -10)$. Find the slope of the line, if possible, or write *not defined*.

10. $\overleftrightarrow{RS}$
11. $\overleftrightarrow{ST}$
12. $\overleftrightarrow{RT}$ *(12-4)*

Suppose line t has slope $-\frac{1}{4}$. Complete each statement.

13. Line t __?__ from left to right.
    a. rises b. falls c. neither
14. If $\overleftrightarrow{AB} \perp t$, then $\overleftrightarrow{AB}$ has slope __?__. *(12-5)*
    a. -4 b. $\frac{1}{4}$ c. 4
15. The line through points $(2, 5)$ and $(10, 3)$ __?__ line t.
    a. is parallel to b. is perpendicular to c. neither

Line j has equation $2x - 3y = 6$.

16. Find the missing coordinates of the following points on line j. *(12-6)*
    a. $P(0,$ __?__$)$ b. $Q($__?__$, 0)$ c. $R($__?__$, 2)$
17. Show that $(2, -1)$ does not lie on line j.
18. Draw the graph of line j.

You are given points $A(0, 0)$, $B(4, 2)$, $C(1, 8)$, and $D(-3, 6)$.

19. *Prove:* $AC = BD$
20. *Prove:* $ABCD$ is a rectangle. *(12-7)*

UNIT F CUMULATIVE REVIEW

Refer to the figure shown. Find each of the following.

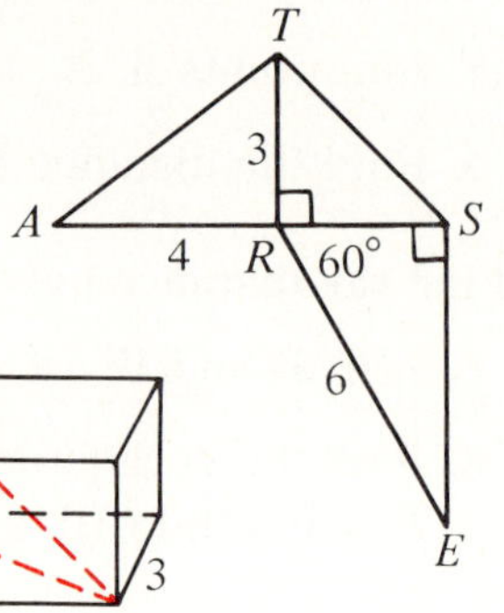

1. *SR* 2. *SE* 3. *TS* 4. *TA*

5. The perimeter of $\triangle TAR$

6. The area of $\triangle STA$

Refer to the rectangular solid shown.

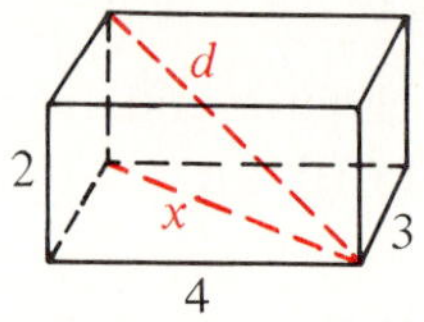

7. Find the value of x.

8. Find the length d of the diagonal.

9. A regular square pyramid has height 15 and slant height 17.
   a. How long is an edge of the base? b. What is the volume?

10. A cone has radius 5 and height 12. Find:
    a. the slant height b. the total area c. the volume

Refer to $\triangle GHI$. Express each ratio in simplest form.

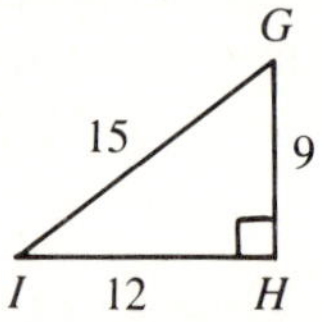

11. $\sin G$ 12. $\cos I$ 13. $\tan G$ 14. $\sin I$

Use the table on page 517 to complete each statement.

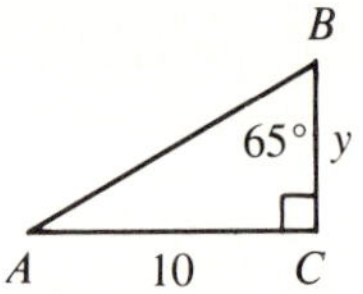

15. $\sin B =$ _?_ 16. $\tan A =$ _?_

17. To the nearest tenth, $y =$ _?_.

Complete each statement.

18. Point D lies in quadrant _?_.

19. The coordinates of point A are (_?_, _?_).

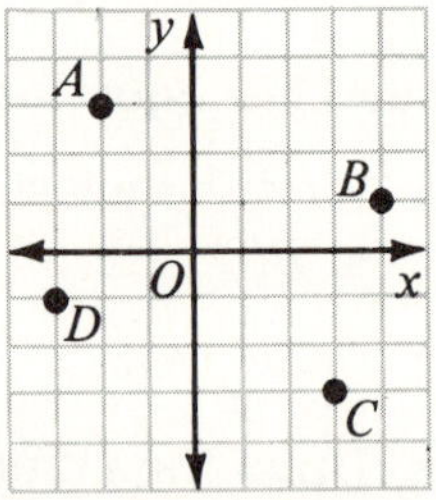

20. The length of $\overline{AD}$ is _?_.

21. The midpoint of $\overline{DC}$ is _?_.

22. The slope of $\overleftrightarrow{AD}$ is _?_.

23. The slope of $\overleftrightarrow{DC}$ is _?_.

24. Any line perpendicular to $\overleftrightarrow{AB}$ has slope _?_.

25. Prove that quadrilateral $ABCD$ is a parallelogram.

Table of Squares

Number n	Square n^2	Number n	Square n^2	Number n	Square n^2	Number n	Square n^2
1	1	**26**	676	**51**	2601	**76**	5776
2	4	**27**	729	**52**	2704	**77**	5929
3	9	**28**	784	**53**	2809	**78**	6084
4	16	**29**	841	**54**	2916	**79**	6241
5	25	**30**	900	**55**	3025	**80**	6400
6	36	**31**	961	**56**	3136	**81**	6561
7	49	**32**	1024	**57**	3249	**82**	6724
8	64	**33**	1089	**58**	3364	**83**	6889
9	81	**34**	1156	**59**	3481	**84**	7056
10	100	**35**	1225	**60**	3600	**85**	7225
11	121	**36**	1296	**61**	3721	**86**	7396
12	144	**37**	1369	**62**	3844	**87**	7569
13	169	**38**	1444	**63**	3969	**88**	7744
14	196	**39**	1521	**64**	4096	**89**	7921
15	225	**40**	1600	**65**	4225	**90**	8100
16	256	**41**	1681	**66**	4356	**91**	8281
17	289	**42**	1764	**67**	4489	**92**	8464
18	324	**43**	1849	**68**	4624	**93**	8649
19	361	**44**	1936	**69**	4761	**94**	8836
20	400	**45**	2025	**70**	4900	**95**	9025
21	441	**46**	2116	**71**	5041	**96**	9216
22	484	**47**	2209	**72**	5184	**97**	9409
23	529	**48**	2304	**73**	5329	**98**	9604
24	576	**49**	2401	**74**	5476	**99**	9801
25	625	**50**	2500	**75**	5625	**100**	10,000

Table of Square Roots

The square roots are given correct to three decimal places.

Number n	Positive Square Root $\sqrt{n}$	Number n	Positive Square Root $\sqrt{n}$	Number n	Positive Square Root $\sqrt{n}$	Number n	Positive Square Root $\sqrt{n}$
1	1	**26**	5.099	**51**	7.141	**76**	8.718
2	1.414	**27**	5.196	**52**	7.211	**77**	8.775
3	1.732	**28**	5.292	**53**	7.280	**78**	8.832
4	2	**29**	5.385	**54**	7.348	**79**	8.888
5	2.236	**30**	5.477	**55**	7.416	**80**	8.944
6	2.449	**31**	5.568	**56**	7.483	**81**	9
7	2.646	**32**	5.657	**57**	7.550	**82**	9.055
8	2.828	**33**	5.745	**58**	7.616	**83**	9.110
9	3	**34**	5.831	**59**	7.681	**84**	9.165
10	3.162	**35**	5.916	**60**	7.746	**85**	9.220
11	3.317	**36**	6	**61**	7.810	**86**	9.274
12	3.464	**37**	6.083	**62**	7.874	**87**	9.327
13	3.606	**38**	6.164	**63**	7.937	**88**	9.381
14	3.742	**39**	6.245	**64**	8	**89**	9.434
15	3.873	**40**	6.325	**65**	8.062	**90**	9.487
16	4	**41**	6.403	**66**	8.124	**91**	9.539
17	4.123	**42**	6.481	**67**	8.185	**92**	9.592
18	4.243	**43**	6.557	**68**	8.246	**93**	9.644
19	4.359	**44**	6.633	**69**	8.307	**94**	9.695
20	4.472	**45**	6.708	**70**	8.367	**95**	9.747
21	4.583	**46**	6.782	**71**	8.426	**96**	9.798
22	4.690	**47**	6.856	**72**	8.485	**97**	9.849
23	4.796	**48**	6.928	**73**	8.544	**98**	9.899
24	4.899	**49**	7	**74**	8.602	**99**	9.950
25	5	**50**	7.071	**75**	8.660	**100**	10

Table of Trigonometric Ratios

Angle	Sine	Cosine	Tangent	Angle	Sine	Cosine	Tangent
1°	.0175	.9998	.0175	**46°**	.7193	.6947	1.0355
2°	.0349	.9994	.0349	**47°**	.7314	.6820	1.0724
3°	.0523	.9986	.0524	**48°**	.7431	.6691	1.1106
4°	.0698	.9976	.0699	**49°**	.7547	.6561	1.1504
5°	.0872	.9962	.0875	**50°**	.7660	.6428	1.1918
6°	.1045	.9945	.1051	**51°**	.7771	.6293	1.2349
7°	.1219	.9925	.1228	**52°**	.7880	.6157	1.2799
8°	.1392	.9903	.1405	**53°**	.7986	.6018	1.3270
9°	.1564	.9877	.1584	**54°**	.8090	.5878	1.3764
10°	.1736	.9848	.1763	**55°**	.8192	.5736	1.4281
11°	.1908	.9816	.1944	**56°**	.8290	.5592	1.4826
12°	.2079	.9781	.2126	**57°**	.8387	.5446	1.5399
13°	.2250	.9744	.2309	**58°**	.8480	.5299	1.6003
14°	.2419	.9703	.2493	**59°**	.8572	.5150	1.6643
15°	.2588	.9659	.2679	**60°**	.8660	.5000	1.7321
16°	.2756	.9613	.2867	**61°**	.8746	.4848	1.8040
17°	.2924	.9563	.3057	**62°**	.8829	.4695	1.8807
18°	.3090	.9511	.3249	**63°**	.8910	.4540	1.9626
19°	.3256	.9455	.3443	**64°**	.8988	.4384	2.0503
20°	.3420	.9397	.3640	**65°**	.9063	.4226	2.1445
21°	.3584	.9336	.3839	**66°**	.9135	.4067	2.2460
22°	.3746	.9272	.4040	**67°**	.9205	.3907	2.3559
23°	.3907	.9205	.4245	**68°**	.9272	.3746	2.4751
24°	.4067	.9135	.4452	**69°**	.9336	.3584	2.6051
25°	.4226	.9063	.4663	**70°**	.9397	.3420	2.7475
26°	.4384	.8988	.4877	**71°**	.9455	.3256	2.9042
27°	.4540	.8910	.5095	**72°**	.9511	.3090	3.0777
28°	.4695	.8829	.5317	**73°**	.9563	.2924	3.2709
29°	.4848	.8746	.5543	**74°**	.9613	.2756	3.4874
30°	.5000	.8660	.5774	**75°**	.9659	.2588	3.7321
31°	.5150	.8572	.6009	**76°**	.9703	.2419	4.0108
32°	.5299	.8480	.6249	**77°**	.9744	.2250	4.3315
33°	.5446	.8387	.6494	**78°**	.9781	.2079	4.7046
34°	.5592	.8290	.6745	**79°**	.9816	.1908	5.1446
35°	.5736	.8192	.7002	**80°**	.9848	.1736	5.6713
36°	.5878	.8090	.7265	**81°**	.9877	.1564	6.3138
37°	.6018	.7986	.7536	**82°**	.9903	.1392	7.1154
38°	.6157	.7880	.7813	**83°**	.9925	.1219	8.1443
39°	.6293	.7771	.8098	**84°**	.9945	.1045	9.5144
40°	.6428	.7660	.8391	**85°**	.9962	.0872	11.4301
41°	.6561	.7547	.8693	**86°**	.9976	.0698	14.3007
42°	.6691	.7431	.9004	**87°**	.9986	.0523	19.0811
43°	.6820	.7314	.9325	**88°**	.9994	.0349	28.6363
44°	.6947	.7193	.9657	**89°**	.9998	.0175	57.2900
45°	.7071	.7071	1.0000	**90°**	1.0000	0.0000	Undefined

Postulates

THE ADDITION POSTULATE
page 29

If $a = b$ and $c = d$, then $a + c = b + d$.

THE SUBTRACTION POSTULATE
page 29

If $a = b$ and $c = d$, then $a - c = b - d$.

THE MULTIPLICATION POSTULATE
page 29

If $a = b$, then $ac = bc$.

THE DIVISION POSTULATE
page 30

If $a = b$ and $c \neq 0$, then $\frac{a}{c} = \frac{b}{c}$.

THE SUBSTITUTION POSTULATE
page 30

If $a = b$, then a can be substituted for b in any equation or inequality.

POSTULATE 1
page 34

Through any two points there is exactly one line.

POSTULATE 2
page 34

Through any three noncollinear points there is exactly one plane.

POSTULATE 3
page 34

If two points lie in a plane, then the line joining them lies in that plane.

POSTULATE 4
page 34

If two planes intersect, then their intersection is a line.

POSTULATE 5 (The Ruler Postulate)
page 34

Each point on a line can be paired with exactly one real number called its coordinate. The distance between two points is the positive difference of their coordinates.

POSTULATE 6 (The Protractor Postulate)
page 35

Suppose O is a point of $\overleftrightarrow{XY}$. Consider all rays with endpoint O which lie on one side of $\overleftrightarrow{XY}$. Each ray can be paired with exactly one real number between 0 and 180.

POSTULATE 7
page 67

If two parallel lines are cut by a transversal, then corresponding angles are equal.

POSTULATE 8
page 72

If two lines and a transversal form equal corresponding angles, then the lines are parallel.

POSTULATE 9 (The SSS Postulate)
page 107

If three sides of one triangle are equal to the corresponding parts of another triangle, the triangles are congruent.

POSTULATE 10 (The SAS Postulate)
page 116

If two sides and the included angle of one triangle are equal to the corresponding parts of another triangle, the triangles are congruent.

POSTULATE 11 (The ASA Postulate)
page 121

If two angles and the included side of one triangle are equal to the corresponding parts of another triangle, the triangles are congruent.

POSTULATE 12
page 238

The area of a rectangle is given by the formula:
Area = base × height

POSTULATE 13 (The AA Postulate)
page 317

If two angles of one triangle are equal to two angles of another triangle, then the triangles are similar.

Theorems

CHAPTER 2 Introducing Proof

THEOREM 1 *page 46* — If two angles are complements of equal angles (or of the same angle), then the two angles are equal.

THEOREM 2 *page 47* — If two angles are supplements of equal angles (or of the same angle), then the two angles are equal.

THEOREM 3 *page 47* — Vertical angles are equal.

THEOREM 4 *page 69* — If two parallel lines are cut by a transversal, then alternate interior angles are equal.

THEOREM 5 *page 69* — If two parallel lines are cut by a transversal, then same-side interior angles are supplementary.

THEOREM 6 *page 72* — If two lines and a transversal form equal alternate interior angles, then the lines are parallel.

THEOREM 7 *page 72* — If two lines and a transversal form supplementary same-side interior angles, then the lines are parallel.

THEOREM 8 *page 72* — In a plane, if two lines are each perpendicular to a third line, then the two lines are parallel.

CHAPTER 3 Triangles

THEOREM 1 *page 92* — The angle sum of a triangle is 180°.

COROLLARY *page 93* — An exterior angle of a triangle is equal to the sum of the two opposite angles of the triangle.

THEOREM 2 *page 126* — (AAS Theorem) If two angles and a non-included side of one triangle are equal to the corresponding parts of another triangle, the triangles are congruent.

THEOREM 3 *page 126* — (HL Theorem) If the hypotenuse and a leg of one right triangle are equal to the corresponding parts of another right triangle, the triangles are congruent.

CHAPTER 4 Using Congruent Triangles

THEOREM 1 *page 154* — Any point on the perpendicular bisector of a segment is equidistant from the endpoints of the segment.

THEOREM 2 *page 155* — Any point that is equidistant from the endpoints of a segment is on the perpendicular bisector of the segment.

THEOREM 3 *page 167* — If two sides of a triangle are equal, then the angles opposite those sides are equal.

COROLLARY *page 167* — An equilateral triangle is also equiangular, and each angle has measure 60°.

THEOREM 4 *page 173* — If two angles of a triangle are equal, then the sides opposite those angles are equal.

COROLLARY *page 173* — If a triangle is equiangular, it is also equilateral.

CHAPTER 5 Polygons

THEOREM 1 *page 194* — If a convex polygon has n sides, then its angle sum is given by the formula:
$S = (n - 2) \times 180°$

THEOREM 2 *page 195* — The exterior angle sum of any convex polygon, one angle at each vertex, is 360°.

THEOREM 3 *page 204* — Opposite sides of a parallelogram are equal.

THEOREM 4 *page 204* — Opposite angles of a parallelogram are equal.

THEOREM 5 *page 204* — Consecutive angles of a parallelogram are supplementary.

THEOREM 6 *page 204* — Diagonals of a parallelogram bisect each other.

THEOREM 7 *page 210* — The diagonals of a rectangle are equal.

THEOREM 8 *page 210* — The diagonals of a rhombus are perpendicular, and they bisect the angles of the rhombus.

THEOREM 9 *page 210* — The midpoint of the hypotenuse of a right triangle is equidistant from the three vertices.

THEOREM 10 *page 214* — If a quadrilateral has one pair of opposite sides that are both parallel and equal, then the quadrilateral is a parallelogram.

THEOREM 11 *page 215* — If a quadrilateral has both pairs of opposite sides equal, then the quadrilateral is a parallelogram.

THEOREM 12 *page 215* — If a quadrilateral has diagonals that bisect each other, then the quadrilateral is a parallelogram.

THEOREM 13 *page 221* — The base angles of an isosceles trapezoid are equal.

THEOREM 14 *page 221* — The median of a trapezoid has two properties:
(1) It is parallel to the bases.
(2) Its length equals half the sum of the base lengths.

THEOREM 15 *page 225* — (The Midpoints Theorem) The segment joining the midpoints of two sides of a triangle is parallel to the third side and half as long.

CHAPTER 6 Areas

THEOREM 1 *page 239* — The area of a square is given by the formula:
Area = side squared

THEOREM 2 *page 244* — The area of a parallelogram is given by the formula:
Area = base × height

THEOREM 3 *page 247* — The area of a triangle is given by the formula:
$$\text{Area} = \frac{1}{2} \times \text{base} \times \text{height}$$

THEOREM 4 *page 251* — The area of a trapezoid is given by the formula:
$$\text{Area} = \frac{1}{2} \times \text{height} \times \text{sum of the bases}$$

THEOREM 5 *page 256* (The Pythagorean Theorem) In a right triangle, the square of the hypotenuse is equal to the sum of the squares of the legs.

THEOREM 6 *page 259* If the square of one side of a triangle is equal to the sum of the squares of the other two sides, then the triangle is a right triangle.

CHAPTER 8 Similar Polygons

THEOREM 1 *page 325* (The Triangle Proportionality Theorem) If a line intersects a triangle and is parallel to one side, then it divides the other two sides proportionally.

COROLLARY *page 326* If three parallel lines intersect two transversals, they divide the transversals proportionally.

THEOREM 2 *page 330* If two similar polygons have a scale factor $a:b$, then:
(1) the ratio of their perimeters is $a:b$;
(2) the ratio of their areas is $a^2:b^2$.

CHAPTER 9 Circles

THEOREM 1 *page 350* A radius drawn to a point of tangency is perpendicular to the tangent.

THEOREM 2 *page 350* If a line lies in the plane of a circle and is perpendicular to a radius at its outer endpoint, the line is tangent to the circle.

THEOREM 3 *page 356* In a circle, equal central angles have equal minor arcs.

THEOREM 4 *page 356* In a circle, equal minor arcs have equal central angles.

THEOREM 5 *page 360* In a circle, equal chords have equal arcs and equal arcs have equal chords.

THEOREM 6 *page 360* In a circle, equal chords are equidistant from the center. Chords that are equidistant from the center are equal.

THEOREM 7 *page 360* A diameter that is perpendicular to a chord bisects the chord and its arc.

THEOREM 8 *page 364* An inscribed angle is equal to half its intercepted arc.

THEOREM 9 *page 365* An angle formed by a chord and a tangent is equal to half its intercepted arc.

THEOREM 10 *page 369* An angle formed by two chords is equal to half the sum of the intercepted arcs.

THEOREM 11 *page 369* An angle formed by two secants is equal to half the difference of the intercepted arcs.

THEOREM 12 *page 373* If two chords intersect inside a circle, the product of the lengths of the segments of one chord equals the product of the lengths of the segments of the other.

CHAPTER 10 Areas and Volumes of Solids

THEOREM 1 *page 414* Suppose two similar solids have scale factor $a:b$. Then:
(1) the ratio of corresponding segments is $a:b$;
(2) the ratio of the areas is $a^2:b^2$;
(3) the ratio of the volumes is $a^3:b^3$.

CHAPTER 11 Right Triangles

THEOREM 1 In a 45° -45° -90° triangle, the hypotenuse is $\sqrt{2}$ times as long as a leg.
page 432

THEOREM 2 In a 30° -60° -90° triangle:
page 432
(1) the hypotenuse is twice as long as the shorter leg;
(2) the longer leg is $\sqrt{3}$ times as long as the shorter leg.

Constructions

CONSTRUCTION 1 *Given:* An angle
page 25 *Construct:* A bisector of the angle

CONSTRUCTION 2 *Given:* A line and a point on the line
page 26 *Construct:* A perpendicular to the line through the point

CONSTRUCTION 3 *Given:* A line and a point not on the line
page 26 *Construct:* A perpendicular to the line through the point

CONSTRUCTION 4 *Given:* An angle
page 77 *Construct:* An angle equal to the given angle

CONSTRUCTION 5 *Given:* A line and a point not on the line
page 78 *Construct:* A line that passes through the point and is parallel to the given line

CONSTRUCTION 6 *Given:* A segment
page 154 *Construct:* A perpendicular bisector of the segment

CONSTRUCTION 7 *Given:* A triangle
page 163 *Construct:* A circle passing through the vertices of the triangle

CONSTRUCTION 8 *Given:* A triangle
page 164 *Construct:* An inscribed circle

CONSTRUCTION 9 *Construct:* A 60° angle
page 174

CONSTRUCTION 10 *Given:* Point A on $\odot O$
page 351 *Construct:* A tangent to $\odot O$ at point A

Review Exercises

CHAPTER 1

In the diagram, $\overline{AD} \perp \overline{FC}$ and $\angle DOE = 30°$.
a. Find the measures of the following angles.
b. Then classify each as acute, right, or obtuse.

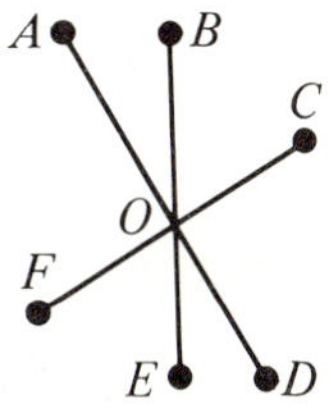

1. $\angle BOA$ 2. $\angle BOF$
3. $\angle COD$ 4. $\angle BOD$
5. $\angle BOA$ and $\angle DOE$ are called ___?___ angles.
6. Name an angle equal to $\angle BOD$.

In the diagram, $\overrightarrow{OX}$ bisects $\angle WOY$ and $\overrightarrow{OY}$ bisects $\angle XOZ$.

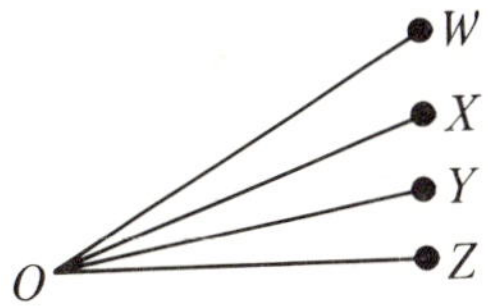

7. $\angle WOX = \angle$ ___?___ 8. $\angle YOZ = \angle$ ___?___
9. Which postulate allows you to conclude that $\angle WOX = \angle YOZ$?

Refer to the rectangular solid shown.

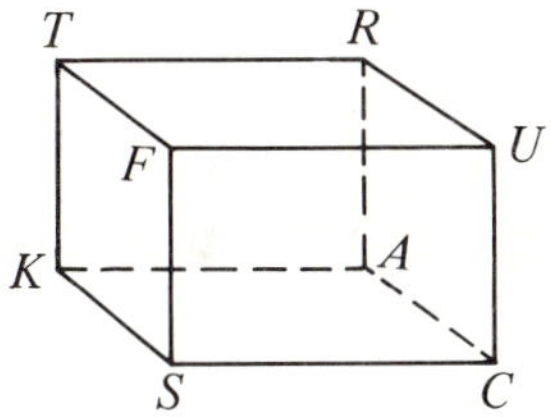

10. K, S, F, and ___?___ are coplanar points.
11. Plane $TFUR \parallel$ plane ___?___
12. Name two planes that intersect in $\overleftrightarrow{FS}$.
13. How many planes contain points R, U, F, and S?
14. Construct a 90° angle.

CHAPTER 2

Given: $\overline{AB} \perp \overline{BC}$ and $\overline{AD} \perp \overline{DC}$.

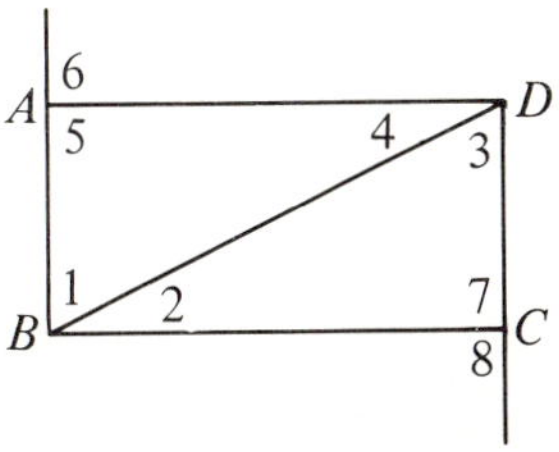

1. $\angle 1$ and $\angle$ ___?___ are called complementary angles.
2. $\angle 5$ and $\angle 6$ are called ___?___ angles.
3. If $\angle 1 = \angle 3$, state the theorem that allows you to conclude that $\angle 2 = \angle 4$.
4. If $\angle 5 = \angle 7$, state the theorem that allows you to conclude that $\angle 6 = \angle 8$.

Exercises 5–11 refer to the figure at the right.

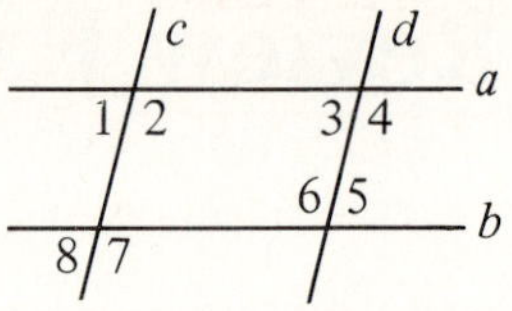

5. $\angle 1$ and $\angle 8$ are called __?__ angles.

6. $\angle 2$ and $\angle$ __?__ are called same-side interior angles.

7. $\angle 6$ and $\angle 7$ are called __?__ angles.

Use the given information to tell which lines must be parallel.

8. $\angle 3 = \angle 5$ 9. $\angle 1 = \angle 3$ 10. $\angle 8 = \angle 5$ 11. $\angle 4 + \angle 5 = 180°$

12. Write a complete proof in two-column form.
Given: $\angle 2 = \angle 3$
$\overrightarrow{QS}$ bisects $\angle PQR$.
Prove: $\overleftrightarrow{PQ} \parallel \overleftrightarrow{SR}$

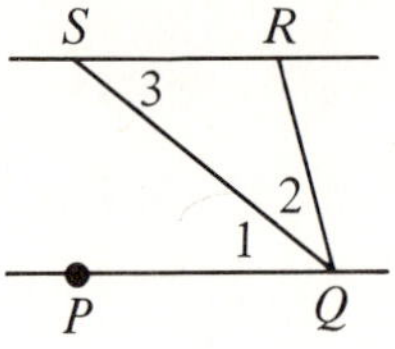

13. Draw an acute angle, $\angle A$.
Then construct an angle equal to $\angle A$.

CHAPTER 3

In each exercise, find the value of *x*.

1.
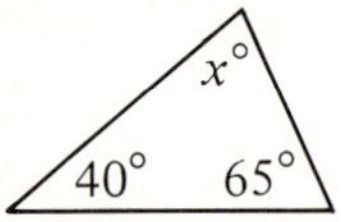

2.
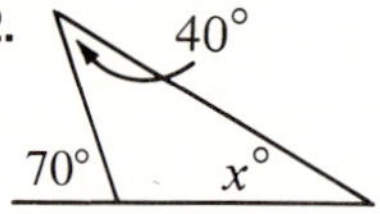

3.
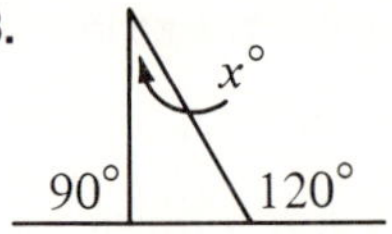

4. In $\triangle ABC$, $AB = AC$.
   a. $\triangle ABC$ must be a(n) __?__ triangle.
   b. Is it possible that $\triangle ABC$ is an equilateral triangle?
   c. Is it possible that $\triangle ABC$ is an obtuse triangle?
   d. Is it possible that $\triangle ABC$ is an acute triangle?

5. Suppose $\triangle ANT \cong \triangle LER$.
   a. $AT =$ __?__ b. __?__ $= ER$ c. $\angle N = \angle$ __?__

Write SSS, SAS, ASA, AAS, or HL to indicate a method you could use to prove two triangles congruent.

6.
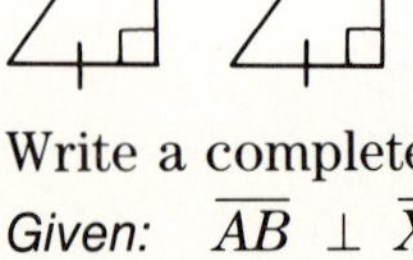

7.
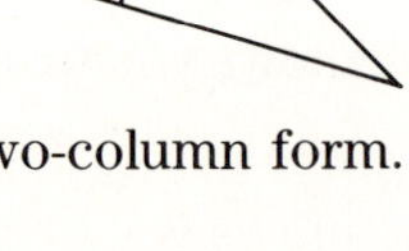

8.
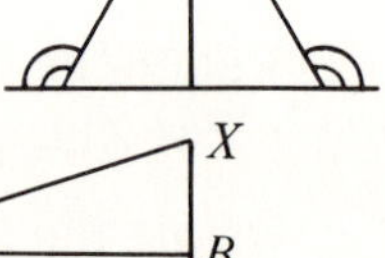

9. Write a complete proof in two-column form.
Given: $\overline{AB} \perp \overline{XY}$
$AX = AY$
Prove: $\triangle ABX \cong \triangle ABY$

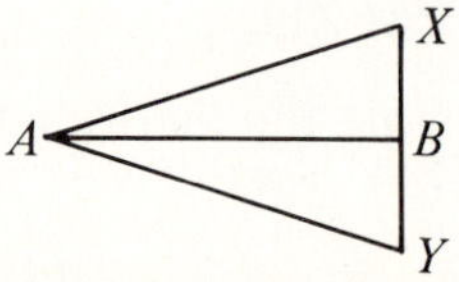

CHAPTER 4

1. Suppose that $\triangle RUN \cong \triangle TEX$. What reason can you write to support the statements: $UN = EX$ and $\angle N = \angle X$?
2. Draw a large $\triangle ABC$ with $\overline{AB}$ clearly longer than $\overline{AC}$. Draw the altitude $\overline{AH}$ from point A and the median $\overline{AM}$ from point A.
3. Draw a figure like the one for Exercise 2, but much larger. Construct the perpendicular bisectors of any two sides of the triangle. Then construct a circle circumscribed about $\triangle ABC$.

In each diagram, equal angles are marked. Find the value of *x*.

4. 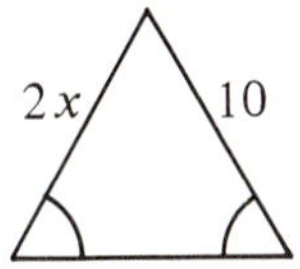

5. 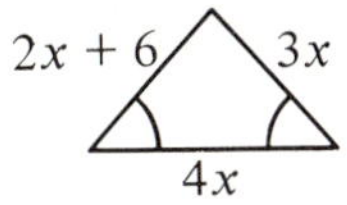

6. 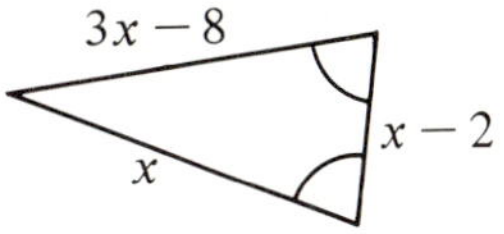

Write complete proofs in two-column form.

7. *Given:* $\angle A = \angle X$
   $AB = XY$
   $AC = XZ$
   Prove: $BC = YZ$

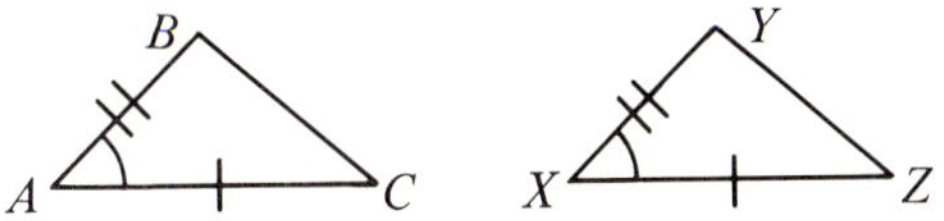

8. *Given:* $RS = TS$
   Prove: $\angle 3 = \angle 2$

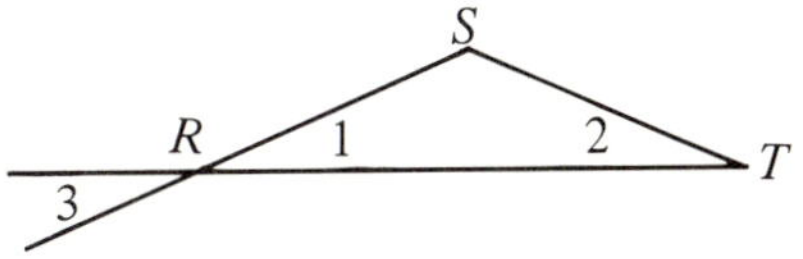

CHAPTER 5

Tell whether each statement is true or false.

1. An octagon has ten sides.
2. A regular polygon is equilateral.
3. Every square is a rectangle.
4. Every rectangle is a square.
5. The interior angle sum of a hexagon is 720°.
6. In a regular pentagon, each interior angle has measure 110°.
7. The diagonals of a rhombus are perpendicular.
8. The diagonals of a trapezoid bisect each other.
9. The diagonals of a parallelogram bisect each other.

Tell whether each statement is true or false.

10. The legs of a trapezoid are parallel.

11. If the opposite angles of a quadrilateral are equal, then it must be a parallelogram.

12. If a quadrilateral has two 80° angles and two 100° angles, then it must be a parallelogram.

13. Base angles of an isosceles trapezoid are supplementary.

14. The diagonals of a rectangle are equal.

15. Write a complete proof in two-column form.
Given: X, Y, and Z are the midpoints of the sides of $\triangle ABC$.
Prove: $AXYZ$ is a $\square$.

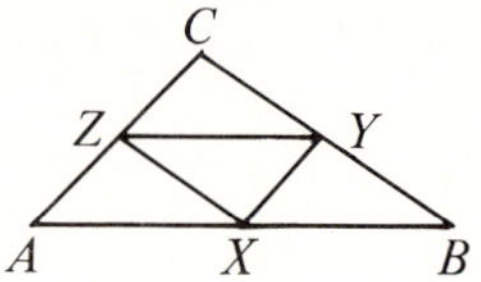

CHAPTER 6

1. A rectangle has sides 7 m and 3 m. Find the perimeter and the area.

2. Find the area of a square whose perimeter is 12 cm.

Find the area of the parallelogram, the triangle, and the trapezoid.

3.
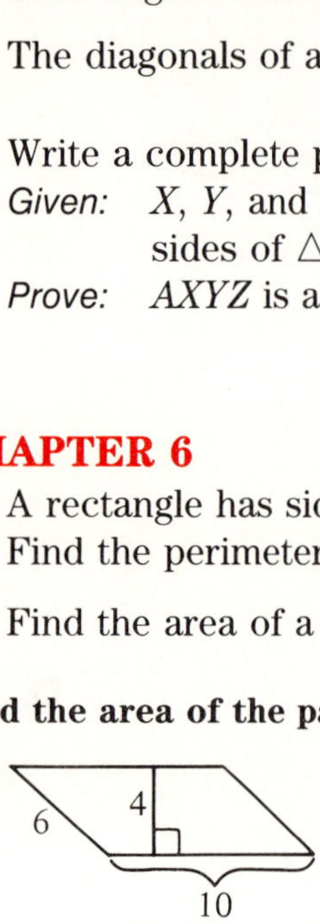

4.
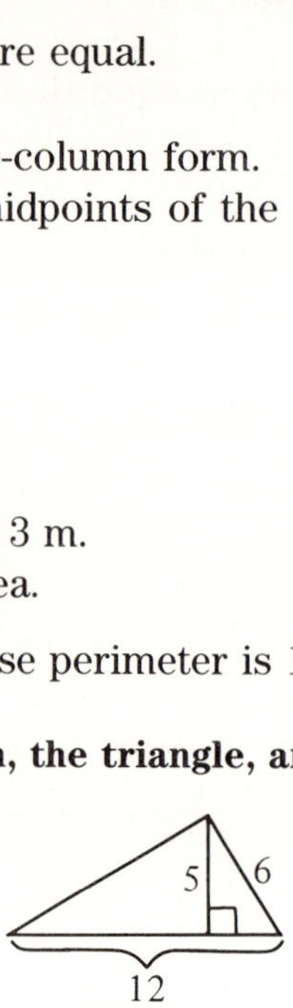

5.
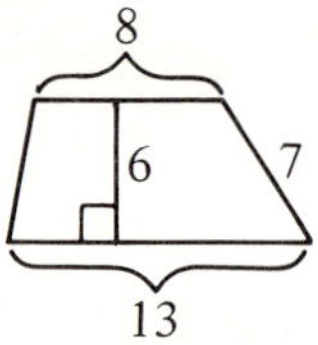

In each diagram, find the value of *x*.

6.
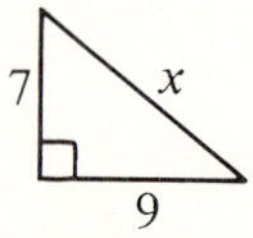

7.
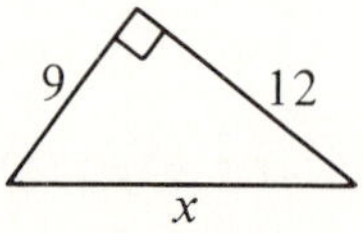

8.
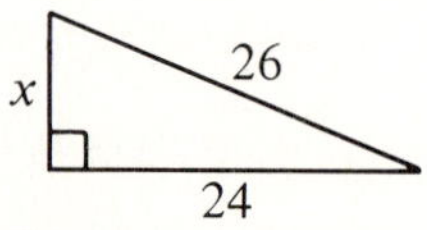

Three sides of a triangle are given. Classify the triangle as acute, right, or obtuse.

9. 3, 4, 5

10. 3, 4, 6

11. 7, 7, 7

Find the circumference of each circle, in terms of π, given:

12. $r = 6$

13. $r = 1.5$

14. $d = 8$

15–17. Use 3.14 for π. Find the area of each circle described above, correct to tenths.

CHAPTER 7

Express each ratio in simplest form.

1. $\frac{15}{9}$
2. $\frac{24}{36}$
3. $\frac{15}{20}$
4. $\frac{3ac}{9ab}(a \neq 0, b \neq 0)$
5. 18:20
6. 9:12:15
7. 2 m:40 cm
8. 3 km:500 m

9. $\frac{SA}{AL}$
10. $\frac{SR}{ST}$
11. $\frac{AR}{LT}$
12. $\frac{TR}{RS}$

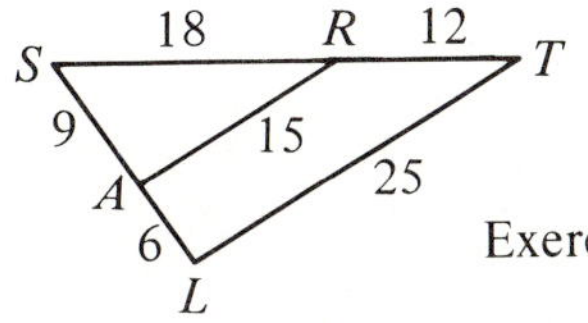

Exercises 9–12

Find the value of *x*.

13. $\frac{4}{x} = \frac{6}{9}$
14. $\frac{18}{x} = \frac{9}{4}$
15. $x:9 = 6:27$
16. $\frac{x-3}{4} = \frac{x}{8}$
17. Two bags of cookies cost $1.78. How much will 5 bags cost?

Exercises 18 and 19 refer to the diagram.

18. Measure the distance on the map between Livonia and Geneva to the nearest centimeter. Then give the scale of the map.
19. Find the actual distance between Raleigh and Geneva.
20. Make a scale drawing of a soccer field 100 m by 60 m.

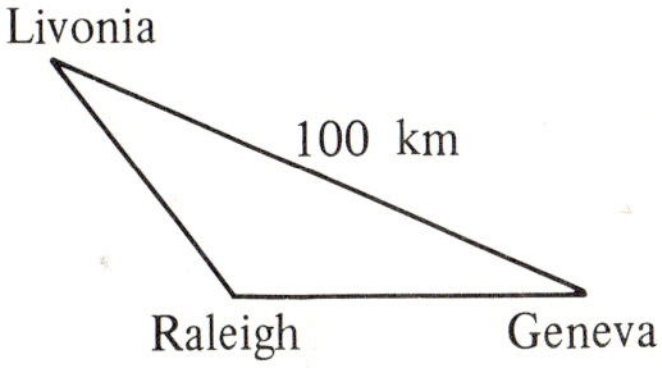

CHAPTER 8

State whether the following polygons must be similar.

1. Any two right triangles
2. Any two regular pentagons
3. Any two congruent polygons
4. Any two isosceles trapezoids

Find the values of *x* and *y* in each diagram.

5.
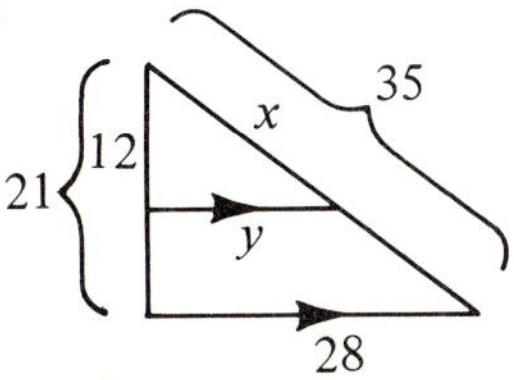

6.
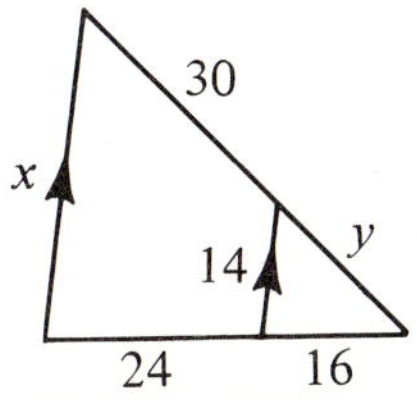

7.
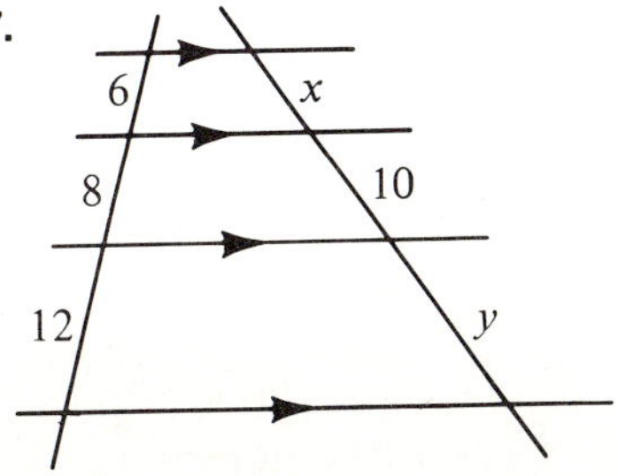

Exercises 8–10 refer to the figure shown.

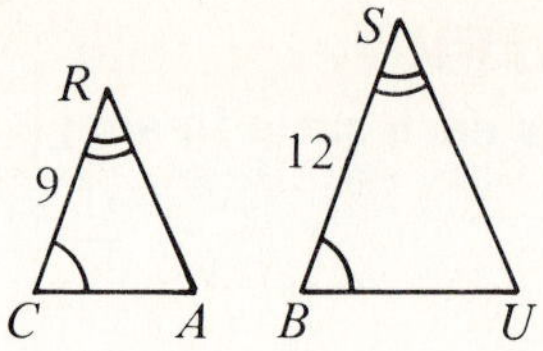

8. Which postulate supports the statement:

$$\triangle RCA \sim \triangle SBU?$$

9. Find the ratio of the perimeters of the triangles.

10. Find the ratio of the areas of the triangles.

11. Write a proof in two-column form.
Given: Trapezoid $ABCD$
Prove: $\frac{AO}{CO} = \frac{BO}{DO}$
(*Hint:* Find two similar triangles.)

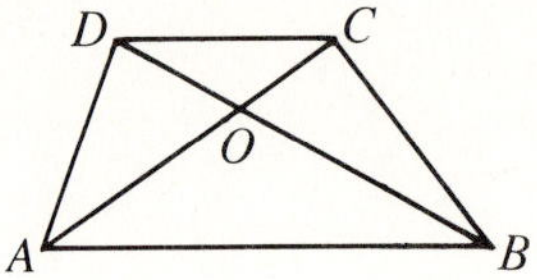

CHAPTER 9

In the diagram, $\overleftrightarrow{UV}$ is tangent to $\odot O$. $\overset{\frown}{AB} = 70°$ and $\overset{\frown}{BC} = 80°$. Find the measure of the arc or angle.

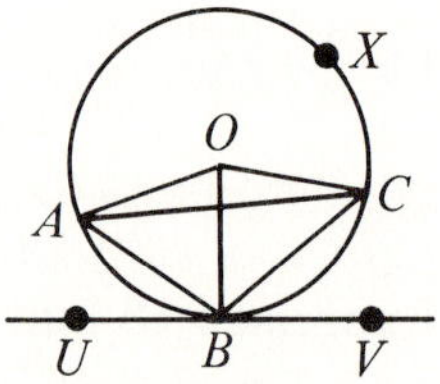

1. $\overset{\frown}{ABC}$
2. $\overset{\frown}{AXC}$
3. $\angle OBU$
4. $\angle AOB$
5. $\angle ACB$
6. $\angle CBV$

$\overline{YZ}$ is a diameter of $\odot P$. $\overline{YZ} \perp \overline{RT}$.

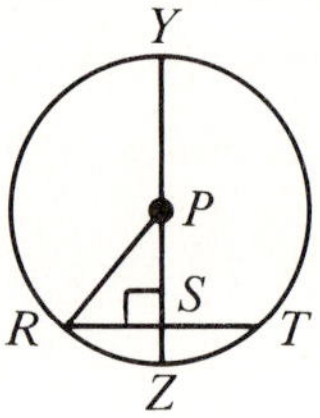

7. If $\overset{\frown}{RT} = 100°$, then $\overset{\frown}{RZ} =$ __?__° and $\overset{\frown}{RY} =$ __?__°.
8. If $RS = 7$, then $ST =$ __?__ and $RT =$ __?__.
9. If $PR = 10$ and $PS = 8$, then $RS =$ __?__.
10. If $YZ = 30$ and $RT = 18$, then $PS =$ __?__.

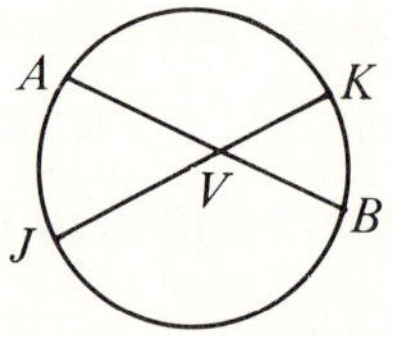

11. If $\overset{\frown}{AJ} = 60°$ and $\overset{\frown}{KB} = 40°$, then $\angle AVJ =$ __?__°.
12. If $\overset{\frown}{JB} = 170°$ and $\angle AVK = 155°$, then $\overset{\frown}{AK} =$ __?__°.
13. Suppose $\overrightarrow{AK}$ and $\overrightarrow{JB}$ intersect at P (not shown). If $\overset{\frown}{AJ} = 70°$ and $\overset{\frown}{KB} = 40°$, then $\angle APJ =$ __?__°.
14. If $AV = 12$, $VB = 9$, and $JV = 18$, then $VK =$ __?__.

15. Construct a circle. Then construct a tangent to the circle.

CHAPTER 10

1. Name three possible positions of two different lines, j and k.
2. Name three possible positions of a line l and a plane X.

3. Can two planes, P and Q, be called *skew* planes?

A rectangular solid is shown. Find:

4. The area of a base
5. The lateral area
6. The total area
7. The volume

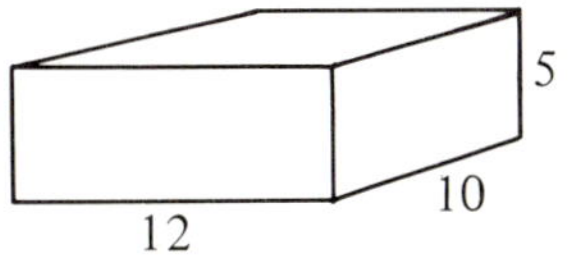

In each exercise, find:
a. the lateral area b. the total area c. the volume.
In Exercises 8 and 9, express your answers in terms of π.

8. 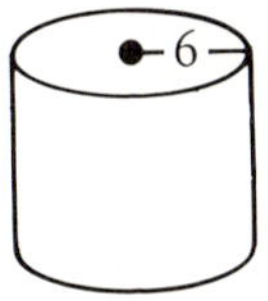

Cylinder

9. 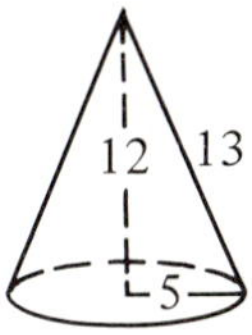

Cone

10. 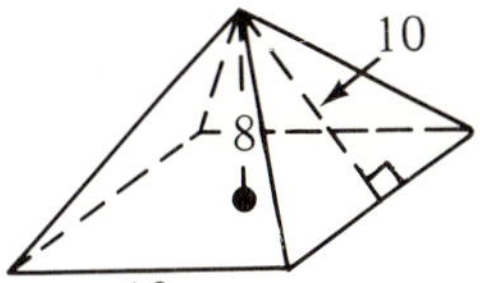

Regular square pyramid

11. In a certain cylinder, the radius is 3 and the height is 2. Find the volume correct to tenths. (Use 3.14 for π.)
12. A sphere has radius 6. Find the area and volume in terms of π.
13. The area of a sphere is 36π. Find the volume in terms of π.
14. The slant height of a cone is 8, and the lateral area is 24π. Find the diameter.

CHAPTER 11

Find the exact values of x and y.

1. 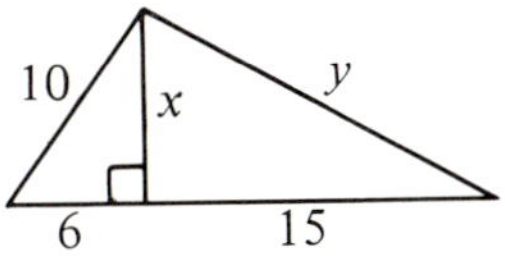

2. 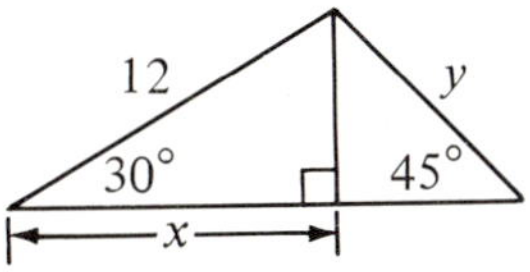

3. 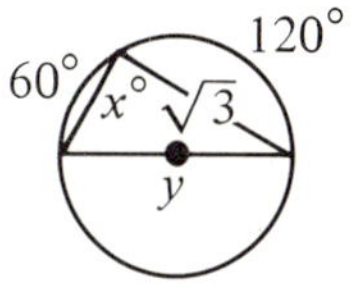

Find the area of each polygon.

4. A square with diagonals of length $8\sqrt{2}$
5. A rectangle with width 5 and diagonals of length 13
6. A $30°$-$60°$-$90°$ triangle with hypotenuse of length 4
7. An equilateral triangle with sides 6 cm long
8. A parallelogram with sides 4 m and 8 m, and with a $60°$ angle
9. An isosceles triangle with sides of length 13, 13, and 10

10. A box is 12 cm by 9 cm by 7 cm. Find the length of a diagonal.

11. Suppose a cone has diameter 12 and height 8. Find:
a. the radius **b.** the slant height **c.** the total area **d.** the volume.

Find x, correct to tenths. Use the table on page 517.

12.

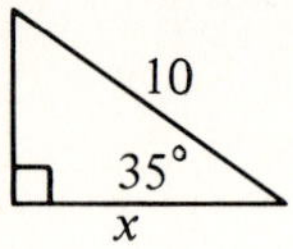

13.

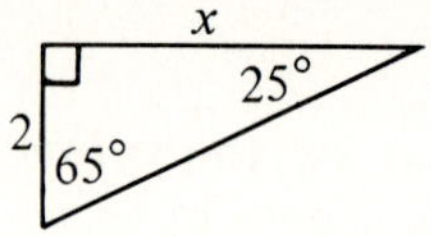

14.

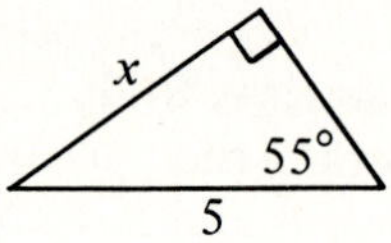

CHAPTER 12

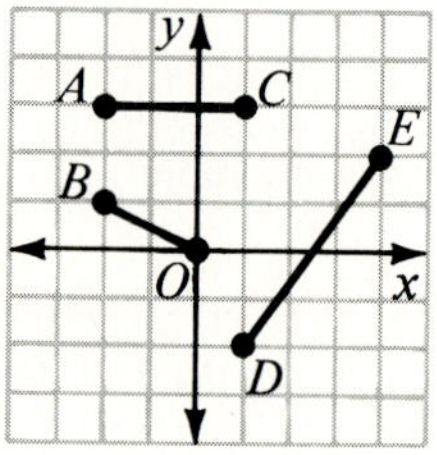

1. Point B lies in quadrant __?__.
2. The coordinates of point C are (__?__, __?__).
3. $AC =$ __?__, and $OB =$ __?__.
4. The midpoint of $\overline{OB}$ has coordinates (__?__, __?__).
5. The slope of $\overline{AC}$ is __?__, and the slope of $\overline{DE}$ is __?__.
6. Does point (3, 1) lie on the graph of the equation $5x - y = 13$?
7. The slope of line j is $\frac{3}{8}$. What is the slope of:
   a. any line parallel to j? **b.** any line perpendicular to j?
8. The vertices of a triangle are $K(-2, 3)$, $E(3, 5)$, and $L(7, -5)$. What angle of $\triangle KEL$ is a right angle?
9. Use the equation $2x + 3y = 12$.
   a. Copy and complete the table of values.
   b. Using a pair of coordinate axes, plot the points. Then draw the graph of $2x + 3y = 12$.

x	y
0	__?__
__?__	0
3	__?__
−3	__?__

10. Draw the graph of the equation $x = 3$.
11. The vertices of a triangle are $A(-1, -1)$, $B(3, -3)$, and $C(3, 2)$. Prove that $\triangle ABC$ is an isosceles triangle.

Appendix: Transformational Geometry

SYMMETRY

Perhaps you have made a design by placing some ink on a folded piece of paper and then pressing the paper together. The figure is said to be **symmetric** with respect to the line of the fold. The two figures on opposite sides of the line of symmetry are identical in shape and size. However, they are reversed in direction.

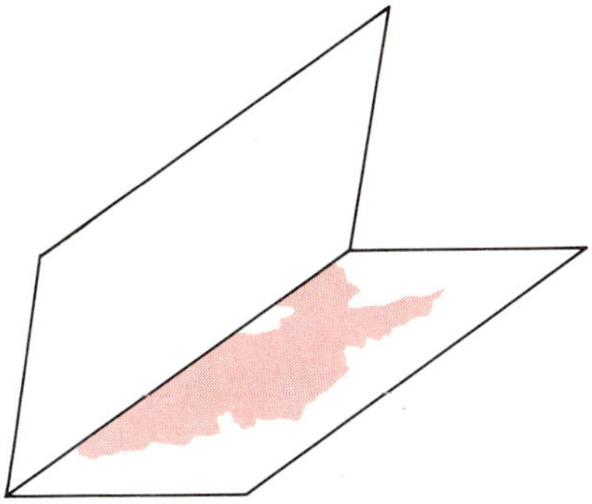

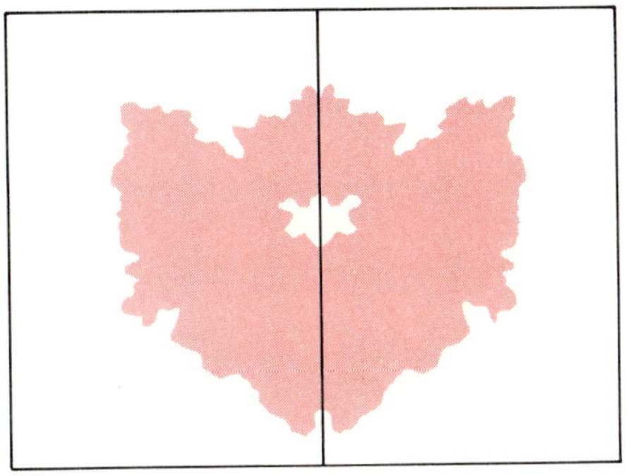

Line symmetry occurs frequently in art and nature as illustrated below. The Mexican art and the butterfly each have one line of symmetry.

The daffodil shows that it is possible to have more than one line of symmetry for the same figure.

Two points are symmetric to one another if the line of symmetry is the perpendicular bisector of the segment joining the two points.

A and *B* are symmetric points with respect to line *m*.

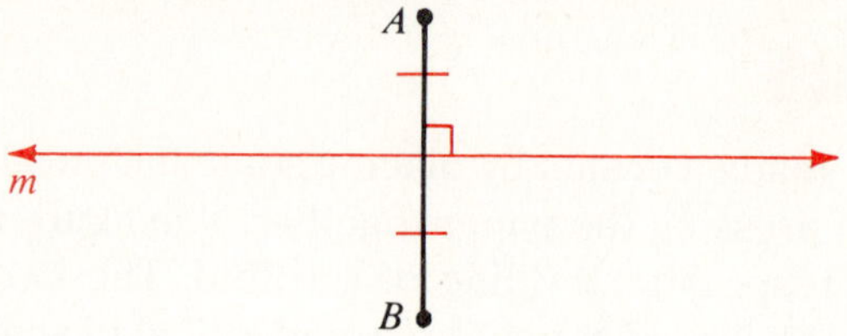

A figure is said to have line symmetry in a geometric sense if *every* point of the figure is symmetric to another point of the figure.

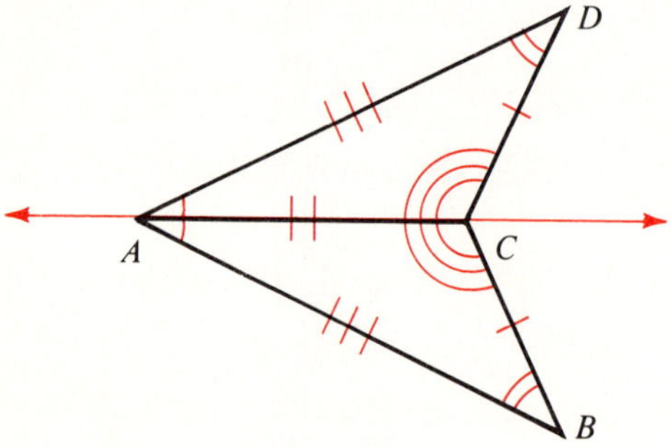

△ *ABC* is symmetric to △ *ADC* with respect to $\overleftrightarrow{AC}$.

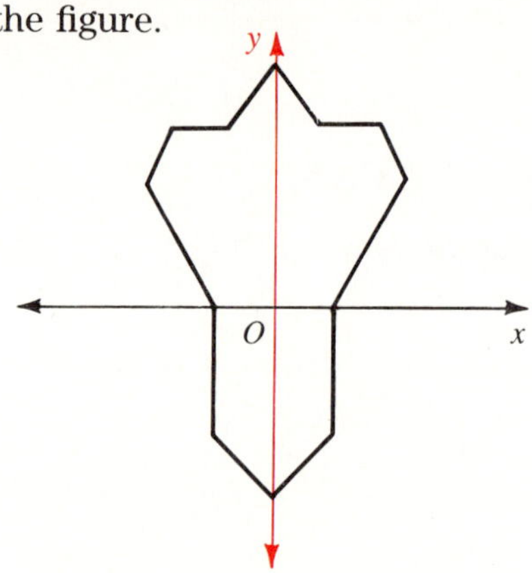

The figure is symmetric with respect to the *y*-axis.

EXPLORATIONS

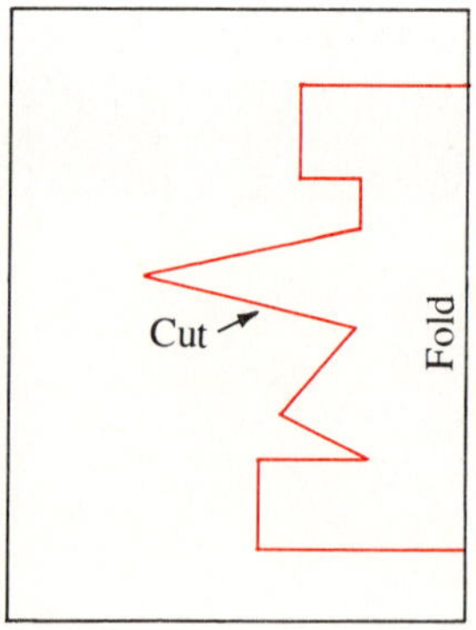

A. Fold a sheet of paper and draw a figure as shown. With scissors cut along the drawn line. Unfold the paper to reveal the symmetric figure.

Create an original piece of symmetric art in a similar manner. Trade your artwork with a classmate and then find each other's symmetries.

B. On a sheet of graph paper, draw axes.

Locate the following points:
$(0, -8)$, $(3, -4)$, $(8, -4)$, $(5, 0)$, $(8, 4)$, $(3, 4)$, and $(0, 8)$.

Use a ruler to connect the points in the given order. Then locate the points needed to complete a figure symmetric to the *y*-axis. Draw the figure.

C. Draw another set of axes on your graph paper.

Draw a dashed line through the points (0, 0), (3, 3), and (−3, −3). It will bisect the first and third quadrants.

Locate the following points:

(7, 7), (−1, 7), (−4, 4), and (−4, −4).

Use a ruler to connect the points in the given order. Then locate the points needed to complete a figure that is symmetric with respect to the dashed line. Draw the figure.

EXERCISES

For each figure, tell whether the dashed line is a line of symmetry.

1. 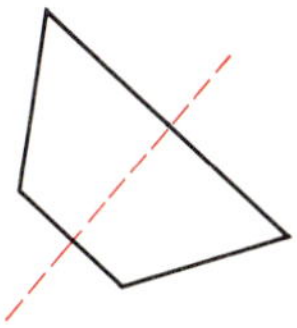

2. 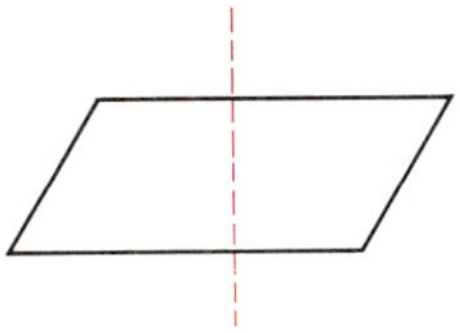

3. 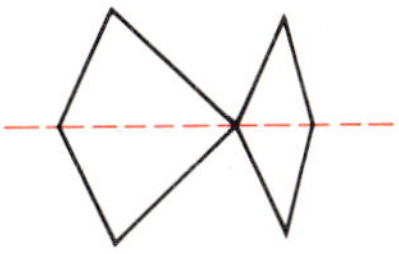

4. 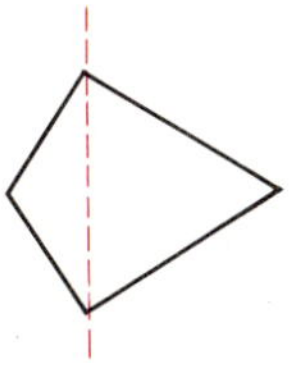

5. 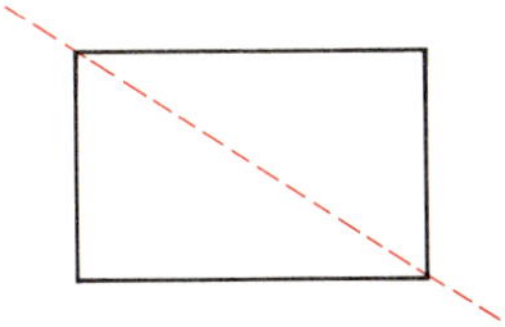

6. 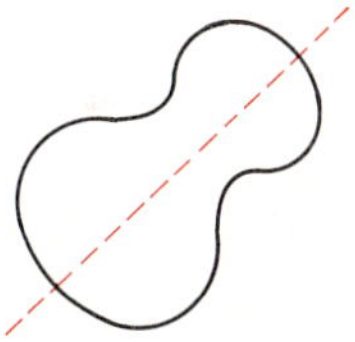

These symbols were used by alchemists during the Middle Ages to represent copper, lead, and tin, respectively. How many lines of symmetry can be drawn for each figure?

7. 

8. 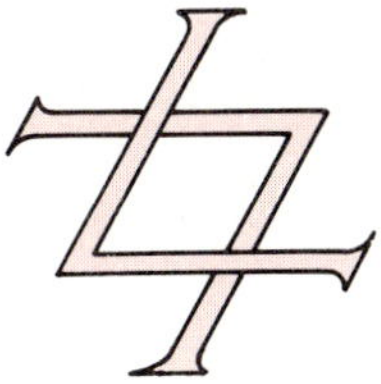

9. 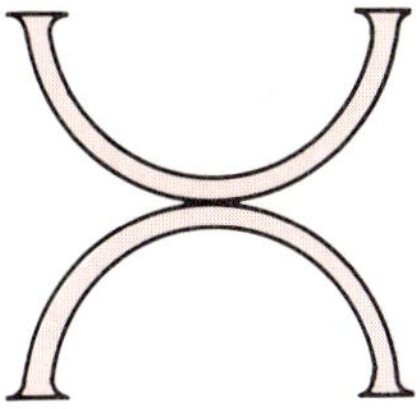

For each figure, state the number of lines of symmetry that can be drawn.

10. An equilateral triangle
11. A circle
12. A square
13. The figure shown at the right

TRANSLATIONS

When the position, size, or shape of a figure is changed according to a given rule, it is called a **geometric transformation.** We will study three types of transformations in which the new figure is congruent to the original figure. Since congruent figures have the same size and shape, we will only be discussing changes in position.

In the game of chess, the knight must move two spaces parallel to one side of the board and one space parallel to the other side. Such a move is a special kind of transformation that is called a **translation.** By the rules of chess, each of the other pieces must move or slide to another square in a special manner which also represents a translation.

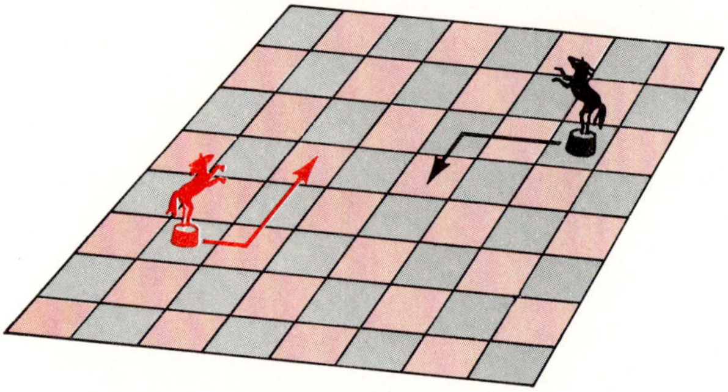

A translation may be thought of as a slide by a figure in any direction without turning. If $\triangle ABC$ slides to a new position, $\triangle A'B'C'$ (read A prime, B prime, C prime), the new triangle, is said to be the **image** of the original triangle.

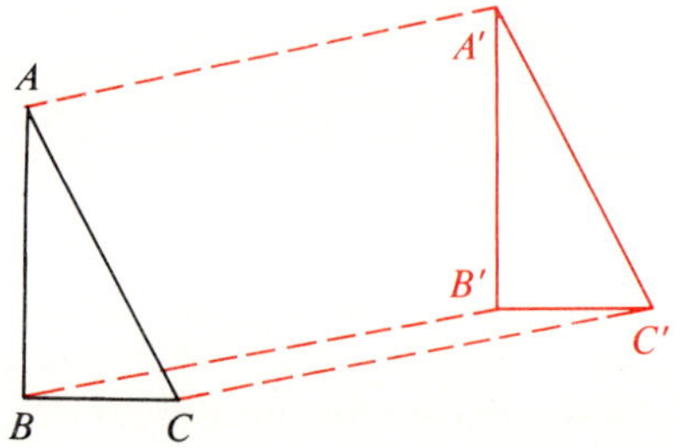

$\triangle RST$ has vertices with coordinates as shown. If the triangle slides 6 units to the right and 2 units up, the coordinates of the vertices of the image triangle are $R'(4, -1)$, $S'(4, 3)$, and $T'(0, 3)$. Notice that these coordinates can be obtained by adding 6 to the x-coordinate and 2 to the y-coordinate of each vertex of $\triangle RST$. This suggests an algebraic representation of a translation as follows.

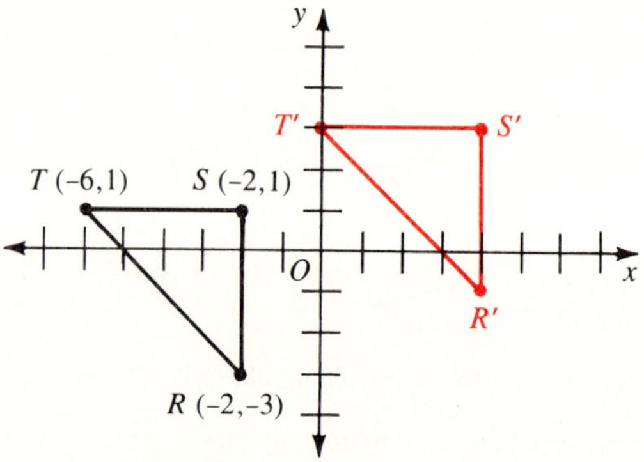

If the coordinates of some point in a given figure are (x, y), then the coordinates of its image point by a translation of a units horizontally and b units vertically are (x + a, y + b).

If a and b are:	*Then the translation is:*
positive	to the right and up
negative	to the left and down

EXAMPLE If the coordinates of one point of a figure are (3, 8) and its image point under a translation is at (3 + 4, 8 − 5), or (7, 3), the entire figure is translated 4 units to the right and 5 units down.

EXPLORATIONS

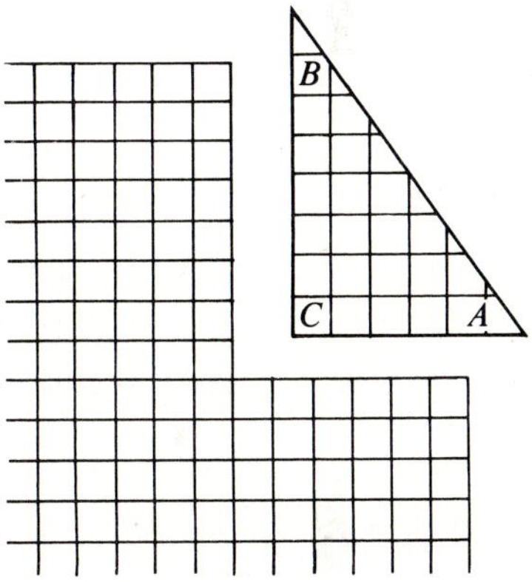

From a sheet of graph paper, carefully cut out a triangle as shown and label the vertices A, B, and C. (Save this triangle for use in the next two sections.)

On another sheet of graph paper, draw and label the x and y axes.

Part A

Step 1 Place the cutout triangle on the graph paper with two vertices at the given points.

Step 2 Write the coordinates of the third vertex.

Step 3 Physically move the triangle as directed without turning.

Step 4 Write the coordinates of the three vertices of the image triangle.

1. $A(2, 2)$, $C(-4, 2)$; 4 units right, 2 units down
2. $B(1, -3)$, $C(9, -3)$; 3 units left, 2 units up
3. $A(-6, 0)$, $B(0, -8)$; 6 units right, 8 units up
4. $C(-6, -1)$, $A(-6, -7)$; 2 units left, 1 unit down
5. $A(-2, 1)$, $B(-8, 9)$; 10 units right
6. $B(10, 5)$, $C(2, 5)$; 6 units down

Part B

Step 1 Given the coordinates of three points, use your cutout triangle to find the coordinates of the other three points.

Step 2 Describe the translation.

	A	B	C	A'	B'	C'	Translation
7.	(7, 2)	?	(1, 2)	(8, 4)	?	?	?
8.	?	(3, 2)	(−5, 2)	?	(2, −1)	?	?
9.	(−6, 3)	(0, −5)	?	?	(6, −5)	?	?
10.	(4, 3)	?	(4, −3)	?	?	(0, 1)	?
11.	?	(−5, 6)	(−5, −2)	?	?	(−5, −3)	?
12.	(3, 6)	(−5, 0)	?	(7, 1)	?	?	?

EXERCISES

State whether each of the following represents a translation.

1. 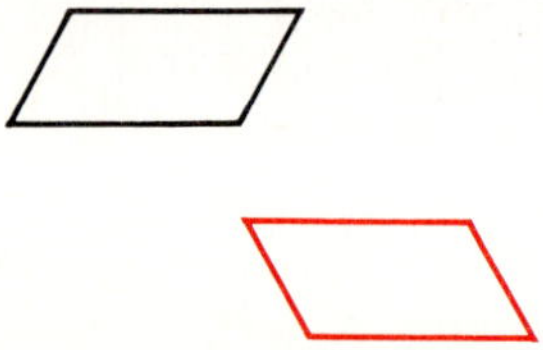

2. 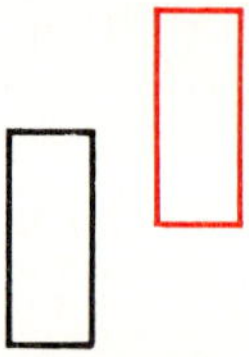

3. 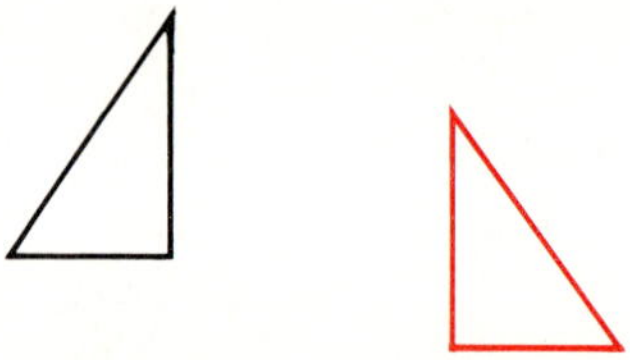

4. 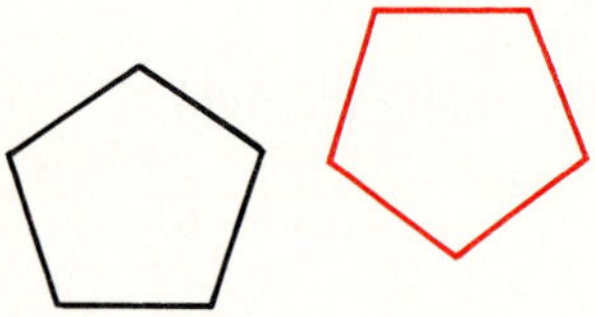

5. 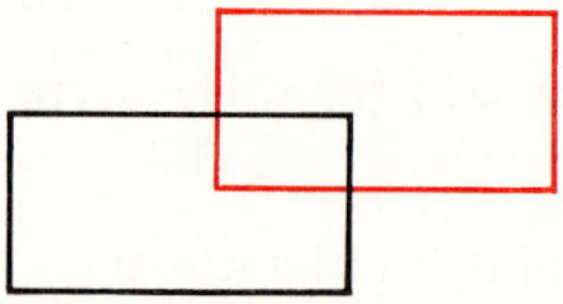

6. 

7. 

8. 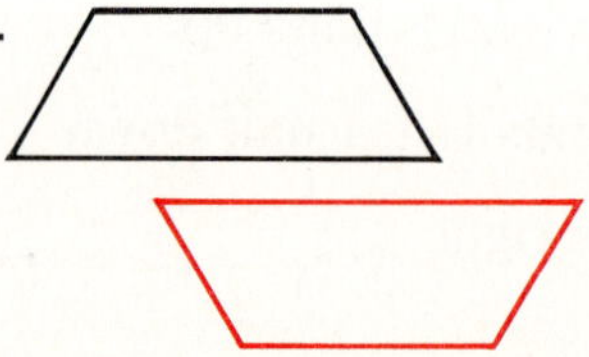

9. 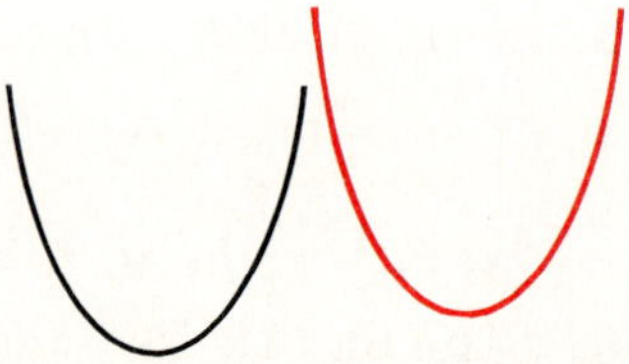

REFLECTIONS

A reflection is another type of geometric transformation in which the image is congruent to the original figure. You are already familiar with reflections in a mirror or still water.

Reflection transformations are closely related to the symmetric figures discussed previously. Symmetry involves a single figure. Each point is symmetric to another point of the figure. Reflection involves a figure and its image. They are symmetric to one another with respect to a line of reflection. This line is the perpendicular bisector of the segments joining the points in the figure below with the corresponding points in the image.

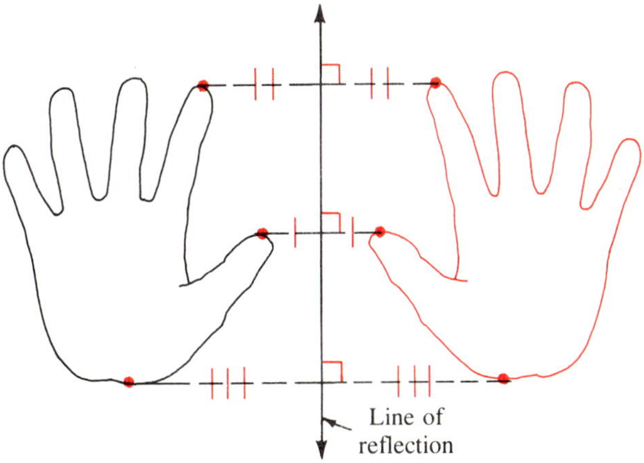

In the coordinate plane, three lines have special significance as lines of reflection: the y-axis, the x-axis, and the line whose equation is $y = x$. The last one bisects the first and third quadrants.

In each case below, observe the relationship between the coordinates of a point and its image.

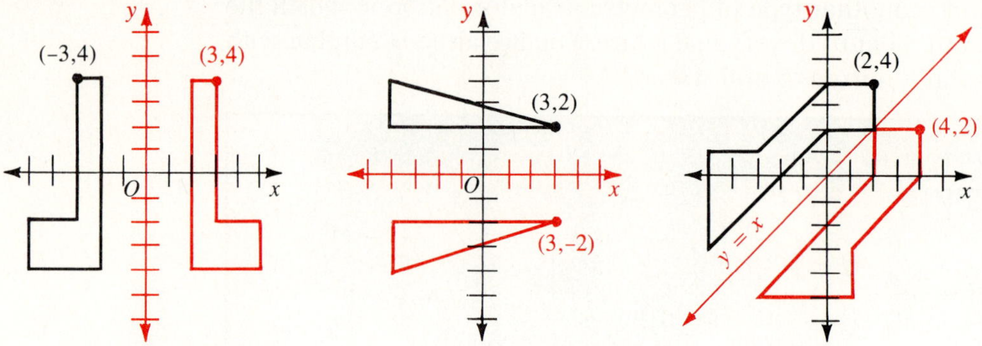

EXPLORATIONS

Use the cutout triangle from the previous section for these exercises. Turn the triangle over and write A', B', and C' on the back of A, B, and C, respectively. A reflection will require you to flip the triangle to the other side of the line of reflection.

On a sheet of graph paper, draw and label the x-axis, the y-axis, and the line whose equation is $y = x$.

Part A

Step 1 Place the cutout triangle on the graph paper, with two vertices at the given points.

Step 2 Write the coordinates of the third vertex.

Step 3 Physically move the triangle as directed.

Step 4 Write the coordinates of the vertices of the image triangle.

For Exercises 1–3, (a) reflect in the y-axis and then
(b) reflect in the x-axis.

1. $A(8, 1)$, $C(2, 1)$ **2.** $B(-2, -2)$, $C(6, -2)$ **3.** $A(-6, 0)$, $B(0, -8)$

For Exercises 4–9, reflect in the line $y = x$.

4. $B(2, 1)$, $C(10, 1)$ **5.** $A(3, 1)$, $B(11, 7)$ **6.** $A(1, -8)$, $C(1, -2)$

7. $B(4, 4)$, $C(-4, 4)$ **8.** $A(5, -4)$, $B(-1, 4)$ **9.** $A(-3, -3)$, $C(-3, 3)$

Part B

Given the coordinates of three points and a line of reflection, use your cutout triangle to find the coordinates of the other three points.

	A	B	C	Reflection	A′	B′	C′
10.	(9, 1)	?	(3, 1)	y-axis	(−9, 1)	?	?
11.	?	(−1, 6)	(−1, −2)	y-axis	?	?	(1, −2)
12.	(−2, 0)	?	(−2, 6)	x-axis	?	?	(−2, −6)
13.	(2, 5)	(8, −3)	?	x-axis	(2, −5)	?	?
14.	?	(8, −7)	(8, 1)	line $y = x$	?	(−7, 8)	?
15.	(4, −4)	(−2, 4)	?	line $y = x$	(−4, 4)	?	?

Your observations should lead you to the following conclusions.

1. If a figure is reflected in the y-axis and the coordinates of a point of the figure are (a, b), the coordinates of its image are $(-a, b)$.
2. If a figure is reflected in the x-axis and the coordinates of a point of the figure are (a, b), the coordinates of its image are $(a, -b)$.
3. If a figure is reflected in the line whose equation is $y = x$ and the coordinates of a point of the figure are (a, b), the coordinates of its image are (b, a).

EXAMPLES

The image of (3, 5) reflected in the y-axis is (−3, 5).
The image of (3, 5) reflected in the x-axis is (3, −5).
The image of (3, 5) reflected in the line $y = x$ is (5, 3).

EXERCISES

Write the coordinates of the image of each point when the given reflection line is used.

1. (7, 1); y-axis
2. (−6, 2); y-axis
3. (4, 6); x-axis
4. (5, −8); x-axis
5. (2, 9); $y = x$
6. (−7, 4); $y = x$

Consider only capital letters for the following exercises.

7. Find all letters of the alphabet that look like the same letter when reflected in the y-axis.

8. Find all letters of the alphabet that look like the same letter when reflected in the x-axis.

9. Which letters look the same under both reflections?

ROTATIONS

The burro in the picture helps make molasses by turning the machinery that squeezes cane. It rotates about a central point, always staying at the same distance from that point.

The last geometric transformation we will consider is the rotation. In a rotation, each point of the given figure moves about a center point as if it were attached to the end of a spoke on a wheel. Each point rotates through the same angle and the image figure is congruent to the original figure. We describe a counterclockwise rotation as being positive and a clockwise rotation as being negative.

To do the rotations that follow, you will need a ruler, a protractor, and a compass.

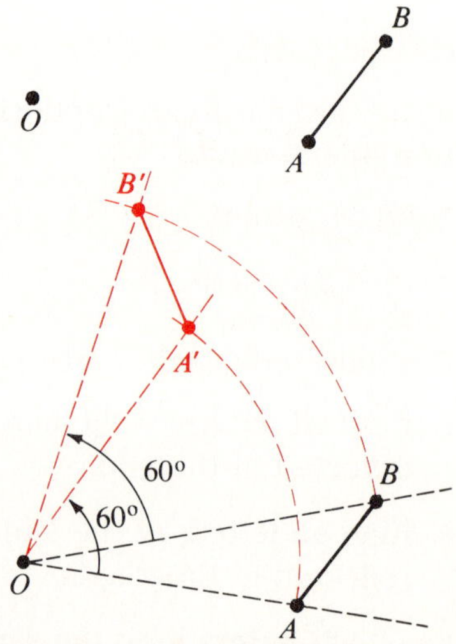

EXAMPLE Rotate $\overline{AB}$ by 60° about center O.

1. Draw $\overrightarrow{OA}$.
2. With a protractor, measure an angle of 60° counterclockwise from $\overrightarrow{OA}$, with O as vertex. Draw the second side of the angle.
3. With compass point at O, strike an arc through A to intersect the other side of the angle. Label the point of intersection A'.
4. Repeat Steps 1–3, starting with $\overrightarrow{OB}$ and locating B.
5. Draw $\overline{A'B'}$. This is the image of $\overline{AB}$ under the rotation.

In the coordinate plane we frequently use the origin as the center of rotation.

EXAMPLE Rotate $\triangle DEF$ by 90° about the origin.

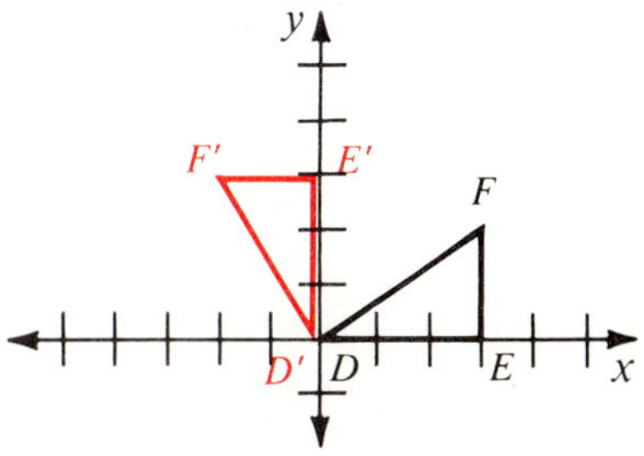

In this case, D remains where it is since it is the center of rotation. E is on the x-axis and a 90° rotation places its image on the y-axis. Since $\overline{EF}$ is vertical, a 90° rotation requires its image to be horizontal. Since $\triangle DEF \cong \triangle D'E'F'$, we can locate F' and draw the sides of $\triangle D'E'F'$.

EXAMPLE Rotate $\triangle RST$ by 90° about the origin.

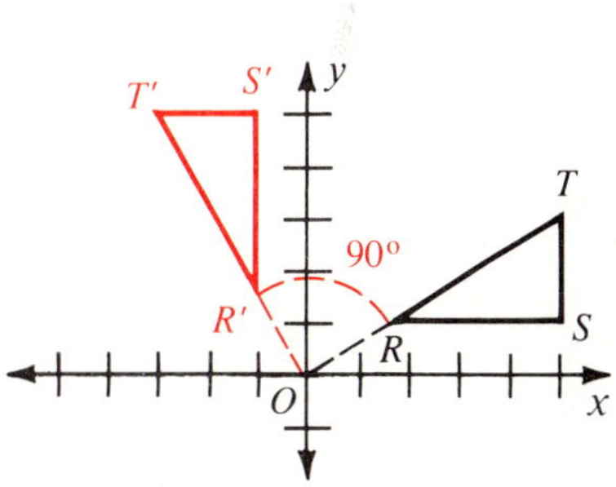

In this case, we must first locate one image point. Since $\angle ROR' = 90°$, $\overline{OR} \perp \overline{O'R'}$. We also know that $OR = O'R'$. Recall your earlier studies of slopes and distances in the coordinate plane. This should lead you to discover that R' must be at $(-1, 2)$ if R is at $(2, 1)$. We may then use the ideas from the first example to complete $\triangle R'S'T'$.

EXPLORATIONS

A. Draw a figure similar to the one shown.

Using a ruler, protractor, and compass, find the image of $\triangle ABC$ under a rotation of 70° about O.

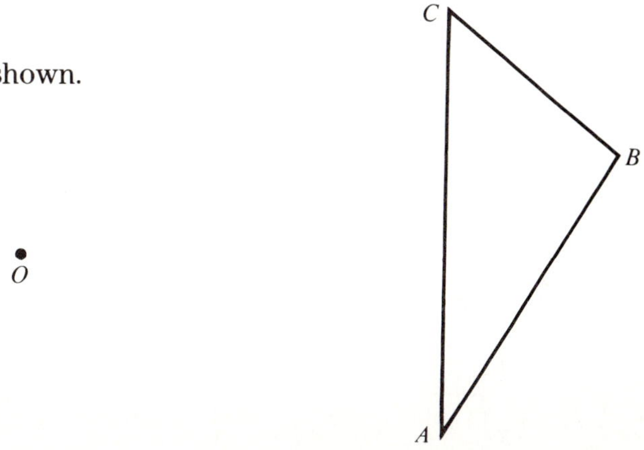

B. Use the cutout triangle from the previous two sections.

Step 1 On a sheet of graph paper, draw and label axes.

Step 2 Place the cutout triangle on the graph paper with two vertices at the given points.

Step 3 Write the coordinates of the third vertex.

Step 4 Rotate the triangle 90° counterclockwise about the indicated point as the center of rotation.

Step 5 Write the coordinates of the three vertices of the image triangle.

1. $A(6, 0)$, $C(0, 0)$; center at the origin

2. $B(2, 0)$, $C(10, 0)$; center at the origin

3. $C(-1, 0)$, $A(-7, 0)$; center at the origin

4. $B(-8, -7)$, $A(0, -1)$; center at point $(0, -1)$

5. $A(8, 1)$, $C(2, 1)$; center at point $(2, 1)$

EXERCISES

Identify each of the following transformations as a translation, reflection, or rotation.

1.

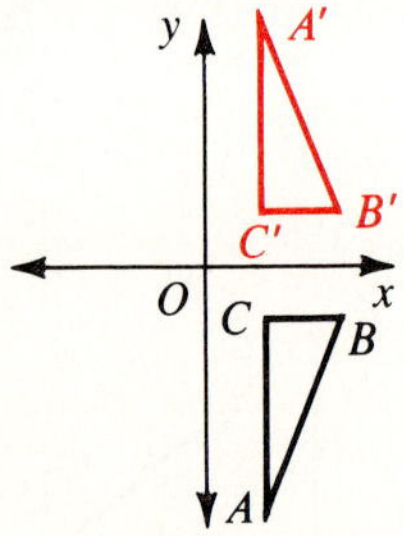

2.

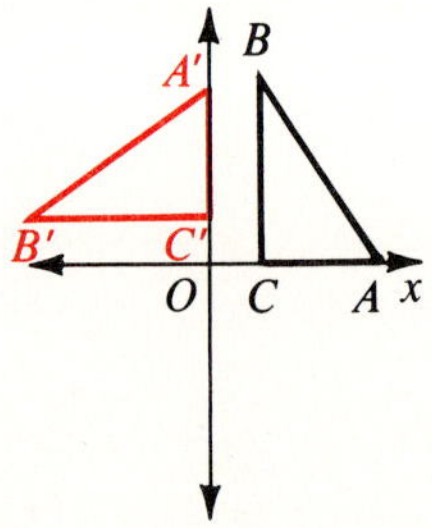

3.

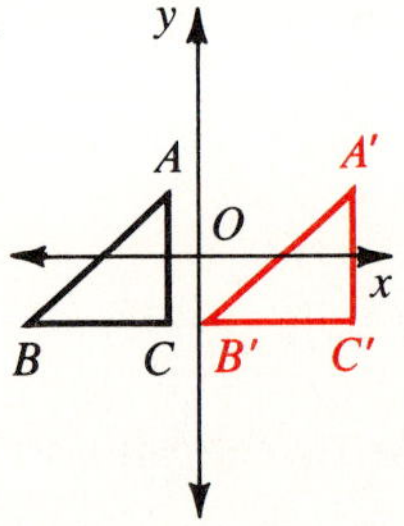

4.

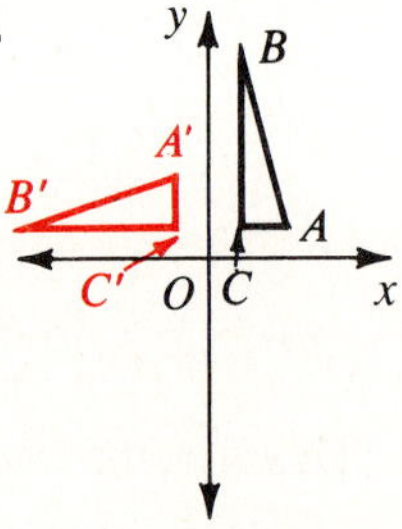

5.

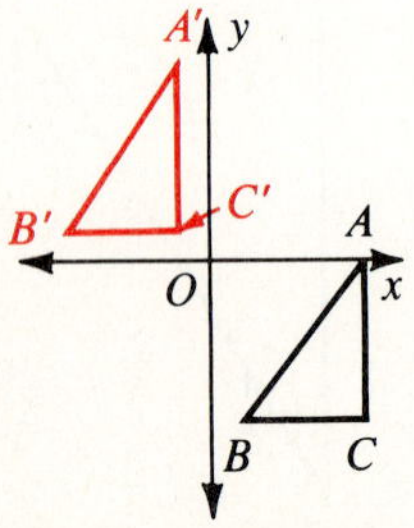

6.

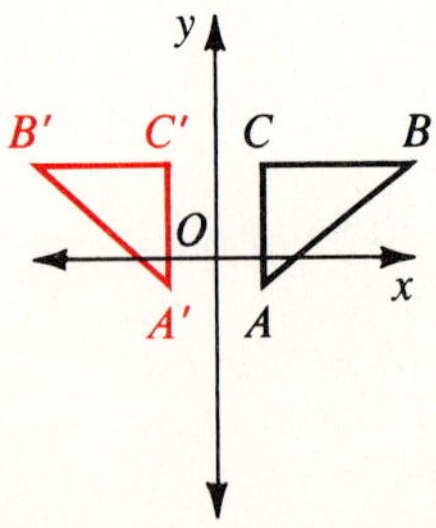

7.

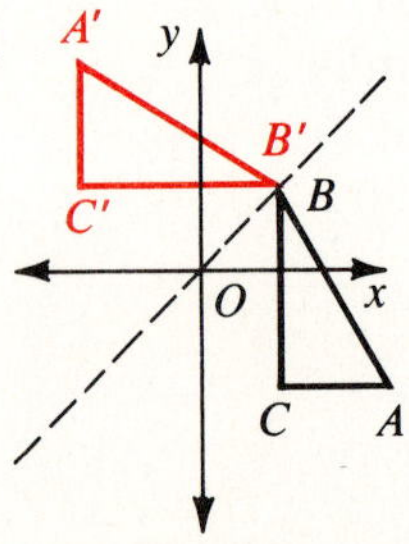

8.

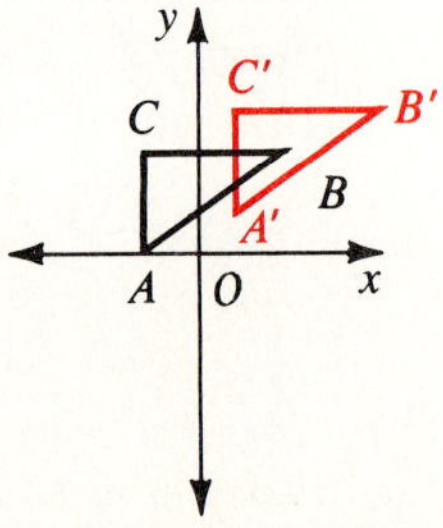

Glossary

A

acute angle (p. 15) An angle whose measure is between 0° and 90°.

acute triangle (p. 97) A triangle with three acute angles.

altitude of a triangle (p. 158) A segment, drawn from any vertex, perpendicular to the line that contains the opposite side. The opposite side is considered to be the *base* of the triangle.

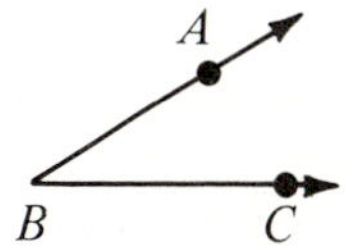

angle (p. 10) A figure formed by two rays or two segments with a common endpoint. B is the *vertex* of $\angle ABC$. $\overrightarrow{BA}$ and $\overrightarrow{BC}$ are the *sides* of $\angle ABC$.

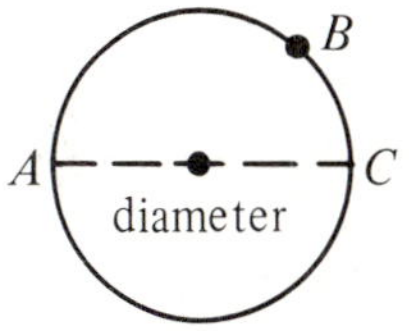

arc (p. 355) Part of a circle. A *semicircle*, such as $\overparen{ABC}$, is exactly half a circle. A *minor arc*, such as $\overparen{AB}$ or $\overparen{BC}$, is less than a semicircle. A *major arc*, such as $\overparen{BAC}$ or $\overparen{ACB}$, is greater than a semicircle.

area (p. 238) The amount of surface in a region. Area is measured in square units.

B

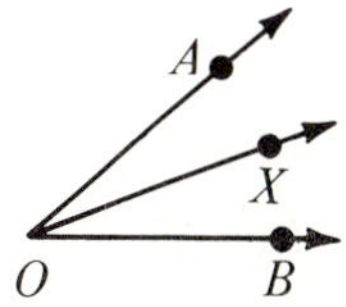

bisector of an angle (p. 20) A ray or a line which divides an angle into two equal angles. If $\angle AOX = \angle BOX$, $\overrightarrow{OX}$ bisects $\angle AOB$.

bisector of an arc (p. 359) A line, ray, or segment that contains the midpoint of an arc.

bisector of a segment (p. 154) A segment, ray, or line that contains the midpoint of a segment.

C

central angle of a circle (p. 355) An angle whose vertex is the center of a circle.

chord of a circle (p. 346) A segment that joins two points on a circle.

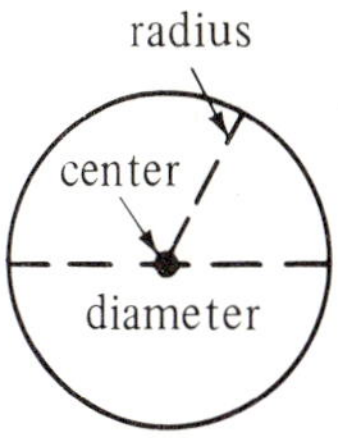

circle (p. 263) A figure, in a plane, whose points are all the same distance from a particular point in the plane. This point is the *center* of the circle.

circumference (p. 263) The distance around a circle.

circumscribed circle (pp. 163, 361) A circle is circumscribed about a polygon when each vertex of the polygon lies on the circle.

circumscribed polygon (p. 361) A polygon is circumscribed about a circle when each side of the polygon is tangent to the circle.

collinear points (p. 6) Points that lie on one line.

common tangent (p. 351) A line that is tangent to each of two coplanar circles. A common *internal* tangent intersects the segment joining the centers of the circles. A common *external* tangent does not intersect the segment joining the centers of the circles.
complementary angles (p. 46) Two angles whose measures total 90°.
conclusion (p. 50) A result reached by reasoning. In a statement of the form "If A, then B," B is the conclusion.

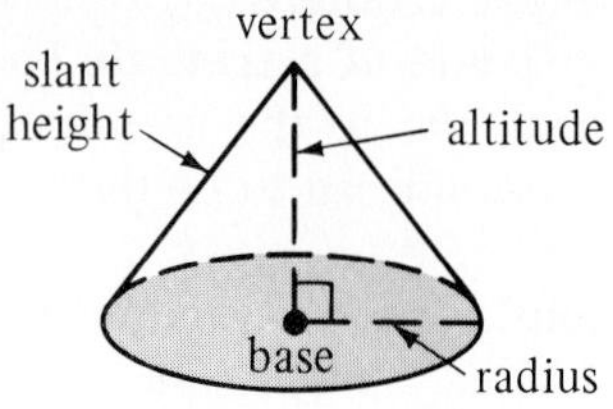

cone (p. 406) The *base* of a right circular cone is circular. Any segment that joins the vertex to a point on the circle is a *slant height* of the cone.

congruent triangles (p. 102) $\triangle ABC \cong \triangle DEF$ if the following statements are true:

$\angle A = \angle D$ $\quad \angle B = \angle E$ $\quad \angle C = \angle F$
$AB = DE$ $\quad BC = EF$ $\quad AC = DF$

consecutive sides of a polygon (p. 190) Two sides which intersect.
consecutive vertices of a polygon (p. 190) The endpoints of a side.
constructing a geometric figure (p. 25) Drawing a figure using only a straightedge and a compass.
converse of a statement (p. 50) A statement formed by exchanging the hypothesis and the conclusion of the given statement.
convex polygon (p. 190) Imagine fitting a rubber band along the edges of a polygon. If the rubber band fits snugly, the polygon is convex.
coordinate(s) (pp. 7, 468) On a number line, the number paired with a point. In the coordinate plane, the numbers which are paired with a point. Point (2, 4) has *x-coordinate* 2 and *y-coordinate* 4.
coordinate axes (p. 468) The horizontal and vertical number lines in the coordinate plane.
coordinate plane (p. 468) A plane which contains a horizontal number line (the *x-axis*) and a vertical number line (the *y-axis*). Every point in the coordinate plane can be named by a pair of numbers.
coplanar lines (p. 66) Lines that lie on one plane.
coplanar points (p. 33) Points that lie on one plane.
corollary of a theorem (p. 93) A statement which can be proved easily by applying a theorem.
corresponding parts of two polygons (p. 102) In $\triangle ABC$ and $\triangle DEF$, if point A is matched with point D, B with E, and C with F, then (1) the *corresponding sides* are $\overline{AB}$ and $\overline{DE}$, $\overline{BC}$ and $\overline{EF}$, and $\overline{AC}$ and $\overline{DF}$; (2) the *corresponding angles* are $\angle A$ and $\angle D$, $\angle B$ and $\angle E$, and $\angle C$ and $\angle F$.

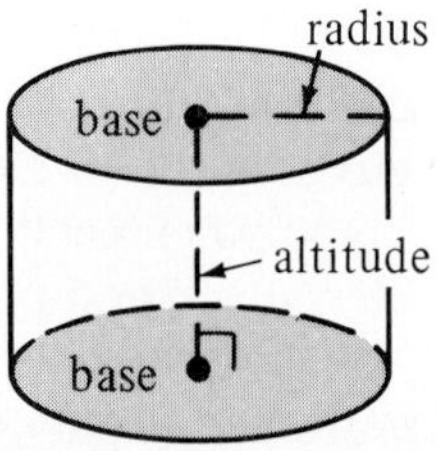

cylinder (p. 398) In a right circular cylinder, the *bases* are congruent circles which are parallel. Any segment that is perpendicular to the bases and has an endpoint in each base is an *altitude* of the right circular cylinder.

D

diagonal of a polygon (p. 190) A segment which joins two nonconsecutive vertices of the polygon.

diameter of a circle (p. 346) A chord that passes through the center of a circle. Also, the length of such a chord.

distance from a point to a line (p. 359) The length of the perpendicular segment from the point to the line.

E

equal angles (p. 15) Two angles whose measures are equal.

equal arcs of a circle (p. 355) Two arcs whose measures are equal.

equal segments (p. 144) Two segments whose lengths are equal.

equal sides (p. 97) Two sides whose lengths are equal.

equiangular polygon (p. 191) A polygon with all angles equal.

equiangular triangle (p. 97) A triangle with all angles equal.

equidistant points (p. 154) If $AB = AC$, then A is equidistant from B and C.

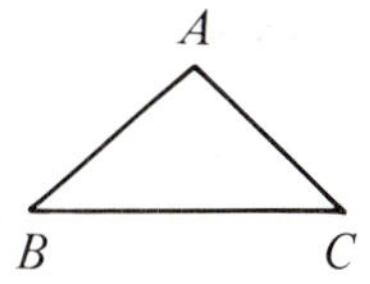

equilateral polygon (p. 191) A polygon with all sides equal.

equilateral triangle (p. 97) A triangle with all sides equal.

G

graph of an equation (p. 496) The geometric figure which contains all the points whose coordinates make the equation a true statement, and no other points.

H

hypotenuse of a right triangle (p. 97) The side opposite the right angle. The longest side of a right triangle.

hypothesis (p. 50) Information that is given. In a statement of the form "If A, then B," A is the hypothesis.

I

if . . . then statement (p. 50) A statement of the form "If A, then B."

inscribed angle (p. 364) An angle whose vertex lies on a circle and whose sides are chords of the circle.

inscribed circle (pp. 164, 361) A circle is inscribed in a polygon when each side of the polygon is tangent to the circle.

inscribed polygon (p. 361) A polygon is inscribed in a circle when each vertex of the polygon lies on the circle.

intercepted arc(s) (pp. 364, 365, 369) In each figure, the angle intercepts the arc or arcs shown in color.

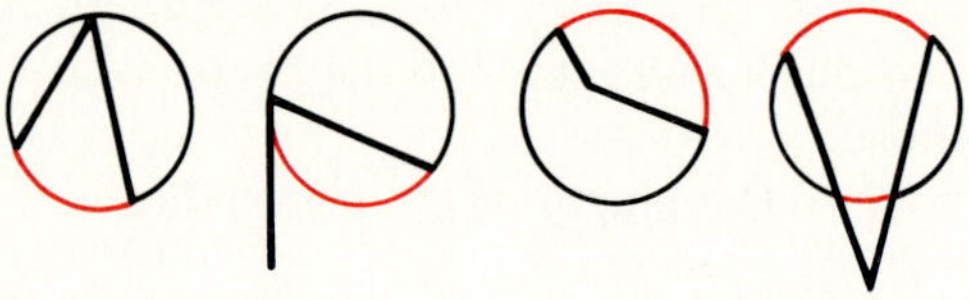

intersecting lines (pp. 18, 388) Two lines that meet in one point. Two intersecting lines are coplanar.

intersecting planes (pp. 34, 389) Two planes that meet in one line.

isosceles trapezoid (p. 220) A trapezoid with equal legs. There are two pairs of *base angles:* ∠ 1 and ∠ 2, ∠ 3 and ∠ 4.

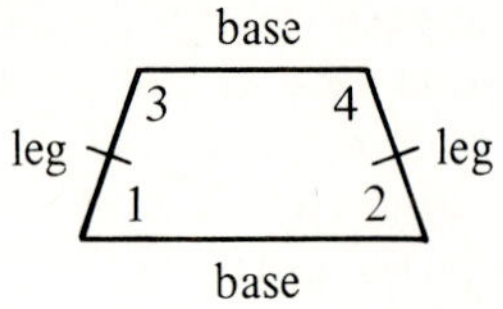

isosceles triangle (p. 97) A triangle with at least two equal sides. ∠ 1 and ∠ 2 are the *base angles.* ∠ 3 is the *vertex angle.*

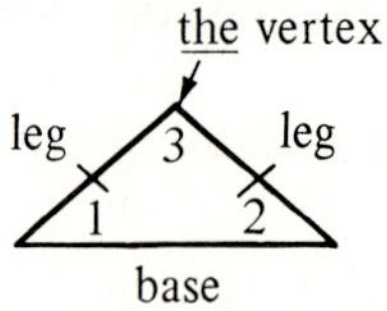

L

lateral area (p. 394) In a right prism or a regular pyramid, the sum of the areas of the lateral faces. In a cylinder or a cone, the area of the lateral surface.

legs of a right triangle (p. 97) The two sides other than the hypotenuse.

legs of a trapezoid (p. 220) The nonparallel sides of a trapezoid.

length of a segment (p. 7) The distance between the endpoints of a segment.

line parallel to a plane (p. 388) A line and a plane which have no point in common.

line perpendicular to a plane (p. 388) A line which is perpendicular to every line in the plane that passes through the point of intersection.

M

measure of an angle (p. 10) The number of degrees of an angle. The number is greater than 0 and less than or equal to 180.

measure of a major arc (p. 355) The difference of 360° and the measure of the minor arc.

measure of a minor arc (p. 355) The measure of its central angle.

measure of a semicircle (p. 355) The measure of a semicircle is 180°.

median of a trapezoid (p. 220) The segment joining the midpoints of the legs of a trapezoid.

median of a triangle (p. 158) A segment that joins a vertex of a triangle to the midpoint of the opposite side.

midpoint of an arc (p. 359) If $\overset{\frown}{CM} = \overset{\frown}{MD}$, then M is the midpoint of $\overset{\frown}{CD}$.

midpoint of a segment (p. 7) Point M is the midpoint of $\overline{AB}$ if M lies on $\overline{AB}$ and $AM = MB$.

N

noncollinear points (p. 33) Points that do not all lie on one line.
number line (p. 7) A line which matches points with the real numbers.

O

obtuse angle (p. 15) An angle whose measure is between 90° and 180°.
obtuse triangle (p. 97) A triangle with one obtuse angle.
origin of the coordinate plane (p. 468) The point in which the coordinate axes intersect. Point (0, 0).

P

parallel line and plane (p. 388) A line and a plane that have no point in common.
parallel lines (p. 66) Two coplanar lines that have no point in common.

parallel lines cut by a transversal (p. 66) The following angles are formed:
alternate interior angles: ∠ 3 and ∠ 6, ∠ 4 and ∠ 5
corresponding angles: ∠ 1 and ∠ 5, ∠ 2 and ∠ 6, ∠ 3 and ∠ 7, ∠ 4 and ∠ 8
interior angles: ∠ 3, ∠ 4, ∠ 5, ∠ 6
same-side interior angles: ∠ 3 and ∠ 5, ∠ 4 and ∠ 6

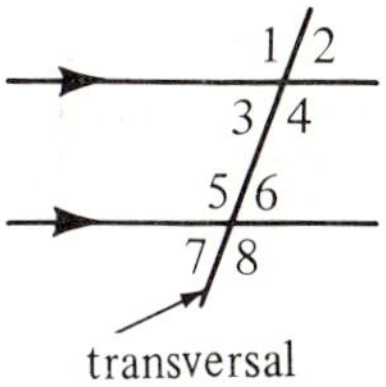

parallel planes (p. 34) Two planes that have no points in common.

parallelogram (p. 198) A quadrilateral with both pairs of opposite sides parallel. Either pair of sides may be considered the *bases*.

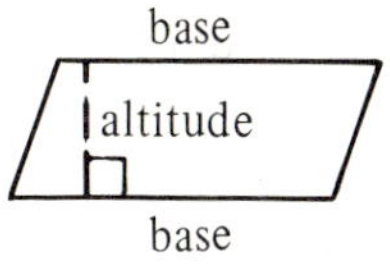

perimeter (p. 240) The distance around a region; the sum of the lengths of the sides of a polygon.
perpendicular bisector of a segment (p. 154) A segment bisector that is perpendicular to the segment.
perpendicular lines (p. 15) Two lines that meet to form four right angles.
perpendicular planes (p. 389) If a plane contains a line that is perpendicular to another plane, the two planes are perpendicular.
pi (p. 264) A Greek letter denoted by π. The quotient $\frac{\text{circumference}}{\text{diameter}}$ for any circle.
plotting a point (p. 468) Locating the point in the coordinate plane.

polygon (p. 190) The figure shows one example of a polygon. Parts of a polygon are named as shown. $\angle 1$ is an *interior angle.* $\angle 2$ is an *exterior angle.*

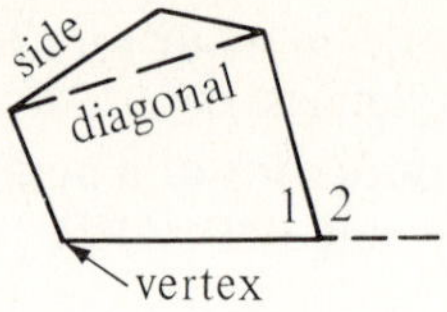

postulate (p. 29) A statement which is accepted without proof.

proportion (p. 289) An equation which states that two ratios are equal. In the proportion $\frac{4}{6} = \frac{2}{3}$, 6 and 2 are the *means,* and 4 and 3 are the *extremes.*

Q

quadrilateral (p. 190) A polygon with four sides.

R

radius of a circle (p. 346) A segment that joins the center and a point on the circle. Also, the length of such a segment.

ratio (p. 286) If a and b are numbers and $b \neq 0$, the ratio of a to b is the quotient $\frac{a}{b}$. This is sometimes denoted by $a:b$.

ray (p. 6) Part of a line. $\overrightarrow{AB}$ starts at *endpoint* A, goes through point B, and continues indefinitely.

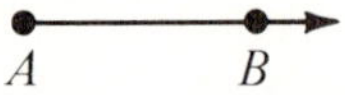

rectangle (p. 198) A parallelogram with four right angles. Any side of a rectangle may be considered a *base.*

rectangular solid (pp. 393, 440) A right prism whose *bases* are rectangular. The segment shown in color is a *diagonal* of the rectangular solid.

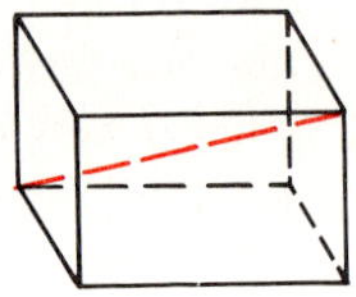

regular polygon (p. 191) A polygon which is both equilateral and equiangular.

regular pyramid (p. 402) The *base* of a regular pyramid is a regular polygon. The *lateral faces* are congruent isosceles triangles. The altitude from the vertex of a face is a *slant height.*

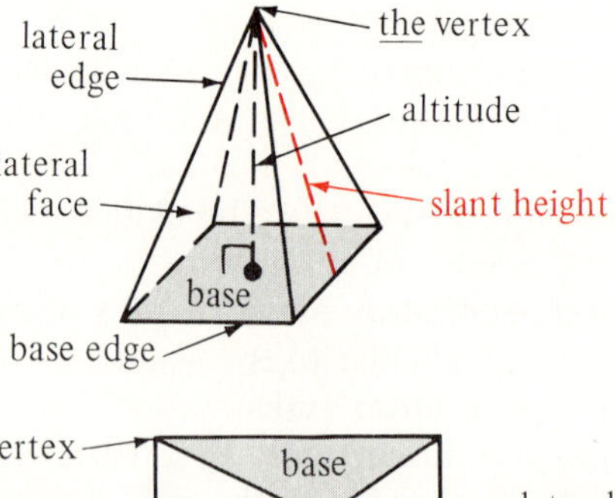

rhombus (p. 198) A parallelogram with four equal sides.

right angle (p. 15) An angle whose measure is 90°.

right prism (p. 393) In a right prism, the *bases* are congruent and parallel. The *lateral faces* are rectangular. The *lateral edges* are perpendicular to the bases. Any segment, such as a lateral edge, that is perpendicular to the bases and has an endpoint on each base, is an *altitude* of the right prism.

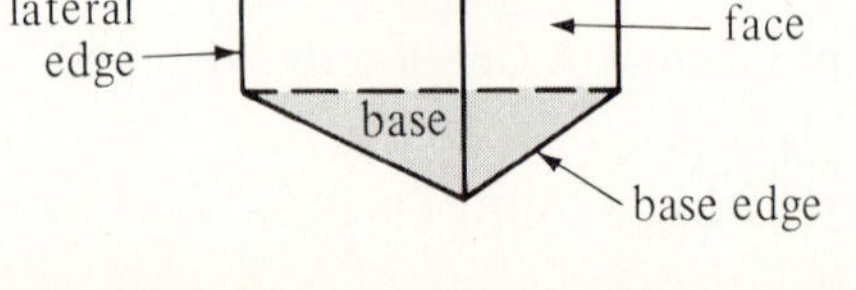

right triangle (p. 97) A triangle with one right angle. The hypotenuse is the longest side of a right triangle.

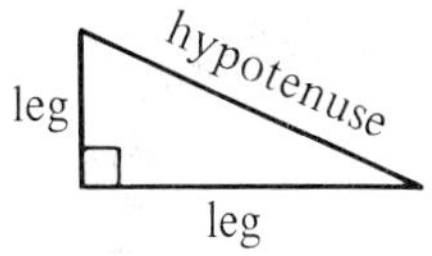

S

scale (p. 296) The ratio of a distance in a drawing to the actual distance represented.

scale factor of two similar polygons (p. 313) The ratio of two corresponding sides.

scalene triangle (p. 97) A triangle with no two sides equal.

secant of a circle (p. 346) A line that contains a chord.

segment (p. 6) Part of a line. A and B are the *endpoints* of $\overline{AB}$.

similar polygons (p. 312) Two polygons with the same shape.
In two similar polygons:
(1) Corresponding angles are equal.
(2) Corresponding sides are in proportion.

similar solids (p. 414) Two solids with the same shape.
In two similar solids:
(1) Corresponding angles are equal.
(2) Corresponding segments are in proportion.

skew lines (p. 66) Two lines that are not coplanar and do not intersect.

slope of a line (pp. 486, 487) $\frac{\text{rise}}{\text{run}}$. Also, if a line contains points (x_1, y_1) and (x_2, y_2), then the slope of the line is $\frac{y_2 - y_1}{x_2 - x_1}$, provided $x_1 \neq x_2$.

sphere (p. 346) A figure in space whose points are the same distance from a particular point. This point is the *center* of the sphere.

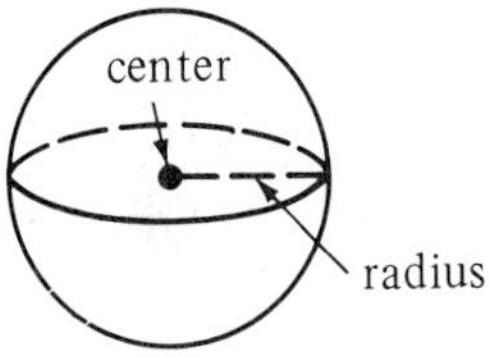

square (p. 198) A parallelogram with four right angles and four equal sides.

straight angle (p. 15) An angle whose measure is 180°.

supplementary angles (p. 47) Two angles whose measures total 180°.

T

tangent to a circle (p. 346) A line or segment, in the plane of a circle, that intersects the circle in exactly one point. This point is called the *point of tangency.*

tangent circles (p. 351) Two coplanar circles that are tangent to a line at one point. ⊙R and ⊙S are *externally* tangent. ⊙S and ⊙T are *internally* tangent.

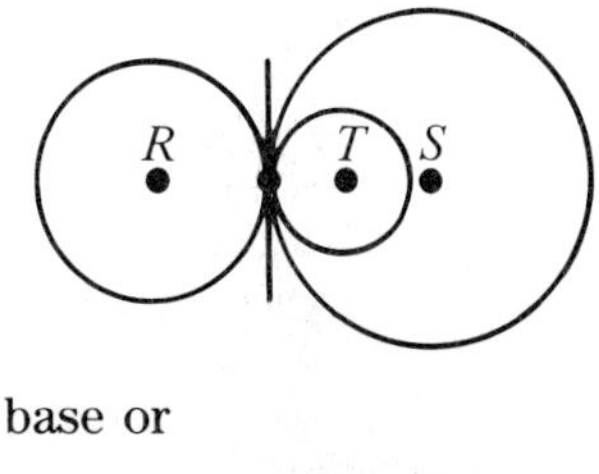

theorem (p. 46) A statement that can be proved.

total area (p. 394) The sum of the lateral area and the area of the base or bases.

trapezoid (p. 198) A quadrilateral with just one pair of opposite sides parallel. The segment joining the midpoints of the legs is the *median* of a trapezoid.

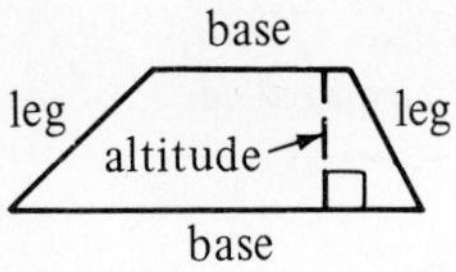

triangle (p. 15) A polygon with three sides. In $\triangle ABC$, each of points A, B, and C is a *vertex* of the triangle. $\overline{AB}$, $\overline{BC}$, and $\overline{AC}$ are the *sides* of the triangle. $\angle 1$, $\angle 2$, and $\angle 3$ are the *angles* of the triangle. $\angle 4$ is an *exterior angle* of $\triangle ABC$. $\angle 1$ and $\angle 2$ are *opposite angles* with respect to exterior $\angle 4$.

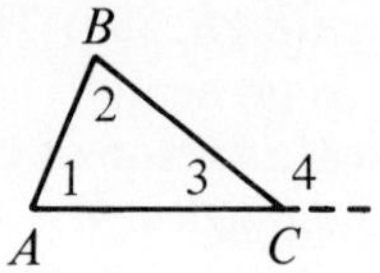

trigonometric ratios (pp. 447, 452)

$$\textit{tangent} \text{ of } \angle A = \frac{\text{leg opposite } \angle A}{\text{leg adjacent to } \angle A}$$

$$\textit{sine} \text{ of } \angle A = \frac{\text{leg opposite } \angle A}{\text{hypotenuse}}$$

$$\textit{cosine} \text{ of } \angle A = \frac{\text{leg adjacent to } \angle A}{\text{hypotenuse}}$$

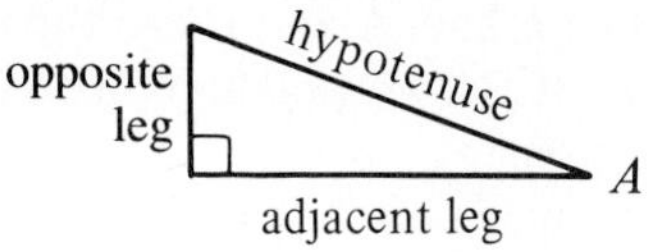

V

vertical angles (p. 18) Two angles formed by intersecting lines. $\angle 1$ and $\angle 2$ are vertical angles.

volume (p. 394) The amount of space in a solid. Volume is measured in cubic units.

Index

A
Acute angle, 15
Acute triangle, 97, 260
Alternate interior angles, 66–68, 72
Altitude
    of cone, 406
    of cylinder, 398
    of parallelogram, 243
    of regular pyramid, 402
    of right prism, 393
    of trapezoid, 251
    of triangle, 158
Angle(s), 10
    acute, 15
    alternate interior, 66–68, 72
    bisector of, 20
        construction of, 25, 149
    central, 355–356
    complementary, 46
    corresponding
        of congruent triangles, 102–103, 144–145
        of parallel lines, 67, 72–73
        of similar polygons, 312–313, 317–318
        of similar solids, 414
    of elevation and depression, 457–458, 475
    equal, 15
        construction of, 77, 151
        ways to prove, 144, 173
    inscribed, 364–365
    interior, 66–67
    measure of, 10–11, 35
    obtuse, 15
    of polygon, 194–195
    right, 15
    same-side interior, 67–68, 72
    sides of, 10
    straight, 15
    supplementary, 47
    of triangle, 15, 92–93
    vertex of, 10
    vertical, 18, 47
Angle sum
    of polygon, 194–195
    of triangle, 92
Answers
    to Selected Exercises, 562–579
    to Self-Tests, 559–561
Applications
    Angles of Elevation and Depression, 457–459
    Art and Geometry, 54–55
    Golden Rectangle, 208
    Indirect Proof, 334–335
    Latitude and Longitude, 23–24
    Locus, 482–485
    Mechanical Drawing, 431
    Mirrors and Billiards, 272–273
    Rigidity of Triangles, 113–114
    Shapes in Nature, 505
    Surface Area and Volume, 413
    Triangle Inequality, 179
    Using Circles to Set Up Schedules, 377
    Using a Compass, 302–303
Arc(s) of circle, 355–356
    bisector of, 359
    of chord, 359
    equal, 355–356, 359–360
    intercepted, 364–365, 369–370
    major, 355
    measure of, 355
    midpoint of, 359
    minor, 355–356
Area(s), 238
    of circle, 268–269
    of cone, 406, 443–444
    of cylinder, 398–399
    of equilateral triangle, 322
    of parallelogram, 243–244
    of rectangle, 238–240
    of regular polygon, 481
    of regular pyramid, 402–403, 443
    of right prism, 394
    of similar polygons, 329–330

Area(s) (*continued*)
of similar solids, 414–415
of sphere, 410
of square, 239
of trapezoid, 250–251
of triangle, 247, 292, 481
Axes, coordinate, 468

B
Base(s)
of cone, 406
of cylinder, 398
of isosceles triangle, 97, 168
of parallelogram, 243
of rectangle, 238
of regular pyramid, 402
of right prism, 393
of trapezoid, 220
of triangle, 247
Bisector(s)
of angle, 20
construction of, 25, 149
of angles of triangle, 164
of arc, 359
perpendicular, of segment, 154–155
construction of, 154
perpendicular, of sides of triangle, 163
of segment, 154

C
Calculator Activities, 71, 292, 322, 332, 409, 436, 475, 481
Career Notes
Actuary, 316
Aircraft Assembly Technician, 451
Astronomer, 32
Auto Mechanic, 262
Carpenter, 500
Marine Dietician, 59
Physical Therapist, 202
Plumber, 172
Realtor, 125
Center
of circle, 263, 346
of sphere, 346
Central angle(s) of circle, 355–356
Chapter reviews. *See* Reviews.
Chapter tests. *See* Tests.
Charts, 180–181
Chord(s) of circle, 346
arc of, 359
diameter perpendicular to, 359–360
equal, 359–360
intersecting within a circle, 369–370, 373
Circle(s), 263, 346
arc(s) of, 355–356
area of, 268–269
center of, 263, 346
central angle of, 355–356
chord of, 346
circumference of, 263–265
circumscribed, 163–164, 361
diameter of, 263–264, 346
externally tangent, 351
inscribed, 164, 361
inscribed angle of, 364–365
internally tangent, 351
major arc of, 355
minor arc of, 355
radius of, 263, 346
secant of, 346
semicircle of, 355
tangent(s) to, 346, 350–351
Circumference of circle, 263–265
Circumscribed circle, 163–164, 361
construction of, 163
Circumscribed polygon, 164, 361
Collinear points, 6, 33–34
Compass, 25
Complementary angles, 46, 134–135
Computer Activities
Best Box, 508–509
Complementary and Supplementary Angles, 134–135
Random Numbers and Pi, 276–277
Ten-Speed Bicycles, 378–379
Conclusions, 50
Conditional, 50
Cone, 406, 443–444
Congruence, methods for proving, 127
Congruent triangles, 102–133, 144–145
Consecutive sides of polygon, 190–191
Consecutive vertices of polygon, 190–191

Constructions, 25
angles, equal, 77
bisector of angle, 25
circumscribed circle, 163
inscribed circle, 164
justifying, 149
list of, 522
parallel lines, 78
perpendicular bisector of segment, 154
perpendicular lines, 26
60° angle, 174
tangent to circle, 351
Consumer Applications
Calling Long Distance, 392
Consumer Price Index, 148
Cost of Driving, 246
Do You Get What You Pay For?, 120
Energy Usage, 76
Life Insurance, 349
Time is Money, 14
Converse, 50–51
Convex polygon, 190
angle sums of, 194–195
Coordinate, 7
Coordinate axes, 468
Coordinate geometry, 468–474, 476–480, 486–504
Coordinate plane, 468
Coordinates, 468
Coplanar lines, 66, 388
Coplanar points, 33–34
Corollary, 93
Corresponding angles of parallel lines, 67, 72–73
Corresponding parts
of congruent triangles, 102–103, 144–145
of similar polygons, 312–313, 317–318
of similar solids, 414–415
Cosine ratio, 452–453
Counterexamples, 418–419
Cumulative reviews. *See* Reviews.
Cylinder, 398–399

D

Diagonal(s)
of parallelogram, 203–204
of polygon, 190–191
of rectangle, 209–210
of rectangular solid, 440–441
of rhombus, 209–210
Diagrams, 81, 180–181
Diameter of circle, 263–264, 346
perpendicular to chord, 359–360
Distance
between points, 7, 34
in coordinate plane, 472
from point to line, 207, 359

E

Edge
of regular pyramid, 402
of right prism, 393
of solid, 405
End-of-course reviews. *See* Reviews.
Endpoint(s)
of ray, 6
of segment, 6, 154–155
Enrichment materials. *See:*
Applications
Calculator Activities
Career Notes
Computer Activities
Consumer Applications
Experiments
Explorations
Problem Solving Strategies
Puzzle Problems
Reading Geometry
Equal angles, 15
construction of, 77, 151
ways to prove, 144, 168
Equal arcs, 355–356, 359–360
Equal chords, 359–360
Equal segments, 97
ways to prove, 144–145, 174
Equality, postulates of, 29–30
Equation of line, 496–498
Equiangular polygon, 191
Equiangular triangle, 97, 167–168, 173–174
Equilateral polygon, 191
Equilateral triangle, 97, 167–168, 173–174
area of, 292, 322
Experiments, 22, 120, 132, 162, 178, 193, 333, 349, 392, 405, 409

Explorations, 167, 203, 209, 214, 215, 220, 243, 250, 255, 317, 350, 359, 364, 369, 414, 476–477, 482, 488, 491–492
Exterior angle
 of polygon, 195
 of triangle, 93
Exterior angle sum of polygon, 195

F
Face
 of regular pyramid, 402
 of right prism, 393
 of solid, 405
45°-45°-90° triangle, 432

G
Golden rectangle, 208
Graph
 in coordinate plane, 468
 of equation, 496–498
 on number line, 7

H
Hexagon, 190
Hypotenuse, 97, 210
Hypothesis, 50

I
If . . . then statement, 50–51, 56–57
Indirect proof, 334–335
Inscribed angle, 364–365
Inscribed circle, 164, 361
 construction of, 164
Inscribed polygon, 163, 361
Intercepted arc, 364–365, 369–370
Interior angles of two parallel lines, 66–68, 72–73
Interior angle sum of polygon, 194–195
Intersecting line and plane, 388
Intersecting lines, 18, 388
Intersecting planes, 34, 389
Isosceles trapezoid, 220–221
Isosceles triangle, 97, 167–168, 173
 parts of, 97, 168

L
Lateral area
 of cone, 406, 444
 of cylinder, 398–399
 of regular pyramid, 402–403, 443
 of right prism, 394
 of similar solids, 414–415
Lateral edge, 393, 402
Lateral face, 393, 402–403
Legs
 of isosceles triangle, 97, 168
 of right triangle, 97
 of trapezoid, 220
Length of segment, 7, 34
Line(s), 6, 388
 auxiliary, 336–337
 coplanar, 66, 388
 equation of, 496–498
 intersecting, 18, 388
 intersecting plane, 388
 parallel, 66–73, 326, 388, 491
  construction of, 78
 parallel to plane, 388
 perpendicular, 15, 72, 491–492
  construction of, 26, 149, 151
 skew, 66, 388
 slope of, 486–488
Locus, 482–483

M
Major arc, 355
Maps and scale drawings, 296–297, 299
Measure
 of angle, 10–11, 35
 of arc, 355
Median
 of trapezoid, 220–222, 501
 of triangle, 157, 158–159
Midpoint
 of arc, 359
 of segment, 7, 154, 476–478
Midpoint formula, 477
Minor arc(s), 355–356
Mixed reviews. *See* Reviews.
Möbius band, 392

N

Noncollinear points, 6, 33–34
Number line, 7

O

Obtuse angle, 15
Obtuse triangle, 97, 260
Opposite angles of a triangle, 93
Origin of coordinate plane, 468

P

Parallel lines, 66–73, 326, 388, 491
    construction of, 78
Parallel planes, 34, 389
Parallelogram, 198, 203–204, 214–216
    area of, 243–244
    properties of, 204, 231
Patterns, 506–507
Pentagon, 190, 193, 194
Perimeter
    of polygon, 98, 240
    of similar polygons, 330
Perpendicular bisector(s)
    of segment, 154–155
        construction of, 154
    of sides of triangle, 163
Perpendicular lines, 15, 72, 491–492
    construction of, 26, 149, 151
Perpendicular planes, 389
Pi (π), 264, 276–277
Point(s), 6
    collinear, 6, 33–34
    coordinate of, 7
    in coordinate plane, 468
    coplanar, 33–34
    distance between, 7, 34, 472
    noncollinear, 6, 33–34
Plane(s), 33
    intersecting, 34, 389
    intersecting line, 388
    parallel, 34, 389
    parallel to line, 388
    perpendicular, 389
Polygon(s), 190–191
    angle of, 194–195
    circumscribed, 164, 361
    consecutive sides of, 190–191
    consecutive vertices of, 190–191
    convex, 190
    diagonal of, 190–191
    equiangular, 191
    equilateral, 191
    inscribed, 163, 361
    perimeter of, 98, 240
    regular, 191, 481
    side of, 190–191
    similar, 312–313, 329–330
    vertex of, 190–191
Postulate(s), 29
    AA, 317–318
    ASA, 121
    of equality, 29–30
    list of, 518
    Protractor, 35
    Ruler, 34
    SAS, 115–116
    SSS, 107–108
Prism, right, 393–395
Problem Solving Strategies
    Asking Questions, 82–83
    Auxiliary Lines, 336–337
    Counterexamples, 418–419
    Diagrams and Charts, 180–181
    Patterns, 506–507
    Working Backwards, 274–275
Proof, 60–61
    indirect, 334–335
    using coordinate geometry, 501
    See also Strategies of proof.
Proportion(s), 289–290, 293
Protractor, 10–11
Puzzle Problems, 17, 101, 112, 132, 157, 162, 224, 254, 261, 295, 324, 354, 368, 376, 401, 405, 442, 446, 456, 459
Pyramid, regular, 402–403, 443
Pythagorean Theorem, 255–256, 428
    converse of, 259–260
    use of, 432–444

Q

Quadrant, 468
Quadrilateral, 190, 194
    See also Parallelogram, Rectangle, Rhombus, Square, Trapezoid.

R
Radius
of circle, 263, 346
of cone, 406
of cylinder, 398
of sphere, 346
Ratio, 286–287, 378–379
Ray, 6
Reading Geometry
Diagrams, 81
How to Read Your Geometry Textbook, xi–xii
Independent Study, 153
Symbols, 133
Three-Dimensional Figures, 417
Words and Meanings, 229
Rectangle, 198, 209–210
area of, 238–240
properties of, 231
Rectangular solid, 393–395, 508–509
diagonal of, 440–441
Reflections, 537–539
Regular polygon, 191
area of, 481
Regular pyramid, 402–403, 443
Reviews
chapter, 39–40, 85–86, 137–138, 183–184, 231–232, 279–280, 305–306, 339–340, 381–382, 421–422, 461–462, 511–512
cumulative, 88, 186, 282, 342, 424, 514
end-of-course, 523–530
mixed, 42, 140, 234, 308, 384, 464
skills
Algebraic Expressions and Equations, 278
Algebraic Fractions, 460, 510
Arithmetic Operations, 38
Coordinates in the Plane, 84
Decimals and Percents, 182
Distributive Property, 338
Fractional and Decimal Equations, 304
Quotients of Square Roots, 380
Squares and Square Roots, 230
Whole Numbers and Fractions, 136
Working with Polynomials, 420
Rhombus, 198, 209–210
properties of, 231
Right angle, 15
Right prism, 393–395
Right triangle(s), 97, 210, 255–260, 428
45°-45°-90°, 432, 437
hypotenuse of, 97, 210
legs of, 97
list of common lengths, 428
30°-60°-90°, 432–433, 437
using, 440–453
Rotations, 540–542

S
Same-side interior angles, 67–68, 72–73
Scale drawings, 296–297, 299
Scalene triangle, 97
Secant(s) of circle, 346, 369–370
Segment(s), 6
bisector of, 154
divided proportionally, 323, 325–326
endpoints of, 6, 155
equal, 97
ways to prove, 144, 174
length of, 7, 34
midpoint of, 7, 154, 476–478
perpendicular bisector of, 154–155
construction of, 154
Self-Tests. *See* Tests.
Semicircle, 355
Side(s)
of angle, 10
corresponding, 102–103, 144, 312, 414
of polygon, 190–191
of triangle, 97
Similar polygons, 312–313, 329–330
Similar solids, 414–415
Similar triangles, 317–333
Sine ratio, 452
Skew lines, 66, 388
Skills reviews. *See* Reviews.
Slant height
of cone, 406
of pyramid, 402
Slope of line, 475, 486–488
of parallel lines, 491
of perpendicular lines, 491–492

Solid(s)
- models of, 417
- rectangular, 393–395
- regular pyramids, 402–403
- right circular cones, 406
- similar, 414–415
- spheres, 410

Sphere, 346
- area and volume of, 410

Square, 198, 209–210
- area of, 239
- properties of, 231

Square roots, table of, 516
Squares, table of, 515
Straight angle, 15
Strategies of proof
- proving a quadrilateral is a parallelogram, 216
- proving two angles equal, 144, 168
- proving two lines parallel, 73
- proving two segments equal, 144, 174
- proving two triangles congruent, 127

Supplementary angles, 47, 134–135
Symbols, 133
- list of, x

Symmetry, 531–533

T

Table
- of square roots, 516
- of squares, 515
- of trigonometric ratios, 517

Tangent(s) to circle, 346, 350–351
- angle formed by chord and, 365
- common, 351
- construction of, 351

Tangent ratio, 447–448, 452
Tests
- chapter, 41, 87, 139, 185, 233, 281, 307, 341, 383, 423, 463, 513
- self-tests
  - answers to, 559–561
  - Ch. 1, 14, 22, 37
  - Ch. 2, 53, 80
  - Ch. 3, 101, 112, 125, 132
  - Ch. 4, 152, 166, 178
  - Ch. 5, 197, 207, 219, 228
  - Ch. 6, 246, 254, 262, 271
  - Ch. 7, 295, 301
  - Ch. 8, 322, 333
  - Ch. 9, 358, 368, 376
  - Ch. 10, 401, 412
  - Ch. 11, 439, 446, 456
  - Ch. 12, 480, 495, 504

Theorem(s), 46
- AAS, 126
- HL, 126–127
- list of, 519–522
- Midpoints, 225–226, 502
- Pythagorean, 255–256
  - converse of, 259–260

30°-60°-90° triangle, 432–433, 437
Three-dimensional figures, 417
Total area
- of cone, 406
- of cylinder, 398
- of regular pyramid, 402
- of right prism, 394
- of similar solids, 414

Transformational geometry
- reflections, 537–539
- rotations, 540–542
- symmetry, 531–533
- translations, 534–536

Translations, 534–536
Transversal, 66–68, 72, 326
Trapezoid, 198, 220–222
- area of, 250–251
- isosceles, 220–221
- median of, 220–221, 501
- parts of, 220–221, 250

Triangle(s), 15
- acute, 97, 260
- altitudes of, 158–159
- angle bisectors of, 164
- angle sum of, 92
- angles of, 15, 92–93
- area of, 247, 292, 481
- circumscribed, 164
- congruent, 102–132, 144–145
- corresponding parts of, 102–103, 144, 317
- equiangular, 97, 167, 173–174
- equilateral, 97, 167, 173–174
- exterior angle of, 93

Triangle(s) (*continued*)
- inscribed, 163
- isosceles, 97, 167–168, 173
- medians of, 158–159, 162
- obtuse, 97, 260
- opposite angles of, 93
- perpendicular bisectors of sides, 163
- right, 97, 210, 255–260, 428, 432
- scalene, 97
- segment joining midpoints of two sides of, 225–226, 502
- side of, 97
- sides divided proportionally, 323, 325–326
- similar, 317–330
- vertex of, 15

Triangle Inequality, 179
Triangle Proportionality, 325
Trigonometry, 447–458, 475
- finding areas with, 481
- table of, 517

V

Vertex,
- of angle, 10
- of cone, 406
- of polygon, 190–191
- of regular pyramid, 402
- of right prism, 393
- of solid, 405
- of triangle, 15

Vertical angles, 18, 47
Volume(s), 394
- of cone, 406, 443–444
- of cylinder, 398–399
- of regular pyramid, 402–403, 443
- of right prism, 394–395
- of similar solids, 414–415
- of sphere, 410

X

x-axis, 468
x-coordinate, 468

Y

y-axis, 468
y-coordinate, 468

Answers to Self-Tests

CHAPTER 1

Page 14 **1.** True **2.** False **3.** True **4.** True **5.** point R **6.** $\angle PQS$ or $\angle SQP$ **7.** $\angle PQR$ or $\angle RQP$ **8.** 180°

Page 22 **1.** right **2.** acute **3.** obtuse **4.** straight **5.** $\angle BED$, $\angle AEF$; $\angle BEA$, $\angle DEF$ **6.** $\angle ABF$, $\angle FBC$

Page 37 **1.**

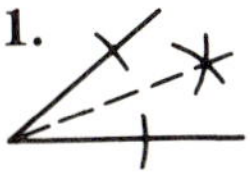

2. The Substitution Postulate **3.** The Addition Postulate **4.** 2; 5 **5.** W **6.** $WXYZ$ **7.** WX

CHAPTER 2

Page 53 **1.** 70° **2.** 142° **3.** $\angle SAL$ **4.** $\angle SAK$ **5.** Hyp.: you read the newspaper every day; Concl.: you are well informed **6.** If you are well informed, then you read the newspaper every day.

Page 80 **1.** A postulate is accepted without proof; a theorem is proved. **2.** $\angle 2$, $\angle 7$; $\angle 3$, $\angle 6$ **3.** $\angle 2$, $\angle 3$; $\angle 6$, $\angle 7$ **4.** $\angle 1 = \angle 3 = \angle 8 = 130°$; $\angle 2 = \angle 4 = \angle 5 = \angle 7 = 50°$ **5.** $x = 70$; $y = 50$ **6.** Show that corresponding angles are equal, alternate interior angles are equal, same-side interior angles are supplementary, or both lines are perpendicular to a third line in the same plane.
7. 1. $l \parallel m$ (Given) 2. $\angle 2 = \angle 3$ (If 2 $\parallel$ lines are cut by a trans., alt. int. angles are equal.) 3. $\angle 1 = \angle 2$ (Given) 4. $\angle 1 = \angle 3$ (Subst. Post.)
8.

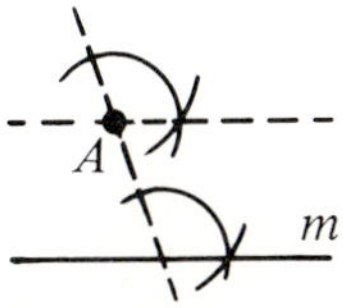

CHAPTER 3

Page 101 **1.** 40 **2.** 120 **3.** 70 **4.** 74 **5.** d or c **6.** a **7.** c **8.** b

Page 112 **1.** BOJ **2.** $\angle O$ **3.** EP
4. 1. Given 2. From algebra 3. Given: D is the midpt. of $\overline{XY}$. 4. SSS Post.

Page 125
1. 1. $\angle 1 = \angle 2$ (Given: $\overrightarrow{TV}$ bisects $\angle ETO$.) 2. $TE = TO$ (Given) 3. $TV = TV$ (From algebra) 4. $\triangle TEV \cong \triangle TOV$ (SAS Post.)
2. 1. $\angle 1 = \angle 2$; $\angle 3 = \angle 4$ (Given: $\overleftrightarrow{TV}$ bisects $\angle ETO$ and $\angle EVO$.) 2. $TV = TV$ (From algebra) 3. $\triangle TEV \cong \triangle TOV$ (ASA Post.)

Page 132 **1.** SAS, ASA, AAS
2. 1. $\angle 1$ and $\angle 2$ are right angles. (Given: $\overline{RT} \perp \overline{SV}$) 2. $RS = TV$; $RM = TM$ (Given) 3. $\triangle RMS \cong \triangle TMV$ (HL Thm.)

CHAPTER 4

Page 152
1. 1. $RM = TM$; $UM = SM$ (Given) 2. $\angle RMU = \angle TMS$ (Vertical angles are equal.) 3. $\triangle RMU \cong \triangle TMS$ (SAS Post.) 4. $RU = TS$ (Corr. parts of $\cong$ $\triangle$s are =.)
2. 1. $\angle X = \angle Y = 90°$ (Given) 2. $AX = AY$ (Given) 3. $AB = AB$ (From algebra) 4. $\triangle AXB \cong \triangle AYB$ (HL Thm.) 5. $\angle 1 = \angle 2$ (Corr. parts of $\cong$ $\triangle$s are =.) **3.** SSS

Page 166
1.

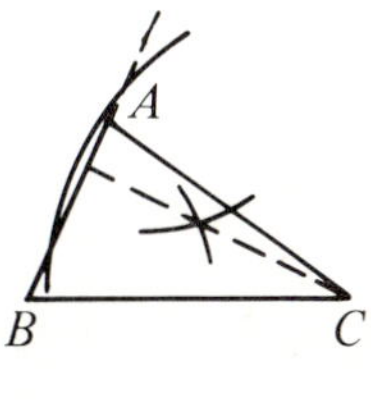

2.

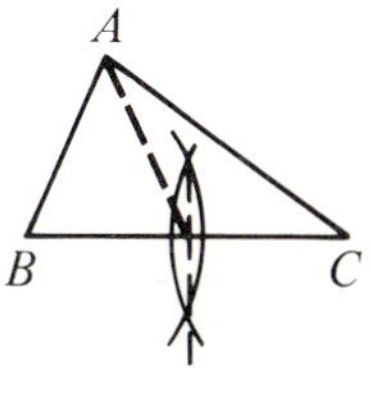

3. 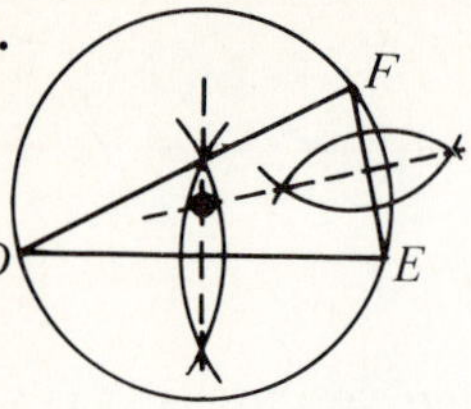

4. 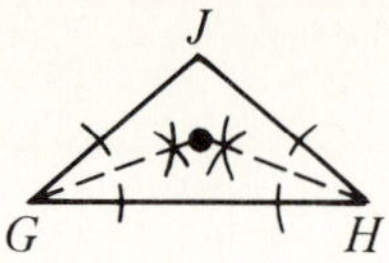

Page 178 **1.** If 2 sides of a triangle are equal, then the angles opposite those sides are equal. **2.** 60 **3.** 40 **4.** 44 **5.** 5 **6.** 4 **7.** 12

CHAPTER 5

Page 197

1. For example: 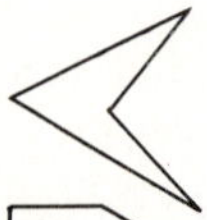

2. For example: 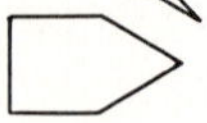

3. For example: 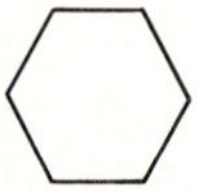

4. 120° **5.** 60°

Page 207 **1.** A quadrilateral with both pairs of opposite sides parallel **2.** A quadrilateral with just one pair of opposite sides parallel **3.** A parallelogram with four equal sides **4.** A parallelogram with four right angles **5.** A parallelogram with four right angles and four equal sides **6.** *CD* **7.** $\overline{AD}$ **8.** *BX* **9.** 5 **10.** *BCD* **11.** 3 **12.** *BCD* or *DAB*

Page 219 **1.** $BD = 12$; $BE = 6$ **2.** $\angle 1 = \angle 4 = 25°$; $\angle 2 = 90°$; $\angle 3 = 65°$ **3.** 28
4. 1. $WZ = XY$; $\angle 1 = \angle 2$ (Given)
2. $\overline{WZ} \parallel \overline{XY}$ (If 2 lines are cut by a trans. so that alt. int. angles are equal, the lines are ∥.)
3. *WXYZ* is a ▱. (If a quad. has 1 pair of opp. sides both ∥ and equal, the quad. is a ▱.)

Page 228 **1.** 9 **2.** $\angle G = \angle O = 60°$; $\angle A = 120°$ **3.** *PTKL*, *SPKT*, *PKAT* **4.** 10 **5.** 9

CHAPTER 6

Page 246 **1. a.** 26 **b.** 30 **2. a.** 5 cm **b.** 20 cm **3.** 48 **4.** 60

Page 254 **1.** 20 **2.** 12 **3.** 84 **4.** 91 **5.** 26

Page 262 **1.** 10 **2.** 4 **3.** 10 **4.** 50 **5.** obtuse **6.** right **7.** acute **8.** right

Page 271 **1.** $\frac{C}{d}$ **2.** 3.14 **3.** $d = 10$; $C = 10\pi$; $A = 25\pi$ **4.** $r = 8$; $C = 16\pi$; $A = 64\pi$ **5.** $r = 10$; $d = 20$; $A = 100\pi$ **6.** $r = 6$; $d = 12$; $C = 12\pi$

CHAPTER 7

Page 295 **1.** $\frac{2}{3}$ **2.** $\frac{5}{3}$ **3.** 5:8 **4.** 1:2:3 **5.** 14 **6.** 15 **7.** 9 **8.** 2 **9.** \$4.45 **10.** \$79.20

Page 301 **1.** about 3 cm; 1 cm: 3 km or $\frac{1}{300{,}000}$ **2.** 12 km **3.** 8.4 km **4.** Suggested scale: 1 cm:5 m or $\frac{1}{5000}$

CHAPTER 8

Page 322 **1.** *JKP* **2.** *KP*; *PJ* **3.** 4:8 or 1:2 **4.** *J* **5.** not similar **6.** similar **7.** not similar **8.** similar

Page 333 **1.** $x = 6$; $y = 15$ **2.** $x = 9$; $y = 2\frac{1}{4}$ **3.** 1:3; 1:9 **4.** 3:7, 9:49 **5.** 5:6; 5:6

CHAPTER 9

Page 358 **1.** $\overline{OX}$ or $\overline{OY}$ **2.** $\overline{XY}$ **3.** $\overleftrightarrow{ST}$ **4.** $\overleftrightarrow{PQ}$ or $\overline{PQ}$ **5.** $\overline{ST}$ **6.** 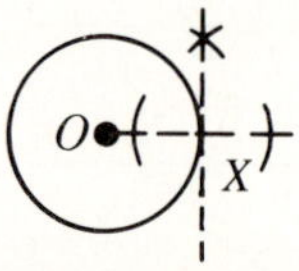

7. 2 **8.** tangent **9.** 80° **10.** 280°

Page 368 **1.** 6 **2.** 32° **3.** 12 **4.** 40° **5.** 76° **6.** 156°; 204° **7.** 30°
8. For example:

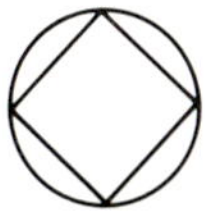

(Note that a rhombus inscribed in a circle must be a square.)

Page 376 **1.** 25° **2.** 17° **3.** 110° **4.** 50° **5.** 126° **6.** 120° **7.** 10 **8.** 6

CHAPTER 10

Page 401 **1.** *DSTE, ETUF* **2.** *CDSR, FETU* **3.** $\overleftrightarrow{CF}$, $\overleftrightarrow{DE}$, $\overleftrightarrow{CR}$, or $\overleftrightarrow{DS}$ **4.** 184 **5.** 160 **6.** 64π **7.** 192π **8.** 256π

Page 412 **1.**

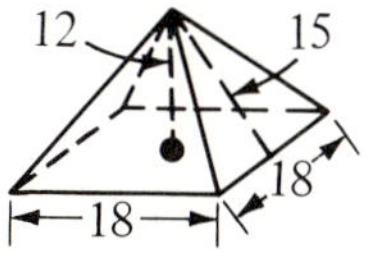

2. 540 **3.** 864 **4.** 1296 **5.** 5 **6.** 15π **7.** 24π **8.** 12π **9.** To the nearest integer, 201 **10.** To the nearest integer, 268

CHAPTER 11

Page 439 **1.** $x = 12; y = 9$ **2.** $x = 2; y = 2\sqrt{2}$ **3.** $x = 5; y = 5\sqrt{3}$ **4.** 18 **5.** $27\sqrt{3}$ **6.** $27\sqrt{3}$

Page 446 **1.** 12 cm **2.** 4 **3.** 16π **4.** 260

Page 456 **1.** $x = 14.3$ **2.** $\frac{8}{15}$ **3.** $\frac{8}{17}$ **4.** $\frac{8}{17}$ **5.** $\frac{15}{17}$ **6.** $y = 16.4; z = 11.5$

CHAPTER 12

Page 480 **1.** $(5, -4)$ **2.** 6 **3.** 10 **4.** 3 **5.** 13 **6.** $\left(-3, -2\frac{1}{2}\right)$ **7.** $(1, 2)$ **8.** $(1, -1)$

Page 495 **1.** horizontal **2.** vertical **3.** $\frac{4}{3}$ **4.** -2 **5.** $\frac{3}{7}$ **6.** $\frac{2}{5}$

Page 504 **1.** no **2.** (0, 8); (8, 0); (3, 5) **3.** $(0, -4); (3, 0); \left(5, 2\frac{2}{3}\right)$

4. For example:

x	y
3	4
4	2
5	0
6	−2

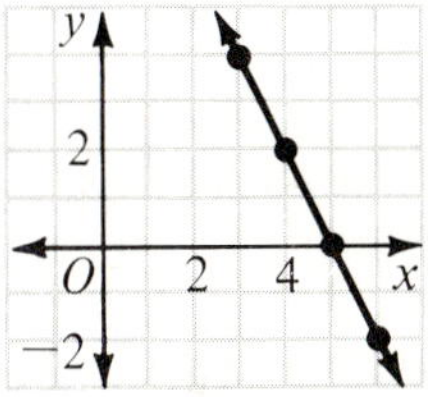

5. Slope of $\overline{AB} = 0 =$ slope of $\overline{DC}$.
Slope of $\overline{AD} = 4 =$ slope of $\overline{BC}$.
Therefore, $\overline{AB} \parallel \overline{DC}$ and $\overline{AD} \parallel \overline{BC}$.
By definition, *ABCD* is a parallelogram.

Answers to Selected Exercises

CHAPTER 1

Exercises, page 5 **1.** the bottom line **3.** The spaces are the same; for any diam.; the dist. will be $\frac{1}{\pi}$m.

Written Exercises, pages 8–9 **1. a.** 7 **b.** yes **3.** −3 **5.** −1 **7.** 4 **9.** 2 **11.** 5 **13.** 13 **15.** 17 **17.** true **19.** false **21.** 4 **23.** 16; −10 **25.** 41; 24 **27.** 6; $a + 3$

Written Exercises, page 13 **1.** $\overrightarrow{IA}, \overrightarrow{IH}$ **3.** *HIO* **5.** *DIO*; 180 **7.** 30° **9.** 100° **11.** 60° **17.** ∠1 = 120°; ∠2 = 65°; ∠3 = 90°; ∠4 = 35° **19.** *LAS* **21.** *BAS*

Consumer Applications, page 14 toaster: 3 h 7 min; television: 40 h; paperback: 42 min

Written Exercises, pages 16–17 **1.** acute **3.** obtuse **5.** straight **7.** ∠s 3, 5, 7, 10 **9.** ∠s 2, 3; 5, 6; 6, 7; 8, 10 **11.** ∠*CRA*, ∠*DRA*, ∠*DRB*, ∠*CRB* **13.** ∠*VQT*, ∠*TQR*, ∠*UQS* **15. a.** 65° **b.** 65° **c.** 25° **17.** 60°

Puzzle Problems, page 17

Written Exercises, page 19 **1.** ∠*DOE* **3.** ∠*FOD* **5.** 40° **7.** 70° **9.** 65° **11.** 25° **13.** 35° **15.** $x = 20$; $y = 9$

Written Exercises, page 21 **1.** 20; 10 **3. a.** 45° **b.** 65° **5.** 4 **7.** 9 (including ∠*LFA*) **9.** $\overline{FE}$ **11.** Answers will vary; the ∠s will be bisected. **13. c.** ∠3 = ∠4 **d.** ∠2 = ∠3 (Vert. ∠s are =.), and ∠1 = ∠4 (Vert. ∠s are =.); since ∠1 = ∠2, ∠3 = ∠4 (Subst. Post.)

Applications, page 24 **1. a.** Arctic **b.** Indian **c.** South Atlantic **d.** North Pacific **e.** North Atlantic **f.** South Pacific **3.** Africa

Written Exercises, pages 27–28 **1.** See Const. 1, p. 25. **3.** See Const. 1, p. 25. **5.** See Const. 2, p. 26. **7.** See Const. 3, p. 26. **9.** See Const. 3, p. 26. **11.** See Const. 1, p. 25.

13–15.

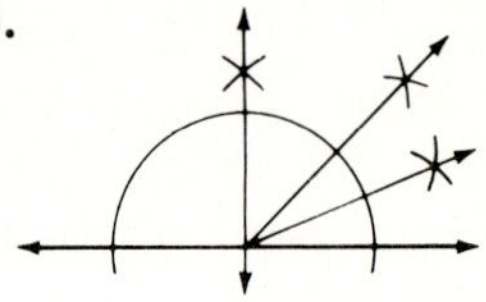

17.

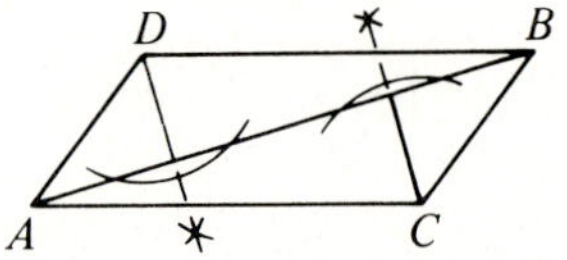

19.

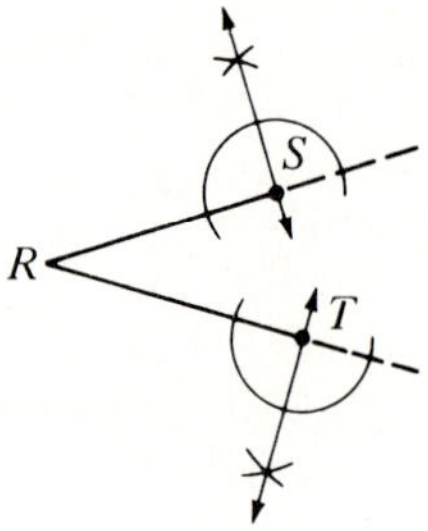

21.

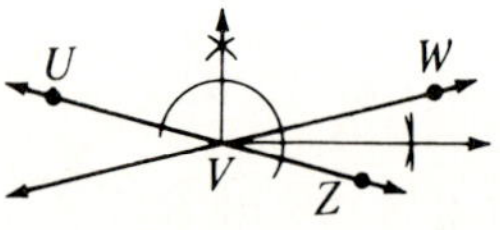

Written Exercises, pages 31–32 **1.** div. **3.** mult. **5.** add. **7.** add. **9.** add. **11.** add. **13.** 2; 3; 5 **15.** ∠1 = ∠2 since $\overrightarrow{AK}$ bisects ∠*RAC*, and ∠3 = ∠4 since ∠3 = 90° − ∠1 = 90° − ∠2 = ∠4. **17.** *AB* = *PQ* or *BM* = *QN*

Written Exercises, pages 36–37 **1.** *EFGH* **3.** *DCGH* **5.** *BF* **7.** *D* **13.** Through any 2 pts. there is exactly 1 line. **15.** more than one plane **17.** one plane **19.** True **21.** True

Skills Review, page 38 **1.** 9 **3.** −12 **5.** −21 **7.** −11 **9.** 8.1 **11.** −16.1 **13.** −4 **15.** 11 **17.** −14 **19.** −7 **21.** −3.4 **23.** 14.4 **25.** 60 **27.** −91 **29.** 0 **31.** −88 **33.** 89.6 **35.** −73.5 **37.** −7 **39.** 13 **41.** 4 **43.** −3 **45.** −23.1 **47.** 90 **49.** $>$ **51.** $=$ **53.** $<$

Chapter Review Exercises, pages 39–40
1. For example: $\overleftrightarrow{AC}$ **2.** $\overrightarrow{BC}$ or $\overrightarrow{BE}$ **3.** 4 **4.** −6 **5.** 11 **6.** *Q* **7.** 180 **8.** $\overline{BN}$ and $\overline{BQ}$ (or $\overline{BK}$) **9.** $\angle KQR$ **10.** 3 or *SQN* **11.** right **12.** acute **13.** obtuse **14.** straight **15.** 5 **16.** *RQB* **17.** 45°

18–19.

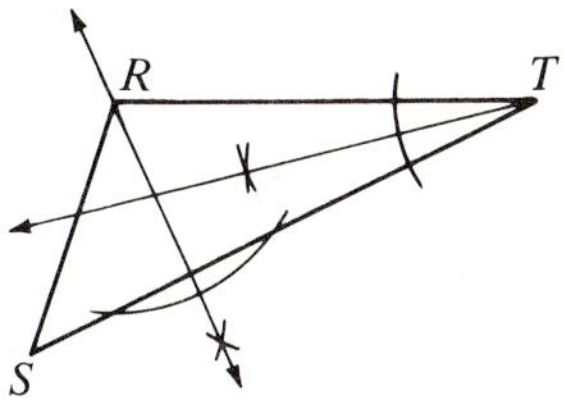

20. div. **21.** add. **22.** subst. **23.** *A* **24.** *S* **25.** *SLQA* **26.** *KL* **27.** exactly one

Mixed Review, pages 42–43
Algebra Problems: **1.** $x = 60$ **3.** $k = 6$ **5.** $x = 20$ **7.** $a = 4$ **9.** $q = -1$ **11.** $j = 3$ **13.** $x = 4$ **15.** $t = -4$ **17.** $w = -4$ **19.** $g = \frac{1}{2}$ **21.** no sol. **23.** $h = \frac{9}{2}$ or $4\frac{1}{2}$ **25.** $k = 3$ **27.** $h = 0$ **29.** All real numbers are solutions.
Geometry Problems: **1.** See Const. 3, p. 26. **3.** 1; Through any 2 pts. there is exactly 1 line. **5.** *ACE*; 180 **7.** True **9.** False **11.** Neither is longer; $RS = XY$. **13.** $x = 6$; $y = 40$ **15.** add. **17.** sometimes

CHAPTER 2

Written Exercises, pages 48–49 **1.** 40°; 130° **3.** 72°; 162° **5.** 42°; 138° **7.** 45°; 45° **9.** $\angle 4$ **11.** $\angle 2$, $\angle 8$ **13.** If 2 ∠s are supplements of = ∠s, then the 2 ∠s are =. **15.** See Const. 2, p. 26 **17.** subtr. **19.** add. **21.** 45°

Written Exercises, pages 52–53
1. *Hyp.*: The sum of two angles is 90°; *Conclu.*: The angles are complements. **3.** *Hyp.*: $5x + 7 = 27$; *Conclu.*: $x = 4$ **5.** *Hyp.*: It's a nice day; *Conclu.*: We can go to the beach Saturday. **7.** If two angles are comp., their sum is 90°. **9.** If $x = 4$, then $5x + 7 = 27$. **11.** If we go to the beach Saturday, then it will be a nice day. **13.** If *X* is a spider, then *X* is an insect. **15.** If two angles are vert. angles, then they are equal. **17.** If *X* is a turtle, then *X* is a reptile. **19.** If a person lives in a glass house, then he or she should not throw stones. **21. a.** False **b.** If two angles are vert. angles, then they are equal. **c.** True **23. a.** True **b.** If the sides of a figure are all the same length, then it is a square. **c.** False **25. a.** False **b.** Every square is a rectangle. **c.** True **27. a.** False **b.** Every number divisible by ten is also divisible by five. **c.** True

Applications, page 55 **1.** Art should resemble that shown in text.
3. Art may vary.

Written Exercises, pages 58–59
1. If *X* is a spider, then *X* has hairy legs. **3. a.** Daddy Long-Legs has hairy legs. (Type 1) **b.** no concl. (Type 3) **c.** no concl. (Type 4) **d.** An octopus is not a spider. (Type 2) **5. a.** Type 1 **b.** correct **7. a.** Type 4 **b.** not correct **9. a.** Type 2 **b.** correct **11.** Nothing can be concluded.

Written Exercises, pages 62–65
1. 1. Given 2. Vert. ∠s are = . 3. Subst. Post. **3.** 1. Given 2. Vert. ∠s are = . 3. Subst. Post. 4. straight 5. Subst. Post. 6. $\angle 3$ and $\angle 2$ are supplements.
5. 1. Def. of ∠ bisector 2. Vert. ∠s are =. 3. 4 (Vert. ∠s are = .) 4. Subst. Post. 5. $\overleftrightarrow{XY}$ bisects $\angle DOB$.
7. 1. *Given:* $\overrightarrow{RO}$ bisects $\angle YRS$. 2. *Given:* $\overrightarrow{SO}$ bisects $\angle YSR$. 3. Given 4. Subst. Post. 5. Add. Post. 6. Def of supp. ∠s 7. Supplements of = ∠s are = .

9. 1. $\angle 3$ and $\angle 4$ are supp. (Def. of supp. $\angle$s) 2. $\angle 5$ and $\angle 6$ are supp. (Def. of supp. $\angle$s) 3. $\angle 3 = \angle 5$ (Given) 4. $\angle 4 = \angle 6$ (Supplements of = $\angle$s are = .)
11. 1. $\angle 3$ and $\angle 6$ are supp. (Given) 2. $\angle 3$ and $\angle 2$ are supp. (Def. of supp. $\angle$s)
3. $\angle 2 = \angle 6$ (Supplements of = $\angle$s are = .)

Written Exercises, pages 69–71 **1.** (3) **3.** (1) **5.** (4) **7.** (2) **9.** $\angle 1 = \angle 3 = \angle 5 = \angle 7 = 110°$, $\angle 2 = \angle 4 = \angle 6 = \angle 8 = 70°$ **11.** *AD, BC, AC; AB, CD, AC* **13.** corresponding; = **15.** alt. int.; = **17.** alt. int.; = **19.** corresponding; = **21.** $\angle 1, \angle 5, \angle 8, \angle 9, \angle 12, \angle 13, \angle 16$ **23.** $x = 48$; $y = 132$ **25.** $x = 75$; $y = 105$ **27.** $x = 50$; $y = 70$ **29.** 1. Given 2. s.-s. int. $\angle$s are supp. 3. Given 4. If 2 $\parallel$ lines are cut by a trans., then s.-s. int. $\angle$s are supp. 5. Supplements of the same $\angle$ are = .

Calculator Activities, page 71 **1.** 64.360 **3.** 153.237 **5.** 109.201 **7.** 63.822 **9.** 4.945 **11.** 2750.019 **13.** 2800.600 **15.** 41.508 **17.** 28.889 **19.** 13.242 **21.** 384.333 **23.** 0.014 **25.** 0.2 **27.** 0.96875 **29.** 0.671875 **31.** 0.5556 **33.** 0.4444 **35.** 0.8333

Written Exercises, pages 74–76 **1.** yes **3.** yes **5.** no

7.

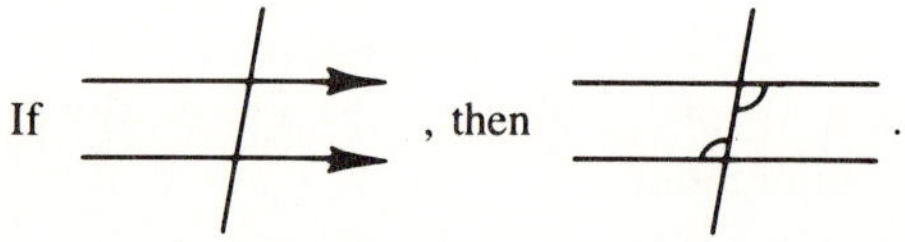

9.

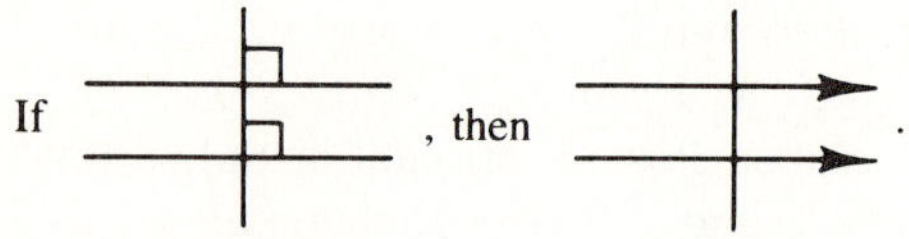

11. $\overleftrightarrow{BC}$ and $\overleftrightarrow{DF}$, $\overleftrightarrow{AB}$ and $\overleftrightarrow{CD}$ **13.** 120 **15.** $e \parallel f$ **17.** $e \parallel f$ **19.** $c \parallel d$ **21.** $c \parallel d$ **23.** $c \parallel d$ **25.** 1. Given 2. supp. $\angle$s 3. the same $\angle$ are = 4. If 2 lines and a trans. form = corr. $\angle$s, then the lines are $\parallel$.
27. It is possible for skew lines, or even $\perp$ lines, to each be $\perp$ to the same line.

Consumer Applications, page 76
1. oven **2.** about 54¢

Written Exercises, pages 79–80 **1.** See Const. 4, p. 77. **3.** See Const. 4, p. 77. **5.** See Const. 5, p. 78.

7.

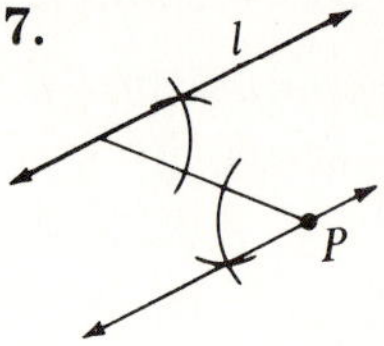

9. a.

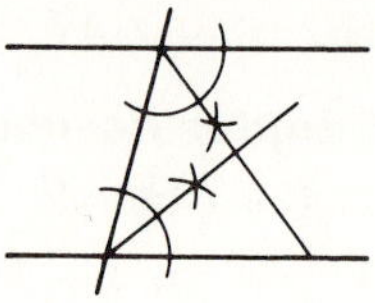

b–c. 90° **d.** right $\angle$s

11.

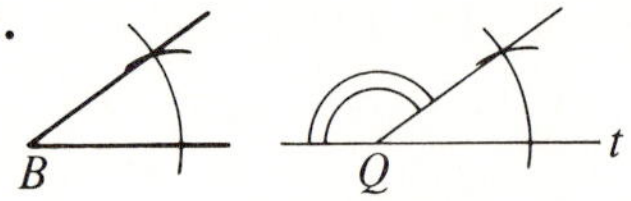

13.

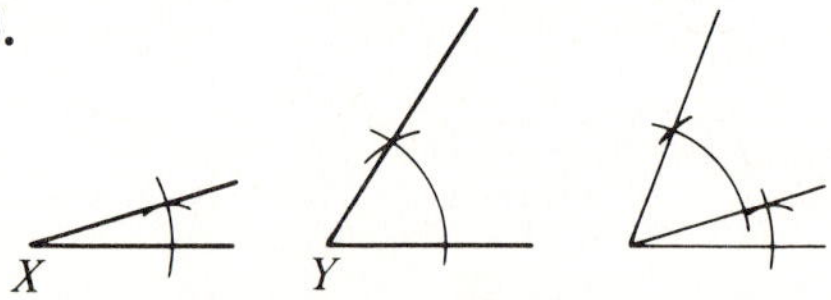

Reading Geometry, page 81 **1.** True **3.** False **5.** False **7.** False

Problem Solving Strategies, pages 82–83 **1. b.** Gail's gross income for the entire day **c.** 33 herbs; $33.00 **d.** $49.50 **3. b.** $\angle EHB = 90°$

Skills Review, page 84 **1.** (6, 2) **3.** (−3, −2) **5.** (5, −6) **7.** (−6, 0) **9.** (−1, 5) **11.** *L* **13.** *V* **15.** *M* **17.** *Z* **19.** *S* **33.** (2, 4) **35.** (3, 2) **37.** (5, −1) **39.** (6, −4) **41.** (1, −6) **43.** (0, −7)

Chapter Review Exercises, pages 85–86
1. 60 **2.** 150 **3.** $\angle 4$ **4.** Complements of = $\angle$s are = . **5.** If an angle is acute, then it has a measure less than 90°. **6.** *Hyp.*: An angle is acute; *Conclu.*: It has a measure less than 90°. **7.** If an angle has a measure less than 90°, then it is acute. **8. a.** no concl. **b.** no concl. **c.** Cinnamon has sharp claws. **d.** Cleo is not a cat.
9. 1. $\angle 1$ and $\angle 2$ are supp.; $\angle 3$ and $\angle 4$ are supp. (Def. of supp. $\angle$s) 2. $\angle 1 = \angle 4$ (Given) 3. $\angle 2 = \angle 3$ (Supplements of = $\angle$s

are = .) 4. $\angle 3 = \angle 5$ (Vert. $\measuredangle$s are = .) 5. $\angle 2 = \angle 5$ (Subst. Post.) **10.** corr. $\measuredangle$s **11.** alt. int. $\measuredangle$s **12.** s.-s. int. $\measuredangle$s **13.** corr. $\measuredangle$s **14.** $x = 58$; $y = 62$ **15.** $x = 125$; $y = 55$ **16.** $x = 55$; $y = 60$ **17.** $a \parallel b$ **18.** $a \parallel b$ **19.** $c \parallel d$ **20.** $c \parallel d$ **21.** See Const. 5, p. 78.

Cumulative Review, page 88 **1.** 4 **3.** 4 **5.** Z **7.** False **9.** True **11.** False **13.** True **15.** If two angles are equal, then they are vert. angles; not true.

CHAPTER 3

Written Exercises, pages 94–96
1. $\angle 1 = 80°$, $\angle 2 = 100°$ **3.** $\angle 5 = 50°$, $\angle 6 = 130°$ **5.** $\angle 1 = 30°$, $\angle 2 = 80°$, $\angle 3 = 70°$ **7.** 90° **9.** 100° **11.** $\angle 1 = 30°$, $\angle 2 = 50°$ **13.** $\angle 1 = 45°$, $\angle 2 = 85°$ **15.** $\angle R = 50°$, $\angle S = 60°$, $\angle T = 70°$ **17.** 30
19. They must also be equal, since each is 180° − (the equal sum of 2 $\measuredangle$s). **21.** 130
23. $\angle D = 40°$, $\angle E = 55°$, $\angle F = 85°$
25. 1. The $\angle$ sum of a $\triangle$ is 180° 2. Subtr. Post. 3. straight 4. Subtr. Post. 5. Subst. Post. **27.** 540 **29.** 90

Written Exercises, pages 99–100
1. isosceles **3.** scalene **5.** acute
7. right **9.** **a.** Any $\triangle$ whose $\measuredangle$s are all less than 90° with no sides =. **b.** Any $\triangle$ whose $\measuredangle$s are all less than 90° with 2 sides =.
11. **a.** Any right $\triangle$ with no sides =.
b. Any right $\triangle$ with = legs. **13.** $\angle C = 60°$, $\angle CAX = 30°$, $\angle AXC = 90°$ **15.** 5 cm, 6 cm, and 7 cm **17.** 10 cm, 10 cm, and 16 cm **19.** **a-b.** Constructions may vary.
c. The sides opp. the = $\measuredangle$s are = .

Written Exercises, pages 104–106
1. **a.** $\angle P$ **b.** $\angle T$ **c.** $\angle I$ **d.** PI **e.** EJ **f.** JM **3.** $\angle R = \angle X$; $\angle S = \angle Y$; $\angle T = \angle Z$; $RS = XY$; $RT = XZ$; $ST = YZ$
5. $\triangle ABC \cong \triangle REW$ **7.** not $\cong$ **9.** no
11. no **13.** yes **15.** no **17.** $\triangle MNP \cong \triangle MNQ$; $\angle PMN = \angle QMN$; $\angle P = \angle Q$; $\angle PNM = \angle QNM$; $MP = MQ$; $PN = QN$; $MN = MN$

Written Exercises, pages 109–111
1. 1. $\overline{CD}$ 2. *Given:* X is the mdpt. of $\overline{AB}$. 3. Given 4. SSS Post. **3.** 1. Given 2. From algebra 3. SSS Post. **5.** 9
7. $\frac{17}{3}$ or $5\frac{2}{3}$ **9.** $\angle B = 60°$; $\angle R = 100°$; $\angle S = 60°$; $\angle T = 20°$ **11.** $\angle A = y°$; $\angle B = 180° - (x + y)°$; $\angle S = 180° - (x + y)°$; $\angle T = x°$ **13.** See exercise instructions.
15. **c.** (3, 2) and (3, −2)

Applications, page 114 Models 3 and 6 are rigid figures.

Written Exercises, pages 117–119
1. Yes; yes; the lengths of each pair of sides are = . **3.** yes **5.** $\angle QXP$; $\angle SXR$ **7.** yes; yes **9.** SSS Post. **11.** SAS Post.
13. 1. *Given:* $\overrightarrow{ZW}$ bisects $\angle XZY$ 2. From algebra 3. Given 4. SAS Post.

15.

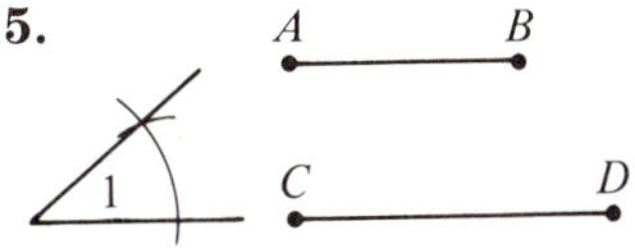

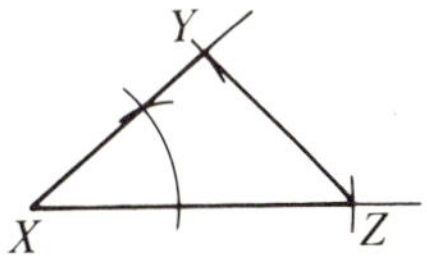

SAS Post.

17. 1. $\overline{RS} \parallel \overline{VT}$ (Given) 2. $\angle 1 = \angle 2$ (If 2 $\parallel$ lines are cut by a trans., then alt. int. $\measuredangle$s are = .) 3. $VS = VS$ (From algebra) 4. $RS = VT$ (Given) 5. $\triangle VRS \cong \triangle STV$ (SAS Post.)

Written Exercises, pages 122–124
1. JXP **3.** CA; PJ
5. 1. Given 2. equals 8 3. ASA Post.

7.

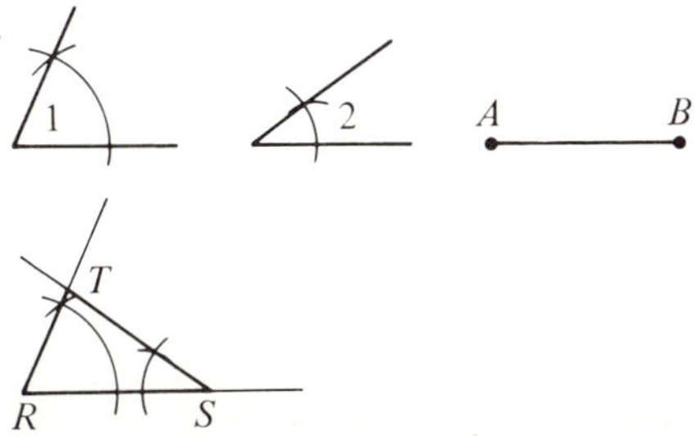

9. a. *ZNE* **b.** ASA Post. **11.** 1. $CB = DB$; $\angle ABC = \angle ABD$ (Given) 2. $AB = AB$ (From algebra) 3. $\triangle ACB \cong \triangle ADB$ (SAS Post.) **15. e.** ASA Post.

Written Exercises, pages 128–131
1. HL **3.** ASA **5.** HL (or AAS) **7.** AAS **9.** SSS **11.** none **13.** SAS, ASA, AAS **15.** 1. Given 2. From algebra 3. Given 4. HL Thm. **17.** 1. Def. of supp. ∠s 2. Given 3. Supplements of = ∠s are =. 4. Given 5. From algebra 6. ASA Post. **19.** $x = 110$, $y = 30$, $z = 12$ **21.** $x = 40$, $y = 13$, $z = 20$ **23.** $\angle X$, $\overline{YM}$ **25.** SAS Post. **27.** no **29.** yes; SAS Post.

Puzzle Problems, page 132 $\triangle AZB \cong \triangle AXB$ since $AB = AB$, $\angle ZAB = \angle XAB$, and $\angle ABZ = \angle ABX$ (ASA Post.). Thus $BX = ZX = 12$ m.

Reading Geometry, page 133 **1.** No; $\angle ABC$ has vertex *B* and $\angle BAC$ has vertex *A*. **3.** Use tick marks to indicate = sides and arcs to indicate = ∠s.

Computer Activities, page 135 **1.** $90 - A$ **3. a.** 55° **b.** 37.5° **c.** 60° **d.** 45° **5. a.** 90° **b.** 135° **c.** 45° **d.** 20°

Skills Review, page 136 *Across* **a.** 686 **h.** 2617 **o.** 361 **q.** 1226 **s.** 2125 **u.** 6590 **x.** 749 **z.** 2004 *Down* **a.** 6314 **c.** 615 **e.** 21,960 **g.** 5432 **i.** 625 **k.** 76,014 **m.** 1936 **s.** 27 **1.** $1\frac{1}{2}$ **3.** $11\frac{1}{6}$ **5.** $\frac{1}{6}$ **7.** $7\frac{7}{10}$ **9.** $\frac{5}{9}$ **11.** $\frac{5}{6}$ **13.** 32 **15.** 20 **17.** $5\frac{3}{8}$ **19.** 176 **21.** $\frac{2}{3}$ **23.** $8\frac{1}{6}$

Chapter Review Exercises, pages 137–138 **1.** 80 **2.** 70 **3.** 35 **4.** 132 **5.** 15 **6.** 40 **7.** 14 cm **8.** 22 cm **9.** scalene **10.** $\triangle SFZ$ **11.** $\triangle GFZ$ **12.** $\triangle SFG$ **13.** *R* **14.** *RD* **15.** *RDI* **16.** SSS **17.** ASA **18.** HL **19.** SAS **20.** 1. $\angle 1 = \angle 2$ (*Given:* $\overrightarrow{GH}$ bisects $\angle RGS$.) 2. $\angle R = \angle S$ (Given) 3. $GH = GH$ (From algebra) 4. $\triangle RHG \cong \triangle SHG$ (AAS Thm.)

Mixed Review, pages 140–141 **1.** If two lines are perpendicular, then they form right angles. **3.** 1. Given 2. Supplements of = ∠s are = . 3. Given 4. ASA Post. **5.** Def. of mdpt. **7.** add. **9.** 4 **11. a.** yes **b.** yes **c.** yes **13.** If 2 lines and a trans. form = corr. ∠s, then the lines are ∥. **15.** $x = 38$, $y = 8$ **17.** True **19.** False **21.** AAS

23.

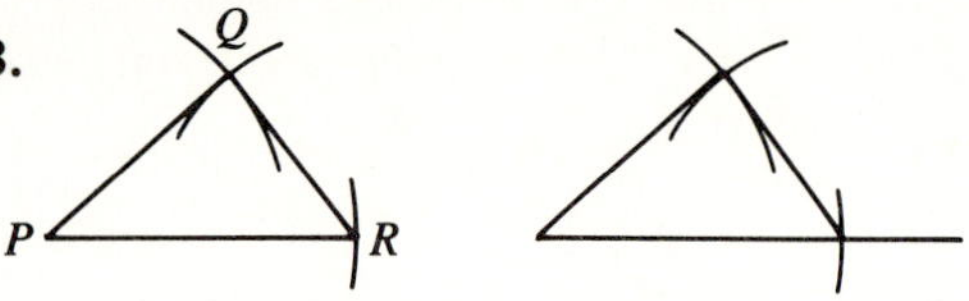

25. 90° **27.** False

CHAPTER 4

Written Exercises, pages 146–148
1. $\angle H = \angle U$ **3.** $\angle J = \angle W$ **5.** $IJ = VW$ **7.** $\triangle MEN \cong \triangle OLN$ **9.** $\triangle OEL \cong \triangle MLE$ **11.** 1. Given 2. *Given:* $\overrightarrow{CD}$ bisects $\angle ACB$. 3. From algebra 4. SAS Post. 5. Corr. parts of ≅ △s are =. **13.** 1. Given 2. From algebra 3. AAS Thm. 4. Corr. parts of ≅ △s are =. **15.** 1. $AX = BX$; $\angle A = \angle B$ (Given) 2. $\angle 1 = \angle 2$ (Vert. ∠s are =.) 3. $\triangle AXC \cong \triangle BXD$ (ASA Post.) 4. $CX = DX$ (Corr. parts of ≅ △s are =.) **17.** $RT = RQ$; $ST = SQ$ (Given) 2. $SR = SR$ (From algebra) 3. $\triangle RTS \cong \triangle RQS$ (SSS Post.) 4. $\angle T = \angle Q$ (Corr. parts of ≅ △s are =.)

Written Exercises, pages 150–152
1. f. ASA Post. **3. g.** SSS Post. **5. f.** SAS Post. **7. a.** SSS **b.** Corr. parts of ≅ △s are =. **c.** construction; algebra; SAS **d.** Corr. parts of ≅ △s are =. **e.** 90 **9.** $AY = OC$, $AX = OB$, and $XY = BC$ by const.; thus $\triangle AXY \cong \triangle OBC$ by the SSS Post.; $\angle A = \angle O$ since corr. parts of ≅ △s are =.
11. a. isosceles, since $DZ = EZ$ by const.

c. Yes, $DZ = EZ$, $\angle DZV = \angle EZV$ (by const.) and $ZV = ZV$, so $\triangle DVZ \cong \triangle EVZ$ by the SAS Post.

Reading Geometry, page 153 **1.** Lesson 4–2 **3.** equidistant **5.** inscribed

Written Exercises, pages 156–157
1. midpoint **3.** ⊥ bisector **5.** Any pt. on the ⊥ bisector of a seg. is equidistant from the endpts. of the seg. **7.** See Const. 6, p. 154. **9.** See Const. 3, p. 26. **11.** See Const. 6, p. 154. **13.** the midpt. of the hypotenuse **15.** the vertex of the ∠ opp. the base

Written Exercises, pages 160–161
1.
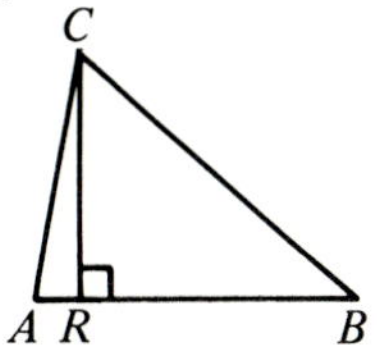

3.
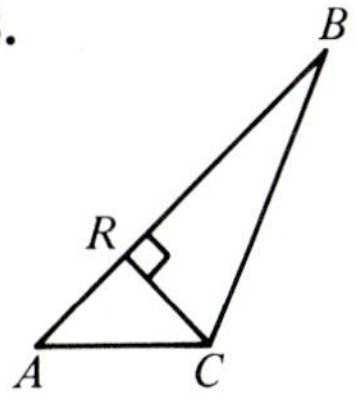

5.
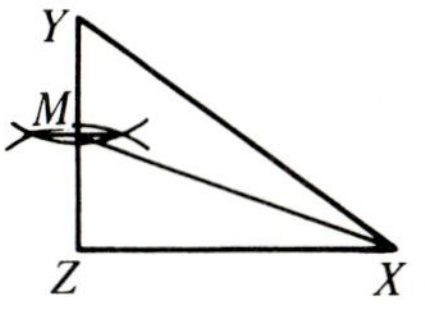

7.
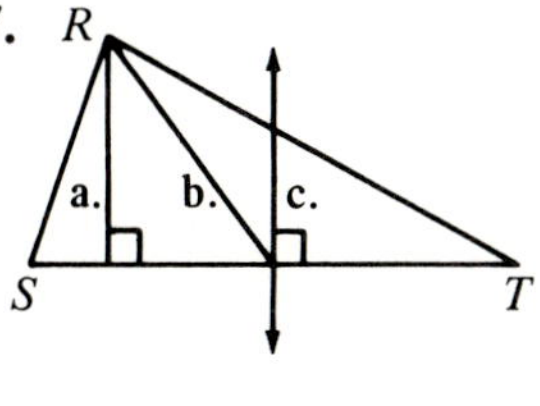

9.
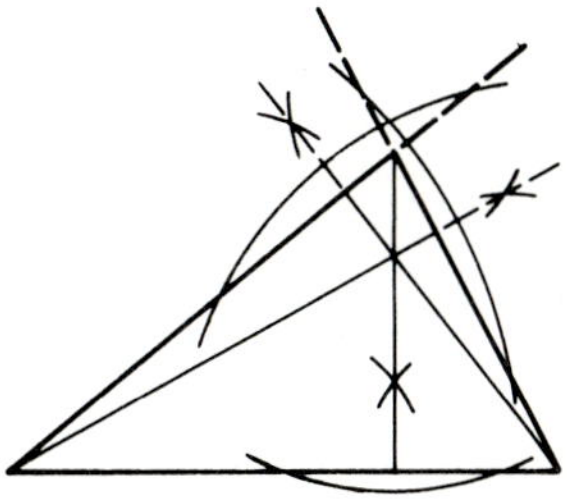

11.

13. d. Actual measurements will vary; in each case, $\frac{2}{3}(AM) = AP$, $\frac{2}{3}(BN) = BP$, and $\frac{2}{3}(CK) = CP$. **15.** The first and third columns are the same; $\frac{2}{3}$. **17. c.** yes

Written Exercises, pages 165–166
1.
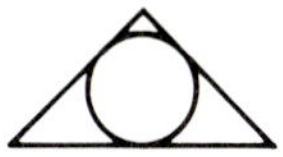

3.

5. See Const. 7, p. 163. **7.** See Const. 7, p. 163. **9.** See Const. 8, p. 164. **11.** Const. the ⊥ bisectors of $\overline{XZ}$ and $\overline{YZ}$; label their intersection O and use $\overline{OX}$ as radius.
13. See Const. 7, p. 163 and Const. 8, p. 164.

Written Exercises, pages 169–171
1. $\angle 2 = 65°$, $\angle 3 = \angle 4 = 115°$, $\angle 5 = 50°$
3. $\angle 1 = \angle 2 = 64°$, $\angle 3 = 116°$, $\angle 5 = 52°$
5. $\angle 2 = 68°$, $\angle 3 = \angle 4 = 112°$, $\angle 5 = 44°$
7. $\angle 1 = \angle 2 = 69°$, $\angle 3 = \angle 4 = 111°$
9. $\angle 2 = j°$, $\angle 3 = \angle 4 = 180° - j°$, $\angle 5 = 180° - 2j°$ **11.** 55 **13.** 9 **15.** 6 **17.** 25
19. 180 **21.** 1. Given 2. If 2 sides of a △ are =, then the ∠s opp. them are =. 3. Subst. Post. 4. The sum of the ∠s in a △ is 180°. 5. Subst. Post. 6. Div. Post. 7. Subst. Post.

Written Exercises, pages 175–177 **1.** 6 **3.** 6 **5.** 9 **7.** 10 **9.** 8 **11.** 1. Given 2. If 2 ∠s of a △ are =, then the sides opp. those ∠s are =. 3. Subst. Post.
13. Bisect the 60° angle from Ex. 12.

15.
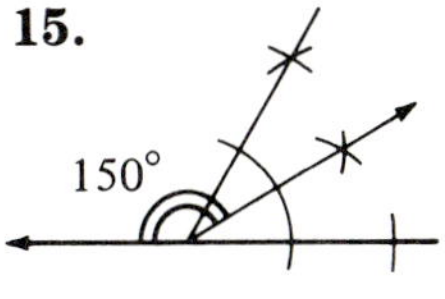

or

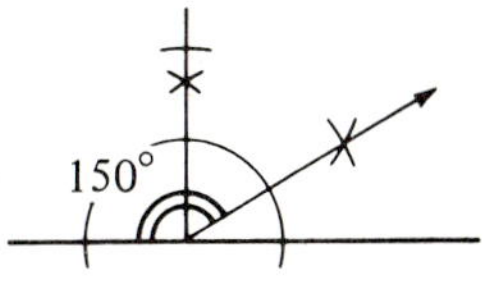

17.
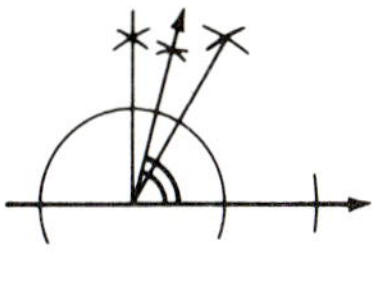

other constructions possible

19. 11 **21.** 1. $\angle J = \angle N$ (Given) 2. $PJ = PN$ (If 2 ∠s of a △ are =, then the sides opp. those ∠s are =.) 3. *JK* = *NM* (Given) 4. △*JPK* ≅ △*NPM* (SAS Post.) 5. ∠5 = ∠6 (Corr. parts of ≅ △s are =.)

Applications, page 179 **1.** yes **3.** no **5.** no

Problem Solving Strategies, pages 180–181 **1.** 15 handshakes **2.** Fang

Skills Review, page 182 **1.** 7.569 **3.** 8.479 **5.** 88.93 **7.** 2.457 **9.** 157.92 **11.** 1.86 **13.** 58.40 **15.** 11.5817 **17.** 90 **19.** 205 **21.** 0.071 ≈ 0.07 **23.** 66.666 ≈ 66.67 **25.** 75% **27.** 60% **29.** 87.5% **31.** 50% **33.** 900.00 **35.** 384 **37.** 20%

Chapter Review Exercises, pages 183–184 **1.** 1. Given 2. Vert. ∠s are =. 3. SAS Post. 4. Corr. parts of ≅ △s are =. 5. If 2 lines and a trans. form = alt. int. ∠s, then the lines are ∥. **2.** 1. Const. 2. Const. 3. SSS Post. 4. Corr. parts of ≅ △s are =. **3.** See Const. 6, p. 154. **4.** on *k* **5.** *Z* is equidistant from *C* and *D* **6.** $\overline{RY}$ **7.** $\overline{TX}$ **8.** Use Const. 3, p. 26, to const. the ⊥ from *M* to $\overline{NO}$. **9.** See Const. 7, p. 163. **10.** See Const. 8, p. 164. **11.** 1. Draw the bisector of ∠*K*, $\overline{KL}$. (Every ∠ has a bisector.) 2. *KX* = *KY* (Given) 3. ∠*XKL* = ∠*YKL* (Def. of ∠ bisector) 4. *KL* = *KL* (From algebra) 5. △*XKL* ≅ △*YKL* (SAS Post.) 6. ∠*X* = ∠*Y* (Corr. parts of ≅ △s are =.) **12.** 9 **13.** 7 **14.** 6

Cumulative Review, page 186 **1.** 1; 3 **3.** isosceles **5.** obtuse **7.** *J* **9.** HL **11.** ASA **13.** AAS **15.** False **17.** True **19.** True **21.**

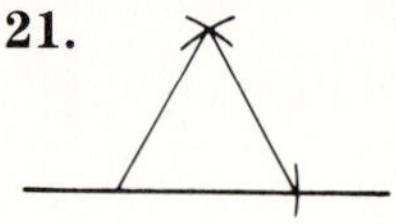

CHAPTER 5

Written Exercises, pages 192–193
1. yes **3.** yes **5.** no **7.** acceptable **9.** not acceptable **11.** 3 **13.** equiangular **15.** equilateral

17. For example:

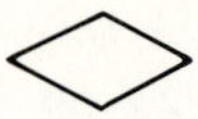

19. For example:

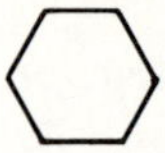

21. For example:

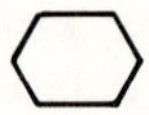

23. 3 **25.** 5 **27.** *n* − 3 **29.** 5 **31.** 14

Written Exercises, pages 196–197
1. 2 × 180°; 360° **3.** 9 × 180°; 360° **5.** 12; 360° **7.** 60°, 120° **9. a.** pentagon **b.** 540°; 108° **c.** 360°; 72° **11. a.** octagon **b.** 1080°; 135° **c.** 360°; 45° **13.** 6; 60° **15.** 18; 160° **17.** No; 3 × 120° = 360°, but no whole number times 135° is 360°. **19. a.** 360 **b.** 180 **c.** supplementary **d.** s.-s. int. ∠s are supp.

Written Exercises, pages 200–202
1. a. 1, 3, 4, 6, 8, 10, 11 **b.** 2, 5, 7, 12 **c.** 3, 6, 10 **d.** 4, 6, 11 **e.** 6 **f.** 6 **3.** False **5.** False **7.** *APTS, STQB, PDVU, UVCQ, APQB, PDCQ, ADCB* **9.** *APTS, UVCQ* **11.** ∠*B*, ∠*PUV*, ∠*UVC*, ∠*UQC*, ∠*QTS*, ∠*AST*, ∠*APT* **13.** 1. *Given: ABCD* is a rhombus. 2. From algebra 3. SSS Post. 4. Corr. parts of ≅ △s are =. 5. If 2 lines and a trans. form = alt. int. ∠s, then the lines are ∥. 6. Def. of ▱ **15.** A square is also a rectangle, so it has all the properties of a rectangle.

17.

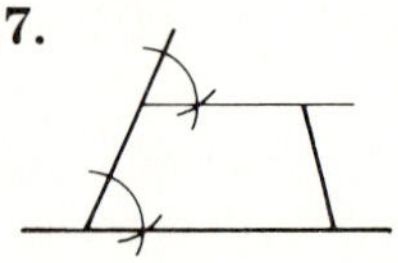

19. rhombus

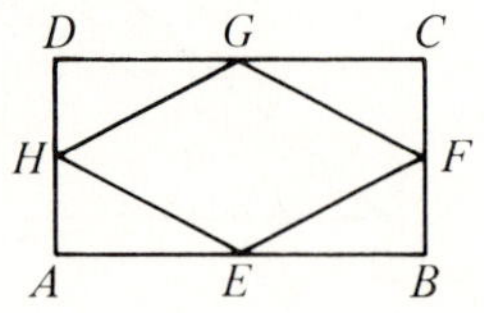

21. square

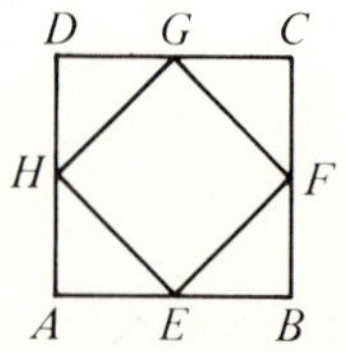

Written Exercises, pages 205–207 **1.** 6 **3.** 115° **5.** 10 **7.** 4 **9.** 110° **11.** yes; no **13.** 4 **15.** 70° **17.** 70° **19.** 86° **21.** 55° **23.** 16 **25.** 36 **27.** 3 **29.** 5
31. 1. *ABCD* is a ▱. (Given) 2. Draw $\overline{AC}$. (2 pts. determine a line.) 3. $\triangle ABC \cong \triangle CDA$ (Ex. 30) 4. $BC = DA$; $AB = CD$ (Corr. parts of $\cong$ $\triangle$s are =.) **33.** 1. Given 2. Def. of ▱ 3. If 2 $\parallel$ lines are cut by a trans., then alt. int. $\angle$s are =. 4. Opp. sides of a ▱ are =. 5. ASA Post. 6. Corr. parts of $\cong$ $\triangle$s are =.

Written Exercises, pages 211–213
1. b, d **3.** b, d **5.** a, b, c, d **7.** c, d **9.** c, d **11.** $\angle 1 = \angle 2 = \angle 5 = 25^\circ$, $\angle 3 = \angle 4 = 65^\circ$, $\angle 6 = 90^\circ$ **13.** $SO = 4$; $TO = 3$ **15.** 34 **17.** $TO = FO = 7$ **19.** $\angle 1 = \angle 4 = 15^\circ$, $\angle 2 = \angle 3 = 75^\circ$, $\angle 5 = 30^\circ$ **21.** $\angle 1 = \angle 2 = \angle 3 = \angle 4 = 45^\circ$, $\angle 5 = 90^\circ$ **23.** 5 **25.** 1. Given 2. Opp. sides of a ▱ are =. 3. Def. of rectangle 4. From algebra 5. SAS Post. 6. Corr. parts of $\cong$ $\triangle$s are =.

Written Exercises, pages 216–219 **1.** If a quad. has diagonals that bisect each other, then the quad. is a ▱. **3.** If a quad. has one pair of opp. sides that are both $\parallel$ and =, then the quad. is a ▱. **5.** 1. Given 2. If 2 $\angle$s of a $\triangle$ are =, then the sides opp. those $\angle$s are =. 3. Given 4. Subst. Post. 5. Given 6. If a quad. has both pairs of opp. sides =, then the quad. is a ▱. **7.** 1. The diagonals of a ▱ bisect each other. 2. Mult. Post. 3. *Given: P* is the mdpt. of $\overline{AO}$ and *Q* is the mdpt. of $\overline{CO}$. 4. Subst. Post. 5. The diagonals of a ▱ bisect each other. 6. If a quad. has diagonals that bisect each other, then the quad. is a ▱. **9.** ▱; if a quad. has diagonals that bisect each other, then the quad. is a ▱. **11.** square; the sides are = (see Ex. 10) and $\perp$. **13.** The jaws are extensions of opp. sides of the rectangle formed by the four bolts on the handle.

Written Exercises, pages 223–224
1. $\angle Q = 50^\circ$, $\angle R = \angle S = 130^\circ$ **3.** $\angle P = \angle Q = 80^\circ$, $\angle R = 100^\circ$ **5.** $\overline{SP}$; $\overline{RQ}$ **7.** 13 **9.** 12 **11.** 13 **13.** 5 **15.** 4 **17.** 4.7 **19.** 4 **21.** 6 **23.** No, the figure would then be a ▱ by Thm. 10, p. 214.

Puzzle Problems, page 224 The winning horse was Slowpoke, owned by Tic.

Written Exercises, pages 226–228
1. $\overline{XY}$ **3.** 10 **5.** 58 **7.** $ST = 4$; $RT = 5$; $RS = 6$; $p(\triangle ABC) = 30$; $p(\triangle RST) = 15$ **9.** $AB = 8$; $BC = 12$; $AC = 10$; $p(\triangle ABC) = 30$; $p(\triangle RST) = 15$ **11.** $BC = 12$; $AC = 8$; $ST = 7$; $p(\triangle ABC) = 34$; $p(\triangle RST) = 17$ **13.** 16 **15.** $y = 5$ **17.** $y = 13$ **19.** 44 **21. a.** *PRZQ* **b.** 50

Reading Geometry, page 229 **2.** Yes; if 2 sides of a $\triangle$ are =, then the $\angle$s opp. those sides are =. **4.** They are =. **5.** No; the second pair of sides must also be $\parallel$ or the first pair of sides must be the same length, in order for the figure to be a ▱.

Skills Review, page 230 **1.** -8 **3.** 7; -7 **5.** -13 **7.** 10; -10 **9.** 15 **11.** $7n$ **13.** $3\sqrt{7}$ **15.** 20 **17.** $8\sqrt{2}$ **19.** $7\sqrt{2}$ **21.** 14 **23.** $10\sqrt{7}$ **25.** 18 **27.** 16 **29.** 6 **31.** $5\sqrt{3}$ **33.** 4 **35.** 2

Chapter Review Exercises, pages 231–232 **1.** yes **2.** yes **3.** no **4.** no **5.** 540° **6.** 360° **7.** 135° **8.** False **9.** True **10.** False **11.** $\angle PRQ = 30^\circ$; $\angle PQR = 132^\circ$ **12.** $OS = 7$; $PR = 24$ **13.** 10 **14.** $\angle RKX = 50^\circ$; $\angle KXN = 50^\circ$ **15.** $\angle FSG = 30^\circ$; $\angle LOF = 90^\circ$ **16.** If a quad. has diagonals that bisect each other, then the quad. is a ▱. **17.** If a quad. has one pair of opp. sides that are both $\parallel$ and =, then the quad. is a ▱. **18.** If a quad. has both pairs of opp. sides =, then the quad. is a ▱. **19.** 11 **20.** $\angle B = 60^\circ$; $\angle X = 120^\circ$; $\angle Y = 120^\circ$ **21.** 6 **22.** 16

Mixed Review, pages 234–235 **1.** 44 **3.** trapezoid **5.** Yes; opp. sides of a ▱ are =. **7.** yes **9.** yes **11.** The altitudes drawn to the base lie on a single line.

13. 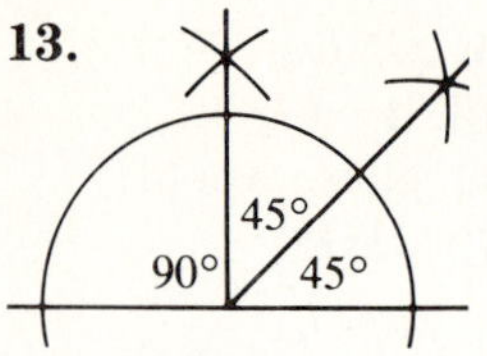

15. If a quad. has equal diagonals, then it is a rectangle; false. 17. never 19. 17 21. -11

23. Sketches may vary. For example, 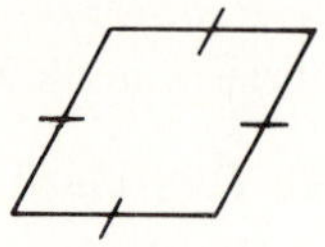

25. 13 27. Use Const. 4 on page 77. 29. (1) Midpoints Thm. (2) Def. of ▱ (3) Opp. ∠s of a ▱ are =.

CHAPTER 6

Written Exercises, pages 241–242 **1.** 6 sq. units **3.** 84 sq. units **5.** 49 sq. units **7.** $p = 16$; $A = 12$ sq. units **9.** $p = 14.4$; $A = 12.8$ sq. units **11.** $p = 26\frac{4}{5}$; $A = 44\frac{2}{5}$ sq. units **13.** $p = 4$; $A = 1$ sq. unit **15.** $p = 12$; $A = 9$ sq. units **17.** $p = 20$; $A = 25$ sq. units **19.** $s = 9$; $A = 81$ **21.** $s = 7$; $p = 28$ **23.** $p = 8k$; $A = 4k^2$ **25. a.** 3600 cm^2 **b.** 0.36 m^2 **27. a.** 15,000 cm^2 **b.** 1.5 m^2 **29.** 35 cm **31.** 260 m^2 **33. a.** 72 cm^2 **b.** 24 cm^2 **c.** 108 cm^2 **35.** $1296

Written Exercises, page 245 **1.** 45 sq. units **3.** 28 sq. units **5.** 56 sq. units **7.** 84 sq. units **9.** 30.4 sq. units **11.** $p = 4k + 14$; $A = k^2 + 5k$ **13.** $p = 16k$; $A = 5k^2 + 20k$

Written Exercises, page 249 **1.** 14 **3.** 32 m^2 **5.** 56 **7.** 8 **9.** 20 m^2 **11.** 24 **13.** $\frac{3t^2 - 5t}{2}$

Written Exercises, pages 252–254 **1.** d **3.** a **5.** 96 **7.** 125 **9.** 87.5 **11.** 28 cm^2 **13.** 54 cm^2 **15.** 88 cm^2 **17.** 49 **19.** 66.5 **21.** 8 **23.** 3 **25.** 20.4 squares

Puzzle Problems, page 254

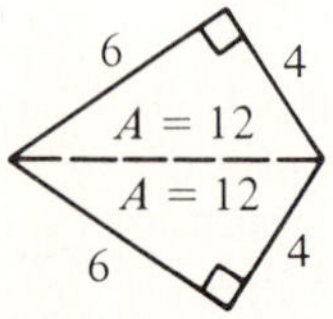

Written Exercises, pages 257–258 **1.** $x^2 = 3^2 + 4^2$; $x = 5$ **3.** $x^2 = 6^2 + 8^2$; $x = 10$ **5.** $17^2 = x^2 + 15^2$; $x = 8$ **7.** 10 **9.** 9 **11.** 10 **13.** 6.4 **15.** 7.8 **17.** 5.2 **19.** 7.1 **21.** 7.2 **23.** $x \approx 7.2$; $A \approx 86.4$ **25.** $x \approx 6.4$; $A = 20$ **27.** $x \approx 3.9$; $A \approx 66.3$

Written Exercises, page 261 **1.** yes **3.** no **5.** yes **7.** no **9.** yes **11.** no **13.** right **15.** acute **17.** right **19.** obtuse **21.** no **23.** $\overline{AC}$

Puzzle Problems, page 261 Luis is winning; George is last.

Written Exercises, pages 266–267 **1.** 14 **3.** 10.2 cm **5.** 8 cm **7.** 10.8 **9.** 10π **11.** 12π **13.** 33π m **15.** 0.8π km **17.** 36.8π km **19.** 1 **21.** 12.5 m **23.** 26.5 m **25.** 94.2 m **27.** 13.2 km **29.** 40.8 m **31.** 50 cm **33.** 3.2 cm **35.** 1.5 km **37.** 6 cm **39.** 1074 cm

Written Exercises, pages 269–271 **1.** 25π **3.** 9π **5.** 1.44π cm^2 **7.** 12.25π cm^2 **9.** $\frac{1225}{36}\pi$ **11.** 5 **13.** 9 cm **15.** 8 cm **17.** 153.9 **19.** 4.5 cm^2 **21.** 226.9 cm^2 **23.** 576π **25.** 30.25π **27.** 7π cm **29.** $25\pi - 50$ **31.** 15π

Applications, page 273 **1.** $\angle 1 = \angle 2 = \angle 6 = 60°$, $\angle 3 = \angle 4 = 30°$, $\angle 5 = 120°$ **3.** If 2 lines and a trans. form supp. s.-s. int. ∠s, then the lines are ∥. **5.** yes **7.** Opp. sides of the table are ∥, so alt. int. ∠s are =. **9.** If 2 lines and a trans. form = alt. int. ∠s, then the lines are ∥.

Problem Solving Strategies, pages 274–275 **1.** A began with $4.50; B began with $7.50 **3. a.** $3.60 **b.** 24 sixths of a mile **c.** 4 miles **5.** 72 days

Computer Activities, pages 276–277 **1.** Answers will vary. **3.** Answers will vary. **5.** 3.14

Skills Review, page 278

1. $\frac{86}{n}$ **3.** $13 + 3x$ **5.** $\frac{1}{2}c - 18$ **7.** $7(a + b)$

9. $n - 27 = 15$ 11. $\frac{1}{5}n - 13 = 27$
13. $3n > 16$ 15. $n + 43 < 75$ 17. $b = 24$
19. $n = 50$ 21. $m = -4$ 23. $x = -13$
25. $t = -24$ 27. $a = 0$ 29. $k = 11$
31. $k = -24$

Chapter Review Exercises, pages 279–280 1. $p = 14; A = 10$
2. $p = 100; A = 600$ 3. $h = 7; p = 28$
4. $p = 22.4; A = 28.8$ 5. 50 6. 42 7. 45
8. 36 9. 44 10. 96 11. 50 12. 34
13. 32 14. 4 15. 10 16. 9 17. 3
18. obtuse 19. right 20. acute 21. 12π
22. 25π 23. 50.2 24. 75.4 25. 9 26. 9

Cumulative Review, page 282
1. always 3. sometimes 5. sometimes
7. sometimes 9. $\perp$ 11. TS; RS
13. MF; ME 15. 78 17. 40 19. 48 m^2
21. no

CHAPTER 7

Written Exercises, pages 287–288 1. $\frac{1}{2}$
3. $\frac{3}{4}$ 5. 3:4 7. 1:2:3 9. $\frac{2}{3}$ 11. $\frac{2}{5}$
13. $\frac{4}{5}$ 15. $\frac{5}{9}$ 17. $\frac{2}{1}$ 19. $\frac{1}{3}$ 21. $\frac{3}{2}$ 23. $\frac{3}{5}$
25. $\frac{5}{8}$ 27. $\frac{5}{8}$ 29. $\frac{3}{2}$ 31. $\frac{1}{1}$ 33. a. $\frac{3}{5}$
b. $\frac{3}{8}$ 35. a. $\frac{1}{3}$ b. $\frac{1}{4}$ 37. $\frac{3}{10}$ 39. $\frac{1}{1500}$

Written Exercises, pages 291–292
1. $3a$ 3. $3r$ 5. $\frac{3}{4}$ 7. $\frac{5}{3}$ 9. $\frac{12}{4}$ or $\frac{3}{1}$
11. $8f$ 13. $\frac{4}{3}$ 15. $\frac{7}{3}$ 17. $\frac{7}{5}$ 19. $\frac{-2}{5}$
21. 3 23. 30 25. 6 27. 8 29. 4
31. 3 33. $\frac{6}{5}$ 35. 3 37. -1

Calculator Activities, page 292 1. 24
3. 15.6 5. 81.3 7. 95.5

Written Exercises, pages 294–295
1. \$85 3. \$9.75 5. 5 books 7. \$16.50
9. \$240 11. 1000 bricks

Puzzle Problems, page 295 Ron is the photographer; Bobbie won the track race; Jo won the math contest.

Written Exercises, page 298 1. 2080 m
3. 2960 m 5. 4880 m 7. 3120 m 9. l = 4 m; w = 3.75 m; p = 15.5 m; A = 15 m^2
11. l = 4.75 m; w = 3.5 m; p = 16.5 m; $A \approx 16.6$ m^2 13. about 21 km 15. about 38 km 17. a. Concord-Manchester: 20 mm; Manchester-Portsmouth: 48 mm; Portsmouth-Concord: 56 mm b. 70 km
c. 25 km

Written Exercises, pages 299–301
1. Using a scale of 1 cm:5 m, the drawing should be 10 cm × 4.2 cm. 3. Drawing should be 4.8 cm × 6 cm.

5.
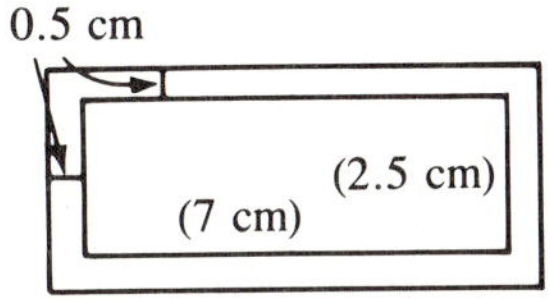

7.
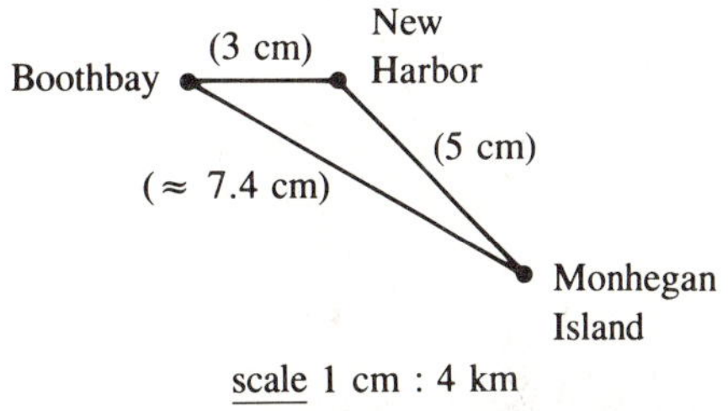

Distance between Boothbay and Monhegan is about 30 km.

9. a. 20 km b. 22 km

Applications, page 303 1. 90° 2. 315°
3. 215° 4. 277°

Skills Review, page 304 1. $a = 90$
3. $b = -34$ 5. $y = 14$ 7. $n = \frac{6}{7}$ 9. $b = 2$
11. $c = \frac{3}{2}$ 13. $a = 5\frac{1}{2}$ 15. $k = -30$
17. 120 cm tall

Chapter Review Exercises, pages 305–306 1. $\frac{3}{5}$ 2. $\frac{7}{2}$ 3. $\frac{8}{3}$ 4. 3:2
5. b:c 6. 8:5:4 7. 20:1 8. 1:100
9. 1:9 10. $\frac{9}{5}$ 11. $\frac{1}{2}$ 12. $\frac{3}{4}$ 13. $\frac{3}{4}$ 14. $\frac{4}{5}$

15. $2m$ **16.** $\frac{7}{5}$ **17.** $\frac{y}{4}$ **18.** 6 **19.** 15
20. 20 **21.** \$2.25 **22.** \$11.00
23. 5 cm; $\frac{1}{1{,}700{,}000}$ **24.** 68 km **25.** 51 km
26.

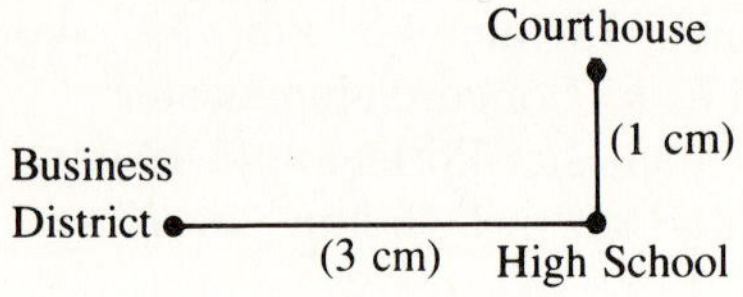

scale 1 cm : 1 km

Mixed Review, pages 308–309 **1.** If a triangle has a 60° angle, then it is equilateral; false. **3.** Each leg is 13 cm. **5.** 105° **7.** 104°, 104°, 152° **9.** 9.5 **11.** $t = 3$ **13.** SSS
15.

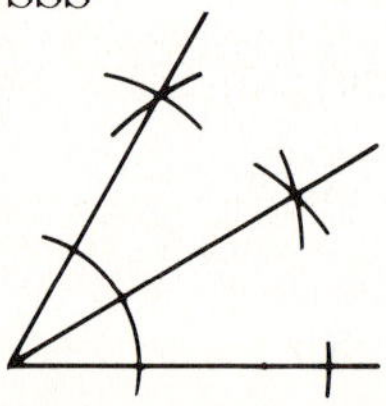

17. $\angle 1 = 75°$; $\angle 2 = \angle 3 = 60°$; $\angle 4 = 135°$ **19.** 168 sq. units
21. 120 sq. units **23.** 5 oranges **25.** 2240 km **27.** 2:5 **29.** HL Thm.

CHAPTER 8

Written Exercises, pages 314–316
1. yes **3.** no **5.** yes **7.** The $\triangle$s are $\cong$, so corr. $\angle$s are equal and corr. sides are in proportion. **9.** Corr. $\angle$s are equal and, since all sides in each $\triangle$ are equal, they are in proportion. **11.** trapezoid *MLKN; LK; KN* **13.** yes **15. a.** $\frac{2+4+2+4}{3+6+3+6} = \frac{12}{18}$ **b.** yes **17.** $x = 5$; $y = 30$; $z = 24$ **19.** $x = 5$; $y = 12$; $z = 20$ **21.** 20 and 22.5 **23.** F is 8 units from E in the horizontal direction, F at (14, 13) or (−2, 13) **25.** yes

Written Exercises, pages 319–321
1. no **3.** yes **5.** no **7.** yes **9.** yes **11. a.** *ADE* **b.** *DE; EA* **13. a.** *THG* **b.** 12; 10 **c.** $x = 8$; $y = 15$ **15. a.** *DOH* **b.** 6 **c.** 8 **17.** 1. Given 2. Opp. sides of a ▱ are $\parallel$. 3. If 2 $\parallel$ lines are cut by a trans., alt. int. $\angle$s are =. 4. AA Post. 5. Corr. sides of $\sim$ $\triangle$s are in proportion. **19.** *BOF* **21.** *FAB; FEC*

Calculator Activities, page 322
1. 6; 9; 12; 15; 18; 21

Written Exercises, pages 323–324
1. $\frac{3}{6} = \frac{x}{8} = \frac{y}{12}$; $x = 4$; $y = 6$
3. $\frac{x}{x+6} = \frac{3}{7} = \frac{y}{8}$; $x = 4\frac{1}{2}$; $y = 3\frac{3}{7}$
5. $x = 18$, $y = 21\frac{2}{3}$ **7.** $x = 9$; $y = 10\frac{2}{3}$
9. $x = 9$; $y = 6\frac{6}{7}$ **11.** 7

Puzzle Problems, page 324
The center figure

Written Exercises, pages 327–328 **1.** $\frac{n}{j}$
3. $\frac{y}{x+y}$ **5.** *YC* **7.** *AC* **9.** $x = 12$; $y = 9$ **11.** $x = 20$; $y = 36$ **13.** $x = 8$; $y = 10$ **15.** $x = 18$ **17.** $x = 13\frac{1}{3}$ **19.** 1. Given 2. exactly one line 3. If a line intersects a $\triangle$ and is $\parallel$ to one side, then it divides the other two sides proportionally. 4. Subst. Post. **21.** Lot A has the greater area by 1800 m^2. **23.** 9

Written Exercises, pages 331–332
1. p: 1:2; A: 1:4 **3.** p: 2:3; A: 4:9 **5.** scale factor: 3:5; A: 9:25 **7.** scale factor: 2:3; p: 2:3 **9.** scale factor: 3:7; p: 3:7 **11. a.** 1:2 **b.** 1:2 **c.** 1:4 **13. a.** 3:5 **b.** 3:5 **c.** 9:25 **15.** p: 3:5; A: 9:25 **17. a.** 20 **b.** 32 **19. a.** $\frac{1}{2}$ **b.** $\frac{1}{4}$ **21.** 25.2 cm^2 and 44.8 cm^2

Calculator Activities, page 332
1. $d = 12.4$ cm; $C = 38.9$ cm; $A = 120.7$ cm^2 **3.** $r = 15.9$ cm; $d = 31.8$ cm; $A = 793.8$ cm^2

Applications, page 335 **1.** Assume $a \parallel b$. Then, since $\angle 1$ and 2 are corr. $\angle$s, $\angle 1 = \angle 2$. This contradicts the given info. that $\angle 1 \neq \angle 2$. Thus, we conclude $a \nparallel b$.

Problem Solving Strategies, pages 336–337 **1.** $6\sqrt{5}$ **5.** $\overline{OT}$ and $\overline{OV}$ **7.** $\overline{CX}$ (or $\overline{OC}$ and $\overline{OX}$) **9.** altitude $\overline{DE}$

Skills Review, page 338 **1.** $8x + 33$ **3.** $a - 9$ **5.** $2 - 9n$ **7.** $1 + 0.4m$ **9.** $8a + 61.3$ **11.** $w = 6$ **13.** $a = -7$ **15.** $b = 2$ **17.** $m = -4$ **19.** $y = -48$ **21.** $d = 32$ **23.** $a = 1$ **25.** $1.45 **27.** $4(3n + 5) = 128$; $n = 9$

Chapter Review Exercises, pages 339–340 **1.** *JIG* **2.** *JG; IG* **3.** *I* **4.** *FLAP* **5.** 2:3 **6.** 27 **7.** *V; U* **8.** *VUT* **9.** 32 **10.** 24 **11.** 21 **12.** $\frac{r}{r+s}$ **13.** $\frac{c}{c+d}$ **14.** 18 **15.** 4 **16.** 6:5 **17.** 6:5 **18.** 36:25

Cumulative Review, page 342 **1.** $\frac{1}{7}$ **3.** $\frac{1}{20}$ **5.** 4 **7.** 11 **9.** $40 **11.** not ~ **13.** ~ **15.** *ROS* **17.** *SE* **19.** *A* **21.** 2 **23.** 4:1

CHAPTER 9

Written Exercises, pages 347–348 **1.** $\odot A$ **3.** $\overline{RB}$, $\overline{CD}$ **5.** $\overline{BC}$ **7.** $\overleftrightarrow{BC}$ **9.** none **11.** 38 **13.** 15 **15.** 13 **17.** $4\frac{1}{4}$ **19.** *J, K, E, F, G, C, D, T* **21.** $\overline{EG}$ **23.** $\overleftrightarrow{JK}$ **25.** A chord that passes through the center of the sphere **27.** possible **29.** possible **31.** 10

Written Exercises, pages 353–354 **1.** internal:2; external:2 **3.** internal:none; external:2 **5.** internal:none; external: none **7.** ext. tangent: Ex. 2; int. tangent: Ex. 6 **9.** True **11.** True **13.** Draw $\overrightarrow{PX}$; use Const. 2, p. 26, to const. the ⊥ to $\overleftrightarrow{PX}$ through *X*. **19.** 6 **21.** 3 **23.** 12

Written Exercises, pages 357–358 **1.** 110° **3.** 58° **5.** $42\frac{1}{2}°$ **7.** 120° **9.** 45° **11.** 49 **13.** $\overset{\frown}{XP}$, $\overset{\frown}{PY}$, $\overset{\frown}{PQ}$, $\overset{\frown}{YQ}$, $\overset{\frown}{QX}$ **15.** $\overset{\frown}{XPY}$, $\overset{\frown}{XQY}$ **17.** 140 **19.** 21 **21.** 56 **23.** 90 **25.** 90

Written Exercises, pages 362–363 **1.** inscribed **3.** circumscribed **5.** neither **7.** 10 **9.** 5 **11.** 8 **13.** yes **15.** no **17.** See text. **19.**

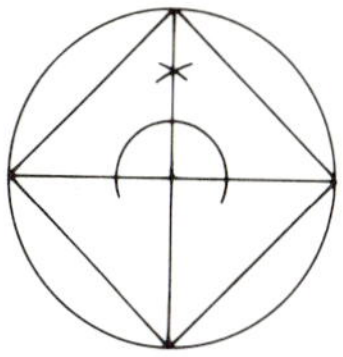

21. 8 **23.** 10 **25.**

Written Exercises, pages 366–367 **1.** 25 **3.** 185 **5.** 56 **7.** 184 **9.** 90 **11.** 60 **13.** 64 **15.** 48; 48 **17.** 88; 44 **19.** yes; yes **21.** 51° **23.** 37° **25.** 94° **27.** 125 **29. a.** 72; 72 **b.** ∠1 = ∠2 = ∠3 = 36°; ∠4 = 108°

Written Exercises, pages 371–372 **1.** 40° **3.** 38° **5.** 80° **7.** 35° **9.** 59° **11.** 45° **13.** 39 **15.** 180 **17.** 84 **19.** 58 **21.** 27 **23.** 40

Written Exercises, pages 374–376 **1.** 2 **3.** 4 **5.** $8\frac{3}{4}$ **7.** $2\frac{3}{4}$ **9.** 6 **11.** 0.9 **13.** $\overset{\frown}{XC}$; $\overset{\frown}{XC}$ **15.** *PD; PX* **17.** 6 **19.** 6

Puzzle Problems, page 376 The construction does not work when the chords are ||, because the ⊥ bisectors are the same line.

Computer Activities, page 379 **1.** 1.43, 1.67, 2.00, 2.35, 2.86, 1.86, 2.17, 2.60, 3.06, 3.71 **3.** Change the following lines:

80 IF G < = 5 THEN R = 36/B
100 IF G > 5 THEN R = 50/B

5. 744 wheel turns per mile **7. a.** 2600 turns **b.** 1000 turns **c.** 2600 turns **d.** 2860 turns

Skills Review, page 380 **1.** $\frac{1}{9}$ **3.** $1\frac{5}{6}$ **5.** $3\frac{1}{4}$ **7.** $\frac{\sqrt{3}}{10}$ **9.** $\frac{\sqrt{5}}{4}$ **11.** $\frac{6x}{5}$ **13.** $\frac{c\sqrt{2}}{10}$ **15.** $\frac{3n\sqrt{2}}{7}$ **17.** $\frac{5x\sqrt{3}}{6}$ **19.** $\frac{d\sqrt{5}}{4}$ **21.** $\frac{\sqrt{21}}{7}$ **23.** $\frac{\sqrt{10}}{4}$ **25.** $\frac{\sqrt{35}}{5}$ **27.** $\frac{2\sqrt{11}}{11}$ **29.** $\frac{\sqrt{6}}{2}$ **31.** $5\sqrt{2}$ **33.** $2\sqrt{15}$ **35.** $2\sqrt{14}$

Chapter Review Exercises, pages 381–382 **1.** $\overline{OB}$ or $\overline{OC}$ **2.** $\overline{CB}$ **3.** $\overleftrightarrow{BC}$ **4.** $\overleftrightarrow{BE}$ **5.** 2 **6.** 2 **7.** $\odot R$ **8.** $\odot S$ **9.** 180 **10.** 50 **11.** 48 **12.** 54

13. 47 **14.** 7 **15.** 70 **16.** 5 **17.** 4 **18.** 50 **19.** 132 **20.** 7 **21.** 10 **22.** 35 **23.** 80 **24.** 6 **25.** 45 **26.** 30 **27.** 55

Mixed Review, pages 384–385 **1.** right **3.** 10 sides **5.** 1. *WXYZ* is a ▱, $\overline{WY}$ and $\overline{XZ}$ intersect at *P*. (Given) 2. $WX = YZ$ (Opp. sides of a ▱ are =.) 3. $XP = ZP$; $WP = YP$ (Diagonals of a ▱ bisect each other.) 4. $\triangle WPX \cong \triangle YPZ$ (SSS Post.) **7.** obtuse **9.** *GHF* **11.** $C = 16\pi$ cm; $A = 64\pi$ cm^2 **13.** 9 **15.** $13\frac{1}{2}$ **17.** See Const. 10, p. 351. **19.** 68.4 sq. units **21.** 119 sq. units **23.** \$2.45 **25.** 1. $\angle QRT = \angle SRT$; $\angle QTR = \angle STR$ (*Given:* $\overline{RT}$ bisects $\angle QRS$; $\overline{TR}$ bisects $\angle QTS$.) 2. $RT = RT$ (From algebra) 3. $\triangle QRT \cong \triangle SRT$ (ASA Post.) 4. $QT = ST$ (Corr. parts of ≅ ⚠ are =.) **27.** none **29.** 10 **31.** 5

CHAPTER 10

Written Exercises, pages 390–391
1, 3. See text examples on p. 388. **5.** $\overleftrightarrow{PQ}$, $\overleftrightarrow{SR}$, $\overleftrightarrow{DC}$ **7.** $\overleftrightarrow{PQ}$, $\overleftrightarrow{SR}$, $\overleftrightarrow{QB}$, $\overleftrightarrow{RC}$ **9.** planes *PQRS, SDCR, ABCD, ABQP* **11.** $\overleftrightarrow{PQ}$ and $\overleftrightarrow{SD}$, or $\overleftrightarrow{SR}$ and $\overleftrightarrow{PA}$ **13.** || **15.** True **17.** False **19.** True **21.** False **23.** False

25.

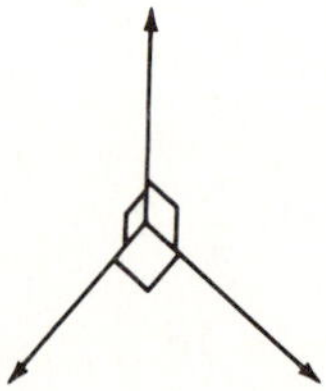

27.

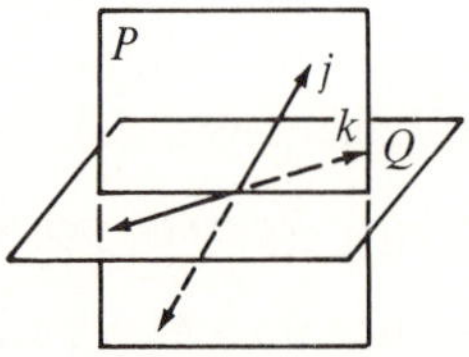

29.

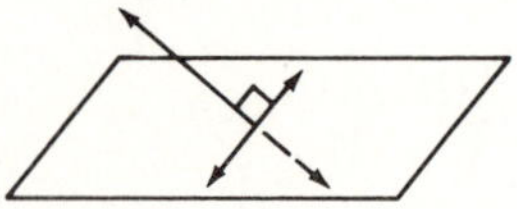

31. one
33. one

Written Exercises, pages 396–397
1. 4 cm **3.** 8 cm **5.** 24 cm **7.** 20 cm^2 **9.** 184 cm^2 **11.** 120 cm^3 **13.** 50 cm^2 **15.** L.A. = 48; T.A. = 72; $V = 36$ **17.** $h = 3$; L.A. = 84; T.A. = 164 **19.** $B = 4$; L.A. = 16; T.A. = 24; $V = 8$ **21.** $B = 25$; L.A. = 100; T.A. = 150; $V = 125$ **23. a.** 4 **b.** 8 **25.** 34.4% **27.** 88,750 L

Written Exercises, pages 399–400 **1.** 2 **3.** 5 **5.** 4π **7.** 28π **9.** $C = 10\pi$; $B = 25\pi$; L.A. = 20π; T.A. = 70π; $V = 50\pi$ **11.** $C = 2\pi$; $B = \pi$; L.A. = 12π; T.A. = 14π; $V = 6\pi$ **13.** $r = 3$; $B = 9\pi$; L.A. = 12π; T.A. = 30π; $V = 18\pi$ **15.** $h = 4$; $C = 6\pi$; $B = 9\pi$; L.A. = 24π; T.A. = 42π **17.** 226 **19.** shorter can

Puzzle Problems, page 401 □ is opposite ●; x is opposite ○; ◇ is opposite 

Written Exercises, page 404 **1.** True **3.** False **5.** True **7.** False **9.** True **17.** 64 **19.** 144 **21.** L.A. = 240 cm^2; T.A. = 384 cm^2; $V = 384$ cm^3 **23.** 120 m^2 **25.** 72 cm^3

Puzzle Problems, page 405
Sam is 1.75 m tall.

Written Exercises, pages 407–408
1. 54π **3.** 30π **5.** 20π **7.** 180π **9.** 48π **11.** $C = 24\pi$; $B = 144\pi$; L.A. = 180π; T.A. = 324π; $V = 432\pi$ **13.** $r = 6$; $B = 36\pi$; L.A. = 60π; T.A. = 96π; $V = 96\pi$ **15.** $C = 10\pi$; $B = 25\pi$; $h = 12$; $l = 13$; L.A. = 65π; T.A. = 90π **17.** 5810 *g* **19.** $\frac{\pi}{4}$

Calculator Activities, page 409
1. ≈ 290 **3.** ≈ 10 **5.** ≈ 10

Written Exercises, pages 411–412
1. $A = 144\pi$; $V = 288\pi$ **3.** $A = 256\pi$; $V = \frac{2048}{3}\pi$ **5.** $A \approx 1017$; $V \approx 3052$ **7.** $A \approx 452$ cm^2; $V \approx 904$ cm^3 **9.** $d = \frac{1}{4}$; $A = \frac{1}{16}\pi$; $V = \frac{1}{384}\pi$ **11.** $r = 5$; $d = 10$; $V = \frac{500}{3}\pi$ **13.** $d = 3$; $A = 9\pi$; $V = \frac{9}{2}\pi$ **15.** $r = 7$; $d = 14$; $A = 196\pi$ **17.** $A = 4\pi r^2 = 4\pi n^2$; L.A. $= 2\pi rh = 2\pi n \cdot 2n = 4\pi n^2$

Applications, page 413 small cube: $\frac{\text{T.A.}}{V} = \frac{3}{1}$; larger cube: $\frac{\text{T.A.}}{V} = \frac{1}{10}$

Written Exercises, pages 415–416
1. 9:16; 27:64 **3.** 1:36; 1:216 **5.** 2:5; 8:125 **7.** 1:4; 1:16 **9.** 10 cm **11.** 1000 cm^3 **13.** 1:9 **15.** 53,333L **17. a.** $3\frac{1}{3}$ **b.** 1500 cm^2; 50,000; $33\frac{1}{3}$

Problem Solving Strategies, pages 418–419

1.
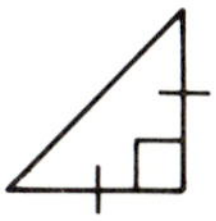

3.
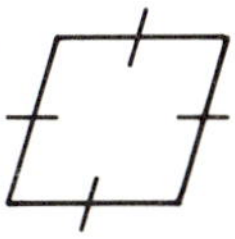

5.
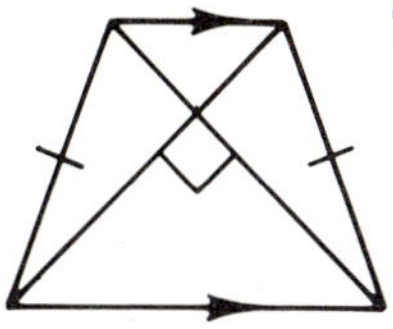

7.

9.
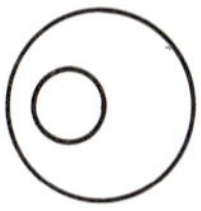

11. The altitudes of any obtuse △ meet outside the △. **13.** Let square A have sides of length 1 and square B have sides of length 2. Then the area of A is 1, the area of B is 4, and their ratio is 1:4. **15.** For example, $7 > 5$ and $7 \neq 6$. **17.** For example, $3^2 = 9 = (-3)^2$ and $3 \neq -3$. **19.** For example, a 30° angle and a 60° angle are comp. **21.** $(-5)^2 = 25 > 16 = 4^2$ and $-5 < 4$ **23.** ∠1 and ∠2 could be the two equal angles of an isosceles △.

Skills Review, page 420 **1.** $3a^5 - a^4b - a^3b^2$ **3.** $r^2 + 8r + 16$ **5.** $100 - 60x + 9x^2$ **7.** $b^2 - 26b + 169$ **9.** $x^2 - 15x + 54$ **11.** $-30hk - 24hk^2 + 6k^5$ **13.** $a^2 - 144b^2$ **15.** $24k^2 + 13hk - 2h^2$ **17.** $9m^2 - 4n^2$ **19.** $5ax(3a - 2x)$ **21.** $(y - 10)(y - 10)$ or $(y - 10)^2$ **23.** $(a + 7)(a - 5)$ **25.** $(k + 7)(k + 7)$ or $(k + 7)^2$ **27.** $(c - 10)(c - 3)$ **29.** $3ab^2(9a^2b + 2c)$ **31.** $(12 + n)(12 + n)$ or $(12 + n)^2$ **33.** $(h + 8)(h - 2)$ **35.** $(n - 9)(n - 3)$ **37.** $(k - 11)(k + 2)$ **39.** $(r + 7)(r + 4)$ **41.** $(d - 19)(d - 1)$

Chapter Review Exercises, page 422
1. True **2.** False **3.** False **4.** False **5.** 90 **6.** 228 **7.** 408 **8.** 540 **9.** 12π **10.** 96π **11.** 168π **12.** 288π **13.** 576 **14.** 720 **15.** 1296 **16.** 1728 **17.** 10π **18.** 65π **19.** 90π **20.** 100π **21.** 100π **22.** $\frac{256\pi}{3}$ **23.** 4:25; 8:125 **24.** 7:9; 343:729 **25.** 4:3; 16:9

Cumulative Review, page 424 **1.** ⊙R **3.** 90 **5.** 47 **7.** RS **9.** 49 **11.** True **13.** True **15.** False **17.** True **19.** 56π **21.** 24π

CHAPTER 11

Written Exercises, pages 429–430
1. 13 **3.** 6 **5.** 15 **7.** $x = 8$; $y = \sqrt{37}$ **9.** $x = 9$; $y = 12$ **11.** $x = 12$; $y = 5$ **13.** $x = 4$; $y = 4$ **15.** $x = 15$; $y = 8$ **17.** 108 sq. units **19.** Yes; $6^2 + 8^2 = 36 + 64 = 100 = 10^2$, so △$ACB$ is a right △. **21.** 25π **23.** $p = 20$; $q = 15$ **25.** yes **27.** no **29.** 12 **31.** Yes; $12^2 + 9^2 = 144 + 81 = 225$ and $15^2 = 225$. **33.** 128 sq. units **35.** $25\pi - 48$ sq. units

Applications, page 431
1.
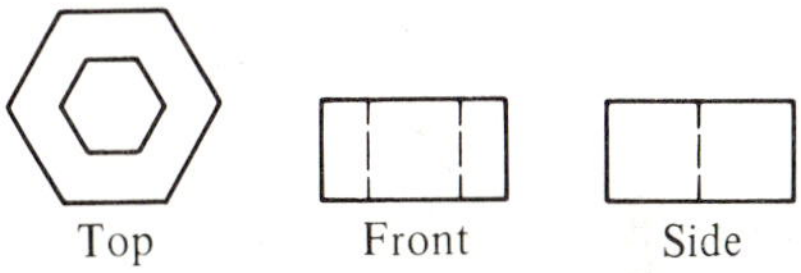

3.

Written Exercises, pages 434–436
1. $x = 8$; $y = 8\sqrt{2}$ **3.** $x = 7$; $y = 7\sqrt{2}$ **5.** $x = 12$; $y = 6\sqrt{3}$ **7.** $x = 10$; $y = 10\sqrt{3}$ **9.** $x = 8$; $y = 8\sqrt{3}$ **11.** 90 **13.** 30 **15.** $3\sqrt{3}$ **17.** $6\sqrt{2}$ **19.** $3\sqrt{2}$ **21.** 8 **23.** $5\sqrt{2}$ **25.** 50 **27.** $5\sqrt{2}$ **29.** Each side has length $6\sqrt{2}$. **31.** $BC = 6$; $AC = 6\sqrt{3}$; $CD = 3\sqrt{3}$; $AD = 9$; $DB = 3$ **33.** $AB = 8$; $BC = 4$; $CD = 2\sqrt{3}$; $AD = 6$; $DB = 2$ **35.** $AB = 28$; $BC = 14$; $AC = 14\sqrt{3}$; $AD = 21$; $DB = 7$

Calculator Activities, page 436
1. ≈ 1.4 g **3.** ≈ 1.3 g

Written Exercises, pages 437–439 **1.** 2 **3.** $50\sqrt{3}$ **5.** $30\sqrt{3}$ **7.** $x = 2\sqrt{3}$; $A = 4\sqrt{3}$ **9.** $x = 8$; $A = 64\sqrt{3}$ **11.** 30° **13.** $4\sqrt{3}$ **15.** 60° **17.** 7 **19.** $\frac{49}{2}\sqrt{3}$ **21.** 90° **23.** 4 **25.** $32\sqrt{3}$ **27.** 5:3 **29.** 25:9 **31.** 32 **33.** 55 **35.** $\sqrt{3}$ cm **37.** $\sqrt{3}$

Written Exercises, page 442 **1.** $x = 2\sqrt{5}$; $d = 3\sqrt{5}$ **3.** $x = \sqrt{5}$; $d = \sqrt{14}$ **5.** $x = \sqrt{29}$; $d = \sqrt{65}$ **7.** $c = 8$; $x = 15$ **9.** $x^2 = a^2 + b^2 = 1^2 + 1^2 = 2$, so $x = \sqrt{2}$ **11.** 30.1 cm **13.** no

Puzzle Problems, page 442

a.

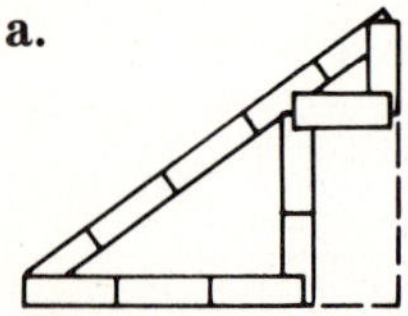

b.

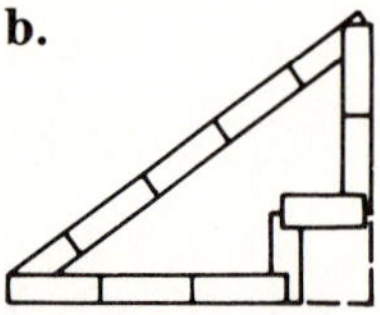

Written Exercises, pages 445–446
1. $l = 5$; L.A. = 80; T.A. = 144; $V = 64$
3. $l = 10$; L.A. = 240; T.A. = 384; $V = 384$
5. $h = 8$; L.A. = 240; T.A. = 384; $V = 384$
7. $l = 10$; L.A. = 80π; T.A. = 144π; $V = 128\pi$
9. $l = 17$; L.A. = 255π; T.A. = 480π; $V = 600\pi$ **11.** $r = 15$; L.A. = 255π; T.A. = 480π; $V = 600\pi$ **13.** $2\sqrt{3}$ **15.** $16\sqrt{3}$ **17.** 480 cm^2 **19.** $\frac{256\sqrt{161}}{3}$ cm^3 **21.** 18 **23.** $\frac{4000}{3}\pi$

Puzzle Problems, page 446 10

Written Exercises, pages 449–451
1. 0.0875 **3.** 2.7475 **5.** 30° **7.** $\tan A = \frac{5}{12}$; $\tan B = \frac{12}{5}$ **9.** $\tan A = 1$; $\tan B = 1$ **11.** 2.7 **13.** 2.5 **15.** 13.7 **17.** 0.8 m **19. a.** $\frac{3}{4}$ **b.** 0.75 **c.** 37° **21. a.** $\frac{2}{3}$ **b.** $0.\overline{6}$ **c.** 34° **23.** 0.4000

Written Exercises, pages 454–455
1. 0.0872 **3.** 1.0000 **5.** 75° **7.** 65° **9.** $\frac{5}{13}$ **11.** $\frac{5}{12}$ **13.** $\frac{5}{13}$ **15.** 67° **17.** $\frac{24}{25}$ **19.** $\frac{24}{25}$ **21.** $\frac{24}{7}$ **23.** $\frac{1}{2}$ **25.** 30° **27.** 0.9 **29.** 5.7 **31.** 6.7 **33.** 10.0 **35.** about 3.8 m **37.** 35° **39.** 76°

Applications, pages 458–459 **1. a.** 15° **b.** 15° **c.** less than 15° **2. a.** $\angle 2$; angle of elev. **b.** $\angle 3$; angle of dep. **c.** $\angle 5$; angle of dep. **d.** $\angle 8$; angle of elev. **7.** $\tan 40° = \frac{16}{x}$ **9.** 92 m **11.** 79 m

Puzzle Problems, page 459

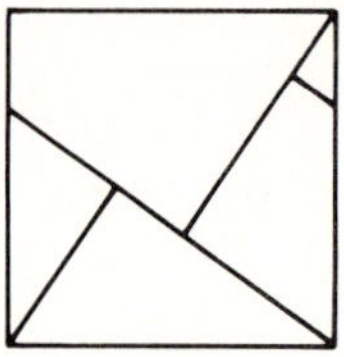

Skills Review, page 460 **1.** $\frac{-a}{4}$ **3.** $\frac{bc^2}{-3}$ **5.** $\frac{2}{a + b}$ **7.** $\frac{2}{a - b}$ **9.** $a - 8$ **11.** $\frac{d - 5}{13}$ **13.** $\frac{5}{2}$ **15.** $\frac{3}{b}$ **17.** $\frac{x - 2}{3}$ **19.** $\frac{y^2}{x + 5}$ **21.** $\frac{x + 2}{4}$ **23.** $\frac{m}{2n^2}$ **25.** $\frac{b}{2}$ **27.** $\frac{4}{7}$ **29.** $\frac{3(1 + d)}{c + 2}$

Chapter Review Exercises, pages 461–462 **1.** $x = 10$; $y = 15$ **2.** $x = 12$, $y = 15$ **3.** $x = 2$; $y = 4\sqrt{2}$ **4.** $x = 18$, $y = 9\sqrt{3}$ **5.** $x = 4$; $y = 4\sqrt{2}$ **6.** $x = 5$, $y = 5\sqrt{3}$ **7.** $18\sqrt{3}$ **8.** 18 **9.** $25\sqrt{3}$ **10.** 10 **11.** $\sqrt{109}$ **12.** 240 **13.** 384 **14.** 8 **15.** 384 **16.** 15 **17.** 135π **18.** 216π **19.** 324π **20.** $\frac{3}{4}$ **21.** $\frac{4}{5}$ **22.** $\frac{4}{5}$ **23.** $\frac{4}{3}$ **24.** 1.5 **25.** 4.7 **26.** 5.6

Mixed Review, pages 464–465 **1.** True **3.** True **5.** False **7.** 1. Given 2. Def. of isos. trap. 3. Base $\angle$s of an isos. trap. are =. 4. Given 5. Def. of $\perp$ lines 6. $\triangle AED \cong \triangle BFC$ (AAS Thm.) 7. $\angle ADE = \angle BCF$ (Corr. parts of $\cong$ $\triangle$s are =.) **9.** The $\triangle$s have the same height and equal bases. **11.** 50; 25 **13.** $x = 2$; $y = 5$ **15.** $x = 17$; $y = 7\sqrt{3}$ **17.** 2940π cubic units **21.** 24 **23.** $3\sqrt{2}$ **25.** 20

CHAPTER 12

Written Exercises, pages 470–471

1. IV; II **3.** III **5.** y **7. b.** C **9. b.** C **11.** $J(2, 2)$; $K(5, 5)$ **13.** $J(-5, 5)$; $K(5, 2)$ **15.** $J(-3, -1)$; $K(5, -4)$ **17.** 5 **19.** 3 **21.** 6 **23.** trapezoid **25.** rectangle **27.** $T(-6, 2)$ **29.** $T(8, 0)$ **31.** $(-3, 1)$, $(3, -1)$, or $(3, 9)$

Written Exercises, pages 473–474

1. $\sqrt{29}$ **3.** $\sqrt{61}$ **5.** $\sqrt{5}$ **7.** $3\sqrt{2}$ **9.** 5 **11.** $\sqrt{61}$ **13.** 7.1 **15.** 4.5 **17.** $p = 12$; $A = 6$ **19.** $p = 30$; $A = 30$ **21.** $p = 20$; $A = 25$ **23.** $p = 34$; $A = 56$

Calculator Activities, page 475

1. a. tangent **b.** sine **3. a.** sine **b.** cosine **5.** 0.2079; 0.9781; 0.2126; 2.1 m; 10.2 m

Written Exercises, pages 479–480

1. $(5, 3)$ **3.** $(-1, 4)$ **5.** $(4, -2)$ **7.** $(0, -3)$ **9.** $(8, 3)$ **11.** $(8, 6)$ **13.** $(5, 3)$ **15.** $(2, -5)$ **17.** $(5, 4\frac{1}{2})$ **19.** $(3, 2.85)$ **21.** $(\frac{3}{4}, -5)$ **23.** $(10, 5)$ **25.** $(3, -1)$

Calculator Activities, page 481

1. 28.19 **3.** 259.81 **5.** 50.23

Applications, pages 484–485

1. 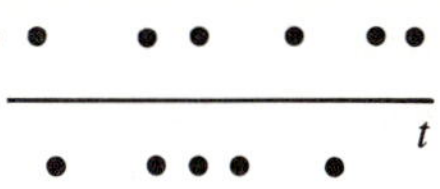

b. 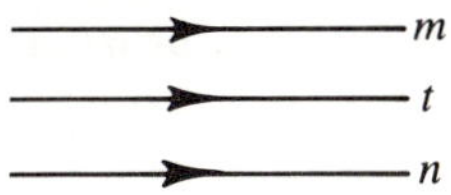

c. Lines m and n, $\parallel$ to t, each 3 cm from t. **3.** A plane $\parallel$ to planes R and S, midway between them **5.** Any pt. on the $\perp$ bisector of a seg. is equidistant from the endpts. of the seg. **7.** At the intersection of the diagonals of the rectangular ceiling **9.** 2 planes, each $\parallel$ to plane P, and 1 m from it. **11.**

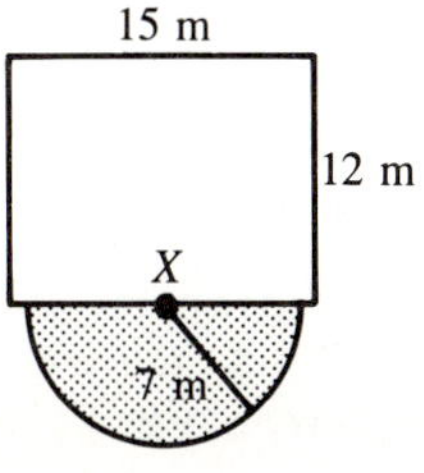

13. At the intersection of the $\perp$ bisectors of the sides of the $\triangle$ formed by the 3 towns **15.** an infinite number **17.** none

Written Exercises, pages 489–490

1. a. a **b.** c **c.** d **d.** b **3.** $\frac{1}{3}$ **5.** $-\frac{7}{3}$ **7.** 6 **9.** -7; -3; $\frac{7}{3}$ **11.** 3; -4; $\frac{3}{-4}$ **13.** -5; 2; $\frac{-5}{2}$ **15.** $\overline{CD}$ **17.** $\overline{AB}$, $\overline{CD}$, and $\overline{AD}$ **19.** For example: $(1, 5)$, $(2, 6)$, and $(-1, 3)$ **21.** For example: $(-1, -4)$, $(2, -4)$, and $(6, -4)$

Written Exercises, pages 493–495

1. a. $-\frac{5}{7}$ **b.** $\frac{7}{5}$ **3. a.** not defined **b.** 0 **5.** $\parallel$ **7.** $\perp$ **9.** 0; 0; yes; no **11.** -2; -2; yes; no **13.** $\overleftrightarrow{EF}$: $\frac{2}{7}$; $\overleftrightarrow{FG}$: $-\frac{7}{2}$ **15.** $\overline{RT} \perp \overline{SV}$ **17.** ▱ **19. a.** $(1, 3)$ **b.** $(6, 3)$ **c.** 0; 0; 0 **d.** Their slopes are =. **e.** 8; 2; 5; yes

Written Exercises, pages 499–500

1. For example: $(0, -1)$, $(3, -3)$, $(-3, 1)$, $(-6, 3)$

3.

x	y
0	6
6	0
2	4

5.

x	y
0	0
0	0
3	9

7.

x	y
0	−4
6	0
12	4

9.

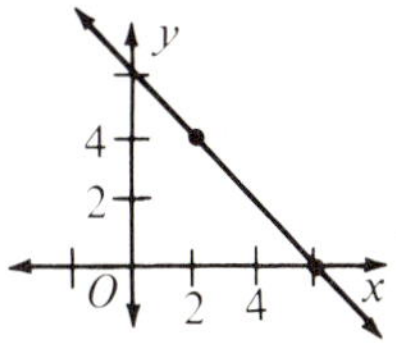

11.

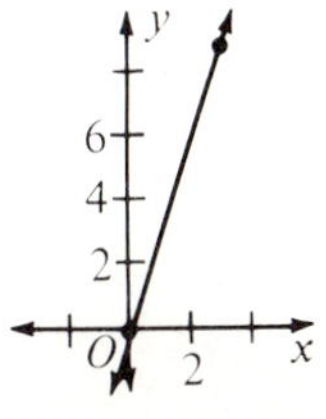

13.

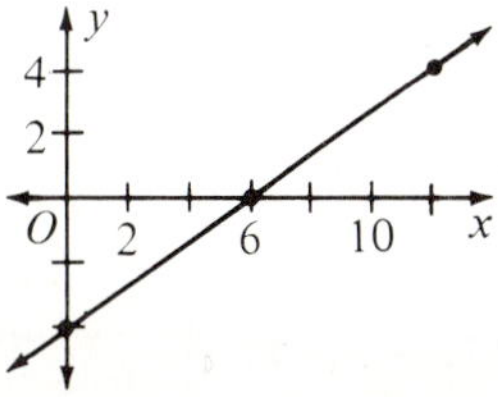

15. 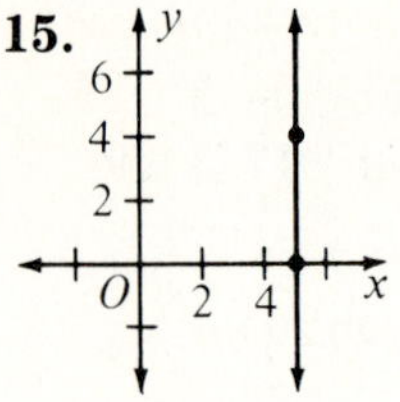

17.

x	y
3	0
1	3
−1	6

19.

x	y
3	−2
0	2
−3	6

21. (4, 2) **23.** (2, 2) **25.** 8 **27.** 3

Written Exercises, pages 502–504
1. a. 5 **b.** 4 **c.** 4; 41 **d.** 5; 4; 41 **e.** $\sqrt{41}$ **3. a.** (2, 3) **b.** See text. **c.** (0, 3); (2, 0) **d.** 2; 13 **e.** 2; 3; 13 **f.** 3; 2; 13 **g.** 13 **5.** T has coord. (a, a). Slope of $\overline{RT} = \frac{a-0}{a-0} = \frac{a}{a} = 1$ and slope of $\overline{VS} = \frac{a-0}{0-a} = \frac{a}{-a} = -1$. Since (slope of $\overline{RT}$) $\cdot$ (slope of $\overline{VS}$) $= 1(-1) = -1$, $\overline{RT} \perp \overline{VS}$.
7. S must have coord. $(4a, 0)$. Let A and B be the mdpts. of $\overline{RT}$ and $\overline{ST}$, respectively. A has coord. $(\frac{2a}{2}, \frac{2b}{2})$, or (a, b) and B has coord. $(\frac{2a+4a}{2}, \frac{2b}{2})$, or $(3a, b)$. $RB = \sqrt{(3a)^2 + b^2} = \sqrt{9a^2 + b^2}$ and $AS = \sqrt{(-3a)^2 + b^2} = \sqrt{9a^2 + b^2}$. Thus $AS = RB$.

Problem Solving Strategies, pages 506–507 **1.** 204 squares; 30 = 16 + 9 + 4 + 1 **2.** 36π; $9 \times 4\pi$ **3.** 4 + 3 + 2 + 1 = 10; 45 lines

Computer Activities, page 509
2. a. 600 cm^2 **b.** 10 cm × 10 cm × 10 cm **3. a.** H = 11.08; TA = 601.54 **b.** H = 9.07; TA = 601.44 **c.** no **4. a.** TA = 952.4592 **b.** 12.6 cm × 12.6 cm × 12.6 cm
5. a. They are all =. **b.** cube

Skills Review, page 510 **1.** $\frac{4x}{7}$ **3.** $-xy$ **5.** $\frac{6b^3 + 2b}{a}$ **7.** $\frac{1}{x+y}$ **9.** $\frac{m}{n}$ **11.** $b - 1$ **13.** $\frac{7a}{10}$ **15.** $\frac{5a}{6}$ **17.** $\frac{7}{2a}$ **19.** $\frac{7 - 5a^2}{ab}$ **21.** $\frac{m}{12}$ **23.** $\frac{14 + x}{2x}$ **25.** $\frac{-2}{3n}$ **27.** $\frac{4x}{3}$

Chapter Review Exercises, pages 511–512 **1.** (3, 0) **2.** (−2, 0) **3.** x **4.** (3, −3) **5.** III **6.** 10 **7.** 7 **8.** 5 **9.** 10 **10.** (4, 0) **11.** (3, −4) **12.** (4, 0) **13.** $(-1, \frac{5}{2})$ **14.** $\frac{1}{2}$ **15.** $\frac{3}{-2}$ **16.** $\frac{-1}{2}$ **17.** 0 **18.** $\frac{3}{4}$ **19.** not defined **20.** True **21.** False **22.** False **23.** True **24.** no **25.** no **26.** 7 **27.** Slope of $\overleftrightarrow{AB} = \frac{6-4}{6-0} = \frac{2}{6} = \frac{1}{3}$; slope of $\overleftrightarrow{CD} = \frac{3-2}{4-1} = \frac{1}{3}$; since their slopes are =, $\overleftrightarrow{AB} \parallel \overleftrightarrow{CD}$. **28.** $CD = \sqrt{1^2 + 3^2} = \sqrt{10}$; $AB = \sqrt{6^2 + 2^2} = \sqrt{40} = 2\sqrt{10}$; $\frac{1}{2} \cdot 2\sqrt{10} = \sqrt{10}$, so $CD = \frac{1}{2} AB$.

Cumulative Review, page 514 **1.** 3 **3.** $3\sqrt{2}$ **5.** 12 **7.** 5 **9. a.** 16 **b.** 1280 **11.** $\frac{4}{5}$ **13.** $\frac{4}{3}$ **15.** 0.9063 **17.** 4.7 **19.** (−2, 3) **21.** (0, −2) **23.** $-\frac{1}{3}$
25. (Slope of $\overline{AB}$) $= -\frac{1}{3} =$ (slope of $\overline{CD}$) and (slope of $\overline{AD}$) = 4 = (slope of $\overline{BC}$), so $\overline{AB} \parallel \overline{CD}$ and $\overline{AD} \parallel \overline{BC}$. Thus $ABCD$ is a ▱.

REVIEW EXERCISES

Chapter 1, page 523 **1. a.** 30° **b.** acute **3. a.** 90° **b.** right **5.** vertical **7.** XOY **9.** Subst. Post. **11.** $KSCA$ **13.** R, U, F, and S are not coplanar points.

Chapter 2, pages 523–524 **1.** 2 **3.** If 2 ∠s are complements of = ∠s, then the 2 ∠s are =. **5.** corresponding **7.** alternate interior **9.** $c \parallel d$ **11.** $a \parallel b$

Chapter 3, page 524 **1.** 75 **3.** 30 **5. a.** LR **b.** NT **c.** E **7.** ASA
9. 1. $\overline{AB} \perp \overline{XY}$ (Given) 2. △ABY and △ABX are rt. △s. (Def. of rt. △) 3. $AX = AY$ (Given) 4. $AB = AB$ (From algebra) 5. △$ABX \cong$ △ABY (HL Theorem)

Chapter 4, page 525 **1.** Corr. parts of ≅ △s are =. **3.** See Const. 7 **5.** 6
7. 1. ∠A = ∠X; $AB = XY$; $AC = XZ$

(Given) 2. $\triangle ABC \cong \triangle XYZ$ (SAS Post.) 3. $BC = YZ$ (Corr. parts of $\cong$ ⚠ are =.)

Chapter 5, pages 525–526 **1.** False **3.** True **5.** True **7.** True **9.** True **11.** True **13.** False **15.** 1. X, Y, and Z are the midpoints of the sides of $\triangle ABC$. (Given) 2. $\overline{ZY} \parallel \overline{AX}$; $\overline{YX} \parallel \overline{ZA}$ (The Midpoints Thm.) 3. $AXYZ$ is a ▱. (Def. of a ▱)

Chapter 6, page 526 **1.** $p = 20$ m; $A = 21$ m^2 **3.** 40 **5.** 63 **7.** 15 **9.** right **11.** acute **13.** 3π **15.** 113.0 **17.** 50.2

Chapter 7, page 527 **1.** $\frac{5}{3}$ **3.** $\frac{3}{4}$ **5.** 9:10 **7.** 5:1 **9.** $\frac{3}{2}$ **11.** $\frac{3}{5}$ **13.** 6 **15.** 2 **17.** \$4.45 **19.** 62.5 km

Chapter 8, pages 527–528 **1.** no **3.** yes **5.** $x = 20$, $y = 16$ **7.** $x = 7.5$, $y = 15$ **9.** $\frac{3}{4}$ **11.** 1. $ABCD$ is a trapezoid. (Given) 2. $\overline{DC} \parallel \overline{AB}$ (Def. of trapezoid) 3. $\angle CAB = \angle DCA$; $\angle DBA = \angle CDB$ (If 2 $\parallel$ lines are cut by a trans., corr. ⚠ are =.) 4. $\triangle DOC \sim \triangle BOA$ (AA Post.) 5. $\frac{AO}{CO} = \frac{BO}{DO}$ (Corr. sides of $\sim$ ⚠ are in proportion.)

Chapter 9, page 528 **1.** 150° **3.** 90° **5.** 35° **7.** 50; 130 **9.** 6 **11.** 50 **13.** 15 **15.** Use Construction 2 to construct a line $\perp$ to the line passing through the center of the constructed ⊙ and any point on the ⊙.

Chapter 10, pages 528–529 **1.** intersecting, parallel, skew **3.** no **5.** 220 **7.** 600 **9. a.** 65π **b.** 90π **c.** 100π **11.** 56.5 **13.** 36π

Chapter 11, pages 529–530 **1.** $x = 8$, $y = 17$ **3.** $x = 90$, $y = 2$ **5.** 60 **7.** $9\sqrt{3}$ cm^2 **9.** 60 **11. a.** 6 **b.** 10 **c.** 96π **d.** 96π **13.** 4.3

Chapter 12, page 530 **1.** II **3.** 3; $\sqrt{5}$ **5.** 0; $\frac{4}{3}$ **7. a.** $\frac{3}{8}$ **b.** $-\frac{8}{3}$

9.

x	y
0	4
6	0
3	2
−3	6

11. $BC = 5$; $AC^2 = 4^2 + 3^2 = 25$; $AC = \sqrt{25} = 5$; $BC = AC$ So $\triangle ABC$ is an isosceles triangle.

APPENDIX: Transformational Geometry

Exercises, page 533 **1.** yes **3.** yes **5.** no **7.** 2 **9.** 2 **11.** an infinite number **13.** 8

Exercises, page 539 **1.** (−7, 1) **3.** (4, −6) **5.** (9, 2) **7.** A, H, I, M, O, T, U, V, W, X, Y **9.** H, I, O, X

Exercises, page 542 **1.** reflection **3.** translation **5.** translation **7.** reflection

CREDITS

Book design and production: TextArt, Inc.
Cover design: Sylvia Steiner
Illustrations: Eva Burg Vagreti
Technical art: Network Graphics and ANCO / Boston

PHOTOGRAPHY

Chapter openers

pages 2–3: Harald Sund / Image Bank
pages 44–45: Peter Gavin / Photo Researchers
pages 90–91: John Maher / EKM Nepenthe
pages 142–143: Chris Jones / The Stock Market
pages 188–189: Frank Siteman / EKM Nepenthe
pages 236–237: Grant Heilman
pages 284–285: L.L.T. Rhodes / Earth Scenes
pages 310–311: Lou Jones / Image Bank
pages 344–345: Georg Gerster / Photo Researchers
pages 386–387: Tim Jewett / EKM Nepenthe
pages 426–427: Hazel Hankin / Stock Boston
pages 466–467: Tom Ballard / EKM Nepenthe

Other photographs

page 23: Shirley Richards / Photo Researchers
page 32: Alan Carey / The Image Works
page 54: Art Resource, N.Y.
page 55: Michael Holford Photographs
page 57: Mark Boulton / Photo Researchers
page 59: Tyrone Hall
page 114: Daniel S. Brody / Stock Boston
page 125: Ken Lax
page 172: Glyn Cloyd / Taurus Photos
page 202: Robert Frerck / Odyssey Productions
page 208: Art Resource, N.Y.
page 258: The Granger Collection
page 262: Ken Lax
page 294: Michal Heron / Woodfin Camp & Assoc.
page 300: George Whiteley / Photo Researchers
page 302: George Bellerose / Stock Boston
page 316: Craig Hammel / The Stock Market
page 334: Ellis Herwig / Stock Boston
page 377: Bob Daemmrich
page 378: Courtesy of Panasonic Co.
page 402: Robert Frerck / Odyssey Productions
page 451: Carl Wolinsky / Stock Boston
page 482: Pierre Berger / Photo Researchers
page 500: Donald C. Dietz / Stock Boston
page 505: Richard B. Hoit / Photo Researchers
Walter G. Hodsdon
Joel E. Arem, Ph. D., F.G.A.
page 531: Richard Parker / Photo Researchers
Courtesy Museo Nacional De Artes E Industrias Populares, Mexico City
Jerome Wexler / Photo Researchers
page 537: J.L. Stage / Image Bank